RADAR AND ARPA MANUAL

RADAR AND ARPA MANUAL

Radar, AIS and Target Tracking for Marine Radar Users

THIRD EDITION

ALAN BOLE

Radar/ARPA nautical consultant and former Principal Lecturer in Navigation Systems at Liverpool John Moores University, UK

ALAN WALL

Head of Nautical Science and Co-director of Liverpool Logistics Offshore & Marine Research Institute, Liverpool John Moores University, UK

ANDY NORRIS

Maritime Consultant and Honorary Professor for Navigation Technology, University of Nottingham, UK

AMSTERDAM • BOSTON • HEIDELBERG • LONDON
NEW YORK • OXFORD • PARIS • SAN DIEGO
SAN FRANCISCO • SINGAPORE • SYDNEY • TOKYO

Butterworth-Heinemann is an imprint of Elsevier

Butterworth-Heinemann is an imprint of Elsevier
225 Wyman Street, Waltham, MA 02451, USA
The Boulevard, Langford Lane, Kidlington, Oxford, OX5 1 GB, UK

First edition 1990
Paperback edition 1992
Reprinted 1997, 1999 (twice), 2000, 2001, 2003
Second edition 2005
Reprinted 2006 (twice), 2007, 2008

Library of Congress Cataloging-in-Publication Data
A catalog record for this book is available from the Library of Congress

British Library Cataloguing-in-Publication Data
A catalogue record for this book is available from the British Library

ISBN: 978-0-08-097752-2

For information on all Butterworth–Heinemann publications
visit our website at http://store.elsevier.com

Printed and bound in the UK

14 15 16 17 18 10 9 8 7 6 5 4 3 2 1

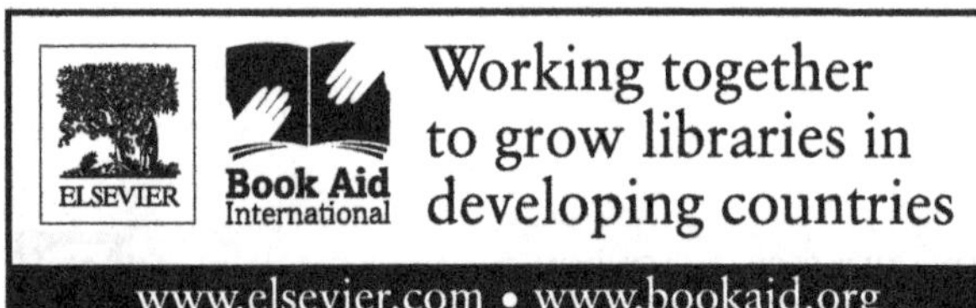

Dedication

To Bill and Keith

Contents

Preface xv
Acknowledgements xvii

1. Basic Radar Principles 1

1.1 Introduction 1
1.2 Principles of Range and Bearing Measurement 2
1.2.1 The Echo Principle 2
1.2.2 Range as a Function of Time 3
1.2.3 Directional Transmission and Reception 4
1.2.4 Display of Radar Information 5
1.3 Principles of Bearing Measurement 10
1.3.1 The Heading Marker 10
1.3.2 Bearing Measurement 11
1.4 Display Modes 12
1.4.1 Orientation Modes 12
1.5 Motion and Stabilization Modes 17
1.5.1 Relative-Motion Presentation 19
1.5.2 The True-Motion Presentation 22
1.5.3 Choice of Presentation 26

2. The Radar System – Technical Principles 29

2.1 Introduction 29
2.2 Basic Functionality 29
2.2.1 Transmitter 29
2.2.2 Antenna 30
2.2.3 Receiver 32
2.2.4 Radar Processor 32
2.2.5 Display and User Interface 33
2.3 The Radar Signal 34
2.3.1 Fundamental Considerations 34
2.3.2 Choice of Frequency 35
2.3.3 Pulse Radar Concepts 36
2.4 The Radar Transmitter 42
2.4.1 Magnetron-Based Transmitter 42
2.4.2 Magnetron Operation 44
2.5 Antenna Principles 45
2.5.1 Antenna Concepts 45
2.5.2 Array Antennas 49
2.5.3 Antenna Rotation Mechanism 60
2.6 Radar Signal Reception and Processing 61
2.6.1 The Radar Equation 61
2.6.2 The Role of the Receiver and Processor 62
2.6.3 The Radar Receiver 64
2.6.4 Receiver Techniques 67
2.6.5 Radar Interference 75
2.7 Radar Processing Principles 77
2.7.1 Conversion to the Digital Domain 77
2.7.2 Digital Processing 79
2.7.3 Range and Azimuth Sampling 81
2.7.4 Clutter Suppression 84
2.7.5 Target Tracking 87
2.8 The Radar Display and User Controls 89
2.8.1 The Radial Scan PPI 90
2.8.2 The Digital Display 92
2.8.3 Flat Panel Display Technology – LCDs 94
2.8.4 Formatting the Display 96
2.8.5 The Display of Radar Echoes 100
2.8.6 The User Interface 105
2.9 Solid-State Radar Principles 108
2.9.1 Coherent Radar Concepts 110
2.9.2 Correlation 114
2.9.3 An Example of Correlation 116
2.9.4 Range Sidelobes 120
2.9.5 Further Considerations on Pulse Compression 126
2.10 The Siting of Units on Board Ship 128
2.10.1 Antenna Siting 128
2.10.2 The Transceiver Unit 133
2.10.3 The Display Unit 134
2.10.4 Compass Safe Distances 134
2.10.5 Exposed and Protected Equipment 135
2.10.6 Power Supplies 135
2.10.7 High-Voltage Hazards 135
2.10.8 Interswitching 135

3. Target Detection 139

3.1 Introduction 139
3.2 Radar Characteristics 140
3.2.1 Transmitter Characteristics 140
3.2.2 Antenna Characteristics 140
3.2.3 Receiver Characteristics 141
3.2.4 Minimum Detection Range 141
3.2.5 Detection Performance Standards 143
3.3 Target Characteristics 143
3.3.1 Aspect 143
3.3.2 Surface Texture 144
3.3.3 Material 144
3.3.4 Shape 145
3.3.5 Size 145
3.3.6 Responses from Specific Targets 151
3.4 Target Enhancement – Passive 152
3.4.1 Corner Reflectors 152
3.4.2 Arrays of Reflectors 154
3.4.3 The Lunenburg Lens 155
3.5 Target Enhancement – Active 157
3.5.1 The Racon Principle 158
3.5.2 The Racon Appearance on the Display 158
3.5.3 Frequency and Polarization 161
3.5.4 Sources of Radar Beacon Information 165
3.5.5 The Radaflare 166
3.5.6 Search and Rescue Transponders 166
3.5.7 Active Radar Reflectors 169
3.6 The Detection of Targets in Sea Clutter 170
3.6.1 The Nature of Sea Clutter Response 170
3.6.2 The Clutter Problem Summarized 175
3.6.3 The Suppression of Displayed Sea Clutter Signals 176
3.7 The Detection of Targets in Precipitation Clutter 183
3.7.1 The Nature of Precipitation Response 183
3.7.2 Attenuation in Precipitation 184
3.7.3 The Effect of Precipitation Type 185
3.7.4 The Suppression of Rain Clutter 186
3.7.5 Combating the Attenuation Caused by Precipitation 190
3.7.6 Exploiting the Ability of Radar to Detect Precipitation 190
3.8 The Radar Horizon 191
3.8.1 The Effect of Standard Atmospheric Conditions 191
3.8.2 Sub-Refraction 194
3.8.3 Super-Refraction 195
3.8.4 Extra Super-Refraction or Ducting 196
3.9 False and Unwanted Radar Responses 197
3.9.1 Introduction 197
3.9.2 Indirect Echoes (Reflected Echoes) 197
3.9.2.1 The Effect of On-Board Obstructions 198
3.9.3 Multiple Echoes 201
3.9.4 Side Echoes 202
3.9.5 Radar-to-Radar Interference 204
3.9.6 Second-Trace Echoes 206
3.9.7 'False' Echoes from Power Cables 210

4. Automatic Radar Target Tracking, Specified Facilities 215

4.1 Introduction 215
4.1.1 Integral and Stand-Alone Displays 216
4.1.2 Carriage Requirements, Standards and Operator Training 216
4.1.3 Compliance with IMO Performance Standards 217
4.2 The Acquisition of Targets 217
4.2.1 The Acquisition Specification 218
4.2.2 Manual Acquisition 218
4.2.3 Fully Automatic Acquisition 218
4.2.4 Automatic Acquisition by Area 219
4.2.5 Guard Zones 219
4.2.6 Guard Rings and Area Rejection Boundaries 219
4.3 The Tracking of Targets 220
4.3.1 The Tracking Specification 220
4.3.2 Rate Aiding 221
4.3.3 The Number of Targets to Be Tracked 221
4.3.4 Target Loss 222
4.3.5 Target Swap 223
4.3.6 The Analysis of Tracks and the Display of Data 223
4.3.7 Target Trails and Past Positions 229
4.4 Vectors 232
4.4.1 Relative Vectors 233
4.4.2 True Vectors 233
4.4.3 Trial Manoeuvre 234
4.5 The ARPA Display 236
4.5.1 The Size of the Operational Display 236
4.5.2 The Range Scales on Which ARPA Facilities Should Be Available 236
4.5.3 The Modes of Display 236
4.5.4 The ARPA Data Brilliance Control 237
4.5.5 The Use of the Screen Marker for Range and Bearing Measurement 237
4.5.6 The Effect of Changing Range Scales 237

4.6 The Display of Alphanumeric Data 237
4.7 Alarms and Warnings 238
4.7.1 Guard Zone Violation 238
4.7.2 Predicted CPA/TCPA Violation 239
4.7.3 Lost Target 239
4.7.4 Time to Manoeuvre 240
4.7.5 Track Change 240
4.7.6 Anchor Watch 240
4.7.7 Tracks Full 240
4.7.8 Wrong or Invalid Request 241
4.7.9 Safe Limit Vector Suppression 241
4.7.10 Trial Alarm 241
4.7.11 Performance Tests and Warnings 241
4.8 Automatic Ground-Stabilization 241
4.9 Navigational Lines and Maps (See Also Section 8.4.6.3) 242
4.10 Target Simulation Facility 244
4.11 The Predicted Point of Collision 245
4.11.1 The Concept of Collision Points 245
4.11.2 The Behaviour of Collision Points If the Observing Ship Maintains Speed 246
4.11.3 The Behaviour of the Collision Point When the Target Ship's Speed Changes 247
4.11.4 The Behaviour of the Collision Point When the Target Changes Course 249
4.12 The PREDICTED AREA of DANGER (PAD) 250
4.12.1 The PAD in Practice 251
4.12.2 Changes in the Shape of the PAD 252
4.12.3 The Movement of the PAD 252
4.12.4 The Future of PADs 253

5. Automatic Identification System (AIS) 255

5.1 Organization of AIS Transmissions 256
5.1.1 Autonomous and Continuous Mode 257
5.1.2 Assigned Mode 257
5.1.3 Polling Mode 258
5.1.4 Class B Transmissions – CSTDMA 258
5.2 AIS Information Transmitted by a Class A Vessel 259
5.3 AIS Messages and Types 261
5.3.1 Class A Transmissions 261
5.3.2 Class B Transmissions 261
5.3.3 SAR Aircraft 261
5.3.4 AIS SART 261
5.3.5 Aids to Navigation 263
5.3.6 AIS Binary Messages 263
5.4 AIS Units and Bridge Displays 264
5.4.1 Stand-Alone Minimum Keyboard Display 264
5.4.2 Stand-Alone Graphical Display 265
5.4.3 Integration of AIS with ARPA and/or ECDIS 265
5.4.4 Pilot Plug and Display 266
5.4.5 Class B Equipment 266
5.4.6 AIS Receivers 266
5.5 AIS Usability 267
5.5.1 Target Swap 267
5.5.2 The Transmission of AIS Signals Around Large Land Features 267
5.5.3 AIS Capacity 267
5.5.4 AIS Dependence on GNSS 268
5.5.5 AIS Participation by Vessels 269
5.5.6 AIS Vulnerability to False Reports, Spoofing and Jamming 269
5.6 Benefits of AIS to Shore Monitoring Stations 270
5.7 Radar/ARPA and AIS Comparison for Collision Avoidance 271
5.8 Other AIS Applications and Applications Associated with AIS 273
5.8.1 AIS Internet-Based Networks 273
5.8.2 Satellite AIS 273
5.8.3 Long Range Identification and Tracking LRIT 274

6. Operational Controls 277

6.1 Use of Controls and Optimum Performance 277
6.2 Setting Up the Radar Display 278
6.2.1 Preliminary Procedure 278
6.2.2 Switching On 278
6.2.3 Setting the Screen Brilliance 279
6.2.4 Default Conditions and Start-Up 281
6.2.5 Setting the Orientation of the Picture 282
6.2.6 Setting the Presentation of the Radar Picture 282
6.2.7 Obtaining the Optimum Picture (Magnetron Radar) 284
6.2.8 Optimum Picture on Coherent Radar Systems 288
6.3 Performance Monitoring 288
6.3.1 The Principle of the Echo Box 288
6.3.2 Echo Box Siting 290
6.3.3 Power Monitors 292

6.3.4 Transponder Performance Monitors 292
6.3.5 Calibration Levels 293
6.3.6 Performance Check Procedure 293
6.3.7 Modern Trends 294
6.4 Change of Range Scale and/or Pulse Length 294
6.5 The Stand-by Condition 295
6.6 Controls for Range and Bearing Measurement 295
6.6.1 Fixed Range Rings 296
6.6.2 Variable Range Marker 297
6.6.3 Parallel Index Lines 297
6.6.4 The Electronic Bearing Line 297
6.6.5 Free Electronic Range and Bearing Line 298
6.6.6 Joystick/Tracker Ball and Screen Marker 298
6.6.7 Range Accuracy 299
6.6.8 Bearing Accuracy 300
6.7 Controls for the Suppression of Unwanted Responses 301
6.7.1 Sea Clutter Suppression 301
6.7.2 Rain Clutter Suppression 304
6.7.3 Interference Suppression 305
6.8 Echo Stretch 305
6.9 Using an Automatic Radar Plotting Display 305
6.9.1 The Input of Radar Data 305
6.9.2 Switching on the Computer 306
6.9.3 Heading and Speed Input Data 306
6.9.4 Setting the Vector Time Control 308
6.9.5 Selecting the Relative or True Vector Mode 308
6.9.6 Selecting Ground- or Sea-Stabilized True Vectors 308
6.9.7 Safe Limits 310
6.9.8 Preparation for Tracking 310
6.10 AIS Operational Controls 311
6.10.1 Stand-Alone AIS Equipment 311
6.10.2 AIS Integrated with ARPA 313

7. Radar Plotting Including Collision Avoidance 317

7.1 Introduction 317
7.2 The Relative Plot 317
7.2.1 The Vector Triangle 319
7.2.2 The Plotting Triangle 319
7.2.3 The Construction of the Plot 320
7.2.4 The Practicalities of Plotting 321
7.2.5 The Need to Extract Numerical Data 323
7.2.6 The Plot in Special Cases Where No Triangle 'Appears' 324
7.3 The True Plot 324
7.4 The Plot When Only the Target Manoeuvres 326
7.4.1 The Construction of the Plot (Figure 7.8) 327
7.4.2 The Danger in Attempting to Guess the Action Taken by a Target 328
7.5 The Plot When the Own ship Manoeuvres 328
7.5.1 The Plot When the Own Ship Alters Course Only 328
7.5.2 The Construction of the Plot (Figure 7.10) 329
7.5.3 The Plot When the Own Ship Alters Speed Only 329
7.5.4 The Construction of the Plot (See Figure 7.11) 330
7.5.5 The Use of 'Stopping Distance' Data in the Form of Tables, Graphs and Formulae 331
7.5.6 The Plot When the Own Ship Combines Course and Speed Alterations 334
7.5.7 The Plot When the Own Ship Resumes Course and/or Speed 335
7.5.8 The Plot When Both Vessels Manoeuvre Simultaneously 336
7.6 The Theory and Construction of PPCs, PADs, SODs and SOPs 338
7.6.1 The Predicted Point of Collision 338
7.6.2 The Construction to Find the PPC 339
7.6.3 The Predicted Area of Danger 339
7.6.4 The Construction of the PAD 340
7.6.5 The Sector of Danger 341
7.6.6 The Construction of a Sector of Danger 342
7.6.7 The Sector of Preference 343
7.6.8 The Construction of an SOP 344
7.7 The Plot in Tide 345
7.7.1 The Construction of the Plot 346
7.7.2 The Course to Steer to Counteract the Tide 346
7.7.3 The Change of Course Needed to Maintain Track When Changing Speed in Tide 346
7.8 Manual Plotting – Accuracy and Errors 347
7.8.1 Accuracy of Bearings as Plotted 348
7.8.2 Accuracy of Ranges as Plotted 348
7.8.3 Accuracy of the Own Ship's Speed 348
7.8.4 Accuracy of the Own Ship's Course 349
7.8.5 Accuracy of the Plotting Interval 349

7.8.6 The Accuracy with Which CPA Can Be Determined 349
7.8.7 The Consequences of Random Errors in the Own Ship's Course and Speed 350
7.8.8 Summary 350
7.9 Errors Associated with the True-Motion Presentation 351
7.9.1 Incorrect Setting of the True-Motion Inputs 351
7.9.2 Tracking Course Errors 353
7.9.3 Tracking Speed Errors 353
7.10 Radar Plotting Aids 353
7.10.1 The Radar Plotting Board 354
7.10.2 Threat Assessment Markers ('Matchsticks' or 'Pins') 354
7.10.3 The Reflection Plotter 354
7.10.4 The 'E' Plot 358
7.10.5 Electronic Plotting Aid (EPA) 359
7.10.6 Auto-Tracking Aid 360
7.11 The Regulations for Preventing Collisions at Sea as Applied to Radar and ARPA 361
7.11.1 Introduction 361
7.11.2 Lookout – Rule 5 362
7.11.3 Safe Speed – Rule 6 362
7.11.4 Risk of Collision – Rule 7 363
7.11.5 Conduct of Vessels in Restricted Visibility – Rule 19 363
7.11.6 Action to Avoid Collision – Rule 8 364
7.11.7 The Cumulative Turn 365
7.11.8 Conclusion 366
7.12 Intelligent Knowledge-Based Systems as Applied to Collision Avoidance 367

8. Navigation Techniques Using Radar and ARPA 371

8.1 Introduction 371
8.2 Identification of Targets and Chart Comparison 372
8.2.1 Long Range Target Identification 372
8.2.2 The Effect of Discrimination 373
8.2.3 Shadow Areas 374
8.2.4 Rise and Fall of Tide 376
8.2.5 Radar-Conspicuous Targets 376
8.2.6 Pilotage Situations 378
8.3 Position Fixing 378
8.3.1 Selection of Targets 379
8.3.2 Types of Position Line 379
8.4 Parallel Indexing 380
8.4.1 Introduction 380
8.4.2 Preparations and Precautions 382
8.4.3 Relative Parallel Indexing: The Technique 385
8.4.4 Progress Monitoring 392
8.4.5 Parallel Indexing on a True-Motion Display 394
8.4.6 Modern Radar Navigation Facilities 396
8.4.7 Unplanned Parallel Indexing 402
8.4.8 Anti-Collision Manoeuvring While Parallel Indexing 402

9. ARPA – Accuracy and Errors 407

9.1 Introduction 407
9.2 The Accuracy of Displayed Data Required by the Performance Standard 407
9.3 The Classification of ARPA Error Sources 408
9.4 Errors That Are Generated in the Radar Installation 408
9.4.1 Glint 409
9.4.2 Errors in Bearing Measurement 409
9.4.3 Errors in Range Measurement 410
9.4.4 The Effect of Random Gyro Compass Errors 411
9.4.5 The Effect of Random Log Errors 412
9.5 Errors in Displayed Data 413
9.5.1 Target Swap 413
9.5.2 Track Errors 413
9.5.3 The Effect on Vectors of Incorrect Course and Speed Input 415
9.5.4 The Effect on the PPC of Incorrect Data Input 418
9.6 Errors of Interpretation 420
9.6.1 Errors with Vector Systems 420
9.6.2 Errors with PPC and PAD Systems 421
9.6.3 The Misleading Effect of Afterglow 422
9.6.4 Accuracy of the Presented Data 422
9.6.5 Missed Targets 423

10. Ancillary Equipment 425

10.1 Global Navigation Satellite Systems 425
10.1.1 Global Positioning System 425

10.1.2 The Measurement of Range and Time from the Satellite 428
10.1.3 The Position Fix 430
10.1.4 User Equipment and Display of Data 430
10.1.5 Accuracy and Errors 432
10.1.6 Differential GPS 437
10.1.7 Improvements to GPS 437
10.1.8 Other Navigation Satellite System Developments 438
10.1.9 Inter-Relationship of GNSS with Radar 442
10.1.10 Inter-Relationship of GNSS with ECDIS (See Sections 10.2 and 11.3.3) 442
10.2 Electronic Charts (ECDIS) 442
10.2.1 Vector Charts 443
10.2.2 Raster Charts 444
10.2.3 Comparison Between Paper and Electronic Charts 444
10.2.4 Comparison Between Vector and Raster Charts 445
10.2.5 Unapproved Electronic Chart Data 446
10.2.6 Publications Associated with Charts 446
10.2.7 Relationship of ECDIS with Radar and Target Tracking 447
10.3 Integrated Systems 447
10.3.1 Integrated Bridge Systems 447
10.3.2 Integrated Navigation Systems 448
10.3.3 Typical Systems That May Be Integrated 448
10.3.4 Connectivity and Interfacing 449
10.3.5 Advantages of Integration 453
10.3.6 Potential Dangers of Data Overload 453
10.3.7 System Cross-Checking, Warnings and Alarms 453
10.3.8 Sensor Errors and Accuracy of Integration 453
10.3.9 Data Monitoring and Loggings 454
10.4 Voyage Data Recorders (SEE ALSO SECTION 11.3.5) 454
10.4.1 Equipment 454
10.4.2 Non-Radar Data Recorded 455
10.4.3 Radar and Radar Tracking Data Recorded 456
10.4.4 Playback Equipment 456
10.4.5 Future of VDRs 457

11. Extracts from Official Publications 459

11.1 Extracts from Regulation 19, Chapter V, Safety of Navigation, of IMO-SOLAS Convention 459
11.2 IMO Performance Standards for Radar Equipment 460
11.2.1 Extracts from IMO Resolution MSC.192 (79), Performance Standards for Radar Equipment for New Ships Constructed After 1 July 2008 460
11.2.2 Extracts from IMO Resolution MSC.191 (79) Performance Standards for the Presentation of Navigation-Related Information on Shipborne Navigational Displays for New Ships Constructed After 2008 477
11.2.3 Extracts from IMO SN/Circ. 243 Guidelines for the Presentation of Navigation-Related Symbols, Terms and Abbreviations (As Amended) 484
11.2.4 Extract from IMO Resolution A.615(15) Marine Uses of Radar Beacons and Transponders 499
11.2.5 Extract from IMO Resolution A.802(19) Performance Standards for Survival Craft Radar Transponders for Use in Search and Rescue Operations (As Amended) 502
11.2.6 Extract from IMO Resolution A.384(X) Performance Standards for Radar Reflectors 503
11.3 IMO Performance Standards for Other Related Equipment 503
11.3.1 Extract from IMO Resolution MSC.74(69) Annex 3 Performance Standards for a Universal Shipborne Automatic Identification System (AIS) 503
11.3.2 Extract from IMO Resolution MSC.112 (73) Performance Standards for Shipborne Global Positioning System (GPS) Receiver Equipment Valid for Equipment Installed on or After 1 July 2003 505
11.3.3 Extract from IMO Resolution MSC.232 (82) Revised Performance Standards for Electronic Chart Display and Information Systems (ECDIS) Adopted on 5 December 2006 507

11.3.4 Extracts from IMO Resolution MSC 252 (83) Performance Standards for Integrated Navigation Systems (INS) 507
11.3.5 Extract from IMO Resolution A.861(20) Performance Standards for Shipborne Voyage Data Recorders (VDRs) (As Amended) 516
11.3.6 Extract from IMO Resolution A.694(17) General Requirements for Shipborne Radio Equipment Forming Part of the Global Maritime Distress and Safety System (GMDSS) and for Electronic Navigational Aids 519
11.4 Extracts from UK Statutory Instrument 1993 No. 69, the Merchant Shipping (Navigational Equipment) Regulations 1993 520
11.4.1 Extract from Part IV, Radar Installation 520
11.4.2 Extract from Part IX, Automatic Radar Plotting Aid Installation 521

Glossary of Acronyms and Abbreviations 523

Index 525

Preface to the Third Edition

There have been considerable advances in technology in recent years which has meant that a major revision has been necessary.

In the past, much of the work of the navigator involved the correct use of the controls in setting up the display and the correct interpretation of the displayed data – in particular, radar plotting to determine risk of collision. These problems have been largely solved by the development of digital techniques which have allowed the data to be electronically processed resulting, among other facilities, in auto-clutter suppression and target tracking (ARPA).

Unfortunately, the advances in technology have brought with them their own problems. The move from analogue to digital techniques has opened up considerable possibilities, in particular, to integrate the displayed outputs from what were independent instruments on to a common display monitor. This can give rise to information overload and/or display congestion if used indiscriminately.

A common failing now is for operators not to input, update or regularly check the data being fed to the systems upon which the output depends (courses, speeds, ship's data, etc.). As a result, for the navigator, the displayed data can be erroneous/misleading. The behaviour of an observer on another vessel will depend on the information being received (e.g. from AIS) and from information determined (e.g. from radar/ARPA). Serious confusion can arise when there are inconsistencies in what the instruments are telling the observer.

In recent years, there have been considerable changes and increases in the technical specifications of all navigational equipment (although rarely retrospective), and also, the Carriage Requirements. These have, to a large extent, been taken into account in this treatment. Also, the basic ideas behind solid-state coherent radars have been included within Chapter 2, as these are being increasingly fitted to vessels.

IMO and national advice on matters of safety and good practice is still included where applicable. The correct use of the equipment is paramount and it is in this area that we have continued to stress the importance of 'good practice' which has been built up over the years.

Although small vessels and pleasure craft are not specifically required to carry this equipment, many of them do and in their interest; it is hoped that many aspects of the material covered here will prove of value for them.

Some material relating to the development of radar has been retained in order to provide a background to understand where today's equipment is coming from and to underpin the theory upon which present-day radars are based. Most of the descriptions which related to specific earlier equipment has been removed, in spite of the fact some of that equipment may still be in use today.

Another significant change is that the latest IMO performance standards for radar on ships no longer refer to the term ARPA and instead use the term Target Tracker, as the equipment now has to integrate and present AIS (Automatic Identification System) data with radar tracked data. This new edition has therefore included a much larger discussion of AIS with the inclusion of the new Chapter 5. This trend away from independent to integrated equipment has meant that, for completeness, the inter-relationship between radar/ARPA, AIS, GPS and ECDIS has had to be included, but not to the same technical depth as the radar and ARPA.

Alan Bole
Radar/ARPA nautical consultant and former Principal Lecturer in Navigation Systems at Liverpool John Moores University, UK
Alan Wall
Head of Nautical Science and Co-director of Liverpool Logistics Offshore & Marine Research Institute, Liverpool John Moores University, UK
Andy Norris
Maritime Consultant and Honorary Professor for Navigation Technology, University of Nottingham, UK

Acknowledgements

First edition

The authors wish to express their gratitude to:

The International Maritime Organization (IMO) for permission to reproduce the various extracts from resolutions adopted by the Assembly.

The Controller of Her Majesty's Stationery Office for permission to reproduce the extracts from M 1158 and Statutory Instrument No. 1203 (1984).

Captain C. E. Nicholls of Liverpool Polytechnic for his major contribution to Chapter 8.

Second edition

We again express our thanks to IMO for permission to reproduce updated extracts from various resolutions adopted by the Assembly.

In this edition we are grateful for the considerable assistance of June Bole and Alison Wall in proofreading the manuscript and for their help, support and encouragement in our completion of this book.

Mr B. Price of Sandown College, Liverpool, for his helpful comments based on a reading of Chapter 2.

Mr Andrew O. Dineley for his assistance in producing the computer printout of the manuscript.

Families and friends without whose assistance, support and understanding, this undertaking would never have been completed.

Third edition

Mr Barry Wade of Kelvin Hughes for his helpful comments on Chapter 2.

We again express our thanks to IMO for permission to reproduce updated extracts from various resolutions adopted by the Assembly.

CHAPTER

1

Basic Radar Principles

1.1 INTRODUCTION

Radar forms an important component of the navigational equipment fitted on virtually all vessels apart from the very smallest. Its display of critical information is easily assimilated by a trained user and has acted as a focus for the presentation of other navigational data, giving it a deserved prominence on the bridge of a vessel. It is poised to retain its central electronic navigational role into the foreseeable future, equalled only in display significance by the rather more recent development, the electronic chart. Together, they will provide the basis of the major displays for marine navigation into an increasingly integrated navigational world.

The word RADAR is an acronym derived from the words *Radio Detection and Ranging*. The scientist Heinrich Hertz, after whom the basic unit of frequency is named, demonstrated in 1886 that radio waves could be reflected from metallic objects. In 1904 a German engineer, Christian Hülsmeyer, obtained a patent in several countries for a radio wave device capable of detecting ships, but it aroused little enthusiasm because of its very limited range. Marconi, delivering a lecture in 1922, drew attention to the work of Hertz and proposed in principle what we know today as marine radar. Although radar was used to determine the height of the ionosphere in the mid-1920s, it was not until 1935 that radar pulses were successfully used to detect and measure the range of an aircraft. In the 1930s there was much simultaneous but independent development of radar techniques in Britain, Germany, France and America. Radar first went to sea in a warship in 1937 and by 1939 considerable improvement in performance had been achieved. By 1944 naval radar had made an appearance on merchant ships and from about the end of the Second World War the growth of civil marine radar began. Progressively it was refined to meet the needs of peacetime navigation and collision avoidance.

The civil marine radars in use today differ markedly from their ancestors of the 1940s in size, appearance and versatility, but the basic data that they offer, namely target range and bearing, are determined by exploiting the same fundamental principles unveiled so long ago. An understanding of such principles is an essential starting point in any study of marine radar, even though recent developments in the use of a technology known as *coherent radar* have somewhat complicated the picture. This latter technology is explained in some detail in Section 2.9, but first it is useful to gain an understanding of the basic principles behind radar.

1.2 PRINCIPLES OF RANGE AND BEARING MEASUREMENT

1.2.1 The Echo Principle

An object (normally referred to as a target) is detected by the transmission of radio energy as a pulse or otherwise, and the subsequent reception of a fraction of such energy (the echo) which is reflected by the target in the direction of the transmitter. The phenomenon is analogous to the reflection of sound waves from land formations and large buildings. Imagine somebody giving a short sharp shout through cupped hands to focus the sound energy. The sound wave travels outwards and some of it may strike, for example, a cliff. Some of the energy which is intercepted will be reflected by the cliff. If the reflected energy returns in the direction of the caller, and is of sufficient strength, it will be heard as an audible echo, resembling the original shout. In considering this analogy, the following points can usefully assist in gaining a preliminary understanding of pulse radar detection:

A. The echo is never as loud as the original shout.
B. The chance of detecting an echo depends on the loudness and duration of the shout.
C. Short shouts are required if echoes from close targets are not to be drowned by the original shout.
D. A sufficiently long interval between shouts is required to allow time for echoes from distant targets to return.
E. It can be more effective to cup one's hands over the mouth when shouting and put a hand to the ear when listening for the echo.

Now considering radar, its basic building blocks are illustrated diagrammatically in Figure 1.1. The antenna is used both to transmit the signal and to receive its reflection. On transmit, the antenna is acting very much like the cupped hand, focussing the energy in a particular direction. On receive it is acting more like a hand to the ear, collecting more received energy from that direction. The transmitter has a similar role to that of the mouth and vocal chords of the shouter, and the radar receiver acts as the ear. The processor clarifies the received signal and judges its distance, perhaps somewhat similar to what a trained human brain can do in identifying and assessing a received sound wave. Finally the radar displays the information to a human operator, perhaps analogous to a human writing down the estimated range and direction of the object producing the echo.

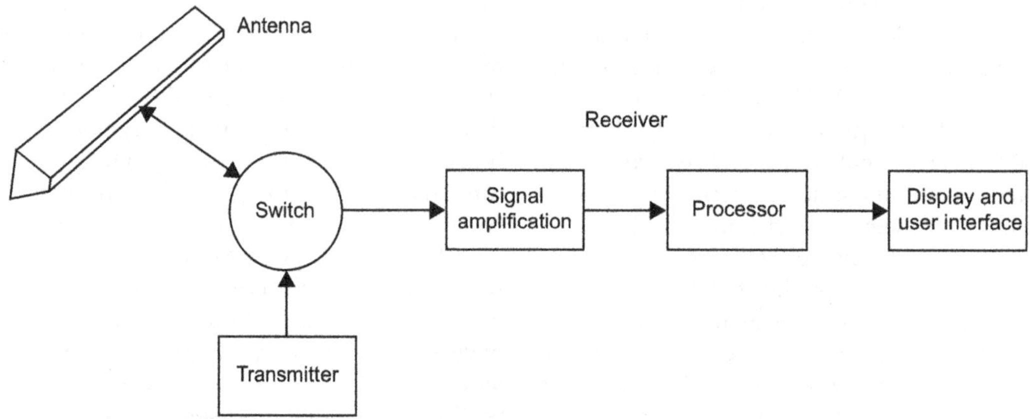

FIGURE 1.1 The basic radar system.

The antenna of a marine radar rotates steadily in the horizontal plane giving a complete rotation about every 2 s. This means that radar pulses consecutively cover all directions over 360° at each rotation of the antenna. The speed of radio waves is so high, about one million times greater than sound waves, that the antenna receives all the reflected energy from a particular transmitted pulse before it has appreciably rotated.

1.2.2 Range as a Function of Time

It is self-evident that the time which elapses between the transmission of a pulse and the reception of the corresponding echo depends on the speed of the pulse and the distance which it has travelled in making its two-way journey. If the speed of the pulse is known and the elapsed time can be measured, the range of the target producing the echo can be calculated.

The velocity of radio waves is dependent on the nature of the medium through which they travel. In fact, within the Earth's atmosphere it is hardly different to that within a space-type vacuum, that is 299,792,458 m/s. In our own minds this is easiest to be considered to be almost precisely 300,000,000 (three hundred million) metres per second, or as 300 metres per microsecond (μs), where 1 μs represents one millionth part of a second (i.e. 10^{-6} s). Using this value it is possible to produce a simple general relationship between target range and the elapsed time which separates the transmission of the pulse and the reception of an echo in any particular case (Figure 1.2).

Let D = the distance travelled by the pulse to and from the target (metres)

R = the range of the target (m)
T = the elapsed time (μs)
S = the speed of radio waves (m/μs)

Then $D = S \times T$
and $R = (S \times T)/2$
hence $R = (300 \times T)/2$
thus $R = 150T$

The application of this relationship can be illustrated by the following example.

EXAMPLE 1.1

Calculate the elapsed time for a pulse to travel to and return from a radar target whose range is (a) 40 m (b) 12 nautical miles (NM).

a. $R = 150T$
thus $40 = 150T$
hence $T = 40/150 \approx 0.27$μs

This value is of particular interest because 40 m represents the minimum detection range that must be achieved to ensure compliance with IMO Performance Standards for Radar Equipment (see Section 11.2.1). While this topic will be fully explored in Section 3.2.4, it is useful at this stage to note the extremely short

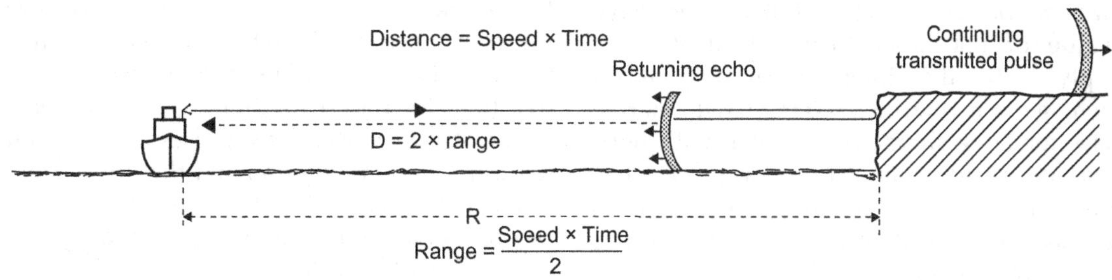

FIGURE 1.2 The echo principle.

time interval within which transmission and reception must be accomplished.

b. $R = 150T$

$$\text{Since } 1\ \text{NM} = 1852\ \text{m},$$
$$12 \times 1852 = 150T$$
$$\text{hence } T = 12 \times 1852/150 = 148.16\ \mu\text{s}$$

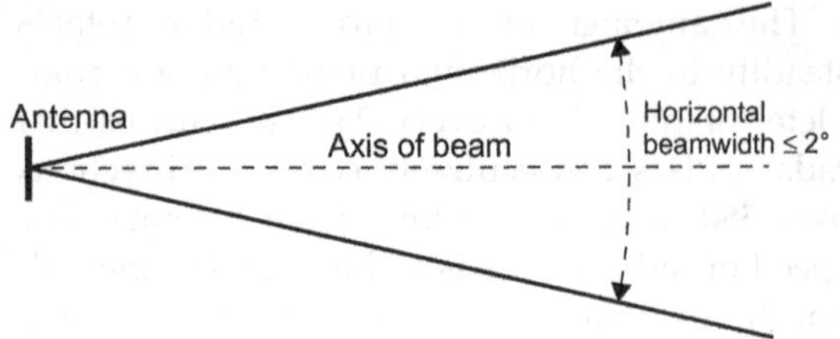

FIGURE 1.3 The horizontal beam width.

This result is noteworthy as it represents the elapsed time for a commonly used marine radar range scale. The elapsed times established in this section are of the order of millionths of a second and therefore need special instrumentation to be able to measure them accurately. In the early days of radar this was cutting-edge technology, but with the advent of quartz timing technology, and fast microelectronics it is no longer a major issue. Such technology is low cost, accurate and ubiquitous, with most humans owning multiple examples of precision timing in their watches, mobile phones, computers, TVs and cars.

1.2.3 Directional Transmission and Reception

In a marine radar system it is cost and space effective to use a single antenna for both transmission and reception. It is designed in such a way (see Section 2.5) as to focus the transmitted energy into a beam which is very narrow in the horizontal plane. The angle within which the energy is constrained is called the *horizontal beamwidth* (Figure 1.3). It must have a value of not more than 2.0° if it is to comply with the international regulations which govern marine radar. Civil marine radars for large ships are available with horizontal beamwidths as narrow as 0.75°. The equivalent reception property of the antenna is such that it will detect energy which has returned from within the angular limits of the horizontal beamwidth; that is from those targets that have been illuminated by the corresponding radar transmission. Its insensitivity to picking up unwanted noise from other directions effectively increases its ability to detect the reflected echoes.

An essential feature of a marine radar is that it should provide continuous coverage over the full 360° of azimuth angle. To achieve this the antenna has to rotate and no part of the vessel should obscure the radar beam, such as masts and other superstructure. Typical antenna rotation rates are 24–45 revolutions per minute, resulting in a complete rotation occurring every 1.3–2.5 s, depending on the system.

The interval between successive transmitted pulses has to at least allow the transmitted signal to travel out to the furthest target of interest and back again, although there are other considerations, which are discussed in Section 2.3.3.2. This interval is normally considered as a pulse repetition frequency (PRF), that is the number of pulses transmitted in 1 s. If we take, as an example, a value of 1500 pulses per second (1500 Hz); this is equivalent to one pulse every 667 μs. Taking a representative time for one revolution of the scanner to be 2 s, it is seen that 3000 pulses are transmitted during one revolution and that the scanner rotates through 0.12° between pulses. The picture is thus 'built up' of approximately 3000 radial lines of reflected echoes.

1.2.4 Display of Radar Information

1.2.4.1 The A-Scan Display

The A-scan is a useful concept to help understand the makeup of a reflected radar signal and how it can be displayed. This basic type of display is sometimes used today by engineers and technicians for special purposes, but is not a display that is available on a marine radar when used as a navigational aid. An A-scan display plots the returned radar signal as a graph, see Figure 1.4. The horizontal axis represents time and the vertical axis represents the strength (amplitude) of the received signal. The plot, sometimes called a trace, commences at the instant each radar pulse is transmitted. This event is indicated by a vertical spike, known as the transmission mark. Returning echoes also generate spikes in the plot. The amplitude of these spikes are related to the strength of the echo. The equivalent 'real-life' situation is also shown in the diagram.

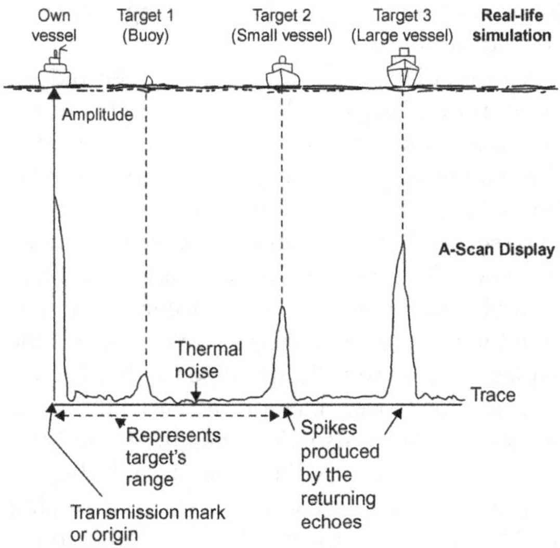

FIGURE 1.4 The A-scan display.

The horizontal distance between the transmission mark and an echo spike is a measure of the range of the target. Using the result from Example 1.1(b), it is evident that if the full extent of the plot is to represent a range of 12 NM (the selected range scale) this is equivalent to a timescale of approximately 148 μs.

1.2.4.2 The Plan Position Indicator Display

The A-scan shows the amplitude of the reflected radar energy as a function of range at a particular azimuth bearing angle of the radar antenna. In principle, this angle could be shown in degrees as an information box on the display, allowing the user to determine the range and azimuth of any target in view, as the antenna rotated. In practice, this would not be a very effective display. With targets only being visible for a short period, once per revolution of the antenna, the human brain would have difficulty in assessing any real situation.

What is perhaps ideally required is a plan view, such that the radar image creates a 'map' of the surrounding area, allowing easy assimilation of the current situation by the user. This is particularly relevant in our modern world as it also allows the radar display to show conventionally charted features as an 'underlay' to the radar image, putting them into geographical context.

The term Plan Position Indicator (PPI) has been used for this type of radar display, since the 1940s. Nowadays the precise image on the display is produced by digital processing technology. This effectively computes the amplitude of the received signal, as shown in Figure 1.4 for the A-scan, at small increments of range. The increment used is known as the *range cell* increment. The process produces a computerized list of signal strengths against range for the particular azimuth angle (bearing) of the antenna. The next radar pulse is transmitted when the antenna has turned through a small angle, known as the *azimuth cell* increment, creating another list of signal strengths for each range cell increment. This ongoing process results in a digitally stored table of signal strengths against range and azimuth angle. This process is illustrated in Figure 1.5.

The main diagram is a plan view of an area with the ship's position at the centre, showing its heading as a vertical arrow. It looks rather similar to a radar display, but in this case it is solely representing the actual geographical situation. Each cell increment in the azimuthal direction is depicted as a radial line and each increment in the radial direction as a circle, centred at the own ship's position. The actual increment size is chosen so that a point target would be detected in a number of adjacent range and azimuth sampling points, taking into account the beamwidth of the antenna and the length of the transmitted pulse. This means, in practice, that there are many more azimuth cells than are depicted in Figure 1.5, typically 1,024 or more covering the full 360°. Also, the length of a range cell is typically measured in tens of metres but depends on the chosen pulse length, see Section 2.3.3.1. The reflected signal strength measurement is centred at the crossing points of these lines and circles. The radar stores them as a table of values of signal strengths (amplitudes), which is also depicted in Figure 1.5. The illustrated table uses realistic values, including the depiction of signal strength. Signal strength, in this example, is based on a scaled value of between 0 and 1,023.

To display the image the radar's digital processor has to convert the ranges and azimuths of the measurements to 'x' and 'y' coordinates, relative to the own ship, using the simple mathematical concept illustrated in Figure 1.6. After scaling the x and y positions to allow them to be represented on the radar display, the received echoes are indicated at their equivalent position by spots of appropriate colour and intensity, depending on the received signal strength. At every revolution of the antenna the stored data, and hence the resultant radar image, is updated. This process produces the conventional image of displayed radar targets, such as that illustrated in Figure 1.7.

Today, the radar display is a conventional 'flat-panel' electronic screen, similar to that used on modern TVs and computer displays. The technology lying behind flat-panel displays is discussed in Section 2.8.3. The radar image is conventionally shown within a circular domain, with a radius equivalent to the selected maximum display range. This is no longer compulsory under the international regulations that govern marine radar but remains a widespread practice, reflecting the fact that original radar cathode ray tube (CRT) displays were circular rather than rectangular. These used to epitomize a radar installation, see Figure 1.8. There is always a bearing scale shown around the periphery of the *operational display area*, whether circular or otherwise. On older radars this used to be engraved on the rim surrounding the CRT. Nowadays, the scale is produced electronically and forms part of the displayed image. The bearing scale is labelled in degrees.

Also displayed are range circles centred on the origin (position of the own ship). These are known as range rings and can be set to convenient values by the operator, see Figure 1.9. This shows the radar on a 12 NM scale, with range rings set at 2 NM spacing. The rings can be switched off, if not required. In addition there are tools that enable a user to accurately determine the range and bearing of any target on the display. These are fully discussed in Section 6.9 and, for instance, include a variable range marker (VRM), as also illustrated in Figure 1.9.

The normally circular operational display area of a radar is a useful means of assessing quickly whether any particular display on a ship's bridge is set up as a radar or as an electronic chart. The latter image is generally

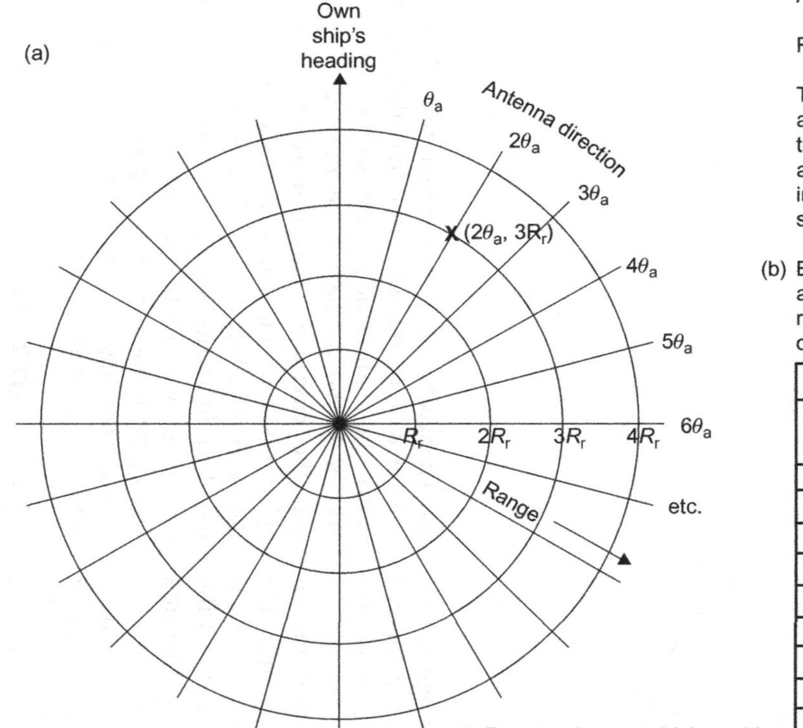

Azimuth cell increment = θ_a, e.g. 0.3°

Range cell increment = R_r, e.g. 10 metres

The amplitude of the received signal at each azimuth and range cell increment, such as at the point X ($2\theta_a$, $3R_r$) in the diagram, is measured and temporarily stored by the radar. This results in a table of values of the amplitude of the received signal at every cell.

(b) Extract from example table, showing possible target at about (12.0° , 3,762 metres) and only small reflections, such as from waves, in the vicinity of (12.3°, 2,400 metres):

Cell value		Amplitude
θ (Degree)	R (Metres)	(e.g., max value 1,024)
---	---	---
12.0	3,750	105
12.0	3,760	904
12.0	3,770	495
12.0	3,780	75
---	---	---
12.3	2,390	20
12.3	2,400	14
12.3	2,410	25
12.3	2,420	16
---	---	---

FIGURE 1.5 Creating a table of received signal strengths: (a) plane view of area with the own ship's position at centre and (b) table of amplitudes.

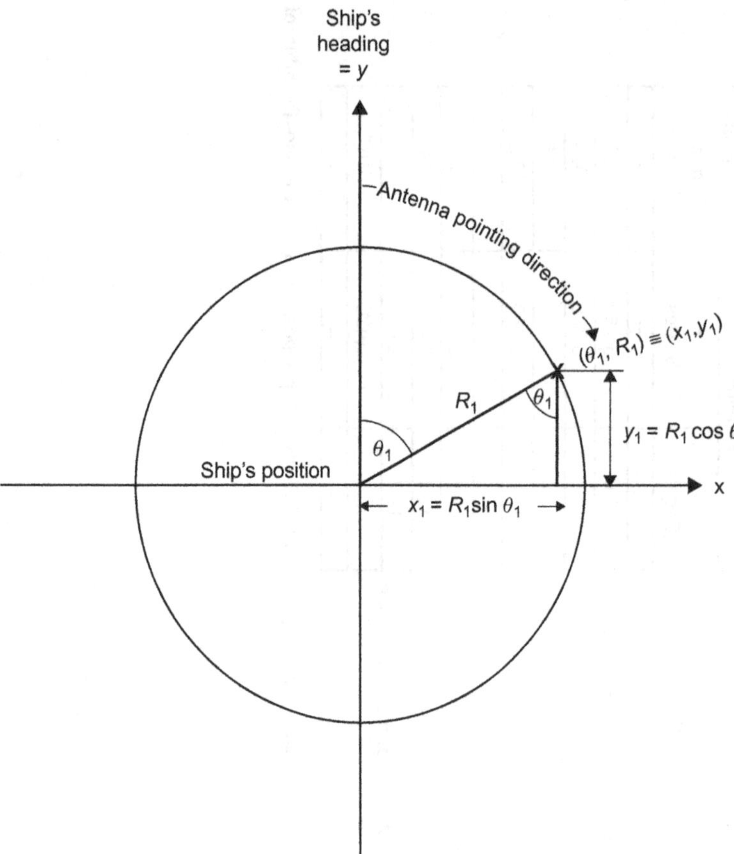

FIGURE 1.6 Conversion of range and azimuths to (x,y) Cartesian coordinates.

displayed as a rectangle, even though it may have radar data included in the displayed image. Section 10.2 explains the significant differences between viewing and using radar-derived data on a radar display (PPI) and on an electronic chart. In general, the simple rule is that fundamental decisions concerning collision avoidance should always be made on the radar display, but the main route monitoring activity should be using the chart display.

The term *PPI* will perhaps cease to be used over time, especially with the increased use of *multifunction displays*, which can be set to be used at any one time as a radar, electronic chart or other navigational display. However, this book will use the terms PPI and radar display interchangeably, as is in common usage at the time of writing. Into the future the likely trend is that the main radar display will increasingly become known as the *collision avoidance display* and the electronic chart display (when not being used for route planning) as the *route monitoring display*. When radar data is being used for other functions, such as position fixing or assistance with route monitoring, these task will be performed on the appropriate display modes, showing relevant radar information as well as other available data.

1.2.4.3 *Target Trails*

It is very often useful on a radar display for the past track of targets to remain visible, at

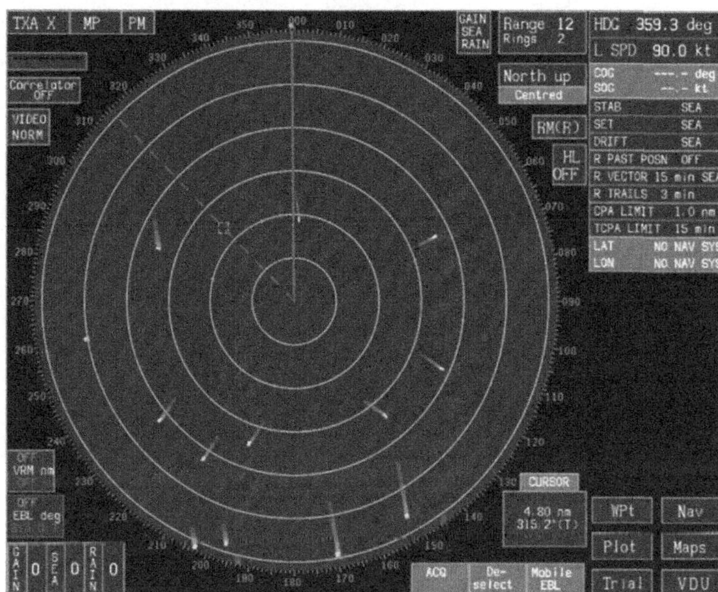

FIGURE 1.7 Displayed radar targets. *Figure courtesy of Kelvin Hughes.*

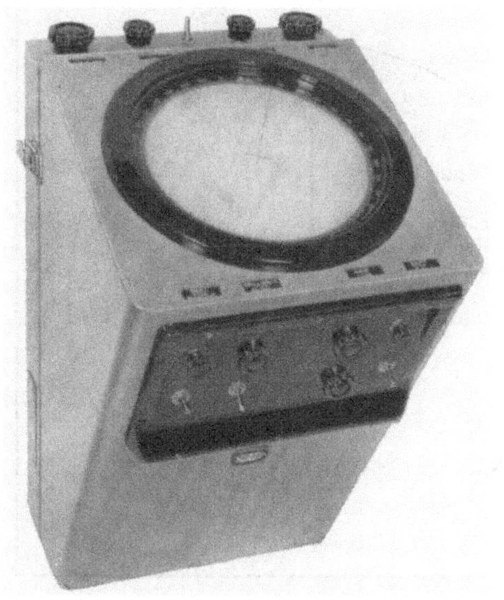

FIGURE 1.8 Older radar display. *Figure courtesy of Kelvin Hughes.*

least for a few minutes. This can give a much clearer visualization of the movement of critical targets. Targets are said to leave a *trail* on the display. Originally this feature was achieved by using CRTs with a very high image persistence. Any instantaneous image on the display only slowly faded because of the specially chosen phosphors used on the

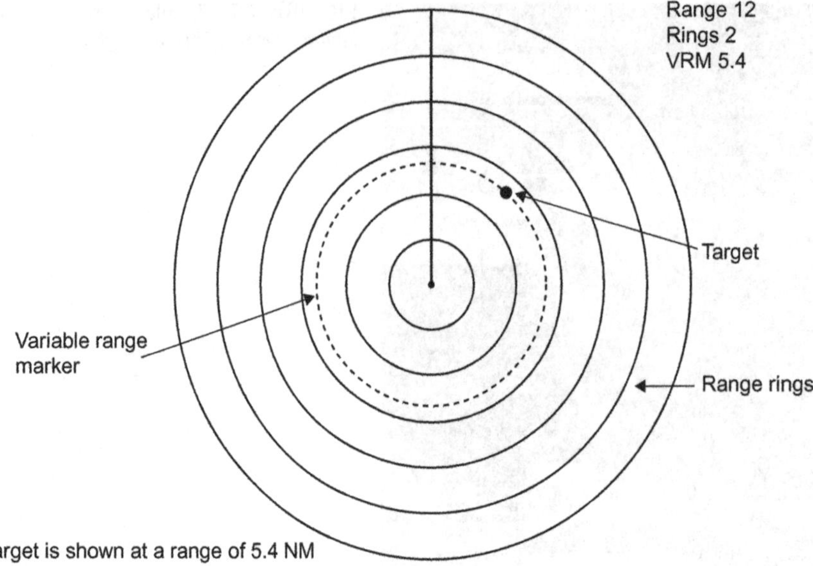
Range 12
Rings 2
VRM 5.4
Target
Variable range marker
Range rings
A target is shown at a range of 5.4 NM

FIGURE 1.9 Radar range rings and VRM.

display surface of the CRT. Consequently, targets would create a line on the display, showing their past positions. Close to the most recent position of the target the trail would be bright and would gradually fade to being invisible further along its length, see Figure 1.10. Nowadays this effect is artificially created by digital processing of the displayed radar image. This allows greater flexibility in the display of trails, such as their time length to extinction and whether or not they are displayed. It also more clearly distinguishes between targets and trials, for instance, by the use of different colours.

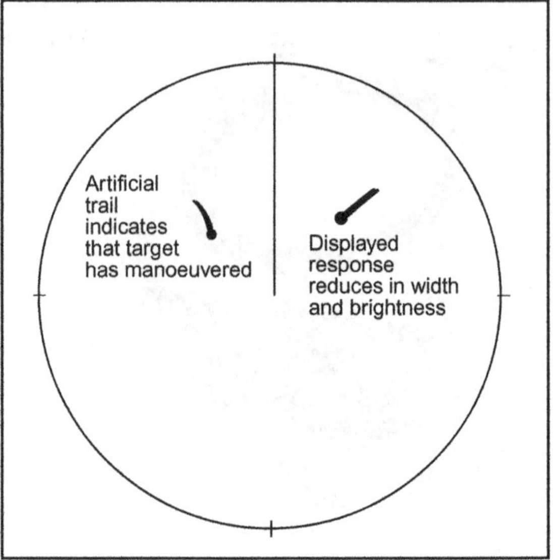
Artificial trail indicates that target has manoeuvered
Displayed response reduces in width and brightness

FIGURE 1.10 Target trails.

1.3 PRINCIPLES OF BEARING MEASUREMENT

1.3.1 The Heading Marker

In general, a bearing is the angle between the direction of a chosen reference and that of an object of interest. On a PPI display the fundamental reference is the instantaneous direction of the observing vessel's heading.

As the axis of the beam of the radar antenna crosses the ship's fore-and-aft line in the forward direction, a sensor within the turning mechanism of the radar antenna is activated and the associated electronics sends a timing pulse to the radar receiver. This pulse is used to synchronize the display electronics to the antenna rotation and, in particular, is used as the reference for the *heading marker* or *heading indicator*. In addition, *azimuth pulses* are generated at regular angular increments as the antenna rotates to take into account its potentially uneven rotation due to wind, vibration and vessel motion effects. Thus all targets are displayed, not only in the correct angular relationship to one another, but also in the correct angular relationship to the own ship's heading (see Figure 1.11).

The angle between the observed vessel's heading and the direction of the horizontal beam is sometimes called the *antenna angle*. IMO Performance Standards (see Section 11.2.1) require that the heading marker is able to be aligned to within 0.1°. The procedure for checking this accuracy is discussed in Section 6.6.8. There is a danger that a target may be masked if it lies in the direction of the heading marker. The specification recognizes this danger by requiring that there is a provision for temporarily switching the marker off. However, it is such an important feature on the radar display that it cannot be permanently switched off. In particular, the appearance of the heading marker confirms the orientation of the display (see Section 1.4).

A modern radar may also be able to display a stern line, drawn on a reciprocal bearing to the heading, which can be very useful when manoeuvring astern. This line can be switched on or off, as required. The heading line remains visible when the stern line is selected.

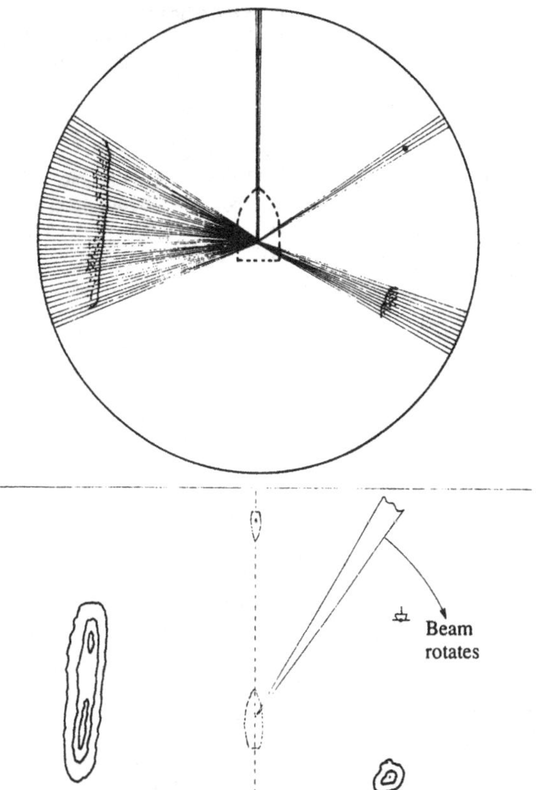

FIGURE 1.11 The build-up of the picture.

1.3.2 Bearing Measurement

IMO Performance Standards (see Section 11.2.1) require that provision be made for quickly obtaining the bearing of any object whose echo appears on the display. Traditionally this was fulfilled by a variety of mechanical and electromechanical devices which enabled the observer to measure the angle between the heading marker and the object of interest. On a modern radar, electronic bearing lines (EBLs) are used for this measurement. In particular, these are designed to be able to quickly determine the bearing of a target with respect to the own ship's heading. In the basic setting of the radar the EBL emanates

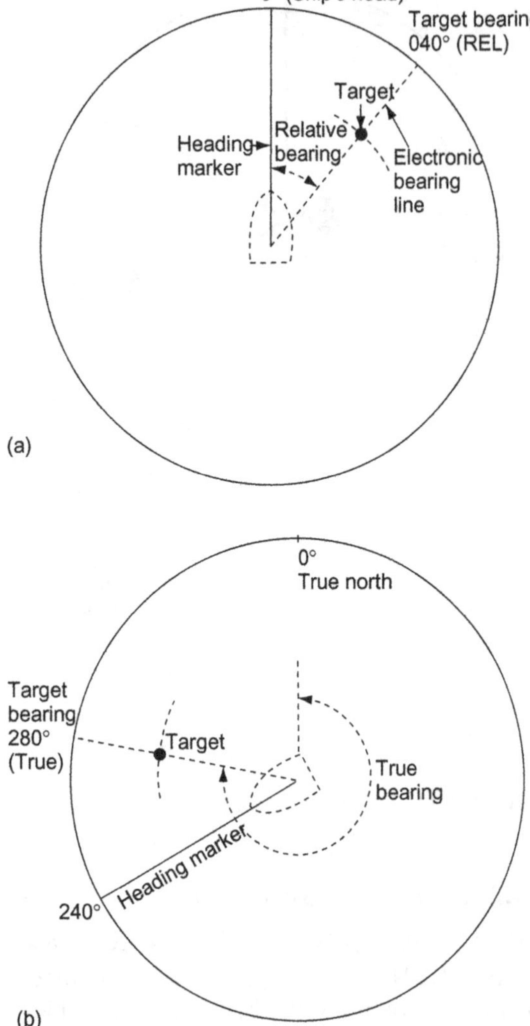

FIGURE 1.12 Measurement of bearing: (a) relative bearing and (b) true bearing.

from the centre (origin) of the display, there will the own ship's position, to the edge of the operational display area, where its angular position can be read off from the *bearing scale* around the periphery of the area. Using the appropriate controls the operator can orientate the EBL such that it passes through the target of interest. This is illustrated in Figure 1.12(a).

In addition to the bearing scale facility there will also be a numerical readout of the bearing on the display, which is more typically used nowadays rather than the bearing scale. The bearing scale mainly helps the user to keep an appropriate awareness of bearing. A variety of bearing measurement facilities and the correct procedure for their use are discussed in Section 6.6.

1.4 DISPLAY MODES

There are a number of display modes on a modern radar for determining exactly how the radar shows information in relation to the outside world. These cover three different areas, which are referred to as orientation, motion and stabilization modes. The *orientation mode* defines how the 'vertical' direction of the display aligns with the outside world horizontal (azimuthal) direction; the *motion mode* defines how the own-vessel moves with respect to the display; and the *stabilization mode* defines how absolute movement is referenced – relative to the ground or relative to the sea. Taking the display to be a conventional graphical representation in x and y coordinates, it is the y-direction that is considered to be vertical and the x-direction as being horizontal.

1.4.1 Orientation Modes

A vessel's radar display provides a choice of orientation modes. The natural mode is the one already described where the heading direction of the ship, and therefore the heading line, is vertically upwards on the display. This is known as the *head-up* mode.

There are two other orientation modes available. One is termed *north-up*, where the vertical direction represents true-north and the other is *course-up*, where the vertical direction of the display represents the desired course of the vessel.

1.4.1.1 *Head-Up Orientation*

This orientation, where the heading marker is always vertical on the display, is illustrated in Figure 1.13. As the vessel's heading changes, so does the orientation of the displayed image – the image is *vessel stabilized*, aligning with the view from the bridge windows, but is unstabilized with respect to true-north. The figure shows the situation just before and after a course change. This was the only orientation mode available on very early marine radars because of cost and technological limitations. However, the only significant attraction of using the basic head-up mode today is that it does not need a working gyro or compass input to the radar, unlike the other orientation modes on a modern radar, north-up and course-up. These modes, described separately in the sections below, stabilize the orientation of the radar image. For this reason, head-up mode is often described as *unstabilized*. If compass problems are encountered its use may be essential and so needs to be fully understood.

The head-up unstabilized mode is superficially attractive because of the very fact that the displayed radar image corresponds directly with the scene as viewed through the wheelhouse window. A well placed display unit, close to the bridge windows and facing forwards, means that irrespective of whether the user is viewing the radar screen or looking forward through the wheelhouse window, objects on the starboard side of the ship will lie on the right of the display and those on the port side will lie on the left.

However, this orientation mode became generally little used after north-up stabilization was introduced on marine radars. This was for a number of reasons. Firstly, the head-up image of earlier radars could become very unclear when in head-up mode. The 'afterglow' trail of static targets, especially of extended targets such as land masses, could obliterate critical small moving targets when the image rotated. This is not such a serious problem on modern radars set to head-up mode because of the digital processing technology now employed. Secondly, small yawing movements of the vessel create corresponding oscillations in the orientation of the radar image, which can make precise target range and bearing measurements difficult. This generally remains an issue, even on a modern radar set to unstabilized head-up mode. The third issue is that the bearing scale on an unstabilized head-up radar is not true-north related, and therefore creates extra work in establishing the true bearing of targets.

A particular reason for north-up mode becoming so frequently used was the general attractiveness of using an orientation which matches that of the paper chart, since it considerably benefits situation awareness. In fact, with the advent of electronic charts, which can also be displayed in head-up mode, the use of a head-up orientation mode potentially becomes more attractive. Before the era of electronic charts the use of head-up mode was mainly confined to special situations, such as when negotiating rivers, estuaries, narrow channels and locks, or when no compass interface was available. While course-up mode, described in Section 1.4.1.3, is a good alternative, many radars have an advanced head-up mode that is generally called *stabilized head-up*. This uses the gyro/compass input to orientate the bearing scale such that the heading direction is referenced relative to true-north, together with any other indications of bearing on the radar display. Smart processing can also prevent small yawing motions of the vessel creating an oscillating image, generally allowing targets, measurements to be easily performed, and also improving the clarity of the display.

It should be borne in mind that in both stabilized and unstabilized head-up mode, an unwary or poorly trained observer can be misled by the angular rotation of the display as

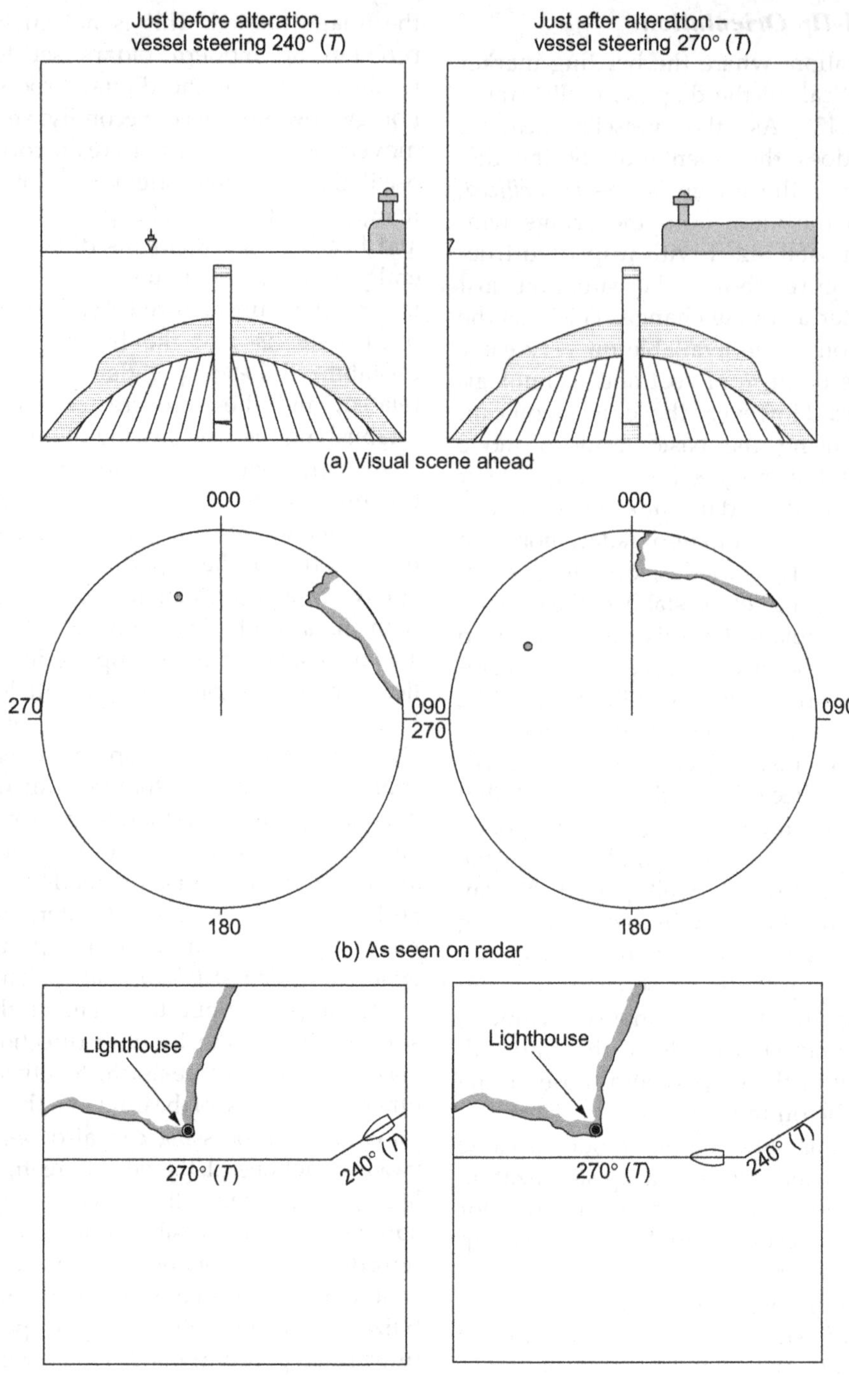

FIGURE 1.13 Head-up orientation (unstabilized).

the own-vessel heading changes. For example, a small change of course by the observing vessel may give the impression that the bearing of a target is changing, while in fact the true bearing is remaining constant. The extremely important topic of systematic observation of target movement is discussed at length in Chapter 7.

1.4.1.2 North-Up Orientation

In north-up orientation, the heading marker is aligned with the graduation on the bearing scale that corresponds with the instantaneous value of the ship's heading relative to true-north. It means that 000° on the bearing scale aligns with true-north. Thus the observer views the picture with north at the 'top' of the screen and it is for this reason that the orientation is so named. Figure 1.12(b) shows the same situation as that displayed in the head-up mode in Figure 1.12(a) but with the system set to north-up, assuming that the own ship is on a heading of 280°. Compass stabilization is essential to maintain north-up orientation, not least when the observing vessel alters course or yaws about its chosen course (Figure 1.14, which compares the cases for head-up, north-up and course-up operation). The stabilization signal can be derived from any transmitting compass, but in practice the signal source is often a gyro compass, which is compulsory for larger vessels. The principles of north-up orientation are illustrated in Figure 1.15.

A major benefit is that the orientation compares directly with that of the paper chart. Also, because the display is stabilized it removed the significant disadvantage of earlier radars that changes in heading caused significant blurring of the radar displayed image when in head-up mode. These two factors have led to north-up mode becoming the most commonly used orientation option on most vessels. It also remains relevant when using electronic charts in north-up mode. Some users find using electronic charts and radar in north-up

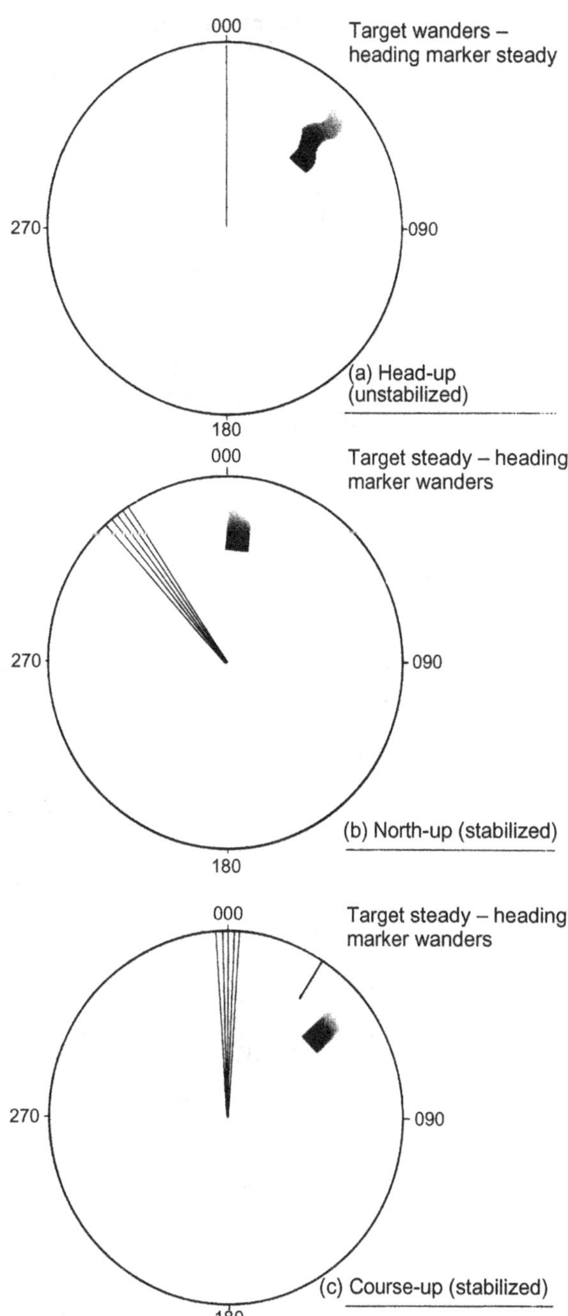

FIGURE 1.14 Target trails and the effect of yaw.

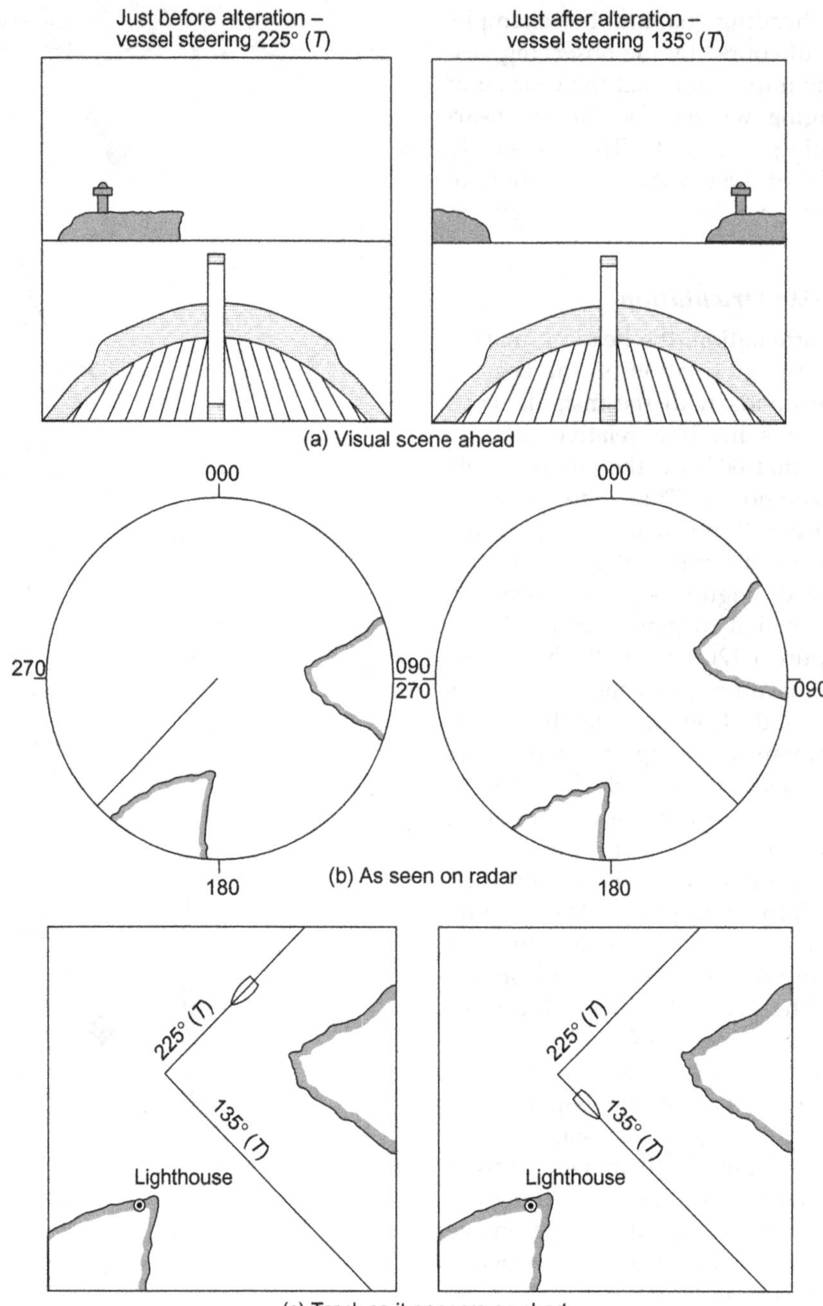

FIGURE 1.15 North-up orientation (stabilized).

preferable, as it aligns both the radar and the chart image with the mind image they have of the area, easing situation awareness. For others, especially when on a southerly course, they find north-up awkward or uncomfortable to view as it appears 'upside down'.

1.4.1.3 Course-Up Orientation

In course-up orientation the vertical direction on the display is aligned to the bearing which represents the desired course of the vessel. This can be obtained either automatically or semi-automatically from route planning information stored within the radar or by the operator selecting a particular course. By virtue of the compass stabilization, changes in the vessel's instantaneous heading are reflected by sympathetic angular movements of the heading marker, thus maintaining the ship's course (the reference course) in alignment with the display's vertical direction. For the same reason, the angular wander of echoes associated with an unstabilized display is eliminated. On modern radars the bearing scale will be relative to true-north, but older radars may have the vertical direction always shown as 000°, representing the desired course. Figure 1.16 illustrates course-up orientation.

Provided that the observing vessel does not stray very far from her chosen course, this orientation can be more effective than a stabilized head-up orientation because it eliminates all angular wander of the picture due to yaw, while maintaining the heading marker approximately vertical on the display. Inevitably a major alteration of course will become necessary either due to the requirements of collision avoidance or to those of general navigation. When the vessel is steadied on the new course the orientation, although not meaningless, will have lost its property of being substantially head-up. The problem is that the orientation is still *previous-course-up* and the picture should be re-oriented to align the heading marker to the vertical direction of the display (see Figure 1.16(d)).

1.4.1.4 Choice of Orientation

The fundamental function of any civil marine radar is to provide a means of measuring the ranges and bearings of targets for collision avoidance and the determination of the observing vessel's position in order to ensure safe navigation. The ease with which these objectives can be achieved is affected by the choice of orientation. Where the various techniques of collision avoidance and navigation are described in this text, appropriate attention will be given to the influence of orientation. The practical use and setting up of orientations is discussed in Chapter 6. Table 1.1 summarizes the essential features of the three described orientations.

Except in emergency situations, when azimuth stabilization has been compromised by equipment failure, head-up unstabilized orientation has nothing to offer other than its subjective appeal, because by its very nature it regularly disrupts the steady-state condition conducive to measurement of bearing and tracking of echo movement (see Figure 1.14(a)). The stabilized north-up and course-up orientations do not exhibit this angular disruption and hence are equally superior in fulfilling the fundamental requirements. Fortunately they are complementary in that while one is north-up, the other is orientated in such a way as not to alienate the user who has a ship's-head-up preference. On some radar systems, stabilized head-up orientation may be included as an alternative to the use of course-up mode.

1.5 MOTION AND STABILIZATION MODES

There are two motion modes, known as relative motion and true motion. Relative means relative to the own ship, while true means

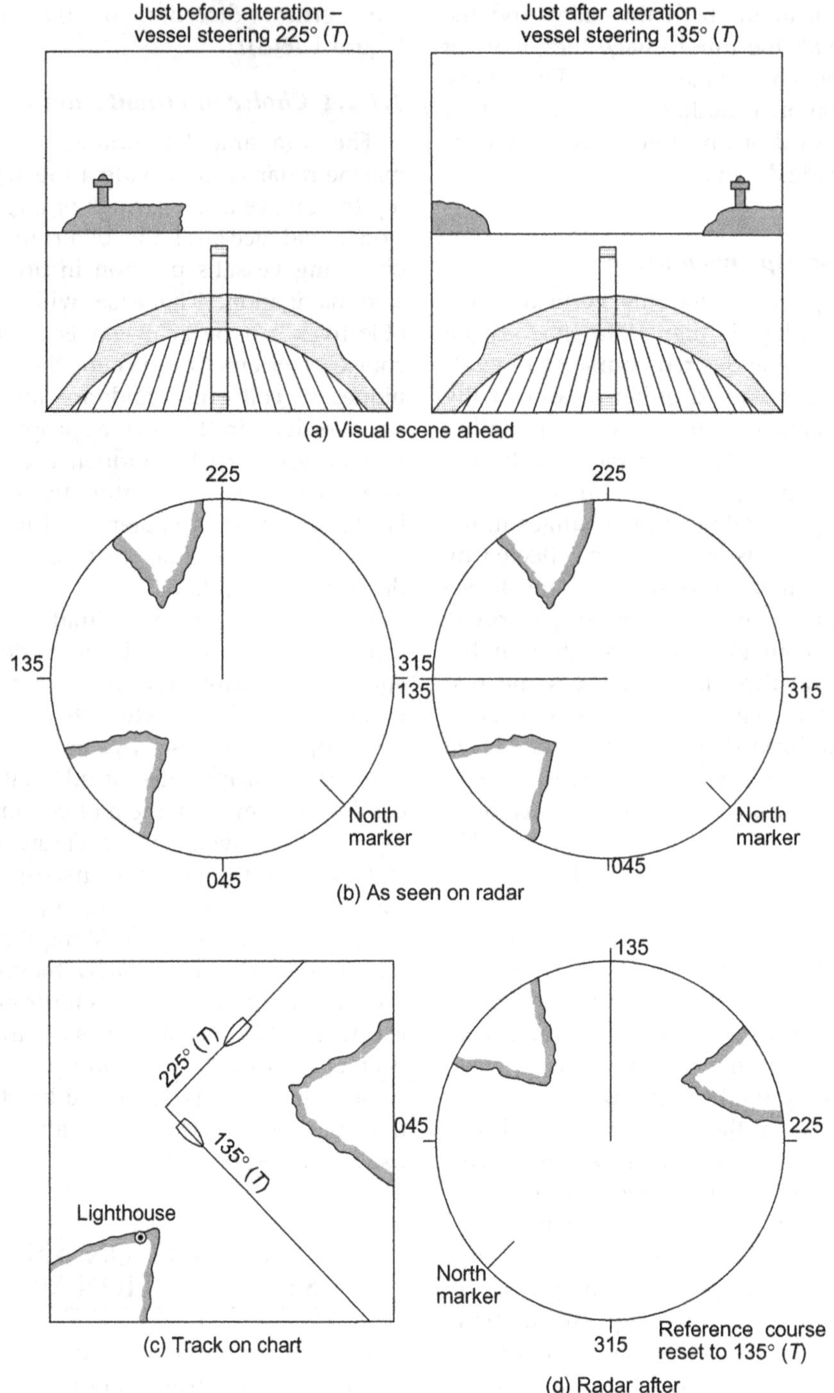

FIGURE 1.16 Course-up orientation (stabilized) – resetting the reference course.

TABLE 1.1 Picture Orientations Compared

Feature	Orientation		
	Head-Up, Unstabilized	**North-Up, Stabilized**	**Course-Up, Stabilized**
Blurring when observing vessel yaws or alters course	Yes: can produce very serious masking	None	None
Measurement of bearings	Awkward and slow	Straightforward	Straightforward
Angular disruption of target trails when observing vessel yaws or alters course	Yes: can be dangerously misleading	None	None
Correspondence with wheelhouse window view	Perfect	Not obvious	Virtually perfect except after large course change
Correspondence with chart	Not obvious	Perfect	Not obvious

relative to an outside reference system. The outside reference is split into two stabilization modes – ground stabilized and sea stabilized. Ground stabilization requires an external sensor signal that at least can determine the speed over ground (SOG) of the observing vessel. In today's world this is typically based on the output of a Global Navigation Satellite System (GNSS) using data from the US Global Positioning System (GPS) and/or Russia's Globalnaya Navigatsionnaya Sputnikovaya Sistema (GLONASS). Other systems can also give this information, as discussed in Section 6.9.6. When sea stabilized, the important sensor is the speed log, which measures the vessel's speed through the water (STW).

1.5.1 Relative-Motion Presentation

In relative-motion presentation the origin of the display, which always represents the effective position of the observing vessel, is stationary. Commonly, the origin is located at the centre of the display circle, but the user can move this to a position of choice to better meet the needs of the actual situation. The essential feature is that the origin is stationary and as a consequence targets exhibit their motion relative to the observing vessel. The setting of relative motion is independent of the chosen orientation mode. Also, in terms of where the targets appear on the display at any one time, it is immaterial whether ground or sea stabilization has been set, simply because all positions are referenced to the observing vessel. The essential features of a relative-motion presentation are best illustrated by an example. In this example it is assumed that any leeway is negligible.

Consider the case of an observing vessel on a steady heading of 000° (*T*) at a speed of 10 knots through the water in a tide (which is uniform throughout the area) setting 270° (*T*) at a rate of 4 knots. For this case the basic PPI view would look very similar, whether the orientation mode was set to head-up (unstabilized or stabilized), course-up or north-up. A simplified chart of the situation is illustrated in Figure 1.17(a), showing four targets:

i. Vessel *A* which is located 7 NM due north of the observing vessel and is stopped in the water heading 045° (*T*).
ii. Vessel *B* which is located 8 NM due east of vessel *A* and is on a steady heading of 270°

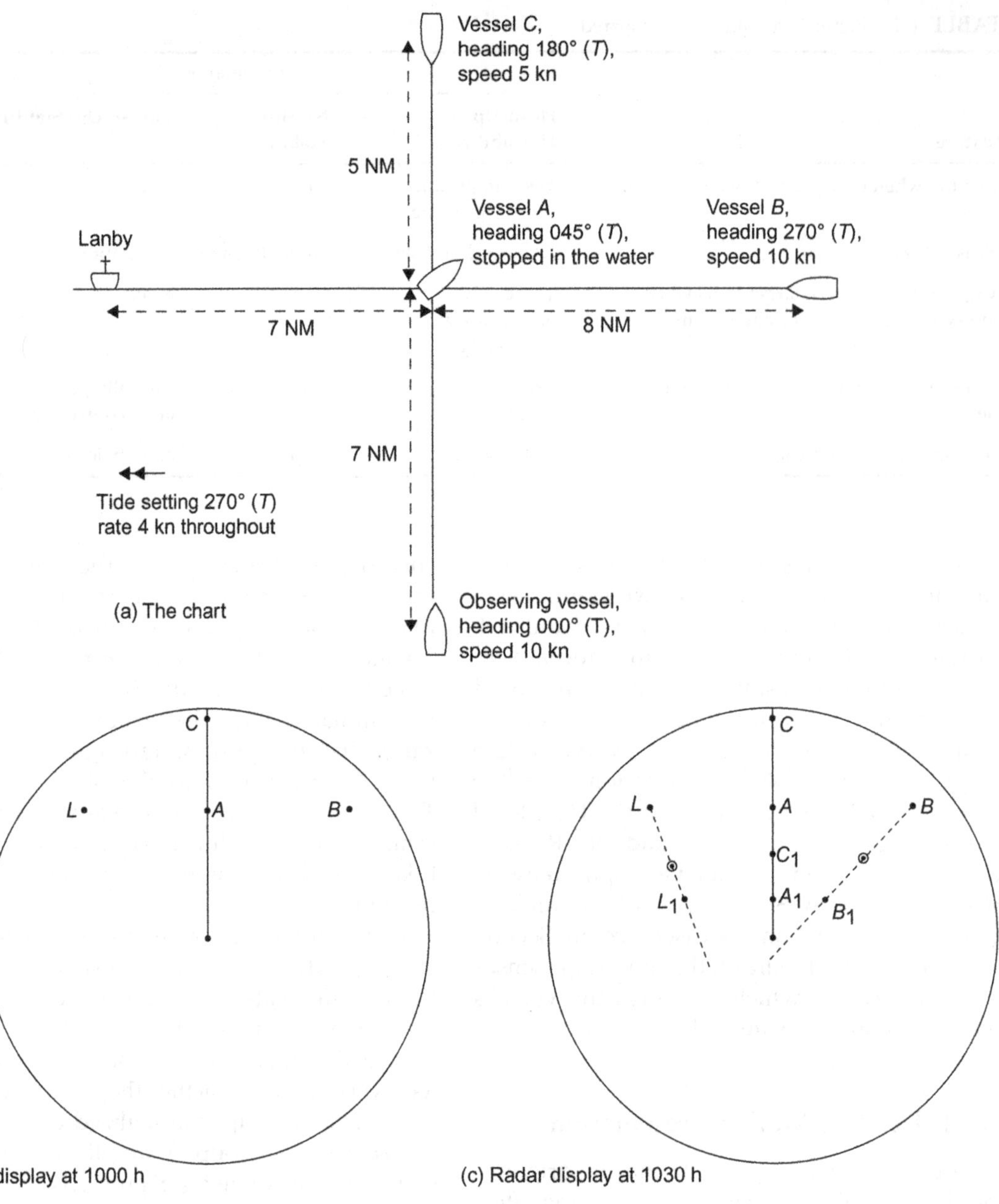

(b) Radar display at 1000 h

(c) Radar display at 1030 h

FIGURE 1.17 Relative-motion presentation.

(*T*) at a speed of 10 knots through the water.

iii. Vessel *C* which is located 5 NM due north of vessel *A* and is on a steady heading of 180° (*T*) at a speed of 5 knots through the water.

iv. A large navigational buoy *L* which is anchored and therefore, for the purposes of this example, can be considered to be stationary over the ground. Its position is 7 NM due west of vessel *A*.

To assist in the understanding of relative motion, Figure 1.17(b) represents the observing vessel's PPI as it would appear at 1000 h. For comparison, Figure 1.17(c) represents the same PPI showing the positions of the echoes as they would appear at 1030 together with a record of their 1000 positions. It will be noticed that the shape of the echoes normally gives little indication of the outline of the targets, as explained in Section 2.8.5. Consider now the movement of each of the four echoes in turn, commencing with that of the water-stationary target *A*, which offers a simple basis on which an understanding of all relative motion can be built. It is important to remember the assumption that the observing vessel is maintaining a steady heading.

In the period 1000–1030 the observing vessel will move north by a distance of 5 NM through the water. Because the origin remains stationary, and the range of target *A* decreases at 10 NM/h (knots), it follows that the echo of *A* will move down the heading marker by a distance of 5 NM in the 30 min interval. This reveals the basic property of the relative-motion presentation which is that the echo of a target which is stationary in the water will move across the screen in a direction reciprocal to that of the observing vessel's heading, at a rate equal to the observing vessel's STW. Importantly, this is not generally the case if heading is replaced by course over ground (COG) and STW is replaced by SOG.

Consider now the movement of the echo of vessel *B* which at 1000 was 8 NM due east of the stationary vessel *A*. As *B* is heading directly towards *A* at 10 knots, it follows that its 1030 position will be 3 NM due east of *A*. Figure 1.17(c) reveals that the trail left by the echo of vessel *B* offers an indication of how far off the target will pass if neither vessel manoeuvres. However, the echo has moved across the screen in a direction and at a rate which is quite different from the target's course and speed. An appreciation of this fact is absolutely essential if the basic presentation is to be interpreted correctly and used in assessing collision avoidance strategy. (In practice, target tracking vectors would be used to make such decisions, as discussed in detail in Section 4.4 and Chapter 7) Further consideration of the figure will show that the relative motion of echo *B* is the resultant of that of a water-stationary target (which is determined by the observing vessel's course and STW) and the true motion of the vessel *B* through the water. An analogous argument can be based on ground referenced motions. The proper use of radar for collision avoidance is based on systematic observation and analysis of both the relative motion and the true motion of the other targets in an encounter (see Chapter 7).

Consider now the movement on the screen of the echo of vessel C. At 1000 its position was 5 NM due north of the water-stationary vessel *A* and heading directly towards it at 5 knots. It follows that at 1030 its position will be 2.5 NM north of vessel *A*. As shown in Figure 1.17(c), because the echo of vessel *A* has itself moved across the screen by 5 NM in a direction of south, the aggregate movement of echo *C* is 7.5 NM in the same direction. Thus, as in the case of vessel *B*, the echo has moved across the screen in a way that is different from the movement of the vessel through the water. However it should be noted that, by coincidence, the track across the screen of echo *C* is in the same direction as that of the water-stationary target *A*. This reveals a further feature of the relative-motion presentation, which is that the echoes of targets which are stopped in the water, targets which are on a reciprocal course to the observing vessel and targets which are on the same course as the observing vessel, but slower, will all move across the screen in the *same* direction (but at different speeds). This feature has the potential to mislead the untrained or unwary observer into confusing, for example, a target that is being overtaken with one that is on a reciprocal course. This further emphasizes the necessity of having a good understanding of these basic principles and a systematic approach when using the radar for collision avoidance (see Chapter 7).

Initially the east/west distance between the buoy and the stationary ship was 7 NM. As the tide is setting the stationary vessel down on to the buoy at 4 knots, it follows that this distance will have reduced to 5 NM by 1030. A study of Figure 1.17(c) will show that the echo of the buoy has moved across the screen in a direction which is the reciprocal of the observing vessel's ground track at a speed equal to the speed of the observing vessel over the ground. This property is exploited in the use of radar for navigation (as opposed to collision avoidance); the various procedures are set out in Chapter 8.

1.5.2 The True-Motion Presentation

It has been shown that in a relative-motion presentation the movement of all echoes across the screen is affected by the course and speed of the observing vessel. In a correctly adjusted true-motion presentation, the echo movement of all targets is rendered independent of the motion of the observing vessel. This is achieved by causing the origin of the picture to track across the screen in a direction and at a rate which corresponds with the motion of the observing vessel. There is clearly a fundamental difference to the actual movement of the origin as to whether ground or sea stabilisation has been set, although the displayed basic geometrical layout of targets with respect to the origin (but not its orientation or absolute position) always remains identical on the PPI. This remains true whatever the orientation, motion or stabilisation mode, simply because the 'world outside' is obviously not affected by the settings of the radar.

It is clear that after a period of time the origin – that is the position of the observing vessel – will move to the edge of the display. It then has to be reset, either by user intervention or by an automatic process set up by the user. Strategies for resetting are discussed in Section 6.2.6.3.

1.5.2.1 *True-Motion Sea-Stabilized Presentation*

To produce a true-motion sea-stabilized presentation, the origin of the picture must be made to track across the screen in a direction and at a rate that corresponds with the observing vessel's course and STW. In the example in the illustration (Figure 1.18) the course is 000° (*T*) and the speed is 10 knots.

Figure 1.18(b) shows the PPI of the observing vessel as it would appear at 1000. The origin of the picture is offset in such a way as to make optimum use of the available screen area (see Section 6.2.6.3). Figure 1.18(c) shows the position of the four echoes as they would appear at 1030, together with an indication of their 1000 positions for the purpose of comparison. The movement of each of the four echoes will now be considered in turn, commencing with target *A* which is stopped in the water.

In the interval 1000–1030 the origin will move due north by a scale distance of 5 NM, while in the same time target *A* will remain on the heading marker but its range will decrease by 5 NM. It follows that the net motion of the echo of target *A* will be zero. Consideration of Figure 1.18 reveals the basic property of a correctly setup true-motion sea-stabilized presentation, which is that the echo of a target which is stationary in the water will maintain a constant position on the screen.

At 1000 the moving target *B* was located 8 NM due east of vessel *A*. As it is heading

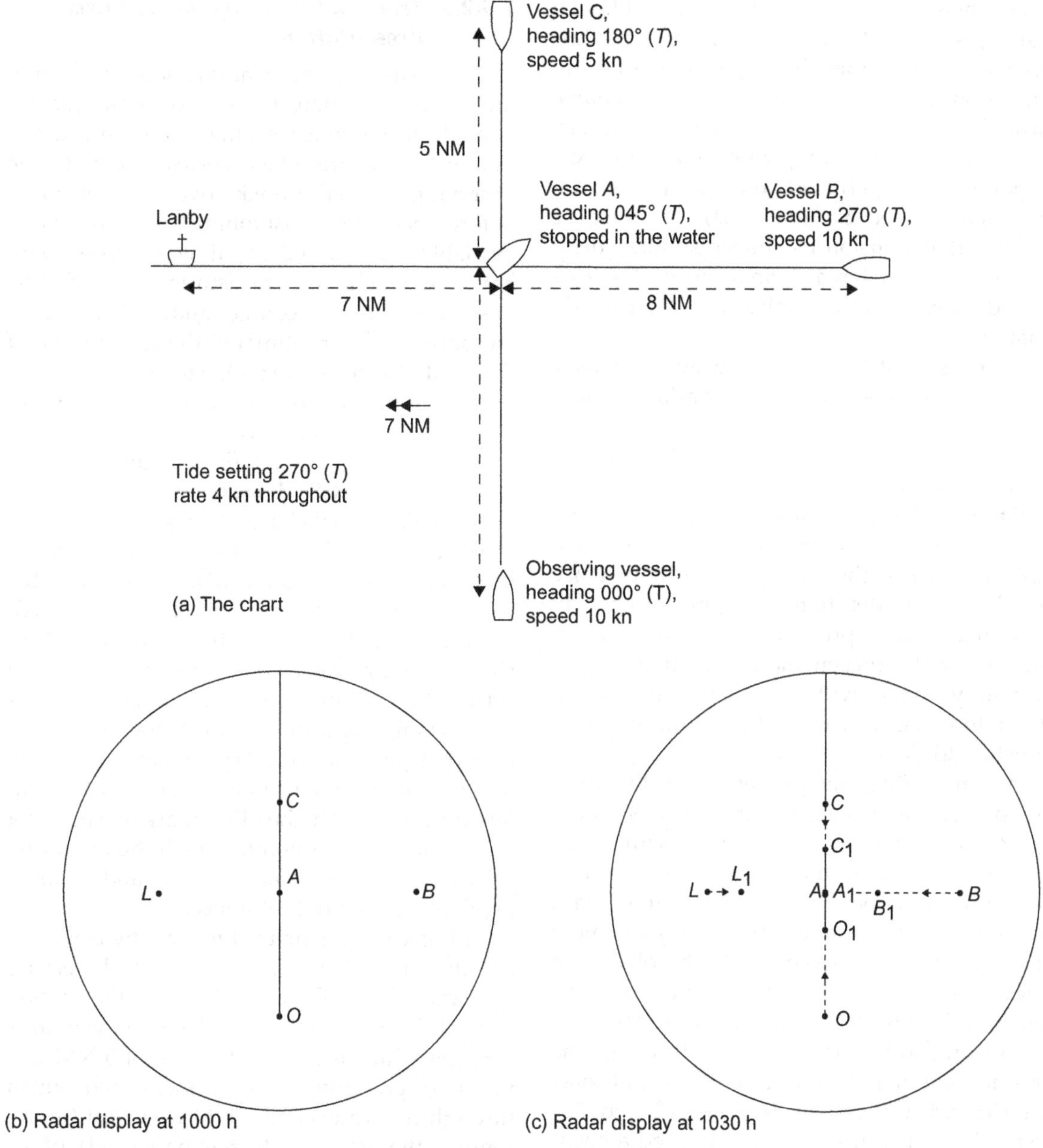

FIGURE 1.18 True-motion sea-stabilized presentation.

directly towards *A* its bearing from *A* will remain steady, but the range will have decreased to 3 NM by 1030. Figure 1.18(c) shows that the echo of target *B* will move across the screen in a direction and at a rate which corresponds with the target vessel's course and STW. A similar argument will reveal that the echo of vessel *C* will move across the screen in a direction of 180° (*T*) at a scale speed of 5 knots. The presentation thus has the property that the target trails offer an indication of the headings (actually course through the water – CTW) of all moving targets. This feature is complementary to the corresponding property of the relative-motion presentation (see Section 1.5.1). It must be stressed that collision avoidance strategy must be based on systematic analysis of the displayed target movements, as detailed in Chapter 7.

As a result of the tide, the water-stationary vessel *A* will be set directly towards the buoy and by 1030 the east/west distance between the two will have reduced to 5 NM. It has been established that echo *A* will maintain its position on the screen, and thus it follows that in the interval from 1000 to 1030, echo *L* will move east across the screen by a scale distance of 2 NM. Consideration of Figure 1.18(c) will show that a third property of the true-motion sea-stabilized presentation is that land-stationary targets will move across the screen at a rate equal to the tide but in the opposite direction to the set.

In considering the properties of the true-motion sea-stabilized presentation it is essential to appreciate that the accuracy with which the displayed target movements are presented is completely dependent on the accuracy with which the direction and rate of the movement of the picture origin represents the observing vessel's course and STW. The true-motion presentation is only as good as the input data.

The practical procedure for setting up the presentation and the effect of errors and inaccuracies are covered in Sections 6.2 and 7.9, respectively. Because the scenario used a heading of north for the observing vessel, the question of which orientation mode is in use is irrelevant. It should be noted that any orientation mode (except head-up unstabilized) can be used with true motion, irrespective of heading.

1.5.2.2 True-Motion Ground-Stabilized Presentation

To create ground stabilization of a true-motion presentation, the origin of the picture is made to move across the screen in a direction and at a rate which correspond with the observing vessel's track over the ground. Before accurate positioning systems were available, such as GNSS, it was necessary to have independent measurements of the observing vessel's course and STW, plus a measurement that estimated the set and rate of the tidal stream or current. This complexity of understanding is now not normally necessary as the PPI can effectively be considered to be referenced to a ground fixed coordinate system, such as WGS-84 used by GPS and electronic navigational charts (ENCs).

For comparison purposes it is convenient to illustrate the presentation with reference to the same scenario as was used in the two preceding examples. Figure 1.19(b) shows the PPI of the observing vessel as it would appear at 1000, while Figure 1.19(c) shows the echoes as they would appear at 1030 together with recorded plots of the 1000 positions. It is seen that this origin moves in a direction which differs from that of the heading marker. The latter represents the direction in which the observing vessel is heading at any instant and is independent of any tidal influence.

In this case it is helpful to start by considering the ground-stationary target *L*. Reference to Figure 1.19(c) shows that in the period 1000–1030 the origin will have moved to a position which is a scale distance of 5 NM due north (representing the vessel's movement through the water) and 2 NM due west (representing the set and drift experienced) of its 1000 screen location. In the same interval, the north/south distance between the observing vessel and the buoy will have decreased by 5 NM due north (representing the vessel's movement through the water) and 2 NM due west

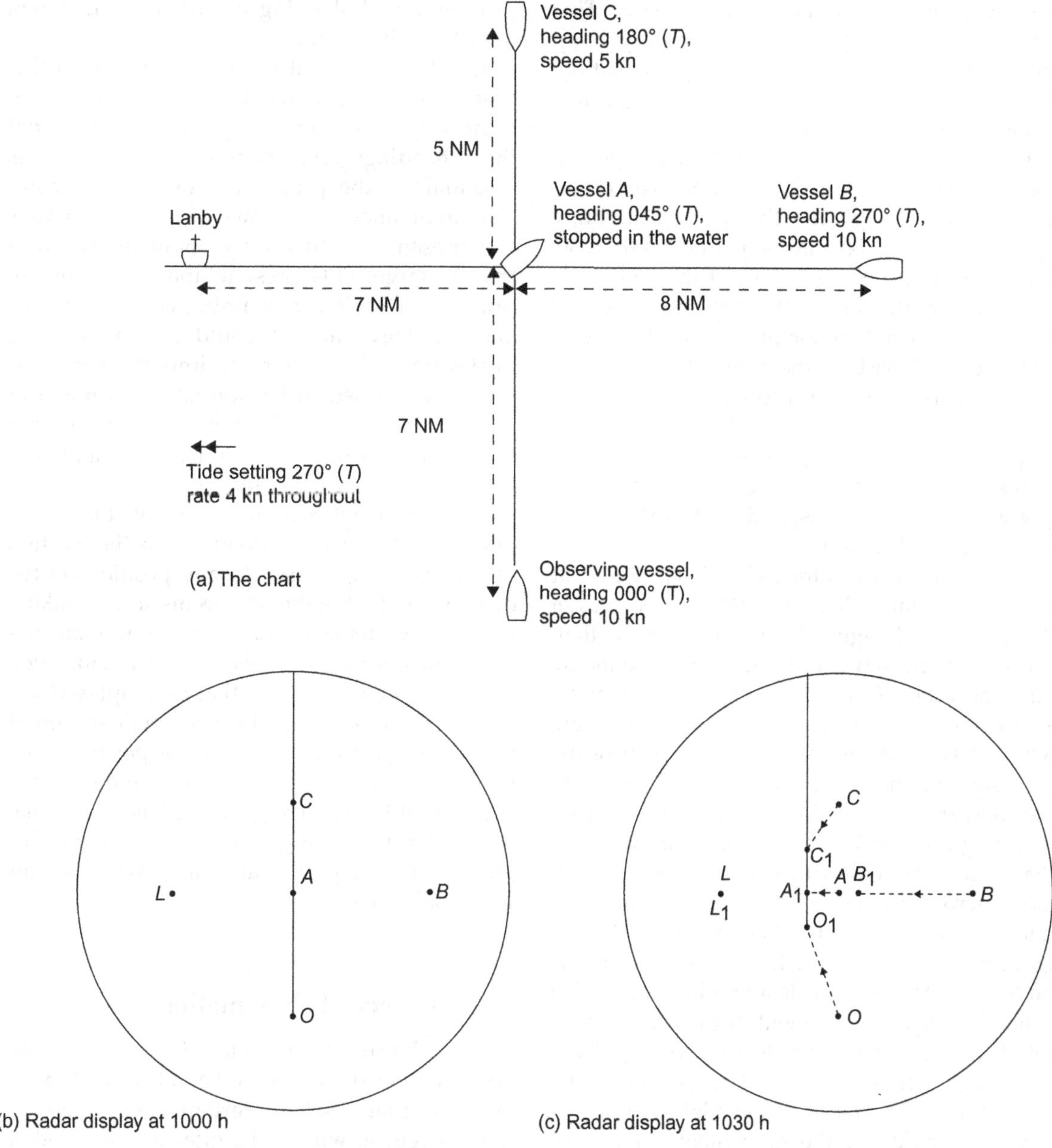

FIGURE 1.19 True-motion ground-stabilized presentation.

(representing the set and drift experienced) of its 1000 screen location. In the same interval, the north/south distance between the observing vessel and the buoy will have decreased by 5 NM while the east/west distance will have decreased by 2 NM. It follows from these two statements that the echo of the buoy L will exhibit neither north/south movement nor east/west movement across the screen. This reveals the key property of the correctly setup true-motion ground-stabilized presentation, which is that the echo of a land target will remain stationary on the screen.

Consider now vessel *A* which is stopped in the water heading 045° (*T*). As previously established, it will be set directly towards the buoy by a distance of 2 NM in the period 1000–1030. It follows from this, and the fact that echo *L* is stationary on the screen, that the echo of vessel *A* will move in a direction of due west by a scale distance of 2 NM in the interval considered. This reveals a second property of the true-motion ground-stabilized presentation, which is that the echo of a target which is stopped in the water will move across the screen in a direction and at a rate that corresponds with the set and rate of the tidal stream.

Having regard to target *C*, it is clear that at 1030 its position will be 2.5 NM north of vessel *A*. A study of Figure 1.19(c) will show that, because the movement of echo *A* represents the tide, the echo of vessel *C* will therefore move across the screen in a direction and at a rate which is the resultant of the set and drift of the tidal stream, and the vessel's course and STW. This shows the general property of the true-motion ground-stabilized presentation, which is that echoes of vessels that are under way will move across the screen in a direction and at a rate which represents their track over the ground. The example of the vessel *A* which is under way but stopped is a special case of this general rule. The movement of the echo of vessel *B* also illustrates this feature, but perhaps not so dramatically because it is heading in the same direction as the set of the tidal stream.

Consideration of the movements of echoes *A* and *C* in particular will emphasize how an untrained or unwary observer might be misled into *erroneously* concluding that vessel *A* was slow-moving and crossing showing a red sidelight (when in fact it is stopped in the path of the observing vessel), and that vessel *C* was a passing vessel showing a red sidelight (when in fact it is head on).

It is thus essential to appreciate that in this presentation the movement of the echoes of vessels that are under way does *not* represent their headings. Information on headings is essential for the proper use of radar in collision avoidance. It follows that, in principle, this presentation, if used without the use of a tracked target (TT) presentation, is not appropriate as a basis for planning collision avoidance strategy since it could be dangerously misleading. This extremely important topic is further discussed in Section 6.9.6. As noted in the previous section, it should be appreciated that either north-up or course-up orientation can be used with true motion.

In this mode it is common to have the option of being able to fix the origin, that is the position of the observing vessel, to any position on the display. At first sight this seems to be making the display identical to the relative-motion presentation discussed in Section 1.5.1. However, an important difference is that the displayed target trails will be shown in their original ground referenced position and not as a position relative to the observing vessel. This removes the problem of the changing 'surveillance area' that occurs when the origin moves across the display, which in particular can cause problems with 'look-ahead'.

1.5.3 Choice of Presentation

The choice of presentation made in any given circumstances will be influenced by a number of factors. To a certain extent, material to the decision will be the question of whether the radar is being used primarily for collision avoidance or for position fixing and progress monitoring. Because of some of the subtleties, it is appropriate to defer more detailed discussion of the factors affecting such a choice until after further consideration has been given to the operating principles of the radar system, the practical procedures for the setting-up and maintaining of the presentations, and the general philosophy of the use of radar for collision avoidance and navigation (see Chapters 6–8). However, at this stage it is useful, for comparison purposes, to summarize the major features of each presentation and comment briefly on its suitability for use in collision avoidance and navigation.

In the relative-motion presentation the echo movement of targets which are under way is that of the target relative to the observing vessel. Systematic observation of this movement readily offers a forecast of the distance off at which a target will pass (the closest point of approach or CPA) and the time at which the target will reach its closest point of approach (TCPA). This information is an effective measure of the risk of a close-quarters situation developing. The presentation, without the addition of target tracking vectors, gives no direct indication of the heading or speed of target vessels. Thus the relative-motion presentation gives a direct indication of some of the information required for collision avoidance, but the remainder must be found by deduction (see Chapter 7).

The echoes of land targets on a relative-motion presentation trace out a trail which is the reciprocal of the observing vessel's track made good over the ground. This feature is

TABLE 1.2 Presentations – Summary of Features (assuming no automatic plotting available)

Feature	Presentation		
	Relative-Motion	**True-Motion Sea-Stabilized**	**True-Motion Ground-Stabilized**
Ease of assessing target's CPA/TCPA from trails or hand plots	Directly available	Resolution required	Resolution required
Ease of assessing target's course, speed and aspect	Resolution required	Directly available	Resolution required – potentially misleading
Need for additional sensor inputs: course and speed	No, but course input desirable	Yes	Yes
Need for data on tide set and rate or ground speed and course	No	No	Yes
Displayed information relative to:	Observer	The water	The ground
Particular application for collision avoidance/ navigation	Partial contribution to collision avoidance data; ideal for parallel indexing	Partial contribution to collision avoidance data	Difficult to achieve without Automatic Radar Plotting Aid (ARPA) but if achieved provides stationary map
Limitations for collision avoidance	Target heading not directly available	CPA not directly available	No collision avoidance data directly available
Limitations for navigation	Movement of land echoes may hinder target identification	Limited movement of land echoes	None if stabilization effective

particularly useful in progress monitoring and position fixing when the radar is being used for navigation as opposed to collision avoidance (see Chapter 8).

The true-motion sea-stabilized presentation makes the headings and speeds of targets available directly. In respect of the use of radar for collision avoidance it can be seen that the relative-motion and the true-motion sea-stabilized presentations are complementary. The true motion does have the added advantage that it makes it very much easier to identify target manoeuvres and, further, the continuity of target motion is not disrupted when the observing vessel manoeuvres or yaws.

Table 1.2 summarizes the features of the presentation described above and is from a traditional hand plotting perspective or from observing the target trails on the screen.

Modern observers usually have the assistance of an ARPA or TT (target tracker) for ascertaining target data. This negates much of the discussion above as the continuous tracking and computing power means that true and relative target data (vectors) can be displayed at the 'push of a button'. Also, if looking at true data (vectors), the display of sea and ground stabilized data (vectors), if available, can be changed at a push of another button. This does not reduce the need for observers to be aware of the different properties of these different vectors, but it does mean the choice of presentation can be selected on modern equipment using less critical criteria.

The choice of presentation in modern ARPA equipment therefore often becomes an option in how the observer wishes to control the area being shown on the display. Relative-motion means that the centre spot stays in one place, with the observer having a choice of the centre spot being at the geometric centre or offset so more is seen ahead of a perceived dangerous or useful direction.

The standard true-motion options mean that the centre spot moves across the display with time so the area being covered ahead reduces. This can be perceived as a disadvantage, although the occasional resetting of centre spot backwards is a quick and straightforward option on modern equipment. An advantage is that a stationary feature on the display such as a conspicuous headland can be maintained on the display as the vessel passes it.

It still should be remembered that target trails are normally controlled by the choice of presentation, and this affects the interpretation of target manoeuvres.

This modern use of presentation controls with ARPA and TT facilities is explored more fully in Section 6.9.

CHAPTER

2

The Radar System – Technical Principles

2.1 INTRODUCTION

In Chapter 1 the fundamental principles of range and bearing measurement which underlie the generation of a marine radar picture were discussed. The way in which the target information is displayed for use in navigation and collision avoidance was also described in general terms. This chapter looks in more detail at the structure of modern marine radar systems. The treatment of the principles is substantially qualitative and does not need a detailed understanding of science or engineering, but some principles do require some basic mathematical treatment.

Much of the chapter describes the principles of conventional magnetron-based pulse radar systems, but Section 2.9 describes the principles of solid-state radar. An increasing number of marine radars are of this type. In general, it is easier to understand such radars after getting a firm understanding of pulse radar concepts, since many of these are directly applicable to 'coherent' solid-state radars.

2.2 BASIC FUNCTIONALITY

The block diagram in Figure 2.1 shows the main functional elements of a marine radar and the following subsections elaborate on each of these. The basic diagram is applicable to both pulse and solid-state radar systems, but only the former are discussed in this section.

2.2.1 Transmitter

The function of the transmitter is to generate the radiated electromagnetic energy. For a standard pulse radar the transmitter provides the correct radio frequency (RF) of the pulses, together with their repetition frequency, length, shape and power. Nowadays, the transmitter is often located upmast, close to the antenna, separated from it by the *rotating joint*. This allows the transmitted RF energy to travel between the typically static transmitter and the rotating antenna. On some systems the transmitter is below-deck and there is a *transmission line*, normally a *coaxial cable* at S-band and a *waveguide* at X-band, connecting it to the rotating joint of the antenna unit. A coaxial cable is one having a central conductor surrounded by insulating material together with an outer screening sheath, as illustrated in Figure 2.2(a). The traditional waveguide is hollow rigid copper tubing with a precise rectangular or circular cross-section (Figure 2.2(b)). Waveguide can also be made flexible, such as

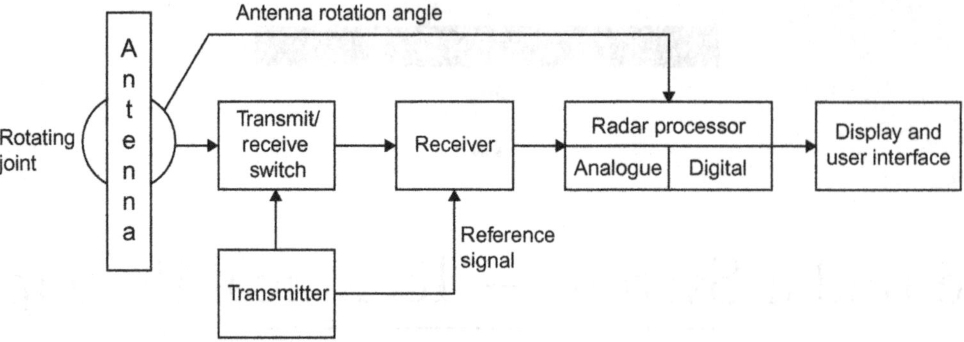

FIGURE 2.1 The main functional elements of a marine radar.

by corrugating the walls and using an elliptical cross-section. Waveguide is a technically good solution because it creates little loss in the transmitted or received signal. Typical runs can be up to 20–30 m in length. However, it is a very expensive and bulky transmission line, especially at S-band.

In both coaxial cable and waveguide the energy travels in the form of electric and magnetic fields, bounded by the conducting metallic surfaces that make up the walls of the transmission line. In a coaxial cable the insulating *dielectric* material between the central conductor and the outer screening sheath absorbs some of the transmitted energy creating inefficiencies. Air forms the insulating material within a waveguide, which is one reason that makes it low loss compared to a coaxial cable. In the earlier days of radar the dielectric loss in coaxial cable was immense, forcing the use of waveguide. Nowadays, the better dielectrics available favour the use of coaxial cable for marine radars, at least at S-band, when compared with the use of lower loss but very expensive and bulky waveguide.

In the block diagram of Figure 2.1 a connection is shown between the transmitter and the receiver. This supplies a common time reference so that, for instance, the time taken by a particular echo to arrive at the receiver can be accurately measured.

2.2.2 Antenna

Antenna, *scanner* and *aerial* are all names that are used to describe the device which radiates the radio energy into space and collects the returning echoes. Its construction defines the shape (power distribution) of the radar beam in both the horizontal and vertical planes. In order to achieve the required directional characteristic (see Section 1.2.3) the horizontal limits of the beam must be narrow, in the order of 1–2°. By contrast, the beam is wide in the vertical plane in order to maintain adequate performance when the vessel is rolling and pitching in a seaway. The International Maritime Organization (IMO) Performance Standards for radar set out certain range performance requirements and these must be achieved when the vessel is rolling or pitching up to ±10°. In principle, the antenna could be mounted on a stabilized platform that compensates for the vessel pitch and roll, as implemented on some naval radars. In practice, this is very expensive and the simple low-cost solution is to ensure that the antenna's vertical beamwidth accommodates the motion, effectively requiring it to be a minimum of 20°.

To achieve the desired 360° of azimuth coverage the antenna is typically rotated continuously and automatically in a clockwise

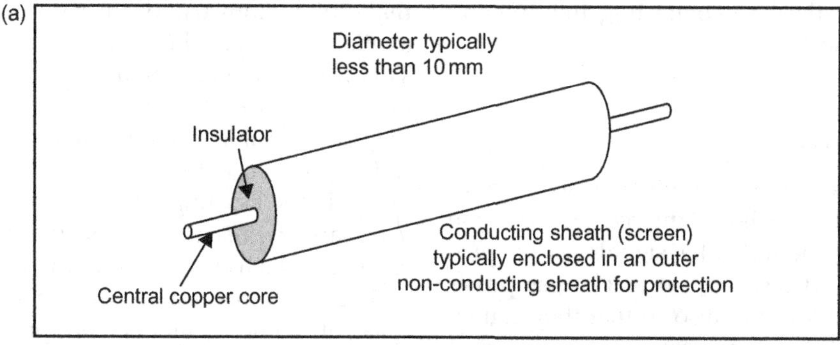

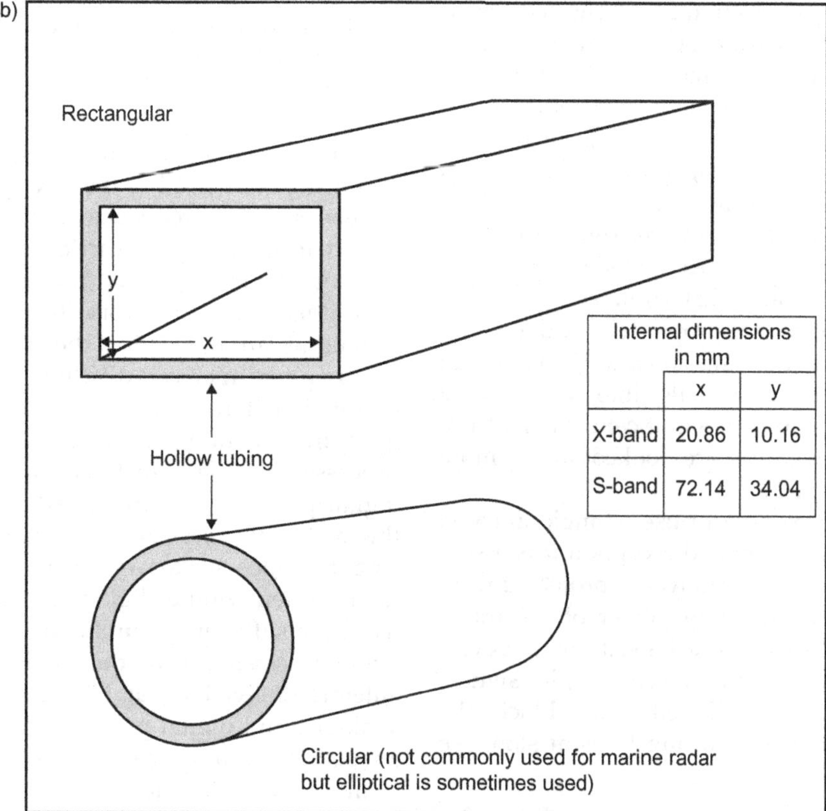

	Internal dimensions in mm	
	x	y
X-band	20.86	10.16
S-band	72.14	34.04

FIGURE 2.2 Typical transmission lines: (a) coaxial cable and (b) waveguide.

direction, when viewed from above, thus rotating the beam to consecutively cover all azimuth angles. A signal measuring the precise pointing angle, either in analogue or in digital form, is relayed to the radar processor to ensure that the received echoes are correctly referenced with the antenna azimuth angle relative to the ship's heading (relative bearing), see Section 1.2.4.2 and Figure 1.5. The transmission of this signal is represented in

Figure 2.1 by the line connecting the antenna to the processor.

2.2.3 Receiver

The function of the receiver is to amplify the very weak echoes intercepted by the antenna. This permits the signals to be processed such that they appear as clear as possible on the display, and also so that they can be used for automated processes, such as target tracking (TT), see Section 2.7.5. The receiver is a very critical module of a radar as it must accurately amplify the received signal without any distortion. The very low levels of the original signal make this a demanding process. In particular, a poorly designed receiver can add unwanted *noise* into the received signal, which can then totally obscure weak signals and can also distort stronger ones. In today's crowded radio spectrum any distortion effects within the receiver can inappropriately add components of other transmissions, such as digital telecommunications signals, into the received radar signal causing significant interference effects. These issues are looked at in more detail in Section 2.6.5.

In order to be able to use a single antenna for both transmission and reception it is essential that the sensitive receiver is protected from the potentially damaging power of the transmitted signal. This is achieved by a device known as the transmit/receive (T/R) switch, sometimes called a T/R cell, which blocks the input to the receiver during transmission, see Section 2.6.3.1.

2.2.4 Radar Processor

The prime function of the radar processor is to turn the received radar signals into an optimized image that can be shown on the operator's display. In particular, the image must be constructed to ensure that a trained user can make maximum use of the information that is available within the returned radar signals. This fundamentally includes a clear indication of the presence of objects by generating on the display a visible response whose angular and radial position with respect to, for instance, the own ship's heading line and indicated position, are representative of the bearing and range at which the corresponding target lies.

The radar processor function typically consists of some analogue pre-processing, followed by extensive digital processing. Analogue processing means that the signal is manipulated before it is digitized and so retains its 'continuous' structure and is not represented as a table of data. It can be depicted by a smooth curve of voltage or current against time. It has the advantage that relatively simple processing functions can be almost instantaneously carried out by a set of low-cost devices such as transistors, resistors, capacitors and inductors (coils). Analogue techniques are used when the speed of processing required makes it difficult or expensive to perform digitally, and nowadays, these techniques are generally confined to the very early processes of radar functionality. The speed of digital processing is always advancing and so this means that less and less analogue processing becomes necessary. A future advanced radar system would digitize the signal once it had been sufficiently amplified – the only analogue processing being the amplification of the antenna-received signal. Figure 2.1 shows both analogue and digital processing phases, which are elaborated in later sections of this chapter.

In general, the digital processing attempts to optimize the detection of wanted targets while reducing the deleterious effects of unwanted echoes, such as reflections from precipitation and the sea, known as clutter. It also provides the tracking function for targets, so that parameters such as the speed and course of selected targets may be shown graphically and numerically on the display and allows automatic collision warnings to be made. The tracking functionality of a modern radar can also be made to include the integration of radar information with that being received from the ship's Automatic Identification System (AIS). AIS is described in Chapter 5.

Digital processing also formats the image for display, scaling it appropriately to fit the screen and putting it into the correct orientation and motion modes (see Section 2.8.4). Finally it ensures that supporting information is visible on the display, such as the heading line, bearing scale and range rings, and that user instigated calculations, such as the positional interrelationships of targets and the own ship, can be determined. All the processing takes into account the specific user settings, which form part of the user interface described in the following section.

2.2.5 Display and User Interface

The user interface, including the display of the radar image and the control of the radar's functionality, is obviously of high importance to the radar operator. It must be appropriately positioned on the ship's bridge, to enable it to be used effectively in conjunction with other systems, while providing a good view from the bridge windows. The display needs to be clear under all operating conditions in day, dusk and night. In particular, at night, it must give clarity without diminishing the night vision of the user and others on the bridge. The controls and operating menus need to be easy to understand, with values that have been set by the user readily observable, helping to ensure that the radar is always set optimally to meet the needs of the immediate situation. In today's terminology, it must be user-friendly.

Although older radar displays are based on rectangular cathode ray tubes (CRTs), similar to those used on old televisions, newer systems are all based on flat panel technology (described in Section 2.8.3). They are very similar to modern good quality computer and TV displays, and will undoubtedly continue to follow the general evolution of mainstream display technology. A radar may have a computer keyboard as part of its user interface, but most of its functionality will normally be accessed via a small number of dedicated buttons and switches, a tracker ball and a screen-accessible menu. Again, mainstream evolution in all aspects of user interfaces will continue to be adopted by radar system manufacturers, whenever applicable, such as the use of touch screens and the possible future use of head-up displays, auxiliary hand-held displays, voice control and three-dimensional displays.

Today's rectangular display is generally orientated such that the display surface is approximately vertical, with its longest dimension in the horizontal plane, similar to a conventional TV or computer display. The orientation is commonly described as *landscape mode*. On some systems, especially river radars, the longest dimension is in the vertical plane, where the display is said to be in *portrait mode*. Larger displays may have their surfaces tilted well away from the vertical to prevent the window view being obstructed. Possibly, the even larger displays that may be used in the future will have their surfaces almost horizontal, although reflections from the bridge windows will have to be taken into consideration.

On most systems the radar 'picture' is bounded by a circle, effectively replicating the historical circular CRT display. Auxiliary information, such as radar settings and radar TT data are placed outside this circle. There is no internationally agreed standard for positioning such auxiliary information and so different radar models can place information in very different positions. The function and position of control buttons and switches is also not highly regulated, neither is the structure of any menu. It is therefore important that before using a particular radar at sea, the user becomes familiar with its specific operation, see Chapter 6.

2.3 THE RADAR SIGNAL

2.3.1 Fundamental Considerations

The radar needs to transmit a signal such that when it is echoed from a target and received, as much relevant information about that target can be collected, especially including its instantaneous range and bearing. There is a need to collect information from targets at all bearings from 0° to 360° relative to own ship, and out to a specified range. In addition, the signal characteristics need to provide sufficiently rapid updates in information so that the successive changes in relative position of own ship and any target are small enough to allow a human operator or any automatic system within the radar to be able to keep track of all individual targets. Excessive positional jumps of targets between refreshes of the information would create the potential for confusion because the human user or the automatic process within the radar would not be able to *correlate* targets between updates.

The conceptually easiest way of meeting these fundamental requirements is to send out short pulses of RF energy at the radar operating frequency. The action of the antenna confines the energy to a small angular sector in azimuth; the antenna steadily rotates such that the radar beam illuminates all azimuth angles in a relatively short space of time, typically about 2–3 s. The beam must rotate slowly enough to allow the echoes from the most distant targets to be received before the beam rotates away from their particular azimuth angle. For a relatively short range radar, such as a commercial marine radar, this is generally not a practical limitation. The important aspect is that the beam must rotate fast enough so that data from any particular azimuth angle is updated at intervals short enough to ensure the correlation of individual targets at every revolution. The time gap between successive pulses needs to at least allow the transmitted pulse to be reflected and received from a target at the current maximum range setting of the radar. The time taken by the pulse to travel from the radar antenna to the target and back is, of course, a measure of the range of the target from the radar. This straightforward conception forms the basis of a conventional pulse radar, even though there are many practical considerations to be taken into account, which are elaborated further in Section 2.3.3.

One important aspect to be determined, properly detailed in Section 2.3.3.1, is the length of the pulse itself. To achieve good range discrimination between targets it has to be extremely short; there then has to be a long gap before the next pulse is transmitted in order to allow the target echoes to be received. The inter-pulse gap for a marine radar is typically 2500–6000 times longer than the pulse itself. In effect, it means that the transmitter spends most of its time not transmitting. When a pulse is transmitted it is therefore not surprising that its magnitude has to be very large – in practice typically 25,000 W (25 kW) for a ships' radar. By comparison, the ship's main very high frequency (VHF) radio typically transmits at 25 W and has an operating range somewhat comparable to a ship's radar.

Generating 25 kW of pulsed power is a very specialist requirement and the magnetron, described in Section 2.4, has evolved to be able to do this. However, if the pulses can be made to be much longer, without compromising radar performance, it would allow more conventional transmitters based on semiconductor technology to be utilized, similar to those used today by base stations for modern telecommunications. Advanced *coherent* radar principles essentially allow this to happen by using special processing techniques known as *pulse compression*, which are described in Section 2.9. In effect, the processing artificially reduces the length of the received echoes so that good target discrimination and range accuracy is maintained. The very stable signal that can be generated by a semiconductor-based transmitter also results in the radar receiver being able to be more sensitive to the received echoes than is the case for a conventional magnetron-based system, further helping to reduce the total power required, including the actual amount of pulse compression. Also, the stable signal enables more information to be extracted from the echoes, such as the radial velocity of the target, greatly helping the detection process, especially in clutter. Today's coherent radars reduce the peak power required to a few hundred watts, compared to the 25 kW typically needed for a magnetron-based marine radar.

However, a good knowledge of basic pulse radar principles is needed to fully understand the action of this newer radar technology. In particular, the concepts surrounding pulse length, pulse repetition frequency (PRF), power and frequency are all very relevant.

2.3.2 Choice of Frequency

An essential consideration for the radar is the frequency at which it operates. This is fundamentally connected to two important issues. The first is the angular resolution that is required and the second is how well the signal travels within the atmosphere of the earth. In Section 2.5.1.4 it is shown that angular resolution is fundamentally controlled by the antenna; the wider the antenna, the narrower its horizontal beamwidth can be made. Practical marine radars for ships need to be able to discriminate between two point targets at the same range that are separated by 2.5° or more; also a target bearing accuracy of 1° is required from the complete system. Both these requirements are embodied within IMO's radar performance standards, see Section 11.2.1. It means that the marine radar antenna needs to have a horizontal beamwidth somewhat less than 2.5°. This only becomes practical at microwave frequencies; otherwise the antenna horizontal dimensions become too unwieldy. It became clear in the early days of marine radar that operation below about 2.5 GHz started to become impractical and that operation well above this frequency allowed even more practically sized antennas.

However, it was also found that operation at higher frequencies increased the attenuation of the signal by the atmosphere and, even worse, by rain and other precipitation. This put a practical limit of around 10 GHz on marine radars, which was also affected by the difficulty in earlier days of producing signals significantly above this frequency. By international agreement two groups of radio frequencies are now allocated for use by civil marine radar systems. One group is at X-band, specifically those frequencies which lie between 9.2 and 9.5 GHz. These frequencies correspond with a wavelength of approximately 3 cm. The second group is at S-band and includes the frequencies lying between 2.9 and 3.1 GHz and corresponds with a wavelength of approximately 10 cm. In fact there was another band allocated at around 5 GHz (C-band), but this band never really caught on for marine radar.

X- and S-band radars have different strengths and weaknesses, and so on appropriately large vessels both are ideally installed. In fact it is an IMO requirement that both X- and S-band systems are fitted on all vessels above 3000 gt. The characteristics of the two systems are compared in Table 2.1. Basically, at S-band there is better performance in rain, but because of the practical limitation on antenna size the horizontal beamwidth is generally larger than optimum. At X-band, the antenna used can either be large enough to give excellent resolution or be smaller, and therefore more practical for smaller vessels, but still large enough to

TABLE 2.1 X-Band and S-Band Compared

Feature	Comparison
Target response	For a target of a given size, the response at X-band is greater than at S-band
Bearing discrimination	For a given antenna width the horizontal beamwidth effect in an S-band system will be approximately 3.3 times that of an X-band system
Vertical beam structure	The vertical lobe pattern produced by an S-band antenna is about 3.3 times as coarse as that from an X-band antenna located at the same height (see Section 2.5.2.5)
Sea clutter response	The unwanted response from sea waves is less at S-band than at X-band, thus the probability of targets being masked is less
Precipitation response	The probability of detection of targets which lie *within* an area of precipitation is higher with S-band transmission than with X-band transmission
Attenuation in precipitation	In any given set of precipitation conditions, S-band transmissions will suffer less attenuation than those at X-band

provide adequate resolution. Performance in precipitation and also at longer ranges is, however, compromised.

It is seen from Table 2.1 that there are many ways in which the characteristics of X- and S-band transmissions are complementary, and a knowledge of these leads to more effective use of the equipment. The various circumstances in which these complementary characteristics can be exploited are discussed in more detail in later sections of this book.

2.3.3 Pulse Radar Concepts

A marine pulse radar, such as that realized by a conventional magnetron-based transmitter, has to produce pulses of defined length, PRF, power and shape that are optimized to achieve the required radar performance defined by IMO. The actual pulse length, PRF and power of each transmitted pulse are of considerable importance in the effective detection of targets. The final pulse shape influences the accurate measurement of range and, perhaps surprisingly, the interference of the radar, both with other radars and with other RF systems, particularly digital telecommunications. Interference issues are discussed in Section 2.6.5.

2.3.3.1 Pulse Length

Pulse length is defined as the duration of a single transmitted radar pulse and is often quoted in microseconds (μs), although pulses rather shorter than 1 μs are sometimes given in nanoseconds (ns), where 1 ns = 10^{-9} s; and so there are 1000 ns to 1 μs. As an example, a typical short pulse length of 0.05 μs can alternatively be stated as 50 ns. In general, the target echoes from longer pulses are easier to detect than those from shorter pulses, simply because long pulses contain more energy. Consider the analogy of an electric heater with a power of 3 kW. If it had been on for only 5 min any rise in the room's temperature may be imperceptible. On the other hand, if it had been on for 50 min the room is likely to have become perceptibly warmer. Energy is typically measured in watt hours or kilowatt hours, with the heater examples being the equivalent of $3 \times 5/60 = 0.25$ kW hours and $3 \times 50/60 = 2.5$ kW hours, respectively. However, short radar pulses generally allow more easy and accurate determination of the range of a target, and also permit two targets closely spaced in range and at the same bearing to be displayed as separate targets. This is known as *range discrimination*, see Figure 2.3. Very short pulses are required when operating at a low maximum displayed range. For instance, a radiated pulse of 1 μs has a length in space of about

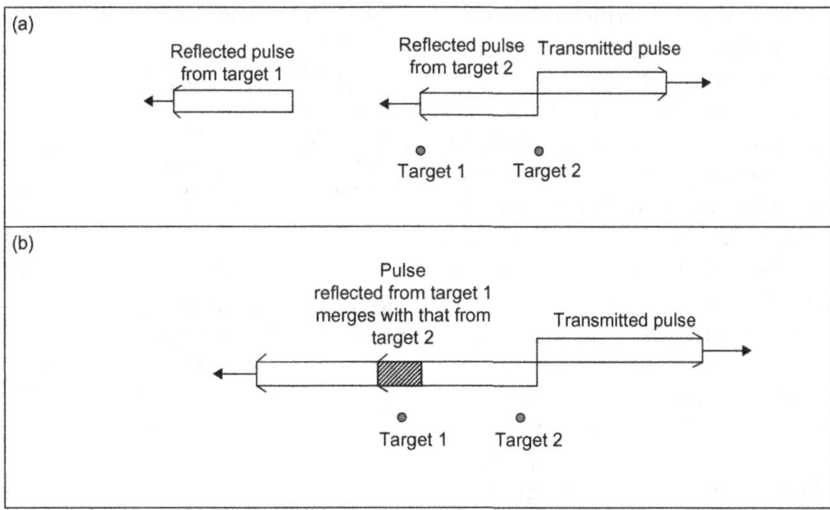

FIGURE 2.3 Range discrimination: (a) reflections from short pulses remain separate – good target, discrimination, and (b) reflections from long pulses merge – poor target discrimination.

300 m, effectively limiting the minimum range to at least half this figure – a conventional radar cannot receive while the pulse is still being transmitted. On the other hand, a pulse of 0.05 μs is only 7.5 m in length, greatly enhancing the minimum detectable range. See Section 3.2.4 for discussion of minimum range.

In certain special circumstances, such as those where targets are difficult to detect against the background of sea or precipitation clutter, the use of a short pulse generally improves the probability of detection of a wanted target. This is because of the better resolution that results, greatly enabling a small 'persistent' target to be discerned from, say, the noisy background of sea clutter. A longer pulse would have a greater amount of clutter embedded in it, making it more difficult to distinguish a real target within its return. Marine radars typically have three or four pulse lengths available, although their selection by the user may be appropriately limited by the radar, according to the selected range scale. On a conventional magnetron-based marine radar, available pulse lengths typically fall within the range 0.05–1.3 μs. If four different pulse lengths are used, they are typically labelled as short, medium, long and extra long.

It is useful to understand the reasons why the user's choice of pulse lengths is restricted according to the selected range. If the radar is being used for general surveillance on the longer range scales, a long pulse is needed to increase the energy received from more distant targets. It is explained in Section 2.6.1 that the radar's ability to detect more distant targets rapidly diminishes and so the highest practical energy must be transmitted to obtain adequate longer range performance; the use of longer pulses helps achieve this. At shorter range scales the returns are generally stronger and the observer is more likely to be concerned with the finer details of the picture. The theoretical minimum radial length of any echo, as eventually displayed on the radar screen, is determined by the pulse length. Hence, at short range the shorter the pulse length, the better the detail. The fact that the wanted targets are closer-in means that the highest possible energy is not required. The ability of the radar to display separately two targets which are on the same bearing and closely spaced in range is known as *range discrimination*, and this is discussed further in Sections 2.7.3.1 and 2.8.5.2.

Thus, in general usage, a short pulse is likely to be more appropriate on the shorter range scales and vice versa for longer pulses. Nevertheless, there are some occasions on which it may be helpful to use longer pulses on the shorter range scales or shorter pulses at longer range. The correct use of pulse length selection requires a good understanding of radar use. However, at this stage it is useful to summarize the various factors as set out in Table 2.2. Typical pulse lengths and PRFs available at selected ranges scales are given in Table 2.3.

TABLE 2.2 Short and Long Pulses – Features Compared

Feature	Short Pulse	Long Pulse
Long range target detection	Poor. Use when short range scales are selected.	Good. Use when long range scales are selected and for poor response targets at short range.
Minimum range	Good. Use when short range scales are selected.	Poor. Use when long range scales are selected and minimum range is not a major consideration.
Range discrimination	Good.	Poor.
Effect on echo paint	Short radial paint. Produces a well-defined picture when short range scales are selected.	Long radial paint when short range scales are selected but the effect is acceptable when long range scales are selected.
Effect on sea clutter	Reduces the probability of the masking of targets due to saturation.	Increases the probability of the masking of targets due to saturation.
Effect in precipitation	Reduces the probability of the masking of targets due to saturation.	Increases the probability of the masking of targets due to saturation. However, the use of long pulse helps to combat the attenuation caused by precipitation and will increase the probability of detecting targets which lie beyond rain.

2.3.3.2 *Pulse Repetition Frequency*

PRF is normally expressed as the number of pulses transmitted in 1 s and is therefore denoted in Hertz or pps (pulses per second). Typical values for a marine radar are 1000–3000 pps. The pulse repetition interval (PRI) is the time interval between pulses. It should be noted that PRF and PRI effectively refer to the same feature and are simply related by the expression $PRF = 1/PRI$.

For any user selected range scale, the PRI must at least be long enough to allow the immediately previous transmitted pulse to travel out and back to a target situated at the maximum displayed range of the radar. If the next pulse was transmitted before this interval the radar receiver would not be able to decide whether a return was from a long range target illuminated by the preceding pulse or a short range target from the subsequent pulse. In fact, this possibility of confusion always remains, even when the PRI is rather longer than the time interval needed for the transmitted pulse to travel to and from a target at the maximum displayed range. This is because the return from a target situated beyond the maximum displayed range may be received after the next pulse has been transmitted, and erroneously displayed as a target at much shorter range. It is generally termed the *second time around* or *second trace echo* effect.

TABLE 2.3 PRF and Pulse Length – Some Representative Values

Range Scale Selected (NM)	Pulse Length Selected			
	Short		Long	
	PRF (Hz)	PL (μs)	PRF (Hz)	PL (μs)
0.25	2000	0.05	2000	0.05
0.5	2000	0.05	1000	0.25
0.75	2000	0.05	1000	0.25
1.5	2000	0.05	1000	0.25
3.0	1000	0.25	500	1.0
6.0	1000	0.25	500	1.0
12	1000	0.25	500	1.0
24	500	1.0	500	1.0
48	500	1.0	500	1.0

To reduce the possibility of such an effect it means that the radar designer has to build in extra 'dead-time' to the PRI, whilst not lowering the PRF to such an extent that it adversely affects the detection of targets at all azimuth angles. A very low PRF would result in appreciable rotation of the antenna before the next pulse, in the limit leaving azimuthal gaps where no targets could be detected. To also help reduce the possibility of such effects, the pulse length is chosen to be suitably short to limit the effective energy being radiated in order that excessive power does not exacerbate the situation (see also Section 2.3.3.3). This is all a bit of a compromise, but by experience suitable PRFs and pulse lengths are readily determined that give adequate performance. Even so, second trace echo effects are not an uncommon experience to the user, especially from very large targets, for example wind turbines located at distances rather longer than the user-set maximum displayed range (see also Section 3.9.6).

Even a point target will generally reflect several pulses back to the radar before the antenna beam no longer illuminates it, because of its rotation. This increases the received energy from any target, thereby increasing the likelihood that it will be detected, as further discussed in the next section. The radar designer carefully determines the PRFs and pulse lengths that are available for the user, with Table 2.3 indicating representative values. This shows that they are both dependent on the user-selected maximum display range. Long pulses are used on the longer range scales and are associated with low PRF, while short pulses are used on the short range scales and are associated with high PRF.

In some circumstances the pulses are best not transmitted at regular intervals but with a small random variation in the length of time between each pulse. This is known as *pulse jitter* and the random process is effected by a digital process known as *pseudo-random* number generation. It is 'pseudo-random' because the numbers are the result of a defined algorithm that generates a sequence of numbers that appears to be random but is actually predictable. This can significantly reduce interference effects with other radars, as described in Section 2.6.5.1, albeit with a small degradation in overall radar performance.

2.3.3.3 *Power Considerations*

It is obvious that the range at which a target can be detected is dependent on the actual power of the transmitted signal. In fact, many other factors affect the range at which detection takes place, but the transmitted signal power is perhaps the most obvious. On a magnetron-based radar, the instantaneous power is measured in kilowatts and may typically be 3 kW for a small craft radar or 25 kW for a ship's radar. As previously noted, this seems large, especially when compared to a 25 W VHF transmitter that has a similar operational range. In fact, as discussed in the previous section, what is more important is the energy that is effectively available to illuminate targets, with energy being the product (multiplication) of power and time.

First of all, consider a conventional 25 kW maritime magnetron-based radar, operating at medium range with a pulse length of 0.25 μs and a PRF of 1000 Hz. A radar with such settings is only radiating power for 0.25 μs every 1/1000th of a second (every 10^{-3} s). This results in the average (mean) power being generated given by:

$$\text{peak power} \times \text{pulse length/PRI}$$
$$= 25{,}000 \times (0.25 \times 10^{-6})/10^{-3} = 6.25\ \text{W}$$

This is a better comparison to the power rating of more continuous systems such as domestic electric heaters and VHF radios. The radar transmitter is developing just 6.25 W hours of energy.

A target is only illuminated as the antenna beam sweeps across it. If the antenna is rotating at 24 revolutions per minute, this is equivalent to $24 \times 360°$ per minute or $24 \times 360/60°$ per second, that is 144° per second. If the antenna azimuth beamwidth is 1° then the beam is only directing appreciable energy towards the target for 1/144 s every revolution of the antenna; that is for about 6.9 ms. This is called the *dwell time* of the radar beam. If pulses are being transmitted every millisecond it is seen, for this example, that a burst of about 7 pulses is directed towards a particular point target at every revolution of the antenna.

It can be useful to consider the concept of 'effective energy'. Conceptually, the effective energy (E_f) is a representation of the energy in watt seconds that is being transmitted by the radar towards any particular point target at every revolution of the antenna. For our example, this is $6.25 \times 6.9 \approx 43$ mW seconds. It is emphasized that E_f is a conceptual quantity, that is useful in explaining certain basic properties of radar, but is not the full story nor is it widely quoted and it certainly does not represent absolute signal strengths.

For a conventional pulsed radar the effective energy being radiated in any particular azimuthal direction at every revolution of the antenna is therefore given by:

$$E_f = P_p \times P_l \times N_p\ \text{W s}$$

where:

P_p is the peak transmitted power in watts
P_l is the pulse length in seconds
N_p is the number of pulses illuminating a point target every antenna revolution, and so:

$$N_p = \frac{B_w \times P_{rf}}{R_{pm} \times 60}$$

where:

P_{rf} is the PRF in pulses per second
B_w is the antenna beamwidth in degrees
R_{pm} is the antenna rotation rate in revolutions per minute

By substituting the expression for N_p in the original equation for E_f, we get the relationship:

$$E_f = \frac{P_p \times P_l \times B_w \times P_{rf}}{R_{pm} \times 60}\ \text{W s}$$

This simple equation, although not commonly quoted, is generally useful in helping to understand some fundamental radar aspects, including the comparison of conventional magnetron radars with solid-state systems, see Section 2.9. In particular, it shows how the basic parameters of a conventional pulsed radar with a rotating antenna influence the effective energy being radiated, and hence the basic ability of the radar to detect targets. It is seen that the effective energy is increased by any of the following:

- A higher peak power
- A longer pulse length
- More pulses illuminating the target, such as by increasing the PRF or decreasing the antenna rotation rate.

The effective energy, as defined here, is only a concept but is useful in bringing out some important aspects that are otherwise only embedded in more difficult to understand equations. Furthermore, it does not include all real effects, such as the vertical beamwidth of the antenna. Also, in practice, the detectability of a target is additionally enhanced by it being illuminated on multiple revolutions of the radar antenna. This is a rather more complex situation, and a detailed analysis is outside the scope of this book as it involves a good knowledge of statistical mathematics. In fact, detailed statistical issues also arise when more accurately analysing the effective power of a radar, even when just considering the pulses received within the beamwidth of the antenna (N_p). These are also outside the scope of this book.

All these additional factors result in a concept known as the probability of detection (P_d), which is effectively a measure of the likelihood of a target being shown on the radar display. This is normally given as a percentage. A good P_d would typically be better than 90%, whilst a poor one would be less than 50%. It is best understood by the fact that, in effect, a target with a P_d of 90% would be visible on the display for 9 out of 10 revolutions of the antenna. There is a connected parameter known as the probability of false alarm (P_{fa}), that is where an apparent target is shown on the display but that it is noise and not 'real'. In practice, a relatively poor P_{fa} would be larger than 10^{-4}, that is more than 1 in every 10,000 apparent targets on the display is false. A relatively good P_{fa} is typically smaller than 10^{-6}, where only less than one target per million is false. The false targets form visual 'noise' on the display. Such noise is actually used to help the operator judge the correct settings of, for instance, the manual gain control, as detailed in Section 6.2.7.1. IMO defines the required basic detection ranges of a radar at a P_d of 80% and a P_{fa} of 10^{-4}.

2.3.3.4 *Pulse Shape*

The pulse shape is of particular significance in the accurate determination of range, although modern processing systems can compensate for non-idealities. Conceptually, the outline (or envelope) of the pulse should be rectangular and, particularly, the leading edge of the pulse should take the form of a vertical rise. The significance of this is illustrated by Figure 2.4, which shows the envelope of an ideal pulse and

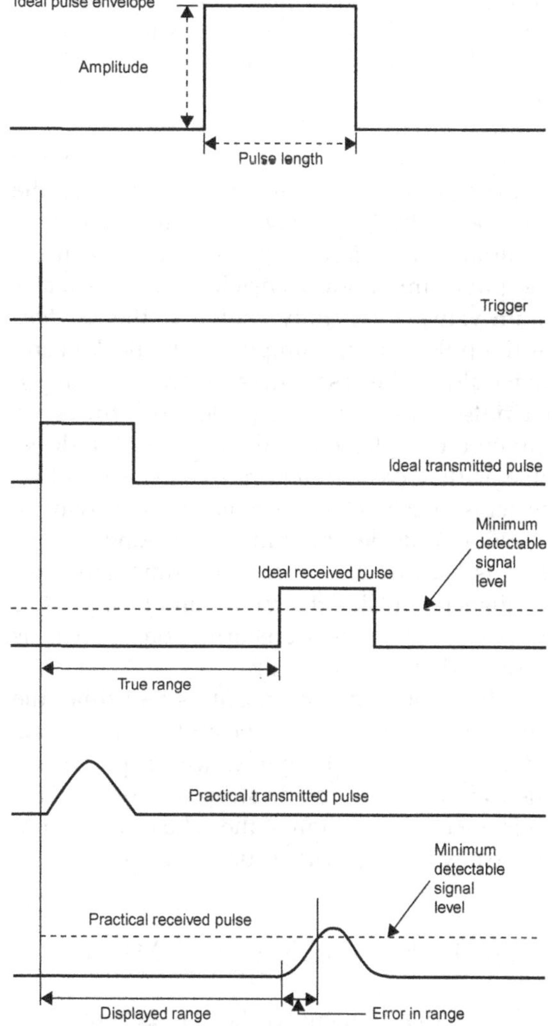

FIGURE 2.4 The significance of pulse shape.

then compares the range measurement obtained using such an ideal pulse shape with that obtained using a pulse shape which differs from the ideal. Because of the great disparity between the strength of the transmitted pulse and that of the returning signals, it is not practical to show their amplitudes to scale. Thus, when considering Figure 2.4, it must be appreciated that the transmitted pulse is of extremely high power, whereas the received echo, even after amplification, is comparatively weak and will only be observed where its amplitude rises to the minimum level at which the receiver can detect signals. That level is called the *minimum detectable signal* and is a receiver characteristic which is discussed further in Section 3.2.3.

Figure 2.4 shows that the range obtained will be in error by an amount which depends on the time taken by the received signal to reach its minimum detection level. The time taken for the pulse amplitude to reach 90% of its maximum value is generally known as the *rise time* of the pulse and for simplicity should be zero, as would be the case with a perfectly rectangular pulse. In practice, very short rise times are difficult to implement and are actually undesirable as they create interference effects on other systems, such as communications networks, working outside the radar operating bands. This is because a very fast rise time generates frequencies outside the radar band by a process known as Fourier transformation, which is described in Section 2.6.4.2.

While not being so crucial as rise time, the forms of the upper envelope and of the trailing edge can also be of significance. In particular, the trailing edge is made steep in order to give a clean-cut termination of the echo so as to minimize any overlap with a subsequent return.

2.4 THE RADAR TRANSMITTER

The radar transmitter has to generate the required transmitted signal of the correct power and detailed characteristics. The detail of the transmitted signal has to take into account many factors, including:

- The target detection performance of the radar as defined by IMO.
- The specific settings of the user, such as the display range.
- The frequency and interference limitation requirements set by the International Telecommunication Union (ITU) – see Section 2.6.5.2.
- Subtleties in the specific design of the radar system that may help a particular manufacturer in improving its overall performance.

This section looks at the operation of a conventional magnetron-based radar transmitter. The transmitter concepts for a solid-state coherent radar are considered in Section 2.9.

2.4.1 Magnetron-Based Transmitter

The basic components of a magnetron-based radar transmitter are shown in Figure 2.5. It consists of the following:

1. The trigger generator, which controls the repetition frequency of the transmitted pulses.
2. The pulse-forming network, which defines the length and shape of the transmitted pulses in conjunction with the magnetron.
3. The magnetron itself, which creates the pulse of RF energy equating to the characteristics set by the pulse-forming network length. This pulse of RF energy is sent via the transmission line to the antenna (see Figure 2.1), which radiates the energy to the outside world.

The trigger generator produces a continuous succession of low-voltage control signals known as *synchronizing* or *trigger pulses.* Commonly they are referred to simply as *triggers.* Each trigger results in the magnetron

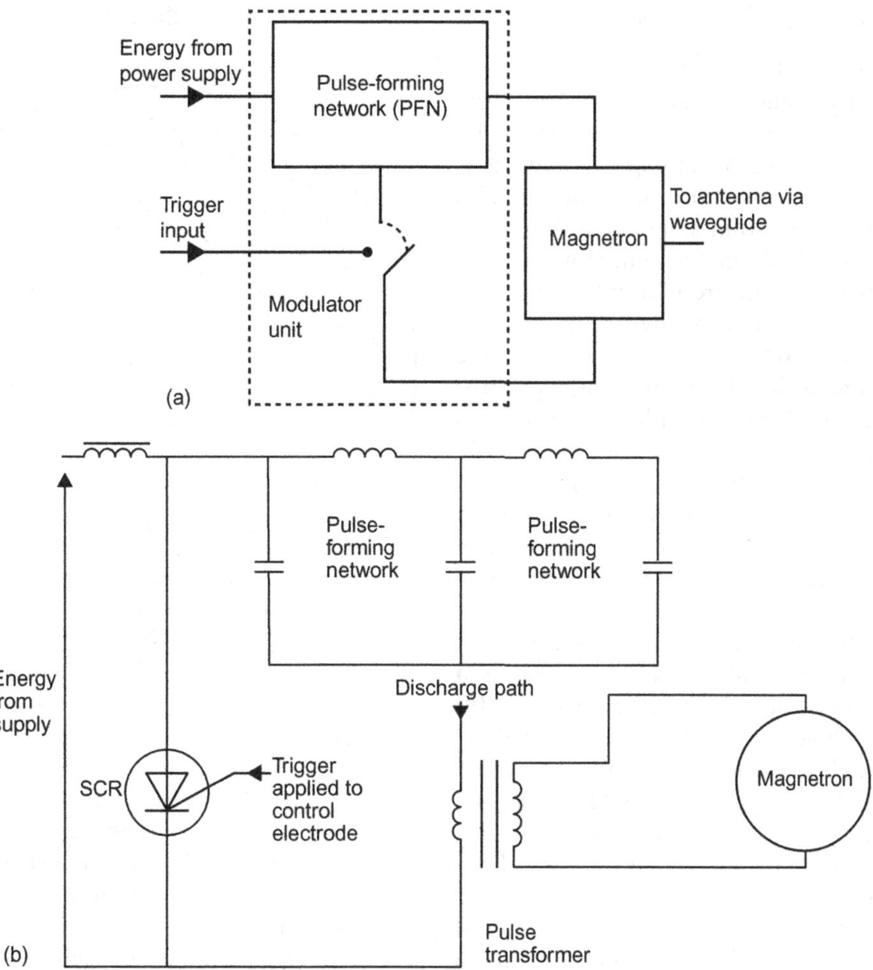

FIGURE 2.5 The modulator unit.

generating an RF pulse which is sent via the transmission line to the antenna. Because the trigger generator controls the PRF, it is sometimes referred to as the PRF generator.

It is normal to consider that the pulse-forming network and associated circuitry *modulates* the transmitting device. This means that the basic RF energy generated by the magnetron is being controlled in the desired manner, in this case forming a pulse of a defined length with controlled rise and fall times. It forms a specific example of *amplitude modulation.* Modern radars typically use designs based on a special type of device, known as a field effect transistor (FET), which has the ability to deal with the very fast rise and fall times required of the modulator and form a pulse modulation signal of several hundred volts, which is needed to drive the magnetron. Intelligence built-in to the FET modulator automatically compensates for the effects of magnetron ageing to ensure that a well-formed signal is transmittable over the working life of a magnetron (typically 10,000 h).

2.4.2 Magnetron Operation

The magnetron (or cavity magnetron as it is more correctly named) was invented in 1939, when electronic systems were dominated by vacuum tube (valve) technology. Today, it is one of the very last vacuum tubes in common use to survive the solid-state revolution. The technology is based on the emission of electrons in a near vacuum from an internal device within the vacuum tube, known as the *cathode*, towards a collecting device, known as the *anode*. The anode has to be at a high positive voltage compared to the cathode for this to happen. Diode vacuum tubes just have a cathode and an anode but amplifying vacuum tubes, which used to do the job that has been dominated by transistors since the 1960s, have other electron controlling devices within them known as *grids*.

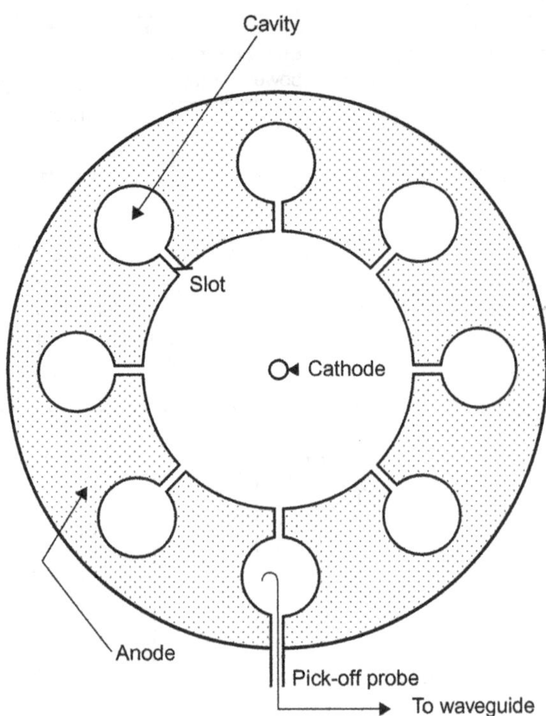

FIGURE 2.6 The cavity magnetron.

The magnetron (Figure 2.6) is essentially a diode vacuum tube in which the anode is a copper cylinder into which are cut cavities (in the form of holes and slots) of very precise dimensions. The cathode is a pillar located along the central axis of the cylinder. A permanent magnet applies an extremely powerful magnetic field which acts along the axis of the cylinder. In the absence of the magnetic field one might expect electrons to flow, in an orderly fashion, radially from the cathode to the anode when a pulse is applied to the cathode. Because the magnetic field created by the permanent magnet is at right angles to the electric field created by the pulse, the electrons are deflected from the path which would take them directly to the anode (an application of the motor principle). Many electrons will eventually reach the anode only after a complex oscillatory journey in which their paths may alternately be directed towards and away from the anode, and their speed increased and decreased. The movement of each electron will be further affected by the electromagnetic influence of the millions of other electrons moving in its vicinity. The effect is extremely complex but it can be summarized by saying that if the change of electron velocity is regular and repetitive then there will be associated with such change, a sympathetic fluctuating electromagnetic field. There are certain limits of steady electric field (provided by the modulator pulse) and a steady magnetic field (provided by the permanent magnet) outside which oscillation will not take place. However, if values are selected which lie within the appropriate limits, the frequency of the oscillations that develop is determined by the physical dimensions of the cavities. The RF energy generated within the magnetron is picked up from one of the cavities by a small probe which couples the radar pulse to the waveguide by means of which it is conducted to the antenna.

Of the energy supplied to the magnetron, less than half contributes to the maintenance of the oscillations during transmission. The remainder generates heat, which is generally dissipated by the fitting of cooling fins to the magnetron assembly. The temperature of the magnetron will increase during the warm-up period (usually considered to be up to half an hour) which follows the initial switching on, with a consequent change in the size of the cavities because of expansion of the copper cylinder. The transmitted RF progressively changes during this period, although good design can minimize such effects. There are implications when consideration is given to maintaining the receiver in tune with the transmitted RF, see Sections 2.6.4.2 and 6.2.7.2. Systems which comply with IMO Performance Standards have a 'stand-by' condition (see Section 6.5) in which the magnetron is kept warm by means of a valve heater while the magnetron is not being used for transmission. This is to enable the equipment to return to peak performance more quickly. When the system is again transmitting, the supplementary heating is reduced or removed as the natural heating returns.

2.5 ANTENNA PRINCIPLES

The antenna system is required to focus the transmitted energy into a beam, to receive reflected signals from targets and to rotate the beam at an appropriate angular speed. It also provides accurate rotation angle and heading marker data to the radar processor. The antenna forms a very important part of the total radar system, contributing greatly to its overall performance, and so it is important to understand its basic concepts.

It is especially difficult to talk about antennas and, in general, about radar performance without using the term *decibel*, abbreviated as dB. This is a mechanism for making very small and very large ratios more comprehensible. In the electronics world it is mainly applied to power ratios. A system that can produce 1000 W compared to one producing 1 W is 1000 times more powerful. If we now take the logarithm to the base 10 of 1000 we get the value 3, which is the power ratio in logarithmic format, with units known as *bels*, named after the telecommunications pioneer Alexander Graham Bell. The more normally used decibel figure is just the number in bels multiplied by 10; this means that a power ratio of 3 bels is more normally referenced as 30 decibels (30 dB). Formally stated:

$$R(\text{dB}) = 10\log_{10}(P_1/P_0)$$

where:

R is the ratio given in decibels
P_1 is the measured power and
P_0 is the reference power

All power ratios tend to be quoted in dB by engineers, even those that are unity (0 dB). Here are some examples:

Power Ratio	dB Equivalent
1	0 dB
2	3 dB (actually 3.01029996...)
10	10 dB
100	20 dB
1000	30 dB
10,000	40 dB

2.5.1 Antenna Concepts

2.5.1.1 The Isotropic Source

An *isotropic source* is a useful concept in antenna theory, even though it cannot actually be realized in practice. It is simply an antenna that radiates equally in all directions or, equivalently, has equal receive sensitivity from all directions. It can be imagined as a point at the centre of an imaginary sphere; at all points on the sphere's surface equal radiation intensity will be present. Equivalently on receive, the isotropic antenna is equally sensitive to radiation coming from any direction. Despite the usefulness and simplicity of the concept, detailed electromagnetic theory shows that such a radiator cannot actually exist. It means that all real antennas will radiate better in some directions than in others. In fact, the essential characteristic of a radar antenna is that it does precisely that, radiating particularly well in one chosen direction. However, a convenient measure of its directional ability can be obtained by comparing the maximum radiation from a real antenna with that which would be obtained from an isotropic source. When given as a ratio this is normally expressed in dBs.

2.5.1.2 *Power Density*

As the power delivered from the transmitter via the antenna moves outwards in space it becomes distributed over a continuously increasing area, and thus the power available to be reflected by any intervening target decreases progressively. Power density is a measure of the power per unit area available at any location in the area of influence of an antenna and is typically measured in watts per square metre (W/m^2).

To illustrate the concept, consider an isotropic radiator situated in 'free' space. If a pulse is transmitted, the power will travel outwards and will be uniformly distributed over the inside surface of an imaginary sphere of ever-increasing radius. If the power delivered by the transmitter to the antenna is W watts, then the power density at a distance r metres from a *lossless* antenna will be given by the power W divided by the surface area of a sphere of radius r. This can be written as:

$$\text{power density} = P_r = \frac{W}{4\pi r^2}$$

This relationship forms part of the radar range equation, which shows the theoretical relationship that exists between transmitter power and target detection range. The radar equation is discussed further in Section 2.6.1 and in Chapter 3. For the case of an isotropic source the power density is the same in all directions, whereas for a real antenna it will be greater in some directions than in others.

2.5.1.3 *Radiation Intensity*

For many purposes *radiation intensity* is a more convenient measure of the radiation from an antenna because it is independent of the range at which it is measured. Radiation intensity is defined as the power per unit solid angle, that is the power incident on that portion of the surface of a sphere which subtends an angle of one radian at the centre of the sphere in both the horizontal and the vertical planes. To illustrate the concept, consider again the case of an isotropic radiator located in free space. If the power delivered to the antenna is W watts then as it travels outwards it will be uniformly distributed over the surface area of a sphere of ever-expanding radius but which subtends a constant solid angle of 4π radians at the centre of the sphere. Thus, at any point in the field of an isotropic radiator:

$$\text{radiation intensity} = U = \frac{W}{4\pi}$$

Radiation intensity can be plotted for different directions from the antenna and this gives rise to what is known as a *radiation pattern*. It is clearly a three-dimensional figure, but in dealing with radar beams it is generally unnecessary to show all three dimensions simultaneously; it is generally more convenient to use two

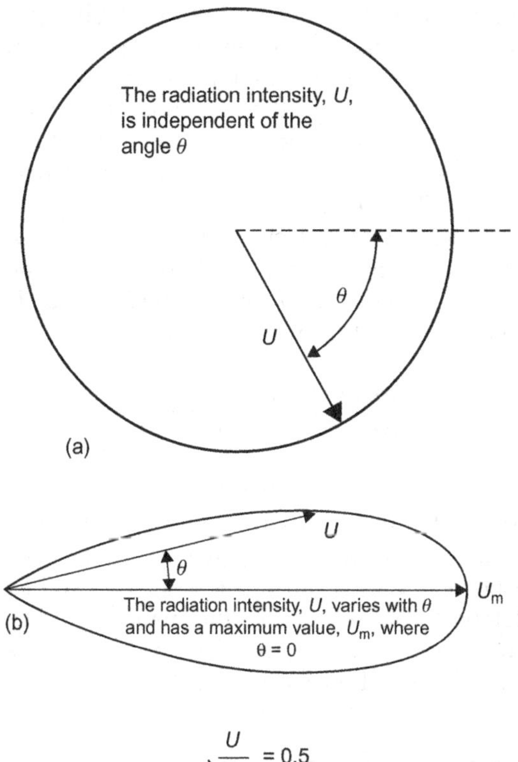

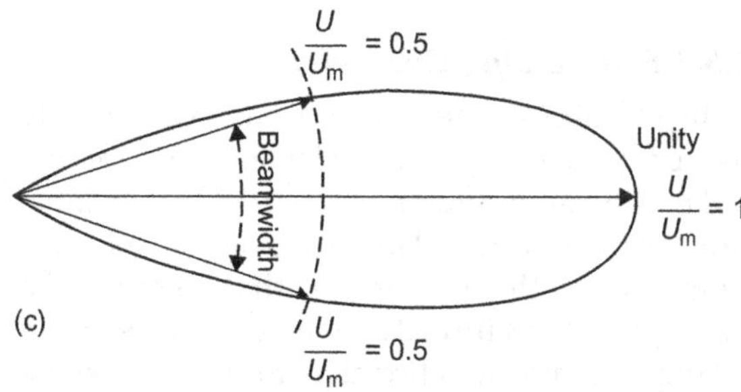

FIGURE 2.7 Examples of radiation patterns: (a) radiation pattern of an isotropic source, (b) directional pattern (absolute units) and (c) directional pattern (normalized).

dimensions to illustrate the beam pattern either horizontally or vertically. Figure 2.7 illustrates the concept, showing the horizontal radiation pattern for an isotropic source and, for comparison purposes, that of a directional antenna.

The radiation in any given direction is represented by the length of a line drawn from the origin in the chosen direction, which terminates at its intersection with the pattern. The line can be measured in watts per unit solid angle, but as we are principally concerned with representation of shape it is more convenient for the representation to be normalized. In the normalized pattern the maximum value is considered to be unity (or 0 dB) and the other values represent the ratio of the radiation intensity in the chosen direction to that of the maximum radiation intensity. When normalized, the shape produced is commonly referred to as the *antenna radiation pattern*.

2.5.1.4 Beamwidth and Sidelobes

In considering the shape of the beam we are particularly concerned with its angular width, normally in both the horizontal and vertical planes. The beamwidth, in either plane, is defined as the angular limits within which the power does not fall to less than half its maximum value. This is readily extracted from the radiation pattern, as illustrated by Figure 2.7(c).

The sidelobes of an antenna are effectively its sensitivity outside of the main beam, as illustrated in Figure 2.8. They arise from unwanted diffraction and other effects, and are more fully described in Section 2.5.2.2. They typically contribute to very little wasted power on transmit but sometimes cause large or close-in targets outside of the main beam to produce spurious received echoes. These effects are examined from an operational view in Section 3.9.4. Sidelobes are generally referenced to the maximum level of the main beam of the antenna. If a particular sidelobe is 1000 times lower than the main beam it would be said to have a level of −30 dB. The negative sign arises because the side lobe power is lower than that of the main beam, and the logarithm of a number less than 1 is negative. However, the negative sign is often omitted in practice. The sidelobe peaks of typical radar

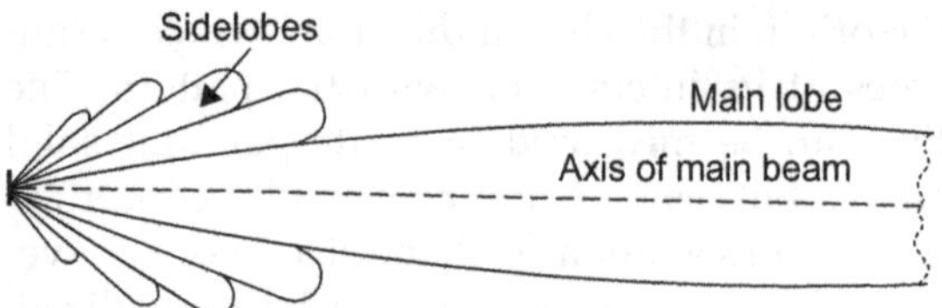

FIGURE 2.8 Radiation pattern showing sidelobes produced by practical antenna.

antennas normally lie in the range −30 to −50 dB and it would be commonly said that the antenna has sidelobes of 30 to 50 dB.

2.5.1.5 *Antenna Gain and Directivity*

Antenna gain is a measure of the maximum effectiveness with which the antenna can radiate the power delivered to it by the transmitter towards a target. Antenna gain is typically given the symbol *G*, and is defined as the maximum radiation intensity produced by the antenna compared to that given by a lossless isotropic radiator supplied with the same level of power. If an antenna's gain is 2 (3 dB), it means that twice the amount of effective power will be sent in the direction of a target than from an isotropic radiator, and so has the equivalent effect of doubling the power of the transmitter in that particular direction. In practice, the gain of a ship's radar antenna will be around 30 dB (1000 times). By definition, the gain of an isotropic radiator is 0 dB (unity).

The antenna gain is a ratio and therefore has no units. Its maximum possible value would be realized in the ideal situation in which none of the energy supplied by the transmitter was wasted and all of it was concentrated within the optimum beam shape. In practice, this ideal cannot be achieved – it is inevitable that some of the transmitter power will be dissipated in overcoming any electrical resistance in the antenna, and also the energy cannot be concentrated in a truly optimal beam shape. However, good design techniques make it possible to reduce the electrical losses in the antenna to small values, and the special antennas used in marine radar can concentrate the power particularly effectively.

An alternative indicator of the measure of power concentration achieved by the antenna is called *directive gain* or *directivity*. This is given the symbol *D* and is defined as:

$$D = \frac{\text{maximum radiation intensity}}{\text{average radiation intensity}}$$

This quantity is independent of the antenna losses and is thus a figure of merit for describing the ability of the antenna to concentrate the power. If there are no losses, then $G = D$. Sidelobe levels relative to the main beam are identical whether directivity or gain is used as the fundamental measurement.

2.5.1.6 *Receiving Characteristics*

In the above, the antenna has been mainly treated as a transmitting element. For marine radar the same antenna is used for transmission and reception. However, advanced electromagnetic theory shows that there is a principle of reciprocity; whatever is said in relation to the directional nature of transmissions applies equally well to reception. This means, for instance, that the transmit and receive gains and directivities of an antenna are identical and if, say, a particular sidelobe, is −30 dB relative to the main beam on transmit, the same would be true on receive.

2.5.1.7 *Polarization*

The energy travelling outward from the antenna is in the form of an electromagnetic wave having electric and magnetic fields which are at right angles to one another and, except very close to the antenna, are also at right angles to the direction of propagation. The fields are said to be *orthogonal* to the direction of propagation. The polarization is defined as the direction of the plane of the electric field. To comply with IMO Performance Standards (see Section 11.2.1), all marine radars using the X-band must be capable of being operated with horizontal polarization. This is to ensure that they properly activate radar beacons and Search and Rescue Transponders (SARTs), which also have horizontally polarized antennas. Other polarizations are allowed, provided there is a clear indication of what is being used and that horizontal polarization can be readily accessed. In principle, circular polarization (see Section 3.7.4.6) offers benefits in some situations, such as in rain, but is currently rarely encountered on commercial marine radars. In practice, current X- and S-band marine radar systems typically only offer horizontal polarization.

2.5.2 Array Antennas

To achieve the directional transmission necessary for the accurate measurement of bearings (see Section 1.3), the horizontal beamwidth of the radar antenna must be appropriately narrow. This also has the effect of producing high antenna gain by concentrating all the available power in one azimuthal direction at a time. Further, such directional transmission is necessary in order to give the system the ability to display separately targets which are at the same range and closely spaced in bearing. This property is known as *bearing discrimination* and is discussed in Section 2.8.5.3.

The wavelengths of the electromagnetic energy contained in radar pulses are much longer than those of visible light but in many ways they exhibit similar characteristics. In the early days of radar, designers exploited this resemblance by using a parabolic reflector antenna to focus the radar beam, much in the same way as a reflecting mirror is used in a car headlight or torch to focus a beam of light. In civil marine radar the use of reflector antennas has been superseded by array antenna technology, but they are still used in some other radar applications such as ground-based air surveillance radars. For this reason, only array technology antennas are discussed in detail within this book.

2.5.2.1 *Multiple Radiating Elements*

An array antenna is comprised of multiple radiating elements. Figure 2.7(a) shows the antenna pattern of a single isotropic radiator. As discussed in Section 2.5.1.1, this is actually a sphere but the figure, for simplicity, is just looking at a single plane passing through the sphere's centre. For a theoretical and lossless isotropic radiator, $D = G = 1$, and therefore the radius of the sphere is unity. Now consider an electromagnetic wave emanating from the isotropic radiator at a certain frequency. This wave travels spherically outwards from the point source at the speed of light. At one particular instant the peaks in the wave could appear in the positions shown in Figure 2.9(a), which similarly to Figure 2.7(a) is depicted in a single plane. It can help to visualize this by comparing it to the water waves emanating on a calm surface after a stone has been thrown into the water. The wave-peaks at a particular time, frozen as in a photograph, correspond to the dashed concentric circles shown in Figure 2.9(a).

If we now plot the amplitude of a wave along a single radius, such as the diametrical line depicted in Figure 2.9(a), we get a sine-like wave but one that is diminishing in amplitude with distance from the isotropic source,

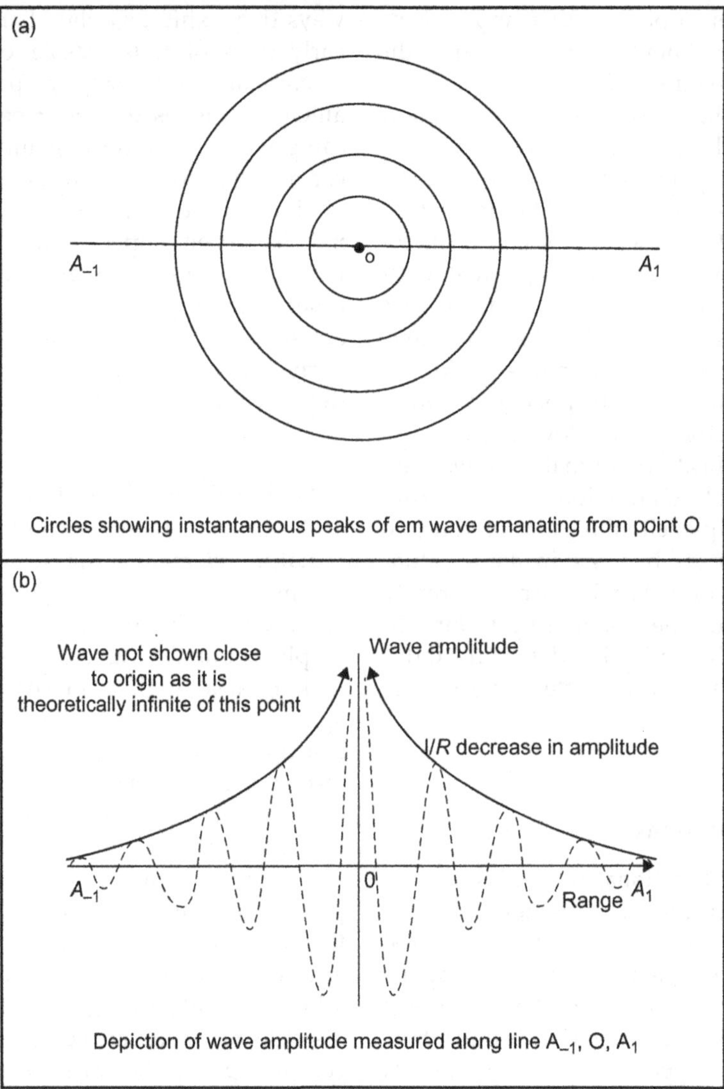

FIGURE 2.9 Isotropic antennas: (a) instantaneous depiction of transmitted wave, and (b) amplitude of wave as distance from source.

as depicted in Figure 2.9(b). The wavelength is that of the radiated signal, which for an X-band radar is about 3 cm. At every doubling of the radial distance the amplitude of the wave diminishes by a factor of 4, simply based on the increase in area of the sphere that is being illuminated as the complete wave travels out from the source. This is often referred to as the $1/R^2$ (one over R-squared) effect. Taking this philosophy in reverse, towards the source, each time halving the distance and therefore increasing the amplitude by four, mathematically ends up with an infinite amplitude of the wave at the point source. This shows that point source theory, although a useful concept, has theoretical problems in extreme circumstances.

Now consider a second isotropic source placed close to the first and radiating the same fixed frequency signal. For good reasons, later explained, it will be assumed that this source is half a wavelength away from the first. This creates an interference pattern, rather like when throwing two stones into the water, perhaps separated by about a metre. Figure 2.10 depicts the geometry in a single plane. Consider what happens along the line that bisects the two sources: the y-axis. The important aspect of this line is that any point along it is exactly the same distance from the two sources. Therefore, if the two sources are transmitting exactly the same signal with identical timing, the signals from each source will be identical in amplitude and phase. Phase is effectively the particular part of the wave's oscillation, whether a peak or trough or somewhere in between. The signals will simply add together along this bisecting line. This is shown in Figure 2.10 where on the y-axis it is seen that the maxima from both sources exactly coincide and so do the minima.

Now consider what happens along the extended line joining the two sources: the x-axis. Because of their half wavelength spacing the signals from the two sources will be exactly out of phase. When one is at a positive peak the other will be at a negative peak and so, when added together, the effect will be to partially cancel each other. At short distances from the 'array' of two sources (or elements), the cancellation will not be complete because each is a slightly different distance away from any point on this line and the $1/R^2$ effect will create slightly different amplitudes. However, when R becomes large, that is at long ranges, the cancellation effectively becomes complete,

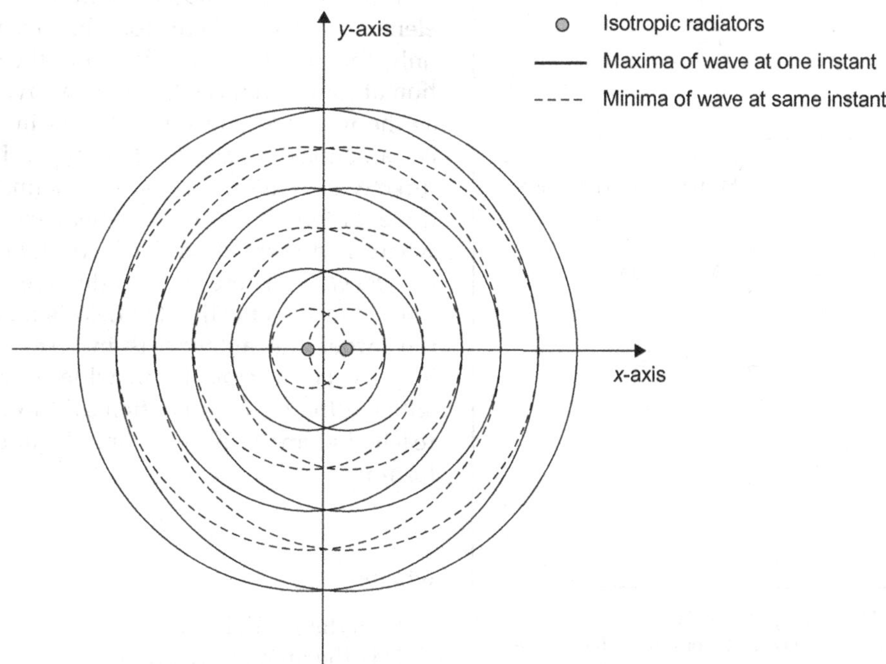

FIGURE 2.10 Two element array – depiction of wave interference.

simply because the very small relative differences in the ranges makes the signal strengths effectively equal (but with opposite *polarity*). This is shown in Figure 2.10, where the maxima from one source and the minima from the other source exactly coincide on the x-axis – and vice versa.

The effect on signal strength measured on every other line that passes through the midpoint between the two array elements is partial but not complete cancellation. The resultant pattern of the two-element array measured at a sufficiently far distance is shown in Figure 2.11(a). It exhibits a 'figure-of-eight' shape, with a peak directivity necessarily greater than unity. The detailed assessment of directivity is actually rather complicated and will not be developed here. In fact, the pattern shown in Figure 2.11 (a) is actually annular in shape because of the spherical beam emanating from each of the isotropic elements. It effectively comes out of the page; forms the lower half of the figure-of-eight shape as it goes back into the page; and then goes below the page to rejoin the upper half of the figure-of-eight. It is seen in Section 2.5.2.3 how this annular shape can be transformed into just a narrow vertical beam suitable for radar applications. Increasing the size of the array to many elements distributed along a line and with half wavelength spacing between each element increases the angular sharpness of the figure-of-eight pattern (i.e. the antenna horizontal beamwidth), as can be seen in Figure 2.11(b), and also introduces a sidelobe effect. The beamwidth of an antenna is normally specified in degrees at the half power (−3 dB) levels.

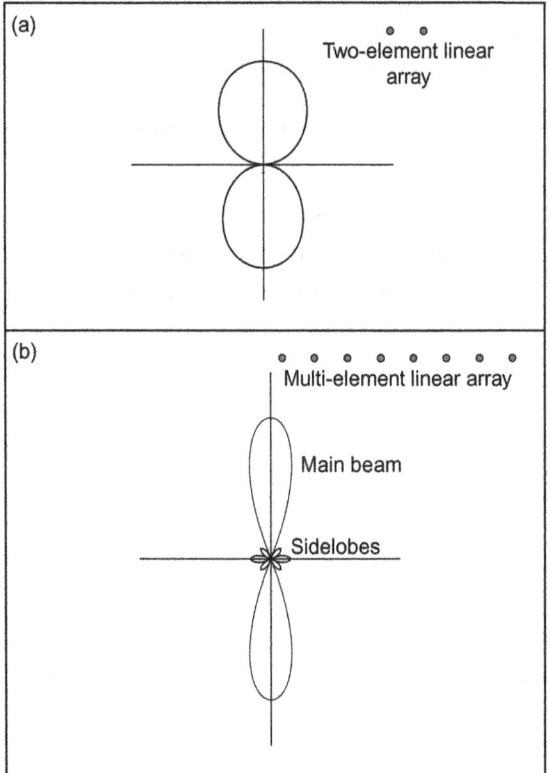

FIGURE 2.11 Array patterns: (a) radiation pattern of two-element cophased array of isotropic radiators and (b) radiation pattern of multi-element cophased array of isotropic radiators.

The 'range-dependent' pattern of the two-element array, where for this example we only get complete cancellation in the 90° direction at long distances from the array, is known as the *near-field* effect and occurs in all practical antennas, whatever their type. The gain/directivity pattern of any antenna can be quite different in the near-field compared to where it has stabilized to a constant shape in the *far-field*. There is no definite transition point, but the effective far-field is more distant the larger the antenna is in terms of wavelengths. For a typical 2 m wide X-band radar antenna the far-field is often said to have been reached at about 270 m, using the approximate formula:

$$F_d = 2D^2/\lambda$$

where:

F_d is the far-field distance
D is the antenna width and
λ is the radiation wavelength

In general, antennas are less effective, especially in terms of gain, beamwidth and sidelobes, when operating at near-field distances. It is rare for the near-field performance of antennas to be quoted.

2.5.2.2 *Sidelobes*

Sidelobes are an unfortunate reality of all directional antennas. They are a consequence of nature and are directly related to diffraction effects in light – and there are similar effects in sound and water waves. Diffraction occurs whenever there is a sharp discontinuity in a radiating or reflecting surface. On the antenna arrays so far described we have a number of elements all radiating and suddenly the antenna ends, which creates diffraction or sidelobe type effects. These effects can be minimized by carefully controlling the power radiated by each element such that, in general, the further an element is away from the centre the less power it should radiate. In a sense this 'softens' the edge effect, lowering the component of diffraction. There have been many complicated algorithms developed over the years in attempts to optimize the performance of antenna arrays by determining the 'optimum' radiated power for each element.

A linear array consisting of elements radiating with equal power has maximum sidelobes of about −13 dB, that is about 20 times lower than the main beam directivity. Good designs can readily lower sidelobes to peak at about −30 dB or even to −40 dB and better. In general, lower than −40 dB peak sidelobes are only essential for military systems, where they are needed to counteract jamming. This is because designing for very low sidelobes creates two significant drawbacks. The first is that as the sidelobes are lowered the beamwidth increases, thus requiring a larger antenna to get the required beamwidth. This increases both cost and causes siting difficulties. The second is that ultra-low sidelobes require very precise control of the power to each element, significantly raising the cost of design and production. It has already been noted that the beamwidth of a linear array is dependent on the number of elements in the array, although in general, the total length of the array is more fundamentally important. Typical array distributions used for civil marine radar antennas yield a horizontal beamwidth given approximately by the formula:

$$\text{HBW (in degrees)} = 70 \times \frac{\text{transmitted wavelength}}{\text{aperture width}}$$

The practical consequences of this are explored in more detail in the following section.

2.5.2.3 *Practical Considerations*

The theoretical linear arrays of isotropic radiators considered so far do not give an antenna pattern usable for marine radar. The directivity patterns shown in Figure 2.11 form figure-of-eight patterns in three dimensions, meaning that the complete pattern has a main beam (and sidelobes) that are annular around the array axis. What is wanted is a vertical pattern that only occupies about 20°, as described in Section 2.5.2.5, and not a full 360°. This can be achieved by using elements that themselves have the desired vertical beamwidth of the complete antenna, rather than being isotropic. This is typically achieved by enclosing the elements within a reflecting surface that constrains the vertical extent of the beam, such as the example illustrated in Figure 2.12(a). Another technique is to use a two-dimensional array of elements that are constrained to radiate in just the forward direction, as conceptually depicted in Figure 2.12(b). Such a *planar* array has defined beamwidths in both the horizontal and vertical directions. The wider beamwidth that is required in the vertical direction means that fewer elements are needed in the vertical dimension than are needed horizontally.

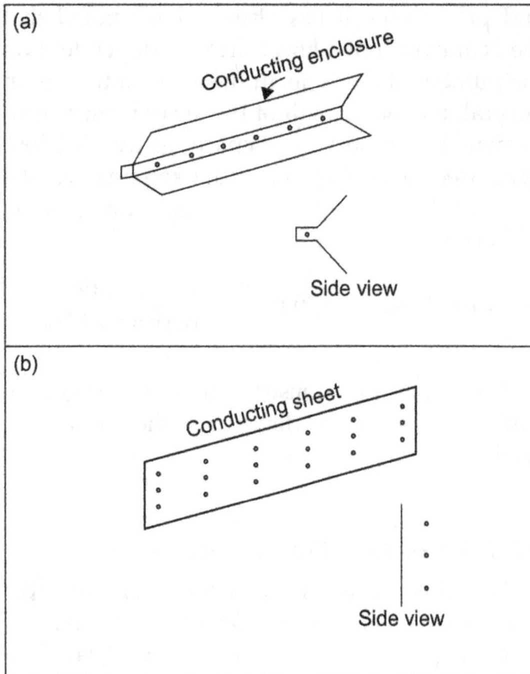

FIGURE 2.12 Vertical pattern shaping: (a) partially enclosed linear array of elements and (b) planar array of isotropic elements backed by conducting sheet.

The important principles are that the horizontal beamwidth is directly proportional to the transmitted wavelength but inversely proportional to the effective width of the antenna array. A practical feeling for the quantities involved is illustrated here by two numerical examples.

EXAMPLE 2.1

Calculate the antenna width necessary to achieve a horizontal beamwidth of 2° for an X-band radar operating at 9.4 GHz:

$$\text{wavelength (m)} = \frac{\text{speed of radio energy (m/s)}}{\text{frequency (Hz)}} = \frac{3.00 \times 10^8}{9.4 \times 10^8}$$

that is wavelength = 0.032 m (3.2 cm).

$$\text{HBW (degrees)} = 70 \times \frac{\text{transmitted wavelength (m)}}{\text{aperture width (m)}}$$

(See Section 2.5.2.2)

$$\text{thus } 2 = \frac{70 \times 0.032\ \text{(m)}}{\text{aperture width (m)}}$$

$$\text{hence aperture width} = \frac{70 \times 0.032}{2}$$

that is aperture width = 1.12 m.

Therefore, to produce a horizontal beamwidth of 2° with X-band transmission at 9.4 GHz, an antenna width of about 1.12 m is necessary.

EXAMPLE 2.2

An S-band marine radar antenna operating with a wavelength of 9.87 cm has a width of 3.8 m. Calculate its approximate horizontal beamwidth:

$$\text{HBW (degrees)} = \frac{70 \times \text{transmitted wavelength (m)}}{\text{aperture width (m)}}$$

thus

$$\text{HBW} = 70 \times \frac{9.87 \times 10^{-2}\ \text{m}}{3.8\ \text{m}}$$

that is HBW = 1.82°.

Therefore, a marine radar antenna of width 3.8 m, transmitting a wavelength of 9.87 cm, will produce a horizontal beamwidth of about 1.8°.

It is clear from the examples why the internationally agreed frequencies (see Section 2.3.2) allocated to civil marine radar needed to be of an order which would give a wavelength of not more than about 10 cm. Longer wavelengths would require an antenna size so large as to preclude practical installation on many merchant vessels.

It is evident that for the same horizontal beamwidth an S-band antenna must be about three times the width of a corresponding X-band antenna. In general, available S-band antennas are not as large as this and hence they normally have wider beamwidths than typical X-band systems. To comply with IMO Performance Standards (see Section 11.2.1), a horizontal beamwidth of less than 2° is required. Radars designed for use on large merchant ships normally offer horizontal beamwidths somewhere in the range 0.65–2° for X-band operation and 1.7–2° in the case of S-band.

2.5.2.4 The Slotted Waveguide Array Antenna

If an electromagnetic wave is imagined impinging on a large conducting sheet having a small aperture within it, then it appears obvious that some energy will 'leak' through the opening, which is indeed the case. The actual energy that can leak through is dependent on a number of parameters, particularly the dimensions of the aperture relative to the wavelength of the incident wave. A very small aperture compared to the wavelength will allow very little energy to get through, but when the dimensions approximate to a half wavelength ($\lambda/2$) there is a marked increase in the energy transmitted. In effect, the aperture is forming a small antenna, radiating energy into the region on the other side of the conducting sheet. Detailed physics can be used to accurately characterize the amount of energy being radiated from the aperture. In particular, if the impinging electromagnetic wave is horizontally polarized, a good radiating aperture is a vertical slot of length $\lambda/2$ with minimal width; the emerging wave is also horizontally polarized.

A waveguide is a conducting tube that is designed to allow electromagnetic energy to travel within it. Typically, the tube has a rectangular, circular or elliptical cross-section. The dimensions of the tube are carefully chosen to suit the band of wavelengths that need to be propagated along the tube. They can be made so that very little energy is dissipated, even in long lengths of waveguide. In fact, circular waveguide formed the basis for the development of fibre-optic cable, which allows signals at optical frequencies to travel along them with little attenuation for very long distances and are extensively used in the telecommunications industry. What is of interest here is that the energy propagating within the waveguide can also be made to 'leak' from slots created within the waveguide walls, similarly to that described for the conducting sheet. These slots can be considered to be the radiating elements of a linear array.

In a rectangular waveguide the slots can be made either in the broad or narrow walls; however, the use of detailed physics is necessary to explain their particular characteristics. Practical slots are approximately a half wavelength long ($\lambda/2$). If in the broad wall of the waveguide, the slots generally need to be cut parallel with the longitudinal axis of the waveguide. Such slots, if cut along the centre line of the wall, emit no energy but when displaced from this line, more and more energy is emitted, peaking when at its edge. For a horizontally placed waveguide array the resulting radiation is vertically polarized.

If cut in the narrow wall of the waveguide the energy radiated by each slot is dependent on its angle with respect to the waveguide's longitudinal axis. When at right angles to this axis no energy is radiated, but as the angle increases so does the radiated energy. In practice, the slots remain nearly at right angles to the waveguide axis. Since the narrow wall of the waveguide has to be rather less than $\lambda/2$ in height to make it an effective waveguide, the slots have to wrap around into the broad wall, simply because they need to be around $\lambda/2$ in length to emit the required energy.

To create a beam that is at right angles to the slotted waveguide array, the slots need to be positioned so that they each emit energy with the same phase. This means that they have to be spaced at intervals of one wavelength. Detailed analysis shows that the wavelength in the waveguide is longer than that in 'free space', typically by a factor of about 1.4 times. Such a spacing would make the elements of the array greater than a wavelength in free space, which has the highly undesirable effect of creating an additional beam in the pattern. Fortunately, advanced theory shows that the phase of the transmission from such slots differs by 180° depending on whether they have been angled in a clockwise or anticlockwise direction. This means that if the slots are spaced at half waveguide wavelength intervals but tilted alternately in the clockwise and anticlockwise directions, they would all radiate with the same phase. In free space the spacing would be rather less than a wavelength, generally removing the possibility of an unwanted additional beam.

The alternate tilting of the slots also has the effect of making the main beam virtually precisely horizontally polarized (for a horizontally oriented array), since the vertically polarized components of the emitted energy effectively cancel each other out. As previously mentioned, this meets IMO polarization requirements for a marine radar, and because of the relative simplicity in its design and manufacture results in the narrow wall slotted array being the basis of many practical marine radar antennas in use today.

Two more detailed effects need to be taken into account in a real system. The first is that the exact half waveguide wavelength spacing of the slots causes a 'resonant' effect that tends to reflect a lot of the energy being supplied by the transmitter back to it from the array and not being radiated. This can be overcome by slightly decreasing (or increasing) the spacing. The effect of this is to put a 'squint' into the main beam direction of maybe 3° or 4° from the broadside direction. This arises from more detailed antenna theory, taking into account that there is a steady change in the radiated phase from each radiating element within the array. This clearly has to be taken into account in the overall design of the radar system. The user needs to appreciate that the axis of the main beam will therefore not be fore-and-aft when the long axis of the antenna is athwartships. The squint is frequency dependent and so this also needs to be borne in mind, especially if a magnetron is replaced with one having a different frequency.

Another issue is that even at the last slot in the array there will still be appreciable energy travelling within the waveguide. This has to be absorbed by a resistive 'load' at the far end of the waveguide. This inefficiency shows in the fact that the gain of such an array is rather less than its directivity, typically by about 5%, which is equivalent to approximately 0.2 dB.

Figure 2.13(a) shows the detail of an antenna waveguide with a typical pattern of slots. Figure 2.13(b) shows a full slotted

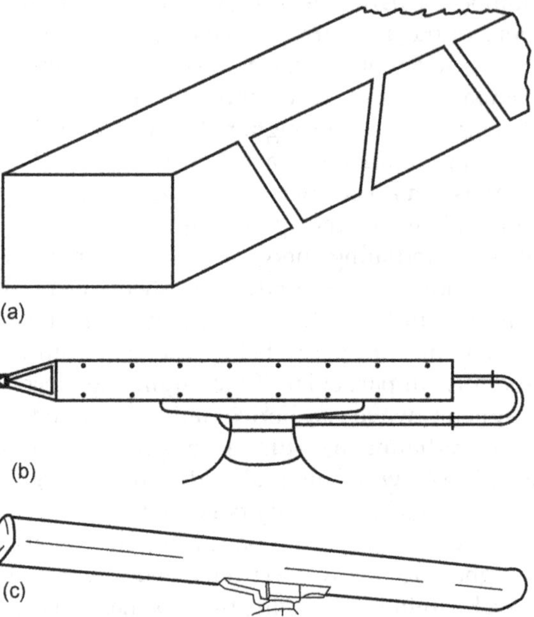

FIGURE 2.13 The slotted waveguide array antenna: (a) detail of waveguide slots, (b) basic antenna assembly and (c) modern enclosed system.

waveguide antenna assembly of an older type which has the advantage that all of the parts can be clearly seen. The width of the aperture is simply determined by the length of the slotted unit. In the vertical plane, as previously described, the aperture is bounded by conducting sheets effectively forming a horn which guides the waves from the narrow vertical dimension of the slotted unit to an aperture of a depth consistent with the correct vertical beamwidth (see Section 2.5.2.5). A sheet of glass-reinforced plastic (GRP – which is often referred to by the name fibreglass) protects the aperture from the ingress of water and dirt. GRP is virtually transparent to radar waves (see Section 3.3.3). Most paints are not transparent to radar energy and so the cover commonly carries the legend 'do not paint'. As illustrated in Figure 2.13(c), common modern practice is to enclose the entire assembly in a GRP envelope.

2.5.2.5 *Vertical Beamwidth*

The minimum vertical beamwidth for compliance with IMO Performance Standards is 20° because of the need to maintain the required performance when rolling and pitching. A further factor in favour of a wide vertical beam is that it reduces the possibility that a target at close range will escape detection by passing under the lower limit of the beam (see Section 3.2.4).

To achieve a 20° beamwidth means that the vertical dimension of the array can be up to 20 times less than the horizontal dimension that would give a 1° beamwidth, characterizing the general outline of a marine radar antenna. The large vertical beamwidth necessary to compensate for the ship's rolling and pitching does, however, create an undesirable effect, known as *vertical lobing*. This is because the radiation directed towards the sea is reflected, forming an interference pattern with the unreflected radiation direct from the antenna.

Figure 2.14 illustrates how the interference pattern is created. Consider two rays of energy which arrive at point *P*, one having travelled directly and the other having been reflected from the surface of the sea at *B*. Let us suppose that the reflected path length is one half wavelength longer than the direct path length. Detailed analysis shows that a 180° phase shift takes place on reflection at the sea surface, and hence the signals arriving at point *P* will be in phase (180° phase shift due to path difference and 180° phase shift due to reflection). Thus, assuming that the signal is perfectly reflected, the signal strength at point *P* will, due to the addition of the reflected signal, be twice what it would have been if the antenna were located in free space. If the reflected path is extended further it must in due course intersect with another direct ray at a point at which the path difference is a full wavelength. At that point there will be 180° phase difference (360° due to path difference and 180° due to phase shift on reflection) between the direct and reflected signals, and the resultant signal strength will be zero. Clearly this pattern of maxima and minima will repeat itself with height.

As a result of the existence of alternate maxima and minima the single main lobe in free space is broken up into a family of lobes of the form shown in Figure 2.14(d) and (e). The diagram shows only a few lobes in order to illustrate the pattern of maxima and zeros. In practice, for X-band transmission, there will be approximately 33 lobes per metre of scanner height. Thus, the lobes are much more numerous and more closely spaced than those illustrated in a diagram of reasonable size to maintain clarity. For S-band signals the figure is approximately 10 lobes per metre of scanner height. Thus the S-band vertical pattern has fewer, more widely spaced lobes. The radiation intensity along the axis of a lobe is twice what it would be in free space and an area of zero radiation lies between each pair of lobes. It can be shown that the elevation of the

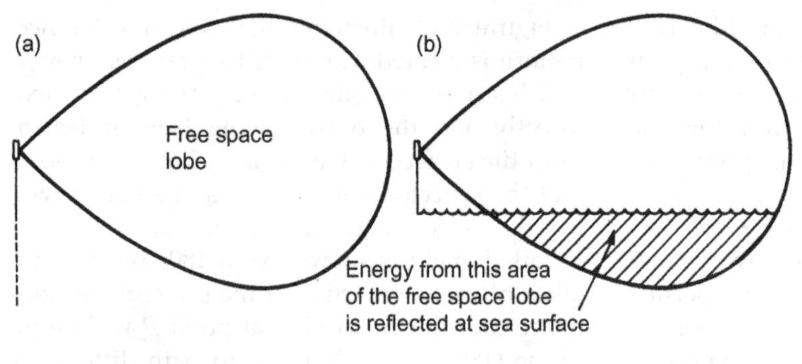

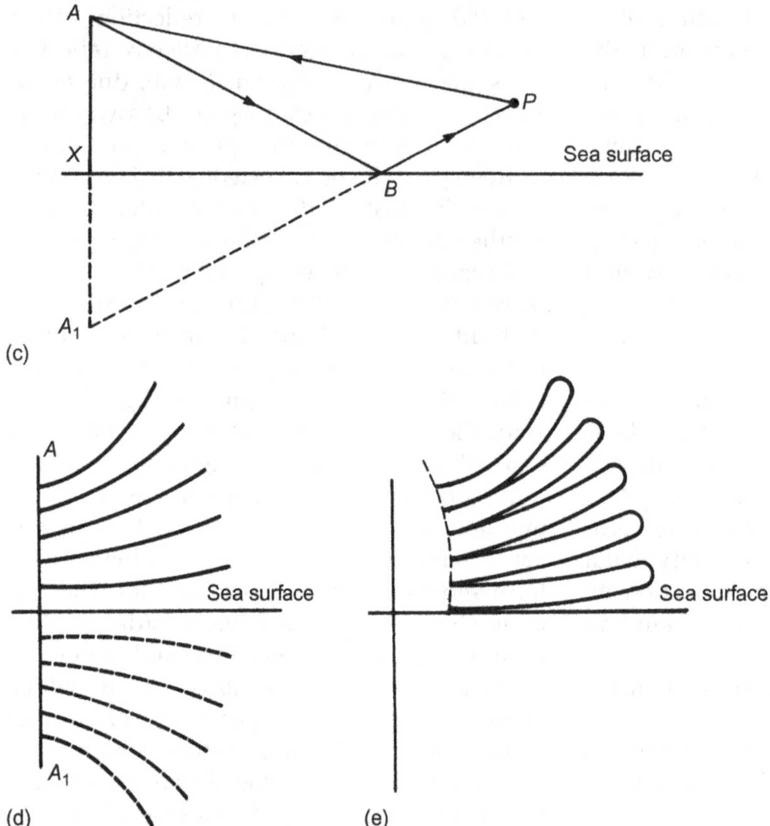

FIGURE 2.14 Vertical lobing pattern. (a) Main vertical lobe in free space; (b) Main vertical lobe close to sea surface; (c) The interference pattern; (d) Hyperbolic pattern; (e) Vertical lobe pattern – only a few lobes are shown. In practice, the number of lobes per metre of antenna height is approximately: At X-band 33; At S-band 10. *Note: The horizontal extent represented in (e) is 2 × that represented in (b).*

lowest maxima in degrees is approximately 14 times the ratio of wavelength to antenna height. For X-band transmission with an antenna height of 20 m the lowest maxima will be approximately 8.5 m above the sea at a range of 12 NM (nautical miles).

Despite the apparently dramatic effect on the vertical radiation pattern of a radar antenna, vertical lobing is not often a real issue. This is because of two main reasons. The first is that the gross effects only occur when the sea is very calm. Waves create multiple reflections that rapidly 'fill in' the nulls of the lobing pattern, maintaining reasonable visibility of a point target at all vertical heights. Also, many real targets are not approximated by a point target, they have appreciable vertical extent. This also generally lessens the effect of vertical nulls. However, when observing a small target in relatively calm conditions, problems can occur. For instance, a radar reflector on a yacht generally forms the largest component of the radar reflection from the vessel and acts as a point target. Under relatively calm conditions a yacht that may have good visual visibility may be unobservable by a radar at specific ranges. Racons (radar beacons, see Section 3.5) also act as point source targets, with the possibility of lobing making them non-reactive to a specific radar at certain ranges.

2.5.2.6 *Parallel Fed and Active Arrays*

An alternative to the slotted waveguide array antenna is using discretely fed radiating elements. The network feeding them just splits up the power from the transmitter so that the required amplitude at every element is produced, see Figure 2.15(a). This is known as a parallel fed array, whilst the slotted waveguide array is known as a serially fed array. The important thing about the parallel fed array is that the path lengths to each element from the transmitter can be made exactly the same. For a serial fed array, as depicted in Figure 2.15(b), the path lengths are all different resulting in a phase progression along the array. As described in Section 2.5.2.4, this causes the beam to squint. A parallel fed array has no squint. The network producing the power division is unsurprisingly known as a *power divider network*. A well-designed parallel fed array can produce very low sidelobes, but is typically more expensive to manufacture than serially fed systems and so they have not been as widely used as marine radar antennas. This is likely to change into the future with improved manufacturing techniques.

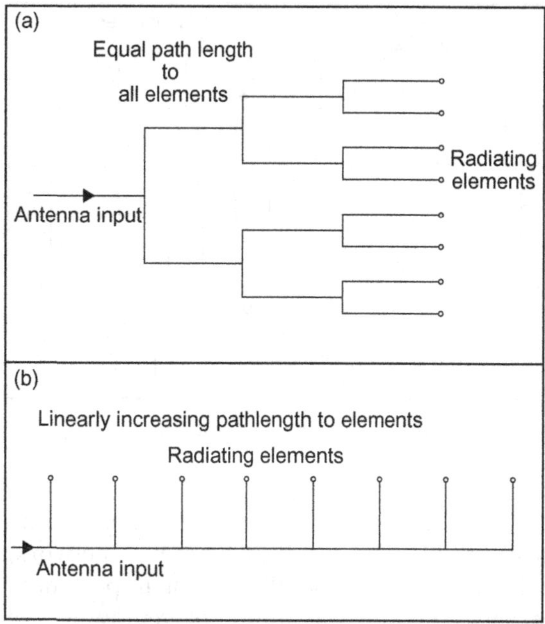

FIGURE 2.15 Array feed networks: (a) parallel fed linear array and (b) serial fed linear array, for example slotted waveguide array.

With the advent of semiconductor-based transmitters for marine radar a totally different type of array becomes feasible, known as the *active array*. In an active array each element has its own mini-transmitter and also a receiver, known as a T/R module, as shown in Figure 2.16, which creates the required signal amplitude and phase for the particular connected element. Such technology is used today on a number of military radars and is especially used for air surveillance. For many years it has been postulated that marine radar will one day take this route. It is tempting to say that this may become feasible for marine radar within the next 10 years, but people have been saying the same for 30 years or more.

Active arrays give a number of benefits. Firstly, the power from each element is low, allowing the use of more affordable RF transistors than are typically needed for a single transmitter. Also, the losses in the RF circuitry

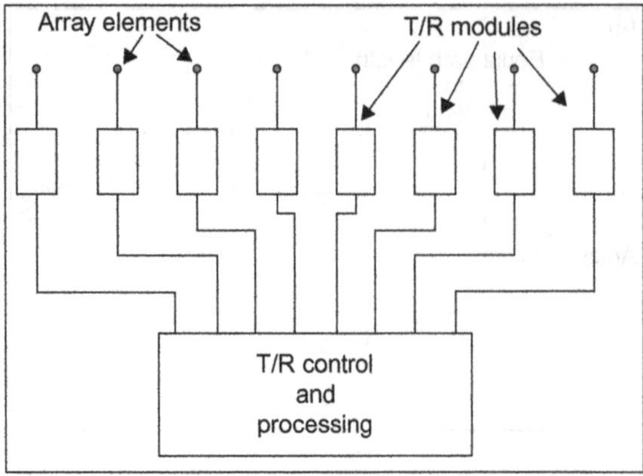

FIGURE 2.16 Active array concept.

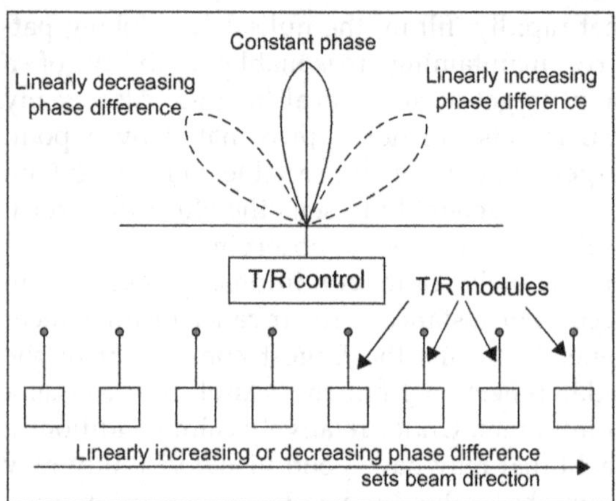

FIGURE 2.17 Beam steering with active array.

can be kept very low because there is no lost power in a beamforming network, whether serial or parallel. Most importantly, it can give the possibility of removing the need for a rotating antenna, saving cost and removing the mechanical issues associated with rotating structures, not least their maintenance. This is because the phase – and even the amplitude – of each radiating element can be electronically adjusted so that it can form an optimized beam emanating from the array in any direction, see Figure 2.17.

The concept would need three or four separate linear or planar arrays, or perhaps even a circular or cylindrical array of elements, as shown in Figure 2.18. Such an antenna would allow a beam to be electronically scanned. Perhaps more than one beam would be active at any one time. The possible future use of active elements in the vertical plane would allow for the beam to also be directed in the vertically to compensate for the vessel's pitch and roll. This would allow a narrower vertical beam, resulting in less reflection from precipitation and lobing effects from the sea. The resultant higher effective gain of the antenna would further reduce the power needed to be produced at each element, helping its cost feasibility. Furthermore, the

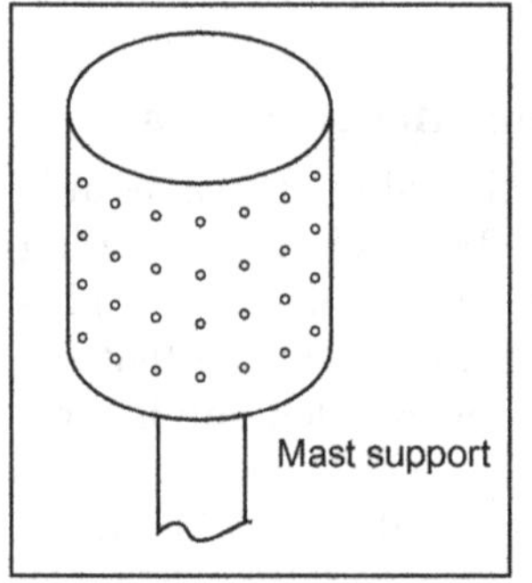

FIGURE 2.18 The active cylindrical array. The marine radar antenna of the future?

antenna structure could be around a mast and so shadowing effects would be greatly alleviated. This is all speculation, but is inevitable at some future time.

2.5.3 Antenna Rotation Mechanism

A significant part of the radar system today is its antenna rotation mechanism, often known as the *turning mechanism* or *turning gear*. Its main active items (Figure 2.19) are an electric motor, for producing the primary turning energy; a gearbox to convert the rotation rate of the motor to the required antenna rotation rate; and a rotating joint (sometimes called a *rojo*) to be able to transfer the transmitted

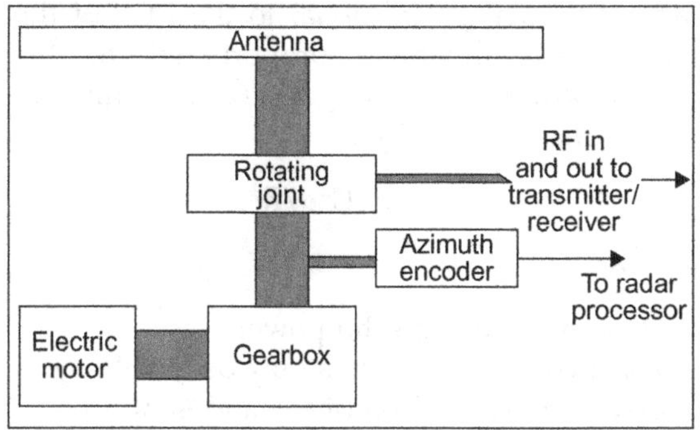

FIGURE 2.19 Radar antenna-turning gear.

and received RF signals to and from the antenna. The structure forms the main mount for the antenna and can also house the transmitter and receiver, when in an upmast configuration. On some newer systems the use of a *direct drive motor* removes the need for a gearbox, with the precise speed of the drive being electronically controlled. This can produce very stable rotation rates, even in extreme conditions of wind and vessel motion.

A small but important part of the turning mechanism is a device known as the *azimuth encoder* or *syncro transmitter*, which provides the accurate angular positional information of the antenna. The encoder generates a signal that allows the precise pointing angle of the antenna to be accurately assessed. It typically sends a precise pulse at a defined azimuth increment, often 1024 pulses per 360° of revolution. It separately supplies a pulse when passing through the heading line, effectively maintaining the calibration of the azimuth pulses. The use of these signals is covered in Section 2.7.3.2.

2.6 RADAR SIGNAL RECEPTION AND PROCESSING

2.6.1 The Radar Equation

Before considering the receiver and processor functionality of a radar in more detail, it is useful to consider what factors fundamentally govern radar performance. The basic list is relatively short, comprising the following: the effective power of the transmitter (see Section 2.3.3.3); the gain of the antenna (assumed to be used for both transmit and receive); the distance of the target from the radar; the ability of the target to reflect signals back to the radar; and the sensitivity of the radar receiver.

Initially ignoring the receiver sensitivity and just considering the level of the reflected signal at the output of the antenna emanating from a particular target at a defined range R, it is possible to make a number of important deductions as certain parameters are varied. If, for instance, the level of the transmitted power is doubled the received energy from the target at the output of the antenna would also double. If the antenna gain is doubled, that is increased by 3 dB, it would result in twice the energy reaching the target, and on reception, a further doubling of received energy. This gives a four times increase in the received power level (6 dB).

Now consider what happens when the target range is altered. It has already been deduced in Section 2.5.1.2 that the transmitted energy follows the relationship $1/R^2$. If the target range was initially at 3 NM and then it moved out to 6 NM, the resultant received energy would be down by the ratio $3^2/6^2 = 0.25$, which is equivalent to −6 dB. This is true for all cases of doubling the range – the signal level at the target becomes 6 dB lower every time the range is doubled. The energy of the target-reflected signal also decreases by 6 dB (equivalent to $1/R^2$) every time the range is doubled. Taking the effect on the transmitted and the received signal together it means that the range effect becomes $1/R^2 \times 1/R^2 = 1/R^4$ – a doubling of range implies a 12 dB reduction in the received signal, that is to a level of 0.063 to what it was previously. This is a very rapid reduction in power as the target moves further from the antenna and epitomizes radar when compared to other radio systems. Radio, TV, wifi and mobile phone communications all have $1/R^2$ characteristics. If we double the transmitted power of a normal radio system the maximum range nominally increases by a factor of 1.41 (the square root of 2); if we double the power of a radar transmitter the effective range only increases by 1.19 (the fourth root of 2).

The ability of a target to reflect incident transmitted energy back towards the radar is also a very important factor in determining the radar's effective range. In general, and not surprisingly, physically small targets are poor reflectors of energy, whilst large ones return a stronger echo. The strength of the reflection is also dependent on the material that comprises a target and its shape. For all targets other than a perfect sphere the amount of energy returned is also highly dependent on their orientation to the radar. It is perhaps not surprising that a ship typically has a lower ability to reflect radar signals when it is head on than when it is broadside on. What is perhaps surprising is that very small changes in their orientation can cause massive fluctuations in the level of the reflected energy. This contributes to a phenomenon known as scintillation, where the observed echo is seen to fluctuate rapidly in intensity.

A useful concept, known as *radar cross-sectional area* (RCS), has been developed to quantify target reflectivity, which is normally measured in square metres. The area can be likened to an electromagnetic 'net' of the specified area that collects the incident radar energy at the equivalent range of the target, with the collected energy being 'retransmitted' towards the radar by a hypothetical isotropic radiator. A target with an RCS of 10 m^2 will hence reflect twice (3 dB) as much power as one with an RCS of 5 m^2. Hence the power available at the receiver is proportional to the power of the transmitted signal; proportional to the square of the gain of the radar antenna; proportional to the RCS of the target; and inversely proportional to the fourth power of the target's range. This can be put mathematically as:

$$P_r = kP_tG^2\sigma/R^4$$

where:

P_r is the received echo power
k is a constant, which is only derivable by advanced considerations – and is of no consequence to a basic understanding
P_t is the effective transmitted power
G is the antenna gain
σ is the RCS of the target
R is the range of the target from the radar

This equation is a form of the *radar equation*, essential for the understanding of fundamental radar principles. It is given some more consideration in a slightly different form in Chapter 3.

Clearly, the larger the value of P_r, the greater is the power available at the receiver. If P_r is very low the receiver will not be able to amplify the reflected power successfully, because its level will be comparable either to the level of natural electromagnetic noise in the vicinity of the radar or to the level of self-noise induced by the receiver itself. Also, if P_r is very high, the receiver may not be able to deal with it successfully, since it would attempt to increase its level beyond that which could be accommodated by the receiver – the receiver would become *saturated*.

2.6.2 The Role of the Receiver and Processor

The primary role of the receiver is to amplify the weak returning echoes intercepted by the antenna and optimize the received signal prior to it being converted into digital format. Once digitized the signal data enters several stages of processing. The first is

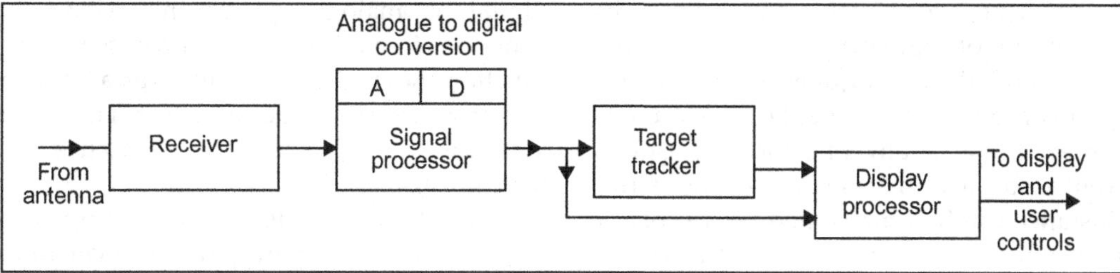

FIGURE 2.20 Receiver processor chain.

generally known as signal processing, which, for instance, helps the wanted echoes to be distinguished from sea and rain clutter. In fact some of this initial processing may be carried out before digitization, depending on the system. The processed data then enters the display processor, which puts it into a format that allows it to be clearly visible to the radar user, for instance on a flat panel display, together with other relevant information, such as range rings, a bearing scale, a heading line and textual data. In addition, there is a TT processor, which takes the data from the digital signal processor and, on a scan-to-scan basis, tracks the movement of echoes from targets, estimating their courses and speeds. The derived data from the tracking processor is sent to the display processor. The process is generally known as the automatic radar plotting aid (ARPA) reflecting its historical background, although it is now referred to as target tracking (TT) by IMO in its Radar Performance Standards. The term ARPA continues to be used in IMO Model Courses defining the training requirements for mariners.

Figure 2.20 shows the interconnection between the receiver and the various processors. The signal processor is further described in Section 2.7; the TT processor in Section 2.7.5; and the display processor in Section 2.8.

2.6.2.1 *Analogue and Digital Processes*

The receiver processes the signal and, in particular, amplifies it by analogue means in order to get it into a form suitable for digitization. Analogue signals are continuously variable, resembling the real-life continuous variations of the actual signals received by the radar antenna, see Figure 2.21(a). In its simplest form a digital signal consists of just on and off states (say, generating 0 and 1 V to distinguish the states), which can only change, or remain the same, at precise timing intervals, see Figure 2.21(b). This enables long sequences, such as 10010100···101, to be generated. If the sequence can then be considered to be subdivided into groups, such as into four, for example 1001, 0100,..., then each of the possible sequences in the group can have an identified meaning. For instance, 0000 could represent the digit zero; 0001 by the digit one; 0010 by two; and 0011 by three, etc. A digital signal

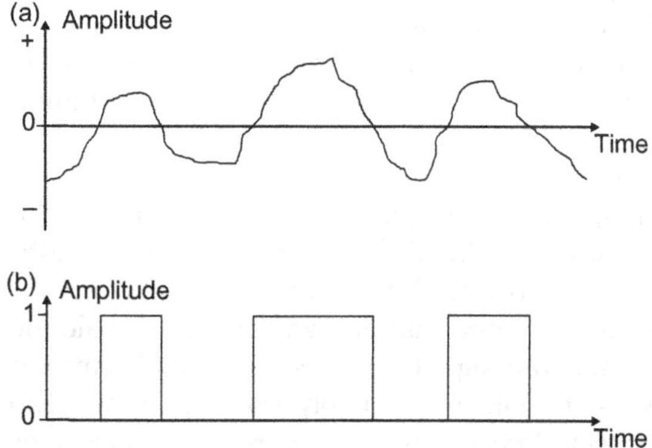

FIGURE 2.21 Analogue and digital signals: (a) analogue signal and (b) basic digital signal.

can represent absolutely anything by suitable coding of regular numbers, letters and symbols into digital sequences; hence the power of computers. In particular, the strength of a radar signal at any one instance can be represented by a digital sequence; the actual time instance can be represented by a separate digital sequence. We call each time-specific measurement of the signal strength (or any other parameter) a *sample*. Once the signal is digitized, it is relatively straightforward to perform very complex processing on it since it becomes a straightforward digital computing process. However, there are several limitations. Firstly, the signal must be of a suitable strength to permit accurate digitization, and so the small signals of a radar return must be amplified by the receiver before digitization is practical. Secondly, if we are to replicate the information within the analogue signal, the signal's successive digital samples must be taken at suitably short intervals. The chosen interval is known as the *sample rate*. The problem with a radar signal is its high transmission frequency. It is currently not economic to digitize received microwave radar signals directly, although this will inevitably change in the future. This means that the receiver also first has to perform some analogue 'processing' in order to help extract the underlying changes to the transmitted signal when it is reflected by a target and collected by the radar antenna. It initially does this by a process known as *down-conversion*, which is described in more detail below.

In particular, the speed limitation of digital processing has to be taken into account; or at least the cost and performance tradeoffs of doing something digitally or by analogue means. In Sections 2.6.3.3 it is explained that the processed signal is best represented logarithmically, since this allows huge variations in signal strength to be more easily accommodated. In principle the conversion to logarithmic representation can be performed by either digital or analogue processing. Many radars continue to use analogue processes for this function, because it can still offer a good compromise in performance and cost. The analogue logarithmic receiver is discussed in Section 2.6.4.5.

In modern systems it is becoming increasingly difficult to separate the receiver function from that of the processing. This is because of the increasing capability of digital processing, for which there are no apparent theoretical limits. For a rising number of RF-based systems the only analogue functionality is the direct amplification of the received signal, and any necessary pre-filtering of the signal to help ensure that unwanted signals, such as interference, are not amplified. Once at an appropriate level of RF amplification the signal is digitized and all subsequent processes are performed in the digital domain. However, for most current magnetron-based radars, the designs follow the 'traditional' route of analogue processing towards a baseband signal (see Section 2.6.4.6), where it is converted into digital format.

2.6.3 The Radar Receiver

A block diagram of a conventional radar receiver is shown in Figure 2.22. Each of the elements is discussed in the following subsections. Section 2.6.4 expands on the design principles of the total receiver and discusses the later stages of the receive process, which merge with that of signal processing and can be carried out in analogue or digital forms.

2.6.3.1 Receiver Protection

The returning pulses collected by the antenna are extremely weak and the receiver must be capable of amplifying signals having a received amplitude of as little as one millionth of a volt. Both the transmitter and the receiver are connected to the same antenna, usually via waveguide. If the powerful transmitted signal

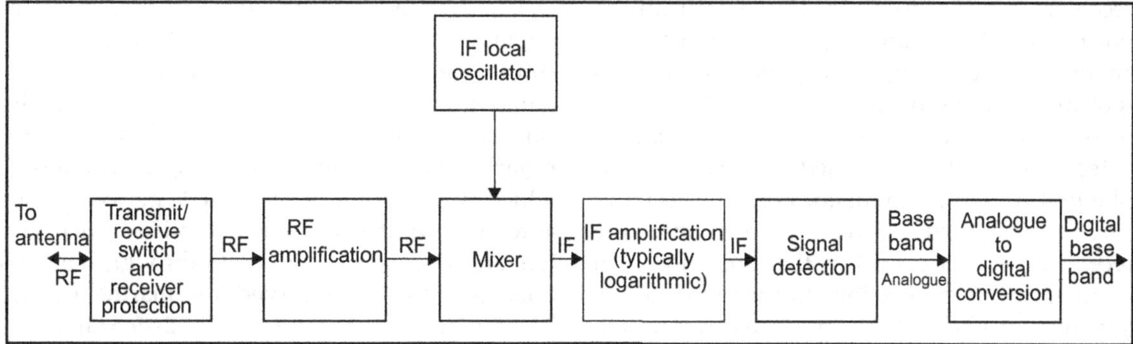

FIGURE 2.22 Block diagram of radar receiver.

was allowed to enter the sensitive receiver, permanent damage would almost certainly occur. At the very least, the receiver would take time to recover its sensitivity to small echoes. For this reason it is normal – and essential for the very high peak powers of magnetron-based radars – to have an automatic switch that consecutively connects the transmitter (for just the duration of the transmitted pulse) and then the receiver to the antenna, so that echoed signals are properly directed and then amplified. This is often called the receiver protector and sometimes the T/R switch. This switch prevents power from the transmitter entering the receiver. On older systems, because of the technology then universally employed, it was known as the T/R cell, and this term is still sometimes used today. On modern systems, the receiver protector typically consists of a *circulator*, based on ferrite technology, which cleverly separates the received and transmitted signals effectively by their direction of travel and a *limiter*.

The circulator cleverly makes use of the field structure of electromagnetic energy when affected by magnetized ferrite material. The physics is surprisingly complicated and so its detailed action is not described here but the ferrite material, working in conjunction with the device geometry, effectively creates a 'non-reciprocal' device, meaning that the behaviour of a wave travelling in one direction is different to one that is travelling in the opposite direction. The functionality of the circulator is shown in Figure 2.23. It is seen that if the radar antenna was connected to Port 3 it would allow transmitted energy to enter from Port 2 and direct received energy to Port 1. Very little energy from Port 2 would enter the receiver at Port 1; also, very little received energy from Port 3 would be directed towards Port 2.

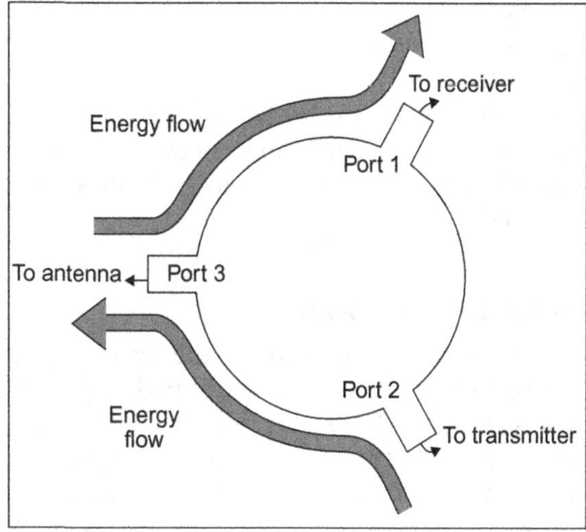

FIGURE 2.23 Energy flow in a three-port circulator.

The limiter, typically based on PIN diode technology, is put between Port 1 and the receiver. For the duration of the transmitted pulse, it blocks any residual transmitted energy from entering the receiver. A signal is sent to the limiter during the transmitted pulse so that its signal blocking functionality is activated. The limiter is needed for two reasons. The first is that the circulator is not perfect and some transmitted power can get into the receiver port. The second is that reflections can occur from the connections to the antenna, the antenna itself and from very close objects, such as the vessel's adjacent superstructure, which can all be large enough to adversely affect the receiver. The limiter is therefore switched in for slightly longer than the duration of the transmitted pulse so that it also blocks such close reflections. These reflections are unlikely to permanently damage the receiver, but they can desensitize it for long enough so that wanted reflections from close targets are obscured.

2.6.3.2 RF Amplification

After receiver protection, the raw signal received by the antenna is amplified, typically by a relatively small factor of about 3. The first stage amplification process is specially designed to introduce very little extra noise into the signal and hence this process is often known as *'low-noise RF amplification'*. On most modern shipborne radar systems the receiver is generally situated upmast, often with the transmitter. Both these actions significantly reduce the RF loss in the system, therefore minimizing the system-added noise, resulting in a greater sensitivity to small targets. Unfortunately, while performance is enhanced by an upmast system it significantly complicates the servicing of the radar.

Today, there are an increasing number of communications services, such as 4G, that are operating at frequencies close to the radar operating bands, particularly at S-band. Deleterious effects can arise if the receiver attempts to amplify a nearby telecoms signal, simply because of its inherent strength compared to radar reflections. For this reason it can be expected that RF filters on marine radars will become increasingly common, which will typically be fitted between the antenna output and the RF amplifier. Such filters are designed to only pass through the frequencies that are reserved for marine radar, 2.7–3.1 GHz at S-band and 9.2–9.5 GHz at X-band, protecting the receiver from transmissions at other nearby frequencies.

2.6.3.3 Mixer and Intermediate Frequency Amplifier

Immediately after the low-noise RF amplifier is a device known as the *mixer*. This is a simple device, employed since the relatively early days of radio communications, which very cleverly produces an equivalent received signal that is modulated at a much lower frequency than the original transmission. On a marine radar, this is typically at 60 MHz, significantly lower than the transmitted radar signal. It does this by *mixing* the signal from an RF oscillator that is tuned to be 60 MHz higher or lower than the transmitted frequency. This oscillator is commonly called the *local oscillator (LO)*. For reasons that will become obvious below, the resultant rate of modulation is known as the *intermediate frequency* (IF). This is a much lower and therefore easier frequency at which to achieve affordable high amplification using conventional transistor semiconductor technology, normally in the form of one or more analogue integrated circuits (IC). An IC is a microminiaturized set of semiconductor-based components. The process of amplification at this point is known as IF amplification. In fact, it is also a frequency at which the signal, after IF amplification, could be directly digitized, although most marine radars at the time of writing do not follow this route.

The problem with radar signals is the huge range of amplitudes that are encountered in the received signal, typically 100 dB or more. To help with this a *logarithmic amplifier* is used, which significantly reduces the dynamic range of the signal to enable easier processing. The *'log amp'* typically operates as an analogue function within the IF amplifier. It is described in Section 2.6.4.5. Nowadays, it can also be digitally realized.

2.6.3.4 Signal Detection and Baseband Processing

From the output of the IF amplifier the signals are passed to a device known as a *detector* that removes the 60 MHz oscillations, just leaving the *envelope* of the received signal. This envelope is a measure of the received signal strength as a function of time. The signal is then often referred to as being at *baseband*. In ideal conditions, on a conventional pulsed radar, the envelope of the return from a point target situated in clutter-free conditions would be in the form of a *rectangular* pulse with no modulating oscillations when at base band. The signal then goes through an analogue-to-digital convertor (A-to-D). Coherent radars have to go through a more complex detection process to preserve the essential 'phase' information. This is described in Section 2.9.

It is evident that the signal path through the receiver can be divided into four stages:

1. Receiver protection and RF amplification
2. Conversion to IF by the mixer and local oscillator.
3. Amplification and analogue (e.g. logarithmic) signal processing at the IF
4. Demodulation to baseband prior to signal digitization.

Stage 3 is particularly relevant to the user in setting up and using the radar system, as settings of the gain and clutter controls can affect the analogue signal processing being performed here.

2.6.4 Receiver Techniques

2.6.4.1 The Mixer Principle

It can be shown that if two sinusoidal signals of differing frequencies are mixed, the resultant complex signal consists of a number of sinusoidal components. One has a frequency which is equal to the difference between the two frequencies that were mixed and which is known as the beat frequency. Other sinusoidal components generated include the sum of the mixed frequencies, twice the higher frequency and twice the lower frequency, and numerous even higher frequency components. The basic concept is more correctly known as the heterodyne principle and the radar receiver is said to be of the superheterodyne type, commonly abbreviated to superhet receivers. The resultant signal is said to have been *down-converted* because the main mixed component within the process is at a lower frequency than the original. This can be considerably lower by a factor of many tens of seconds.

The principle is applied in the radar receiver by mixing the incoming signal, consisting of bursts of electromagnetic energy at the transmitted frequency, with a continuous low power RF signal generated by a device known as the local oscillator. The resulting waveform at the output of the mixer will contain, among others, the component whose frequency is equal to the difference between that of the transmitted signal and that of the local oscillator. This signal is used as the input to the IF amplifier. This amplifier is designed so that it will respond only to that component of the mixer output which lies at the chosen 'beat' frequency. As a result, all the other higher frequency components generated by the mixing process are rejected. Thus the IF section of the receiver deals with pulses whose envelope resembles the shape originally imparted by the transmitter, as modified by targets, and which encloses bursts of oscillations at a frequency which is sufficiently low to be easily amplified and further processed.

2.6.4.2 Tuning and Bandwidth

Since the IF amplifier is pre-tuned to the chosen IF frequency it follows that the local oscillator must be carefully adjusted so that the frequency which it generates differs from that of the transmitter, such as the magnetron (and thus the incoming wanted signals), by an amount equal to the chosen IF. This can be performed automatically by the system but generally magnetron-based systems also have a manual tuning facility, which can be operated by the user from the radar console. This position allows the user to monitor the radar display whilst making adjustments to the tuning of the receiver. It should be borne in mind that in almost all circumstances the automatic circuitry on most modern radars generally matches or even exceeds the capability of a skilled human operator in optimizing this task, but it can be useful for the operator to vary this control in special circumstances.

A magnetron is not highly stable in frequency and can vary in the short, medium and long terms (ranging from minutes – especially immediately after switch-on, whilst the magnetron warms up – to years, as the magnetron ages). This is because the magnetron relies on precisely engineered physical cavities for its control of frequency, which can vary in their precise dimensions, for instance, with temperature, air pressure, shock, vibration and contamination.

It is important to understand the concept of signal bandwidth, when considering receiver and other design aspects concerning radar. It is a feature that affects all radio systems and is most easily understood by considering the information content contained within the signal. On a simple radio system, such as a marine VHF receiver or even the voice channel of a smartphone, very little bandwidth is needed. Speech can be understandable even if only the audio frequencies up to about 5000 Hz are 'modulated' (i.e. superimposed) onto the RF of the transmission. Good quality visual information with stereo sound, such as is used for wide-screen high-definition TV, requires a much greater bandwidth, typically many megahertz. This is a bandwidth perhaps more than a thousand times greater than that used by a simple radio system.

The need for marine radar to identify the presence and accurately measure the range of quite small targets in sea and rain clutter requires a surprisingly large bandwidth – around 20 MHz. It may be thought that a magnetron-based radar is only transmitting a single frequency, just pulsed on and off. However, it is this pulsing that gives the transmitted signal its high bandwidth. Furthermore, the returned signal is even slightly higher in bandwidth because it includes the variations produced by the targets and clutter, the very information that is needed to be 'decoded' to be put onto the radar display. As already emphasized, on a magnetron-based system it is the fast on and off switch times of the pulse, together with the actual pulse length, that fundamentally sets the ability of the radar to detect targets and differentiate them from other targets and clutter.

A mathematical concept known as Fourier analysis, named after its French discoverer Joseph Fourier – born in 1768 – can be used to translate the shape of a signal into its frequency components. It shows that a radar pulse that will meet the short range requirements of a marine radar will of necessity have

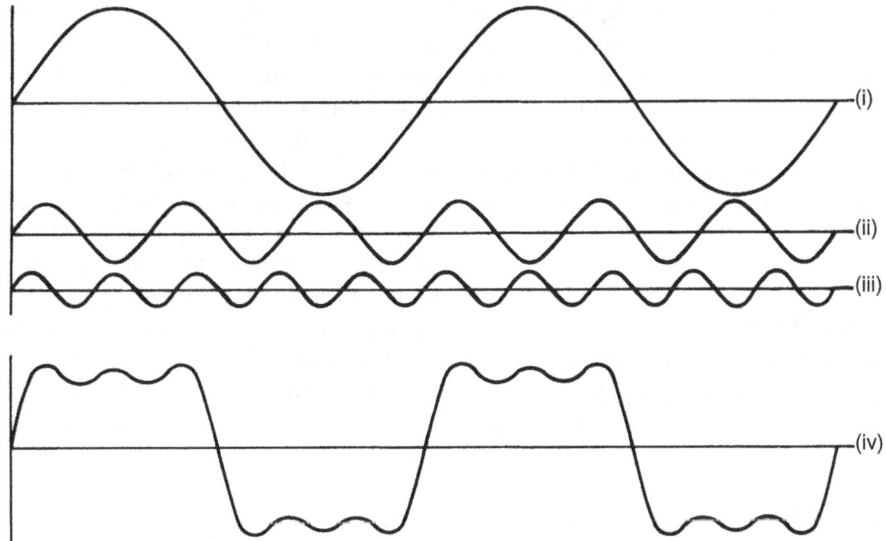

FIGURE 2.24 Forming a near rectangular pulse from three sine waves. (i) Sine wave at amplitude *A* units, frequency *f*. (ii) Sine wave at amplitude *A*/3 units, frequency 3*f*. (iii) Sine wave at amplitude *A*/5 units, frequency 5*f*. (iv) Sum of (i)+(ii)+(iii) → tends toward a square wave.

a bandwidth of about 20 MHz. Unfortunately, the mathematics behind this are too complicated to include in this book. However, Figure 2.24(a) shows how a simple pulse-like structure can be formed out of just three sine waves. The use of more than just three waves makes the resultant wave more and more square, with faster rise and fall times. Fourier analysis effectively starts with the shape of the resultant wave and calculates the single frequency waves needed to produce it. The highest frequency wave needed to adequately replicate the required signal represents the bandwidth of the transmission.

This basic *information* bandwidth must be retained in all the preliminary processing of the radar signal, including the RF, IF and baseband amplification stages, and the 'down-conversion' in the mixer. In most magnetron-based marine radar systems the bandwidth, when long pulse is selected, normally lies in the range 3 to 5 MHz, and from 15 to 25 MHz in the case of short pulse selection.

2.6.4.3 *Signal Amplification*

At all stages in the process signal amplification is achieved by the use of transistor technology. In the earlier stages, particularly in the RF amplifier, these may be discrete devices. At the lower frequency stages, many amplifying transistors will be built into a single 'substrate', giving large amplification with very high stability and needing few external components. This keeps costs down and increases the overall reliability of the system. When designing the amplifier stages care has to be taken to avoid the output from *limiting*. This is when further increase in the input signal will produce no further increase at the output. The amplifier is then said to be saturated. All input signals which exceed the limiting input produce the same level output but are highly distorted. This characteristic is of particular importance in the interpretation of a marine radar picture as a badly set-up radar can result in small echoes, for example from regular sea waves, having the same resultant strength as echoes from strong targets such as ships, creating an overall bright picture with very little detail being discernible. This can happen when the user-set manual gain control has been inappropriately set. The importance of this serious wrongful setting of marine radar is dealt with in detail in Sections 3.6 and 3.7.

2.6.4.4 *Thermal Noise*

In any electrical conductor or device at temperatures other than absolute zero there is random movement of electrons caused by thermal agitation. This results in the random movement of electrical charge, since electrons are negatively charged. This produces an effect known as thermal noise, which is often also called white noise, because like white light, it is equally distributed across a wide frequency spectrum.

The movement of the charge manifests itself as a minute random electrical signal which in the case of an amplifier stage will appear at its output. Although when initially generated the noise signal is of very low level, when amplified by a number of amplifier stages its amplitude could increase by a factor of as much as 1 million (Figure 2.25), rendering it strong enough to produce a very detectable signal on the radar display.

The term 'noise' was adopted because, when the phenomenon was first noticed in audio systems, it could be heard as background hissing in earphones and loudspeakers making it difficult to hear weak signals. In a radar display it appears visually as a close-grained background speckling on the display. As the observer must attempt to discern the echoes of targets against this background, the level of noise generated in the first stage of the amplifier is of fundamental importance in determining the system's ability to display weak echoes. If the returning echo is weaker than the noise produced in this first stage it will not be capable of detection, as successive stages of amplification will merely increase the amount by which the noise exceeds the required signal. However, if initially the signal is stronger than the noise, successive stages of amplification will magnify this difference and improve the probability of detection. Echoes which are close to noise level will always be difficult to detect.

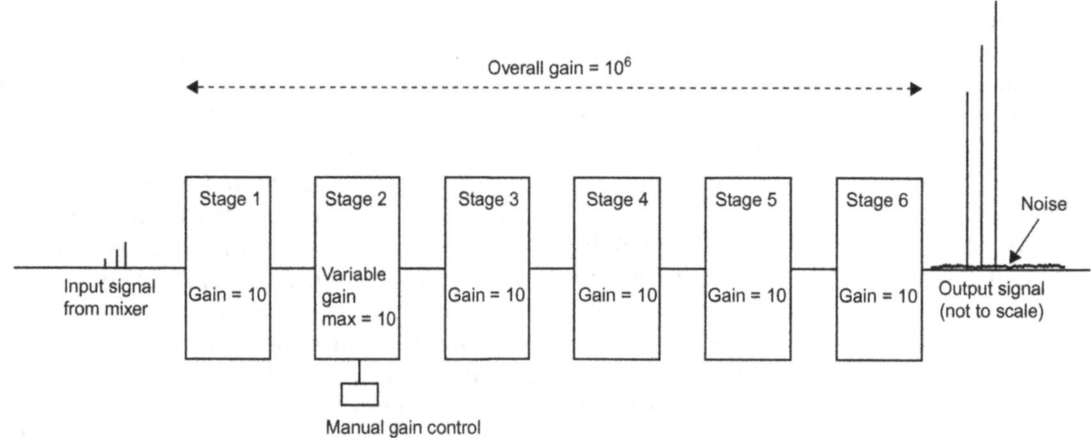

FIGURE 2.25 A multi-stage amplifier.

The level of noise displayed on the screen, and hence the probability of detecting weak targets, will depend on the overall setting of the gain control within the radar. If the gain is set too low, weak echoes will be lost because they will fail to display. If the gain is set too high, weak echoes will be lost, not because they will not be displayed, but because they will be masked by the background noise. This is a different effect from that of amplifier limiting, discussed in the previous section. The correct practical procedure for setting the user gain control is discussed in detail in Chapter 6.

The greater the bandwidth of the overall radar receiver the more noise will be generated and amplified. Increasing the bandwidth allows more noise to be accepted, and hence reduces the probability of detection of weak echoes. To get the best possible performance, the actual IF bandwidth employed is optimized to suit the particular pulse length being transmitted – a narrower bandwidth for longer pulses and a wider bandwidth for shorter pulses. This increases the ability of the radar to see smaller targets at long range settings but increases the radar's accuracy and discrimination of targets at shorter range settings.

2.6.4.5 *The Logarithmic Amplifier and Sensitivity Time Control*

If the complete radar power amplification was say 60 dB (1 million), a signal with a power of 1 μW from the antenna would result in a 1 W signal at the output of the receiver, similarly a 2 μW signal would produce 2 W. The trouble with radar returns is the huge range that has to be accommodated – more than 100 dB (ten thousand million). This is caused by two separate factors. One is the $1/R^4$ range factor that rapidly makes more distant targets give very low returns. The other is that the echo received from targets of interest can have a huge variation in strength. This depends, for instance, if it is from a small target, such as a dangerous but partially submerged object, for example debris or an iceberg; or from a large target, such as a supertanker. In practice, the range and target size effects are about equal, both equivalent to more than 50 dB each, making at least 100 dB in total.

The range effect is countered by a technique known as sensitivity time control (STC), which adjusts the radar receiver gain to being very low immediately after the pulse is transmitted and then climbing at a rate approximately proportional to R^4 until it reaches its maximum gain. STC is also sometimes known as swept time constant, a term that accentuates the time-varying nature of the gain. The changing amplifier gain (also called sensitivity) can be applied at any amplification stage: RF, IF, video – and also after digitization of the signal. There are various advantages and disadvantages as to where it is best applied. In fact, it is often optimized by being separately applied at several different stages of the receiver. The process is described in more detail in Section 2.7.4.1, as it is intimately connected with obtaining the best suppression of unwanted targets, especially sea clutter.

The huge variations in the signals of interest are particularly countered by the use of a logarithmic amplifier, generally included within the IF amplifier chain. This reduces the dynamic range of the signal that has to be processed within further stages of the radar by having a response proportional to the logarithm of the input signal. For instance, a system that converted input signal powers of 10^{-4}, 10^{-3}, 10^{-2} and 10^{-1} (max) watts to strengths of 1/4, 1/3, 1/2 and 1 W, would be a logarithmic amplifier; the 1000:1 ratio of the smallest to largest figures being reduced to just a 4:1 ratio. In this case the equation linking the output (O_p) value with the input (I_p) can be simply written as:

$$O_p = -1/(\log_{10} I_p), \text{ with } I_p \text{ limited to } 10^{-1} \text{ max}$$

The 1000:1 ratio is equivalent to a 30 dB range; it follows that a 50 dB, 100,000:1 ratio will be translated by a log amplifier into an easily managed power ratio of 6:1, assuming that the STC process was carried out prior to log amplification. (In actuality, this is not often the case.)

The huge decrease in dynamic range makes it much easier to decide how the amplitude of the signal should be displayed on the radar display. For instance, a straightforward but hypothetical algorithm for a signal emerging from a log amplifier with a 50 dB range would be:

Brightness level 5 for returns with a magnitude range of 1/2 to 1
Brightness level 4 for returns with a magnitude range of 1/3 to 1/2
Brightness level 3 for returns with a magnitude range of 1/4 to 1/3
Brightness level 2 for returns with a magnitude range of 1/5 to 1/4
Brightness level 1 for returns with a magnitude range of 1/6 to 1/5
Brightness level 'off' for returns with a magnitude below 1/6

Logarithmic amplifiers started being introduced into marine radars in the late 1960s, when the technology became affordably available. Up to then, the huge dynamic ranges that were having to be handled by the complete radar system required very careful user set-up with continual changes to the manual gain and STC setting to optimize target detection and avoid amplifier limiting in the immediate region of interest.

In fact the traditional log amplifier actually relies on the limiting effect inherent in all linear amplifiers and consists of a serially connected set of the latter, as shown in Figure 2.26 (a). In this configuration it is also known as a successive detection log amplifier and the figure illustrates a typical six-stage system. The essential feature of the amplifier is that the gain and limiting output of all stages must be identical. The figure shows that an output is taken from each individual stage, which is then summed by a special circuit.

Consider the situation where there is no saturation. For this case at the output of Amplifier 6, the total amplification would be equal to A^6, where A is the amplification of each stage. The signal from the summing circuit would be the addition of the outputs from all the stages and would total to $A^6 + A^5 + A^4 + A^3 + A^2 + A$. If the signal was large enough to cause the last amplifier to saturate, but not the second to last, the signal output would be $A^5 + A^4 + A^3 + A^2 + A$, with the circuitry ensuring that saturated stage signals are not being summed. For increasing levels of signal, the total gain would progress through the following levels:

$A^6 + A^5 + A^4 + A^3 + A^2 + A$ (for very small signals that do not saturate the last stage)
$A^5 + A^4 + A^3 + A^2 + A$
$A^4 + A^3 + A^2 + A$
$A^3 + A^2 + A$
$A^2 + A$
A (for signals that saturate all but the first stage).

Providing A is reasonably large, say 10 or greater, the gain at each of these transitions is mainly dominated by the first term of each row in the above sequence and so, as the signal level linearly increases, the gain decreases approximately according to the power series A^6, A^5, A^4, A^3, A^2, A – that is, logarithmically. Figure 2.26(b) illustrates this. For signal levels lying in between each of the critical values that just cause adjacent stages to saturate, the amplification process is linear. In the graph of

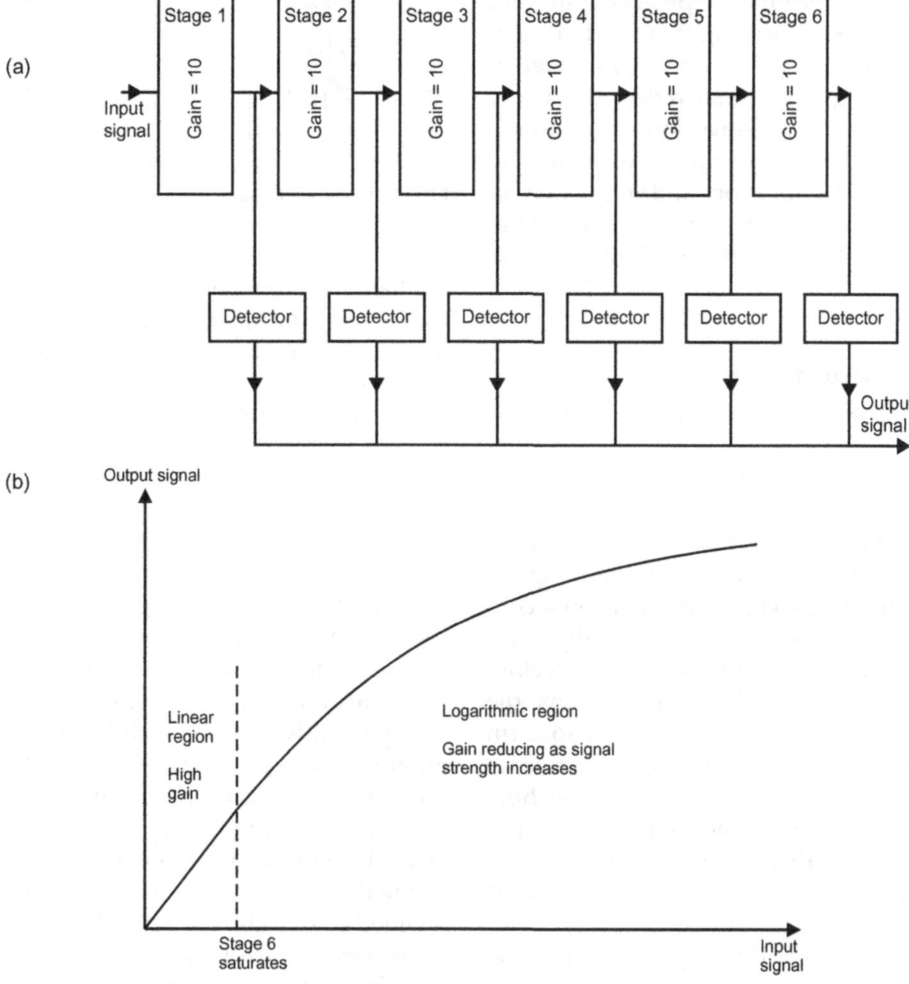

FIGURE 2.26 The logarithmic amplifier: (a) successive detection and (b) gain curve.

Figure 2.26(b), which is a logarithmic plot, the truly linear segments come out as curves. In fact, a well-designed log amplifier would hardly show this effect as compensatory measures can be implemented. The total amplifier will not saturate until the input signal is sufficiently large to saturate the first stage.

The logarithmic amplifier increases the range of signals which can be handled without saturating the receiver. In practice, the radar designer has to use the characteristics of the log amplifier and the generally multiple stages of STC to handle the vast range of the received radar signals. It is not simply that the logarithmic amplifier deals with target size effects and the STC with range sensitivity effects. However, when using a logarithmic receiver with a wide dynamic range all the subtleties of the STC can potentially be carried out in the digital domain.

The radar reception process so far described has been ensuring that no detrimental transmitted energy can enter the receiver chain; collecting the radar-reflected energy from targets by the antenna; low-noise amplification of the RF signal; down-conversion to IF; and then amplification and reduction of the dynamic range of the signal. At this point the signal is still modulated, typically at an IF of 60 MHz.

2.6.4.6 The Second Detector

The final phase of the conventional radar receiver is to eliminate the IF modulation from the received signal, and just be left with its envelope, which can then be digitized. It will feature multiple range (and therefore time) separated copies of the transmitted signal, approximating to a rectangular pulse on a conventional magnetron radar, but generally modified by the details of the specific reflecting objects. It is this signal which contains the essential radar information. It is used to form the displayed image and so is often known as the *video* signal. It is also known as the *baseband* signal, as it contains no frequency components other than that needed to portray the information contained within the reflected signal.

The function of the detector is to extract the envelope from the amplified IF signals (Figure 2.27). In engineering texts it is sometimes referred to as the *second detector*, the mixer being considered to be the first detector; the process involved is fundamentally the same. Also the term *demodulator* is used. This arises from the general terminology of radio communication in which the term modulation is used to describe the process whereby a signal representing information (e.g. speech, music or data) is superimposed, for transmission onto an RF *carrier*. For radar, the modulating signal is the rectangular pulse envelope (see Figure 2.4), and the carrier is the RF

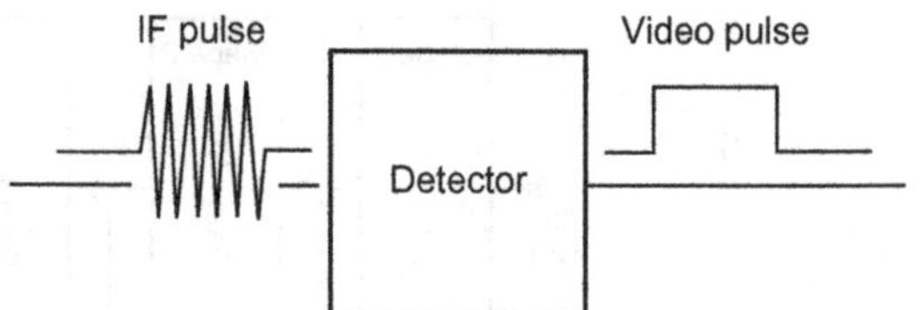

FIGURE 2.27 Pulse detection.

oscillations produced by the magnetron. The function of the demodulator or detector is to recover the modulation, as modified by the target reflections.

The characteristic of the detector must be such that it will respond to the *relatively* low frequencies that are effectively contained (by Fourier analysis) within the pulse envelope – generally less than 20 MHz – but reject the relatively high IF oscillations, typically at 60 MHz. In fact, if the targets are moving they also put a very subtle form of additional frequency modulation onto the transmitted signal, simply because the movement introduces a very slightly varying path length from transmission to reception. If this can be detected by the radar it can give an instantaneous velocity measurement of the target, which greatly helps identification of whether it is clutter or a wanted target. This information can be retained at detection if a special *IQ* demodulator is used, which is described in Section 2.9.1. The frequency instability of marine magnetrons is such that they mask such effects and so *IQ* demodulation is generally only used in coherent radars. In fact the term *coherent* is intimately connected to the concept of *IQ* demodulation and the improved potential performance of such radars is a direct consequence of it.

Once at baseband the signal is typically further (linearly) amplified and then converted from analogue into digital form. The processes often go by the names video amplification and analogue-to-digital conversion. The latter is variously abbreviated as A-to-D, A/D or ADC.

2.6.5 Radar Interference

2.6.5.1 In-Band Interference

Interference issues for radars have become increasingly more complex as time has advanced. When first introduced the main problem was interference with other radars, not necessarily maritime. Radar-to-radar interference tends to cause a characteristic set of discontinuous lines across the display, see Figure 2.28. This is caused mainly by the direct transmissions of other radars being picked up by the main beam and sidelobes of the own-ship radar antenna. It can often make the radar display unusable. For this reason the marine radar bands were made wide enough to allow different radars to be operating on different frequencies, but all within the internationally agreed limits (nowadays, 2.9–3.1 GHz at S-band and 9.2–9.5 GHz at X-band).

This did not remove all interference but a way of significantly reducing the on-screen effect of radar-to-radar effects was developed. This was to introduce a random jitter into the timing of each transmitted pulse, resulting in a non-constant PRF but with the average value equalling the desired one. Of course, this effect has to be appropriately taken into account within the radar receiver. By doing this, it significantly reduces the possibility of interference 'arcs' being produced on the displays of other radars, simply because there is reduced probability that the timing of the own ship and interfering transmissions remain in any kind of step. Instead there are just random instances of interference across the display, creating intermittent rather bright false target effects. Furthermore, these can be eradicated by some clever but simple processing, illustrated in Figure 2.29. This shows a particularly simplified example but is the basic principle of radar *interference rejection* used on magnetron-based marine radars.

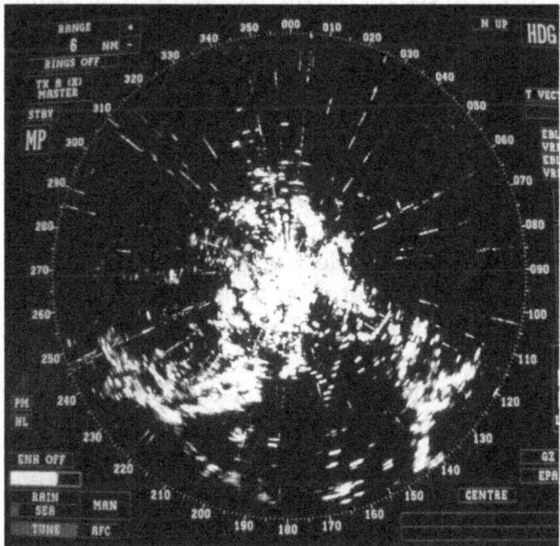

FIGURE 2.28 Interference patterns on display. Reproduced couretsy of Northrop Grumman Sperry Marine.

The returns from two consecutive pulses are shown. For the first (Pulse n), three targets are visible and there is one interference 'spike'. For the second pulse (Pulse $n+1$), it is almost certain that there will not be another interference pulse appearing at exactly the same instant – and therefore equivalent to a particular range. This is because both the interfering and own ship radars are jittering their pulse transmissions to different times. Only real targets are therefore visible in the second trace. The radar then performs a digital process known as correlation. If a detection matches (in range) over the two pulses transmitted then we assume that the target is real, but if we only get a detection at a particular range for just one of the two transmitted pulses we assume that it is interference and it is rejected (deleted).

This was first introduced on marine radars in the early 1980s and proved to be very effective, even though it does reduce real target detectability by a small amount. This is because very small targets will not necessarily produce a detectable echo from every

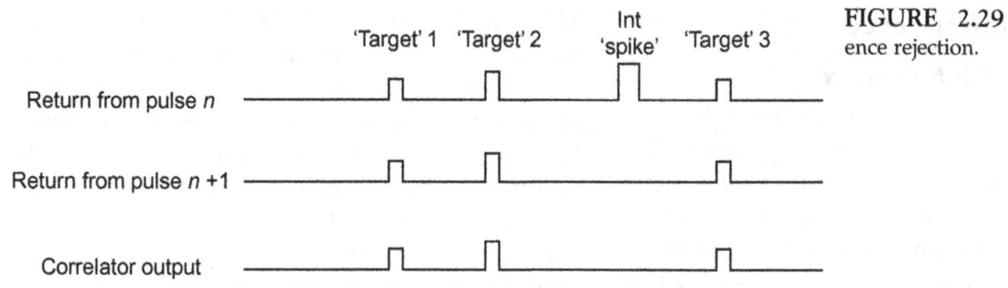

FIGURE 2.29 Radar interference rejection.

transmitted pulse, and hence will be totally rejected by the interference rejection system. For this reason interference rejection should not be applied if no disconcerting interference is visible when it is off.

2.6.5.2 Out-of-Band Interference

The problem with all radar and radio receivers is that they can be susceptible to interference that is not at the frequency at which they operate. Equipment using a particular band must not only take into account any interference possibilities of similar equipment operating within the same band, but also quite different equipment operating in other bands, which may be adjacent or much further apart in frequency. The possible interference mechanisms are quite complex and beyond the scope of this book. A main function of the International Telecommunication Union (ITU) is to agree internationally as to what band-specific equipment uses and what the maximum interference levels can be without significantly affecting the performance of other systems. Like IMO, ITU is an agency of the United Nations.

Over the years ITU has tightened the requirements of use of the marine radar bands, particularly on the allowable out-of-band interference levels. The easiest way for many manufacturers to meet the stricter criteria was to make their radars generally all operate at virtually the same mid-band frequency. This keeps the transmissions furthest in frequency from other bands and so generally reduces the interference issues to non-marine radar equipment. The built-in radar-to-radar in-band frequency rejection had become so good (ignoring the small level of lost target detection performance) that this was an affordable and practical approach.

What is now happening is that with the advent of 4G mobile phones and other developments, more and more systems are needing to operate at frequencies close to the radar bands (at present, especially at S-band) and this is further tightening the requirements to prevent interference. It is worth remembering that the use of a 4G mobile phone in the vicinity of the antenna of an older radar may produce interference. The actual effect may be difficult to see on the display, but it may result in lost targets. This is a potential concern for the future and so expect further information concerning this to be promulgated.

As we learn to live with 4G and its consequences it may be found that some radars are affected by it and other modern transmissions, especially from coastal masts. If this is suspected then the relevant authorities and the radar manufacturer should be alerted. If it does occur at all, it is likely to be observed as a loss of some targets along the bearing connecting the ship to the telecommunications transmitter. Such an effect could possibly only occur if the ship is relatively close to the transmitter, possibly (but not necessarily) closer than 1 NM.

2.6.5.3 Solid-State Coherent Radars

These are described in Section 2.9. The problem with these is that the standard interference rejection techniques do not work as well. It generally means that such radars are designed to work at frequencies that are not occupied by conventional magnetron radars. It is therefore convenient that the latter are centred towards mid-band, as described above. By implication, such radars work closer to the edges of the radar bands and so their transmitters have to be particularly well designed to prevent interfering with 'out-of-band' systems. Some may automatically select their operating frequency at any one time to operate at a frequency not occupied by any nearby radar, or there may be a need for the operator to select a frequency with little or no interference effects.

2.7 RADAR PROCESSING PRINCIPLES

Radar has benefited enormously from digital processing. Although analogue processing can be very fast it is extremely limited in what it can do. In fact, the radar log amplifier is one of the few examples where analogue technology can still just hold its own against digital techniques. However, today's processing power now allows even this function to be affordably performed. In general, the earliest conversion of the received signal to the digital domain is desirable in order to get the best visibility of the wanted targets from the raw received signal. The advent of the microprocessor in the 1970s started this revolution, but it is only since around the start of the twenty-first century that affordable processing power has allowed digitization to be employed prior to the second detection process. Many standard magnetron systems still continue to digitize the signal at this point.

2.7.1 Conversion to the Digital Domain

Section 2.6.2.1 gave some initial insight into digitization and this section further elaborates on the basic concepts of working in the digital domain. We have seen that analogue-to-digital conversion is a process that at very short intervals measures the strength of a signal and assigns a digital value to it and is therefore known as a sampling process. The rate at which the sampling takes place is dependent on the maximum frequency of the information of interest embedded within the video signal. A mathematical concept, known as the Nyquist sampling theorem, shows that sampling has to take place at intervals of $2B$ samples per second, where B is the highest frequency of interest (the bandwidth). For a conventional magnetron radar we have to at least adequately replicate the received pulse. The signal bandwidth typically needed to do this on a marine radar is up to about 20 MHz, as discussed in Section 2.7.3.1. The minimum sampling interval for this case is therefore given by 2×20 million $= 40$ million samples per second, equivalently 40 MHz.

At each sample point we need to adequately represent the received signal amplitude. The amplitude values have typically been compressed by the logarithmic amplifier and some front-end STC but, in general, use of the finest affordable resolution brings benefits to the processing further down the chain. The resolution of this process is generally referenced to the number of digital bits used in representing the signal. The actual device which performs the conversion is known as the analogue-to-digital convertor (A-to-D, A/D or ADC). Early digital radar processors simply used '0' to represent a signal that at the sample point had a level below a defined threshold, and '1' for signals above. In effect, a target was denoted by a 1 and the lack of a target by a 0. By utilizing digital strings many more levels can be represented. Such strings are generally known as *words* and each binary digit within the word is known as a *bit* (from binary digit).

A 2-bit word can represent four levels – 00, 01, 10 and 11. Typically, 00 is used to represent the lowest level signals, with 11 representing the highest. The A/D convertor will assign the appropriate word according to the level of the sampled signal strength, as shown in Figure 2.30. The use of 3 bits gives the possibility of representing eight separate levels; 000, 001, 010, 011, 100, 101, 110 and 111. Electronic processors use binary numbers simply because they can be easily represented by two simple states, typically 'if there is a voltage or current present then the state equates to 1'; and 'if there is no discernible voltage or current then the state equates to 0'. This simple logic is the basis lying behind digital processors. However, the processing of the digital data should not be confused with the digital sampling of a signal. As with all digital processes, we are representing our data as a digital string, which the processor can then manipulate, for instance by performing comparisons or calculations with other digital strings.

Although the word length of the digital sample has to be fully taken into account, the digital processor can operate with an entirely different word length. For instance, samples from a 10-bit A/D converter may be easily processed by a 16, 32 or even a 64-bit processor. In fact an 8-bit processor could also accurately deal with 10-bit samples, although the required 'firmware' would need to take into account the fact that each sample of the data would occupy two 8-bit words within the processor.

Table 2.4 gives the number of possible signal level representations for various word lengths of the A/D convertor. In fact, the relatively fast A/D conversion needed for marine radar can get comparatively expensive above

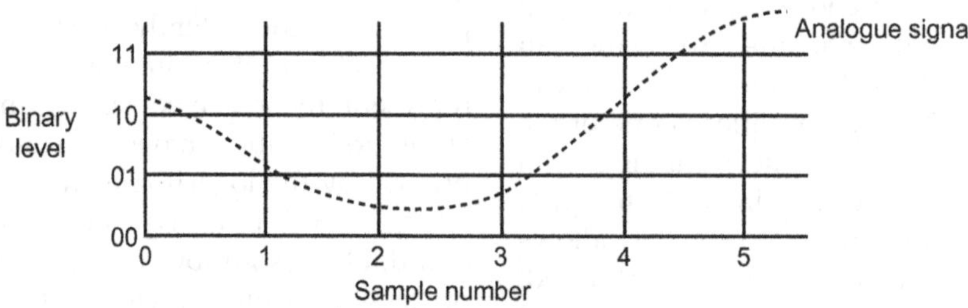

Sample number	Binary level
0	10
1	01
2	00
3	00
4	10
5	11

FIGURE 2.30 Analogue-to-digital conversion.

TABLE 2.4 Strength Levels for Computer Words of Varying Length

Word Length (number of bits)	Number of Levels
1	2
2	4
3	8
4	16
8	256
16	65,536
32	4,294,967,296
64	1.84467×10^{19}

about 16 bits and therefore current marine systems typically use 10–16 bits for their A/D.

A particularly interesting observation is that a 16-bit sample gives a workable dynamic range equivalent to $20 \times \log_{10}(65{,}536) = 96$ dB. Not only is that range suitable for audio – it is the word length used for CDs and for many other recording formats – but it closely approaches the total signal range required for marine radar (100 dB or more). The big difference between audio and marine radar is that the sampling rate for CDs is 44.1 kHz, but for radar is typically 40 MHz – nearly 1000 times faster. It is readily seen that with appropriately fast 16-bit sampling and some minimal RF STC, the logarithmic processing can be performed digitally rather than by passing the signal through a logarithmic amplifier.

2.7.2 Digital Processing

Radar processing begins to become more simple to understand once conceptualized in the digital domain. For instance, if the user adjusts the gain control, it can simply be processed as a multiplication of all signal levels by a fixed number that represents the setting of the control. It is worth looking at this in more detail as it demonstrates some of the things that have to be considered by the radar engineer when designing a system, and by the user when operating it.

As an example, take a system that uses an 8-bit A/D and an 8-bit processor. Assume that the sampled signal level at one instant is at the equivalent decimal level of 25 (0001 1001 binary). Now assume that the gain control has been set at a level that results in a five times multiplication of all the sampled values. The resultant signal is boosted to $5 \times 25 = 125$ decimal (0111 1101 binary) and so the signal level is increased, leading, in general, to a brighter spot on the display that corresponds to that particular sample. If the gain setting is set to 12, the decimal equivalent of the amplified signal would be $12 \times 25 = 300$, but this exceeds the highest level that can be represented by a simple 8-bit word and so the resultant level would be at the maximum value – 1111 1111 binary, which is 255 decimal. This means that all signals that have been processed, for instance by the user gain setting, such that their levels would have exceeded the maximum possible will be shown at the maximum level. This state is called *saturation*. On the display it would result as an increased preponderance of very high-brightness signals being shown, potentially masking significant targets.

In fact, this is no different to analogue processing. As previously mentioned, such processes also have a top limit, normally arising when a component such as an amplifier saturates – it cannot reach a higher voltage because it is limited by the particular DC power supply within the radar. It is therefore important that any user controls that can effectively saturate the system are used with care. This includes parameters such as radar gain and display brightness. In practice, there is a lot that the radar designer can do to mitigate poor setting by the user, not least by having options for automatic control. Despite this, a skilful and knowledgeable user will naturally get far better performance out of a radar than one being used in ignorance.

The given example has really been fundamentally limited by the word length of the assumed radar processor and the basic simplicity of the process. A more sophisticated set-up could, for instance, put the multiplied result into two 8-bit words or more easily utilize a 16-, 32- or even a 64-bit processor. As previously emphasized, the word length of the processor and the word length of the A/D process are entirely separate concepts.

To further illustrate the processing options it is perhaps useful to consider another simple example. Take the case where we multiply two 8-bit numbers together, for instance. This can result in up to a 16-bit answer. A further multiplication with an 8-bit number will produce up to 24 bits – and so on. There are a number of ways of tackling this fundamental issue without generally leading to the calculation saturating. These all lead to some approximation but a good design will be unaffected in any practical sense.

A generally good way of tackling this issue makes use of *floating point* representation. This is very similar to representing a decimal number in *scientific notation* such as 1.2×10^2 (=120). If we square this number we get 1.44×10^4, which, if working to 1 decimal place of accuracy, will be represented as 1.4×10^4. This has maintained the magnitude of the number – and maintained a definable accuracy. By such means, digital processors working in binary can perform lengthy calculations to a defined 'percentage' accuracy without incurring practical limiting issues.

Once in logarithmic form and in the digital domain there is the capability for almost unlimited processing of the received signal, which to a limited extent, is under the control of the operator. In particular, the processing will attempt to optimize the performance taking into account the actual signal format being transmitted, which for a magnetron transmitter is dominated by the pulse length and PRF. To a large extent, these parameters are governed by the particular display range selected by the user, as discussed in Section 2.3.3.2.

IMO radar performance standards require user controls to be able to optimize the performance in both sea and rain clutter. Also a gain control function needs to be provided to set the 'signal threshold gain'. The requirements mention manual and automatic controls, but the term 'manual' is far removed from that applied to earlier radars, which were highly simplified compared to today's systems. For instance, in earlier days the manual gain control would simply raise or lower the gain of the video amplifier, simply altering the brightness of the targets shown on the display. A modern 'manual gain' control may intelligently react to the observed situation giving somewhat different thresholds to different areas of the display dependent on the actual received signal. These would increase or decrease according to the user setting but not necessarily uniformly, such that the optimum display could be more subtly set. A threshold is the level of the signal needed for it to be displayed, for instance, at a particular brightness. The different approaches by different manufacturers do lead to observed differences, such that an individual user may prefer one radar to another.

The blurring between automatic and manual controls is effectively taken onboard by IMO, particularly for the functions optimizing the performance in sea and rain clutter, by the sentence in the Radar Performance Standards: 'a combination of automatic and manual anti-clutter functions is permitted', which follows the requirement that 'effective manual and automatic anti-clutter functions is permitted' (see Section 11.2.1). On some equipment, 'Automatic' implies there is no user adjustment, whereas on other equipment, manual adjustment may still be available. However, the range of adjustment or its total effect may be different to when the equipment is in manual mode.

2.7.3 Range and Azimuth Sampling

By definition of the concept 'cell', the received signal is sampled at every range cell for every discrete azimuth (relative bearing) 'step' of the rotating antenna. These samples are then used individually and in appropriate groups as values for the various algorithms that the radar uses to improve the detection and discrimination of targets. In addition, some data will be processed on a scan-to-scan basis, comparing and enhancing the resultant display at every rotation of the antenna. The range cell samples are typically taken after second detection and video amplification of the signal, as described in Section 2.7.1. If digitization has taken place before this point, the sampling rate will need to have been far greater than the rate needed when just digitizing the amplitude of the signal at every range/azimuth cell. Section 2.7.3.1 shows that the cell sampling of a marine radar when at short rage is perhaps at about 40 MHz, whereas if sampled at IF the rate will need to be at least double the IF frequency, typically $2 \times 60 = 120$ MHz. If such fast sampling is undertaken for the initial digital detection process, they will subsequently be numerically processed to form the received signal strength at the slower sample rate needed to represent the amplitude at each range and azimuth cell.

2.7.3.1 Range Sampling Example

It is useful to look at a specific example of digital sampling when applied to radar range discrimination. It is a fairly straightforward example and was one of the first used on marine radars. It assumes the normal case of signal digitization immediately after video amplification. It is aimed at recovering the essential target information, which is the amplitude of the received signal at every range/azimuth cell. This fundamental thinking forms the basis of many systems in use today.

Example: A marine radar system needs to discriminate two targets separated by 40 m in range on the 1.5 NM range scale. Calculate a suitable pulse length and sampling rate, giving due regard to a required range accuracy of 30 m at short range.

An electromagnetic wave takes $40 \div (3 \times 10^8)$ seconds to travel 40 m, which is equal to 133.3 ns. The situation is illustrated in Figure 2.31. The leading edge of the pulse from the closest target (Target 1) will be

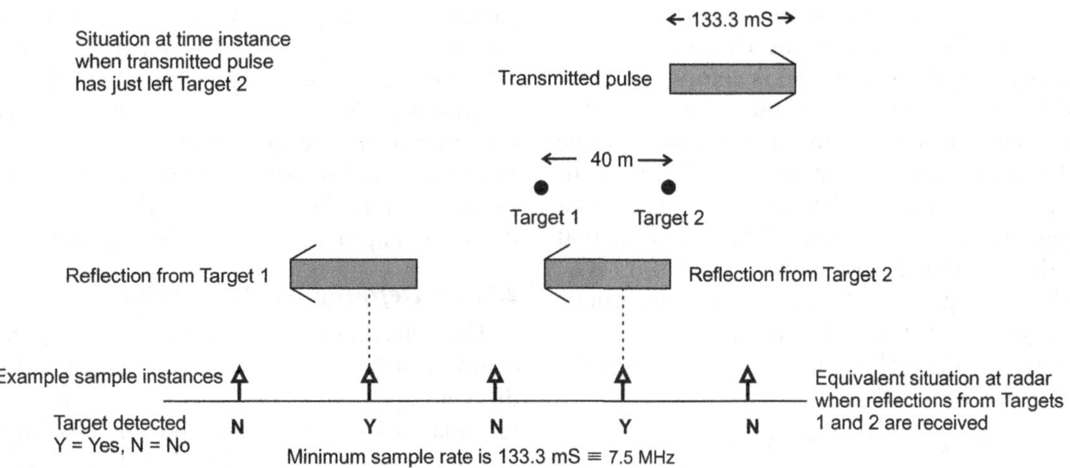

FIGURE 2.31 Pulse length considerations.

received at $133.3 \times 2 = 266.6$ ns before the leading edge of the response from Target 2. This is because the pulse has to travel an extra 133.3 ns both to and from Target 2, compared to the case for the reflection from Target 1. If a pulse length of 133.3 ns was used it would ensure that there is a gap equal to the pulse length between the reflections from Target 1 and Target 2. The sampling rate would then need to ensure that the responses from two such targets were always visibly separate. This would need the sample rate to be equivalent to measuring the received signal at least once every 133.3 ns, that is at 7.5 MHz. Looking at the best range accuracy that could be achieved, assuming perfect timing measurements, it is important to remember that for a radar, the range (R) of a target is given by the formula:

$$R = cT/2$$

where:

c = the velocity of electromagnetic energy
T = the total time between the transmission and reception of a reflected pulse

If time is being quantized to 133.3 ns because of the sampling rate, this equation also shows that the equivalent range quantization is 20 m. This means that the range accuracy degradation due to the chosen sampling rate is at worst 20 m and so is compatible with IMO requirements for a minimum of 30 m accuracy. It is also worth relating this example to the ideas discussed in Section 2.7.1, particularly concerning the Nyquist rate. The sampling rate of 7.5 MHz implies a signal bandwidth of half that amount, just 3.75 MHz, with c being the velocity of electromagnetic radiation and f_b being the bandwidth.

This is equivalent to an electromagnetic wave of wavelength:

$$\lambda = c/f_b = 3 \times 10^8/(3.75 \times 10^6)$$
$$= 80 \text{ m}$$

This is exactly the length of the pulse in free space plus the 40 m minimum needed between targets to be able to distinguish them. It is effectively portraying the minimum needed system bandwidth (3.75 MHz) that is needed to be able to discriminate such targets in range.

A typical real radar operating at short range would actually utilize a considerably shorter pulse length of around 50 ns (0.05 μs), rather than the 133 ns (0.113 μs) pulse used in the example. This is because a shorter pulse gives appreciable benefits in separating wanted targets from clutter with its better range discrimination. The sample rate, following the logic in the above example, would then equate to 20 MHz and be able to distinguish point targets 15 m apart.

In reality, for a well-formed pulse, even faster sampling will give target discrimination benefits. A reasonably well-formed pulse is considered to have an equivalent bandwidth given by the reciprocal of the pulse length. For a 50 ns pulse this equates to 20 MHz. To get the most information from the received signal this would need to be sampled at the Nyquist rate of 40 MHz, and this would improve the range discrimination to about 7.5 m. In effect, what is happening is that the system, with increased sampling, is able to detect shorter gaps between targets. The reason why the simple approach initially illustrated in this example does not come up naturally with the Nyquist rate is because the time gap between the returns from targets at different ranges are effectively doubled (by the fundamental nature of radar), but the reflected pulses remain of the same length as the transmitted pulse.

2.7.3.2 *Defining the Azimuth Cell*

The selection of the azimuth cell is quite a complex task. Radar designers are likely to take different approaches because of the various factors that need to be considered. In this digital age it is easiest to consider that they all effectively amount to creating a table of samples that is defined in azimuth by the PRF. This means that the effective azimuth increment of the table is dependent on the PRF. Samples are taken and effectively stored at every range cell and for every pulse. Take the case of a radar antenna rotating at 24 revolutions per minute, equivalent to 0.4 revolutions per second or 144° per second. If the example radar is operating at short range with a PRF of 3000 Hz, it effectively creates a table such that each azimuth increment is given by 144° ÷ 3000 = 0.048°. At a very long range setting the PRF may be around 375 Hz, with an equivalent azimuth cell increment of 0.384°.

Even a point target will appear at the same range in a number of azimuth cells. This is because of the antenna horizontal beamwidth. It means that processing on several adjacent cells can be used to enhance the appearance of the target on the display. In particular, the probability of detection can be enhanced and the probability of false alarm reduced by the use of suitable algorithms working on the samples of a number of adjacent cells, in both azimuth and range. Also, algorithms can be used to sharpen the displayed images of targets and numerous other tasks to improve the image quality and the overall performance of the radar. This includes the rejection of interfering pulses from other radars, see Section 2.6.5.

In any particular system, the reality of how the processing is actually done may be rather different to this simplified description. It gets even more complicated when considering the possibilities of scan-to-scan processing. It is unlikely that all the detailed samples will be stored over many scans, or even over a complete scan. Certain processed information is likely to be stored and used over several scans but possibly not at the same cell resolution as originally used. This is all in the hands of the radar designer.

One of the practical considerations to be taken into account is the measurement of the actual angle of the antenna horizontal beam when the pulse is transmitted. Because of mechanical influences such as wind and vessel stability the antenna is unlikely to be rotating at a precise angular rate. Therefore, its actual rotation angle has to be constantly measured. In certain circumstances, the rotation could even be momentarily reversed, especially in very high wind conditions. On a modern radar, the instantaneous bearing of the antenna is given by a digital signal, generated from within the turning mechanism. This is used at the reference angle for the discrete set of range cells measured at each 'azimuth' sample. It allows the received data to be associated with the correct bearing relative to the heading line of the vessel at each azimuth cell increment. In reality the azimuth cells previously described are really cells associated with each pulse. Associating the actual pointing angle of the antenna at each radar pulse makes them into azimuth samples. In general, the samples will not be taken at totally fixed azimuth increments – and may even briefly reverse.

The relative bearing of the antenna is an angle which generally increases progressively from 0° to 360° relative to the ship's heading, as it rotates. It can, for instance, be represented by a binary number which increases from zero to some maximum, where that maximum is one azimuth increment below 360°. This is unconnected to the increment that is used for the 'azimuth' cell size of the range/azimuth signal amplitude sample. In principle, the antenna sensor could directly give the bearing as a digitally encoded decimal number representing the rotation angle in degrees or radians. In practice, many systems use a more basic concept, which links to the historic development of marine radars.

Instead of a digital word being transmitted from the azimuth sensors on the antenna rotation mechanism, simple digital pulses are generally output. Two separate pulse outputs are provided. One gives a pulse every time the antenna beam is aligned with the heading of the vessel. The other provides a pulse as the antenna rotates through successive defined angular increments. The angular increment is typically related to a binary 'round' number such as 1024 or 4096. Thus using the 1024 example a pulse is given every time the antenna rotates through another $360/1024 = 0.35°$. The system resets the relative azimuth of the antenna rotation to zero every time a heading pulse is received. For the following 360° it determines the precise angle at any time instant by the number of azimuth pulses it has received. From this information it can then interpolate the angle of the antenna when each pulse is transmitted. It is this value that forms the angle of the particular azimuth cell. Such a simple system can create small errors in very high winds if the antenna rotation momentarily changes directions. However, these errors are corrected when the next heading line pulse is received. More sophisticated systems than that described here can instantaneously cope with such circumstances and are used on some marine radars.

Together with the value of the range cell, this angle is used to identify the stored measurement of signal strength. The process can be considered to form a table of signal strengths against range and relative bearing, as shown in Figure 2.32. The amplitude of the returns will be stored as a digital number. On older radars this could have been simply a 1-bit number, 0 or 1, depending on whether a target was detected or not, based on previous analogue processing of the signal. On modern radars the amplitude is stored in multi-bit form, typically 4–16 bits, allowing the target detection process to become part of the digital processing function.

Relative bearing	Range: ΔR	$2\Delta R$	$3\Delta R$	etc.
O	A_{10}	A_{20}	A_{30}	
$\Delta\varphi$	A_{11}	A_{21}	A_{31}	
$2\Delta\varphi$	A_{12}	A_{22}	A_{32}	
etc.				

ΔR = Maximum range/number of range samples
$\Delta\varphi$ = 360°/number of azimuth samples
A_{XY} = Amplitude of return at (R,φ)

FIGURE 2.32 Table of radar return signal strength.

2.7.4 Clutter Suppression

A great deal of processing is used to help detect targets in clutter. Nowadays, much of this takes place in the digital domain, but the important things to understand are the fundamental concepts, as described in this section.

2.7.4.1 *Sea Clutter*

The amelioration of sea clutter effects is intimately tied up with STC. For normal targets, as underlined by the radar equation, we have seen that the returned signal power varies according to $1/R^4$. However, the sea does not act like a normal target, as its effective RCS varies according to range. If at first we imagine a sea area on a totally flat earth, the effective illuminated area of sea at any one instant would be bounded by the beamwidth of the radar antenna and the length of the range cell, see Figure 2.33. Mathematics readily shows that the area A_s of the sea being 'illuminated' by the radar is given by:

$$A_s = K \times \theta \times R \times \Delta R$$

where:

K is a constant irrelevant to this discussion
θ is the beamwidth of the antenna
R is the range of the range cell and
ΔR is the range cell increment

The important fact is that the area of the sea being illuminated is proportional to the range, R, resulting in the equivalent RCS of the sea also being proportional to R. The sea's RCS increases with range, unlike regular targets

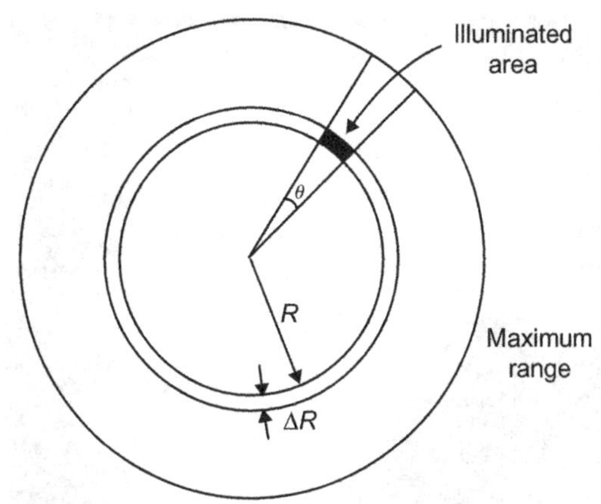

FIGURE 2.33 Area of range/azimuth cell.

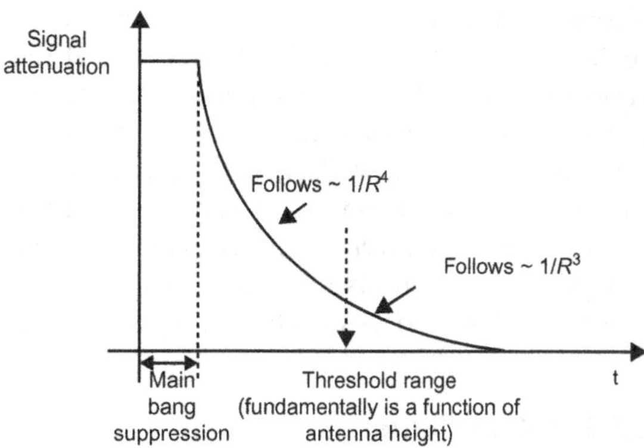

FIGURE 2.34 Sensitivity time control.

which have a range independent of RCS. However, it is also worth recognizing that the RCS of the other major component of clutter (precipitation, such as rain) also tends to increase with range. This time, because the volume of precipitation illuminated also increases with range – and hence with the increase proportional to the square of the range, assuming that the precipitation continues to occupy all the vertical angle of the radar antenna, which in reality is unlikely over any appreciable distance.

For sea clutter, and assuming a flat earth, the fact that its RCS is proportional to range effectively means that the R^4 term in the radar equation reduces to R^3, simply because the RCS (σ) should now be represented by $R \times \sigma_s$, where σ_s is a fixed quantity, effectively representing the RCS of the particular seastate at $R = 1$. Mathematically we have from the representation of the radar equation in Section 2.6.1:

$$P_r = kP_tG^2\sigma/R^4 = kP_tG^2R\sigma_s/R^4 = kP_tG^2\sigma_s/R^3$$

If we applied STC following the standard R^4 law on such sea clutter (see Section 2.6.4.5) it would mean that the clutter would become increasingly dominant with range. Therefore, the applied STC should vary according to R^3 in order to compensate for this, whenever sea clutter is dominant. When we switch on the sea clutter control for a marine radar this is effectively what happens. However, one particular practicality has to be taken into account, which is that we are not operating on a flat earth. At distances beyond the radar horizon there is no sea clutter as the radar beam does not illuminate the sea at such distances, and therefore the STC needs to change to R^4 at this point. The actual horizon is dependent on the height of the radar antenna and so this is generally set by the engineer when the equipment is installed. The standard STC curve when sea clutter is switched on is illustrated in Figure 2.34. The manual sea clutter control adjusts the exact shape of this curve, according to how the equipment has been specifically designed. Very low settings would approximate to an R^4 curve and higher settings to an R^3 curve.

Switching to automatic sea clutter reduction would normally introduce additional processing that effectively altered the shape of the curve according to the actual level of returns. The user may also retain some manual control as to how intensely the automatic system is effectively 'allowed' to distort the curve from the standard R^3 through to R^4 settings.

The use of scan-to-scan correlation, also called rotation-to-rotation correlation, is an additional way of reducing sea clutter effects. Peaks in sea clutter are unlikely to occur at the same range and bearing for consecutive scans and so a digital process that attempts to eliminate the display of reflected signals that are only visible during one scan can significantly reduce such clutter. Unfortunately weak wanted signals can also be eliminated by such a process. The process is described in Section 3.6.3.4.

2.7.4.2 Rain Clutter

Rain clutter, and other types of precipitation clutter such as from hail and snow, is typified by having a continuous return over a long range and at wide angles. Unlike the returns from sea clutter, which tend to be very 'spiky' – the spikes resulting from particular instantaneous sea waves – rain clutter has a very smooth overall response. It is a problem for the user of the radar because the generally increased levels of the total radar return caused by precipitation clutter can mask other targets, as shown in Figure 2.35. On a large area of rain clutter, falling from a well-defined rain cloud, for instance, the reflected signal would rise suddenly and then remain high over a large range, until it would fall suddenly. The effects of such clutter can be mitigated by ensuring that the gain thresholds in such affected areas are appropriately reduced.

Before digital signal processing this was typically performed by having an analogue circuit that performed a differential process. By differentiating the signal with respect to time (i.e. range), the resultant signal will be large at the start and end of the rain area, where the signal changes amplitude suddenly, and near zero where the signal is virtually constant at ranges where the rain was falling. Since echoes of wanted targets rise and fall sharply with range, the differentiating process keeps these very visible, but now with much reduced contribution from the rain. This is illustrated in Figure 2.36.

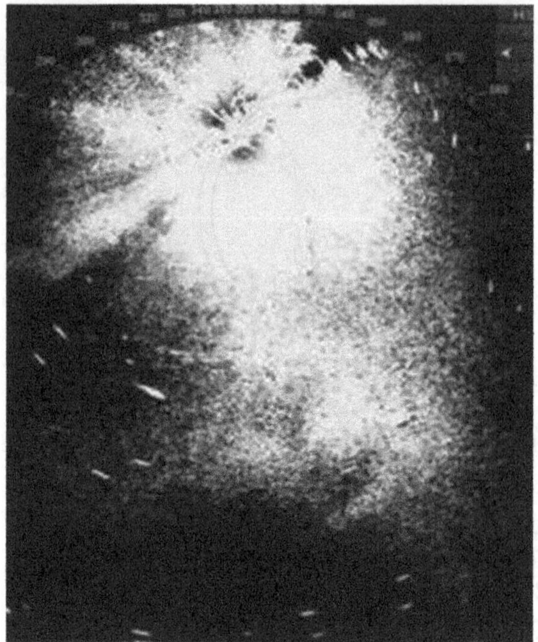

FIGURE 2.35 Precipitation clutter (with no FTC). Reproduced courtesy of Northrop Grumman Sperry Marine.

Today, modern radars can use various digital methods to optimally set the threshold to be able to see small targets in rain, including those based on differentiation. The user's rain control adjusts the level of the resultant threshold up and down, assisted by manufacturer-specific algorithms aimed at getting the best performance. Reflecting the original analogue processing techniques used for differentiation, the rain clutter control process is sometimes known as the *fast time constant* (FTC) control. This reflects the fact that the fast changing elements of the signal in time, for instance the edges of rain clutter and normal target reflections, create a larger processed signal than the slower changing elements, such as that from a large area of rain.

STC is sometimes understood to mean *slow time constant*, as the effective gain applied to the received signal is moving relatively slowly

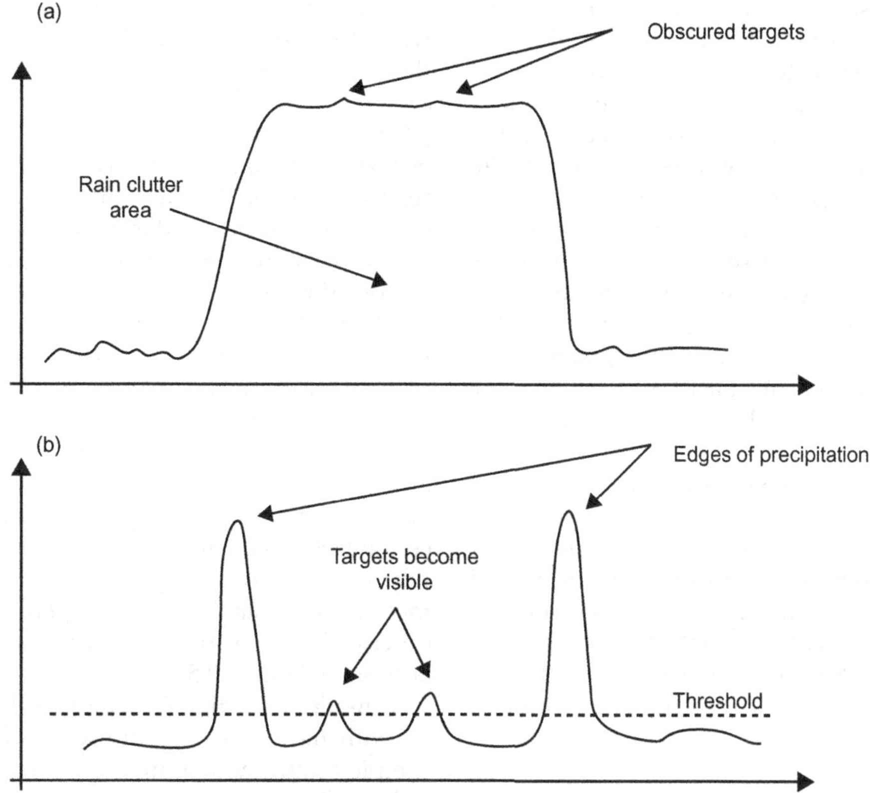

FIGURE 2.36 Differential (FTC) processing: (a) signal before differentiation and (b) signal after differentiation.

in time – at least compared to that implied by the use of FTC. The safe and successful operation of the main manual controls of a radar, gain, rain (FTC) and sea (STC) are dealt with in Sections 3.6, 3.7 and 6.7.

2.7.5 Target Tracking

A very important function of a modern radar is its TT capability. This is still commonly called, albeit somewhat archaically, the Automatic Radar Plotting Aid. In the early days of radar, target plotting was carried out by hand, often assisted by the use of a mechanical plotting aid, which either overlaid the radar display or was immediately adjacent to it. The more modern processor-based systems that started to appear in the 1970s were understandably called 'automatic' radar plotting aids and the term has stuck, although it is not used in the current IMO Radar Performance Standards, where TT is used throughout. It used to be considered a very special and separate function, but on a modern radar it forms just part of the total digital processing system. Its functionality and use is explained in rather more detail in Chapter 7. Only the basics of TT functionality are outlined here to enable it to be put into context with other radar processing functions.

Given a specific radar return the TT process attempts to track the target over time. Either the operator initially selects specific targets, generally using the display cursor and by pressing a selection button, or else allowing automatic criteria to be used for target selection. The critical parameters of the automatic criteria are also in the control of the operator. A simple example would be for the TT to acquire and track all targets within a defined radius of the own ship. For automatic selection the radar processor also has to take decisions on what constitutes a target. This could be, for instance, a return of significant size that appears over several scans of the antenna and exhibiting movement consistent with floating targets, including those targets that are stationary.

From the movement of a tracked target from scan-to-scan its direction and speed is automatically estimated by the processor, allowing it to be shown in numerical form and as a vector on the display, emanating from the target position with a depiction that it is a target being tracked, as shown in Figure 2.37. In particular, the TT processes the information to estimate whether there is a risk of collision with the own vessel if both vessels continue at their current speeds and courses. If operator-set limits on safe passing distances are compromised the radar will produce an alarm. The estimate of a collision risk is entirely based on relative motion, the fundamental working reference for a marine radar, and hence is generally a very reliable estimate. However, the vector information (course and speed) concerning tracked targets can be usefully displayed to the user relative to a number of different references. In principle, these form part of the functionality of the display processor, which drives the display and ensures that the set-up of the final image reflects the settings of the operator. For this reason the discussion concerning reference modes is left to Section 2.8.4, where the processing for all orientation, motion and stabilization display modes is described. Use of these modes is described in Sections 1.4 and 6.2.5.

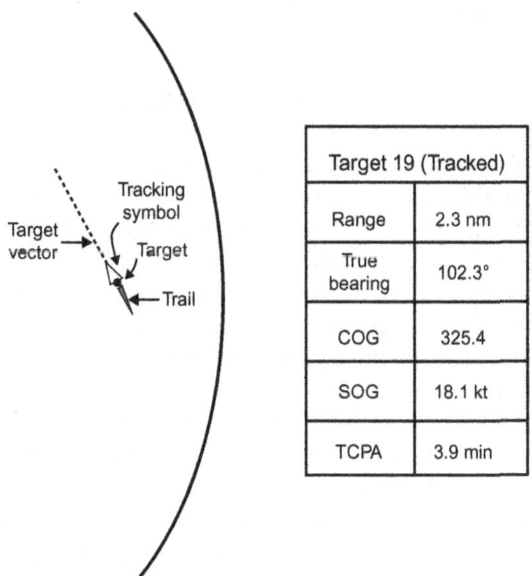

Target 19 (Tracked)	
Range	2.3 nm
True bearing	102.3°
COG	325.4
SOG	18.1 kt
TCPA	3.9 min

FIGURE 2.37 Tracked target.

A modern TT digital system works in parallel with the main radar processes. It uses the sampled information in each range/azimuth cell and from the scan-to-scan correlation process. It also has to work closely with the processes forming the display and those interfacing with the operator settings. To ensure optimum performance it is likely that it will be using basic radar information specifically processed for its own use. For instance, it may not rely on the user-set gain, rain and clutter controls but use automatic settings, optimized for tracking. This will not affect the main radar display as set by the user because it is solely aimed at improving the tracking process. It also may not be constrained by the user range settings, generally allowing more distant targets to continue to be tracked when the user has selected a shorter range scale. This enables the target track information to be instantly available when the user reselects a longer scale. In principle, it may even form 'hidden' tracks on potential targets not yet selected by the operator, so that there is minimal delay when the operator selects a new target to be tracked.

2.8 THE RADAR DISPLAY AND USER CONTROLS

The primary function of the display is to present the radar picture in such a way that it enhances the situational awareness of the user. This includes being able to readily detect all targets that have any significance to safety and for the protection of the environment; to be able to measure those target parameters necessary to make good collision avoidance decisions, such as their range, bearing, past track and present course and speed; and to be able to easily observe those own-ship navigational parameters considered essential for situational awareness. The user controls are to optimize the displayed image for the particular tasks being performed at any one time.

The display and the physical controls form the *user interface* of the radar system. The actual physical controls may be very minimal, such as a tracker ball and its associated button-type switches together with a few dedicated controls, potentially limited to just an on/off power switch. In general, the display and its *menus* give the visual information to enable the user to set up the radar optimally for the immediate task in hand and to show the precise way that the radar has been configured. Dedicated off-screen indicating lights may also show specific settings. Modern radar displays are identical to good quality industrial grade computer displays, fed by a digital signal from the display processor. Since the late 1990s new marine radar displays have been of the flat panel type, generally using liquid crystal display (LCD) technology, which is described in Section 2.8.3.

Displays used until around the mid-1980s were a specific form of CRT known as a radial scan plan position indicator (PPI). They were epitomized by their circular (rather than rectangular) outline and their inherent reliance on the use of long persistence phosphors on the inside surface of the screen to produce the radar image. During the 1980s the use of rectangular format CRTs with standard persistence phosphors became the norm in the marine radar industry, not least because of the increased ability and cost effectiveness of digital processing techniques. During the 1990s colour CRTs were adopted for many marine systems. Rectangular CRTs had formed the basis of TV displays since the 1950s, initially in monochrome and then colour. They were used as the basis for computer displays before being replaced by flat panel technology. CRTs are made of an evacuated glass tube with a near flat viewing face. Unlike flat panel technology, they have a considerable depth making them relatively bulky and heavy, and are therefore not as practical as flat panel technology.

From the earliest days of marine radar until about 1970, the radial CRT display contributed to a significant part of the analogue processing of the received radar signal. In particular, its specific action was effectively used to provide the timing mechanism that enabled the picture to be in synchronism with the real world layout of the actual targets, as detailed in the next section. From the early 1970s the change from a display-processed *real-time* picture to a *synthetic* picture began, although the basic radial scan PPI was initially retained until the mid-1980s, when rectangular CRT raster displays started to become the norm. The radial scan origins of marine radar are still epitomized on most modern systems by the circular outline of the radar displayed area, although the advent of Electronic Chart Display and Information System (ECDIS) is perhaps starting to contribute to a change away from this. The display no longer performs any radar processing function. Its internal processing is dedicated to presenting, as far as possible, a clear and sharp image under all lighting conditions.

2.8.1 The Radial Scan PPI

Since the final 'look and feel' of a modern marine radar is still influenced by its radial scan PPI origins it is useful to briefly look at that technology. It used (Figure 2.38) an air-evacuated narrow glass tube which funnelled out at one end to produce a circular screen. Electrons were emitted from an *electron gun*. This consisted of a *cathode* at the narrow end of the tube, which generated the electrons. The amplified radar-received signal was applied to the cathode, leading to the number of emitted electrons varying according to the instantaneous strength of the signal. The emitted electrons were then accelerated and focused towards the display by appropriately placed *anodes*, all at a very high voltage and also forming part of the electron gun. A *grid* between the cathode and the first anode was used to control the overall strength of the electron steam and therefore the overall brightness of the display. This was simply set by the voltage on the grid, which was directly applied by the operator by means of the *brilliance* control.

The electron beam was positioned on the display according to the rotation of the antenna and the equivalent range of the instantaneously received signal. Electron beams can be deflected by electrostatic and electromagnetic forces and so these were utilized in radial scan PPIs to dynamically position the beam. Marine radar CRTs generally used the electromagnetic force generated when a coil of wire is energized. The larger the current applied to the coil the greater the magnetic force, and hence the greater the deflection of the electron beam. The system is depicted in Figure 2.39. At zero range no current was applied, resulting in a centrally focused beam. At increasing ranges the current had to be appropriately increased. The scan coil was physically rotated in precise synchronism with the antenna to ensure the correct representation of bearing.

The display had a special phosphor coating on its inside surface that produced light when impacted by electrons. The light intensity, typically orange in colour, was governed by the instantaneous strength of the electron beam – which corresponds to the strength of the received radar signal by the action of the electron gun. However, the phosphor was specially chosen such that the light intensity decayed relatively slowly, giving the property

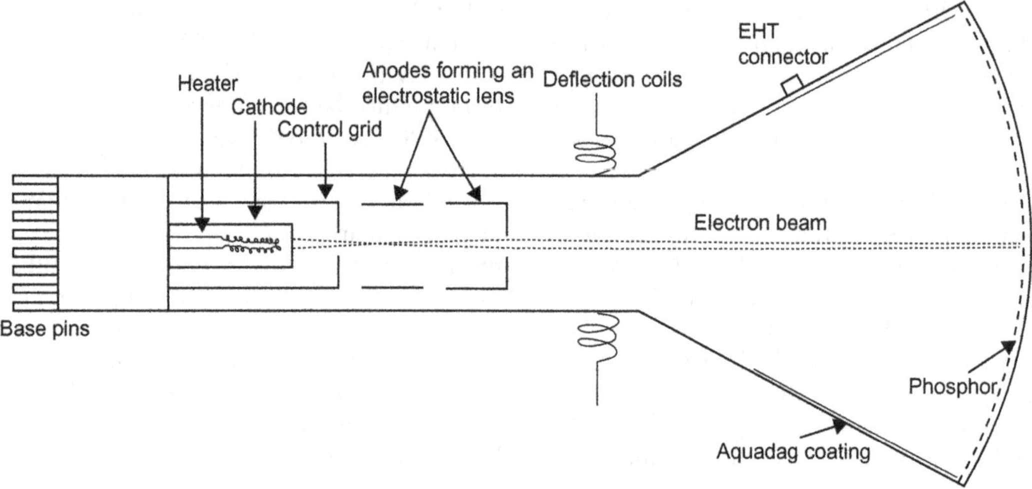

FIGURE 2.38 The radial scan CRT PPI.

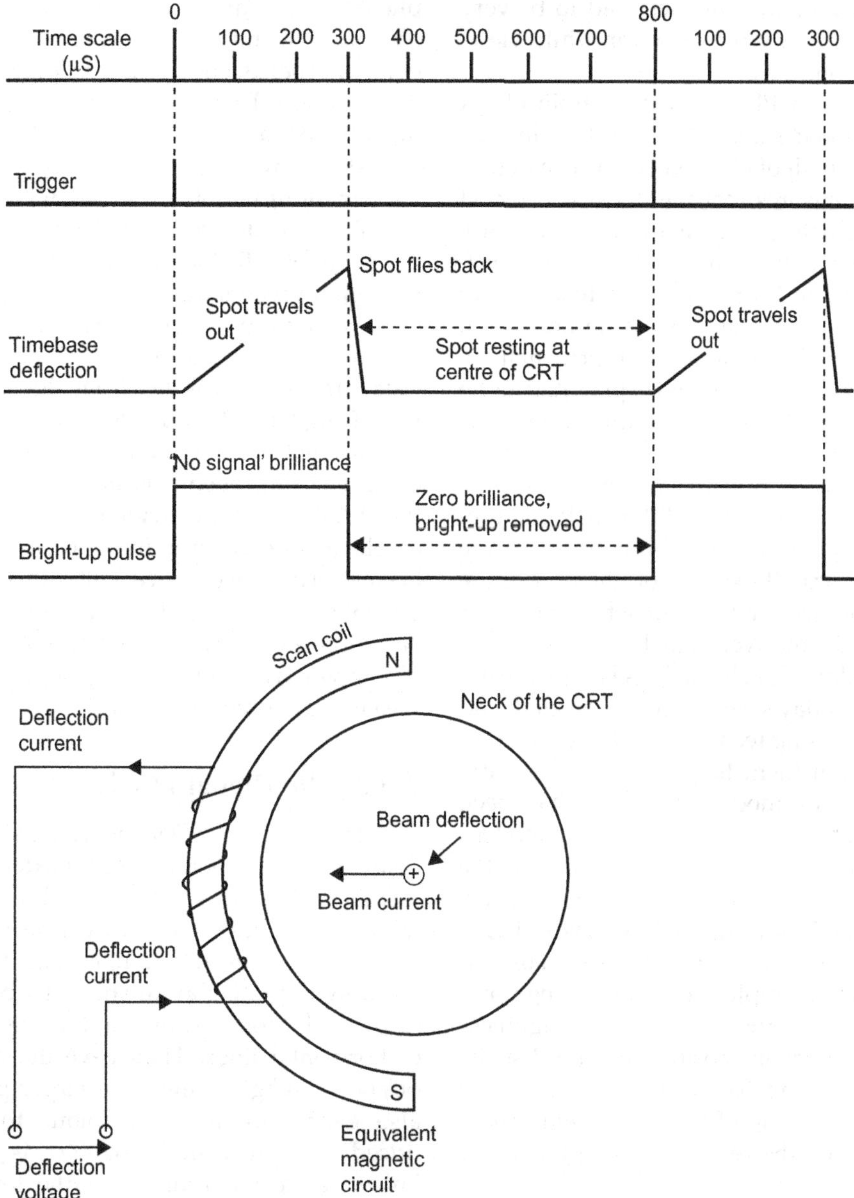

FIGURE 2.39 Rotating-scan coil deflection: time sequence diagram for PRF = 1250, range scale 24 NM.

known as *persistence* or *afterglow*. This kept a target visible while the antenna rotated a full 360°, but was also chosen to be long enough for a moving target to leave a trail across the display, showing its past track over a few minutes. The dynamic range of signals which can be displayed by the phosphor was very limited, typically around 10 and so the settings, particularly gain and brilliance, had to be very carefully adjusted by the operator while viewing the situation.

The radial scan PPI turns the amplitude of the received radar signal at any instant into an equivalent strength of electron beam. The deflection coil then ensures that the beam is directed to the appropriate point on the display. Since the integral geometry of a radial scan PPI, measured in range and bearing (r,φ), matches that of the received radar signal, there is no need for a conversion to a Cartesian (x,y) representation. The timing of the radial scan display had to be kept in good synchronization with the antenna beam pointing angle and also with the pulse transmission times, but it was able to operate satisfactorily on signals with little analogue preprocessing and with no digital processing. Large targets were naturally shown brighter since the strength of the electron beam matched the larger amplitude of the received signal.

The capability of such displays is now greatly exceeded by today's sophisticated digital and analogue processing techniques. However, comparison between them helps towards a greater understanding of modern radar. It has been explained previously in this book that modern radars build a table of amplitudes of the received signal for all cells defined by their discrete range and bearing. This enables digital processing to be carried out on the signals and for the resultant amplitude at any range and bearing to be used, for instance, as a brightness level at the equivalent position on the display. The final processing has to include converting the range and bearing of the cells to equivalent (x,y) positions on the rectangular display, using simple geometry.

The persistence of the phosphor on the old CRTs had several important advantages, which on a modern system have to be replicated by processing. A highly important one is that of scan-to-scan integration. A weak target may not generate a visible return at every scan of the antenna. However, the persistence of the phosphor can allow such weak targets to be observed. If missed for a small number of scans the persistence of the display still shows its presence. This action has to be replicated digitally on a modern radar. Also, the same persistence over many scans shows the relative movement of targets as a trail, which is highly useful radar-derived information, providing the basis of TT. This too has to be replicated by digital processing.

This may sound as if digital processing is just attempting to replicate the action of an old radar. In fact this is far from the truth as the use of digital techniques has very substantially transformed marine radar – and we can expect continued improvements into the future. What the radial scan PPI provided was the basis of an excellent human–machine interface, which has been evolving ever since but still bears fundamental similarities. In fact the electronic chart is based on a very similar human–machine interface, giving rise to the concept of a chart radar, which is explored in Section 10.2.

2.8.2 The Digital Display

From the late 1970s the original radar PPI display was generally superseded by CRT displays similar to those used on domestic televisions. These worked by a beam scanning in the horizontal direction, consecutively moving down the display in small discrete vertical steps and thus forming a picture composed of horizontal lines. They gave the advantages of good daylight viewing, a larger picture and also eventually allowed colour to be introduced. A picture made of lines is commonly known as a raster image – the term is connected with the word 'rake'. The analogue signal modulated the intensity of the beam, just as on the original PPI. The advent of colour tended to pixellate the image as the colour phosphors were arranged on a regular grid across the screen. At each grid-point phosphors for red, green and blue (RGB) were applied during the manufacturing process of the colour CRT, just as for colour TV CRTs.

Although earlier monochrome CRTs used long persistence phosphors, later versions and especially the colour CRTs used standard phosphors, and so analogue processing on the signal prior to its display had to emulate the effects of a long persistence phosphor. Increasingly, the processing needed to do this was digitized, even though the final signal to the CRT had to be converted back to analogue form. Modern flat panel displays used for radar work almost entirely digitally and it is nowadays best to consider that the final step of displaying a radar image is to define the brightness and colour on a closely spaced grid of elements across the display, each element being known as a pixel.

A set of memory locations within the display processor effectively have a one-to-one relationship with each of the pixels on the display. The memory locations contain the calculated amplitudes of the RGB amplitudes of each display pixel. These have been determined by the processed radar image, and by other digital processes which form overlays, such as graphics and text symbols, as shown in Figure 2.40. We have seen that the radar image is basically stored in terms of range (R) and relative bearing (φ) in discrete cells. These have to be transformed into the Cartesian (x,y) coordinates of the display screen, by use of the basic equations (see Figure 2.41):

$$x = R\sin(\varphi), \quad y = R\cos(\varphi)$$

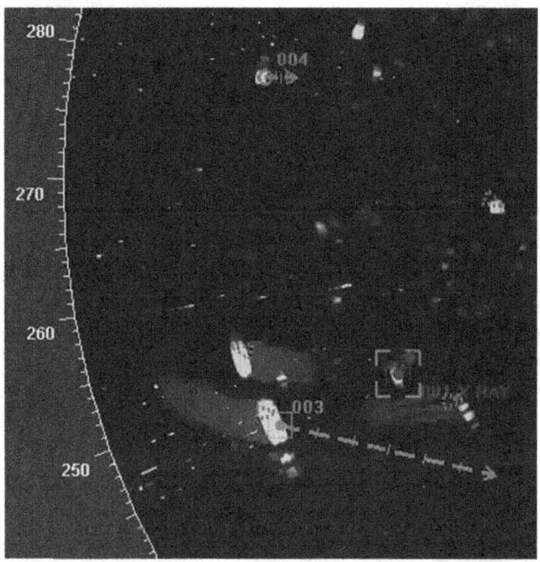

FIGURE 2.40 Graphics overlays. Reproduced courtesy of Kelvin Hughes.

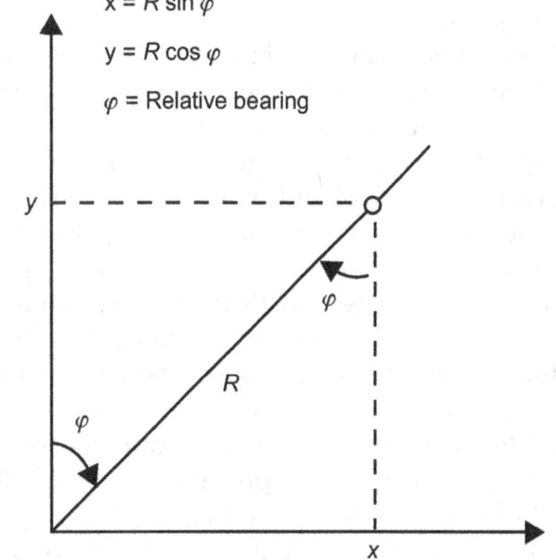

FIGURE 2.41 Cartesian transformation.

2.8.2.1 *The Use of Colour*

Colour is widely used in modern radar displays as it helps clarity. Over time, more and more uniformity over the use of colour is mandated but at present much is left to the discretion of the equipment manufacturer. To illustrate this, it is worth noting IMO requirements for basic radar video images, which replicates Para 6.3.1 of MSC 191(79):2004 – Performance standards for the presentation of navigation related information on shipborne navigational displays:

> *Radar images should be displayed by using a basic colour that provides optimum contrast. Radar echoes should be clearly visible when presented on top of a chart background. The relative strength of echoes may be differentiated by tones of the same basic colour. The basic colour may be different for operation under different ambient light conditions.*

To meet the additional requirements of the International Electrotechnical Commission's performance standards IEC 62288 Presentation of navigation-related information on shipborne navigational displays (Ed 1.0, 2008), the following additional clause (and others) must also be met for radar video images:

> *The colours may be different for operation under different ambient light conditions (day, dusk and night) likely to be experienced on the bridge of a ship, and with due consideration to the night vision of the officer of the watch.*

In particular, the IEC document also points out that *if the colour red is used for the radar video image, then it shall be distinguishable from other uses of the colour red, for example, alarms including dangerous targets.*

In principle, colour can be used to indicate signal strength but oversimplified ways of doing this can be confusing to the user, for instance leading to continuous colour changes for some targets. Typically, a single colour (e.g. yellow or amber) is used for the present position of all targets, irrespective of signal strength. A second colour (possibly brown or green) is then used to indicate more random responses, such as those from sea clutter, while persistent echoes retain the main echo colour. Such a feature can be user selectable, but must be used with care and with an understanding of the principles involved. It has to be appreciated, for example, that buoys, boats and similar small floating targets could return intermittent responses. However, with good design the increasing sophistication in signal processing algorithms certainly can provide very useful extra clarity to targets by using colour.

An important decision is the choice of background colour, which is that which represents the absence of signal (or data). There has been some earlier debate on this, but the current requirements are that a dark non-reflecting background should be used, such that it provides a contrast against the chosen colours for targets. Modern radars typically offer some user selection on this, even if limited to a choice of manufacturer-chosen day, dusk and night settings.

A different colour contrasting with the background and the target itself must be used for target trails. In particular, assimilation of display information is assisted by using a separate colour (or more) for the additional display graphics, such as those associated with range and bearing measurement, for example range rings, variable range markers, electronic bearing lines, heading markers, bearing scales and data readouts. Red is required to be used as a warning colour for selected graphics and data.

2.8.3 Flat Panel Display Technology – LCDs

LCDs are now the standard screen technology used for radar and other bridge displays. Other display technologies will inevitably supersede them, but it is difficult to forecast what or by when. Current candidates include quantum dot light-emitting diode displays (QD-LED), organic light-emitting diode or transistor displays (OLED, OLET) and field emission display (FED). However, it is worth examining LCDs in a little depth as they epitomize much of the thinking behind modern display technology.

Liquid crystals are unusual in that they have a state that does not fall into one of three standard matter states of solid, liquid or gas. This extra state is nearer liquid than solid, but the molecules line themselves up in a fixed orientation, which is normally the property of a solid. This is called the nematic phase. Some liquid crystals are called twisted nematics, because in their nematic phase, the orientation of the molecules naturally twists with respect to each other. The amount of the twist can be controlled by applying a voltage across the liquid crystal.

Another property of liquid crystals is that they are transparent substances that allow light to pass through them. In the case of twisted nematics, light penetrating one side of the crystal will be twisted as it goes through the crystal (i.e. the polarization of the light will be rotated, henceforth referred to as twisted in this text). This property is exploited for LCD, by putting the liquid crystal (layer 3 in Figure 2.42) and its controlling transparent electrodes (layers 2 and 4), between two pieces of glass (layers 1 and 5) each with a polarizing film coating on one side. Finally a mirror (layer 6) is added to the bottom of the second piece of glass. The polarizing film only allows light that has a certain polarization to pass through the film. Light with incorrect polarization simply does not pass through the film. The top glass layer (layer 1) therefore only allows through external light polarized in one direction. The plane of the polarized light is twisted by a controlled amount through the liquid crystal, depending on the voltage on the

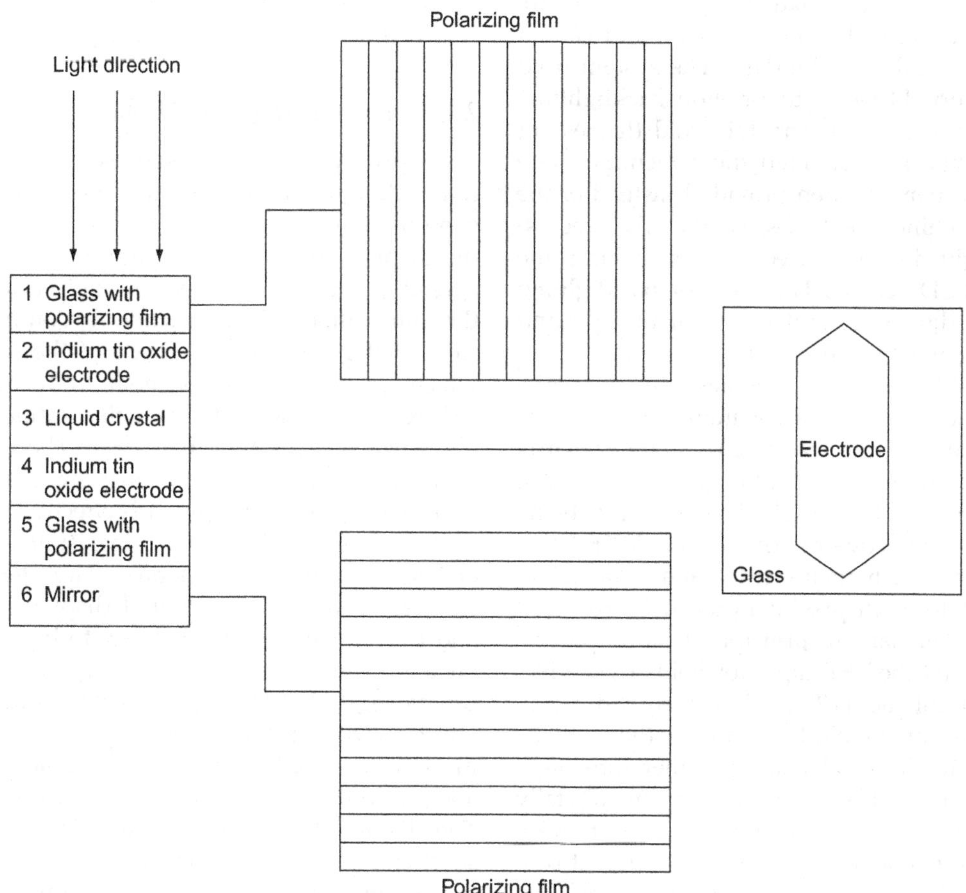

FIGURE 2.42 Principles of the LCD.

electrodes (layers 2 and 4). In one condition, the twist will be such that the plane of the polarized light will match the polarization of the film on the second piece of glass. The light will therefore pass through and be reflected back through the same path by the mirror. The LCD appears blank to an observer in this case. In the second condition, the amount of twist is adjusted by the voltage across the electrodes so that the plane is polarized at 90° to the polarized film on the second piece of glass and no light gets through. The LCD appears black in this case.

The simple LCD with a mirror background, described above, depends on external light to illuminate it and make it visible. These are sometimes termed reflective LCDs and cannot be seen unaided in the dark. The system used on modern LCDs is to provide backlighting. This may consist of very thin cold fluorescent strip lights that surround the screen. A white diffusion panel is then provided instead of the mirror behind the LCDs, so that the fluorescent light is distributed evenly behind the whole LCD screen. The level of backlighting can be adjusted to suit the ambient light and the convenience of the user.

Each LCD element becomes one of a large number of pixels on a rectangular screen. The LCD pixels are arranged in a rectangular matrix and typically use a thin film transistor (TFT) to control each LCD pixel; the technique being known as an active matrix system. A TFT consists of a switching transistor and a capacitor. The TFT for each pixel is updated by row and column (similar in principle to the passive matrix) and the TFT capacitor holds the correct charge until the TFT pixel is refreshed again. This system precisely controls the voltage applied to the LCD pixel and therefore more precisely controls the twist. A very fine digitally controlled graduation of twist can be employed, each graduation letting through a different proportion of light. If a white diffusion panel (see *backlighting* above) is employed behind the LCD, the effect is to obtain different shades of grey (greyscale) from pure white to pure black. This provides a good monochrome picture. To obtain a colour picture, now normal for marine radars, each pixel is split up into three separate sub-pixels fitted with RGB filters, respectively. Any colour can then be achieved by mixing the appropriate amount of each of the three primary colours.

In some systems the display surface also has a touch-sensitive characteristic and the user can operate certain functions simply by touching the appropriate screen location. This is achieved by means of a thin overlay consisting of two layers of plastic between which lies a matrix of a very large number of elemental areas whose conductivity is sensitive to finger pressure.

2.8.4 Formatting the Display

In principle modern technology gives the radar designer almost infinite possibilities in choosing how to represent radar information on the display. In reality, there are internationally agreed practices approved through IMO and the International Electrotechnical Commission (IEC) that govern many aspects of the display, which provides some essential uniformity to reduce the time needed for familiarization training when using a different radar. The display processor deals with the formatting of all the information on the display. It is effectively identical to the display processing functionality embedded within any personal computer (PC). Software controls the required display layouts and functionality, which is likely to be running on a processor that is also doing other radar processing tasks. The display data is output in one of several possible industry standard formats, such as XGA (Extended Graphics Array) and fed to the display itself, via the video cable. The display itself processes this signal further so that the required voltages/currents at the correct time are applied to the individual screen pixels that actually generate the picture.

2.8.4.1 *Display Requirements*

IMO defines the main requirements of the display. In particular, these include the requirements that the operational display area of the radar presentation shall be at least a circle of diameter of:

- 180 mm for ships smaller than 500 gross tonnage
- 250 mm for ships larger than 500 gross tonnage and high-speed craft less than 10,000 gross tonnage
- 320 mm for ships larger than 10,000 gross tonnage.

It is this definition that allows the radar area to be other than a circle, provided that the display area covers at least the required diameter. The requirements given in IEC 62388, the technical specifications for marine radar issued by the International Electrotechnical Commission, which expand on IMO requirements, explicitly state that 'circular, square or rectangular presentations' are those that are permitted.

There is an important IMO requirement that 'the presentation of alphanumeric data, information and text, symbols and other graphical information (e.g. radar image or echoes) shall support readability from typical user positions (i.e. reading distance and viewing angles) under all ambient light conditions (day, dusk and night), likely to be experienced on the bridge of a ship, and with due consideration to the night vision of the watch' (from IMO Resolution MSC.191(79)). Preservation of night vision is a particularly important consideration. On today's bridges the electronic instrumentation can produce very bright images. If incorrectly set by the user at night, their brightness will contract the irises of the eyes of those on watch, greatly affecting their night vision. The view from the bridge windows is a primary navigational aid and it must not be compromised by incorrect setting of the brightness of any display on the bridge.

The typically circular image on a current radar display allows much data to be shown in the area between the radar image and the extremities of the rectangular display. The user will typically have some choice in exactly what data is being displayed, although some data has to be displayed permanently. User menus will typically appear in this area, although start-up menus, when the radar is not fully operational could also occupy the area normally used by the target display area. The convergence between the 'look and feel' of radar and ECDIS displays (but not their functionality, see Section 10.2) will probably mean that the typical radar display area becomes rectangular, rather than circular – but there is an argument that suggests both the ECDIS and radar area displays should be circular, since it would maximize the space for additional data to be displayed whilst generally maintaining the practical effectiveness of the displayed area.

An example of a modern radar display is shown in Figure 2.43. On all modern displays the user has control of the following:

- The effective scale of the display, generally selected as a 'maximum range'
- The position of the own ship on the display, generally known as off-centring
- Whether range rings should be displayed and if displayed, which ones?
- The orientation of the radar display area, such as north-up or course-up, known as the azimuth stabilization or orientation modes, see Section 2.8.4.2
- The display motion modes – true- or relative-motion modes, see Section 2.8.4.3
- The horizontal reference mode of the radar – ground or sea stabilization, see Section 2.8.4.4
- Whether trails should be displayed and the time length for the trails

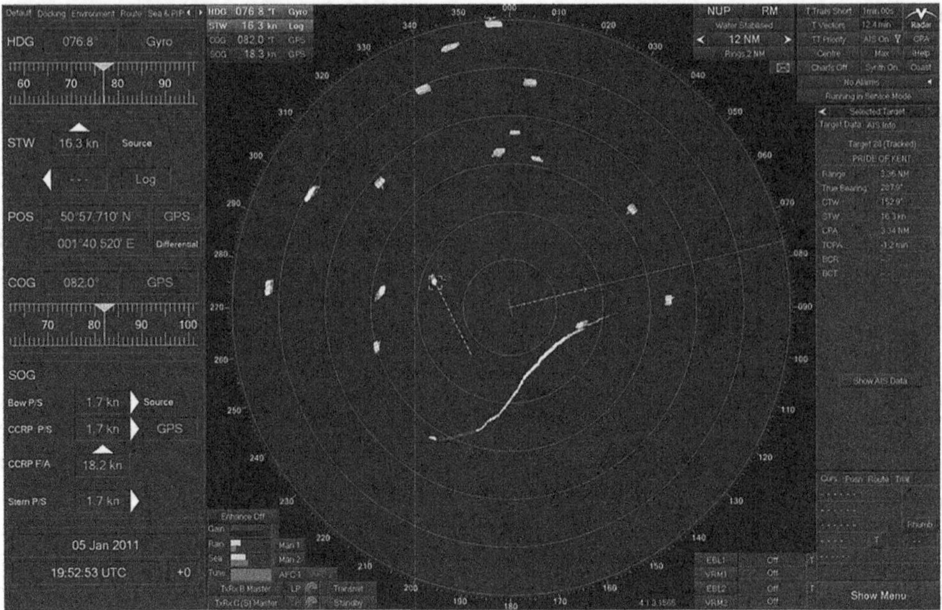

FIGURE 2.43 Modern radar display. Reproduced courtesy of Northrop Grumman Sperry Marine.

- Whether TT should be activated and if so which targets should be tracked, see Sections 4.2 and 4.3
- Whether AIS targets should also be displayed and if so which ones, together with their interaction with radar displayed data (such as correlation), see Section 5.7.

In addition, the user will have control of at least the following functions, with any appropriate indications of the settings being shown on the display:

- Radar gain
- Rain clutter setting (manual level and setting to automatic)
- Sea clutter setting (manual level and setting to automatic)
- User map creation and display
- Selection of radar stand-by or transmit
- Alarm setting and acknowledgement.

The control of these settings will typically be by a combination of dedicated and menu control, which can vary considerably from system to system. This underlines the need for proper familiarization with the particular radars on a vessel before they are used on a formal watch.

2.8.4.2 Display Orientation Modes

There are three orientation modes that are provided on a radar, known as north-up, course-up and head-up (see Section 1.4). Head-up is the 'natural' mode of a marine radar, with the 'vertical' direction on the display being aligned to the heading of the vessel. Nowadays, this mode is generally restricted to the case when the compass input to the radar has failed and should be considered to be an emergency mode. This is because the displayed radar image tends to oscillate in azimuth with small variations in the heading of the vessel, often making it difficult to make accurate radar observations.

If the radar picture is required to be generally aligned with the heading of the vessel, which can be useful in keeping a fast mental comparison with the view from the bridge windows, then course-up is IMO recommended display setting. The user selects the course, which may also be a selection that is automatically linked with other navigational aids, such as the autopilot and ECDIS. For this mode to operate an automatic input from the vessel's compass is required, ideally a gyro compass, which is compulsory on larger vessels. The digital interface between the compass and the radar is defined and follows standards issued by the IEC, in a series known as IEC 61162. Such digital interconnections between navigational equipment are often also called 'NMEA' (see Section 10.3.4), indicating standards issued by the US National Marine Electronics Association, which have appropriate alignment with IEC 61162 standards and are commonly used on smaller vessels. Some compass inputs are still based on an internationally defined analogue signal for their interface.

The most commonly used orientation mode is that of north-up, which aligns the radars 'vertical' direction to north. This matches the radar display to the orientation of paper charts and the usual 'mind image' of an area and so is a very useful mode. The increasing use of ECDIS in course-up mode may possibly lead to north-up mode being less used in the future on both ECDIS and radar. Since the image is set this way to align with the charted view it is true-north that is needed and so the correct variation should be applied, when using magnetic 'transmitting' compasses.

For both course-up and north-up modes the radar image is required to be aligned within 0.5° of the signal being applied by the compass. However, the compass may be in error by more than this amount. For all orientation modes the heading of the own ship is shown by where the heading line intersects the bearing scale, which is generally shown around the edge of the circular radar area. For north-up mode true-north (000°) will be shown at the top of the bearing scale and the heading line will intersect the scale at the true heading. If the own vessel's position is not in the centre of the display the bearings around the display will automatically adjust and will not appear to be uniformly spaced. The vessel's heading will almost certainly also be displayed numerically in a box outside of the radar circle.

If the display is in course-up mode it is likely that the bearing scale can be user-set such that it either displays relative to true-north or relative to the chosen course. If the vessel is on course the heading line will be vertical in both cases but the value indicated on the scale will either be 000° or the actually chosen course, such as 030°. The user must remain fully aware of what convention has been set. Course changes will be indicated by a non-vertical heading line. Sections 1.4 and 6.2.5 explain the use of each of the orientation modes.

2.8.4.3 Display Motion Modes

There are two motion modes – true-motion and relative-motion. When set to true-motion the position of the own ship moves across the display at the scale speed of the vessel's actual motion and in accordance with whether the radar has been set to ground or sea stabilization, see Section 2.8.4.4. When own vessel nears the edge of the display it will reset to a user-specified position. The reset will occur at a preset distance to the edge of the display, also preset by the user. IMO calls the position of the own vessel on the display the CCRP, the Consistent Common Reference Point, see Section 2.10.1.1. True-motion requires the own vessel's changing (relative) position to be assessed by the radar. This may be from compass and log inputs or from a positional input, such as GPS. In fact, IMO's radar requirements for SOLAS vessels always state the use of Speed and Distance Measuring Equipment (SDME), meeting certain minimum requirements, rather than the term 'log'.

When set to relative-motion, own vessel will remain stationary on the display. This is traditionally at the centre of the display but can be set by the user to any position. The latter is often known as off-centring. It may even be possible to set this position so that it is off the display, but great care needs to be taken when so set, simply because close-in and therefore potentially dangerous targets will also not necessarily be shown. On a modern radar, the bearing scale will typically adjust to take into account the displayed position of own ship. Depending on specific circumstances this can look quite distorted and so the digital readout should always be referenced.

2.8.4.4 Horizontal Reference Mode

Ground and sea stabilization form the two horizontal reference modes. The vessel's motion is naturally based around a sea referenced framework. Neglecting wind and wave effects an unpowered vessel would lie stationary with respect to the current flow of the water and all powered motion would be relative to the actual current. In general, vessels nearby will be experiencing the same current. Therefore, correct judgements on collision avoidance can then be made without any reference to ground fixed coordinates, provided there are no (charted) underwater hazards to be avoided. Since radar is primarily a collision avoidance aid, working in sea stabilization mode can be very useful in the right circumstances. Furthermore, it can be easily implemented as the ship's SDME and compass inputs are the only ones necessary to reflect the own vessel's sea stabilized movements on the radar.

For practical reasons, ground referenced motion is often preferred in many situations, especially to maintain safe distances from charted hazards but also to generally ease situational awareness. Ground referencing at sea used to be quite a difficult task, requiring estimates of the actual currents in terms of set and drift and/or by using a Doppler log referenced to the sea bed, with many associated problems. Nowadays, GNSS generally provides a very good solution, allowing good ground referencing to be automatically applied to the radar through the digital interface between the GNSS and the radar. However, the potential weaknesses of GNSS should always be understood when using such a source for ground stabilization, see Section 10.1.

It is important to keep aware of the selected horizontal reference mode. The chosen orientation mode and the display motion modes are generally self-evident, just by observing the display. However, the changes between ground and sea stabilization are not so obvious, except by close examination of the tracks of targets, if when selected, or by the vectors of any targets that are being tracked. There should be an indication of the mode in use somewhere on the display. The practical use of all modes is examined in detail in Section 6.9.

2.8.5 The Display of Radar Echoes

'Echo paint' is a term sometimes used to describe the size and shape of an echo as it appears on the screen. In general, the echo paint is highly influenced by certain characteristics of the radar system, including its processing algorithms, and somewhat less by the actual size and shape of the target. However, it is useful to consider the fundamental issues that affect the display of the echo from a point target. As discussed previously, a point target is a theoretical concept which supposes a target that has the ability to reflect radar energy but which has no dimensions. Clearly this cannot be realized in practice, but it is a useful concept because it simplifies conceptualization and, in practice, many practical targets behave in a manner closely resembling that of a theoretical point target. Figure 2.44 shows a situation in which a radar beam sweeps across two point targets and illustrates how the echo paint is built up.

The principle characteristics of the radar system which affect the size and shape of the displayed echo are the transmitted pulse

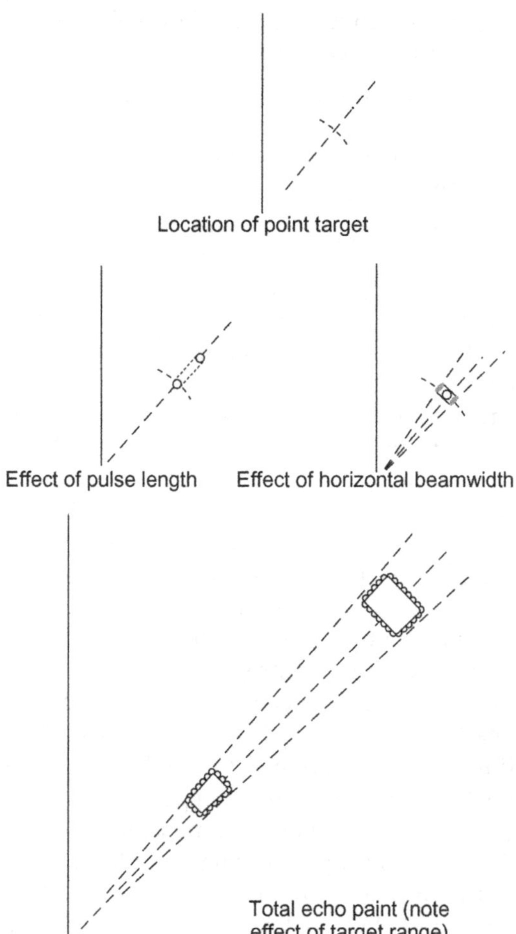

FIGURE 2.44 The echo paint of a point target.

length and the horizontal beamwidth of the antenna. As discussed in Section 2.7.3.1, the pulse length is the major factor to be taken into account when choosing the range cell size for radar target processing, not least because of the limitations that the pulse length put on the discrimination accuracy of the display. The azimuth cell size is dictated by the PRF and the antenna rotation rate. These do not have a major influence on the size of the displayed image and hence the radar's fundamental resolution. It is the horizontal beamwidth of the antenna that is the dominant factor.

Consider a pulse length of 1 μs. As previously discussed, the typical sample rate for such a pulse would be half of this, that is, a sample every 0.5 μs. For a perfect point target two adjacent range cells would indicate the target. (In principle, the target could appear in three adjacent cells if the sampling happened to occur exactly at the start of the received pulse – one sample at the exact beginning, middle and end of a pulse – but this has, in our simple model, a precisely zero chance of happening.) The received pulse thus creates an exact 1 μs 'trace' on the display, just as it would have done in the original analogue-based radars of the past. A length of 1 μs on the display equates to 150 m in range, taking into account that a radar pulse has to travel to and from any target – an electromagnetic wave travels a total of 300 m in 1 μs. For this case 150 m is the fundamental apparent length of the received signal, provided the display resolution (number of pixels) can match or better this figure.

As the beam sweeps across the point target, it will return a number of echoes according to the PRI, the antenna rotation rate and its horizontal beamwidth. The echo paint of the point target is therefore built up of a number of azimuthal elements each of which has a radial length, as defined in the preceding paragraph. This is illustrated by Figure 2.44. The angular width of the paint of a point target will depend on the horizontal beamwidth, while the number of azimuthal elements is dependent on the PRF and the antenna rotation rate. In general, the digital display processing within the radar effectively fills in the azimuthal gaps. The angular elongation of the echo paint is sometimes referred to as half-beamwidth distortion because it has the effect of extending the paint by half the beamwidth on either side of the nominal echo position. The angle contained between the limits of the main lobe of the horizontal beam is constant, but the length of the arc which subtends this angle at the antenna increases in proportion to the range at which the arc is measured. Thus the length which this angular distortion produces increases with the range of the target (see Figure 2.44).

In the limit, the pixel size of the display and/or any limitations within the display processing finally dictate the displayed size of target. Nowadays, this is rarely a practical issue. With early CRT technology the minimum spot size that could be produced did create an additional practical limitation.

2.8.5.1 *Practical Considerations*

In the case of a point target, the aggregate of the effects described above is a paint whose size and shape depends on the pulse length and horizontal beamwidth of the radar system and also on the set maximum range of the display. To appreciate the practical significance of these effects it is helpful to consider the following numerical example.

Example: A marine radar system has a radar display area (PPI) of diameter 250 mm. The horizontal beamwidth is 2°, the pulse length is 1 μs and the display is switched to the 12 NM range scale. A point target X is detected at a range of 10 NM. Calculate the radial length and the angular width of the echo paint as it appears on the screen, and hence determine the distances, in NM, which these dimensions represent.

Let the total dimensions on the display of the echo paint be given by T_R (radial length) and T_A (angular width). The 12 NM range equates to an equivalent time interval of:

$$12 \times 1852 \div (3 \times 10^8) \div 2\ \text{s} = 148.16\ \mu s$$

with the division by 2 being a consequence of the 'to and from' nature of the radar signal. The received pulse is 1 μs and so its equivalent length on the display is given by:

$$T_R = (1 \div 148.16) \times 250 \div 2 = 1.2\ \text{mm on the display}$$

A 10 NM range ring on the display would have a circumference of:

$$\pi \times 250 \times 10 \div 12 = 654.5\ \text{mm}$$

A 2° segment of this would therefore have a length given by:

$$T_A = 654.5 \times 2 \div 360 = 3.6\ \text{mm}$$

Thus the echo paint has a radial length of 1.2 mm and an angular width of 3.6 mm. This is well in excess of the minimum resolution of a modern display and so therefore is a good approximation as to what would be shown on it. In fact, on long range scales the pixel size could dominate the radial length of the paint, which has to be a factor considered by the manufacturer in the design of the display processing of the radar to avoid any unintended suppression of such targets.

In the example chosen, the radius of the screen (125 mm) represents 12 NM, from which it can be deduced that the natural scale of the PPI is such that 1 mm represents 178 m. If this were to be applied to the dimensions of the echo paint it would suggest a target of approximate radial dimension 214 m (1.2 × 178) and approximate angular dimension 641 m (3.6 × 178). This we know is likely to be a distortion of target size, especially the 'angular' width. The area which is bounded by these dimensions is the effective *resolution cell* of the radar. Thus while it is possible to determine the range and bearing of a target with the necessary degree of accuracy by taking the measurement to the centre of the nearer edge of the echo paint (this represents the location of the echoing surface, see Figure 2.44), the size and shape of the paint of all but very large targets bear little or no relationship to the size and shape of the target which produced it. A radar reflector (see Section 3.4) has almost negligible dimensions, but if located at the position of the point target suggested in the example above, it would produce an echo paint which covers a screen area equivalent to that of several large ships. Hence, only when a target has reflecting surfaces which extend beyond the resolution cell does it begin to contribute significantly to the size and shape of the echo on the display.

Except at extremely close ranges it is evident that a ship will in general produce an echo approaching that of a point target. If the calculation in the example is performed for the same radar system, but with the point target located at a range of 2 NM, it will be found that the radial dimension remains unchanged at 1.2 mm (as it is not a function of range) while the angular dimension decreases to 0.5 mm. Thus, if a small target closes from a range of 10 NM to a range of 2 NM, its echo shape will change from one in which the angular width is the greater dimension to one in which the reverse is the case. It is noteworthy that radar characteristics selected in the above examples are representative and do not illustrate the worst possible case.

In the use of radar for collision avoidance the principal significance of the above limitation is that the shape of the displayed echo of a ship gives no indication of how the vessel is heading. Heading inference drawn from the shape of the echo is likely to be dangerously misleading. For example, to the unwary or untrained observer, targets towards the edge of the screen may *appear* to be broadside-on (because of the dominance of the angular distortion), whereas targets towards the centre of the screen may *appear* to be end-on. The importance of understanding the specious nature of the impression created by the screen shape of the echo cannot be overstressed. The techniques for deducing a reliable indication of the heading of other vessels are set out in Chapter 7.

2.8.5.2 Range Discrimination

Range discrimination describes the ability of the radar system to display separately the echoes of two targets which lie on the same bearing but which are closely spaced in range. This was examined from a sampling point of view in Section 2.7.3.1. Obviously, the size of the resolution cell presents limitations to the navigational use of the radar, and this is looked at in detail in Section 8.2.2. Range discrimination is expressed in terms of the number of metres in range by which the targets must be separated in order to prevent their echoes overlapping on the display. In discussing discrimination it must be assumed that both targets are illuminated by the radar beam. If the more distant target is in a blind area caused by the nearer target, no echo will be displayed, irrespective of the ability of the radar to discriminate. IMO Performance Standards set out the requirement for range discrimination in terms of two small targets which lie on the same azimuth and which are separated by 40 m in range. To comply with the standard, the equipment must, on range scales of 1.5 NM and less, be capable of displaying the echoes of two such targets as separate indications when the pair lies at a range of between 50% and 100% of this range scale.

2.8.5.3 Bearing Discrimination

Bearing discrimination, which may also be referred to as bearing resolution, describes the ability of the radar system to display separately the echoes of two targets which lie at the same range but are closely spaced in bearing. The discrimination is normally expressed as the angular separation that two targets at the same range must be separated in azimuth such that their echoes will appear separately on the screen. IMO Performance Standards set out the requirement for bearing discrimination in terms of two small similar targets both situated at the same range between 50% and 100% of the 1.5 NM range scales and less. To comply with the standard, the equipment must be capable of displaying the echoes separately

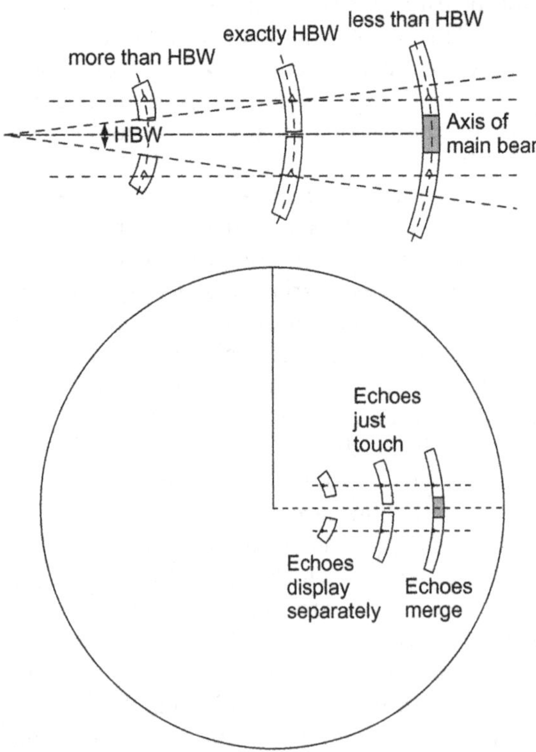

FIGURE 2.45 Bearing discrimination.

when the targets are separated by not more than 2.5° in azimuth.

If two targets lie at the same range and are closely spaced in bearing, a minimum requirement for angular discrimination is that the trailing edge of the rotating beam must leave the first target before the leading edge illuminates the second target. This is illustrated by Figure 2.45. This figure shows the limiting case of discrimination and two other cases for comparison purposes. It follows from the illustration that, in theory, the radar can only discriminate between targets which are separated by at least one horizontal beamwidth, providing the azimuth cell size is small in comparison to this. Section 2.7.3.2 shows that for short range settings this is generally true, with a typical azimuth cell being about 0.05°. The simple theory is also complicated by the fact that the effective antenna beamwidth for discrimination determination is unlikely to be its 3 dB figure, especially for relatively close-in targets – its 10 or even 20 dB beamwidth may be more appropriate to be used. As for range discrimination, the pixel size of the display is unlikely to have a limiting effect.

It is important to remember that the distance represented by the horizontal beamwidth increases in direct proportion to the range at which the discrimination is being considered. The influence of spot size will be less on larger screens and on shorter range scales. Some representative values are illustrated by the following numerical example.

Example: A radar display has a screen of diameter 250 mm. If the effective horizontal beamwidth is 1.5°, calculate the bearing discrimination at a range of 10 NM when the 12 NM range scale is selected and 1 NM when the 1.5 NM range scale is selected.

At 10 NM

The length of the arc of the horizontal beamwidth at 10 NM is given by:

$$A_B = \text{range} \times \text{horizontal beamwidth in radians}$$

$$= (10 \times 1852) \times 1.5 \times \frac{\pi}{180}\,\text{m}$$

$$= 10 \times 1852 \times 1.5 \times \frac{22}{7 \times 180}\,\text{m}$$

$$= 485\ \text{m}$$

At 1 NM

$$A_B = 1 \times 1852 \times 1.5 \times \frac{22}{7 \times 180}\,\text{m}$$

$$= 48.5\ \text{m}$$

Care should be taken that this theoretical 'best' is not marred by excessive brilliance, gain or contrast settings (see Sections 6.2.3 and 6.2.7.1).

A serious shortcoming of the use of civil marine radar for collision avoidance is its inability to offer direct indication of the heading of other vessels. Beamwidth distortion is a major contributor to this shortcoming. It also limits the usefulness of the radar for navigation by merging coastline details, such as bays and lock entrances which lie within one horizontal beamwidth. The significance of this is discussed in Section 8.2.2. Importantly, digital signal processing techniques can be used to help improve angular resolution to better the limits discussed in this section. This is because the small size of the azimuth cell relative to the beamwidth, at least at shorter range settings, gives extra information that can indicate that two targets are present, such as distinctive changes in the amplitudes of the return at a constant range as the antenna rotates.

2.8.6 The User Interface

The traditional radial CRT display had a large number of analogue and push-button controls surrounding the circular screen. Over time the number of features and the corresponding controls/buttons offered on sets tended to increase. The first raster-scan displays followed this traditional layout and sets (particularly ARPA/radar sets) with over 80 controls or buttons were common.

This trend has now reversed itself by the use of on-screen menus which have been encouraged by three interlinked factors:

a. The increased computer literacy of the general population means that most mariners are familiar with the concept of pull-down menus.

b. The increased use of pointing devices/ cursors (see below) enables a flexible arrangement on screen to operate the pull-down menus.

c. The use of rectangular raster-scan screens means that there are convenient spaces to the side of the circular radar area to display both information and the drop-down menus (Figure 2.46).

In the extreme case, the display may only have two buttons and the pointing device. A major advantage is that this considerably reduces the cost and maintenance of the radar system. Hardware is effectively replaced by software that does not wear out. The use of drop-down menus enables the manufacturer to provide many functions without cluttering up the display and radar set. The skill of the manufacturer is arranging the various commands into logical groupings so that an observer can intuitively find the correct information/command through several layers of menus without continual reference to the manual.

Pointing devices. Pointing devices enable the user to control the position of a moving symbol on the display. Two or more 'click' buttons are provided. When pressed they will be programmed to perform an action or display more information.

On a home computer the most common pointing device used is the 'mouse'. Different designs exist, but the principle of a computer mouse is that mechanical, optical or potentially acceleration sensors in its base detect its movement over a surface. Radar displays on vessels are not usually provided with a mouse, but with a tracker ball (or trackball). A tracker ball acts as an inverted mouse with a mechanical ball as the sensor detecting the required motion of the cursor. The user simply spins the ball in the required direction. There are several reasons why a tracker ball is preferable to a mouse for a marine radar display. Firstly, on a rolling ship, the mouse may move of its own volition, the tracker ball is fixed to one place. Secondly,

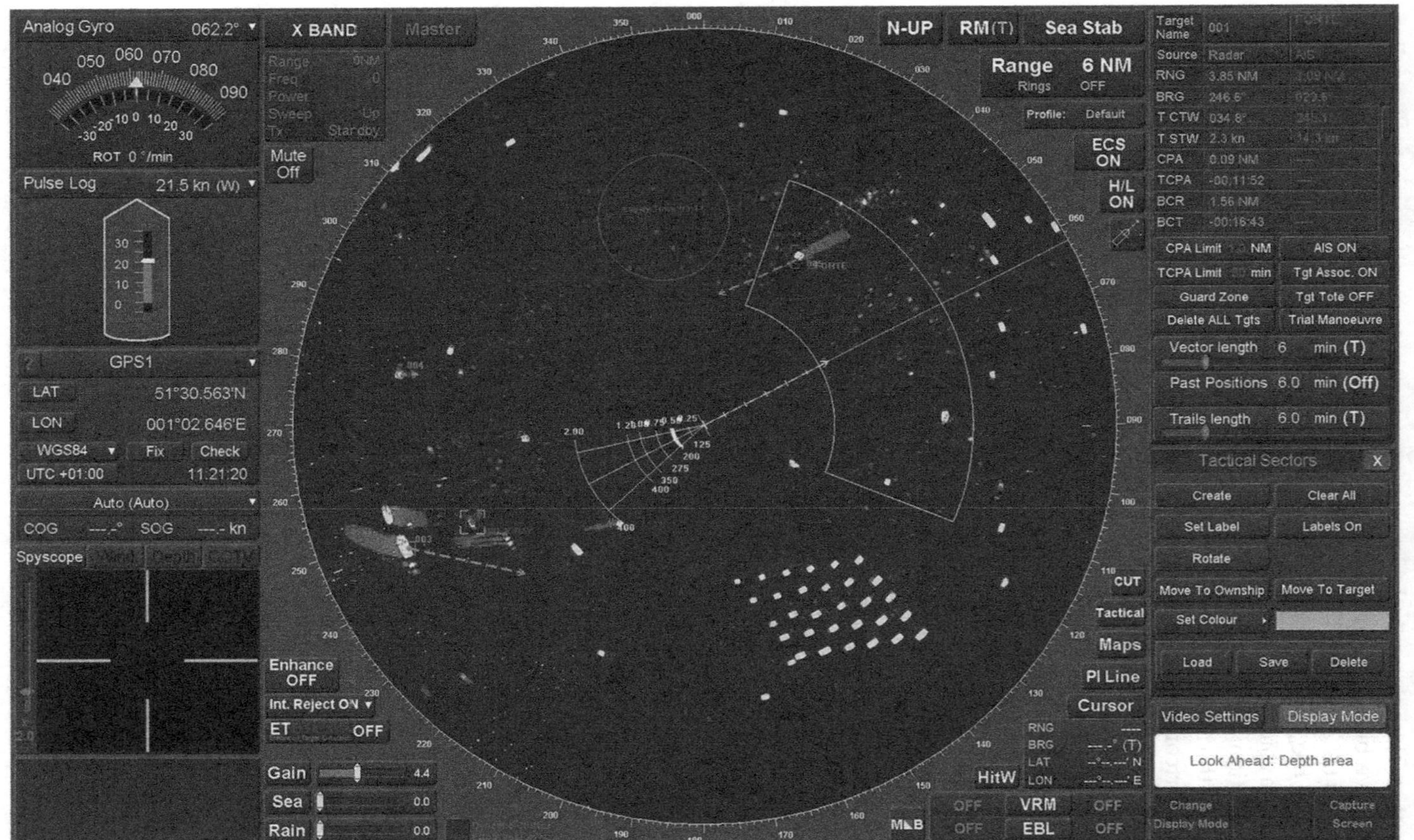

FIGURE 2.46 Example of larger vessel display. Reproduced courtesy of Kelvin Hughes.

FIGURE 2.47 View of a modern bridge. Reproduced courtesy of Northrop Grumman Sperry Marine.

the mouse requires a convenient flat surface to operate and this makes the equipment larger. Thirdly, many users believe that the tracker ball is easier to use from a standing position, which is often the case on a ship's bridge.

Display arrangements. Dedicated screens for marine navigation equipment are increasingly giving way to multifunctional screens, which may be used to display data from a variety of sensors, for example radar or ECDIS as required or integrated from a number of sources, such as radar, ECDIS, log, gyro and echo sounder. One major advantage of the common standardization of display screens across equipment is that in the event of the failure of one screen, the display can be transferred to another. This facility is often referred to as redundancy. A typical example is shown in Figure 2.47.

These advances in shipboard electronics have been very quick to be incorporated into the equipment on smaller vessels where acceptance has in many ways been faster than in larger vessels, where there can be a reluctance to embrace new technology. Smaller multifunction screens display a variety of data, save space, power and weight as well as reduce overall costs and are very popular on small well-equipped craft, such as racing craft. The screens can usually be switched to show standalone data (such as in Figure 2.48(a) and (b)) or the screen can be split (as in Figure 2.48(c)).

For many years raster screen displays were fixed in 4:3 ratio. Different ratios are now commonly in use, including the 16:9 ratio commonly used for domestic televisions. The screen size ratio is selected by the manufacturer to best show the data required within the confines of typical bridges layouts, whilst meeting minimum IMO requirements on the radar display area.

(a)

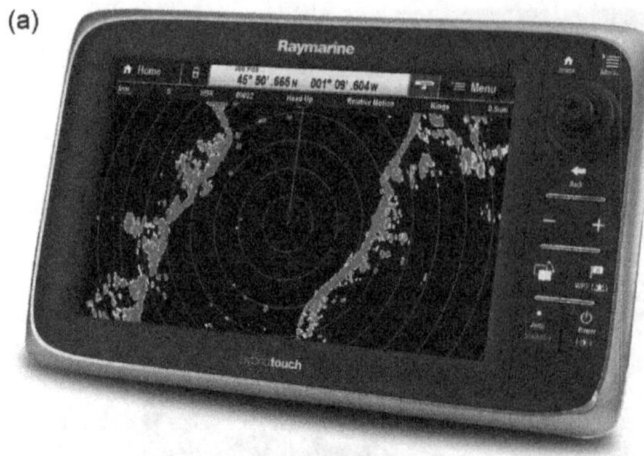

(b)

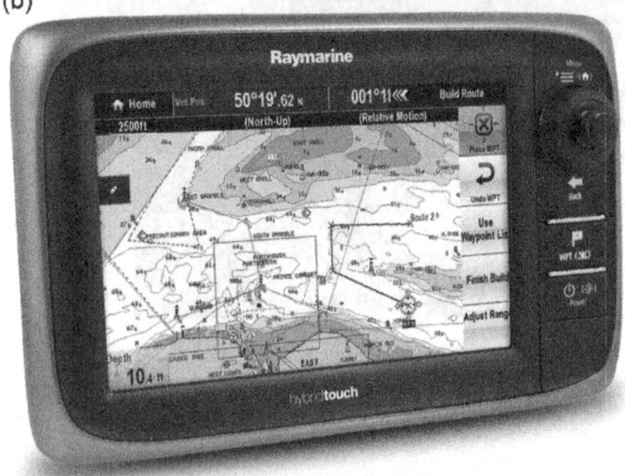

(c)

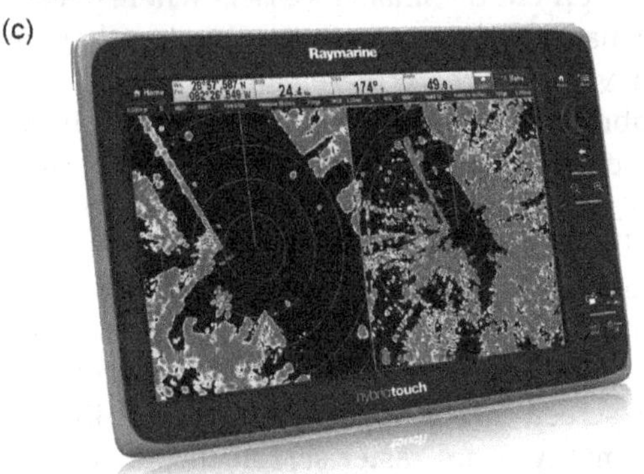

FIGURE 2.48 Multifunction display for smaller vessels: (a) Set to radar display. (b) Set to electronic chart display. (c) Set to 'split screen' radar. Reproduced courtesy of Raymarine.

2.9 SOLID-STATE RADAR PRINCIPLES

The magnetron-based marine radar has served the shipping industry well for over 60 years, but its dominance will decline into the future because of the benefits offered by newer technology based on solid-state transmitters. Such radars can be termed as solid-state, coherent, pulse compressed, digital or even New Technology (NT) radars. The terms all attempt to refer to the same concept but concentrate on a particular feature of its implementation.

The ever-continuing growth in the telecommunications market is effectively promoting the development of the necessary technology, which is then being exploited innovatively by marine radar manufacturers. In developed regions of the world, telecoms users are demanding instantaneous mobile access with ever-increasing data speeds. In less developed regions it is difficult and expensive to put in wired communications and so wireless mobile communications often provide the only affordable option for phone calls, text and Internet access. All this implies an ever-increasing demand for radio communications bandwidth and for its more efficient use. In particular, the main growth area for telecommunications bandwidth is at the microwave frequencies. Newer 4G telecommunications networks are currently operating at S-band and it can be expected that growth will continue to beyond 10 GHz, despite the inherent problems that come with using such high frequencies, not least the increased necessity for the base station and the user to be in direct line of sight.

Marine radar can make use of at least two particular technology improvements arising out of this growth in telecommunications.

The first is the affordable availability of power transistors working at microwave frequencies that have been developed primarily for use in telecommunications base stations. The second is from the development of devices that allow very sophisticated modulation to be put onto the transmitted microwave signal, together with the corresponding devices that can perform the demodulation. These allow a process known as *pulse compression* to be used that decreases the peak power requirement of a radar, generally to levels that can be generated by solid-state transmitters. In essence, solid-state radars use very long pulses to increase the effective energy being radiated, which is otherwise limited by their lower peak power compared to magnetron-based radars. It is perhaps useful to reconsider the discussion on radar transmitted power considerations given in Section 2.3.3.3 to fully understand this important aspect.

Of significance, a fully transistor-based radar transmitter will have a higher reliability than a magnetron-based system simply because it would only utilize low voltages, less than about 50 V compared to a magnetron-based system requiring many thousands of volts. Furthermore, unlike a magnetron the power transistors in a solid-state design do not have to be replaced regularly, saving on servicing needs. Power transistors at S-band suitable for use in civil marine radars started to become affordably available from about 2005, leading to their relatively recent introduction by a number of manufacturers. Transistors for X-band use are becoming increasingly powerful and more affordable, but at the time of writing an X-band solid-state radar would cost rather more than its magnetron-based equivalent and so they are not commonly used in civil marine applications.

The pulse compression process is described in the following subsections and is basically a way of using a long pulse to give the same range accuracy and target discrimination as a very short pulse. A highly important consequence of adopting such technology is that, unlike a standard magnetron-based radar, they become capable of detecting the instantaneous radial velocity of the target returns, which can greatly help in increasing the discrimination of wanted targets from sea and precipitation clutter. It is an example of use of the Doppler effect, named after the nineteenth-century Austrian scientist Christian Doppler. In sound waves, it is this effect that causes the horn of an advancing or retarding locomotive to sound respectively higher or lower in frequency to the listener. A similar effect occurs in electromagnetic waves, simply because relative movement increases or reduces the transmitted wavelength. This is further explored in Section 2.9.5.

The use of Doppler in marine radar requires very sophisticated and still-developing principles, but it potentially offers considerable advantages, which will be increasingly exploited. With the performance improvements expected, it should allow IMO to steadily enhance the fundamental performance requirements for radar, while the use of ever-advancing technology should keep the equipment affordable. Present-day radars are far from meeting what users (and safety and environmental protection pressures) really desire, which is 100% detection of all targets of relevance under all possible conditions, and so there is a fundamental demand for improved performance; the move to solid-state technology opens up these opportunities.

A more minor point is that a particular problem facing future magnetron-based radars is the special design skills needed for their future development – it is a very niche area. Today's electronic engineering requires primarily a mixture of solid-state technology and digital/software skills. The design skills necessary for future magnetron transmitters are becoming increasingly difficult to find and nurture.

All the above considerations contribute to speculation that within a few years most new S-band radars will be solid-state and within 10 years or so it is also likely to become the case at X-band. This is why a broad knowledge of solid-state radar principles is becoming essential for a good user understanding of marine radar.

2.9.1 Coherent Radar Concepts

In a very basic sense a signal propagating as an electromagnetic wave, whether in free space, in a transmission line, or within an active device such as a transistor, has just one property – its instantaneous amplitude with respect to time at any given point in space. By 'time' we here mean very precise time, such that the basic oscillations of the wave (the carrier) are measurable and not just the amplitude modulation on the wave, as on a conventional magnetron-based system. There are huge complications, such as how the wave is divided into electric and magnetic components and how the propagating medium itself affects the signal (including, in free space, the divergence of the beam), but in the end what is the only important factor is its amplitude with respect to precise time, at any one point in space. A radar sends out a signal, modified by the antenna and the ship's characteristics including its velocity, and then receives back a highly modified version. It is modified by many things but particularly by the properties of the various reflecting objects, for instance their distance, size, shape and speed, and also by the atmosphere and extraneous noise components. The receiving process within the radar itself also modifies the signal, such as the movement of the vessel, including the rotation of the antenna, the precise action of the receiver, the effects of internally generated noise and the specific processing that takes place within the radar.

When dealing with a conventional pulse radar we normally just focus on the envelope of the transmitted signal. However, a solid-state radar generally has to transmit a signal that is far from being a simple pulse. A magnetron-based pulse radar transmits a waveform that approximates to the one shown in Figure 2.49(a), where we see a representation of the RF oscillations forming the pulse. In reality, a 1 μs pulse at X-band would comprise about 9300 complete oscillations of the RF energy and 2900 at S-band. A conventional magnetron transmitter is not very good at forming such a precise looking wave. In reality it would take many oscillations to build up to the full level; the actual frequency and amplitude of the wave would slightly 'wander' during the pulse; and it would also die away over many 'cycles' of the oscillating frequency and not be a precise cut-off. Figure 2.49 (b) illustrates these effects. Furthermore, the precise features of each pulse emanating from the magnetron differ.

This instability leads to a conventional magnetron-based marine system being termed as a *non-coherent* radar. (In English, *incoherent* has an entirely different meaning and should not be used in this context.) The stability of the transmitted signal in marine solid-state radars causes these to be known as *coherent* radars. In our use of the terms coherent and non-coherent we are implicitly meaning *phase* coherency. If two signals are compared, does the phase of their oscillations tie up or does it uncontrollably wander about, as shown in Figure 2.49(c). The transmitter of a coherent radar accurately reproduces the desired waveform, which allows us to perform sophisticated processing on the received signal that is not possible on a non-coherent system.

Imagine a very long pulse that is varying precisely linearly in frequency. At a specified pulse start time t_0, let its frequency be f_0; and at time t_1, where the pulse terminates, the frequency is f_1. Furthermore, assume that the amplitude of the pulse is unity between the times t_0 and t_1 and is zero in amplitude, that is not present, outside of these times, see Figure 2.50(a). We say that the pulse has been

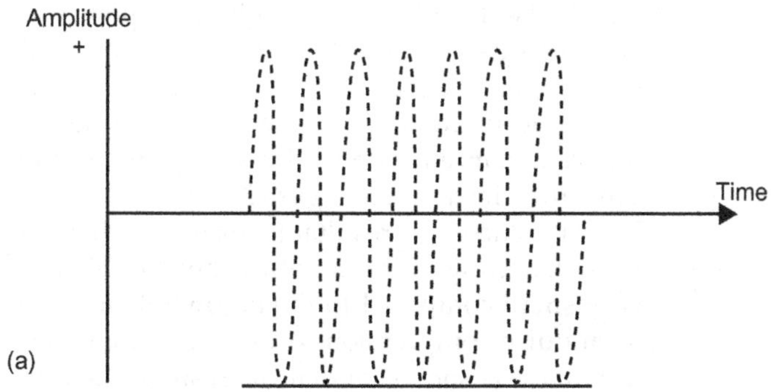

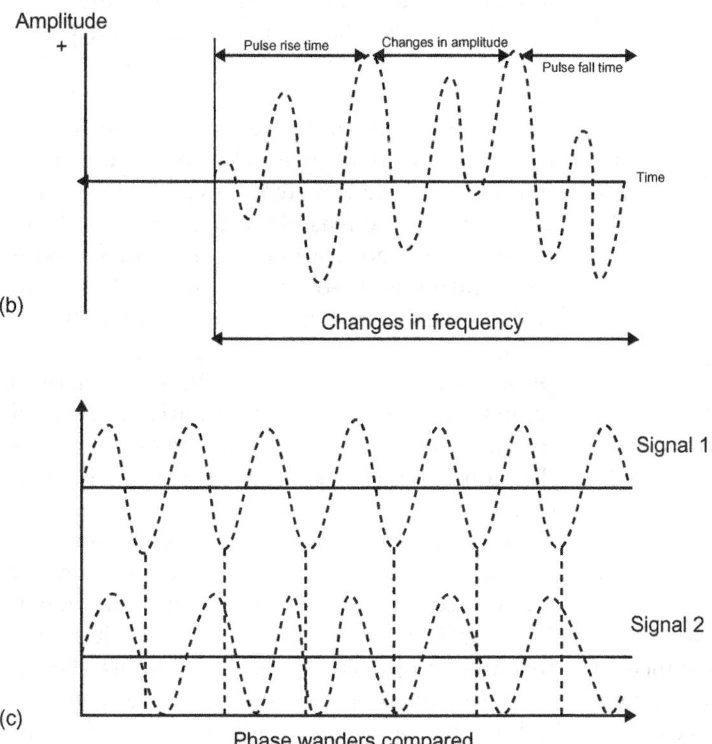

FIGURE 2.49 Showing non-idealities of a magnetron pulse: (a) ideal pulse of magnetron radar, (b) common defects of magnetron pulse (exaggerated) and (c) showing phase non-coherency.

frequency modulated, which just means that during the pulse its amplitude is constant, but the frequency of the signal is varying with time in some defined manner, in this example in a simple linear fashion.

The returned echo from a simple point target would look very similar to this transmitted pulse, except its amplitude would be much reduced. In principle, we could examine the returned pulse in detail. We could identify the precise time that the pulse was at a particular frequency when transmitted. We could then measure the total time taken for the received pulse to be at the identical frequency when

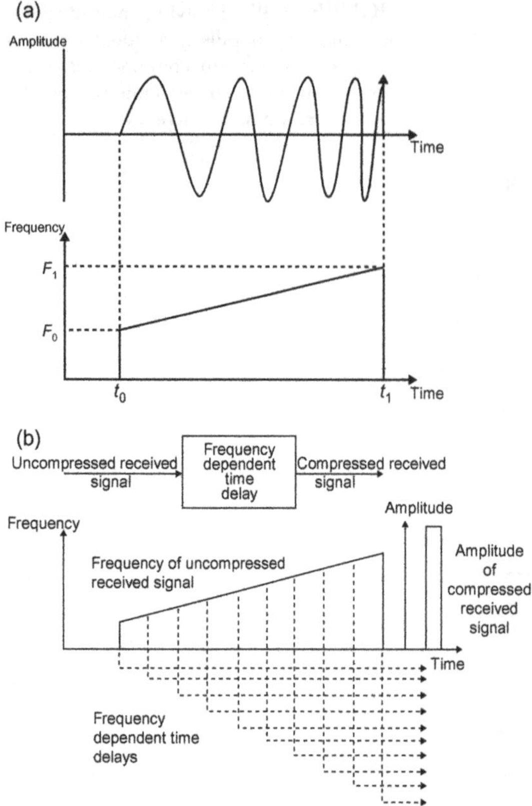

FIGURE 2.50 Pulse compression: (a) frequency modulated pulse; (b) pulse compression of received frequency modulated signal.

reflected from a target. In fact, for a totally static system, for whatever frequency we chose we would get the same time delay – which, of course, corresponds to the actual distance of the target from the radar, following normal principles. The clever additional process that can happen in a coherent radar is through the application of pulse compression. For the particular *frequency modulated* coherent radar example that we are initially looking at, imagine a special circuit that applies a time delay to the received signal that is dependent on the instantaneous frequency of the signal being received. At the lowest (start) transmitted frequency, for example, the added time delay would equal the total pulse length. At the highest frequency it would be zero, and at other frequencies would be such that all the energy of the signal would be concentrated by being appropriately delayed, concentrating all the energy to the same time instant (see Figure 2.50(b)).

In practice, the resultant pulse does have a short length associated with it but one that is very small compared to the original; hence the term pulse compression. A detailed mathematical analysis shows us that the resultant length of the compressed pulse is controlled by the total bandwidth of the signal; the larger the bandwidth, the shorter the compressed pulse. In fact the compressed pulse length is given approximately by the reciprocal of the bandwidth. For a linear frequency modulated signal, the bandwidth is effectively the total frequency variation that occurs across the pulse. It is quite feasible to use analogue techniques to compress a linearly frequency modulated pulse and so the technique has been widely used for some time, particularly for military radars. However, digital pulse compression is now more often used in modern radars and seems to be particularly well suited to meet the requirements of a ship's radar. It is this concept that is developed in the following sections.

For any coherent radar the pulse compression ratio, which may vary according to radar range setting, is an important parameter. If the ratio is given as 50 (i.e. 50:1), it means that the compressed pulse is 50 times shorter than the uncompressed pulse. If a coherent radar with a pulse compression of 50 was required to end up with a compressed pulse length of 1 μs, matching a typical magnetron-based marine radar at the longer range settings, the transmitted uncompressed pulse would be 50 μs in length. Since the energy of the echo from a target is effectively increased by this factor, the peak power of the coherent radar could then be nominally 50 times smaller than that needed for a non-coherent system.

Pulse compression radar necessarily needs a coherent detection process, maintaining both the phase and amplitude characteristic of received signal through to the subsequent correlation process. This is achieved by using a method known as *IQ* demodulation – standing for in-phase and quadrature-phase demodulation. Two separate mixers have to be used for this and the concept is illustrated in Figure 2.51. On standard magnetron-based radars half of the available received energy cannot be 'decoded' due to the non-coherence of the transmitted signal and is therefore left unused. It means that all coherent radars have a 3 dB advantage in their receiver gain compared to non-coherent systems, such as standard magnetron radars. This effectively allows the radar transmitter of a coherent system to radiate half the mean power of a non-coherent system and still get the same detection performance.

Together, it means that a coherent radar with an effective 50:1 pulse compression ratio would have a peak power requirement nominally 100 times less than a conventional pulse radar. For example, the detection performance of a 25 kW peak power magnetron radar would be nominally matched by a 250 W peak power coherent radar with a 50:1 pulse compression ratio. It is this significant reduction in the peak power requirement that enables the use of a solid-state transmitter. The length of the compressed pulse is equivalent to the pulse length of a non-coherent (magnetron) radar, not least when considering its range resolution.

In principle, the transmission from a coherent radar can be continuous, and not divided into pulses at all. For our linear frequency modulated example the transmission would just continually scan in frequency, for example from the lowest to highest, and then immediately repeat with no gaps between the pulses, as shown in Figure 2.52. In fact this forms the basis of frequency modulated continuous wave (FMCW) pulse compression radars, first used on military systems in the mid-twentieth century. It has the advantage that the peak power of the transmitter is equal to the mean power and so, in principle, a marine radar would only need a transmitter capable of giving a signal power of less than 10 W. In fact, this concept has been recently used for some small boat radars but not, so far, on commercial shipborne

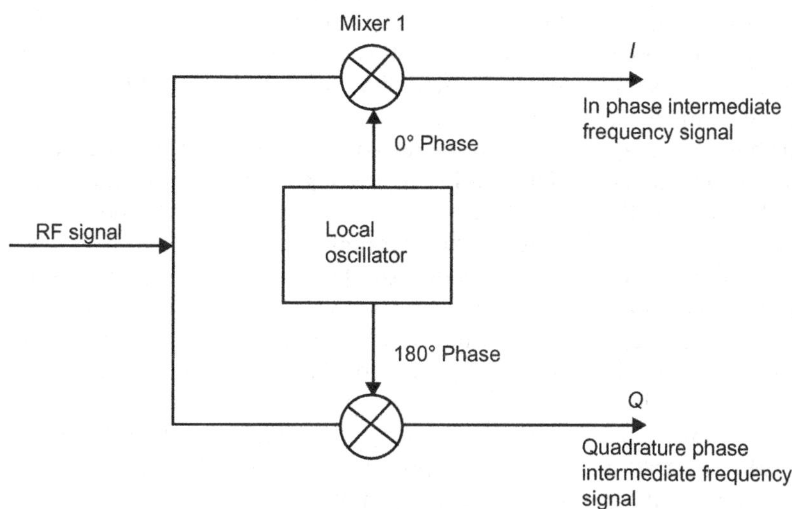

FIGURE 2.51 Demodulation of RF-received signal into *I* and *Q* components at IF.

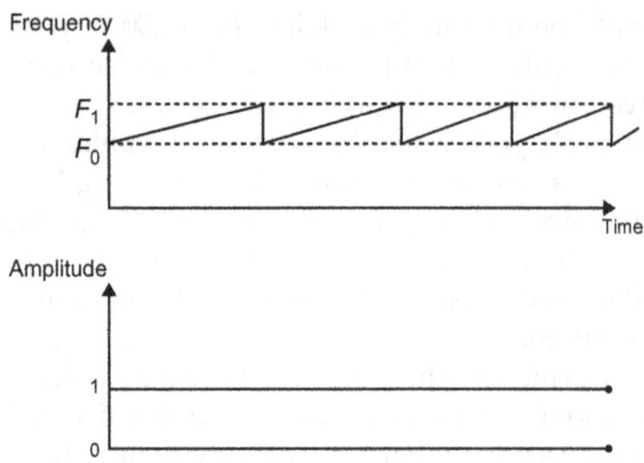

FIGURE 2.52 FMCW signal.

radars. An important consideration on such systems is in effectively isolating the receiver from the directly transmitted signal. However, a more difficult issue for a ship's radar is optimizing the signal to give good performance at both short and long ranges.

2.9.2 Correlation

In the section above we looked at the possibility of using a precisely varying frequency signal as the modulation on a pulse compression radar. The received signal was precisely delayed according to its instantaneous frequency, allowing all the energy of the uncompressed pulse reflected from a particular point target to be concentrated at virtually the same time instant, effectively forming the compressed pulse. There is another particular but equivalent way which this can be looked at, which can be described as *correlation theory*. This allows the application of powerful statistical mathematics and can be applied to any form of transmitted signal. For instance, imagine the received signal of the frequency scanned radar plotted out as a graph of frequency versus time, as shown in Figure 2.53. In general, the signal would look very noise-like, except where there was a distinct target. The transmitted signal is portrayed in the inset to the figure. Now imagine cutting out this inset and moving it along the time axis of the graph until it exactly matches the shape of the received signal. At that point the left-hand edge of the inset is exactly at the time instant that marks the range of the target, as referenced to the start of the transmitted pulse.

This is an example of *autocorrelation* – we have correlated the received signal with a replica of the transmitted signal in order to determine the precise position of the echoing source of the target. This is a special case of correlation, which in the general case, looks at the 'level of match' between one signal and another. In fact correlation, also sometimes called *cross-correlation*, can be applied to many disciplines such as finance, medicine and political science, as well as numerous other areas of engineering and conventional science, simply because correlation is such a fundamental statistical principle. There are several other related concepts and terms used in the literature concerning this complex subject, including convolution, covariance, autocovariance and matched filtering, but these are not used here. For this very simplified treatment we use 'correlation' to denote the general principles and 'autocorrelation' when we wish to stress the correlation of a signal with a replica of the one originally transmitted. In principle, we can perform a correlation process on any transmitted signal that is varying in some way with time. In actuality, amplitude variation is not generally a very useful option, as lower levels of the transmitted signal will not get the same detectability as the higher levels, and therefore the effective range capability of the radar would be compromised.

In practice, digital encoding of the transmitted signal, known as digital modulation, has been found to be an excellent method of generating pulse compression. Both frequency and phase variations can be used for digital encoding. For instance, the digital symbols 0 and 1 could be represented by two different frequencies f_0 and f_1, with the amplitude of their

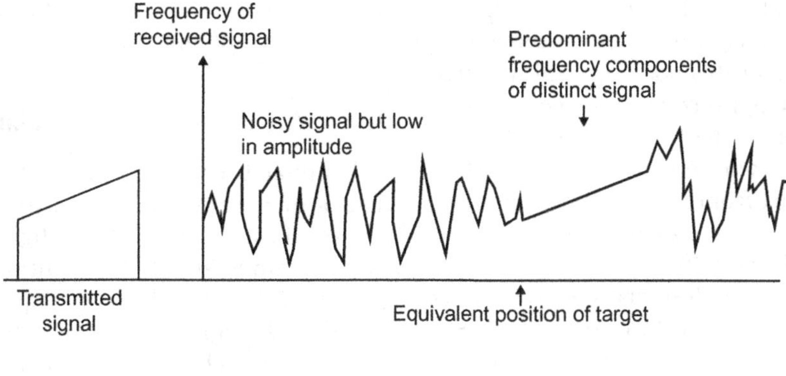

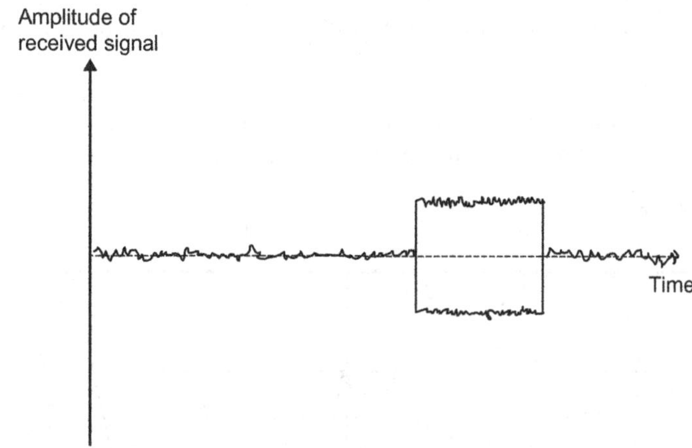

FIGURE 2.53 Correlation of received signal with transmitted signal.

transmissions being equal. This is an example of frequency shift keying (FSK). More complex encoding can allow a faster transmission of a digital sequence, effectively sending multiple digits at the same time instant. For example, four different frequencies can be chosen to represent the digital codes 00, 01, 10 and 11, effectively halving the time to send a code compared to using two frequencies. Importantly, this increases the bandwidth of the signal and therefore its compression factor, similarly to that discussed in Section 2.9.1 concerning frequency modulation. The basic receiver function is effectively to measure the frequency of the received signal at any one time and assign the correct digital code.

An important alternative method is known as phase modulation, which is illustrated in Figure 2.54. A constant amplitude, constant frequency wave is transmitted with discontinuities in its phase. Typically, the phase is limited to jumping between 0° and 180° or between 0°, 90°, 180° and 270°, known respectively as biphase (or binary phase) and quadraphase shift keying (BPSK and QPSK). For the binary case the phases can be considered to represent 0 and 1; for the quadraphase case the phases can represent 00, 01, 10 and 11, very similar to the cases for FSK. The good points about both digital frequency and phase modulation are that, using modern technology, they are reasonably easy to generate for transmission and to decode on reception, although the detailed processes employed lie outside the scope of this book. Digital modulation allows digital processing concepts to be employed at the earliest stages of the radar receiving process, with no analogue processing of the signal apart from receiver protection, RF and IF amplification stages and *IQ* signal detection, see Figure 2.55.

In general, recently designed coherent radar systems, not just for maritime use, are based on digital frequency or phase modulation techniques. The precise choice of modulation depends on the specific design philosophy of the manufacturer. What is also very manufacturer-dependent is the precise sequence of digital 0s and 1s used in the formation of a particular pulse. It is this sequence that the correlator needs to act upon, as described in the next section.

2.9.3 An Example of Correlation

Manufacturers are unwilling to divulge details of the precise sequences that they use, but the basic principles of digital correlation can be understood by considering simple examples. Assume that the frequency or phase modulated signal can be represented by the specific 11-digit sequence 11100010010. The reasons behind using this particular sequence will be explained in the next section. A complete pulse consists of the 11 digits represented by an appropriate change in phase or frequency of the *carrier* signal, as discussed in the previous section. For this particular

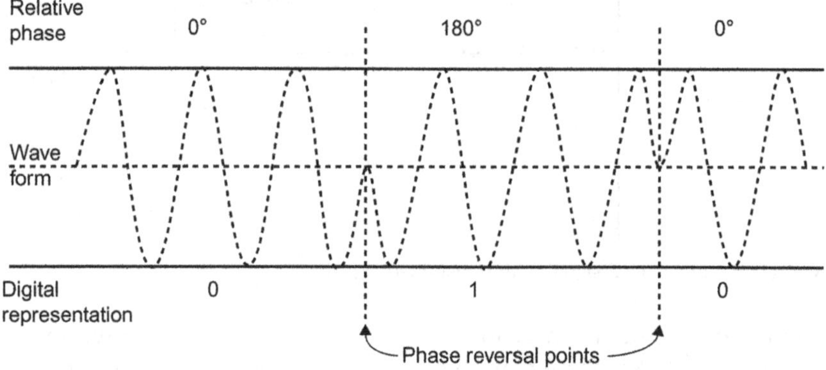

FIGURE 2.54 Phase modulation. *Note: The phase reversals are not necessarily restricted to points where the wave amplitude is zero.*

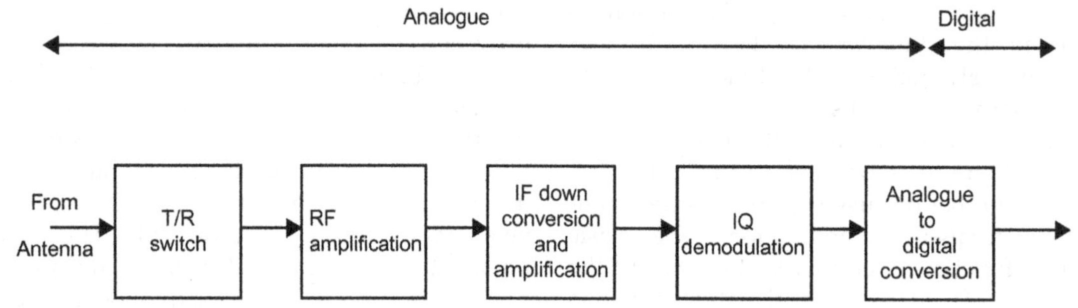

FIGURE 2.55 Analogue stages of typical coherent radar.

example, the correlation process effectively compresses the pulse by a factor of 11, equivalent to the number of digits in the pulse compression sequence, sometimes called a code.

How this happens is given in Table 2.5(a). The top line of the table labels the received time intervals, starting at the time $t = L_t$, which coincides with the total time taken for transmitted energy from the radar to reach and be reflected back from a particular target. Each column represents a time that is equal to the compressed pulse length (L_c) and the total extent of the 11 time cells is the length of the uncompressed pulse (L_u). The table effectively represents time advancing to the right, and also assumes that the first digit of the transmitted sequence is sent at time $t = 0$. The second line shows the received signal, demodulated into a sequence of binary numbers. The first digit of the received uncompressed pulse was received at $t = L_t$, with the final digit of the 11-bit sequence being received at $t = L_t + 10L_c$. The third line shows the binary sequence modulation on the originally transmitted signal, which forms the reference sequence with which we are correlating the received signal. For Table 2.5(a) we are calculating the correlation sum when we have 'fortuitously' time-aligned the transmitted reference code with the received code of the signal reflected from the point target. The received signal is fully correlated, as denoted by the sequences in lines 2 and 3 of Table 2.5(a) being identical.

The fourth line in Table 2.5(a) gives a simple method of calculating the correlation of the 11-digit received sequence. The correlation value +1 has been assigned when the received digit ties up (correlates) with the reference digit; −1 is assigned to those that differ. We then sum the result of all these single digit correlations. In this instance this sum comes to +11, representing the highest correlation possible for this sequence, confirming that there is a target at the assumed range equivalent to the T/R time interval of L_t. Table 2.5(b) gives the case when we perform the correlation with the reference sequence shifted earlier in time by one digit (a time interval equal to T_c). The actually transmitted signal has not changed in time, nor the received sequence; we are just calculating the correlation when the reference sequence is shifted to start at $t = L_t - L_c$. At this particular time cell, there will be no received signal so we have represented this case in the table by N/S (no signal). Also, if there is no received signal, we have applied the number 0 to the sum. The final line sums to zero, showing that there is not a target at the assumed range equivalent to $t = L_t - L_c$.

Table 2.5(c) gives the case when the correlation process is performed commencing at $t = L_t - 2L_c$. For this case the sum of the single digit correlations comes to −1, showing that there is no target at the equivalent range given by $t = L_t - 2L_c$. Table 2.5(d) gives the case when we perform the correlation calculation with the reference sequence starting at $t = L_t + L_c$, where the sum is 0. If we plot the correlation for all possible positions of the reference sequence we get the graph depicted in Figure 2.56. It is seen that only when the sequences exactly match in time do we get a sum of +11, all the other possible situations give a sum of either 0 or −1.

If we select an unrelated sequence we also tend to get a low correlation. Such a sequence could be created by an interfering signal from another radar or a telecommunications transmission, or even from natural noise. Because dissimilar interfering codes generally create low levels of correlation they are therefore less likely to cause visible interference, a further benefit of the correlation process. As an example, the sequence 01010101010 generates a sum of −1 when correlated with the code that has been used for the examples given in Table 2.5.

A final example illustrates the case of perfect anti-correlation or negative correlation, where the 'received' sequence has each of its

TABLE 2.5 11-Bit Barker Code Example

(a) Fully Correlated – Sample Sequence Fully Matched to Received Signal

Time	L_t	L_t+L_c	L_t+2L_c	L_t+3L_c	L_t+4L_c	L_t+5L_c	L_t+6L_c	L_t+7L_c	L_t+8L_c	L_t+9L_c	L_t+10L_c	Sum
Received sequence	1	1	1	0	0	0	1	0	0	1	0	
Sample sequence	1	1	1	0	0	0	1	0	0	1	0	
Correlation	+1	+1	+1	+1	+1	+1	+1	+1	+1	+1	+1	+11

(b) Uncorrelated – Sample Sequence One Range Cell 'Too Early'

Time	L_t-L_c	L_t	L_t+L_c	L_t+2L_c	L_t+3L_c	L_t+4L_c	L_t+5L_c	L_t+6L_c	L_t+7L_c	L_t+8L_c	L_t+9L_c	Sum
Received sequence	N/S	1	1	1	0	0	0	1	0	0	1	
Sample sequence	1	1	1	0	0	0	1	0	0	1	0	
Correlation	0	+1	+1	−1	+1	+1	−1	−1	+1	−1	−1	0

(c) Uncorrelated – with Sample Sequence Two Range Cells 'Too Early'

Time	L_t-2L_c	L_t-L_c	L_t	L_t+L_c	L_t+2L_c	L_t+3L_c	L_t+4L_c	L_t+5L_c	L_t+6L_c	L_t+7L_c	L_t+8L_c	Sum
Received sequence	N/S	N/S	1	1	1	0	0	0	1	0	0	
Sample sequence	1	1	1	0	0	0	1	0	0	1	0	
Correlation	0	0	+1	−1	−1	+1	−1	+1	−1	−1	+1	−1

(d) Uncorrelated – with Sample (Transmitted) Sequence One Range Cell 'Too Late'

Time	L_t+L_c	L_t+2L_c	L_t+3L_c	L_t+4L_c	L_t+5L_c	L_t+6L_c	L_t+7L_c	L_t+8L_c	L_t+9L_c	L_t+10L_c	L_t+11L_c	Sum
Received sequence	1	1	0	0	0	1	0	0	1	0	N/S	
Sample sequence	1	1	1	0	0	0	1	0	0	1	0	
Correlation	+1	+1	−1	+1	+1	−1	−1	+1	−1	−1	0	0

(e) Anti-Correlation – with Received Sequence the Exact Opposite of the Sample

Time	L_t	L_t+L_c	L_t+2L_c	L_t+3L_c	L_t+4L_c	L_t+5L_c	L_t+6L_c	L_t+7L_c	L_t+8L_c	L_t+9L_c	L_t+10L_c	Sum
Received sequence	0	0	0	1	1	0	0	1	1	0	1	
Sample sequence	1	1	1	0	0	0	1	0	0	1	0	
Correlation	−1	−1	−1	−1	−1	−1	−1	−1	−1	−1	−1	−11

L_t = total time that electromagnetic energy takes to travel from radar to target and back to radar.
L_c = length of compressed pulse = duration of single digit frequency or phase encoding on signal.
$11L_c$ = total length of uncompressed pulse.
N/S = no received signal.

digital components reversed, that is a 0 becomes a 1, and a 1 becomes 0. This case is illustrated in Table 2.5(e), where it sums to −11. It is fully correlated in the sense that all 1s in the transmitted code are matched by all 0s in the received code, and vice versa. It is explained in Section 2.9.4 why this, for radar, is identically equivalent to the standard correlation case and therefore positively indicates a target at the equivalent range.

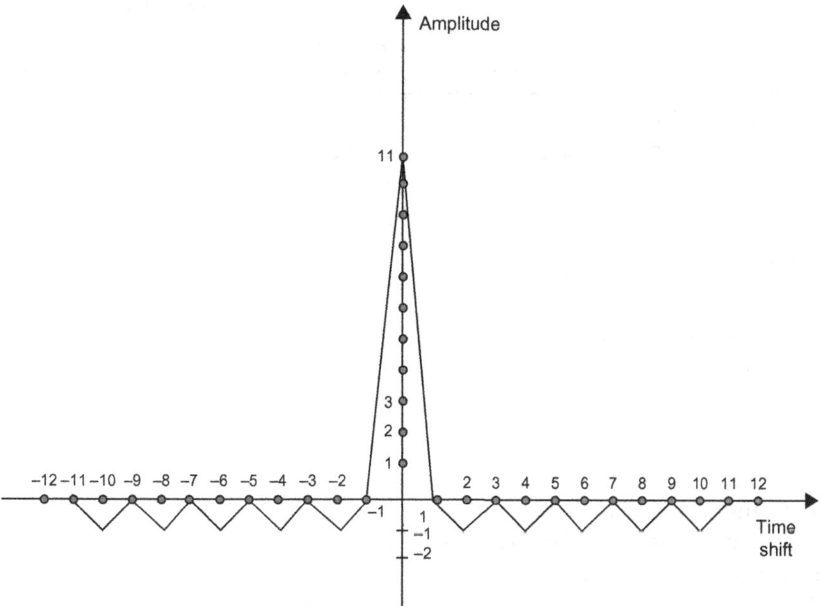

FIGURE 2.56 Autocorrelation for 11-bit Barker code.

The summation process in the correlator generates a 'large' positive number (+11 in the example 11-digit transmitted sequence) when the received signal exactly matches the transmitted signal, when the target is at the sample range. If a target is not at the assumed range then the summation in the correlation process provides a low number (−1, 0, or +1 for the above example). In addition, for a dissimilar received signal, such as noise or interference, it also generally provides a low summation. However, any noise-like effect could momentarily produce a signal with some correlation for a brief instant, resulting in a momentarily significant value of correlation. This is somewhat similar to the characteristics of noise in an analogue system – at any one instant the noise level can be unusually large, but with very low probability. It appears only briefly on the display and is unlikely to persist or produce any scan-to-scan correlation.

Correlation provides real gain to the receive process; the wanted signal has been amplified but, in general, not the unwanted noise or interfering signals. In this simplified treatment for the given example, it appears that the detection 'effectiveness' of the radar has improved by a factor of 11. This is backed up by applying a rigorous mathematical treatment, which shows that the effective range performance of the radar is approximated by the equivalent range of a conventional radar with a pulse length the same as the uncompressed pulse length L_u, but with the range resolution given by the compressed pulse length L_c. This is explored in further detail in the next section. In reality, most pulse compression radars do not fully achieve a 'compression gain' that is exactly equal to L_u/L_c. This is tied up with an important effect known as range sidelobes, which in the worst case can be discernible by the user.

It is worth looking at a few practical numbers to illustrate digital pulse compression. Assume that each digit of the 11 pulse code is 0.05 μs long and so the total pulse length is

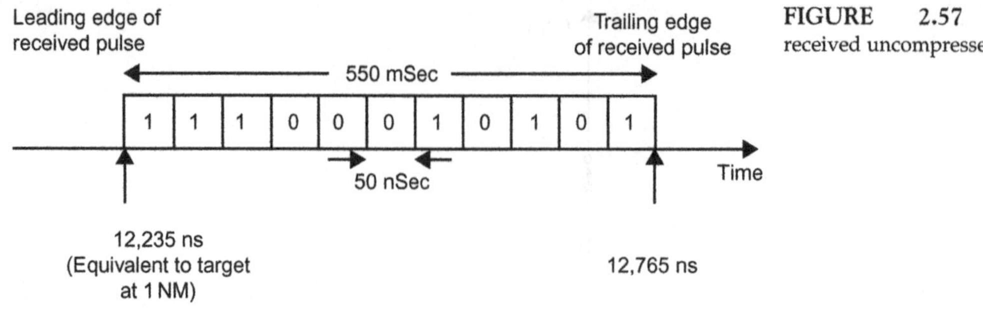

FIGURE 2.57 Depiction of received uncompressed pulse.

0.55 μs. Imagine a point target at 1 NM, which is equivalent to a range of 1852 m. The approximate speed of electromagnetic energy is 2.9979×10^8 m/s. This means that the leading edge of the 0.55 μs composite pulse will be received at a time T_P after it was transmitted, given by:

$$T_P = 2 \times 1852/(2.9979 \times 10^8)\text{s}$$
$$= 12.235\ \mu s$$

The received signal is conceptualized in Figure 2.57. The pulse is shown divided into its 11 segments, with the trailing edge of the composite pulse occurring 0.55 μs later. The timebase of the receive process in the figure has been artificially adjusted to allow each digit of the pulse to exactly fill one range cell of length equivalent to 0.05 μs. The target has been assumed to be at a range of exactly 1 NM (Figure 2.58).

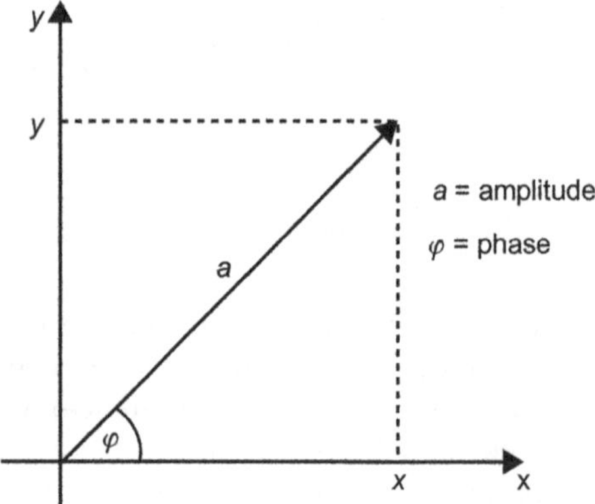

FIGURE 2.58 Phase and amplitude representation of received radar signal.

2.9.4 Range Sidelobes

The digital pulse compression sequence used in the preceding section is, in fact, an example of a very special sequence, known as a *Barker code*. No other 11-bit sequence exists that would give such an emphatic performance for autocorrelation; a sum of +11 when correlated and either −1 or 0 when uncorrelated, as illustrated in Table 2.5. In fact such codes appear to be limited to just nine examples, which all are depicted in Table 2.6. They all have a length of 13 digits or fewer. Table 2.6 gives the codes values as being −1 or +1, which is the more common way that Barker codes are notated rather than the 0 and 1 used in Section 2.9.3, which was used to emphasize their relationship to the usual binary notation. However, using the +1/ − 1 nomenclature as used in this section has other advantages, as will be seen later.

Unfortunately, the shortness of the Barker code sequences gives two unrelated problems

TABLE 2.6 List of Barker Codes

Length	Barker Code
2	+1 −1 and +1 +1
3	+1 +1 −1
4	+1 +1 −1 +1, and +1 +1 +1 −1
5	+1 +1 +1 −1 +1
7	+1 +1 +1 −1 −1 +1 −1
11	+1 +1 +1 −1 −1 −1 +1 −1 −1 +1 −1
13	+1 +1 +1 +1 +1 −1 −1 +1 +1 −1 +1 −1 +1

for many real radar applications, including maritime. The first is that the levels of compression are relatively low, such as a factor of 13, at best. From the discussion in Section 2.9.3, in order to match the range performance of a 25 kW peak power magnetron radar, a pulse compressed radar with a peak power of around 1–2 kW would be needed if based on a 13-bit Barker code. This is not particularly low – and is currently expensive to achieve when using solid-state transmitters. However, a more fundamental problem is the phenomenon known as *range sidelobes*. In fact Figure 2.56 from Section 2.9.3 is an illustration of them. Totally excluding the effects of noise, if the autocorrelation is performed at even-numbered range cells from the correlated position, then we get the values of −1 for the correlation coefficients. These are named range sidelobes, with a relative level of −1, compared to the correlated sum. Because they do not have the value zero, they significantly affect the performance of the radar as now elaborated.

First of all it is necessary to look more closely at the correlation process. For illustration purposes it is convenient to continue to use the 11-bit Barker code. In a coherent radar based on such a code what is effectively happening is that the correlation calculation is being performed at every range cell and at all azimuthal increments. This leads to a correlation amplitude at every cell. The calculation, as described in the previous section, involves the particular range cell and the 10 following it – all at the same azimuth angle. The process can be looked at in a slightly different way if we use the ±1 nomenclature to represent the Barker code and the received signals, rather than the 1/0 nomenclature used in Section 2.9.3. Consider Table 2.7(a), which exactly replicates the example in Table 2.5(a). Table 2.7(b) gives the identical case using the ±1 nomenclature. We can consider that the +1 or −1 represents signal amplitudes measured as a voltage. Therefore, if we have no received signal at any range cell, instead of representing it as N/S (no signal) as we have done in Table 2.5, it would have the value 0. The correlation process then can be described as simply multiplying the received value with the comparison value of the transmitted sequence, as shown in the bottom line of Table 2.7(b), and then summing the result over the 11 range cells. In fact this exactly follows the true mathematical definition of correlation, whereas Tables 2.5(a)–(e) and 2.7(a) are just an equivalent way of looking at things. Table 2.7(c) replicates the case for Table 2.5(c).

For real radar signals the measured signal is defined by an amplitude and a phase value; let these be denoted by a and φ, see Figure 2.56. As the figure shows, they can also be represented by the Cartesian coordinates (X,Y). From Figure 2.56 it can be seen that if $\varphi = 0$, then $X = a$ (and $Y = 0$). If $\varphi = 180°$, then $X = -a$, also with $Y = 0$. This directly corresponds with the ±1 nomenclature used above. This is also related to the concept introduced in the introductory paragraphs of Section 2.9.1 where a signal at any one point can be considered to be simply an amplitude with respect to time. Here time has been greatly simplified to being compared to the phase of a 'reference' signal. It is far from being a complete time reference but is a useful simplification over short intervals.

TABLE 2.7 11-Bit Barker Code Example Using $-1/+1$ nomenclature

(a) Fully Correlated, 0/1 Binary Representation – Replicating Table 2.5(a)

Time	L_t	L_t+L_c	L_t+2L_c	L_t+3L_c	L_t+4L_c	L_t+5L_c	L_t+6L_c	L_t+7L_c	L_t+8L_c	L_t+9L_c	L_t+10L_c	Sum
Received sequence	1	1	1	0	0	0	1	0	0	1	0	
Sample sequence	1	1	1	0	0	0	1	0	0	1	0	
Correlation	+1	+1	+1	+1	+1	+1	+1	+1	+1	+1	+1	+11

(b) Fully Correlated, $-1/+1$ Representation of Table 2.5(a)

Time	L_t	L_t+L_c	L_t+2L_c	L_t+3L_c	L_t+4L_c	L_t+5L_c	L_t+6L_c	L_t+7L_c	L_t+8L_c	L_t+9L_c	L_t+10L_c	Sum
Received sequence	+1	+1	+1	−1	−1	−1	+1	−1	−1	+1	−1	
Sample sequence	+1	+1	+1	−1	−1	−1	+1	−1	−1	+1	−1	
Correlation	+1	+1	+1	+1	+1	+1	+1	+1	+1	+1	+1	+11

(c) Uncorrelated, $-1/+1$ Representation of Table 2.5(c)

Time	L_t-2L_c	L_t-L_c	L_t	L_t+L_c	L_t+2L_c	L_t+3L_c	L_t+4L_c	L_t+5L_c	L_t+6L_c	L_t+7L_c	L_t+8L_c	Sum
Received sequence	0	0	+1	+1	+1	−1	−1	−1	+1	−1	−1	
Sample sequence	+1	+1	+1	−1	−1	−1	+1	−1	−1	+1	−1	
Correlation	0	0	+1	−1	−1	+1	−1	+1	−1	−1	+1	−1

The correlation value is simply the multiplication of the two numbers in the cells above it.

In this nomenclature the amplitude -1 represents a signal with amplitude 1 at a phase of 180°, whereas the amplitude $+1$ is equivalent to an amplitude of 1 with a phase of 0°. As previously mentioned, the *IQ* demodulation illustrated in Figure 2.51 is just a particular way of measuring the amplitude and phase of the signal, but presenting the quantities in a different but equivalent form. Figure 2.51 also explains the relationship between amplitude and phase with the *I* and *Q* components. Therefore, the receive process of the radar provides the received signal at each range cell in amplitude and phase, typically represented in *I* and *Q* components. The autocorrelation process then multiplies these by the appropriate digit of the correlation code.

What is significant to detection is the power of the signal arising from the autocorrelation process in each cell. This is obtained by squaring the amplitudes of the *I* and *Q* components, and is therefore always a positive number. (For those with a knowledge of complex number theory we are really multiplying the complex number representation of the signal by its complex conjugate.) If we do this for the peak (correlated) value of the 11-bit Barker code we get $11 \times 11 = 121$ and a power of either 0 or 1 for all the uncorrelated cases.

The greatest relative signal level for these other cases is therefore given by the ratio 1:121, which in decibels is $10\log_{10}(1/121) \approx -21$ dB. It means that the autocorrelation process for an 11-bit Barker code on a perfectly received signal produces −21 dB peak 'spurious' signals in the 10 cells before and 10 cells after the actual range cell occupied by the target. This seems reasonably good until it is realized that these would appear as small but close range spurious targets on the radar display. In practice, wanted targets have a relative size that can be significantly greater than the 21 dB (121 times) difference between the actual and spurious responses of the 11-bit Barker code process – and even the 22 dB (169) times difference of the longest (13-bit) Barker code.

These spurious responses in range are remarkably similar to the effect on a radar display that radar antenna sidelobes can give in the azimuthal direction, as previously described in Section 2.5.2.2. In fact, the mathematics behind the effects is very similar. This is because the formation of an antenna's mainbeam and sidelobes can be looked at as arising from a full or partial correlation in phase length of the possible routes of electromagnetic radiation from the antenna structure to any point in space. This results in local signal peaks with the main beam being at the highest level of correlation. These similarities, in both the appearance and the mathematics to antenna sidelobes, have given the name 'range sidelobes' to such features in a pulse compression radar. They also arise in analogue methods of pulse compression, such as frequency modulated systems. The effects are also often called *time sidelobes*, especially when correlation techniques are used in communications systems.

It is worth looking at how range sidelobes actually affect the radar image of two targets that are close in range. We can illustrate this by using the 11-bit Barker code, while also showing the applicability of digital correlation to multi-target examples. Consider Table 2.8(a). It shows how the received signal from two identically sized targets separated by just one range cell can be combined. For this example, to simplify the calculation, we have assumed that the received signal from both targets is fortuitously in phase. We then just add the amplitudes in each 'column' to get the combined signal. (It is interesting to note that any assumed phase difference would result in the same final 'answer', but would be more complex to illustrate.) Once combined we can then apply the correlation algorithm (several examples are given in Table 2.8(b)–(e)).

We end up with the graph of correlation against range, shown in Figure 2.59. The peak amplitudes have been reduced from 11 to 10, with the peak range sidelobes all at an amplitude of −2. This shows that there has been a marginal reduction in the signal detection of the targets, and that the peak range sidelobes are now at a level of −14 dB (4/100) compared to the targets. Figure 2.60 shows exactly the same case when Target 2 is taken to have a relative amplitude of 0.316 (−10 dB) compared to Target 1. The peak power of the displayed Target 2 is now −12.5 dB relative to the power of Target 1, whereas it was at −10 dB before pulse compression. It is seen that the peak sidelobes are now −18.2 relative to Target 1 and −5.5 dB (12.7–18.2) relative to Target 2.

The precise results are actually of little significance; what is really being shown is that pulse compression does put some artefacts into the display. The two targets remain clear for both cases but the spurious responses, especially when one target is a lot smaller, start to become prominent. Of course, the resultant effects become even more complicated when there are more than two targets involved. A practical marine radar with a good performance perhaps needs a 50 times compression, with range sidelobes typically rather better than −60 dB. Such a radar would have sufficient compression to enable the use of a solid-state transmitter, and would also generally be able to distinguish targets with about 60 dB range in their radar reflectivity. This ratio does not have to include the huge differences in received signal level resulting from targets at different ranges ($1/R^4$ effect) as it is only applicable to those targets encompassed within the uncompressed pulse at any one instant. Translating to a possible range of targets of interest, a 60 dB range would cover

TABLE 2.8 11-Bit Barker Code – Two Targets

(a) Calculation of Combined Signal by Simple Addition

Target 1	1	1	1	−1	−1	−1	1	−1	−1	1	−1		
Target 2			1	1	1	−1	−1	−1	1	−1	−1	1	−1
Combined (received sequence)	1	1	2	0	0	−2	0	−2	0	0	−2	1	−1

(b) Calculation of Correlation – Sample Sequence Start Aligned with Target 1

Received sequence	1	1	2	0	0	−2	0	−2	0	0	−2	1	−1	0	0	Sum
Sample sequence	1	1	1	−1	−1	−1	1	−1	−1	1	−1	0	0	0	0	
Correlation	1	1	2	0	0	2	0	2	0	0	2	0	0	0	0	10

(c) Calculation of Correlation – Sample Sequence Start Aligned with Target 1/Target 2 Gap

Received sequence	1	1	2	0	0	−2	0	−2	0	0	−2	1	−1	0	0	Sum
Sample sequence	0	1	1	1	−1	−1	−1	1	−1	−1	1	−1	0	0	0	
Correlation	0	1	2	0	0	2	0	−2	0	0	−2	−1	0	0	0	0

(d) Calculation of Correlation – Sample Sequence Start Aligned with Target 2

Received sequence	1	1	2	0	0	−2	0	−2	0	0	−2	1	−1			Sum
Sample sequence	0	0	1	1	1	−1	−1	−1	1	−1	−1	1	−1			
Correlation	0	0	2	0	0	2	0	2	0	0	2	1	1			10

(e) Calculation of Correlation – Sample Sequence Start Aligned Two Cells After Target 2

Received sequence	1	1	2	0	0	−2	0	−2	0	0	−2	1	−1	0	0	Sum
Sample sequence	0	0	0	0	1	1	1	−1	−1	−1	1	−1	−1	1	−1	
Correlation	0	0	0	0	0	−2	0	2	0	0	−2	−1	1	0	0	−2

from say 0.05 to 50,000 m^2, enabling a very small vessel or buoy, both without a radar reflector, to be visible close to a large SOLAS vessel, provided the vessel did not physically shield the buoy. As previously discussed, a 50 times compression allows the real transmitted power to be approximately 50 times lower than that of a non-compressed radar and a further two times (3 dB) when comparing with a magnetron-based non-coherent radar. With such a 100:1 total advantage the basic detection performance of a 25 kW magnetron radar would be obtainable with a 250 W peak power pulse compressed radar.

Unfortunately, there is no Barker code that can give a 50 times compression or 60 dB range sidelobes. Fortunately, the similarity in the mathematics of range and antenna sidelobes means that somewhat similar techniques can be used to design for low sidelobes in both areas. The mathematics gets really complicated and so is not explored here. However, it means that suitable codes for pulse compression can be developed that enable high levels of compression and low range sidelobes. However, getting low sidelobes, as is identically the case for antennas, means that the width of the uncompressed pulse (compare with the width

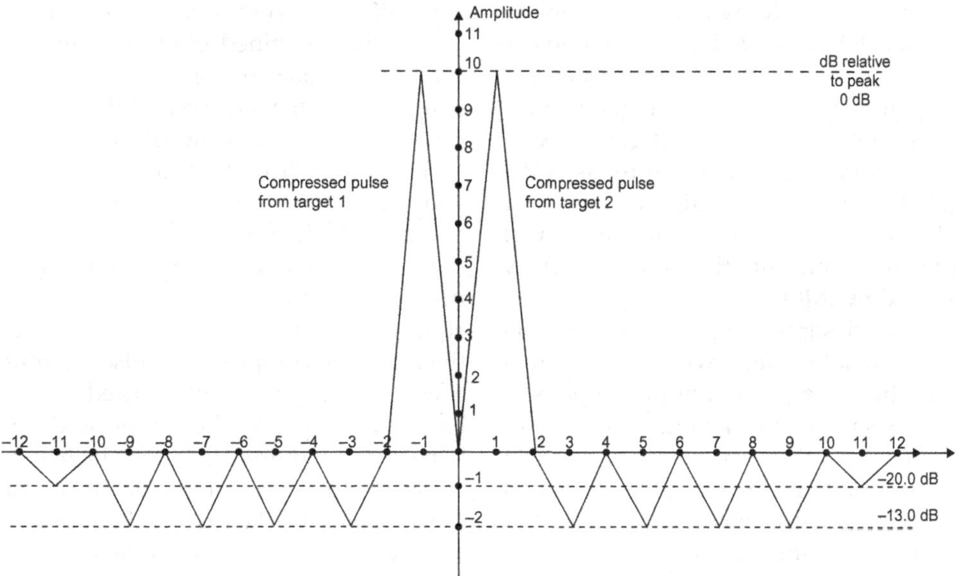

FIGURE 2.59 11-bit Barker code, 2 targets of equal size.

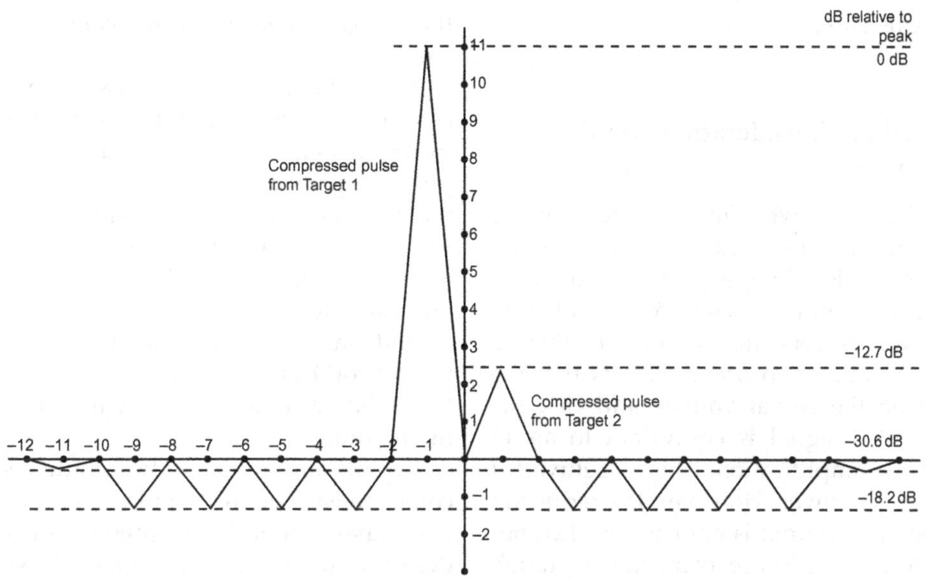

FIGURE 2.60 11-bit Barker code, two targets with Target 2 10 dB smaller than Target 1.

of an antenna) broadens with lower levels of the required sidelobes.

In practice, it means that the effective reduction in peak power is less than the ratio between the uncompressed and the compressed pulse length, L_u/L_c. Therefore, it means that codes rather longer than 50 digits have to be used to get the required compression and the desired range sidelobes. The range sidelobes needed are rather lower than can be achieved on practical antennas. Fortunately, it is easier to produce good range sidelobes as they rely on precise digital processing, rather than a physical structure, as is the case with the antenna. Achieved levels of range sidelobes are currently kept confidential by manufacturers, but into the future are likely to be specified by IMO.

On a well-designed pulse compression radar, range sidelobe effects will be very rarely experienced by a user, certainly in comparison with those experienced in azimuth because of antenna sidelobes. If a range sidelobe effect occurs it will be relatively close-in to the actual target, showing a false smaller target or targets. The danger range is equivalent to the uncompressed pulse length. In practice, a well-designed system should rarely exhibit such effects, but the user needs to be aware of their potential existence.

2.9.5 Further Considerations on Pulse Compression

In Section 2.9.3 we effectively took it for granted that an 11-bit Barker code enabled an 11 times reduction in the peak power when compared to a non-pulse compressed coherent radar. If we explore this a little further, a conundrum appears to occur. This is that the code acts on the signal voltage and that the fully correlated signal is equivalent to an 11 times voltage amplification. This is equivalent to 121 times in power. However, the reduction in peak power required is not equal to 121, but is equal to 11 (the pulse compression ratio). This is explained by the fact that the correlation process has two different effects on the noise component. The first is that there is an increase in the total noise in the system because we are now dealing with 11 cells, instead of one; the second is that the correlation process acts to reduce the total noise. The combined effects equate to a noise level power increase of 11 times, compared to a signal power increase of 121. This produces an overall increase in signal-to-noise ratio of 11 (121/11), which is the true gain of the system and is equal to the pulse compression ratio.

An equivalent way of looking at this is that a coherent radar which did not use pulse compression would need a receiver bandwidth equal to the inverse of the pulse length, effectively the uncompressed pulse. A pulse compression radar would need a receiver bandwidth equal to the compressed pulse. The bandwidth required therefore increases by the pulse compression ratio, that is 11 times for the example Barker code. This increases the noise by 11 times simply because it is a greater range of noise frequencies that will be accepted by the system because of the need to detect the modulation on the wanted signal that varies according to the compressed pulse length.

The radar designer can really exploit pulse compression to get the most out of a system. In particular, it gives the ability to design pulse sequences that effectively cover short, medium and long ranges simultaneously, as well as potentially being able to give special modes for increased detection in particular areas selected by the user. This is simplified on a solid-state transmitter compared to that of a magnetron-based system, simply because of the relative ease of applying any sort of modulation.

For instance, a basic transmitted sequence could consist of three pulses. The first pulse may have very little compression on it and be designed to be very short, so that very close targets are detectable as well as those out to a certain distance. Immediately after the necessary gap to allow signals from the furthest point of (short range) interest to be received, a longer pulse is transmitted with more compression, designed to cover intermediate distances. This medium length pulse could then be followed by a long pulse to cover the longer ranges, perhaps with even greater compression, with the sequence then repeated, see Figure 2.61. Suitable gaps must be left in between pulses to ensure that the returns from the shorter range targets are received. In practice, even more complex sequences are necessary to get the best out of a particular radar design and will undoubtedly be kept confidential by manufacturers. Of course, the system has to be able to cope with the different autocorrelation codes perhaps needed on each pulse.

The optimization of such sequences becomes quite complex and is intimately connected with the antenna beamwidth and its rotation rate and, of course, the minimum performance requirements of the radar as defined by IMO. An important advantage over conventional magnetron-based systems is that such complex pulse sequences are able to cover the whole range of interest of a marine radar, and not just that being displayed at any one time. Therefore, the operator can switch ranges without there being a delay before a full screen update is available. Also, the tracking process (see Section 2.7.5) is not disrupted by a change in the viewing range of the radar display.

A really important aspect, which can be more and more exploited into the future as increased processing power becomes affordably available, is the fact that the compressed pulse retains all its Doppler information. This is because the processed signal retains amplitude and phase information in its *IQ* format. It is an important advantage of coherent radars that has been exploited in other radar sectors for many years, not least since the 1960s by air traffic control (ATC) radars. (All coherent radars preserve the Doppler effect on the received signals but pulse compression is a later refinement.) ATC radars use the Doppler content of the signal to suppress reflections from all static and slowly moving targets. This greatly enhances the view of aircraft, suppressing signals from the ground (ground clutter) – and also from precipitation. These are commonly known as Moving Target Indication (MTI) radars. Earlier ATC radars were based on coherent magnetron technology, which was too expensive to be adopted for marine systems. However, a greater issue is that marine radars need to see static targets as well as moving ones. The Doppler content in the signals can therefore only be exploited by complex processing techniques, which are now just becoming affordable for civil marine applications.

Basically, what can happen is that the amplitude and phase of the returned signal over a small area can be analysed to see if it has the characteristics of a wanted target or is that of clutter. By this means the display of clutter can be rejected but that of wanted targets enhanced. The mathematics of doing this is tied up with the issues of *pattern matching*,

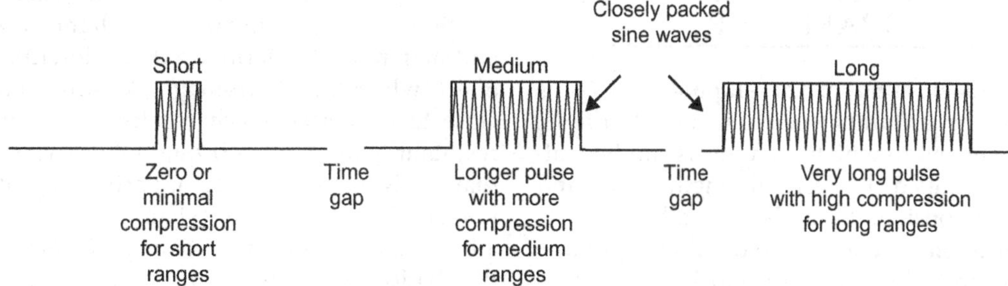

FIGURE 2.61 Example 3-pulse sequence to optimize detection at short, medium and long ranges.

which is used in very diverse areas, including video processing, medicine, finance, policing, robotics and position fixing. There appear to be no limits to the possible reduction for clutter that has a component of its movement in the radial direction – that is, the vast majority. Reductions of 30 to perhaps 50 dB may be feasible, but at present are certainly unaffordable to the marine market.

In fact non-coherent magnetron transmitter technology could potentially use Doppler processing. In principle, each transmitted pulse could be separately analysed in phase and amplitude and autocorrelated with the received signal. It would not give pulse compression but could give the potential for Doppler processing. However, the other previously described advantages of using solid-state technology will probably mean that such a route will never be exploited for commercial marine radar.

In all, the advent of coherent solid-state technology in marine radars gives the possibility of significant user enhancements. In particular, it is highly likely that future radars will be able to give a greatly improved performance especially in detecting small targets in heavy clutter. Current non-coherent magnetron systems are at the peak of their capability in doing this and in many situations they do not meet real user requirements. This technology, in time, could resolve this issue and give real improvements in maritime safety.

2.10 THE SITING OF UNITS ON BOARD SHIP

The ideal siting of the shipboard units is dictated by a number of navigational criteria, but the extent to which such ideals can be realized is limited by certain engineering and practical considerations. In general, the units should be sited so as to avoid failure or undue maintenance difficulties due to heat or fumes and they should be mounted appropriately to prevent the performance and reliability of the system being adversely affected by shock and vibration. Particular navigational and engineering aspects of the siting of each of the physical units are discussed in the following sections, together with a description of interswitching duplicate units.

2.10.1 Antenna Siting

In general, the antenna must be sited with a clear view on a rigid structure which will not twist and give rise to bearing errors. A poorly sited antenna can severely limit the overall performance of the radar in a number of ways. Unfortunately, the siting has to take into account a number of conflicting factors and can be a compromise. In most cases users will have had no say in the acceptance of any compromise, but it is important that the factors which may affect optimum performance are understood.

2.10.1.1 Blind and Shadow Sectors

Blind and shadow sectors occur when the radiation from the antenna is intercepted by obstructions such as masts, the vessel's superstructure and cargo on board the vessel. They can also temporarily occur when the vessel is close to other structures, such as other vessels. The effect is illustrated in Figure 2.62 and it can be seen that it is similar to that of optical shadows cast when a source of light is obstructed. Because of the longer wavelength, radar waves experience more significant diffraction effects compared to light waves. Diffraction is the term used to describe the effect when an electromagnetic wave passes close to an object which results in a complex resultant pattern of varying intensity, rather than a distinct shadow area. Particularly at the longer wavelengths of S-band, it effectively allows a radar to have some performance in the shadow area if the width of the obstruction is small compared to the width of the antenna.

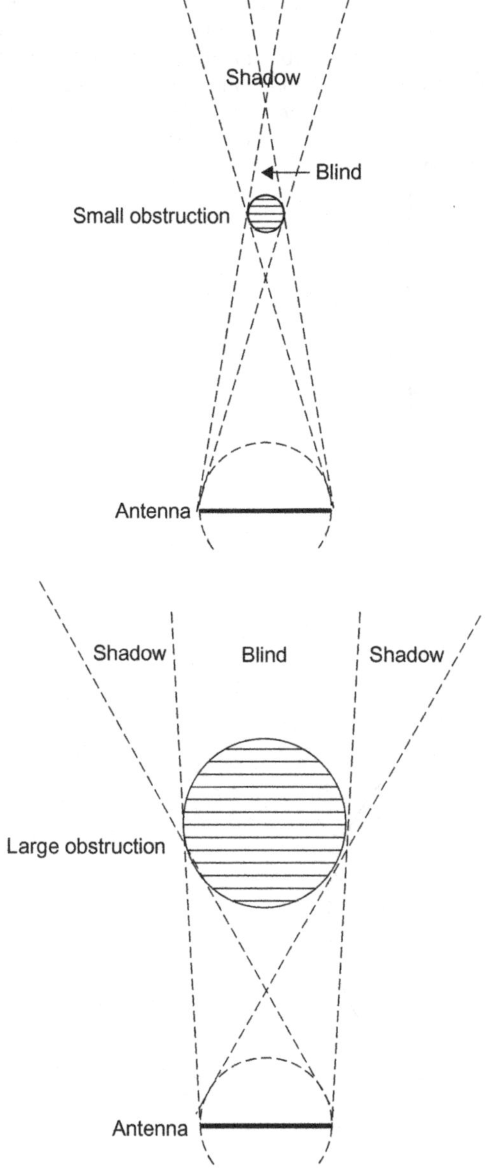

FIGURE 2.62 Blind and shadow sectors.

The size of the antenna, the size of the obstruction and its proximity to the antenna will determine the angular width of the sector and whether it is totally blind or partially shadowed in character.

IMO Radar Performance Standards (see Section 11.2.1) give the following requirements on antenna siting:

> *Blind sectors should be kept to a minimum, and should not be placed in an arc of the horizon from the right ahead direction to 22.5° abaft the beam and especially should avoid the right ahead direction (relative bearing 000°). The installation of the antenna should be in such a manner that the performance of the radar system is not substantially degraded. The antenna should be mounted clear of any structure that may cause signal reflections, including other antenna and deck structure or cargo. In addition, the height of the antenna should take account of target detection performance relating to range of first detection and target visibility in sea clutter.*

It might appear at first sight that the best approach is to locate the antenna unit sufficiently high as to be above all obstructions. This may be possible on some ships but there are also disadvantages for excessive antenna height (see Section 2.10.1.3). Although it may not be possible to use the vertical location of the antenna to eliminate blind and shadow sectors, suitable attention to the horizontal siting may reduce the seriousness of their effects. This is illustrated in Figure 2.63.

Figure 2.63(a) shows a fairly traditional siting with impaired sectors ahead and on both bows. Figure 2.63(b) and (c) shows how these can be relocated astern or on one side, which could have the advantage of improving the detection of targets which are often the 'stand-on' vessels in collision encounters. In considering these solutions it should be remembered that, although in clear weather the collision regulations allocate responsibilities to vessels which relate to the crossing and overtaking situations when vessels are not in sight of one another, the give way/stand on allocation of responsibilities no longer apply. Both vessels now have a responsibility to take action to avoid a close-quarters situation.

There are practical limits to the physical separation between the antenna unit and the transceiver because of the cost and/or losses

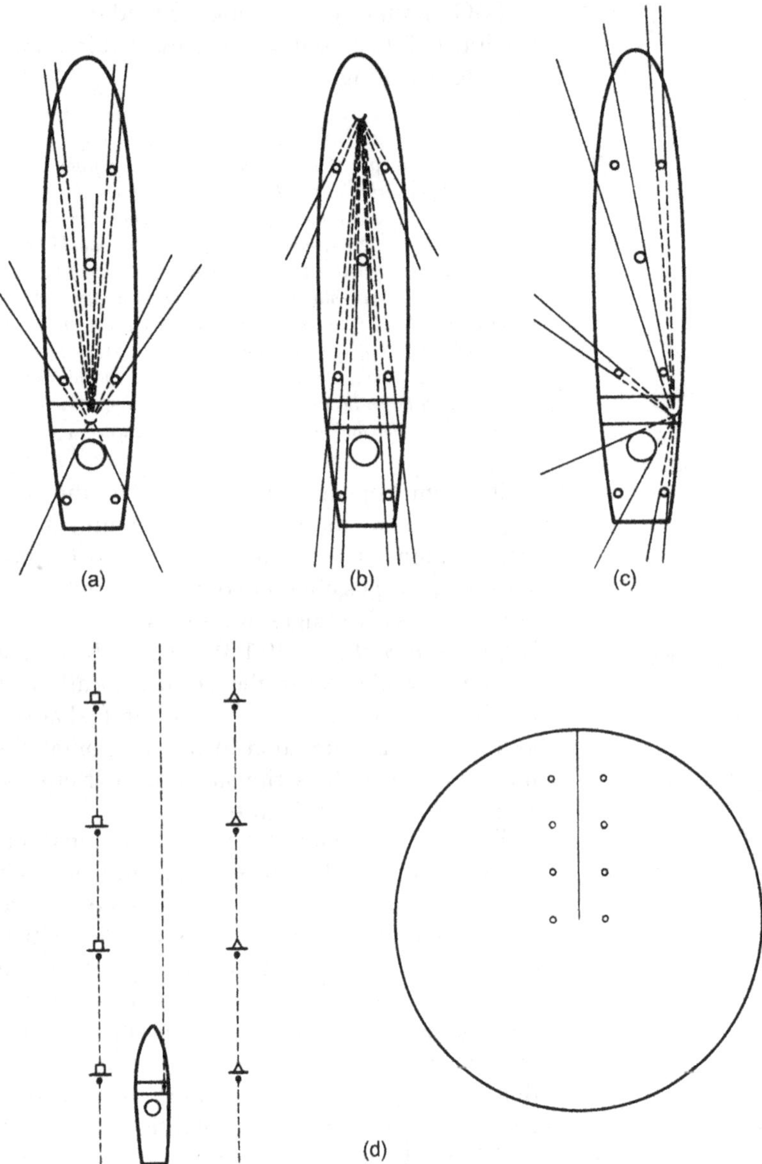

FIGURE 2.63 Antenna siting. (a) Traditional bridge mounting, (b) Mounting forwards, (c) Bridge mounting to starboard, (d) Showing potential positional offset (on radars with no CCRP setting).

in the transmission line between the two. As stated in Section 2.2.1 it is therefore nowadays often the case that the transceiver is mounted close to the antenna-turning gear. If the antenna is sited off the vessel's fore-and-aft centreline, the primary origin of the picture represents the location of the antenna. This was particularly a problem on older systems, especially in pilotage situations because, as illustrated by Figure 2.63(d), failure to appreciate it could result in the observer being seriously misled as to the position of the vessel with respect to mid-channel. However, a more recent IMO requirement on radars (for all fitted since 2008) has been the need for compatibility with working to a Consistent Common Reference Point (CCRP). This is a defined point on the ship to which all navigation sensors are referenced, not least, all radars. The CCRP is typically at the normal conning position. On a correctly set-up radar, it is the position of the CCRP that is at the centre of the bearing scale. On small range settings this can make a significance difference and so it is important that the CCRP has been correctly set on every radar on board.

All radar observers should be aware of the nature and extent of the shadow sectors on board their vessel. The prudent method of ensuring this is to measure the sectors and illustrate the results on a suitable plan-view diagram posted near the radar display (Figure 2.64 and Section 11.4.1). In addition, the information should be recorded in the radar log. In some vessels the trim or draught may affect the shadow pattern and if this is believed to be the case it will be necessary to measure the sectors in more than one condition of loading.

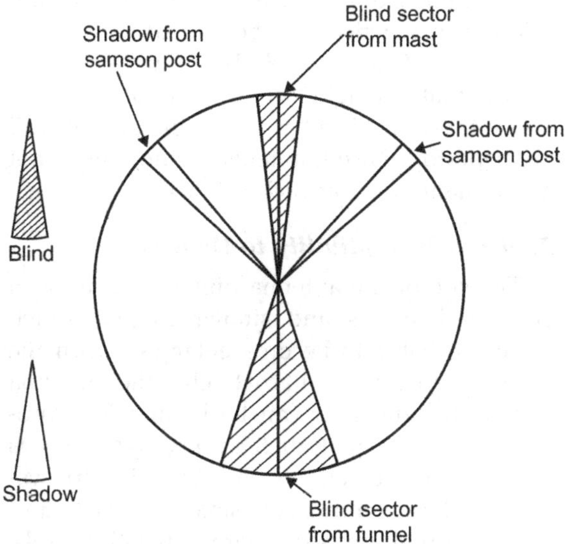

FIGURE 2.64 Plan of shadow sectors.

The angular extent of the sectors can be measured by observing the response of a weak isolated target such as a buoy which is not fitted with a radar reflector (see Section 3.4) and is located at a range of about 1 NM. The ship should be swung through 360° and the limits of the sectors determined by measuring the bearings on which the echo disappears and subsequently reappears. It is important that a weak target is chosen to ensure that the areas of reduced sensitivity are effectively detected. A good target, such as a buoy with a radar reflector, may return an echo which is so strong that a response is obtained despite the loss in energy caused by the offending obstruction. The measurement procedure should be carried out in calm conditions so that the echo is not masked by clutter (see Section 3.6) and to avoid an intermittent response due to the rolling motion of the buoy.

An alternative approach is to measure the sectors against the background of the sea clutter pattern (see Section 3.6). For the same reasons given in the preceding paragraph for selecting a weak target, strong clutter conditions should be avoided. The measurements should not be taken when the vessel is in confined waters, as the sectors may be obscured by indirect echoes (see Section 3.9.2) due to reflections from the obstructions which produce the shadow sectors.

Having measured the shadow pattern, it is important to appreciate the danger of a target approaching on a steady bearing which lies within one of the sectors. Collisions have resulted from this cause. In poor visibility this danger must be addressed by ensuring that from time to time the vessel is *briefly* swung either side of the course by at least half the width of the largest sector in an attempt to detect the presence of any targets within the blind and shadow sectors. In some situations the natural yawing motion of the vessel may adequately fulfil this function.

In ships which frequently navigate astern, for example short sea ferries, the antenna siting must be such as to avoid shadow sectors astern. In many cases such vessels are fitted with a dedicated docking antenna at the stern.

2.10.1.2 *Indirect Echoes*

Obstructions close to the site of the antenna may reflect some of the radar energy on both its outward and return journeys. This gives rise to false echoes known as *indirect* (or *reflected*) echoes. These are discussed in detail in Section 3.9.2.

2.10.1.3 *Antenna Height*

As suggested in Section 2.10.1.1, the location of the antenna above all obstructions is not a simple matter. The height at which it is sited is inevitably a compromise. There are benefits and drawbacks to siting the antenna as high as possible. In respect of shadow sectors, it is worth bearing in mind that if a vessel has cross-trees or similar reflecting areas on the masts, an antenna sited above the level of such an obstruction will cast a shadow on the water, whereas with a lower siting it may cast the shadow in space.

In favour of a high antenna siting is the fact that the distance to the radar horizon, and hence the theoretical maximum detection range of targets, increases with antenna height. This topic is discussed in detail in Section 3.8. The minimum range at which targets can be detected also depends on antenna height. Excessively high and low antenna sitings may increase or decrease this important limit and the implications of this, including the effect on the accuracy with which very short ranges can be measured, are discussed in Section 3.2.4. In general, very short range performance is improved with a low antenna. Some vessels have two interswitched antennas (see Section 2.10.8), one high and one low, to obtain the advantages of each.

A serious penalty incurred by the siting of the antenna at a great height is the effect on the amplitude and range of sea clutter response. The amplitude of sea clutter echoes and the maximum range at which they are detectable increases with antenna height. These responses produce a pattern of echoes centred around the origin which can mask even very strong targets. The danger of targets being lost in sea clutter returns represents one of the most serious limitations of marine radar and is discussed in detail in Section 3.6. Excessive antenna height will exacerbate this fundamental problem and serious consideration must be given to this factor when a shipboard radar installation is planned.

For downmast transmitters, a further problem associated with a high antenna site is the loss of energy which occurs when the transmitted pulse and received signals travel along the necessary lengths of waveguide (see Section 2.2.1).

It is evident from this section that choosing an optimum antenna height involves reconciling a number of conflicting factors. The height selected in practice may be limited by the size and type of the vessel, but experience has shown that a height of between 13 and 20 m above the water level can offer the best overall radar performance if antenna shadowing is not a dominant factor at such heights.

2.10.1.4 *Susceptibility to Damage*

The site of the antenna unit should be well clear of halyards and similar rigging which might be liable to become entangled with the rotating antenna. Alternatively, the antenna and its turning gear can be located in a protective arrangement such as a *radome.* This is rarely implemented on commercial ship systems, but is common on smaller vessels and is also found on some larger naval vessels. A radome is a moulding made of GRP within which the antenna is free to rotate and is typically cylindrical or spherical. It must be kept clean because while GRP (more commonly referred to as fibreglass) is transparent to radar energy (see Section 3.3.3), salt, dirt and similar deposits which may accumulate on the surface of the radome can attenuate both the transmitted and returning signal.

2.10.1.5 Radiation Hazard

In general, there is no radiation hazard to shipboard personnel provided that the antenna is rotating. However, *harmful* effects, particularly to the eyes, can be experienced at very short distances from a stationary antenna if energy is being radiated. The distance will depend on the transmitted power, but in civil marine radars designed for merchant ships it is usually in the region of 0.3–0.6 m. Many administrations require, and good practice dictates, that manufacturers should make radiation level measurements from which it is possible to determine the safe distance. This distance should form part of a warning contained in the operator's manual and should also be printed on the antenna unit. Safe distance warning information is typically available on the antenna unit itself. If it is necessary for personnel to approach within the safe distance of a stationary antenna, it is important to ensure that the equipment is not transmitting and to take steps to prevent the equipment being switched to the transmit condition. An equally serious problem is the possibility that the antenna may start rotating and injure the person. It is an IMO requirement that there is a means to prevent antenna rotation and radiation during servicing or while personnel are in the vicinity of upmast units. On older systems it may be best to have the person qualified to service the equipment remove the fuses from the transceiver and antenna-turning equipment when this work is being carried out, and to place a prominent warning notice on the display.

Service personnel may have to carry out adjustments which require transmission from a stationary antenna, to undertake work on a waveguide while the transmitter is operating and to disconnect the waveguide from the transmitter. All such operations incur a risk of exposure to dangerous levels of radiation. It is essential that such work is only undertaken by fully trained and experienced service personnel. It must always be remembered that *permanent damage to the eyes* is likely to result from looking down a waveguide from which radar energy is being radiated. Some appreciation of the possible harmful effects of radiation can be obtained by recognizing that a domestic microwave oven uses frequencies similar to radar waves to cook meals and is therefore why the radiation is made to automatically cease, immediately the door is opened.

It is also important to note that some cardiac pacemakers can be affected by intense radar-frequency radiation.

2.10.2 The Transceiver Unit

The transceiver unit is so called because it houses the transmitter and all or part of the receiver. Ideally, the transmitter and at least the early stages of the receiver should be located as close to the antenna unit as possible. This reduces the attenuation of the powerful transmitted pulse on its journey up the waveguide to the antenna and, of equal importance, the attenuation of the weak returning signals as they travel down the waveguide to the receiver. Once the signal is at IF, or perhaps even effectively at base band frequencies, the path length between the antenna and the display is no longer an issue.

The waveguide is expensive and thus siting the transceiver in close proximity to the antenna reduces the overall cost of the installation. If long waveguide runs are used, maintaining the cross-section of the waveguide is of critical importance for correct propagation and so must be *plumbed in* with great care. Particular attention must be given to ensure that bends and twists are formed in such a way as to prevent distortion of the waveguide dimensions, and so it is usual to have specially preformed sections for them. In an ideal situation the antenna unit would be mounted vertically above the transceiver so as to allow the use of a straight waveguide run. This is seldom possible in practice. Long waveguide runs are always at risk of accidental damage. The ingress of water or dirt will seriously affect the performance of the guide. Some waveguide systems are filled with a suitable gas in an attempt to maintain the guide cavity in a dry condition and, by pressure indication, to serve as an indicator of any accidental puncture in the wall of the guide.

In many systems the transceiver is fitted within the antenna unit. This may even dispense with the need for a rotating waveguide joint (see Section 2.2.2), eliminates the waveguide run and reduces signal attenuation to a minimum. This approach suffers from the disadvantage that there may be serious difficulties in servicing the unit *in situ*, especially in bad weather.

In general, civil marine radar equipment should not produce undue interference in other radio equipment. However, good practice dictates that radar units in general, and the transmitter in particular, should not be sited close to radio communications equipment. As discussed in Section 2.6.5 some 4G phone systems may produce radar interference at S-band. If this is suspected to be a problem for any particular radar installation, the service agent should be contacted.

While modern transceivers are quiet by comparison with older types, the possibility of acoustic noise being a nuisance to personnel on or off watch is a factor which might influence siting of the unit.

There may be dangerous voltages present and the risk of radiation if the transceiver unit is opened. Only qualified service personnel should therefore perform such a task. In some cases there may also be risk from X-radiation (X-rays). If this is the case many administrations require that the transceiver or other unit should carry a warning notice.

2.10.3 The Display Unit

The display unit will contain the units necessary to perform the display functions described in Section 2.8. On modern equipment, as discussed in Section 1.2.4.2, the display unit is typically equivalent to a computer display, with some additional modules that interconnect to the rest of the radar system. These may even be sited somewhat away from the actual flat panel display and user control panel. Because of the split nature of the radar set-up, some service procedures may require transceiver adjustments whose effect is best judged by viewing the display. If the two units are separated by a great distance, a communication link may greatly facilitate such servicing. This need may well be fulfilled by the use of portable VHF or UHF (ultrahigh frequency) radios. Modern electronics makes the provision of multiple displays a simple matter.

The light level produced by the display can be high and this factor has to be considered in choosing suitable sites for them. Good use of the display controls should be made to ensure that information is visible but, at night, minimally interfering with night vision. IMO requires that 'the orientation of the display unit should be such that the user is looking ahead, the lookout view is not obscured and there is minimum ambient light on the display'.

2.10.4 Compass Safe Distances

Because of their potential magnetic influence, the radar bridge equipment will have been tested to establish the safe distance by which they should be separated from magnetic compasses. This may be indicated on a plate attached to the particular unit. Some radar spares, in particular magnetrons (see Section 2.4.2), have very strong magnetic effects. They may be marked with a magnetic safe distance, but if not should be stowed at least 7 m from magnetic compasses.

2.10.5 Exposed and Protected Equipment

Some radar equipment, such as the antenna, must be designed to withstand the elements while others are intended to be protected from the weather. Certain administrations require the class of equipment to be indicated on each unit.

2.10.6 Power Supplies

Radars for commercial ships normally operate from the standard 110/220 V single-phase mains supply. They may be connected to the emergency power supply to ensure continued operation during a temporary fault in the primary power. Small boat systems are usually designed to run off voltages between 12 and 30 V.

2.10.7 High-Voltage Hazards

Lethal voltages may exist at various points within shipboard radar equipment. The charge held by high-voltage capacitors may take some minutes to discharge after the equipment has been switched off. Maintenance adjustments which require the removal of the outside cover of units should only be attempted by fully qualified personnel.

2.10.8 Interswitching

IMO Performance Standards (see Section 11.2.1) require vessels greater than 3000 tons gross to be fitted with two radar installations, generally one X-band and the other an S-band system. Many vessels below this tonnage limit also carry two radars because the benefits of such a dual installation are recognized. In order to maintain a high degree of reliability and flexibility where a dual installation is fitted, some or all of the units may be made interswitchable and capable of serving either or both displays. This allows unserviceable units to be isolated and also permits the servicing of units requiring maintenance, without affecting the availability of the overall system. IMO Performance Standards (see Section 11.2.1) require that, where interswitching is fitted, the failure of either radar must not cause the supply of electrical energy to the other radar to be interrupted or adversely affected. A typical fully interswitched system is illustrated in Figure 2.65.

Suitable selection of the characteristics of individual units will extend the capability of the system as a whole (see hereafter) to deal with a variety of conditions and circumstances.

On a modern integrated bridge system using integrated navigation concepts (INS) it is increasingly the case that multiple displays can access any radar or any other navigational system, such as ECDIS. For radar there should be some control as to which particular display at any one time has the primary control of a particular radar sensor (master and slave operation, see Section 2.10.8.6). This is because many of the chosen display settings influence the signal being transmitted and possibly to some extent the receiver processing. These settings include display range, gain, sea and rain clutter controls.

2.10.8.1 Choice of Frequency

The characteristics of S- and X-band systems are to some extent complementary (see Section 2.3.2) and an interswitched system which offers both options has much to

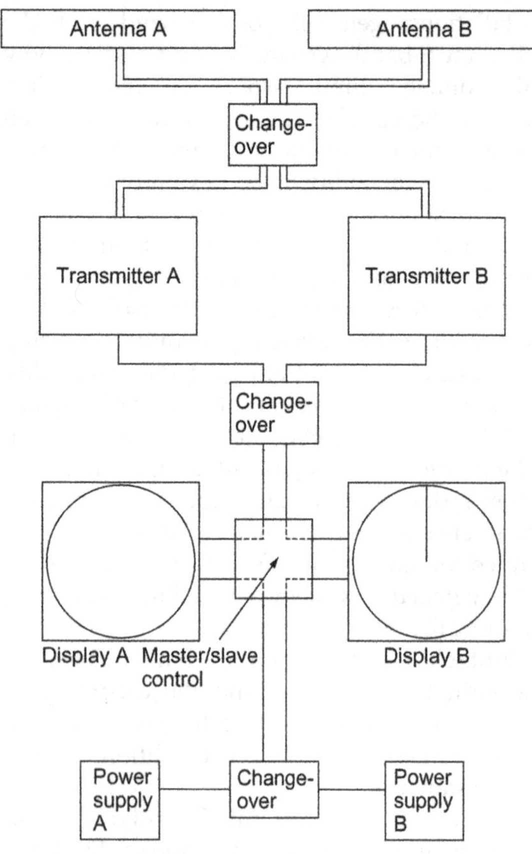

FIGURE 2.65 Interswitched system.

recommend it. It allows the observer to select the frequency which is most appropriate to the particular task in hand. For example, the masking response from sea and rain clutter (see Sections 2.7.4.1 and 2.7.4.2) is less with S-band whereas, with comparable antenna size, X-band generally offers better bearing discrimination (see Section 2.5.1.4).

2.10.8.2 *Choice of Picture Orientation and Presentation*

An interswitched system makes possible the simultaneous availability of two different orientations or presentations. In particular, the relative-motion and true-motion presentations are complementary (see Section 2.8.4.3), and a dual system allows the observer to exploit this.

2.10.8.3 *Choice of Antenna Location*

It is apparent that in the case of a single antenna installation, the choice of antenna siting is a compromise. With an interswitched dual installation, it is possible to ensure that the siting of each antenna gives particular benefits and that these can be exploited by the observer according to the prevailing circumstances. For example, the installation might have one high and one low antenna, one centred and one offset antenna, or one forward and one midships antenna. The particular trade in which a vessel is involved might influence the choice of antenna sites. Each radar transceiver should have the correct settings applied to be consistent with the Consistent Common Reference Position.

2.10.8.4 *Partial Interswitching*

A smaller degree of flexibility with lower cost may be achieved by partial interswitching. For example, one antenna and one power unit might serve two transceivers and two displays. Alternatively, a single antenna, transceiver and power supply might be provided with two displays, one as master and one as a slave.

2.10.8.5 *Switching Protection*

Normally the equipment is protected during switching by the *stand-by* condition (see Section 6.5). In some systems this has to be done manually while in others access to the interswitching control is protected by an interlock.

2.10.8.6 *Master and Slave Controls*

Where two displays are driven from the same transceiver, one display must be designated as master and the other as slave. This has significance in that it affects the use of the operational controls. The master display will control the receiver tuning (see Section 2.6.4.2). Normally the master display will also control PRF and the pulse length (see Sections 2.3.3.1 and 2.3.3.2), whereas the selected range scale is controlled by each display. As a result the pulse length and PRF could be inappropriate to the range scale selected at the slave display. To overcome this problem, at the slave display, an indication of the only available range scales may be given. The display will be protected by suitable circuitry from any technical incompatibility but the observer must be aware of any operational limitations, such as minimum range, discrimination and saturation, which may result (see Sections 3.2.4 and 2.3.3). Pure display functions such as brilliance, range rings, range markers, EBL, centring and rain clutter are available at each display (see Chapter 6).

To overcome problems of mutual interference (see Section 3.9.5), a transmission synchronization circuit may be incorporated in the interswitching unit.

2.10.8.7 Combination of Radar Data from Multiple Antennas

The ability to combine the data from more than one antenna position has been available on some systems installations for some time. It has the advantage of presenting complete radar coverage of an area in one display with the major advantage of eliminating blind and shadow sectors (if they exist) due to obstructions on the vessel. The combination can cause some targets to appear twice due to small calibration errors between the overlapping data coverage from two antennas. Such systems are more common ashore where extended area coverage is required with the combined display at a single control centre. It is usual to use adapted marine radars for this purpose.

CHAPTER

3

Target Detection

3.1 INTRODUCTION

The ability of the radar system to detect and display a given target depends on a large number of factors, some of which are constant and others which may vary in quite a complex manner. The significance of some of these factors may be almost self-evident while that of others may be much less obvious. The *radar equation* is an expression which attempts to formalize the relationship between the range at which a target can be detected and the parameters on which that range depends. In the form in which the equation is used by radar engineers to predict radar performance by taking into account all conceivable factors, it is indeed lengthy and complex, containing statistical and empirical components. However, in its conventional basic form as set out below, it provides an indication of the significance of many of the factors which are of concern to the radar observer (an alternative form of this equation was introduced in Section 2.6.1):

$$R_{max} = \sqrt{\left(\frac{P \times G_0 \times A \times \sigma}{4 \times \pi^2 \times S_{min}}\right)}$$

where:

R_{max} = maximum detection range
P = peak transmitter power
G_0 = antenna gain
A = antenna aperture area (m^2)
σ = target radar cross-section (m^2)
S_{min} = minimxum detectable signal

All but one of the variables appearing in the equation relates to the characteristics of the radar system and each of these is discussed in turn in Section 3.2. The sole representative of the target is its radar cross-section which is a measure of the size of the target as 'seen' by the radar. Except in the case of certain simple shapes, this is a complex quantity, often having a statistical nature; the implications of this are discussed in Section 3.3. This simple form of the radar equation assumes that the radar and the target are in free space. In practice they are both located on a curved earth having a surface of varying character and surrounded by an atmosphere in which various weather effects may be manifest. The simple equation takes no account of the surface over which, or the medium through which, the radar pulse travels, because many of the effects are extremely difficult to quantify. However, a suitable qualitative treatment of these factors is presented in Sections 3.6–3.8.

While each of the various factors will be discussed in turn, it is essential to appreciate that in assessing the ability of the radar to detect a target *all* factors will have to be considered and it must be recognized that the relative significance of individual factors will vary with circumstances.

A further result which follows from the equation is that, in general, the signal strength received varies inversely as the *fourth* power of the range of the reflecting surface. (This can be shown by replacing S_{min} with the more general term S_r representing received signal strength, and then making S_r the subject of the equation.) It is thus important to appreciate that, for distant targets, a small increase in range may produce a comparatively large decrease in response. The response from almost all wanted targets follows this law, but some unwanted responses such as those from the sea and from precipitation do not and these will be discussed in Sections 3.6 and 3.7, respectively. In general, this chapter assumes the use of a conventional pulse magnetron-based system.

3.2 RADAR CHARACTERISTICS

The radar characteristics which appear in the radar range equation as set out above relate directly to the system units described in Sections 2.3–2.5. One (the power) relates to the transmitter, two (antenna gain and aperture) relate to the antenna and one (S_{min}) relates to the receiver. It is thus appropriate to consider the characteristics under these headings.

3.2.1 Transmitter Characteristics

As might be expected, the ability to detect distant targets can be improved by using a more powerful transmitter (see Section 2.2.2.1). In a single radar installation the transmitter power will be a factor which is beyond the control of the observer. However, in dual or interswitched systems (see Section 2.10.8), it may be possible to make a choice between two transmitters of differing powers. The important factor which the radar equation reveals is that the maximum detection range varies as the fourth root of transmitter power. The power of transmitters designed for fitting to large vessels varies with manufacturer, but 10 and 50 KW are representative of low and high values. However, it must be noted that a fivefold increase in power only yields an improvement in predicted detection range of approximately 50%. This relationship must also be considered when contemplating the use of a lower power transmitter to reduce unwanted responses (see Section 6.7.2).

In the conventional form of the radar equation, which uses the peak transmitted power, it initially appears strange that the pulse length (see Section 2.3.3.1), which is clearly a transmitter characteristic, does not appear in the expression for detection range. The effect of pulse length is in fact included in the quantity S_{min}, the minimum detectable signal, which is a function of pulse length and is discussed in Section 3.2.3.

3.2.2 Antenna Characteristics

The radar equation shows that maximum detection range is a function of antenna gain and aperture area. Clearly these two quantities are related, as a study of Sections 2.5.1.5 and 2.5.2 will reveal. For a given rectangular aperture area, an increase in one dimension at the expense of the other will increase the antenna gain in the plane of the larger dimension. It can be shown that G_0 is directly proportional to A and can thus be replaced in the equation by G_0^2. Hence the predicted maximum detection range varies as the square root of antenna gain. Thus antenna gain has a greater influence than transmitter power on long range performance.

As in the case of transmitter power, where a single radar installation is fitted, the observer will have no choice in the matter of antenna gain. In a dual system or interswitched system (see Section 2.10.8), the observer may benefit from the ability to select the more suitable antenna. It should be remembered that bearing discrimination will also be improved by selection of the higher gain antenna.

It may seem strange that wavelength does not appear in the equation. It is in fact implicit in the antenna gain and the target radar cross-section, both of which are functions of wavelength (see Section 3.3.5.4).

3.2.3 Receiver Characteristics

In the absence of unwanted echoes the criterion for the detection of an echo is that the target response must exceed that of the thermal noise generated at the first stage in the receiver (see Section 2.6.4.4). Thus S_{min}, the minimum detectable signal, is a function of receiver sensitivity, while the theoretical maximum sensitivity is of course limited by the extent to which the radar design can minimize the amplitude of the noise generated in the first receiver stage. It cannot be stressed too strongly that the observer will degrade this sensitivity if the tuning and gain controls are set incorrectly (see Section 6.2.4).

Whatever the level of noise generated in the first stage of the receiver, the amount of noise which reaches the display depends on the bandwidth of the receiver (see Section 2.6.4.2). This bandwidth must be wider when shorter pulses are selected and hence the receiver sensitivity is inherently poorer in that condition. The receiver sensitivity, and hence the maximum range at which targets can be detected, is thus a function of the pulse length selected by the observer.

Further study of the radar range equation shows that the transmitter power appears as a large quantity on the numerator of the expression while the receiver sensitivity is represented by a small number in the denominator. Radar designers recognize that small improvements in the noise performance are as effective as massive increases in transmitter power. This principle is at least as important to the radar observer because it must be borne in mind that minor maladjustments of the tuning or gain controls can have the same effect as a large loss in transmitter power.

Where wanted echoes are present among unwanted signals such as clutter and spurious echoes, it may not be possible to exploit fully the sensitivity of the receiver, because the minimum detectable signal level will be determined not by the amplitude of the noise but by the amplitude of unwanted signals. The difficulties posed by such circumstances are discussed in Sections 3.6, 3.7 and 3.9.

It must be appreciated that the available receiver sensitivity can only be fully exploited if the radar display is set up correctly and the controls maintained in correct adjustment to suit changing conditions and requirements. Some adjustments are routine while others, such as searching for targets in clutter (see Section 6.7), require considerable skill and practice. In this connection particular attention is drawn to the various practical procedures set out in Chapter 6. There is little doubt in traditional radar, that but for the difficulty of quantifying it, the skill of the radar observer is a factor which would appear in the radar range equation.

3.2.4 Minimum Detection Range

The radar equation is especially concerned with the prediction of maximum detection ranges and in many circumstances the radar observer is concerned with this aspect of target detection. However, there are certain circumstances, for example pilotage situations, in which the minimum detection range is of particular importance. The importance of a small minimum detection range is recognized by IMO Performance Standards (see Section 11.2.1) which require that specific target types be clearly displayed down to a minimum range of 40 m. The various factors affecting minimum detection range are illustrated in Figure 3.1.

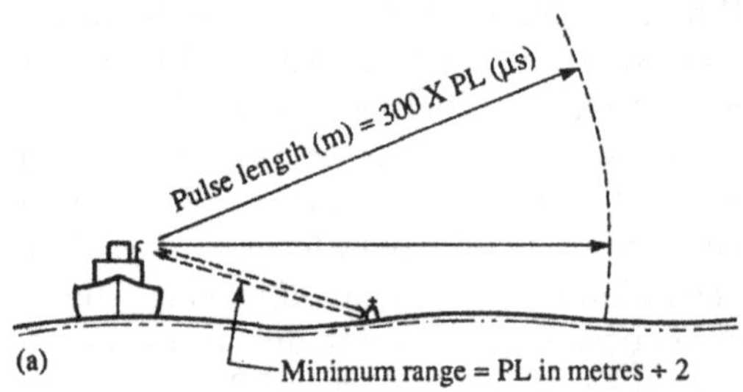

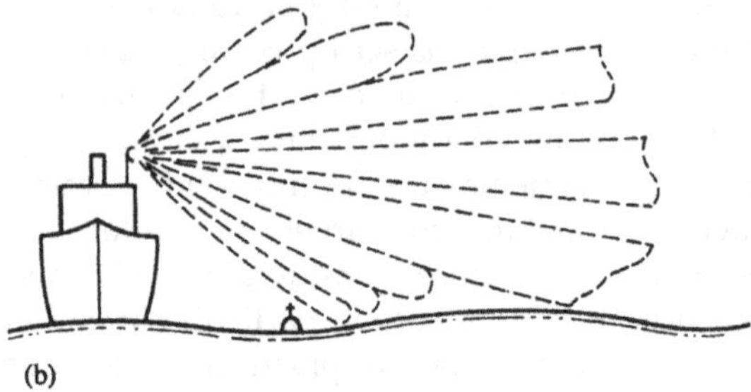

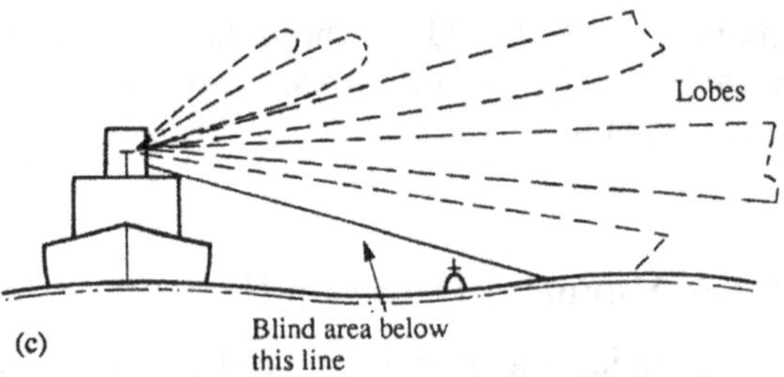

FIGURE 3.1 **Factors affecting minimum detection range.** (a) Pulse length. (b) Vertical beam structure. (c) Ship structure shadowing.

Clearly the radar cannot receive echoes while the pulse is being transmitted and this limitation determines the *theoretical minimum detection range.* The trailing edge of the transmitted pulse will clear the antenna after the start of the timebase by an elapsed time equal to the duration of one pulse length. This instant will coincide with the arrival of an echo from any reflecting surface which lies at a distance of one-half pulse length from the antenna (see Figure 3.1(a)). The theoretical minimum detection range can thus be determined by calculating the distance travelled by radar energy in one-half the duration of the transmitted pulse.

EXAMPLE 3.1

Calculate the theoretical minimum detection range for a radar which has three pulse lengths, the durations of which are 1.0, 0.5 and 0.1 µs.

To achieve the minimum detection range, the observer should select the shortest available pulse length. Half the duration of the shortest pulse length (PL) is given by:

$$\frac{PL}{2} = \frac{0.1}{2}\mu s$$

The distance travelled by radar energy in this time is given by:

$$D_{\min} = \frac{0.1}{2} \times 300\ m(\text{see Section 1.2.2}) = 15\ \text{m}$$

The theoretical minimum detection range = 15 m.

It is evident that the theoretical minimum range illustrated by this example (which uses a typical short pulse length) is considerably less than that required by IMO Performance Standards. In practice, the theoretical minimum range cannot be achieved for three principal reasons, namely:

1. At close range the target may be below the lower extent of the vertical beam though

of course it may be illuminated by the vertical sidelobes. Assuming that the beam is not intercepted by the ship's structure (see below), the range at which the lower edge intersects the sea increases with antenna height (see Figure 3.1(b)).

2. At close range the target may be shadowed in the athwartships direction by the ship's side or in the fore-and-aft direction by the ship's bow or stern. The likelihood of this shadowing decreases with antenna height. The question of whether the minimum range is limited by this factor, or by the lower limit of the beam described in item 1, depends on the height at which the antenna is sited, its location with respect to the ship's fore-and-aft line and the dimensions of the vessel (see Figure 3.1(c)).
3. The transmit/receive (T/R) cell takes a finite time to de-energize.

3.2.5 Detection Performance Standards

Within IMO Radar Performance Standards (see Section 11.2.1), there are requirements covering minimum ranges for detection of coastlines and larger ships, as well as buoys, small boats and their radar reflectors. Table 3.1 gives these detection ranges in clutter-free conditions. The standards have different requirements for S- and X-band systems, taking into account their different attributes.

The standards recognize that performance is much reduced in clutter conditions and the manufacturer is required to state in the manual the reduction in performance for distinct clutter conditions:

- light rain (4 mm per hour) and heavy rain (16 mm per hour);
- sea state 2 and sea state 5;
- a combination of these.

This performance should be considered at different radar ranges and for different target speeds.

TABLE 3.1 Minimum Detection Ranges in Clutter-Free Conditions (for more details see Section 11.2.1)

Target Description	**Target Feature**	**Detection Range in NM**	
Target description	Height above sea	X-band	S-band
	level in metres	NM	NM
Shorelines	Rising to 60	20	20
Shorelines	Rising to 6	8	8
Shorelines	Rising to 3	6	6
SOLAS ships (>5000 gross tonnage)	10	11	11
SOLAS ships (>500 gross tonnage)	5.0	8	8
Small vessel with radar reflector meeting IMO Performance Standards	4.0	5.0	3.7
Navigation buoy with corner reflector	3.5	4.9	3.6
Typical navigation buoy	3.5	4.6	3.0
Small vessel of length 10 m with no radar reflector	2.0	3.4	3.0

3.3 TARGET CHARACTERISTICS

Energy in the pulse which is intercepted by the target is then available for return towards the antenna and hence to the receiver which is now in a receptive state. The amount of energy which is returned towards the antenna, as opposed to that energy which is absorbed and scattered by the target, is dependent upon the following five prime characteristics of the target.

3.3.1 Aspect

The simplest approach to target response is to consider that the energy suffers 'specular'

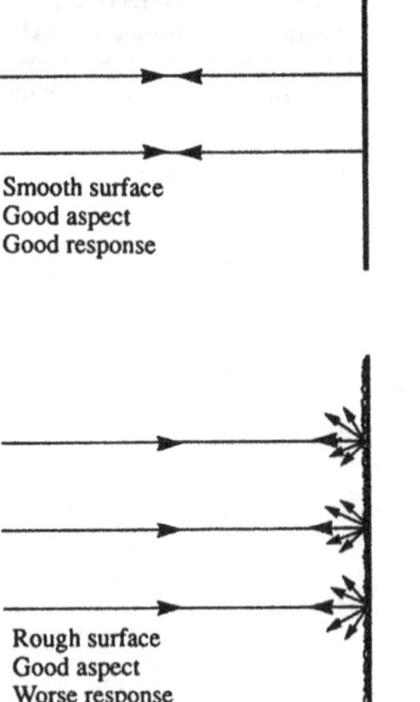

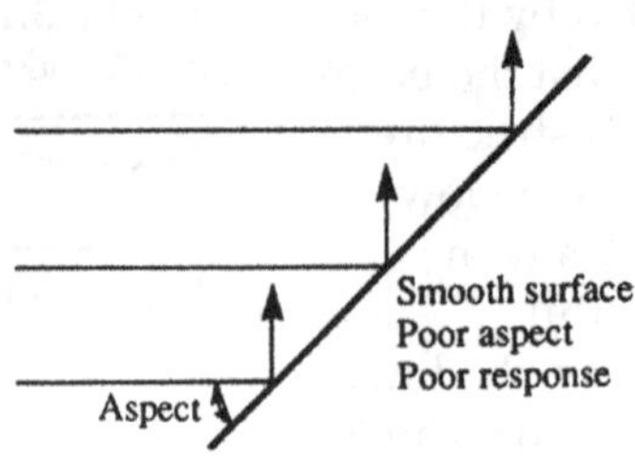

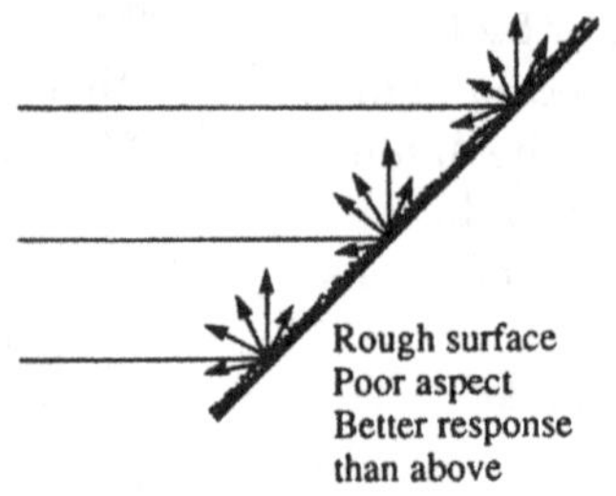

FIGURE 3.2 The effects on target response of aspect and surface texture.

reflection, that is the sort of reflection that occurs when light strikes a plain mirror. Aspect is the angle which the radar rays make with the plane of the mirror and, as can be seen in Figure 3.2, the response will be good when the aspect is 90° and poor at virtually all other angles.

3.3.2 Surface Texture

The extent to which reflection is specular is dependent upon the surface texture of the target, that is whether the surface is 'rough' or 'smooth'. Whether a surface is rough or smooth has to be related to the wavelength of the waves which are striking it. A surface which reflects light poorly because the indentations or facets in the surface are of the same order as the wavelength of the light (approximately 0.001 mm) will appear smooth to radar waves whose wavelength (some 3–10 cm) is very much longer, and specular reflection will result. It should be noted that objects which appear rough to radar waves, and therefore scatter a large proportion of the waves, can occasionally improve the response from a target which has an intrinsically poor aspect (see Figure 3.2).

3.3.3 Material

In general, materials which are good conductors of electricity also return good radar responses. This occurs as a result of absorption and re-radiation of the waves at the same wavelength as those received, rather than from simple specular-type reflection.

Some bodies absorb radiation but, when they re-radiate, the wavelength is different from that at which it was received; still other bodies absorb radiation and re-radiate very little of the energy (this results in the temperature of the body rising, that is the received radiation is converted to heat). Some materials are simply transparent to radar energy. glass

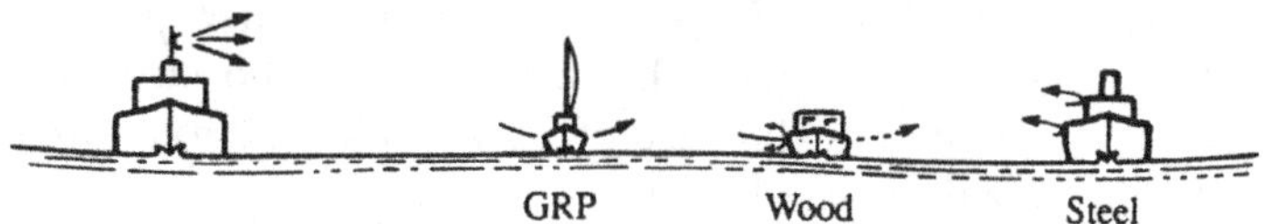

FIGURE 3.3 Some effects of material on target response.

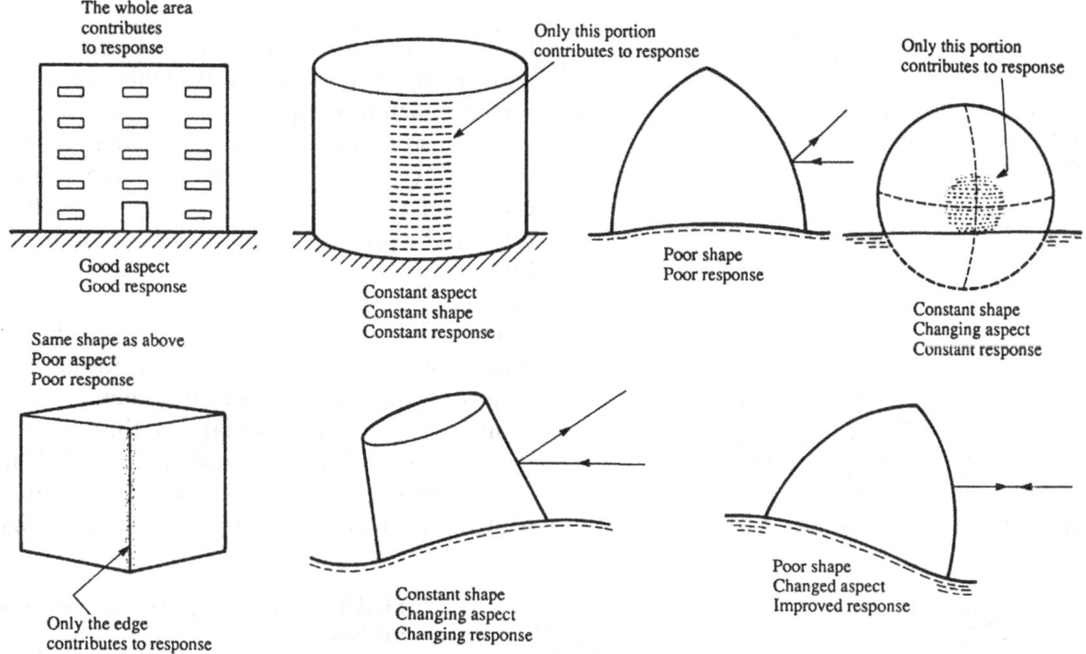

FIGURE 3.4 The effect of target shape on target response.

reinforced plastic (GRP) behaves to a large extent in this way; steel will return good responses, while wooden boats generally produce poor responses (Figures 3.3–3.8).

3.3.4 Shape

It is frequently suggested that shape might be considered as the variable aspect of an object. This is true, but when attempting to assess the degree of response which one might expect from a particular target, it is often convenient to consider the target as of fixed shape but able to change its aspect (e.g. a conical buoy rolling in a seaway).

3.3.5 Size

In general, the more energy intercepted by the target, the better the response is likely to be, i.e. the response is related to the area of the target irradiated by the beam (at any instant). This is not necessarily the same as the intrinsic size of the target. Since the radar beam is angularly wider in the vertical plane than in the horizontal, tall targets will in general produce stronger responses (all other factors being equal).

Consider two targets presenting the same area to the radar; if the linear width of the horizontal beam at the range of the targets is equal to the linear width of target A then, in

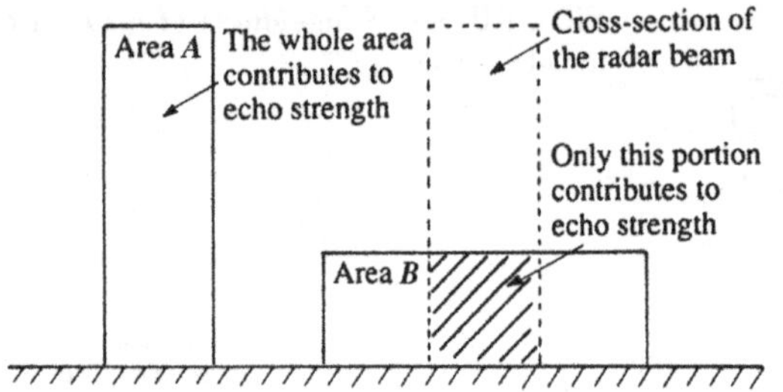

FIGURE 3.5 The effect of irradiated area on echo strength. Area A = Area B = cross-section of the radar beam.

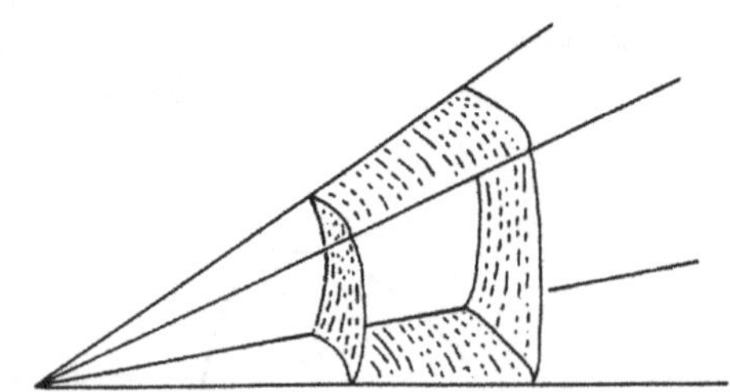

FIGURE 3.6 The resolution cell.

the case of B, only the small irradiated portion of the target will contribute to echo strength, while, in the case of A, virtually the total area will be irradiated.

It is essential to understand clearly that what is being considered here is target response as opposed to actual target size (and therefore its potential *brightness* on the screen) and not its displayed size (i.e. the *area* of the screen which it occupies).

It is also necessary here to consider the 'resolution cell' in relation to the three-dimensional size of the target. The cell is defined by the transmitted pulse length, the horizontal beamwidth and the vertical beamwidth.

Only that portion of the target which falls within the resolution cell at any instant can contribute to the response. In general, as the beam sweeps over small targets, they will fall completely within the resolution cell. With larger targets, all the energy may be intercepted when the target is at short range,

FIGURE 3.7 The portion of the vessel irradiated.

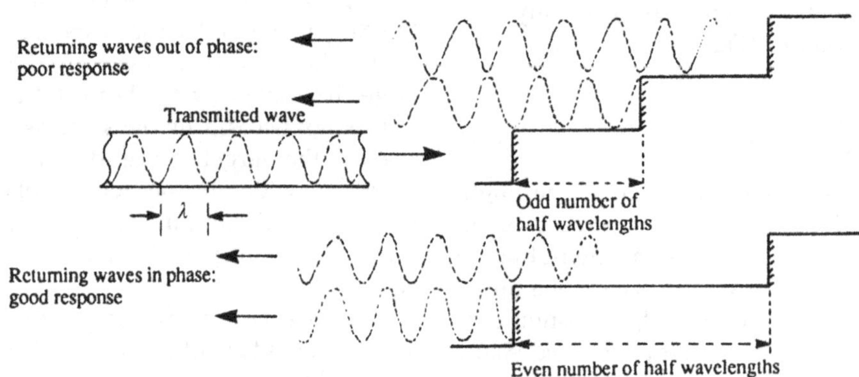

FIGURE 3.8 The effect of coherence on echo strength.

whereas at longer ranges, as the cell 'expands', the target may still exceed the horizontal beamwidth but is unlikely to fill the vertical beam.

Where, within the cell, the target is sloping, for example a coastline, two consequences of the slope need to be considered.

3.3.5.1 Sloping Surfaces – The Effect of Coherence on Response Amplitude

Consider the face of the slope to be made up of small steps. The radial length of the step will determine the phase relationship of the elements of the returning composite wave. If the elements are out of phase on their return to the antenna then the echo strength will be poor, while if they are in phase the response will be good.

Note: The radial length of the steps being considered here is of the order of a fraction of a wavelength (and will take the form of an odd number of quarter wavelengths).

3.3.5.2 Sloping Surfaces – The Effect of Signal Integration on Echo Strength

In this case, the steps which are to be considered are of the order of one-half pulse length, that is half of 0.1–1 μs, which equates to a radial distance of some 15–150 m.

Consider a stepped sloping surface (Figure 3.9) where each step is 10 m in length and the radar pulse length is 0.2 μs, that is the spatial length of the pulse is $0.2 \times 300 = 60$ m.

It can be seen from Figure 3.9 that the responses from the individual steps on the target can integrate (i.e. add up) to a maximum from that portion of the target which lies within any one-half pulse length, but cannot increase indefinitely as the target increases in height. In the above example the one-half pulse length is 30 m.

If, with the same overall size of target, the step height and length are varied so that each step height is halved and the step length is halved, the same aggregate response will be obtained, which is the integration of a larger

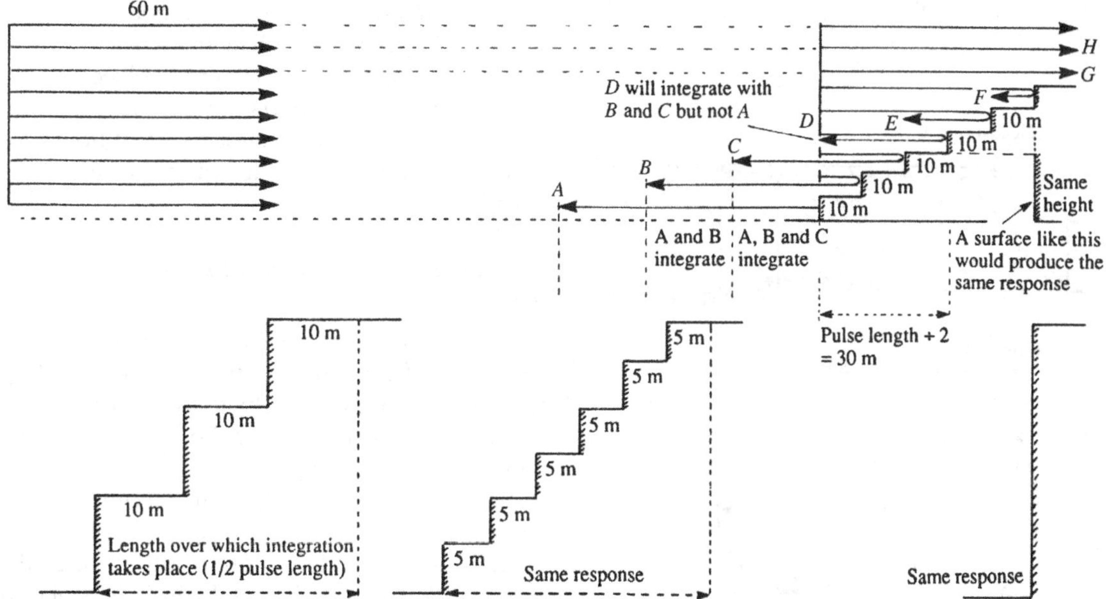

FIGURE 3.9 Integration of signals from sloping surfaces.

number of smaller responses. It is evident that the land still rises to the same height over half the pulse length. This suggests an alternative approach, which is to consider only the vertical height over the one-half pulse length, that is *projected height* (or, in three dimensions, the *projected area* (Figure 3.10).

Note:

a. If the step lengths are not precisely the same and if the difference is of the order of one-half the wavelength, then the effects of coherence will also have to be considered when deciding on whether or not the elements of the response will integrate.
b. If the facets are not true steps, random scattering will also have to be considered.
c. As a result of the most favourable integration of the elements of the response, the maximum echo *strength* would be the same as if a flat plate were placed in the path of the pulse. The plate would have a vertical height equivalent to the projected vertical height of the target which lies within half the radial length of the transmitted pulse.

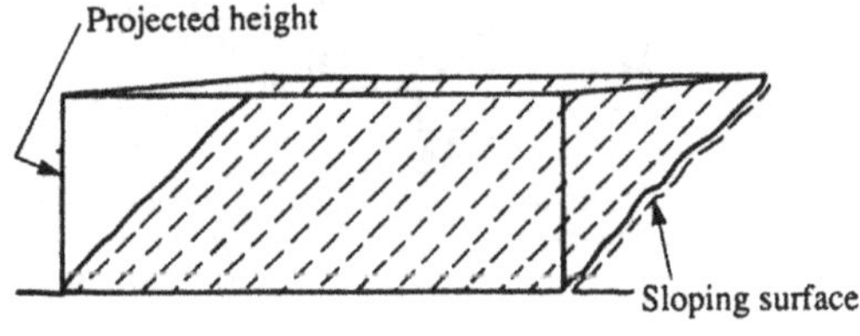

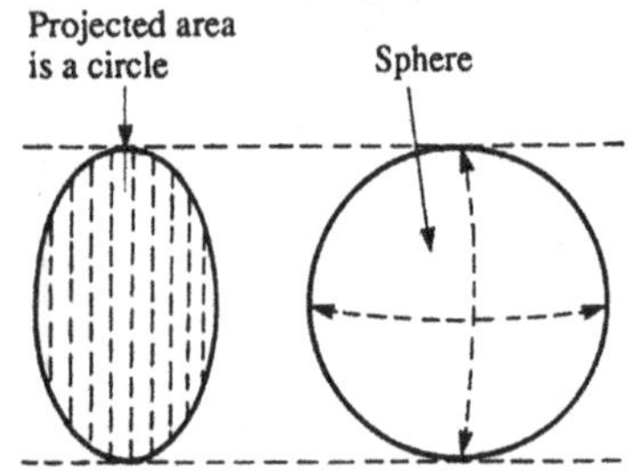

FIGURE 3.10 Projected area.

Where a target is small in relation to the resolution cell, it is convenient to consider the projected area of the target (see also Section 3.3.5.4), but its surface texture, material, aspect and shape must not be ignored.

3.3.5.3 The Equivalent Flat Plate Area (A_f) of a Target

This is the area of a flat plate orientated at right angles to the incident radiation which would return the same energy as the object. The equivalent flat plate areas for some common shapes are given in Figure 3.11.

Note:

a. The values given in Figure 3.11 are maximum values, having been simplified by the removal of terms relating to aspect. If the aspect in either or both planes is changed, the response may decrease. For example, if the flat plate were tilted there would be a marked drop in response, whereas, with the cylinder, a change about a vertical axis would result in no change in response while a change about a horizontal axis would result in a marked drop in response.
b. In deriving the formulae, there is an assumption that the size of the object is very large in comparison with the wavelength. (The formulae should not be applied, for instance, to spherical rain drops.)

3.3.5.4 The Radar Cross-Section of a Target (σ)

The radar cross-section (or equivalent echoing area) of an object is the area intercepting that amount of power which, *when scattered isotropically*, produces an echo equal to that received from the object.

Since a sphere is generally accepted as a typical isotropic radiator (i.e. it radiates in all directions), this definition is indicating the cross-section of a sphere which, if it replaced the target, would produce the same response.

Target	Dimensions	Max. equivalent flat plate area A_t	Max. radar cross-section σ
Sphere	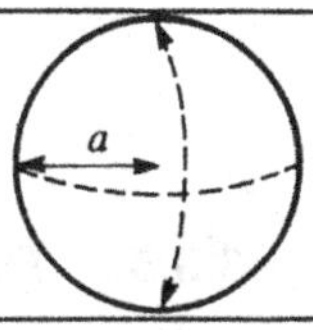 a	$\frac{a\lambda}{2}$	πa^2
Cylinder	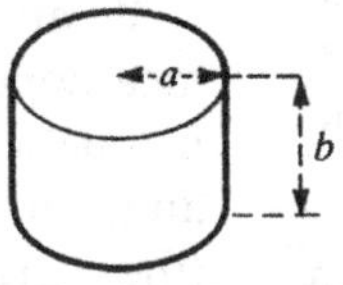 a, b	$b\sqrt{\frac{a\lambda}{2}}$	$\frac{2\pi ab^2}{\lambda}$
Flat plate	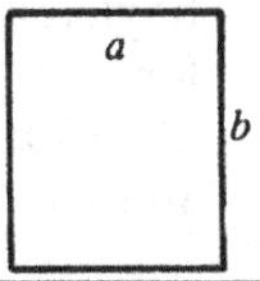 a, b	ab	$\frac{4\pi a^2b^2}{\lambda^2}$
Dihedral corner	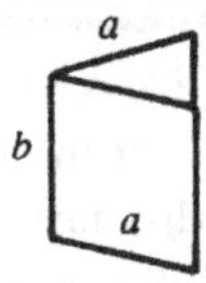 a, b, a	$ab\sqrt{2}$	$\frac{8\pi a^2b^2}{\lambda^2}$
Triangular trihedral	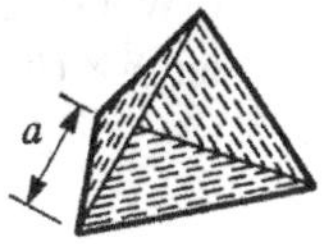 a	$\frac{a^2}{\sqrt{3}}$	$\frac{4\pi a^4}{3\lambda^2}$
Rectangular trihedral	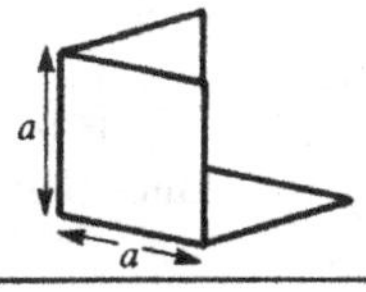 a, a	$a^2\sqrt{3}$	$\frac{12\pi a^4}{\lambda^2}$
Cone	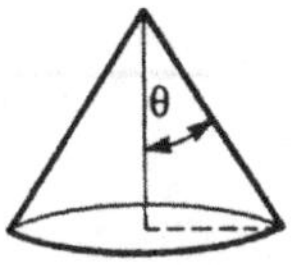 θ	$\frac{\lambda^2 \tan^2\theta}{8\pi}$	$\frac{\lambda^2 \tan^4\theta}{16\pi}$
Lunenburg lens	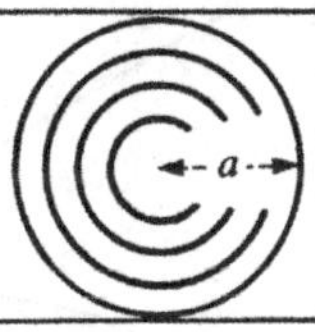 a	πa^2	$\frac{4\pi^3 a^4}{\lambda^2}$

FIGURE 3.11 Radar cross-sections and equivalent flat plate areas for some common targets.

It can be seen that if there is some directivity in the target's response, its radar cross-section will be much greater than its actual physical cross-section or projected area.

EXAMPLE 3.2

Compare the radar cross-section of a flat plate with that of a sphere if the plate is at right angles to the radiation; both plate and sphere have projected areas of 1 m^2 and the wavelength in use is 3.2 cm.

$$\frac{\text{Rader cross}-\text{section of a flat plate}}{\text{Radar cross}-\text{section of a sphere}} = \frac{4\pi a^2b^2/\lambda^2}{\pi c^2}$$

$$\begin{aligned}\text{but } a \times b = 1 &= \pi c^2 \\ &= \frac{4\pi ab}{\lambda^2} \\ &= \frac{4\pi A}{\lambda^2} \\ &= \frac{4 \times 3.142 \times 10^4}{3.2^2} \\ &= 12{,}273 \text{ times}\end{aligned}$$

This means that the flat plate would produce a response some 12,000 times greater than the sphere or, alternatively, a sphere having some 12,000 times the projected area of the plate would have to be placed at the same range as the plate to produce the same response. It is interesting to note that any slight change in the aspect of thc plate will result in a massive drop in the response, whereas no change will occur if the sphere is re-orientated.

One of the surface objects referred to in IMO Performance Standards, in specifying range performance (see Sections 11.2.1), is a navigational buoy having an effective echoing area of approximately 10 m^2. This means one that produces the same strength of echo as a sphere having a projected area of 10 m^2, that is a spherical buoy having a diameter of some 4 m.

The relationship between radar cross-section and equivalent flat plate area is given by the formula:

$$\sigma = \frac{4\pi A_t^2}{\lambda^2}$$

3.3.5.5 *The Rayleigh Roughness Criterion*

It can be shown that the roughness of a surface depends on the size of the discontinuities in relation to the incident wavelength and the angle at which the radiation strikes the surface (the grazing angle). Application of the Rayleigh criterion indicates that the surfaces for all practical purposes are smooth if $(8 \times \delta h \times \sin\theta) < \lambda$, where λ is the wavelength and θ is the grazing angle (Figure 3.12).

EXAMPLE 3.3

If the antenna is at a height of 15 m, at what distance, *d*, from the ship, will the sea surface appear smooth if wave height is 1 m (the radar is working in the X-band)? (At the limit of clutter, $8 \times \delta h \times \sin\theta = \lambda$.)

$$\frac{8 \times 1 \times 15}{\sqrt{(225 + d^2)}} = 0.03$$

$$d^2 = 4000^2 - 225$$

The sea surface will appear smooth at a distance (Figure 3.13).

$$d \geq 4\text{ km}(2.16\text{ NM})$$

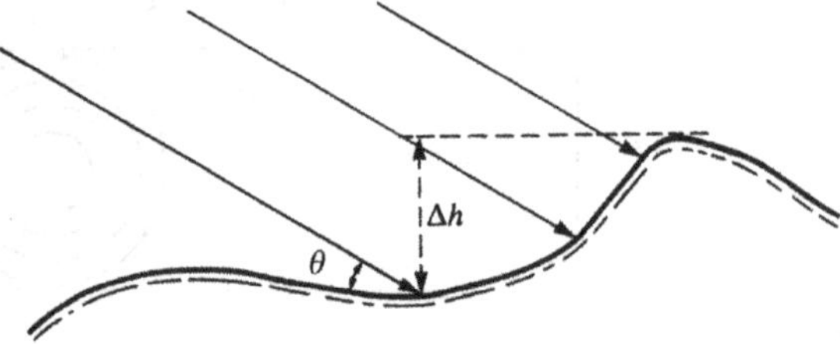

FIGURE 3.12 Grazing angle and roughness.

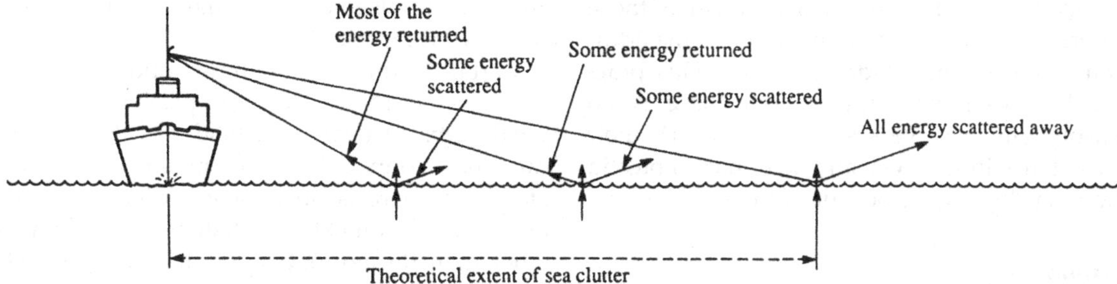

FIGURE 3.13 The theoretical extent of sea clutter.

3.3.6 Responses from Specific Targets

3.3.6.1 *Ice*

Large icebergs, such as those which are formed on the east coast of Greenland and drift down towards the North Atlantic shipping routes, have been found to give greatly varying radar responses. Detection ranges as great as 11 NM have been experienced while, on the other hand, quite large icebergs have approached to within 2 NM without being detected. Even the same iceberg may give greatly differing responses when viewed from different directions.

Particular concern has been expressed because of radar's apparent difficulty in 'detecting ice'. The point should be made that the ice which causes most concern is that form which comprises what are known as growlers. Many theories have been put forward and there have been many practical experiments carried out to determine the best way to use the radar in areas where growlers are expected: two major points emerge.

1. Growlers are intrinsically poor targets because, as melting ice, the surface tends to be smooth and this does nothing to improve their essentially poor shape and aspect. They present a small projected area above the water and the signal returned is frequently of strength comparable to or weaker than that returned from the sea waves which are prevalent in the area in which the growlers are commonly found.
   Theories relating to the effects of high-temperature gradients in the immediate vicinity of the growler and also to the re-radiation characteristic of the ice have not proved conclusive.
2. In areas where growlers are to be expected, radar should not be relied upon to give adequate or even any warning of their presence, especially at night or in poor visibility. Speed and the visual lookout should be set with this in mind.
   The best use of radar under these circumstances is:
   a. a dedicated radar watch by one observer;
   b. regular searching with the anti-clutter control on the short ranges (see Section 6.7.1), remembering to check at frequent intervals on the longer ranges for larger targets;
   c. use of the long pulse in weak clutter;
   d. use of the longer wavelength of the S-band radar.

3.3.6.2 *Radar-Conspicuous Targets*

Targets which are designated *radar-conspicuous* should be those which are known to provide good radar responses and are readily identifiable (see Section 8.2.5). In the past, particular land features which satisfied those criteria were highlighted on charts by the addition of the legend 'Radar Conspic'. This practice has been discontinued, but there is no reason why observers should not mark their own charts in this way and also make notes in the radar log for the benefit of their successors.

3.3.6.3 *Ships*

The structure of ships is such that there are many natural 'corner reflectors' (see Section 3.4.1) and hence, when a target vessel is rolling and pitching in a seaway, its echo strength does not vary quite as much as might be expected. Some rather peculiar effects can be observed, although they are usually quite easily explained by reverting to first principles. For example, long vessels may appear as two or three individual echoes (each of which when tracked by Automatic Radar Plotting Aid (ARPA) might appear to be going in a slightly different direction). They may also be confused for a tug-and-tow or vice versa. Supertankers, because of their low freeboard, may not be detected at inordinately great ranges, when loaded.

It is important to realize that some naval vessels are designed to have a very low radar cross-section, not least by avoiding 'corner reflector effects' in their design. It is to be hoped that in peacetime these vessels carry sufficient passive or active artificial radar enhancers, such as those described in the following section, so as not to appear radar-invisible to other vessels.

3.3.6.4 *Offshore Wind Farms*

The introduction of large offshore wind farms has been a recent addition to the coastal waters around Europe and other parts of the world. The towers and moving turbines can be over 100 m in height with blades over 50 m long. There has been some concern that the radar performance might be reduced or affected by false echoes in the vicinity of offshore wind farms.

Certainly this is sometimes quoted in the media as a reason for not placing onshore wind turbines near to airfields, as the echoes can be confused with moving aircraft. However, there is no known evidence of this in a marine context for offshore wind farms, other than the normal slight shadowing effect caused by targets of a similar size.

On the positive side the regular pattern of wind turbine installations within an offshore field can make good radar-conspicuous navigation marks in areas as the pattern on the radar is easy to compare with the chart. These offshore wind farms are sometimes in navigational areas devoid of other suitable land-based targets, so the wind turbines can provide a good source of bearings and ranges for radar-based navigation. Issues such as the restriction in navigation caused by the offshore wind farms are not in the remit of this book.

3.4 TARGET ENHANCEMENT – PASSIVE

It is essential that some targets which would normally provide poor radar responses, for example buoys, glass fibre and wooden boats, are detected at an adequate range by radar. It was recognized at a very early stage in the development of radar that some form of echo enhancement was needed. IMO have Performance Standards for radar reflectors (see Section 11.2.6).

3.4.1 Corner Reflectors

A corner reflector was seen as a simple device which would return virtually all of the energy which entered it, that is the energy would be returned in the direction from which it had come almost irrespective of the angle at which it had entered the corner (Figure 3.14). In its simplest form, such a corner consists of two metal plates placed at an angle of 90° to each other, attached to a navigation mark which then returns an improved radar response and consequently is detectable at a greater range.

An essential principle of all echo enhancement is that it provides *reserve gain*, which means that, in adverse conditions of sea or weather clutter, the target's improved response can be expected to be greater than the clutter responses and therefore will be detectable (see Sections 6.7.1 and 6.7.2) despite the reduced gain necessary for the suppression of the clutter.

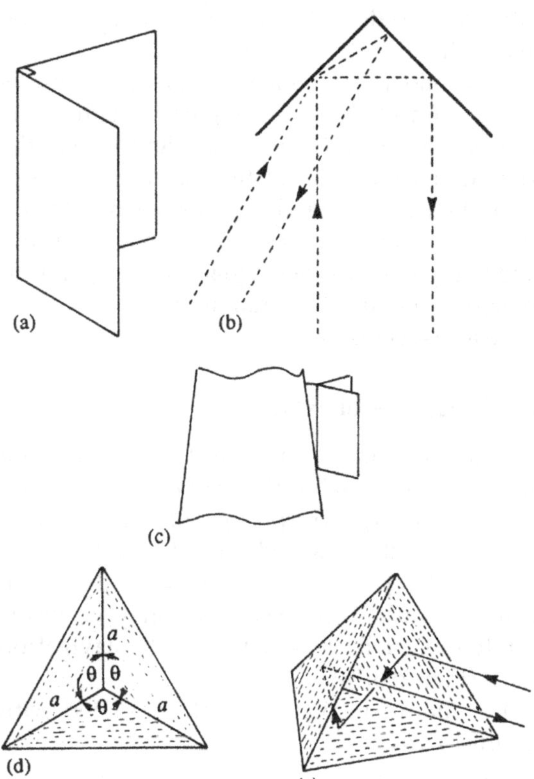

FIGURE 3.14 Corner reflectors.

While a simple corner is adequate on non-floating objects, if the attitude of the target is likely to change as it moves in a seaway, it is necessary to use the now familiar *closed corner* or triangular trihedral (see Figure 3.14(d)) rather than the dihedral corner. The closed corner will ensure that the energy is returned even though the corner is moving through quite a large angle in both the horizontal and vertical axes.

3.4.1.1 *The Accuracy of Construction*

The angle between each pair of planes must be 90° and very little error can be tolerated without affecting the efficient operation of the corner. Where, as a result of damage or careless construction, the angle is other than 90°, even by a very small amount, the response from the corner can be seriously impaired.

Figure 3.15 shows that, with a wavelength of 3 cm and a side length of 1.05 m (i.e. side length to wavelength ratio of 35); if the angle between the planes is 88° the gain is only 10 dB instead of the 30 dB that might be expected – this difference of 20 dB represents a reduction in response by a factor of 100. Note that with smaller sides but still working at the same wavelength, the loss would not be so severe.

Over the years, many designs for the 'home' construction of the corner reflectors and in particular 'collapsible' octahedral arrays (see Section 3.4.2.2) have been published, but it is never made clear how important is the precision required in the construction and assembly of the corners, in order for the full potential of the reflector to be achieved.

The suggestion that 'crumpled' aluminium foil (in a plastic container!) will provide adequate radar responses is a serious over-simplification. The physical principles of adequate enhancement require precise corners and that the size of each facet is at least greater than and preferably a function of wavelength.

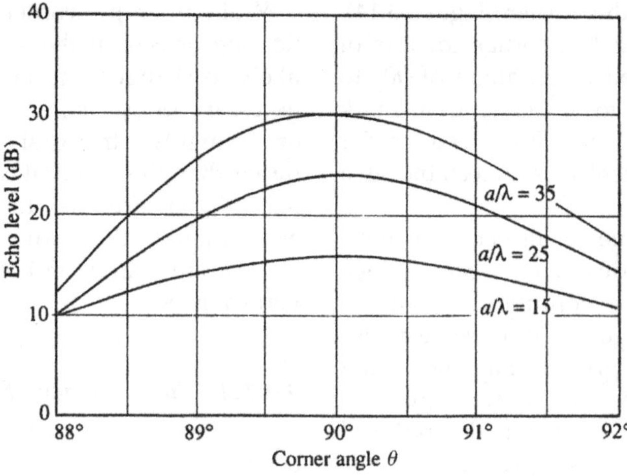

FIGURE 3.15 The effect of an error in all three corner angles of a corner reflector.

3.4.1.2 *The Response from a Corner Reflector*

The three forms of the corner reflector may be compared by referring to Figure 3.11. The following points should be noted:

1. Values shown are maximum values. Where aspect in either plane is changed, additional terms will need to be included in the equations.
2. A_t indicates the area of a flat plate, orientated so as to be perpendicular to the direction of the incident radiation. For example, a triangular trihedral reflector which has a side length *a* of 50 cm reflects as if it were a flat plate having an area of 1443.4 cm^2 and a 90° aspect.

This would appear to indicate that a corner reflector is not as good at returning a response as a flat plate. While this is true, the very important proviso is that a flat plate will only respond at optimum level when it is perpendicular to the incident radiation. If there is only a very slight change in aspect of the flat plate, the response, however, will drop theoretically to zero; with the corner, the level will suffer very little reduction over some ±50° of change in attitude.

A second point which is worthy of note is that the dihedral and rectangular trihedral corners both have better response figures than the triangular trihedral, but the former's response is markedly susceptible to change of attitude while the latter does not have such a wide coverage angle since the square plates tend to intercept the incident radiation at an earlier stage as aspect is varied.

3.4.2 Arrays of Reflectors

A single triangular trihedral corner does not provide 360° of azimuth coverage. By use of a group or array, irrespective of the direction from which it is viewed and of its attitude in the vertical plane, the response will remain virtually unchanged. The array may be incorporated in the basic structure of the navigation mark (Figure 3.16).

The following points should be noted with regard to arrays:

1. Reflectors which have surfaces of wire mesh, especially if *a* is greater than 20 times

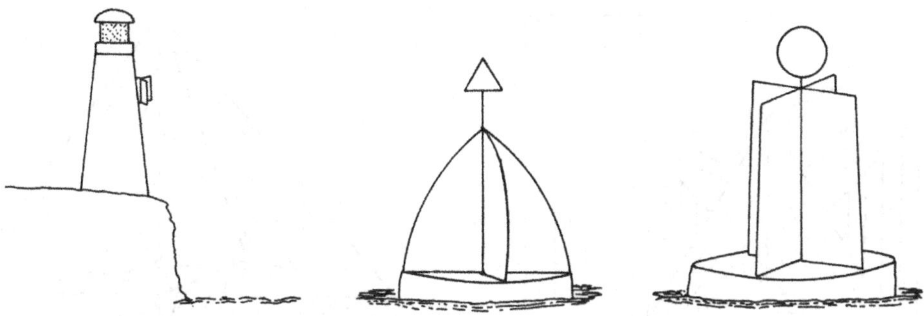

FIGURE 3.16 Arrays of reflectors incorporated into the structures of navigation marks.

the wavelength, tend to warp and so produce poor results.

2. Deterioration of response is likely to result from: (a) imprecise construction or damage, (b) poor intrinsic orientation, and (c) changes in coherence.

3.4.2.1 *Types of Array*

The *dihedral array* has been found to be most appropriate for shore-based navigation marks because the rapid fall-off in response associated with change of attitude in the vertical plane was not a factor.

The *octahedral array* (Figure 3.17) was until recently regarded as the standard for most situations.

In the *pentagonal array* (Figure 3.18), the polar diagram in the horizontal plane shows quite serious arcs of reduced response where the beam passes between the elements.

In the vertical plane, as the navigation mark heels, the response falls off rapidly with angle of heel.

In the *double pentagonal array*, as can be seen from Figures 3.18(b) and 3.19, the interspaced corners fill out the polar diagram in the horizontal plane, thus improving the 360° coverage pattern.

3.4.2.2 *Stacked Arrays (Figure 3.20)*

In order to overcome the effect of the gaps in the polar diagrams for the arrays described above, the individual triangular trihedral corners can be arranged in such a way that, as the response from one corner decreases, the response from the next corner starts to increase and so compensates for that drop. In this way, uniformity of response is maintained over 360° of azimuth.

An advantage of the many commercial forms of stacked arrays is that they are enclosed in a glass fibre housing which protects the metallic element from damage.

3.4.3 The Lunenburg Lens

The Lunenburg lens has been in use for quite a number of years, both on navigation marks and on small craft, but frequently goes unrecognized. It comprises a series of concentric shells of differing refractive index.

3.4.3.1 *Principle of Operation (Figure 3.21)*

Paraxial rays, impinging on the outer spherical surface, are refracted by different amounts at each shell interface so that they are focused at a point on the opposite surface of the sphere. Here the energy is reflected by a metal band in such a way that the exit path of the returning energy is in a direction reciprocal to that of the incident energy.

3.4.3.2 *The Polar Diagram*

Figure 3.22 gives the theoretical polar diagram, and Figure 3.23 shows its practical effect

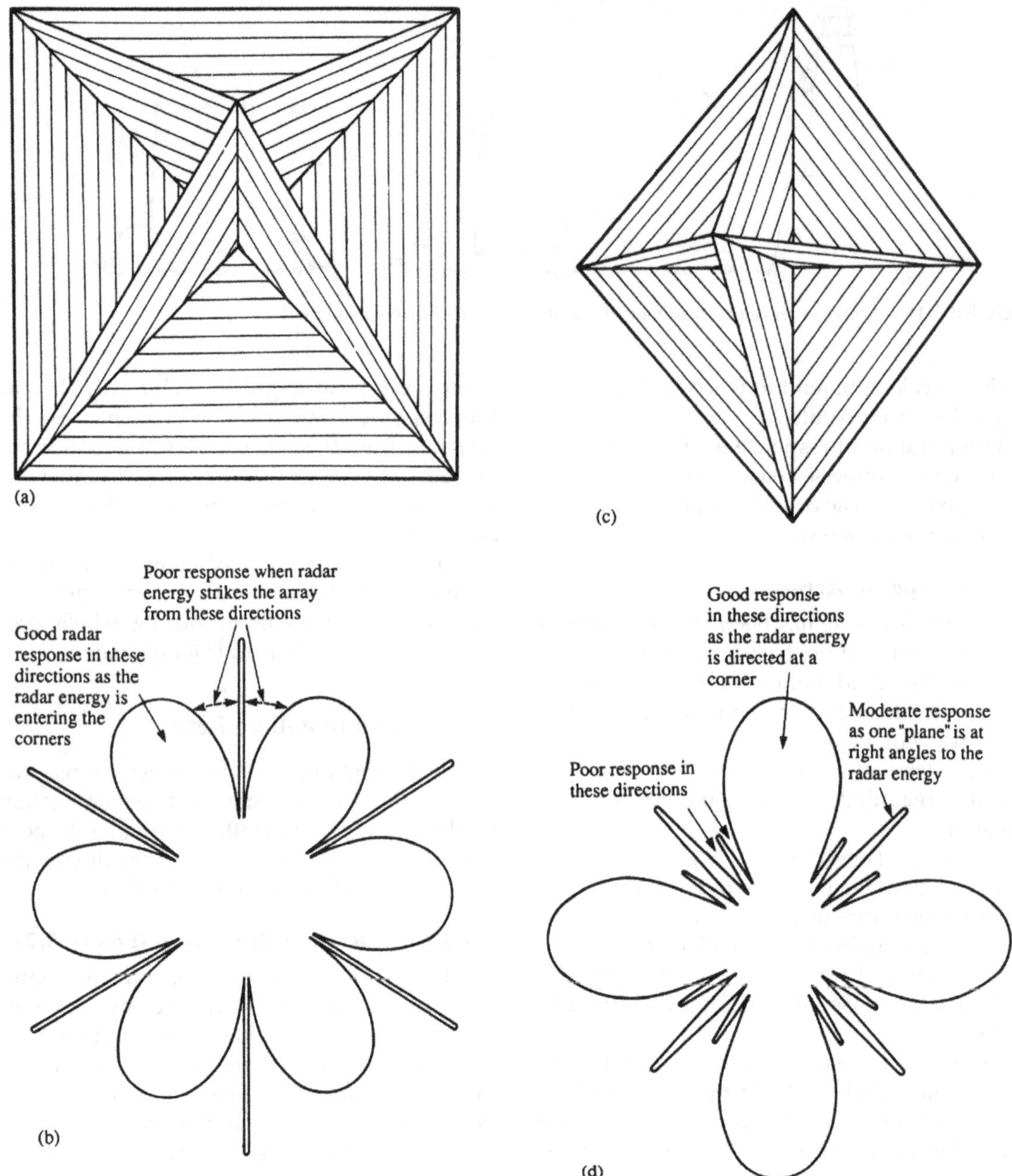

FIGURE 3.17 The octahedral array. (a) Correct orientation. (b) Polar diagram corresponding to (a). (d) Incorrect orientation. (e) Polar diagram corresponding to (c). The 'point-up' orientation gives reduced coverage.

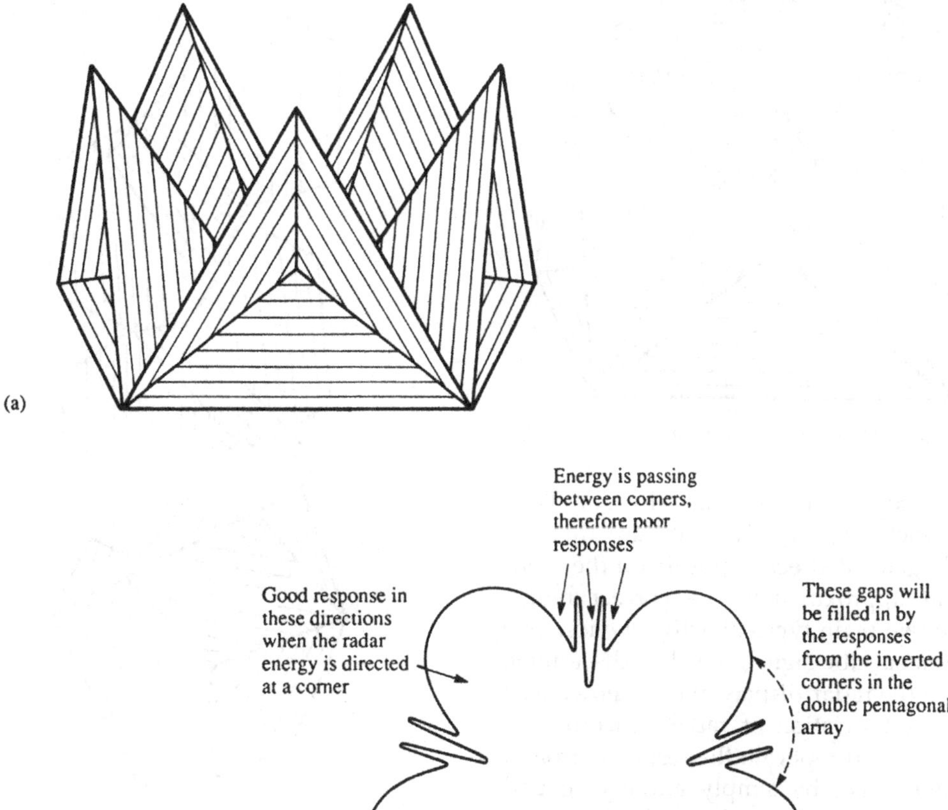
(a)
Energy is passing between corners, therefore poor responses
Good response in these directions when the radar energy is directed at a corner
These gaps will be filled in by the responses from the inverted corners in the double pentagonal array
(b)

FIGURE 3.18 **The pentagonal array.** (a) The array. (b) Polar diagram.

when applied to a buoy. It should be noted that in the horizontal plane, there are no significant 'dropout' sectors. In the vertical plane the coverage is also good, allowing the buoy an angle of heel of some 35° or, conversely, the response will be enhanced even when the buoy is close to a vessel with quite a large antenna height.

3.5 TARGET ENHANCEMENT – ACTIVE

A good alternative method of increasing the radar detectability of a target is by the use of an *active reflector*. Such devices are triggered by the reception of a transmitted radar pulse and immediately generate a radar-like signal which

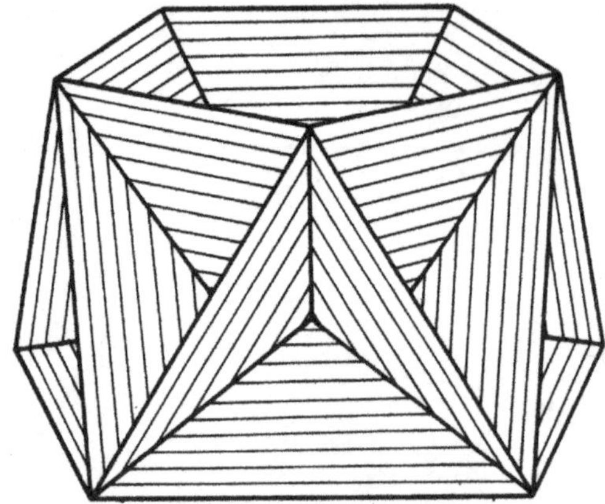

FIGURE 3.19 The double pentagonal array.

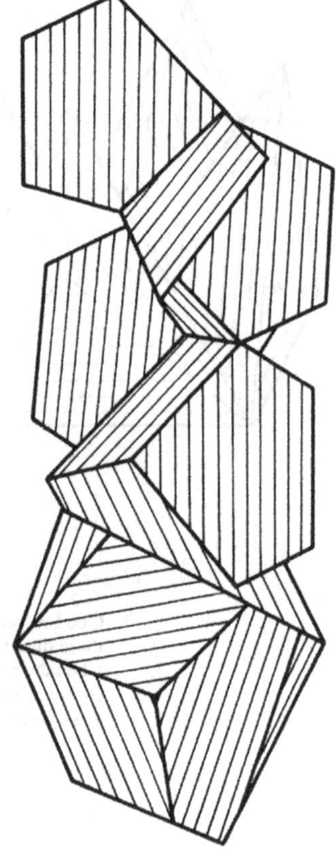

FIGURE 3.20 The stacked array. The polar diagram varies according to how the stack is constructed: in general, there are few blind spots even with 20–30° of heeling.

is transmitted back to the original radar. They have to react very rapidly to ensure that the artificially generated echo appears on the radar display at approximately the correct range. There are two main types of active device. One generates a specific signal, which is discernible as such on the radar display and is known as a racon (an abbreviation of radar beacon). The other attempts to replicate the response from a large point target by simply amplifying and retransmitting received pulses. The latter are generally just referred to as *active radar reflectors*.

Racons are typically used on buoys and beacons and nowadays meet strict requirements issued by the International Association of Lighthouse Authorities (IALA). Active radar reflectors are typically used on small craft and well-designed examples can be very effective at making such craft visible on a radar display.

3.5.1 The Racon Principle

A simple explanation of the operation is that the ship's radar pulse triggers the racon transmitter on the navigation mark, which then responds by transmitting a particular pulse or pulses, virtually instantaneously. The pulses are designed to be much more powerful than the equivalent power contained within the normal radar echo from the buoy or beacon on which the racon is mounted (see Figure 3.24).

3.5.2 The Racon Appearance on the Display

The racon signal appears on the radar screen as depicted in Figure 3.25, that is as a radial 'flash' some 1–2° in angular width. Its radial length depends on the duration of the racon pulse, which is specified by IMO (see Section 11.2.4) as being approximately 20% of maximum range but not more than 5 miles.

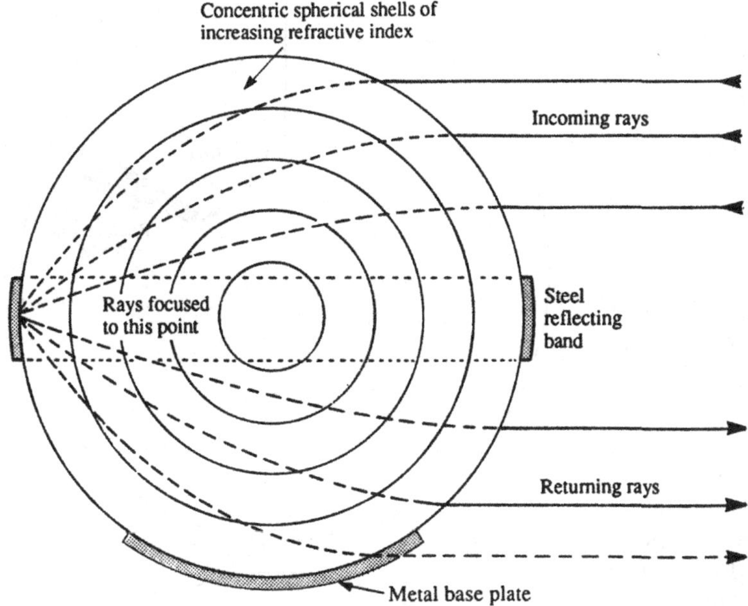

FIGURE 3.21 The Lunenburg lens – the principle.

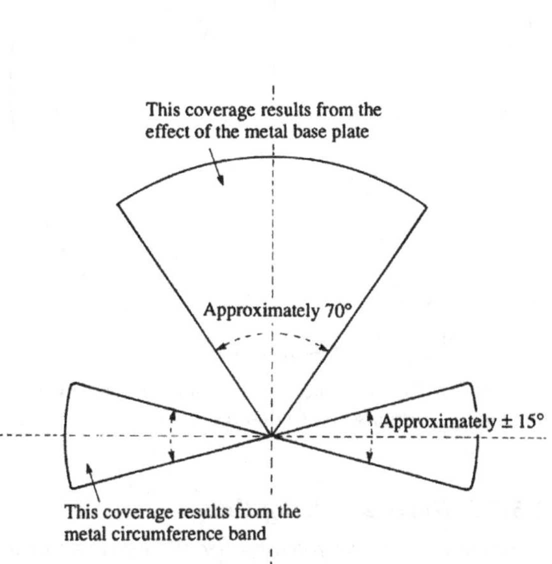

FIGURE 3.22 Theoretical vertical polar diagram of the Lunenburg lens. The horizontal polar diagram is virtually a circle, that is there are no 'preferential' directions.

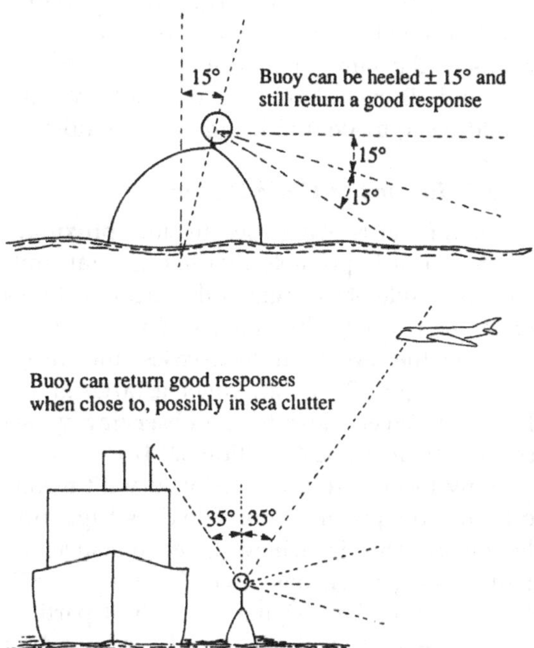

FIGURE 3.23 The vertical plane polar diagram applied to a buoy.

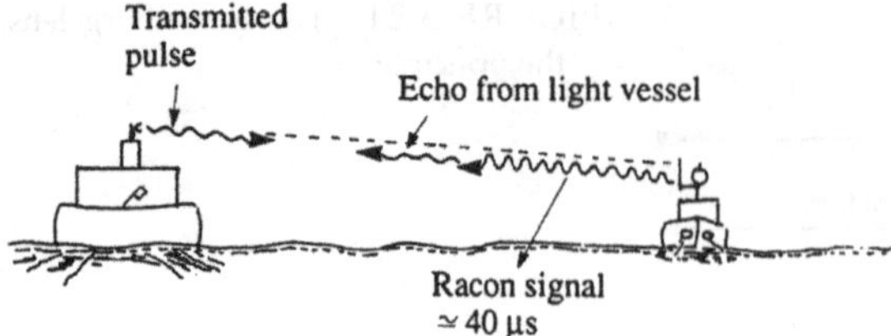

FIGURE 3.24 The racon principle.

Values are likely to lie in the range of 20–40 μs, which represents a length on the radar screen of some 1.5–3 NM.

3.5.2.1 Coding

The racon transmitter may be interrupted by on/off keying in such a way as to give a displayed racon signal of a specific appearance. This normally takes the form of an appropriate morse coding; for instance, a harbour authority might mark a shoal called 'King's Bank' by the morse letter K, as in Figure 3.25(b). This is of particular importance in aiding identification where there are several racons in the same area. The spacing between dots and dashes should be one dot. A dash should be equivalent in length to three dots.

3.5.2.2 Racons at Close Range

When radar is being used in close proximity to a racon, it is possible that energy transmitted in the sidelobes triggers the racon at times other than when the radar's main beam is scanning the target. In such cases, the 'racon clutter' (Figure 3.26) can cover a large arc of the radar screen close to the observing vessel (see also 'side echoes', Section 3.9.4).

Many racons are now equipped with means to reduce the possibility of sidelobes triggering the racon. This is achieved, for instance, by frequency-agile racons (see Section 3.5.3.2) storing the received signal strength at particular frequencies and then only responding when the signal strength at a particular frequency is close to the stored maximum value.

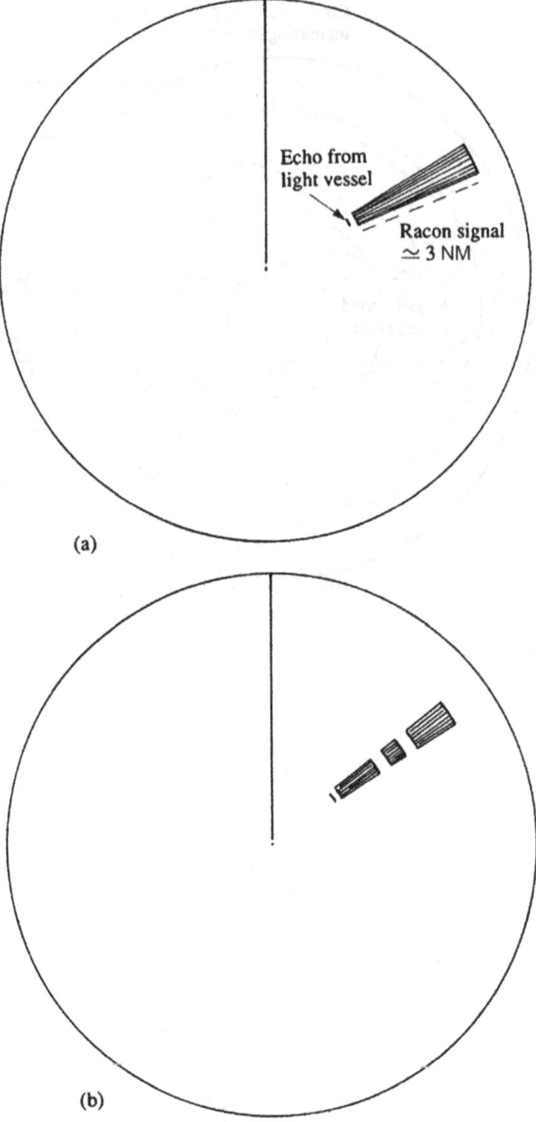

FIGURE 3.25 The racon appearance on the display. (a) Standard response. (b) Coded response.

3.5.2.3 Racons at Long Range

Since one of the reasons for fitting a racon is to improve the detection range of what might otherwise be a poor response target, at longer ranges the response from the actual target may

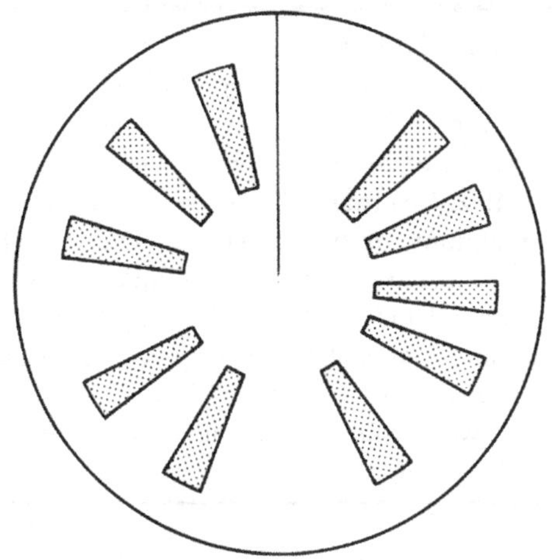

FIGURE 3.26 Racon clutter.

not be detected/displayed although the racon flash appears quite clearly. In this case, measuring the range to the nearest edge of the racon flash is acceptable as it will result in an error of only a hundred metres (the maximum racon receive/transmit delay time) and at long range such an error should not be significant. However, additional care must be exercised when navigating in close proximity to off-lying hazards while using a distant racon to fix position.

3.5.2.4 *The Danger in Using Racons to Assess Correct Tuning of the Radar Receiver*

It is advocated (see Section 6.2.7.2) that the radar receiver be accurately tuned by observing the change in display brightness of certain received signals while adjusting the tuning control. Racon signals *must not* be used in this way since:

a. The racon signal is frequently above saturation level and will not therefore show any marked change as the tuning control is varied.

b. The racon response will be displayed irrespective of the frequency to which the receiver is tuned (see Section 3.5.1), that is the racon will be displayed even when the receiver is off tune.

3.5.3 Frequency and Polarization

It is evident that if the racon signal is to appear on the ship's radar, then the beacon response must be transmitted at the transmission frequency of the ship's radar. During periods of restricted visibility in areas of high-density traffic, the demand on the racon can be extremely heavy, both in the number of interrogations and also the range of frequencies at which a response is required.

For example, at X-band the marine radar transmissions will fall within the band 9200–9500 MHz. The radar receiver bandwidth, that is the band of frequencies within which an individual receiver can detect signals, is in the approximate range 2–20 MHz (see Section 2.6.4.2) and so it can be seen that some form of tunable oscillator is required within the racon transmitter if its responses to all interrogations are to be detectable.

One concession in the demand for a response is that it is not normal for the racon to be 'observed' on every radar scan, i.e. every 3 s or so. This can also be beneficial in that there will be less likelihood of an observer failing to detect a real target which might be obscured by the racon flash.

Racons are used at both X- and S-band. However, IMO Performance Standards (see Section 11.2.1) only require that X-band radars produce a pulse of sufficient power to trigger a marine racon. This is to allow solid-state coherent radars of low peak power to be operated at S-band. It is likely that future S-band (and X-band) racons will be designed to be compatible with such radars.

X-band racon signals have to be horizontally polarized, but S-band racons should respond to both horizontal and vertically polarized radar signals.

3.5.3.1 *Slow-Sweep Racons*

Rather than attempt to respond to each ship's radar pulse every time it is received, the racon oscillator is swept comparatively slowly across the marine radar band. Racon sweep periods of some 60–120 s are typical (Figure 3.27).

By careful consideration of radar receiver bandwidths, scanner rotation rates and also the racon sweep time, some estimate can be made of duration and interval between racon appearances on an individual observer's radar display. This will in general be for sufficient time to allow the identification of the racon-fitted navigation mark (or to obtain a range and bearing of the racon itself). Typically, the racon flash will appear on some two to four consecutive scans of the ship's radar in each sweep period of the racon.

Although the frequency is shown as being continuously swept, an alternative technique which had similar results was to increase the frequency in discrete steps which were held for a short period of time, with the overall 'sweep' period being as before.

EXAMPLE 3.4

A racon with a sweep period of 100 s is triggered by a ship's radar which has an antenna rotation rate of 20 rpm and a receiver bandwidth of 10 MHz. Determine the theoretical maximum and minimum number of responses which can be expected during a single racon frequency sweep if the sweep band was 9300–9500 MHz. Find also the nominal interval between responses from consecutive racon sweeps (Figure 3.28).

Slow-sweep racons are no longer recommended by IMO.

3.5.3.2 *Frequency-Agile Racons*

In such a system, the frequency of the interrogating radar is measured, the racon transmitter is then tuned to that frequency and a (coded) response is transmitted. The time taken for this operation is some 0.4 μs, which corresponds to a ranging error of some 60 m.

The block diagram of a frequency-agile racon is shown in Figure 3.29.

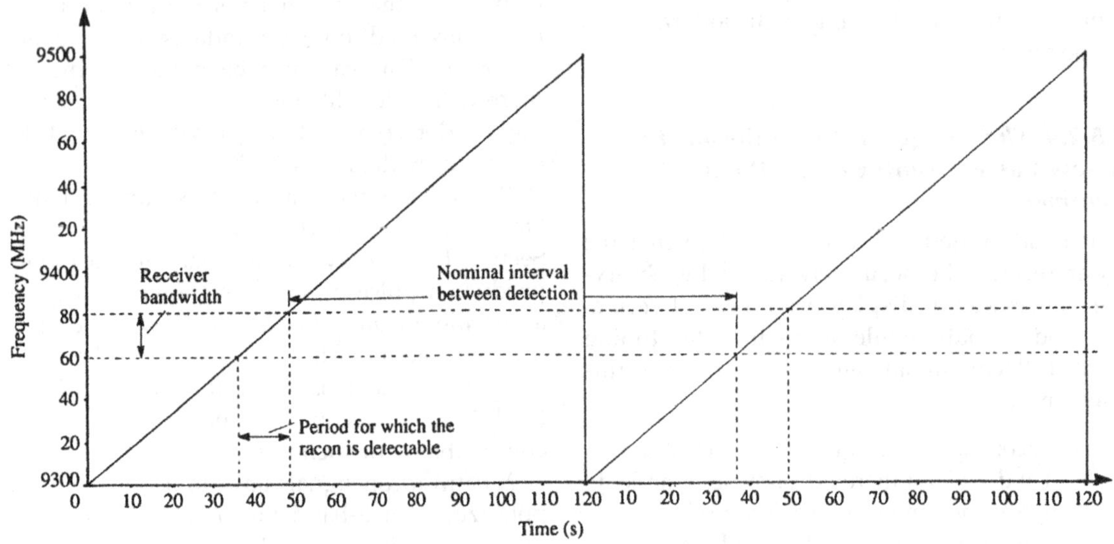

FIGURE 3.27 Slow time/frequency racon sweep.

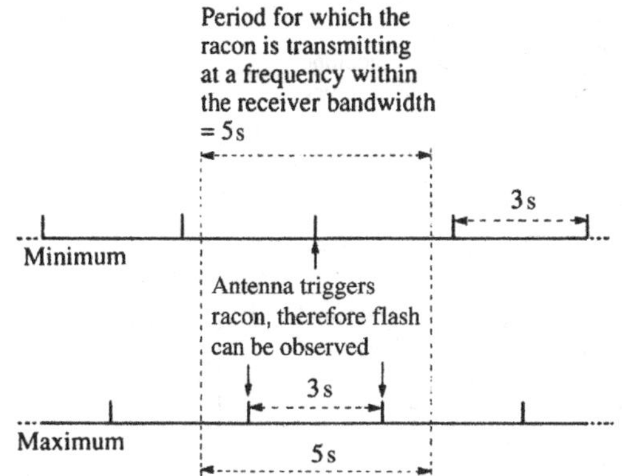

FIGURE 3.28 **Calculation for Example 3.4** 200 MHz in 100 s
therefore 10 MHz in 5 s
Time for 1 antenna rotation = 3 s
maximum 2 flashes
minimum 1 flash
Nominal interval between pulses = 100 s

The following points should be noted:

a. Racons which appear on each scan could mask other targets and so the racon is made to be 'silent' for a short period in each minute.
b. The application of fast time constant (FTC) (see Section 3.7.4) and some video processing techniques including 'interference suppression' (de-fruiting, see Section 3.9.5.1) can mutilate racon signals and should be switched off.
c. The claimed 'mean time between failures' (MTBF) is some 10 years.
d. With frequency-agile racons it is essential that the receiver is on tune, especially if the receiver bandwidth is narrow (i.e. on longer range scales).
e. The ON periods of the racon should be not less than 15 s and there should be an ON period at least once every 60 s.

The reduced ON period conserves battery power (important for offshore navigational aids) as well as reducing clutter. The US coast guard recommends that racons are on 50% of the time (20 s ON, 20 s OFF) for buoys and a maximum of 75% of the time for onshore installations where power supplies are not a problem.

3.5.3.3 *Racon 'Clutter'*

It has always been feared that with the proliferation of radar beacons, racon clutter could become a problem and mask targets. Alternatively, beacon responses can be masked by strong land and rain/sea clutter echoes.

Three solutions have evolved:

1. *Beacon band operation.* This requires all radar beacons to transmit at a particular fixed frequency within a band at the edge of the marine radar frequency band, for example 9300–9320 MHz. A switch on the radar console sets the radar receiver to this 'fixed' frequency and only the radar beacons, but no targets, are displayed. The extra circuitry for all potential users could be expensive.
2. *Interrogated time-offset frequency-agile racons* (ITOFAR). The ITOFAR system requires specially modified beacons and radar sets to operate. When this frequency-agile racon is triggered at a very precise pulse repetition frequency (PRF) (1343.1 pulses per second), it recognizes it as an ITOFAR interrogation and delays its response by a very precise amount (374 μs). If the trigger to the display is delayed by the same amount, the racon signal will be displayed in what was previously the 'clear' outer portion of the radar screen on long range but is now at the correct range on the displayed range scale on a modified set (Figure 3.30). The circuitry required for this solution is less costly than for beacon band operation. On unmodified sets the ITOFAR appear as normal frequency beacons.
3. *Sidelobe suppression.* Circuitry is included in the racon to recognize the sidelobes of a radar. Basically the main beam is assumed to be the most powerful signal received and the racon will ignore all lesser signals.

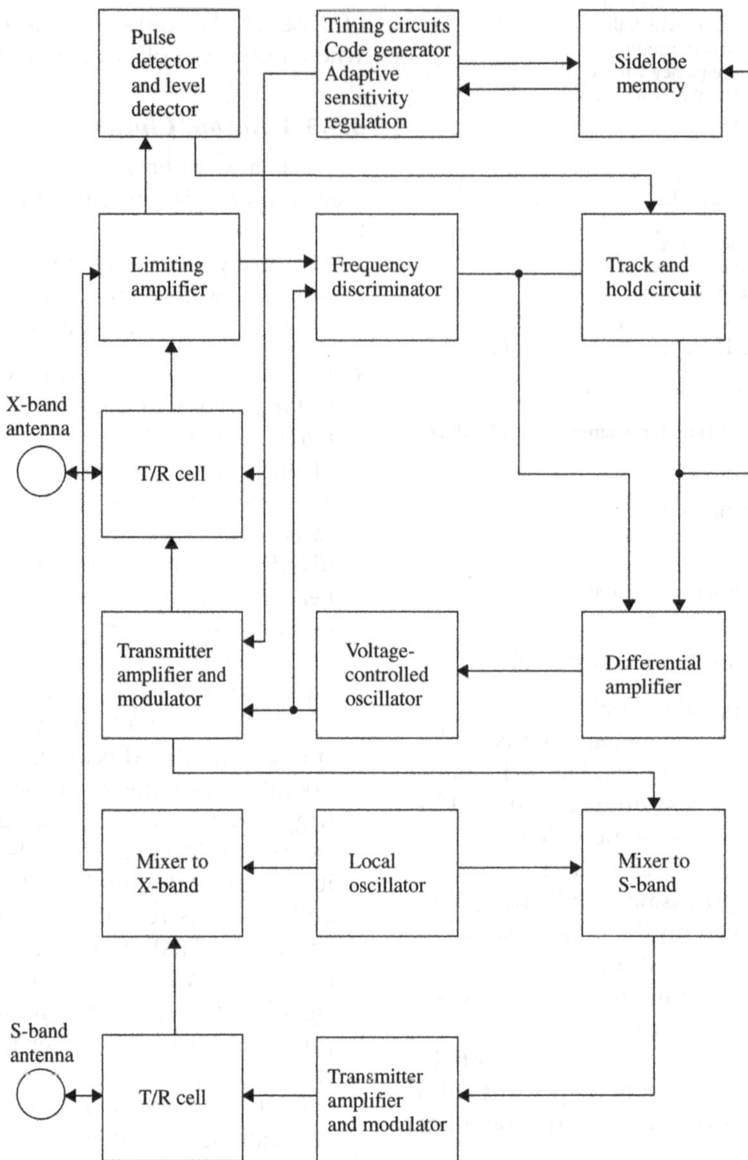

FIGURE 3.29 Block diagram of a frequency-agile racon (based on a diagram of AGA-Ericsson Radio Systems of Sweden).

3.5.3.4 *The Application of Racons*

IMO Performance Standards (see Section 11.2.4) make it very clear that racons are to be used for improving the detection of fixed objects and not moving ones. However, the interpretation of this standard is up to individual administrations. For instance, some administrations use racons to mark turning points (way points) and leading lines, but the U.S. Coastguard does not.

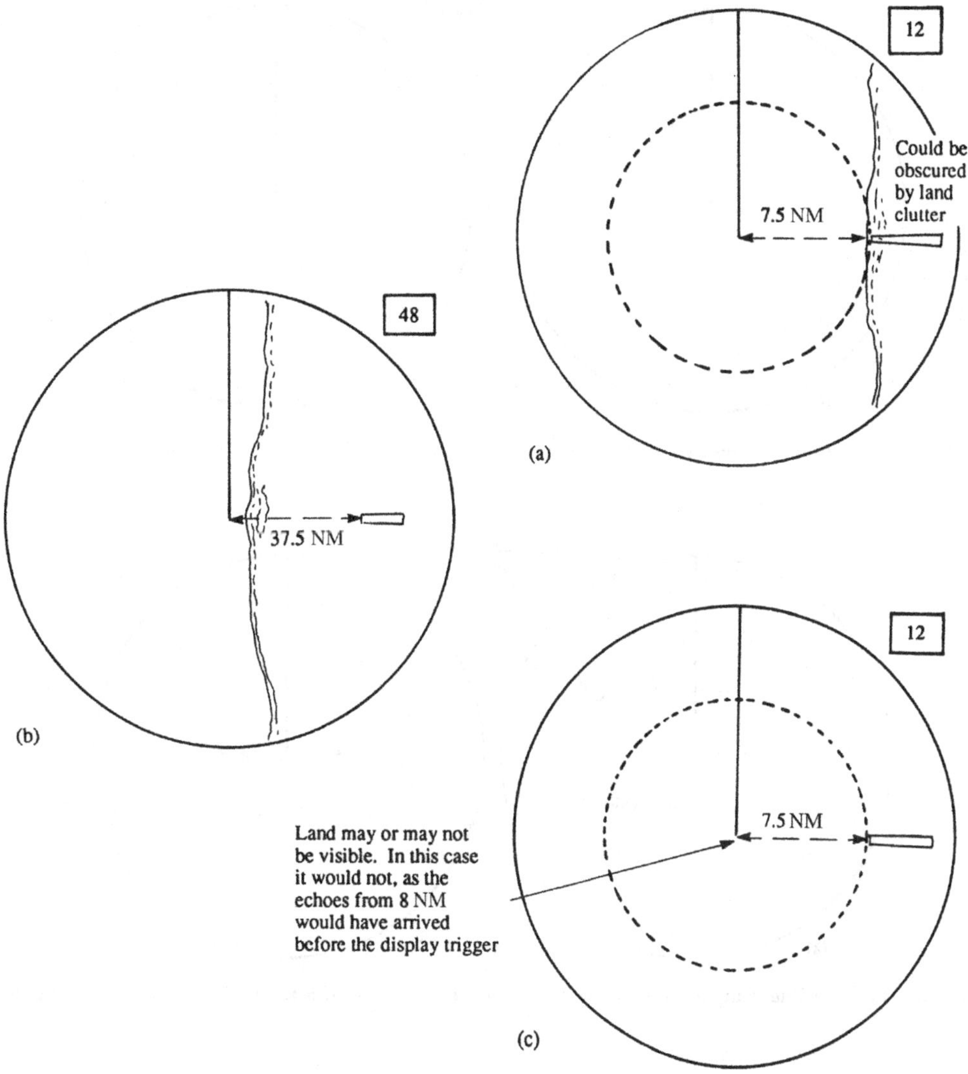

FIGURE 3.30 ITOFAR response. (a) ITOFAR beacon response on all radars. (b) Appearance of the ITOFAR beacon response when display trigger has not been delayed but beacon response has. (c) Display of ITOFAR beacon response if ITOFAR operation is selected at the radar display.

3.5.3.5 Leading Lines Using Racons

Figure 3.31 shows a leading line using two racons. Figure 3.32 shows a leading line using one racon and a buoy with a radar reflector.

3.5.4 Sources of Radar Beacon Information

The *Admiralty List of Radio Signals* volume 2 contains information relating to racons working in both the S- and X-band. Figure 3.33(a)

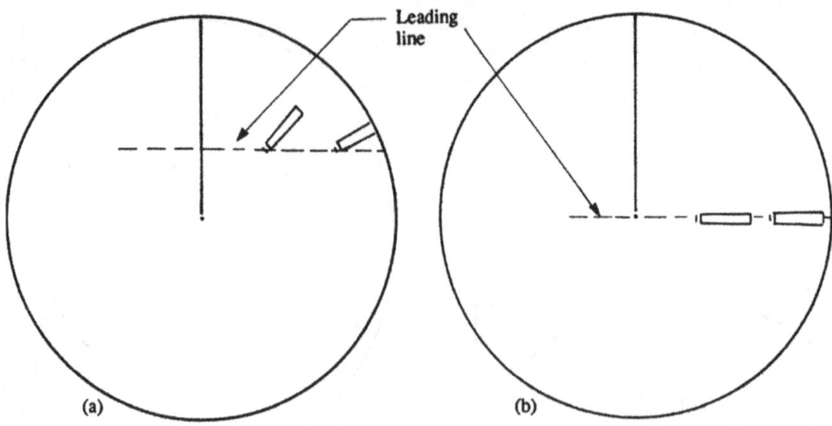

FIGURE 3.31 **Leading line using two racons.** (a) Approaching the line. (b) On the leading line.

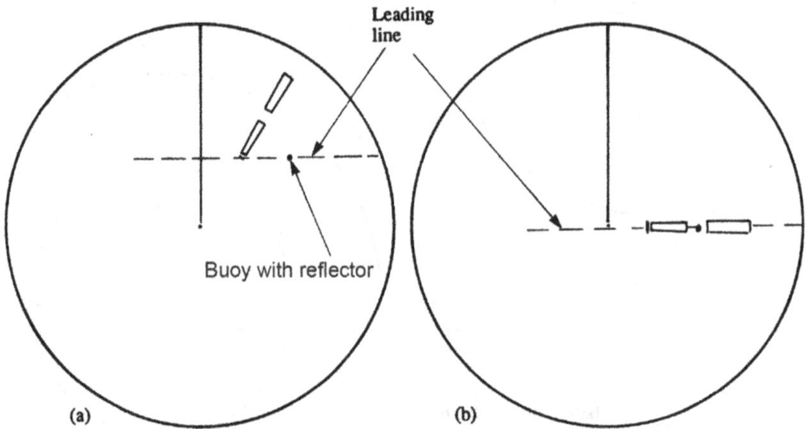

FIGURE 3.32 **Leading line using a coded racon and buoy fitted with a reflector.** (a) Approaching the line. (b) On the leading line.

gives typical extracts. Figure 3.33(b) shows the international chart symbols used for racons.

3.5.5 The Radaflare

This is a rocket which is fired from a pistol. At some 300–400 m altitude, the rocket ejects a quantity of dipoles which respond strongly to 3 cm radar waves and at the same time the rocket gives out a very bright white light. The radar response has a maximum detection range of 12 NM and will last for some 15 min, depending upon weather conditions. It is intended for use as a distress signal for small craft.

3.5.6 Search and Rescue Transponders

The detection of survival craft is a particularly important issue. Traditionally this function

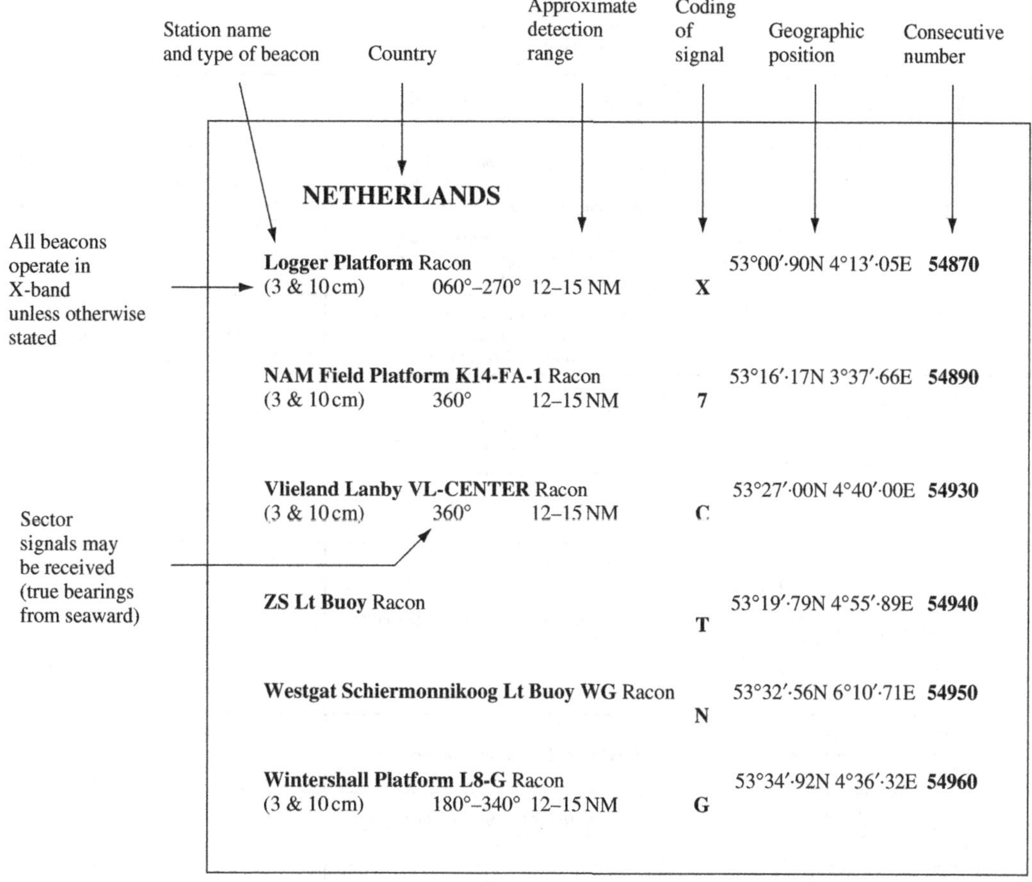

NETHERLANDS

Logger Platform Racon 53°00′·90N 4°13′·05E **54870**
(3 & 10 cm) 060°–270° 12–15 NM X

NAM Field Platform K14-FA-1 Racon 53°16′·17N 3°37′·66E **54890**
(3 & 10 cm) 360° 12–15 NM 7

Vlieland Lanby VL-CENTER Racon 53°27′·00N 4°40′·00E **54930**
(3 & 10 cm) 360° 12–15 NM C

ZS Lt Buoy Racon 53°19′·79N 4°55′·89E **54940**
T

Westgat Schiermonnikoog Lt Buoy WG Racon 53°32′·56N 6°10′·71E **54950**
N

Wintershall Platform L8-G Racon 53°34′·92N 4°36′·32E **54960**
(3 & 10 cm) 180°–340° 12–15 NM G

FIGURE 3.33 (a) Examples of radar beacon information in Admiralty List of Radio Signals Vol. 2, (b) Extract from Admiralty chart symbols.

was carried out on radio frequencies by the RDF (Radio Direction Finder), but this responsibility has now been transferred to the radar display and the RDF is obsolete and no longer a current IMO requirement. Instead, all vessels that come under the Global Maritime Distress and Safety Service (GMDSS) regulations are required to carry a search and rescue transponder (SART) to use in survival craft, so that there is an active indication of its presence on a radar screen. The SART is normally mounted on a short pole on the survival craft. It is a transponder and only transmits when interrogated by a radar signal.

The SART works at X-band frequencies only. It is activated by a radar pulse from a vessel and transmits a series of 12 short radar pulses on the same frequency. The SART appears on the screen as a series of 12 dots (Figure 3.34(a)). Even if the target (survival craft) is not radar conspicuous, the target will be slightly closer than the first dot. However, unlike beacons, sidelobe suppression is not included in a SART, so the dots become arcs

1	Ra	*Coast radar station providing range and bearing from station on request*		485.1 *M11*
2	Ramark	*Ramark, radar beacon transmitting continuously*		486.1 *M14a*
3.1	Racon(Z) (3 cm)	*Radar transponder beacon, with morse identification, responding within the 3 cm (X) band*	† Racon(Z)	486.2 486.3 *M12*
3.2	Racon(Z) (10 cm)	*Radar transponder beacon, with morse identification, responding within the 10 cm (S) band*		486.3
3.3	Racon(Z)	*Radar transponder beacon, with morse identification, responding within the 3 cm (X) and the 10 cm (S) bands*	† Racon(Z) (3 & 10 cm)	
3.4	Racon Obscd Racon(P)	*Radar transponder beacon with sector of obscured reception*		486.4
	Racon(Z) Racon(Z)	*Radar transponder beacon with sector of reception*		
3.5	Racon Racon Racons≠270°	*Leading radar transponder beacons*		486.5
	Racon Racon Lts≠270° Racons≠270°	*Leading radar transponder beacons coincident with leading lights*		
3.6	*Racon* *Racon*	*Floating marks with radar transponder beacons*		486.2 *M12*
4		*Radar reflector (not usually charted on IALA System buoys)*	† *Ra.Refl.*	460.3 465 *M13*
5		*Radar-conspicuous feature*	† Ra conspic	485.2 *M14*

FIGURE 3.33 (Continued).

(see Figure 3.34(b)), as the target gets closer. At very close range, the arcs can extend to be complete circles (see Figure 3.34(c)).

The SART has a battery life on standby (listening for radar signals) of at least 96 h and an operational life transmitting of at least 8 h. The case should be brightly coloured (yellow or orange) and should be capable of floating, unless it is built into a craft. Its construction is very rugged (see Section 11.2.5). It should also

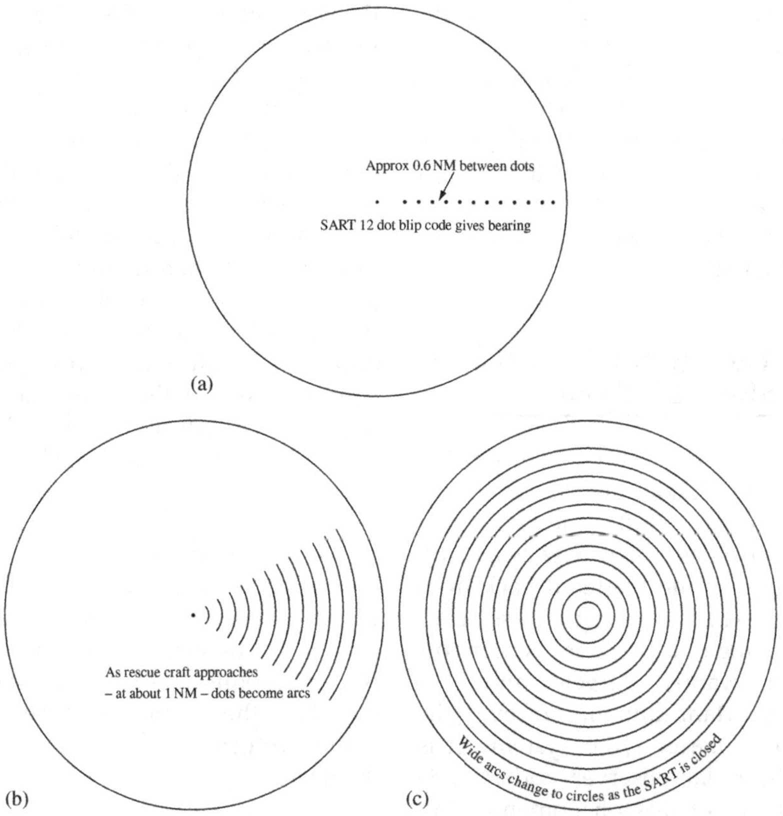
Approx 0.6 NM between dots
SART 12 dot blip code gives bearing
(a)
As rescue craft approaches
– at about 1 NM – dots become arcs
(b)
(c)
Wide arcs change to circles as the SART is closed

FIGURE 3.34 The appearance of a SART on a rescuing ship display.

survive immersion to a depth of 10 m for 5 min as well as being dropped into the water from a height of 20 m.

The SART indicates by audible or visual means (or both) that radar has triggered it into operation, which should be a valuable stimulus for those about to be rescued. The operation of the SART is simple and it should be operable by unskilled personnel. There is also an audible or visual means of ascertaining that it is operating correctly.

Testing of SARTs at sea is allowed, but the transmission should be limited to a few seconds to avoid attracting the unnecessary attention of other radar users.

An AIS version of SART, which works on different principles, is covered in Section 5.3.4.

3.5.7 Active Radar Reflectors

Active radar reflectors are increasingly fitted to small vessels to replace passive radar reflectors. They have the advantage of being small in size and of low weight, and so are easily positioned towards the highest point of the vessel, including the mast-top of a yacht. However, they need to be powered by a suitable source of electric current. They consist of a small (azimuthally) omni-directional antenna attached to an electronics module. They are basically a very simple concept – received radar signals are detected, amplified and immediately retransmitted. The effective delay in response is generally only marginally greater than the transmitted pulse length and so their positional accuracy on the radar display is approximately consistent with the displayed range of target.

3.6 THE DETECTION OF TARGETS IN SEA CLUTTER

In general, the term 'clutter' is used to describe the accumulation on the screen of unwanted echoes. Sea clutter describes the particular case of echoes which arise as a result of the radar energy being scattered back from the surface of the sea. The presence of sea clutter echoes may make it difficult or even impossible to detect wanted targets and this represents an important limitation of even the most modern civil marine radar systems. It is essential that the radar observer understands not only the nature of this phenomenon, but also the principle of the various arrangements available to assist in combating the effect. Since the mid-1970s, with the advent of digital processing and storage (see Section 2.7), the complexity of anti-clutter arrangements has increased progressively and the limitations of these must be appreciated.

The ever-changing surface of the sea represents an immensely complex population of varying radar targets. The radar cross-section (see Section 3.3.5.4) of any element of the sea surface fluctuates in a largely random fashion and can thus only be evaluated in terms of averages or probabilities. In an effort to identify the radar and sea surface characteristics which determine the magnitude of the sea clutter response, eminent radar engineers have spent much time and effort in using statistical techniques to obtain a mathematical model of the sea clutter mechanism. Because of the complexity of the problem, and in particular the difficulty of carrying out controlled experiments with the sea, it is not possible to reconcile completely the detailed mathematical predictions with the observed phenomena, and to this extent it is true to say that the mechanism is not yet fully understood. However, a discussion of such quantitative considerations is beyond the scope of this manual and the reader who wishes to pursue the mathematical approach is referred to any good radar engineering text. To the extent of understanding required for effective radar observation, the mechanism can be explained in the fairly simple qualitative terms used in the following section.

From the point of view of the radar observer, the essence of the clutter problem is that targets weaker than the clutter returns cannot be detected and even targets which return responses which are very much stronger than the clutter may be masked if the clutter returns saturate the receiver or the display.

3.6.1 The Nature of Sea Clutter Response

In absolutely flat calm sea conditions, the sea surface will behave like a mirror and specular reflection will take place (see Section 3.3.1). All radar energy which strikes the surface of the sea will be reflected away and none will be scattered back towards the antenna (some of the energy which strikes targets will travel via the sea surface, as described in Section 2.7.4.1).

If there is any wind at all it will tend to ruffle the surface of the sea producing small wavelets and as the wind strength increases so too will the height of the waves which are generated. In deep water the height and period of the waves depends on the wind speed, the fetch and the length of time for which the wind has been blowing. In shallow water the waves tend to steepen and the distance between successive wave crests decreases. When the wind is blowing against the tide, wave height and steepness will be greater than the observed wind would perhaps suggest. Where two wave patterns are present, as for example may occur in an area where a large wind shift has taken place with the passage of a frontal system, constructive interference between the patterns will produce, from time to time, wave heights larger than the individual component heights. As wave height increases for whatever reason, a point will be reached at which the waves break, producing confused tumbling water and windblown spray. A more complex and less predictable target is difficult to conceive.

Despite the complexity of the problem, one can consider there to be three elements which contribute to the aggregate response:

1. *Specular reflection from those sloping parts of the wavefront which present a good aspect to the incident radar energy.* This would be the sole contribution if the wind could produce a completely smooth sloping wavefront. It is well represented by the situation which arises when swell is present in calm conditions. (*Swell* is the term used to describe waves which have been generated by wind in some distant area and have travelled across the ocean.) Swell may be present in isolation or it may have locally generated *sea* waves superimposed upon it.
2. *Scattering of energy from the surface of the wavefront.* Except when swell is present in flat calm conditions, the sloping surface of the wavefront will be disturbed by the wind and will thus produce a surface texture which is *rough* in terms of the wavelengths used in civil marine radar (see Section 3.3.5.5). The face of the wave can be thought of as being composed of a large number of facets whose aspects change at random with the ruffling and tumbling of the surface water. The resulting backscatter of radar energy will yield responses whenever a favourably placed facet lies within the resolution cell (see Section 3.3.5). It is essential to appreciate that although these facets may be quite small, possibly a few square centimetres, they will produce echoes whose displayed size is determined by the resolution cell. Thus if a facet remains favourable for the time taken by the radar beam to sweep across it (say 10 to 20 ms), it will produce an echo which is one-half pulse long, one beamwidth wide and subject to spot size effect.
3. *Reflection from water droplets close to the surface of the wave.* In higher wind conditions water from the crests tends to be blown off the wave and the droplets so formed will reflect radar energy if sufficiently large (see also raindrops in Section 3.7.3.1).

The relative significance of each contribution depends on the actual conditions. Of particular note is that a large number of independent targets capable of producing backscatter exist on and around the sloping surfaces of sea waves. The observed response from this backscatter depends on characteristics of both the waves and the radar installation, and these will be considered in turn.

3.6.1.1 The Effect of Wave Height

Although, as indicated in the previous section, various factors can affect wave height, in general as the wind force rises so too does the wave height. In attempting to translate the growth in wave height into an increase in clutter response, one must consider not only the increase in the strength of signals received from nearby waves, but also the increases in the range from which detectable signals are received.

The response from an individual wave increases with wave height up to a certain level and then tends to flatten off. The precise nature of the law is a matter for the scientist rather than the radar observer who is more concerned with the effects which can be observed on the display. As wave height increases, the clutter returns close to the observing vessel will in due course saturate the receiver or display; a further increase in wave height will be observed to extend the radius to which this saturation is manifest. Beyond the radius of saturation the amplitude of the displayed clutter signals will decay progressively with range (see Section 3.6.1.2) towards the maximum radial extent of visible clutter, at which the grazing angle has reduced to that at which the radar sees the surface as smooth (see Section 3.3.5.5). In conclusion, the range within which the observer must combat the masking effects of saturation will increase with wave height, as will the range within which targets are undetectable if their response is less than that of the ambient clutter.

3.6.1.2 *The Variation of Sea Clutter Response with Range*

In the introduction to this chapter it was shown that, in general, received echo strength varies inversely as the fourth power of range. However, it was emphasized in Section 2.7.4.1 that the response from sea clutter does not obey this law and that it varies inversely as the third power (cube) of range, simply because the area of sea illuminated is a function increasing with range.

In the absence of saturation, the variation of sea clutter with range will be as illustrated in Figure 3.35(a). If the gain setting of the radar has been set so that saturation can occur, the situations as illustrated in Figure 3.35(b) result. Comparison of sections of the figure shows the masking effect of the presence of sea clutter. The difficulties which arise from the masking and the benefit of employing a logarithmic receiver (see Section 2.6.4.5) will be discussed further in Section 3.6.3.2.

3.6.1.3 *The Effect of Wind Direction*

Even casual observation of the waves generated on the sea surface will reveal that the upwind wavefronts are steeper than their more gently sloping downwind counterparts. The steeper wavefronts present a more effective echoing surface and thus the radial extent of sea clutter echoes will be found to be greater to windward than it is to leeward, as illustrated by Figure 3.36.

The outer limit of the pattern represents the minimum detectable signal (see Figure. 3.13) and hence the radial extent of the pattern offers a suitable criterion by which to judge the setting of the tuning control (see Section 6.2.7.1).

In coastal and estuarial areas the pattern may be modified by seas breaking on banks or in areas of shallow water and the waves may be steepened where the wind blows against the tide.

3.6.1.4 *The Effect of Antenna Height*

Increased antenna height extends the radial limit of the overall sea clutter pattern which appears on the screen and also the range within which the clutter echoes saturate the receiver or the display. This is illustrated by Figure 3.37.

Sloping wavefronts close to the observing vessel present a more favourable aspect for the reflection of the radar energy than those at a distance. This arises because at large angles of depression, the beam *looks down* at nearby wavefronts and there is, as a consequence, a high probability that some energy will be normal to the general surface of the wavefront. If the surface is smooth, specular reflection will take place but in most cases the surface will be

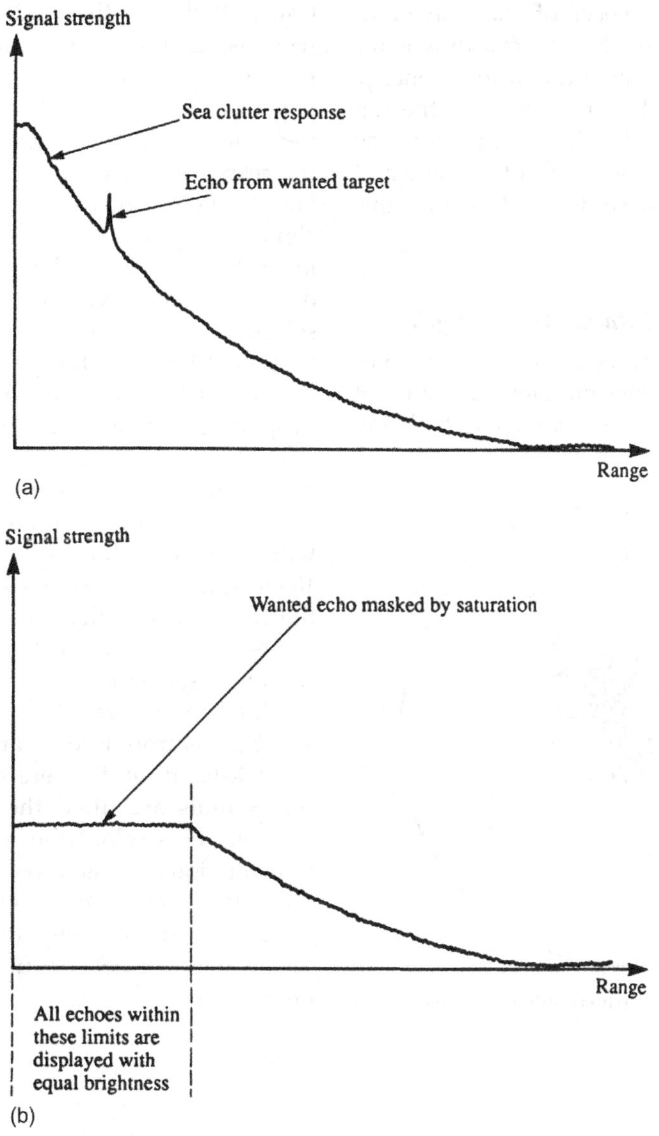

FIGURE 3.35 The variation of clutter response with range. (a) Unsaturated response. (b) Saturated response.

sufficiently rough to ensure scattering and thus a response from surfaces which are not exactly normal to the beam (see Section 3.3.1). It should not be assumed, with the ship upright, that the maximum depression angle of interest is only about 10°, i.e. half the vertical beamwidth (see Section 2.5.2.5). This is because, in practice, there can be sufficient antenna sensitivity well beyond this point, including from the vertical sidelobes of the antenna, that will considerably increase the effective angle. As the range from the observing vessel increases, the aspect of the wavefront will become progressively less favourable for reflection as the direction of the incident energy approaches that which is tangential to the top of the wave. At about this distance, most of the energy will just graze the crest of the wave and clutter echoes will not be detected beyond this limit.

3.6.1.5 *The Effect of Radar Wavelength*

Like most aspects of the sea clutter phenomenon, experimental determination of the exact relationship between the wavelength being transmitted by the radar and the clutter response is difficult to achieve, as is an explanation of the results in terms of the necessarily complex theoretical mathematical model of the mechanism. There is general agreement that the response decreases with increase in wavelength, though there is some debate as to the significant power – $(\text{wavelength})^2$ or $(\text{wavelength})^3$. At best, the civil marine radar observer will have the choice between the selection of 3 or 10 cm wavelength (see Section 2.2.2) and for practical purposes can consider the clutter response to be inversely proportional to wavelength. This is consistent with the concept of surface texture as set out in Section 3.3.2 which suggests that the *roughness* of the surface of the sloping wavefront will be greater for 3 cm than it will be for 10 cm radiation. It is similarly consistent with the theory that reflection from water droplets forms one component of the aggregate response (see Section 3.7.1).

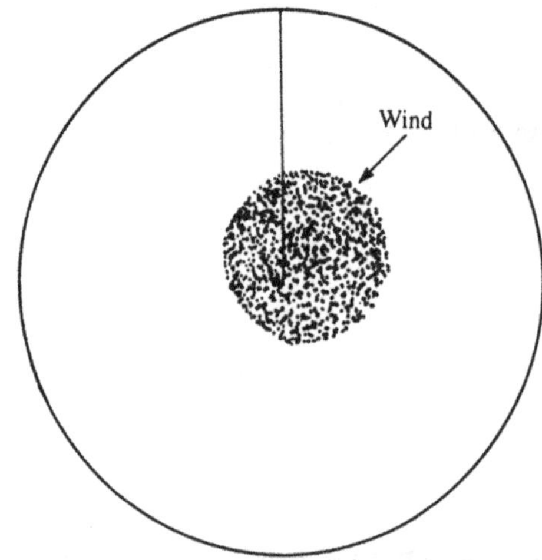

FIGURE 3.36 Sea clutter pattern in the open sea.

Thus, in general, the sea clutter response will be less troublesome at S-band than it will be at X-band, and where dual antenna/transceiver units are fitted the observer may well be able to exploit the potential of S-band transmissions in adverse sea clutter conditions. It is also a factor which should be considered when planning the radar installation for a new vessel or updating an existing installation.

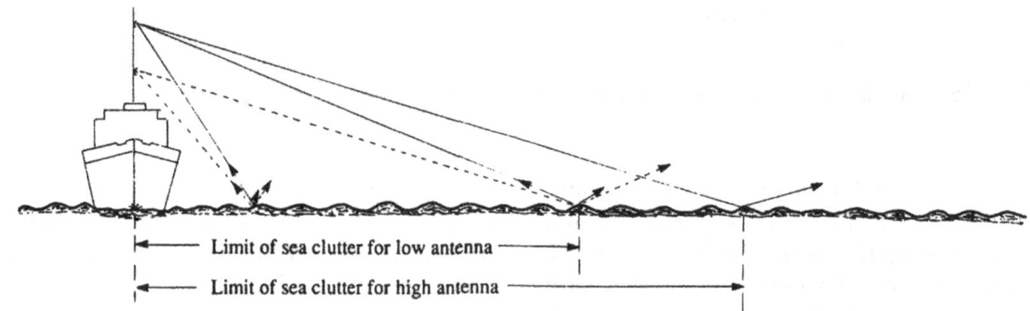

FIGURE 3.37 The effect of antenna height on sea clutter response.

3.6.1.6 The Effect of Pulse Length

In general, subject to any saturation, selection of long pulse rather than short pulse will improve the displayed response of any given target (see Section 2.3.3.1), and sea clutter is no exception to this rule. Thus if wanted targets are masked as a result of saturation, it may be possible to combat the effect by judicious choice of pulse length. Selection of short pulse will reduce the response of both clutter and wanted targets and may bring the clutter response below saturation level, or at least to a level which will allow the observer to apply smaller amounts of clutter suppression (see Section 3.6.3).

Selection of short pulse will also reduce the radial extent of the resolution cell (see Section 2.8.5). This effect can only begin to show if the distance between the farthest scattering element of a wavefront and the nearest scattering element (including any associated spray) of the next wave beyond exceeds half the transmitted pulse length. In certain sea conditions, it is possible to recognize the line of approaching wavefronts, especially in ocean conditions where a long swell is present.

3.6.1.7 The Effect of Polarization (See Section 2.5.1.7)

In clutter conditions there is little if any difference between the response from horizontally polarized radiation and that which is vertically polarized. In almost all cases any difference is in any event academic, since to comply with the traditional IMO Performance Standards relating to the ability to detect radar beacons (see Section 3.5) all systems operating in the 3 cm band must be capable of operating in a horizontally polarized mode. There was thus no stimulus for designers to produce an X-band system with vertical polarization. Although vertical polarization offers some improvement in target response in calm conditions, particularly at S-band, vertically polarized S-band systems are very much the exception rather than the rule. Thus, in general, polarization is not a system characteristic which is under the control of the observer.

In the past, a system offering circular polarization was offered in an attempt to reduce the response from droplets of precipitation. This is discussed under that heading in Section 3.7.4.6; at this stage it is relevant to observe that any reduction so gained would have an effect on that element of sea clutter response originating in reflection from spray. It must, however, be borne in mind that there is some debate as to the significance of the contribution of the droplet component to the aggregate sea clutter response.

3.6.2 The Clutter Problem Summarized

Sea clutter echoes may make it impossible to detect some targets, while the presence of others may only be revealed by skilful adjustment of the controls or with the assistance of some form of signal processing. Before discussing the techniques available for the suppression of sea clutter, it is appropriate to summarize the clutter problem in terms of three distinct cases of target amplitude which are set out below and illustrated by Figure 3.38:

1. *Targets weaker than the ambient clutter.* A target which returns a signal that is weaker than the clutter surrounding it cannot be detected, because any technique used to suppress the clutter will also suppress the echo of the wanted target. Marginal targets may be recognized by the regularity of their paint.
2. *Targets stronger than ambient clutter that does not saturate the receiver or display.* Such a target is capable of detection, but it may not be obvious because of the poor contrast

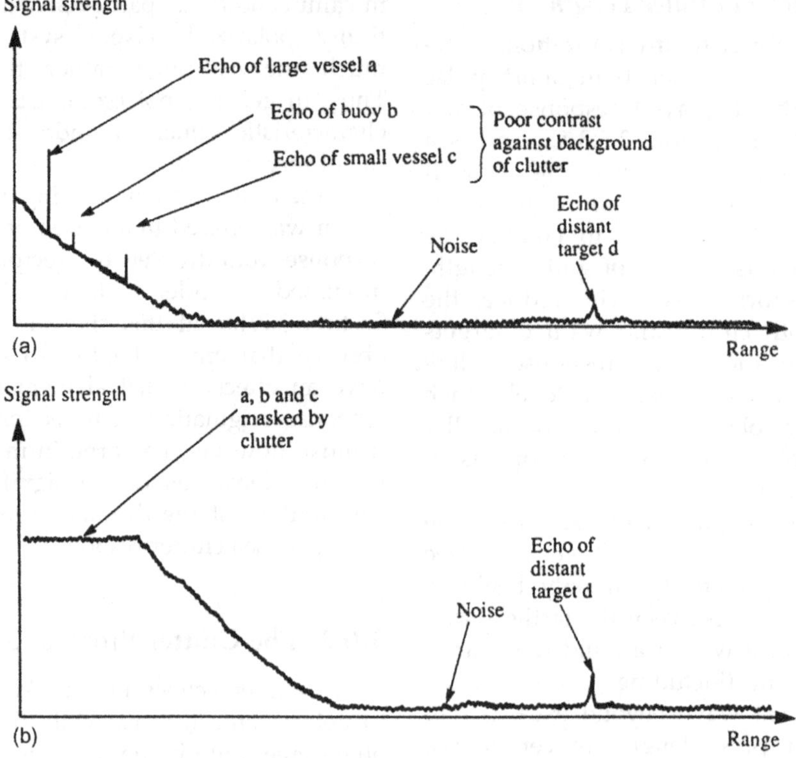

FIGURE 3.38 The masking effect of sea clutter returns. (a) Clutter does not saturate receiver or display. (b) Clutter saturates receiver or display.

when viewed against the background of the clutter echoes. This is likely to be a particular problem if the targets are only marginally stronger than the ambient clutter. The target may be made more obvious by suppression of the clutter returns.

3. *Targets stronger than ambient clutter that saturates the receiver or display.* Such a target will be masked because its echo will have exactly the same brightness on the display as all the sea clutter echoes surrounding it. It is capable of detection subject to suppression of the clutter signal. Skill is required to reveal targets which are only marginally stronger than the ambient clutter.

3.6.3 The Suppression of Displayed Sea Clutter Signals

In Sections 3.6.1.4–3.6.1.6 it was indicated that a judicious choice of antenna height, wavelength and pulse length would reduce sea clutter responses. Except in very light sea conditions it is unlikely that such reduction alone will be sufficient to combat the sea clutter problem, and it will normally have to be supplemented by some form of active suppression applied in the receiver or the display.

3.6.3.1 The Sea Clutter Control

The probability of detecting targets in clutter will be maximized by adjusting the controls so that the echoes of the targets paint against a background of light residual clutter speckle similar to the noise speckle used as the criterion for setting the gain control. This can be achieved for a target at any given range by reducing the setting of the gain control. However, it can be seen from Figure 3.39 that this may only be achieved at the expense of losing more *distant targets,* and without achieving the desired degree of suppression for nearer targets.

Essentially, the problem is that the gain reduces the amplification uniformly across the radius of the timebase, whereas the clutter signals are strongest near the origin and decay with range (see Section 3.6.1.2). The sea clutter control is a specialized type of gain control which reduces the gain at the origin of each radial line and restores it with time as each radial line is written. The control thus allows a varying amount of suppression to be applied. The amount of initial suppression is variable and can be adjusted by the observer by the setting of the front panel control, whereas the rate at which the suppression decays is preset by the manufacturer (Figure 3.40). The joint effect determines the radius to which suppression is effected.

The control is sometimes referred to as *swept gain* or *sensitivity time control* (STC) because of the time-related variation of the gain. Ideally the decay of the suppression curve should obey the same law as that governing the decay of the sea clutter response with range (see Section 3.6.1.2). As discussed in Section 2.7.4.1, the precise characteristics of the STC law on any radar are never made publicly available and are nowadays normally reactive to the actual conditions being experienced – not only when switched to fully automatic sea clutter settings. In earlier days the suppression decayed exponentially with time for the very practical reason that it is the waveform generated by discharging a capacitor through a resistor and hence had the merit of simplicity, reliability and cheapness.

Ideally, the observer wishes to optimize the suppression to match the clutter response, thus leaving wanted targets painting against a light residual clutter speckle. This is illustrated in Figure 3.40 which shows three (excluding the zero setting) particular levels of suppression which have been chosen in such a way that while one level is optimum, the other two levels are representative of the cases of insufficient and excessive suppression *for that particular line* of responses on the display. However, the clutter response varies from line to line (see Section 3.6.1.3) and thus it follows that if the control is set correctly for one radial line, it is likely to be set incorrectly for most, if not all, other radial lines. This is why a modern radar may attempt to provide some subtle aid to such settings, even when set to manual (see Section 3.6.3.3 below, which covers *adaptive gain*). However, there is no single *correct* manual setting for the sea clutter control; to maximize the probability of detecting targets in clutter, the observer must, at regular intervals, systematically adjust the control in small steps through the extent of its travel from full clutter suppression to zero effect. This operation is known as *searching* and considerable skill and practice are required to perform it successfully. It is one of the most important skills associated with marine radar operation; the appropriate practical procedure is described in detail in Section 6.7.1.

3.6.3.2 The Effect of Receiver Characteristic

In a modern system the use of a logarithmic amplifier (see Section 2.6.4.5) will ensure that in almost all cases an unsaturated output is available for input to the video section of the receiver. Thus saturation due to sea clutter returns will not normally take effect until later stages of processing (see Section 2.7.2). Logarithmic amplification and further processing in general make the masking effect of sea responses less troublesome than in the earlier

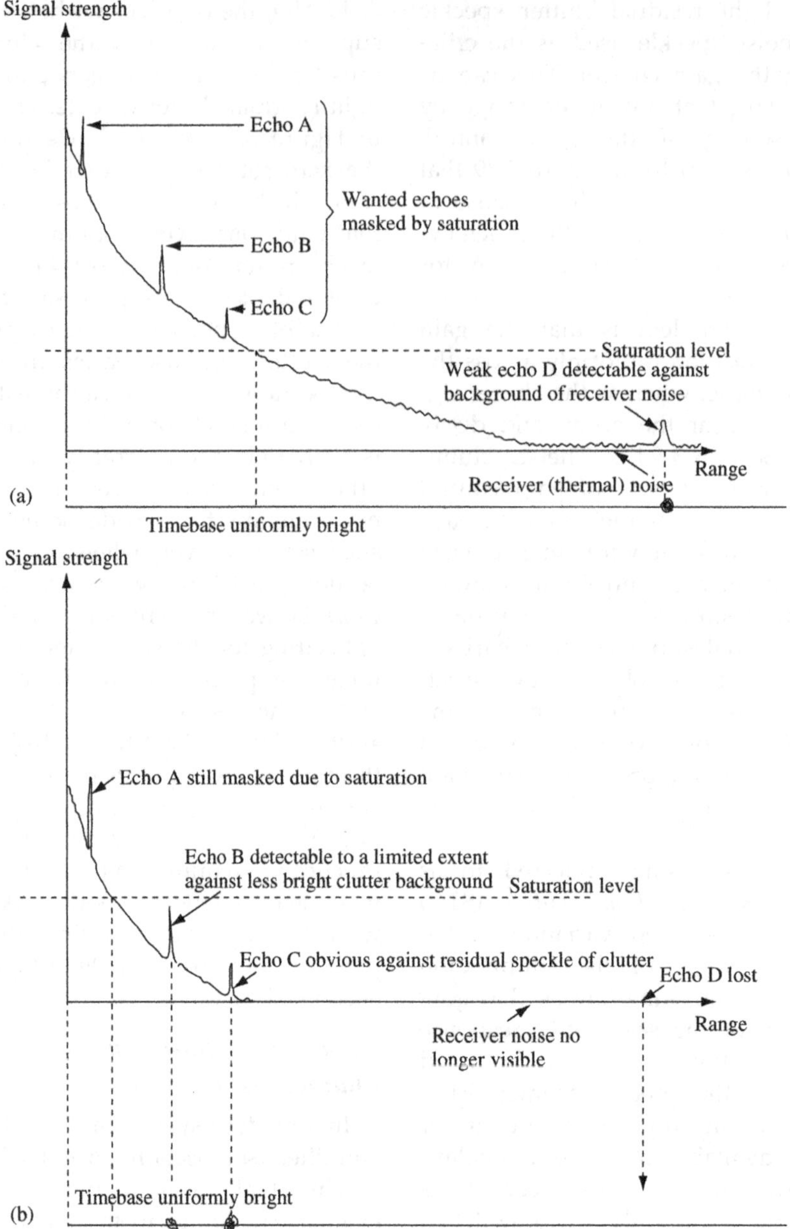

FIGURE 3.39 Suppression of clutter by use of the gain control. (a) Normal gain setting. (b) Gain suppressed.

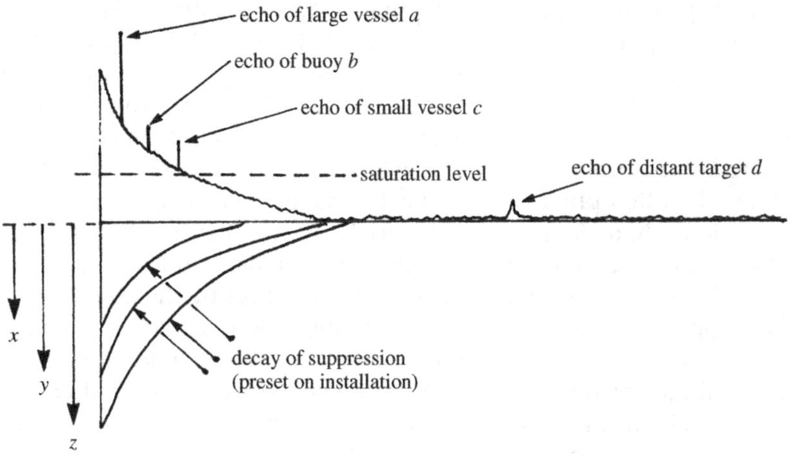
echo of large vessel a
echo of buoy b
echo of small vessel c
saturation level
echo of distant target d
decay of suppression
(preset on installation)
x
y
z

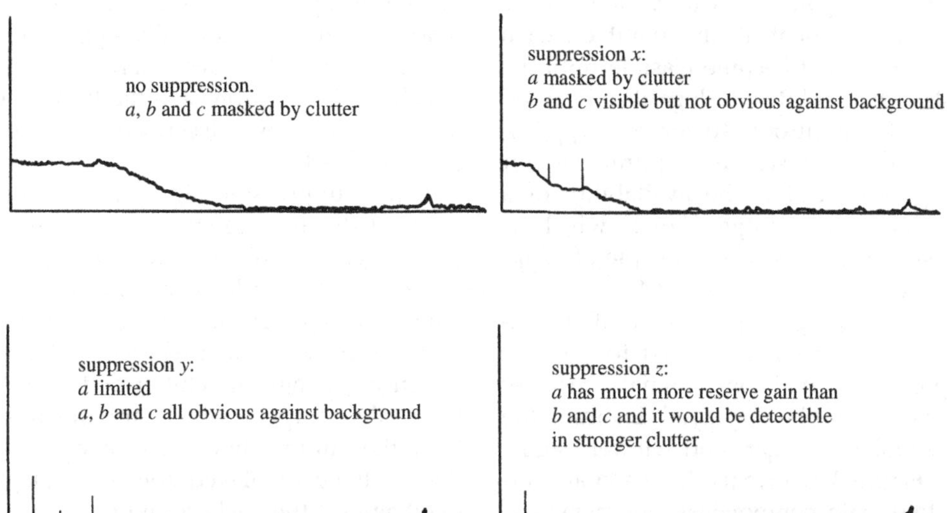
x, y, z represent varying degrees of initial suppression set by sea clutter control
no suppression.
a, b and c masked by clutter
suppression x:
a masked by clutter
b and c visible but not obvious against background
suppression y:
a limited
a, b and c all obvious against background
suppression z:
a has much more reserve gain than
b and c and it would be detectable
in stronger clutter

FIGURE 3.40 Varying degrees of sea clutter suppression (echo of *a* visible throughout).

days of radar, when only linear receivers and analogue processing were in use.

From the treatment presented in Section 2.6.4.5 it is fairly evident that the use of a logarithmic amplifier reduces the risk of targets being masked by sea clutter because it extends the range of signals that will not saturate the intermediate frequency (IF) amplifier. It can also be shown that the logarithmic relationship between receiver output and input also produces a swept gain effect on the clutter signals, thus further improving the potential performance of the system in heavy clutter conditions. Proof of this requires a mathematical treatment beyond the scope of this text. The reader who wishes to pursue this further is referred to a good radar engineering text or to the various technical papers published on the subject of logarithmic receivers.

3.6.3.3 Adaptive Gain

Before digital processing became ubiquitous the operator was continually attempting to alter the clutter suppression characteristics even within a scan. Ideally it should be set correctly for each radial line which makes up the picture. Clearly no human operator possesses the speed or dexterity to perform such a task, but it can be achieved with a considerable degree of success by electronic means. The general name for the technique is *adaptive gain* and it is so called because the associated processing is designed to adjust continually the effective receiver gain so as to *adapt it* to the level of clutter present at any point on each radial line. It did not become feasible until the mid-1970s, when solid-state logarithmic receivers started to be introduced for civil applications at sea. An essential requirement for adaptive gain control is the availability of a very wide dynamic range signal, which is given by such receivers. The principle of adaptive gain is illustrated by Figure 3.41.

The theory of adaptive gain is that if an analogue of the clutter signal is used to produce the suppression signal, only signals stronger than the clutter will be displayed. Examination of the combined target and clutter signal shown in Figure 3.41 reveals that its characteristic has two main components, one being the relatively slow general decay with range (see Section 3.6.1.2), from maximum at the origin to zero at the radial limit of the sea clutter response. Superimposed on this is the second component which has the form of relatively rapid spikes which represent the presence of targets and clutter peaks. The total signal can be thought of as a slowly changing average signal about which rapid excursions occur. The reaction time of the sensing section (sometimes referred to as the integrator) of the adaptive gain circuitry (Figure 3.44) is designed to be too slow to respond to the rapid excursions, but sufficiently fast to react to the slowly changing average signal level. The sensing section thus effectively filters out the high-frequency fluctuations and generates a slowly changing average signal level which is inverted and used as a suppression signal to modify, continuously and instantaneously, the overall gain level set by the observer.

Clearly the slow reaction time of the sensing section will not allow it to respond immediately to a change from zero clutter at the end of one line to maximum clutter at the beginning of the next. This difficulty can be overcome by using the initial clutter signal from the previous line to produce the suppression for the first part of the current line. The adaptive gain effect is normally enhanced by the use of other techniques, such as *differentiation* which tends to emphasize the leading edge of echoes. This is described in detail in Section 3.7.4.4.

If the suppression signal is a faithful analogue of the average signal level, the resultant video signal made available for display (see Figure 3.41) should comprise those signals which are rapid excursions from that level and which represent the presence of targets that are stronger than the clutter and also the clutter peaks. Targets which are much stronger than the clutter should be easy to detect as they will be displayed (on an analogue display) against the background of the light residual speckle of clutter peaks. Targets very close to clutter level will be difficult to detect and the observer may have to rely on an ability to recognize their regular paint against the background of random clutter echoes.

It is possible that there are ship-specific settings for the adaptive gain control. If so, the setting up of the adaptive gain circuitry must be carried out by qualified service personnel, but it should really be performed in consultation with experienced radar observers who, on

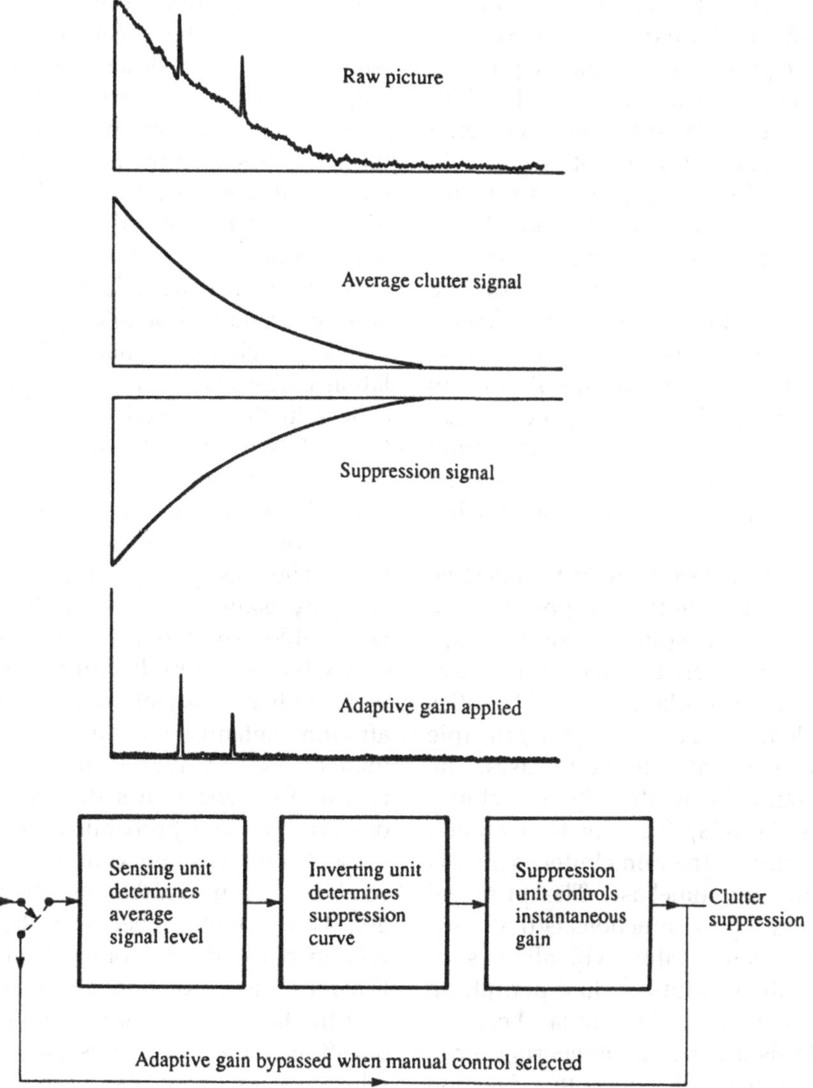

FIGURE 3.41 The principle of adaptive gain.

the basis of systematic use, are in a position to comment on the performance of the system in a wide variety of sea conditions. It has to be recognized that it is virtually impossible to adjust the adaptive gain circuitry so that it will give optimum response over the full range of wind and sea conditions between calm and gale force winds. For this reason, exclusive reliance *should not* be placed on the ability of such circuitry to remove clutter and to display only targets. The practical procedure for the use of the facility is set out in Sections 6.7.1 and 6.7.2.

Because of the averaging effect and the differentiation (see Section 3.7.4.4), adaptive gain circuitry can suppress radar beacon signals (see

Section 3.5) and it is for this reason, among others, that IMO Performance Standards (see Section 11.2.1) require that it should be possible to switch off those signal processing facilities which might prevent a radar beacon from being shown on the radar display. Adaptive gain will have a similar effect on radial performance monitor signals (see Section 6.3) and clearly must be switched off, or minimized, when a performance check is carried out. The effect of adaptive gain on land echoes will vary depending on whether the beam is parallel with or normal to the coastline; in general, it will tend to display the leading edge and suppress what lies behind. For this reason, identification of coastal features (see Section 8.2) is best performed with the adaptive gain facility switched off.

The adaptive gain facility may be effective in suppressing rain clutter responses (see Section 3.7.4.5). In some systems a single adaptive gain facility is offered, whereas in others separate sea and rain adaptive gain facilities may be provided. The adaptive gain principle described above is valid in both cases, the essential difference being that the sea clutter facility applies the adaptive gain to a limited preset range, whereas the rain clutter facility is effective for the full timebase. The principal reason for separating the functions is that a sea clutter adaptive gain facility will allow suppression of proximate clutter while permitting the detection of more distant radar beacons. 'Adaptive gain' is a general engineering term; the display control which effects this function may be labelled in a variety of ways by different radar manufacturers. Normally, when the adaptive gain facility is selected the manual sea and rain clutter controls (see Sections 6.7.1 and 6.7.2) are disabled.

3.6.3.4 Rotation-to-Rotation Correlation

This is a further approach to automatic clutter suppression and is only available in systems which can store the range and bearing data for a full antenna rotation (see Section 2.7.3). The other methods of clutter suppression so far described use difference of amplitude (i.e. signal strength) in their attempt to distinguish between clutter and targets, though it has been mentioned that an observer may identify a weak target in clutter by virtue of the regularity of its paint when compared with random sea clutter responses. Rotation-to-rotation correlation sets out to perform this latter operation automatically.

The principle is similar to line-to-line correlation which is described in Section 3.9.5.1, but differs in that, instead of comparing the contents of two *different* range cells at the same range on successive lines, the comparison is made between the content of the *same* range cell on two (or more) successive antenna rotations. The theory of operation can be understood by asking the question: 'If a wave with a favourable aspect is present in a given range cell, what is the probability that a favourable wave will be present in the same range cell after one antenna rotation, that is after approximately 3 s?' Evaluation of this probability is clearly an exercise in statistics, but it is fairly obvious that the probability of a wave being present in the same cell for two successive rotations is considerably less than that of a target. The difference between the probabilities will increase if the comparison is made for three or more rotations, provided of course that the target does not move out of the cell in question. Thus it is possible, to an extent which depends on the number of rotations over which the comparison is made, to distinguish between regular target returns and random clutter returns.

The way in which the probability theory is exploited and the name given to the facility (see Section 6.7.1.3) vary considerably with manufacturer. In a simple application of the principle, echoes which are present on successive rotations may be displayed at a higher level or in a different colour to those which appear at random from rotation to rotation. In a more complex approach, the comparison may be made over several successive rotations, and multi-levels of display symbology may be used to indicate the consistency of echo returns by promoting them to higher levels of echo prominence (e.g. brightness) as the number of successive rotations on which they appear increases. Demotion to a lower prominence level will occur if a target ceases to be consistent, though probably on the basis of a more tolerant law. The general effect of the correlation will be to favour consistent targets at the expense of clutter, but it must be borne in mind that targets which return an intermittent response, such as a buoy or a small boat in a seaway, will be penalized.

It is essential for radar observers using such a facility on any particular system to be aware of the logical rules on which that version operates. It is also important to appreciate that echo prominence modified in this way is not a measure of echo strength.

Normally in radial-scan synthetic displays, only one, or possibly two, radial lines are stored and it is for this reason that rotation-to-rotation correlation first became available only with the development of raster-scan displays suitable for use on large merchant vessels.

Rotation-to-rotation correlation tends to eliminate some receiver noise and it is thus important to switch the facility off when setting the gain control (see Section 6.2.7.1).

3.7 THE DETECTION OF TARGETS IN PRECIPITATION CLUTTER

Precipitation is the general term used to describe collectively various states in which water can manifest itself in the atmosphere, of which rain, snow and hail are examples. Reflections from precipitation can produce unwanted echoes on the screen and these are, in practice (perhaps somewhat loosely), referred to as *rain clutter* though it is recognized that they may originate from other forms of precipitation. The effect of rain clutter in some ways resembles that of sea clutter, but it has other characteristics which are quite different. Like the responses from the sea, those from rain can mask even strong target echoes by causing saturation within the receiver. They can also make it difficult to detect unsaturated responses against the background of clutter due to poor contrast and impossible to detect targets whose response is weaker than that of the rain. However, precipitation echoes can occur anywhere on the screen and may well change their position quite rapidly. Additionally, the transmitted pulse and the returning echo suffer scattering and attenuation (see Section 3.7.2) on their journey through the precipitation, thus reducing the received echo strength below what it would be in the absence of the precipitation. This has the effect of exacerbating the masking effect of the clutter and reducing the probability of detecting targets located beyond the area of precipitation. It is important to recognize that the solutions to the problems of detecting targets in and beyond rain are distinctly different (see Sections 3.7.5.1 and 3.7.5.2).

3.7.1 The Nature of Precipitation Response

The area of precipitation can be considered to be an aggregation of a large number of randomly distributed individual particles which may take the form of water droplets, snowflakes or ice. Energy from the radar beam may strike these particles and will as a consequence be scattered in a multitude of directions. The aspect of some of the particles will favour backscatter, i.e. reflection in the direction of the antenna, and energy so reflected may produce detectable responses on the radar screen of the observing vessel. Like the backscatter from the multiple facets of sea waves (see Section 3.6.1), the mechanism is extremely complicated and quantitative analysis of the response involves complex statistical theory beyond the scope of this text. However, it is not altogether surprising that it can be shown that the amplitude of the response depends on the size and material of the particles, the number present per unit volume of space illuminated by the radar beam at any time and the range of the precipitation (also the energy in the pulse and the transmission frequency, although, for a particular radar at a particular time, these may be considered fixed: see Section 2.3 on pulse length, transmitted power/PRF and wavelength for the effects of deliberately varying each one).

3.7.1.1 *The Effect of Particle Size*

As is evident from the treatment presented in Section 3.3.5, particle size is only significant when considered in relation to transmitted wavelength. Theoretical calculations show that, provided the particle diameter is small when compared to wavelength, for a given wavelength the scattering response increases in direct proportion to the square of particle diameter. Thus a doubling of particle size will produce a fourfold increase in response.

Conversely, for a given particle size, the response is inversely proportional to the square of transmitted wavelength. Thus in given precipitation conditions selection of an S-band (see Section 2.3.2) system will reduce the rain clutter responses to approximately one-tenth of that on an X-band radar.

3.7.1.2 *The Effect of Particle Intensity*

Particle intensity is the number of particles present per unit volume of space illuminated by the radar beam at any time. Clearly, for any given particle size, the response will increase with the particle intensity.

3.7.1.3 *Precipitation Rate*

It is evident from the two preceding sections that the response from precipitation increases with both the intensity of the particle population and the size of the individual members. Both of these factors determine the precipitation rate which is a measurement meteorologists use to quantify precipitation in units of the volume falling per unit time. The term *precipitation* or *rainfall rate* is quite commonly used in discussing precipitation response. In general, the rain clutter response increases in proportion to the rainfall rate.

3.7.1.4 *The Effect of Range*

It was shown in Section 3.1 that in general, target response decays in proportion to the fourth power of range. By contrast, the response from precipitation falls off in proportion to the square of range. This arises because the radar cross-section of precipitation is range dependent, simply because the volume of rain in a particular cell is range dependent (as is the area of sea clutter). The relationship between the two decay rates favours the detection of clutter.

The remote limits of the rain clutter pattern represent the minimum detectable signal (see Section 3.2.3) and hence the extent of the pattern offers a suitable criterion by which to judge the setting of the tuning control (see Section 6.2.7.2).

3.7.2 Attenuation in Precipitation

Even when the radar waves travel through clear atmosphere, some attenuation takes place. It occurs because some of the energy is absorbed by the gaseous constituents of the atmosphere, in particular oxygen and water vapour. The effect starts to become of major practical significance only at wavelengths shorter than that of X-band. However, the presence of precipitation considerably increases the amount of attenuation that takes place. The extent to which the energy is absorbed depends on particle size, particle form (e.g. water or ice) and rainfall rate. It also increases exponentially with the distance travelled by the energy through the attenuating medium. In any given set of precipitation conditions, S-band transmissions will suffer less attenuation than those at X-band.

3.7.3 The Effect of Precipitation Type

In the preceding sections, the strength of the response, the extent of scattering and the amount of attenuation to be expected from precipitation in general have been considered in terms of the physical characteristics that affect them. Such characteristics will vary with precipitation type and it is therefore appropriate to discuss the effects in terms of various forms in which the precipitation may be manifest.

3.7.3.1 *Rain*

In the case of rain the particles which effect the scattering and attenuation take the form of water droplets. The droplet size cannot exceed a diameter of about 5.5 mm because at this limit the surface tension which holds the water in droplet form is overcome and the droplet subdivides.

Large droplets tend to be found in tropical rainstorms and in general in the rain associated with vigorous convection such as occurs, for example, at and after the passage of a cold front. In such cases the droplet size is an appreciable proportion of the X-band wavelength, very strong clutter echoes will be produced and there will be serious loss of energy due to scattering and absorption. The detection ranges of strongly responding targets within the rain area will be reduced and their echoes may be severely masked by saturation within the receiver. Weaker target responses, for example those from small vessels and buoys, will be rendered undetectable if their echoes are not stronger than that of the rain. While targets in clear areas beyond the rain will not be subject to masking, scattering and absorption will significantly reduce the energy reaching them and therefore reduce their detection ranges. As the droplet size is a much smaller proportion of the 10 cm wavelength, and bearing in mind the square law relationship (see Section 3.7.1.1), it is evident that the problems will be significantly reduced by the selection of S-band under such circumstances.

The droplet size of rain associated with stratiform clouds, such as those which might be expected at a warm front, tends to be smaller than those so far described. It must be remembered that the response and attenuation depend not only on the size of the droplets, but also on the number falling, that is on the rainfall rate.

Drizzle is a form of light rain characterized by low rainfall rate and small droplet size of less than about 0.25 mm, and as a consequence is unlikely to cause serious problems other than in the case of very weak targets. However, the higher relative humidity normally associated with drizzle can result in greater absorption.

3.7.3.2 *Clouds*

The water droplets which form clouds are too small (less than 0.1 mm) to produce detectable responses even at X-band. However, if there is rain or other precipitation within the cloud, it may well be detected.

3.7.3.3 *Fog*

Because of the very small particle size and density, it is most unlikely that fog and mist will return detectable echoes. In most circumstances attenuation results in only a slight reduction in maximum radar detection range. However, in the special case of the intense fogs which arise in polar regions, a significant reduction in detection range may occur.

Smog is a portmanteau word formed from smoke and fog, and describes the effect of the condensation of water droplets on the dirt particles in areas of industrial smoke. It is likely to produce a somewhat higher degree of attenuation than a clean sea fog.

3.7.3.4 *Snow*

Where no melting is taking place, snowflakes are single ice crystals or a conglomeration of such crystals. When precipitation rates are compared, snowfall rates are in general less than rainfall rates. If this is considered in parallel with the lesser response of ice, it is evident that the echoes from snow are likely to be less troublesome than those from rain. *Sleet* describes the condition in which the snow is partially melted and the response will tend towards that of water. The attenuation produced by snow is similarly less than in the case of rain.

The fact that snow reflects radar energy less effectively than rain may be considered to be fortunate, but it must be borne in mind that where snow lies on the surface of wanted targets its relatively poorer reflecting property and its albeit limited absorption characteristic may reduce the detection range of good targets such as land and also render undetectable poor targets such as growlers (see Section 3.3.6.1).

3.7.3.5 *Hail*

Hail is essentially composed of frozen raindrops. Ice reflects radar energy less effectively than water and hence, size for size, hailstones will produce a lesser response than raindrops. Melting on the surface of the hailstone will improve its response towards that of water. In general, the precipitation rates associated with hail are lower than those experienced with rain, and so clutter and attenuation from hail are likely to prove less troublesome than those from rain.

3.7.3.6 *Dry Sand and Dust Haze*

Clearly such phenomena cannot be described as precipitation but, as the effects are similar, it is appropriate to consider them under this general heading. On the basis of particle size and distribution, theory suggests that the effects should be similar to those of fog; practical experience substantiates this. Detectable responses, though not impossible, are extremely unlikely, and a fairly low level of attenuation is to be expected. These characteristics may prove very useful in conditions where visual sighting is severely hampered.

3.7.4 The Suppression of Rain Clutter

The principle effects of rain clutter have been introduced in Section 2.7.4.2. This section covers the issues and the optimization of a radar's performance in the presence of precipitation in rather more detail. As suggested in Section 3.6, in some ways the rain clutter problem resembles that of sea clutter and it is thus not surprising that some suppression techniques may be effective in both cases.

3.7.4.1 *The Choice of Pulse Length*

As in the case of sea clutter, selection of a shorter pulse length will reduce the amplitude of all received echoes and hence may well assist in the detection of targets within an area of rain by bringing rain clutter echoes below saturation and in general improving the contrast between clutter echoes and those of wanted targets. It will of course not assist in combating the attenuation caused by the presence of the precipitation (see Section 3.7.5).

3.7.4.2 *The Choice of Wavelength*

Selection of S-band as opposed to X-band transmission will increase the probability of detection of targets within an area of precipitation clutter by reducing the response from the precipitation and the associated attenuation.

3.7.4.3 Searching Techniques

The probability of detection of targets within areas of precipitation may be improved by using the gain control or the sea clutter control to systematically suppress and restore the clutter signals in a searching fashion similar to that described in Section 3.6.3.1. Where the clutter surrounds the observing vessel, the sea clutter control is likely to prove helpful but it will not be effective for more distant precipitation. The practical procedures for carrying out such searches are described in Section 6.7.1.

3.7.4.4 The Rain Clutter Circuit

This is an alternative approach to the problem of saturation and employs a technique known as differentiation which is performed in the video section of the receiver. The signal is conditioned by circuitry which responds only to increases in signal strength. The term 'differentiation' is mathematical in origin and essentially means to measure the rate of change of some quantity. The theoretical principle of operation is illustrated by Figure 3.42. Figure 3.42(a) and (b) shows the effect of varying degrees of differentiation on a single idealized rectangular pulse, while Figure 3.42(c) shows three idealized rectangular echoes which overlap one another, saturate the receiver and hence are representative of the masking effect of rain clutter. The first pulse represents the rain echo which returns immediately ahead of the target response and the third pulse represents the rain return immediately following that of the target.

Because the circuitry responds only to those sections of the waveform where the signal strength is increasing, its effect is to remove the trailing edges of the echoes, thus making it possible to display the leading edge of the rain and that of the target as separate returns. Clearly the displayed echoes must not be made so short that they become difficult to discern. The finite radial length of the differentiated echo depends on the *time constant* of the circuit which is a measure of the time taken by the circuitry to respond to a levelling off or decrease in signal strength (Figure 3.42(a) and (b)). For this reason the rain clutter control may sometimes be referred to as the FTC. In some systems this is preset by the manufacturer, in which case the rain clutter control is an on/off FTC switch; in other systems it can be set by the observer over a continuous range of values. The effect of the rain clutter control on a practical set of echoes is illustrated by Figure 3.43.

Reference to Figure 3.43 shows that the two principal features of the differentiated output are the leading edges of the rain and the target which is stronger than the rain. It should also be noted that the raw rain signal shows a slow decrease with range, on which is superimposed a small high-frequency fluctuation of the instantaneous signal level. Depending on the setting of the rain clutter control, the differentiated output will also tend to show to a greater or lesser extent the rising edges of the fluctuating peaks; targets close to rain level will have to compete with these for attention. Targets weaker than the rain response cannot be detected. Traditionally differentiation was carried out before any digitization (see Section 2.7.4) and the output illustrated in Figure 3.43 maybe analogue or digital.

It is important to appreciate that although the rain clutter control is so called, it is not necessarily the most effective way of dealing with rain echoes. Many users find the use of the gain control in a searching action much more effective. This is discussed further in Section 6.2.7.1, which sets out the practical procedures for detecting targets in rain. It is worth mentioning that the differentiating effect may be found particularly useful for improving discrimination on short range pictures.

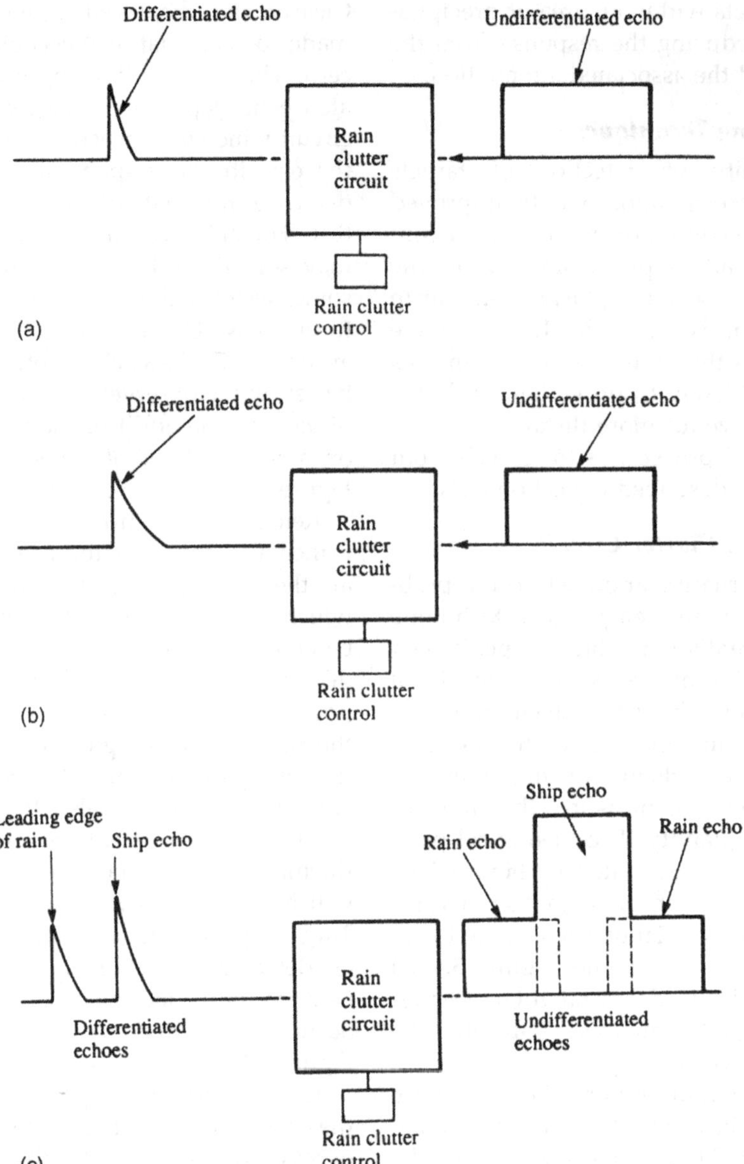

FIGURE 3.42 **The principle of differentiation.** (a) High degree of differentiation. (b) Low degree of differentiation. (c) Separation of echoes by differentiation.

Targets within the rain clutter area can sometimes be detected by observing their shadowing effect in the rain clutter beyond them.

3.7.4.5 *Adaptive Gain*

This technique was described in detail in Section 3.6.3.3 and represents an automatic approach which may be found effective in the

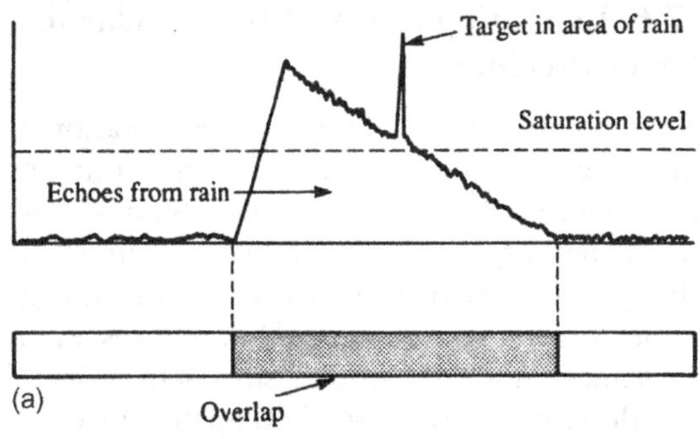

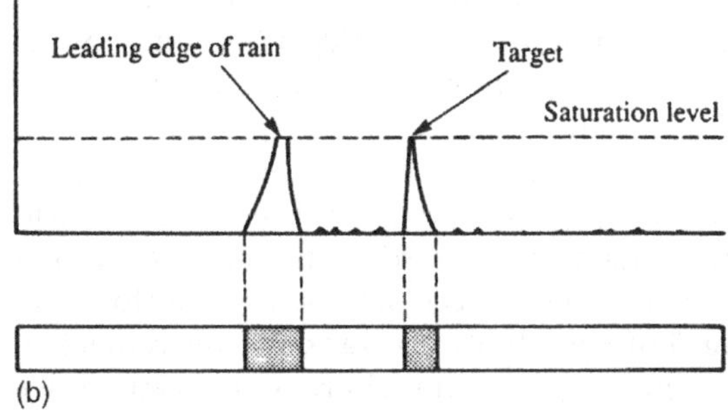

FIGURE 3.43 The effect of the rain clutter control.

suppression of rain clutter echoes. In most systems, when adaptive gain is selected, a fixed amount of differentiation is also applied.

3.7.4.6 Circular Polarization

For the reasons set out in Section 3.6.1.7 there is no stimulus for designers to produce X-band equipment other than with horizontal polarization. Hence, in civil marine applications, vertical polarization is very much the exception rather than the rule. In Section 2.5.1.7 it was explained that the terms 'vertical' and 'horizontal' refer to the plane of the fluctuating electric field associated with the electromagnetic wave.

The visualization of a circularly polarized wave may require some mental effort. Its effect is best explained by considering it to be produced by generating a horizontally polarized wave, and adding to it a vertically polarized wave which has the same amplitude and frequency but is a quarter-cycle out of phase with it. The polarization of the resultant wave thus rotates and hence the use of the term 'circular polarization'. On the return of echoes, the two components of the wave are separated and a further quarter-cycle phase shift is applied to the vertically polarized wave. If the two components of the wave have suffered identical reflection, they will be equal in amplitude, out of phase by one half-cycle and thus will cancel one another. Such symmetrical reflection will be experienced if the target is a small smooth sphere (radar cross-section is a function of polarization). In many circumstances a raindrop comes close to this ideal. Thus the response from raindrops will tend to be cancelled or at least considerably reduced. If the reflection of both components is not symmetrical, then the degree of cancellation will be reduced to a greater or lesser extent, depending on the lack of symmetry. The cancelling achieved in the case of other types of precipitation will not be as effective if the particle is not spherical, as in wet snowflakes. When compared with linear polarization, the response of wanted targets will also be reduced by the use of circular polarization. In general, cancellation can be expected where the energy has suffered an odd number of reflections, but not with an even number. A ship target is a complex arrangement of many reflecting surfaces. It thus behaves as a large scatterer and the energy returned to the observing vessel will contain components which have suffered an even number of reflections.

The effectiveness of circular polarization for the suppression of precipitation thus depends on the reduction in response of wanted targets being less than that suffered by the precipitation. This hope may well not be realized in the case of small targets and this limitation must be firmly borne in mind should such a facility be available. In the past some civil marine radar systems offering this facility were produced, but few, if any, are still at sea and there is no sign at present that they will be produced again. One might speculate that the additional cost and complication was not justified by the benefit gained. However, other radar applications, not least air traffic control systems, make good use of circular polarization.

3.7.5 Combating the Attenuation Caused by Precipitation

The nature of the problem differs depending on whether the target lies within or beyond the area of the precipitation.

3.7.5.1 The Attenuation of the Signal from a Target Beyond Precipitation

The response from a target beyond an area of precipitation will be reduced as a result of scattering and attenuation of the radar signal on its two-way journey through the precipitation (see Section 3.7.2). This attenuation can be combated by the selection of the S-band and the use of a longer pulse length. While searching for specific targets, it may be useful temporarily to turn the gain above the normal setting or to use an echo-stretching facility (see Section 6.8) if the latter is available.

3.7.5.2 The Attenuation of the Signal from a Target Within Precipitation

This is a more difficult problem; the only way in which this attenuation can be effectively reduced is by the use of S-band in preference to X-band transmission (see Section 3.6.1.5). The other techniques suggested in Section 3.7.5.1 will certainly increase the target response but they will also increase the response from the precipitation, thus exacerbating the already serious problem of saturation.

3.7.6 Exploiting the Ability of Radar to Detect Precipitation

The ability of radar to detect precipitation at long range has been found useful in a number of meteorological applications. Shore-based radars having a range of several hundred miles have been used to track the movement of the large areas of precipitation present in hurricanes and other similar storms in order to provide warning of possible danger and damage from the associated weather conditions. The detection range offered by shipboard radar cannot provide sufficient warning to permit avoiding action. The presence of precipitation in other less dramatic weather systems has led to the use of long range surveillance radars in day-to-day general weather forecasting. Again, shipboard equipment does not offer the range necessary for weather forecasting, but the ability to track the movement of precipitation in the short term can prove useful.

Early warning of the approach of precipitation should make it possible to ensure that the vessel is adequately prepared to deal with the additional problems this may pose in the use of radar for collision avoidance. The Master can be called in adequate time, the engine room staff can be prepared for manoeuvres and arrangements can be made for such additional bridge and lookout personnel as may be required. It may be possible to commence the plotting of targets before they enter areas of precipitation. This may well assist in the detection of targets even in the presence of masking, as it is always easier to find an echo if its approximate location is known. The data extracted should also prove useful in planning collision avoidance strategy, although the observer must always be alert to the possibility of a target manoeuvring while masked.

In port, the ability to track approaching rain showers may make it possible to close cargo hatches in sufficient time to prevent cargo

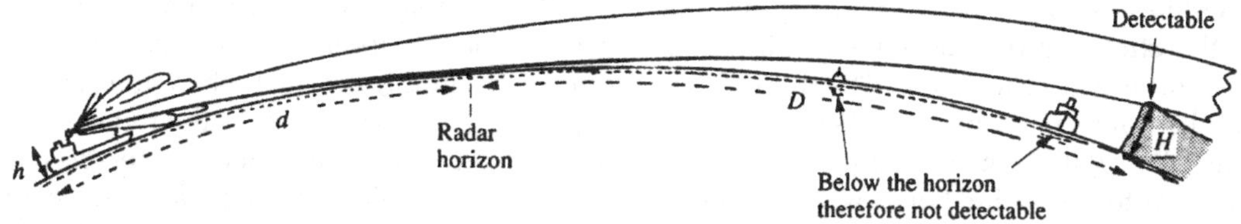

FIGURE 3.44 The radar horizon under standard atmospheric conditions.

damage. At sea, the ability to avoid isolated showers might benefit a freshly painted deck or perhaps an outdoor event on a passenger vessel.

3.8 THE RADAR HORIZON

At marine radar transmission frequencies (nominally 10,000 and 3,000 MHz), the paths followed by the signals may be considered as 'line of sight'. This means that even though the radar is delivering a powerful pulse and the target is capable, if irradiated, of returning a detectable response, the target will not be detected if it is below the *radar horizon*. This is analogous to the visual observation of objects in the vicinity of the horizon.

The effect of the atmosphere on the horizon is a further factor which must be taken into account when assessing the likelihood of detecting a particular target and especially when considering the expected appearance of coastlines.

3.8.1 The Effect of Standard Atmospheric Conditions

Under standard atmospheric conditions, the radar beam tends to bend slightly downward, the distance to the radar horizon being given by the formula:

$$d_{NM} = 1.22\sqrt{h_{ft}} \text{ or } d_{NM} = 2.21\sqrt{h_m}$$

where h is the height of the antenna in feet or metres.

It can be seen from Figure 3.44 that the possibility of detecting targets beyond the radar horizon will, in addition to all the other factors discussed in this chapter, depend upon the height of the target (i.e. whether or not a responsive part of it extends above the horizon). Thus the theoretical detection range based purely on the antenna and target heights is given by the formula:

$$R_d = d + D$$
$$R_d = 1.22\sqrt{h_{ft}} + 1.22\sqrt{H_{ft}} \text{or}$$
$$R_d = 1.21\sqrt{h_m} + 2.21\sqrt{H_{mt}}$$

where h and H are heights of the antenna and the target, respectively, in feet or metres. In both cases, R_d is the theoretical detection range in NM.

This relationship is of course theoretical since it assumes that:

a. Standard atmospheric conditions prevail.
b. The radar pulses are sufficiently powerful.
c. The target response characteristics are such as to return detectable responses.
d. The weather conditions, such as precipitation, through which the pulses have to travel, will not unduly attenuate the signals.

3.8.1.1 The Increase in Detection Range with Increased Antenna Height

While in theory it is correct to say that detection ranges can be increased by increasing the antenna height, in practice there is a limit to the extent to which it is worth pursuing this increase. Above moderate height, the increase in detection range is minimal and not worth the effort (or the consequent additional expense of waveguide, etc.). The reason for this can be appreciated by considering the theoretical relationship set out above: to double the distance to the radar horizon, the antenna will have to be four times as high. If the antenna is at a height of 16 m, the distance to the radar horizon is 8.84 NM. To double this to some 17.7 NM would require an antenna height of some $(17.68/2.21)^2$ NM = 64 m. This point should be borne in mind when considering the extent to which radars offering 96 NM range scales can be effective.

It can be seen from Figure 3.44 that if targets are observed beyond the horizon, it must *not* be assumed that nearer targets will necessarily also be detected. Note that, although the cliff may be detected, there is no chance of receiving an echo from the buoy at this stage.

Consider a vessel having its radar antenna at a height of 16 m above the water. The distance to the radar horizon will be $2.21 \times 4 = 8.84$ NM. This means that when assessing the possibility of detecting a target at a range greater than some 8.5 NM, one must first consider whether or not the target has sufficient height.

The theoretical range at which a sheer cliff face rising to a height of 64 m should be detected on a vessel whose antenna is at a height of 16 m is:

$$\begin{aligned} &2.21\sqrt{(64)} + 2.21\sqrt{(16)} \\ &= 2.21 \times 8 + 2.21 \times 4 \\ &= 26.5 \text{ NM} \end{aligned}$$

Note: Although sheer cliffs immediately suggest a good response target, aspect must be borne in mind, i.e. whether the cliffs are 'square on' or at an angle to the radar rays.

Consider the case where a vessel has its antenna at a height of 16 m above the water and first detects land at a range of 23 NM:

$$23 = 2.21\sqrt{(16)} \times 2.21\sqrt{H}$$

$$\text{therefore } H = \left(\frac{23 - (2.21 \times 4)}{2.21}\right)^2$$

$$= 41.05 \text{ m}$$

The indication here is that, if standard atmospheric conditions prevail, the land which is being detected must be at least 41 m high. When attempting to relate what is being observed on the radar to the chart, only land having a height of at least 41 m should be considered. As can be seen in Figure 3.45, it would be completely incorrect to assume that the vessel is 23 NM from the coastline.

Great care must be exercised when attempting to identify targets which are beyond the radar horizon. Using the example above (and Figure 3.45) it is evident that the land detected is unlikely to be the coastline but is probably higher ground lying inland. It is important to appreciate this principle as failure to do so gives the impression that the vessel is farther offshore than is in fact the case.

It should be noted that, when making in towards a sloping coastline such as that in Figure 3.46, the coastline as displayed on the radar screen can have a disconcerting effect of 'appearing' to come towards the vessel as

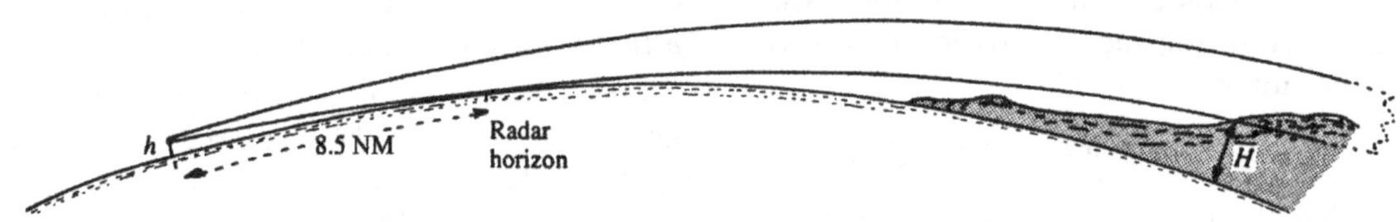

FIGURE 3.45 The effects of target and antenna height.

more of the low-lying land comes above the horizon. This is especially so if it has been assumed that it was the coastline that was being observed in the first instance and positions have been laid off as if this were the case.

Considering a vessel with an antenna at a height of 16 m and assuming standard atmospheric conditions, if echoes from land are observed at a range of 26 NM, then whatever is being observed must be at least some 60 m high (Table 3.2).

Table 3.2 gives the least theoretical height (in metres) of a target detected beyond the radar horizon for various antenna heights. Since the distance to the radar horizon in the above example is some 8.8 NM, any target observed at a greater range than 8.8 NM should be checked against the table to ensure that it has adequate height for detection (Figure 3.45).

3.8.1.2 Standard Atmospheric Conditions

'Standard' conditions are precisely defined as:

Pressure = 1013 mb decreasing at 36 mb/ 1000 ft of height
Temperature = 15°C decreasing at 2°C/ 1000 ft of height
Relative humidity = 60% and constant with height

TABLE 3.2 Least Height of a Target Detected Beyond the Radar Horizon (metres)

	Antenna Height (m)											
Detection range of target (NM)	**8**	**10**	**12**	**14**	**16**	**18**	**20**	**22**	**24**	**26**	**28**	**30**
	6.8	5.1	3.9	2.9	2.0	1.4	0.9	0.5	0.3	0.1	0.0	
12	12.3	10.1	8.2	6.7	5.5	4.4	3.5	2.7	2.1	1.5	1.1	0.7
14	19.5	16.6	14.3	12.2	10.5	9.0	7.7	6.5	5.5	4.6	3.8	3.1
16	28.3	24.8	21.9	19.4	17.2	15.2	13.5	11.9	10.5	9.3	8.1	7.1
18	38.7	34.7	31.2	28.2	25.5	23.1	21.0	19.0	17.2	15.6	14.1	12.8
20	50.8	46.1	42.1	38.6	35.5	32.6	30.1	27.7	25.6	23.6	21.7	20.0
22	64.5	59.3	54.7	50.7	47.1	43.8	40.8	38.1	35.5	33.2	31.0	29.0
24	79.9	74.0	68.9	64.4	**60.3**	56.6	53.2	50.0	47.1	44.4	41.9	39.5
26	96.9	90.4	84.7	79.7	75.2	71.0	67.2	63.7	60.4	57.3	54.4	51.7
28	115.5	108.4	102.2	96.7	91.7	87.1	82.9	78.9	75.3	71.8	68.6	65.6
30	135.8	128.1	121.3	115.3	109.8	104.8	100.2	95.8	91.8	88.0	84.4	81.0
32	157.7	149.4	142.1	135.6	129.6	124.1	119.1	114.4	109.9	105.8	101.9	98.2
34	181.2	172.3	164.5	157.5	151.0	145.1	139.7	134.5	129.7	125.2	121.0	116.9
36	206.4	196.9	188.5	181.0	174.1	167.8	161.9	156.4	151.2	146.3	141.7	137.3
38	233.2	223.1	214.2	206.1	198.8	192.0	185.7	179.8	174.3	169.0	164.0	159.3
40												

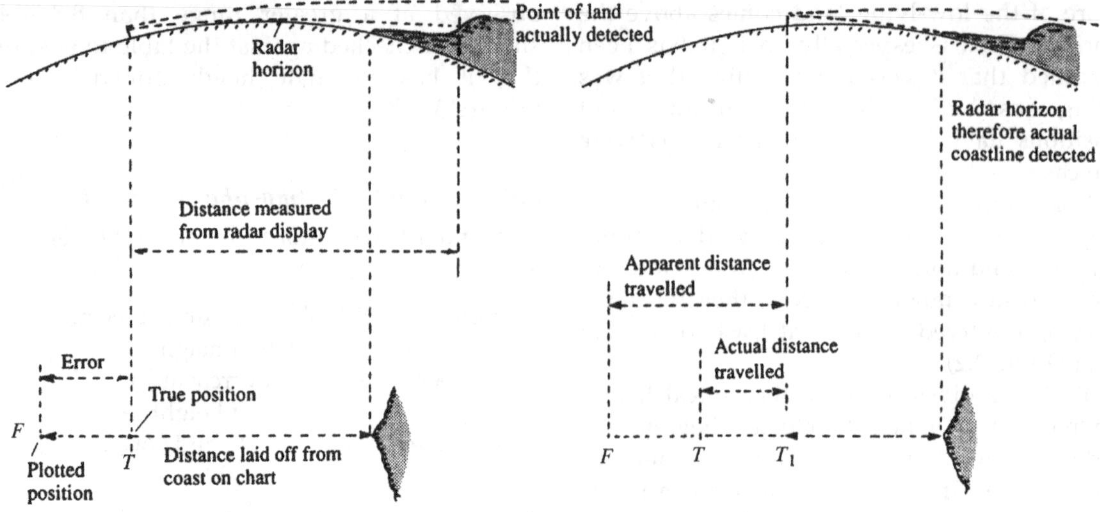

FIGURE 3.46 The apparent movement of a coastline as it comes above the coastline.

These conditions give a refractive index of 1.003 25 which decreases at 0.000 13 units/1000 ft of height.

The term 'standard' should not be confused with the term 'normal', which is extremely subjective and imprecise. Standard conditions are rarely likely to equate to what might be considered normal at any particular location on the earth.

It can be seen that the definition of 'standard' conditions relates to the vertical composition of the atmosphere. The mariner is unlikely to be able to obtain a precise knowledge of this and so must rely on a more general appreciation of the weather conditions, the area of the world and the time of the year, as well as on past experience for guidance as to the likelihood of the existence of standard or non-standard atmospheric conditions and of the effects that they can have on the detection ranges of targets in the area.

In general, 'normal' conditions in the more highly frequented sea areas of the world tend to be super-refractive.

Note: The attenuating and clutter effects of the atmosphere are considered in Section 3.7.2.

3.8.2 Sub-Refraction

3.8.2.1 The Effects of Sub-Refraction on Detection Ranges (Figure 3.47)

Sub-refraction occurs when the refractive index of the atmosphere decreases less rapidly with height than under standard conditions. As a result, the radar beam is bent downward slightly less than under standard conditions. This means that, with all other factors constant, the same target will be detected at a slightly reduced range. In practice, this is likely to mean something of the order of 80% of the detection range under standard conditions, but will obviously depend on the severity of the conditions prevailing at the time.

3.8.2.2 Atmospheric Conditions That Give Rise to Sub-Refraction (Figure 3.48)

These can be specified precisely, but since the mariner is unlikely to have access to data

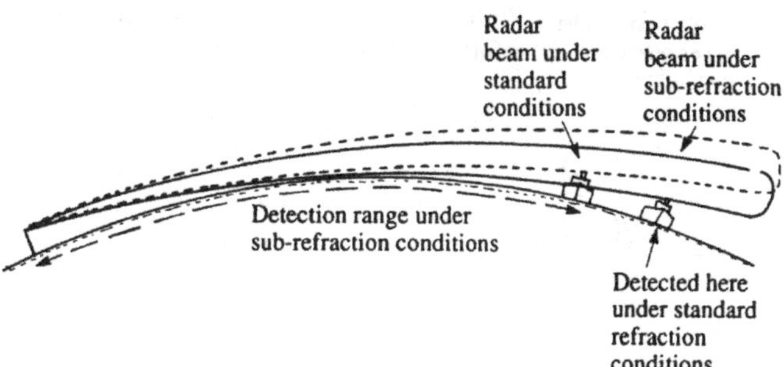

FIGURE 3.47 The effect of sub-refraction on detection ranges.

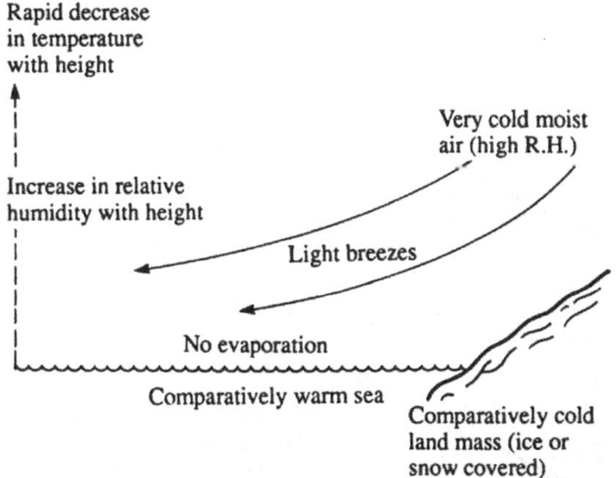

FIGURE 3.48 Atmospheric conditions associated with sub-refraction. The critical factor is land/sea (or sea/sea) temperature difference. When the sea is cold, sub-refraction can occur provided that the sea is even colder.

relating to the vertical composition of the atmosphere, it is perhaps more appropriate to consider the conditions which, from experience, are likely to accompany this phenomenon. These are:

a. An increase in relative humidity with height.
b. A rapid fall in temperature with height.

Both of these conditions do not have to be satisfied at any one time but, obviously, the greater the changes in temperature and relative humidity with height, the greater will be the effect that they will have on the radar beam. This does not mean that, under extreme conditions, the whole beam will rise above the sea surface or that no targets will be detected at all.

These conditions, to a greater or lesser extent, may be encountered virtually anywhere but are most likely in the polar regions, the Grand Banks (especially in the Gulf Stream with northerly winds), east of Japan in winter and in the Mediterranean in winter.

Sub-refraction is generally associated with bad weather conditions, that is with low-pressure systems, which are typically the time when radar is most likely to be needed.

3.8.3 Super-Refraction

3.8.3.1 The Effects of Super-Refraction on Detection Ranges (Figure 3.49)

Super-refraction occurs when the rate of decrease in refractive index with height is greater than under standard conditions. When super-refraction occurs, the radar beam tends to be bent down slightly more and so targets may be detected at ranges which are slightly greater than standard. Increases of some 40% are not uncommon.

Again, the mariner has no means of knowing exactly what sort of atmospheric conditions, are being experienced and so has to rely

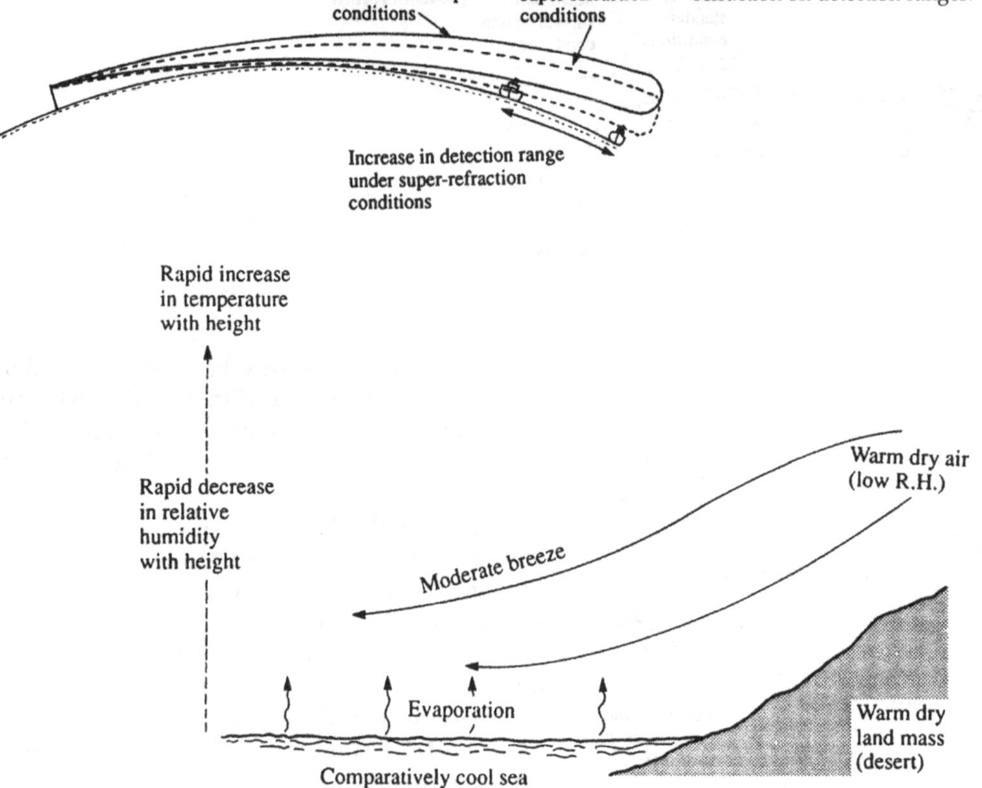

FIGURE 3.49 The effects of super-refraction on detection ranges.

FIGURE 3.50 Atmospheric conditions associated with super-refraction.

on some form of subjective assessment; it is always best to err on the cautious side.

3.8.3.2 *Atmospheric Conditions Associated with Super-Refraction (Figure 3.50)*

These are:

a. A decrease in relative humidity with height.
b. Temperature falling more slowly than standard, or even increasing with height.

These conditions tend to be more frequently encountered in the maritime trading areas of the world, in particular the tropics; other areas include the Red Sea, the Arabian Gulf and the Mediterranean in summer.

Super-refraction is generally associated with fine settled weather, i.e. with high-pressure weather systems.

3.8.4 Extra Super-Refraction or Ducting

3.8.4.1 *The Effects of Extra Super-Refraction (Figure 3.51)*

Under these conditions, the radar energy is, in effect, trapped in a 'duct' formed by the Earth's surface and a highly refractive layer which may be as little as a 100 ft (30 m) above the ground. The effect is to concentrate energy which would otherwise have been lost into space together with the energy which would

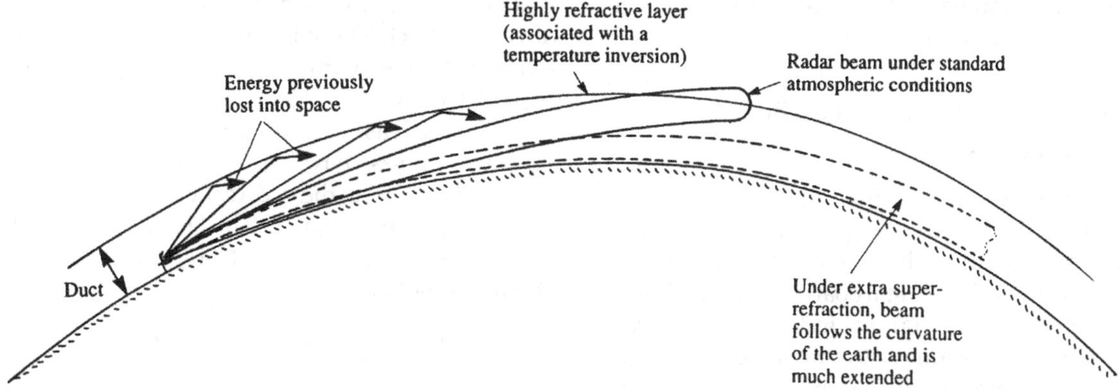

FIGURE 3.51 The effect of extra super-refraction.

normally travel in the direction of the targets. This increased energy now follows the Earth's surface, thus reducing the constraint of the radar horizon and considerably extending the detection ranges of targets. This can result in the detection of unwanted 'second-trace echoes' (see Section 3.9.6).

3.8.4.2 Atmospheric Conditions That Give Rise to Extra Super-Refraction

If the rate of change of refractive index is of the order of four times the 'standard' rate, initially horizontal rays will follow the curvature of the Earth. It is more effective, though, for the energy to be trapped between the Earth and a highly refractive layer (such as a temperature inversion), thus forming a duct in which the energy will travel in a similar manner to propagation within a waveguide with a 'leaky' upper wall.

Note:

a. A temperature inversion by itself must be very pronounced to produce a duct.
b. Relative humidity gradients are more effective than temperature gradients alone.

The areas which are normally associated with extra super-refraction are the Red Sea, the Arabian Gulf, the Mediterranean in the summer with the wind from the south and the area off the west coast of Africa in the vicinity of the Canary Islands. However, extra super-refraction can occur anywhere if the conditions are right. The usual indication that these conditions exist is the appearance of second-trace echoes (see Section 3.9.6).

3.9 FALSE AND UNWANTED RADAR RESPONSES

3.9.1 Introduction

These responses may also be referred to as spurious echoes and are the result of a number of specific causes. In all cases, echoes are displayed on the screen in positions where no genuine targets exist. In general, these are likely when genuine targets are close, but this is not always the case (see Section 3.9.6).

3.9.2 Indirect Echoes (Reflected Echoes)

These can occur when radar energy is deflected in the direction of an object by some obstruction in the path of the radiated energy, either on board the ship or ashore.

The returning energy follows a reciprocal path and so causes an echo to be displayed in the direction of the obstruction. This form of false echo can be particularly misleading when the obstruction is on board the ship.

3.9.2.1 The Effect of On-Board Obstructions

Consider the situation, depicted in Figure 3.52, where energy strikes the foremast and some of it is deflected in the direction of the target on the starboard quarter. Energy reflected by the target follows a reciprocal path and so returns to the antenna from a direction fine on the starboard bow (i.e. the reflection point on the foremast). As a result, a false echo will be displayed at a range which is similar to that of the genuine target (this assumes that the distance to the on-board obstruction is small by comparison with the range of the target), but on the bearing of the obstruction (i.e. the foremast).

Typical of on-board obstructions which can give rise to this form of false echo are a mast, a funnel and cross-trees. In fact, any structure which intercepts energy in the beam can produce the effect. There is a tendency to associate indirect echoes with blind arcs and shadow sectors (see Section 2.10.1.1); while this is frequently the case, in recent times other less obvious causes have been identified. Most notable have been the indirect echoes resulting from reflections from the aft-facing surface of containers (Figure 3.53). Here, three points should be noted:

a. The traditional blind/shadow sector theory need not apply.
b. There is a likelihood of other true echoes appearing on the radar display on the same bearing as the false echoes.
c. The cause of the problem may change each time the stowage of the containers on deck is changed.

It is advisable, when containers are stowed in isolated stacks, that a warning of the possibility of indirect echoes is displayed alongside the radar.

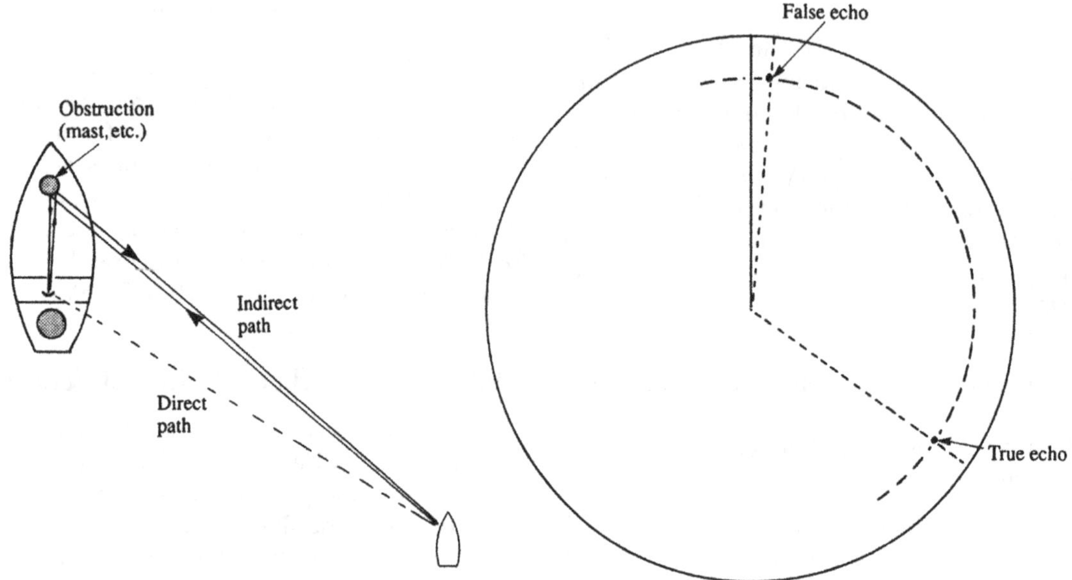

FIGURE 3.52 Indirect echoes.

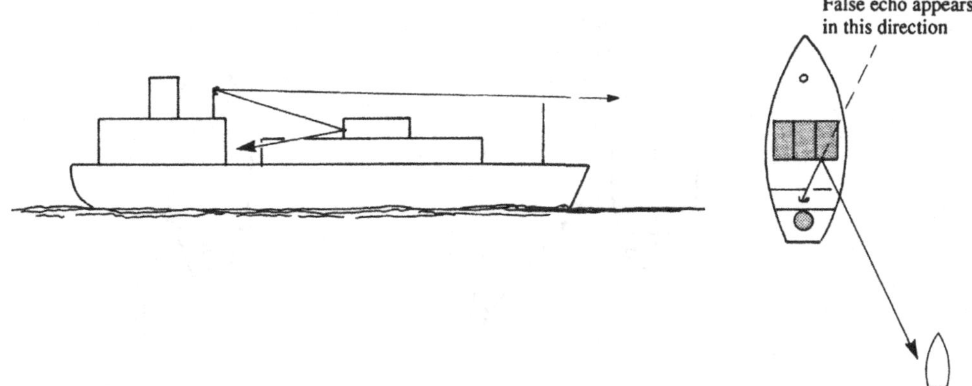

FIGURE 3.53 Reflections from aft-facing surfaces of containers.

3.9.2.2 *The Recognition of Indirect Echoes*

The more obvious causes (blind and shadow sectors) should be studied in clear weather when there are a number of other vessels in the vicinity of the observing ship. Blind and shadow sectors should have been logged (see Section 2.10.1.1) and any echo which appears in these sectors should be regarded as suspect. Visual comparison with the radar should confirm whether the echo is genuine or indirect. Where it is decided that the displayed echo is indirect, it is reasonable to expect to find a genuine target at about the same range, although even this search may be inconclusive.

Many unusual causes of indirect echoes have been reported but, without a careful study of the conditions obtaining at the time, it is very difficult to come to reliable conclusions and to take steps to prevent their recurrence. Deck cargo, deck cranes, breakwaters, ventilators and a host of other on-board structures have aroused suspicion at one time or another.

Indirect echoes can give rise to serious concern when they appear ahead of the vessel, particularly at night or in restricted visibility. At night, it may be suspected that the 'vessel' ahead is not showing the necessary navigation lights, or that the onset of restricted visibility is imminent.

If this phenomenon occurs in restricted visibility and it is considered necessary to manoeuvre to avoid what one must assume is a real target, the resulting apparent behaviour of the echo (if false) is quite unpredictable. For example, the 'target' may appear to make a simultaneous countermanoeuvre or may even disappear (Figure 3.54).

The following safety precautions should be observed:

1. Where echoes are suspected of being false, they should be assumed to be real until proved false beyond all reasonable doubt.
2. Potential causes of indirect echoes such as blind and shadow sectors should be constantly borne in mind and any suspect arc investigated, preferably before arrival in areas of known restricted visibility.
3. Unexplained 'target' behaviour should be logged.
4. Where the cause of indirect echoes has been identified, steps to prevent their recurrence should be taken immediately.

3.9.2.3 *The Prevention of Indirect Echoes*

Where shipboard structures have been identified as the cause of indirect echoes, steps should be taken to prevent their future recurrence. In

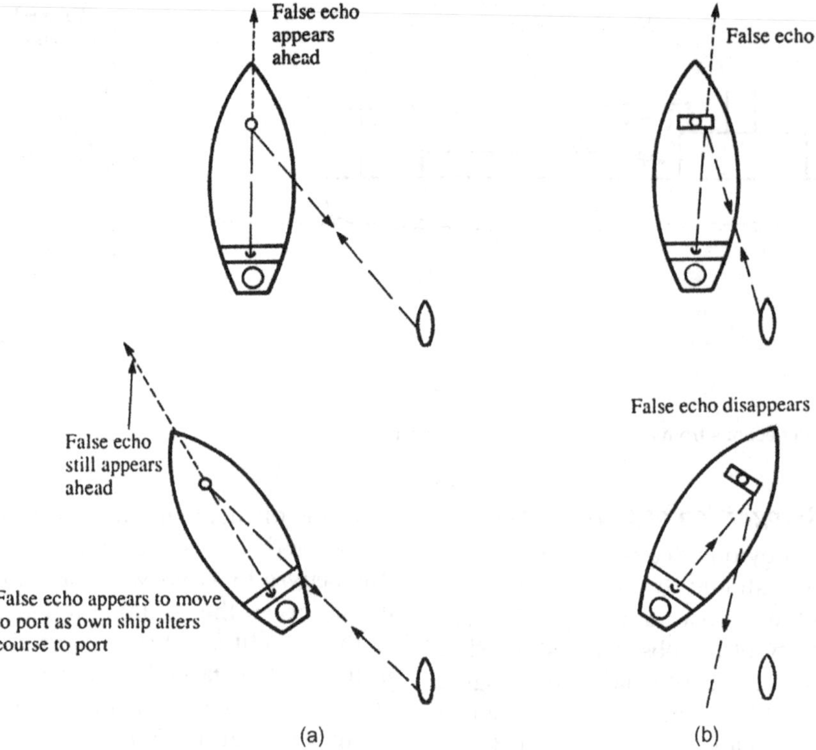

FIGURE 3.54 The effect of a manoeuvre by own ship on an indirect echo.

principle, this means that the energy striking the deflecting obstruction must be dissipated by redirection, scattering or absorption.

Each cause has to be treated individually; typical solutions in the past have included attaching plates angled upwards to the cross-trees, welding angle iron to the after surfaces of the cross-trees and surfacing the obstruction with radar absorbent material (RAM).

Note: Where attempts are made to scatter the energy by 'roughening' the surface, it must be remembered that facets must be of the order of wavelength (see Section 3.3.2).

3.9.2.4 *Indirect Echoes from Objects External to the Ship (Figure 3.55)*

These can frequently be seen when vessels are in built-up areas and other vessels are moving in the vicinity. Because of the conditions under which they occur, they are frequently not recognized.

In general, the obstructions which cause the deflection of energy tend to be large and the false echoes will display on the farther side of them. In many cases they will appear to move over the land, making it easy to recognize them and dismiss them as false. In the open sea, this recognition might not be quite so positive, but there is always the consolation that the nearer of the targets on the same bearing is invariably genuine.

One special form of false echo arising from this cause can be extremely misleading. This is where the indirect echoes occur as a result of a bridge structure spanning the waterway. Consider a vessel approaching a bridge, as in

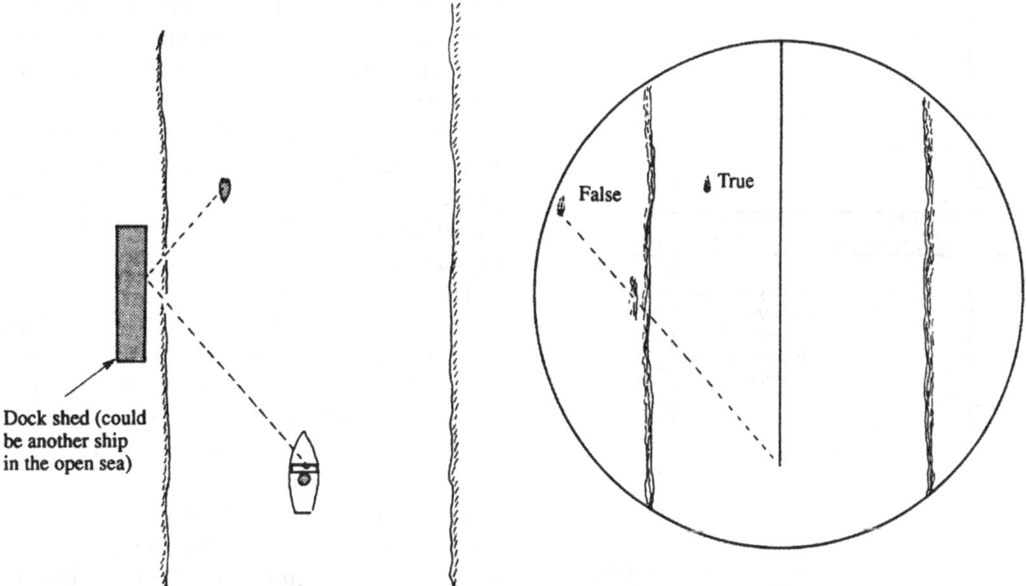

FIGURE 3.55 Indirect echoes which occur as a result of objects external to the ship.

Figure 3.56. Energy reflected back and forth between the bridge and the vessel (in a similar manner to that described in Section 3.9.3) will result in a false echo appearing about as far beyond the bridge as the observer's vessel is from it. As the vessel approaches the bridge, so the false echo will appear to approach the bridge from the opposite direction. In most cases, the bridge will cross the waterway at right angles which will mean that, however the vessel manoeuvres in an attempt to avoid the *'other ship'*, the observed echo will appear to have undertaken a complementary counter-manoeuvre, that is the false echo's position will be on the perpendicular from the observing vessel to the bridge, and its distance from the bridge will be about the same as that of the observing vessel.

Indirect echoes from objects ashore which result from reflections at the bridge can result in a considerable amount of 'clutter' appearing in the water area beyond the bridge. It should be noted that this clutter will be moving as the observing vessel moves and so changes its position in relation to the objects ashore. This clutter can also mask the real responses from other vessels using the waterway.

3.9.3 Multiple Echoes

Multiple echoes are likely when a target is close and energy bounces back and forth between the hulls of the target and the observing ship, with some of the energy entering the antenna at each return (Figure 3.57). The features of this form of response are that the echoes:

a. Lie along a single direction.
b. Are consistently spaced.
c. Tend to move in accord.
d. Tend to diminish in strength with the increase of range.

It should be noted that, while it would seem logical from Figure 3.57 for this form of

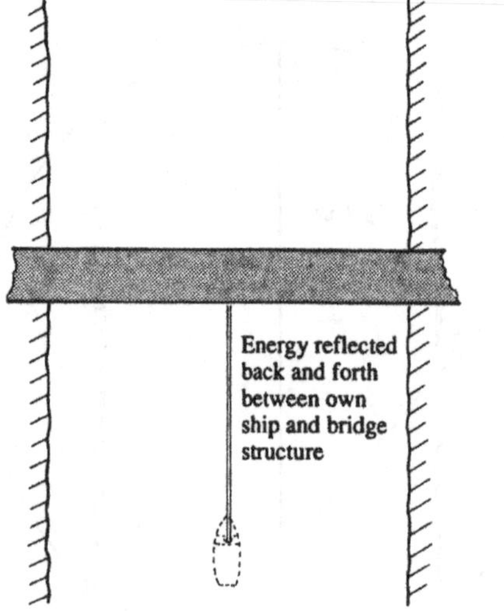

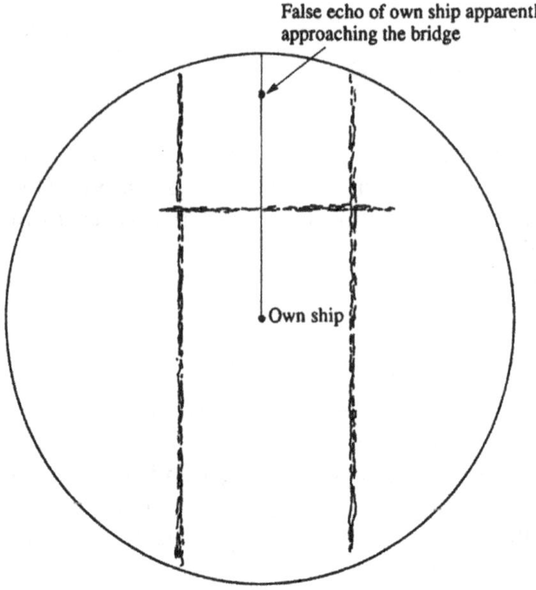

FIGURE 3.56 False echoes from bridges.

response to occur when vessels are abeam, it has been observed to occur on virtually all relative bearings.

Since it is most likely to occur when vessels are close (maybe as much as 4 NM but usually less), if the vessels are constrained within channels or rivers, etc. the false echoes will be seen to be navigating outside the channel or even on the shore.

Note: Because the outer echoes must keep on the same bearing line as the nearer ones, if plotted, their movement can be found to be higher than or contrary to that which might be anticipated.

The chances of multiple echoes can be reduced by ensuring that shorter pulse lengths are selected when using the lower range scales. It may be possible to produce this phenomenon deliberately in order to observe it, by selecting the longer pulse when on a lower range scale and with a close target.

In general, this form of false echo should not give real cause for concern since the genuine target producing the effect is always the nearest. Nonetheless, it is worthwhile to watch out for it in clear weather, when the absence of real targets beyond the nearest one can be confirmed.

Multiple echoes can also be produced by targets ashore, but the multiples will generally be lost amidst the confusion of other land echoes.

3.9.4 Side Echoes

Side echoes are again associated with targets that are at close range, and result from the radar beam being surrounded by smaller beams or lobes (see Section 2.5.2.2).

In the case of the farther target in Figure 3.58, the radar beam sweeping over it will result in the display of an echo of angular width which is approximately equal to the horizontal beamwidth (see Section 2.8.5).

In the case of the nearer target, if energy is returned as each lobe sweeps over it, the angular width of the displayed echo will be of the order of 100°. Some of the echoes will appear to be separate but all will be at the same range,

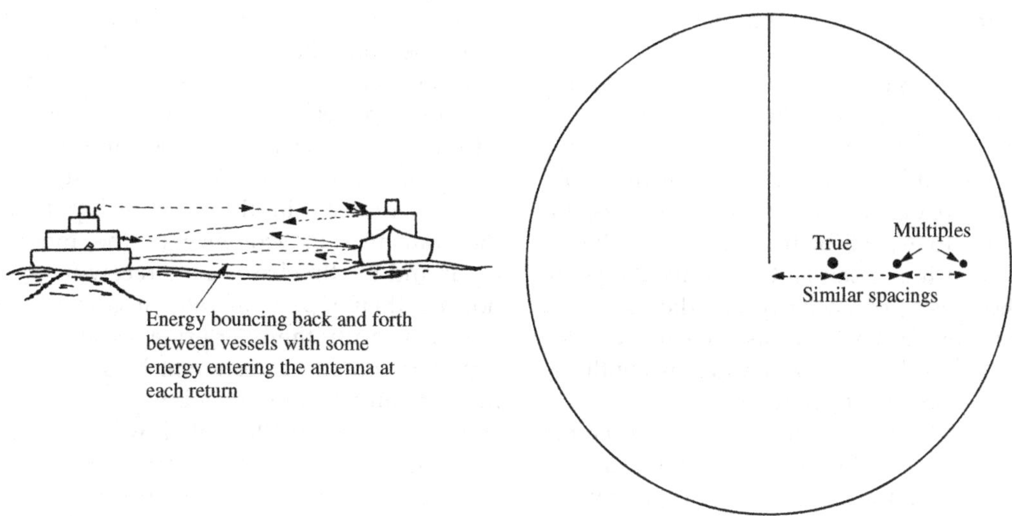

FIGURE 3.57 Multiple echoes.

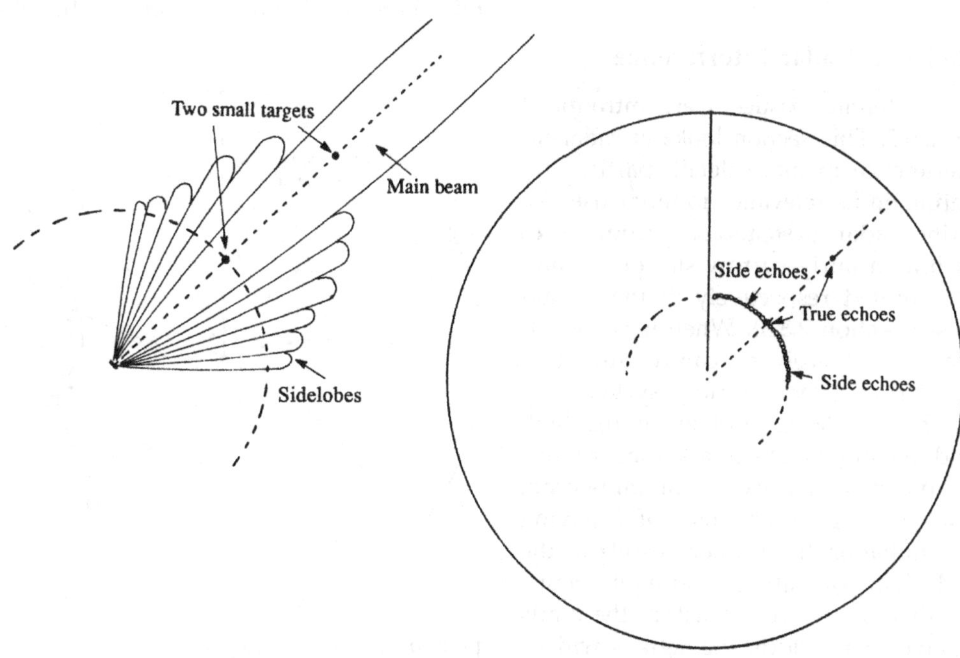

FIGURE 3.58 Side echoe.

i.e. as if all were lying on the same range circle.

This phenomenon is generally associated with smaller antennae and those which are dirty or damaged. While it is normal for the side echoes to be symmetrical about the genuine response, this may not be so in all cases; it is reasonable to expect the true response to be the strongest. If it is essential to identify the genuine echo, this can be achieved by the use of the gain or anti-clutter controls to suppress the weaker side echoes, as when dealing with sea clutter (see Sections 6.2.7 and 6.7).

The effect of side echoes on sea clutter is to intensify the displayed clutter by integration of the echoes and this can be a definite disadvantage.

When side echoes are observed where they have not been seen before, their cause should be determined (a damaged or dirty antenna should be the first suspect) and rectified at the first possible opportunity.

3.9.5 Radar-to-Radar Interference

Radar interference issues were introduced in Section 2.6.5. This section looks at radar-to-radar interference in more detail, particularly concentrating on its relevance to radar use. All civil marine radar systems are required to operate within a fairly narrow slot of 200 and 300 MHz allocated respectively in the S- and X-bands (see Section 2.3.3). When it is considered that the receiver bandwidth (see Section 2.6.4.2) of a marine radar system may be as much as 20 MHz, and given the high power and antenna height of a shipboard system, it is obvious that, except in mid-ocean, there is a very high probability of receiving interfering radiation from other vessels in the vicinity which are operating radar equipment.

If the radiation received is within the limits of the receiver bandwidth, the signals will be amplified in the same way as those reflected from targets and will produce a visible response on the display. In the case of returned echoes, there is a strict time sequence relationship between transmission and timebase cycles and both have the same repetition rate (see Section 2.3.3.2). By contrast, there is no synchronism in the relationship between the transmission sequence of the interfering radar and the timebase of the receiving system, nor are their repetition rates necessarily similar. As a result, the interfering signals produce a pattern on the screen which is totally random. It can take any form but frequently has a spiral character, as illustrated by Figure 3.59.

At best, the presence of interference may be a source of slight irritation which the practised observer may be able to cope with by mental filtering. In severe cases it may seriously impair the ability of the observer to detect targets.

If observers make prudent use of the standby condition (see Section 6.5), the general level of radar-to-radar interference present in the atmosphere can be reduced at source. Despite this, there is still a need to provide the observer

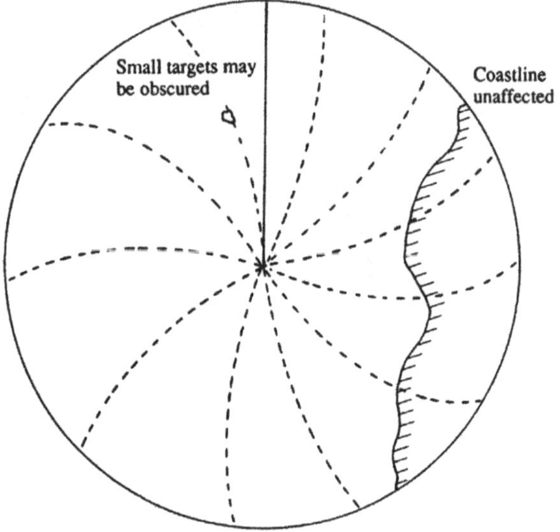

FIGURE 3.59 Radar-to-radar interference – appearance on the display.

with a means of suppressing displayed interference.

3.9.5.1 Interference Rejection

Where picture data is stored, interference can be filtered out by comparing the signals present in successive range cells (see Section 2.7.3) and rejecting those which do not appear on successive azimuth increments (lines). The principle of operation is sometimes known as *line-to-line correlation* and is illustrated by Figure 3.60.

The underlying logic is that, because of the random nature of interference, the probability of two signals appearing in the same range cell on successive range words is negligible. By contrast, as a point target is swept by the radar beam, it will be struck by a number of

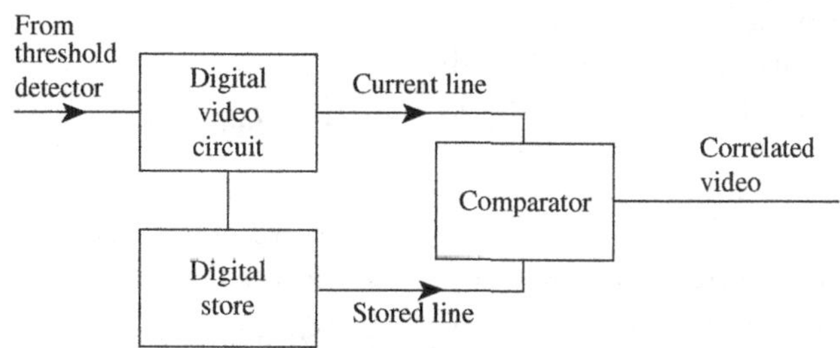

		Input line								Displayed line							
	S	0	0	1	0	1	0	0	0								
S	C	1	0	1	0	1	0	0	1	0	0	1	0	1	0	0	0
C	S	0	0	1	0	1	0	0	0	0	0	1	0	1	0	0	0
S	C	0	1	1	0	0	0	0	0	0	0	1	0	0	0	0	0
C	S	0	0	1	0	0	0	1	0	0	0	1	0	0	0	0	0
S	C	0	0	1	0	0	0	1	0	0	0	1	0	0	0	1	0
C	S	0	0	1	1	0	0	1	0	0	0	1	0	0	0	1	0
S	C	0	0	1	0	0	0	1	0	0	0	1	0	0	0	1	0
C	S	0	0	1	0	0	0	1	1	0	0	1	0	0	0	1	0
S	C	0	0	1	0	0	0	1	0	0	0	1	0	0	0	1	0
C	S	0	1	0	0	0	0	1	0	0	0	0	0	0	0	1	0
S	C	0	0	0	0	0	0	1	0	0	0	0	0	0	0	1	0
C	S	1	0	0	1	0	0	1	0	0	0	0	0	0	0	1	0
S	C	0	0	1	0	0	0	1	0	0	0	0	0	0	0	1	0

FIGURE 3.60 Line-to-line correlation. S = stored line, C = current line. Note. A target is only displayed on the radar screen when a '1'appears on successive displayed lines.

successive pulses all of which will register in the same range cell in successive lines. Thus interference will be removed by the correlation process, whereas the echoes of targets will be unaffected other than the rejection of the first strike. This loss is generally acceptable since even a point target will experience multiple strikes. The output from the interference rejection circuit is known as *correlated video*.

In Figure 3.60 successive lines of digital radar information are shown 'line by line'. The previous line of information is stored (S). This is compared with the current line (C) and only if the target appears on both lines is it displayed. The current line (C) then becomes the new stored (S) line in preparation for comparison with the next new current line.

Such a facility will also reduce the amount of displayed receiver noise and should therefore be switched off when setting the gain control (see Section 6.2.7). Correlation may also result in the rejection of the signals from some radar beacons (see Section 3.5). IMO Performance Standards (see Section 11.2.1 and 11.2.4) state that it should be possible to switch off *any* signal processing facility that might prevent a radar beacon from being shown on the display (see also adaptive gain in Section 3.6.3.3).

Some modern radar systems offer a more sophisticated facility described as *beam processing* by which all hits within the antenna beamwidth are compared with criteria for probability of detection. The information so derived is used to eliminate interference and reduce noise and, additionally, to reduce the angular width of the echo in an attempt to reduce the distorting effect of the horizontal beamwidth (see Sections 2.5.1.4 and 3.9.4).

Mutual interference may be particularly troublesome where two X-band or two S-band systems are fitted on the same vessel. Where the two systems are interswitched (see Section 2.10.8), a pulse synchronization unit may be incorporated. This unit will synchronize the PRFs of both units and time-shift the transmissions in an attempt to ensure that neither system will transmit during the timebase of the other. This will not necessarily eliminate all interference from the adjacent system. Ho/wever, the strongest interfering signals are likely to be those reflected from the ship's structure and nearby targets, and these will be not be displayed. A pulse synchronization circuit will make no contribution to the reduction of interference generated by other vessels.

3.9.6 Second-Trace Echoes

The conditions which must exist before this form of false echo will appear are fairly exceptional:

a. Extra super-refractive atmospheric conditions must exist.
b. Targets must be in the vicinity of the ship and in areas which are precisely defined by the PRF selected by the observer at the time.

Nonetheless, false echoes of this type are regularly reported in some meteorological journals, while on many more occasions observations go unreported as they are not recognized for what they are.

Under conditions of extra super-refraction (see Section 3.8.4), the radar energy follows closely the surface of the Earth and travels to greater distances than under standard conditions. This means that echoes from distant targets can arrive back at the receiver one trace late (i.e. on the 'second trace') or even later, be accepted by the receiver and so be displayed but obviously at an incorrect range.

Consider a radar system operating with the following parameters:

PRF = 1250 pulses/second (pps) and therefore the PRP = 800 μs (an echo which returns after an interval of 800 μs from transmission will be from a target which is at a range of 64 NM).

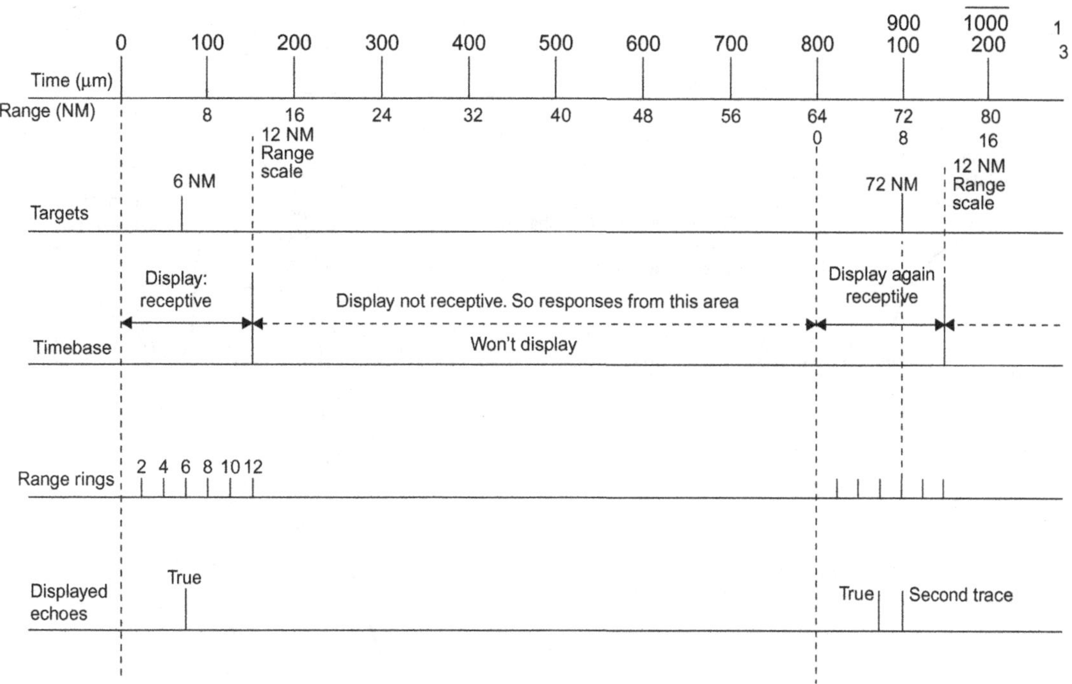

FIGURE 3.61 Time sequence graph.

Long pulse is being used on the 12 NM range scale (timebase = 150 μs) and there are objects at 6 and 72 NM which will respond well to radar pulses if struck.

Note: Under standard atmospheric conditions, the object at 72 NM would be well below the radar horizon; also, the energy in the transmitted pulse would be insufficient to travel to and from targets at that range.

Figure 3.61 shows the time sequence graph for the above data. Consider how this can be translated to the area around the vessel. Figure 3.62 gives the standing area coverage pattern.

By allowing a fairly long 'resting' period while no radar data is collected, radar designers are generally able to ensure that the energy in the pulse is sufficient to detect most targets out to the range of the scale in use (12 NM in this case), but insufficient to return echoes from targets in the 64–76 NM band. Targets in this latter band would otherwise arrive while the receiver was again receptive and the radar data collection period has restarted with a new pulse. So if targets in the 64–76 NM are strong enough to be detected, they would display together with those coming from within the range of the scale in use at the time.

Under extra super-refraction conditions, the situation changes in that the energy trapped in the duct is capable of returning echoes from much greater ranges and so second (or even third and fourth) trace echoes become a real possibility.

As can be seen from Figures 3.61 and 3.62, the positioning of the bands from which second-trace reception is possible is fixed by the PRF in use at the time, while the width of the band is equal to the range of the scale in use.

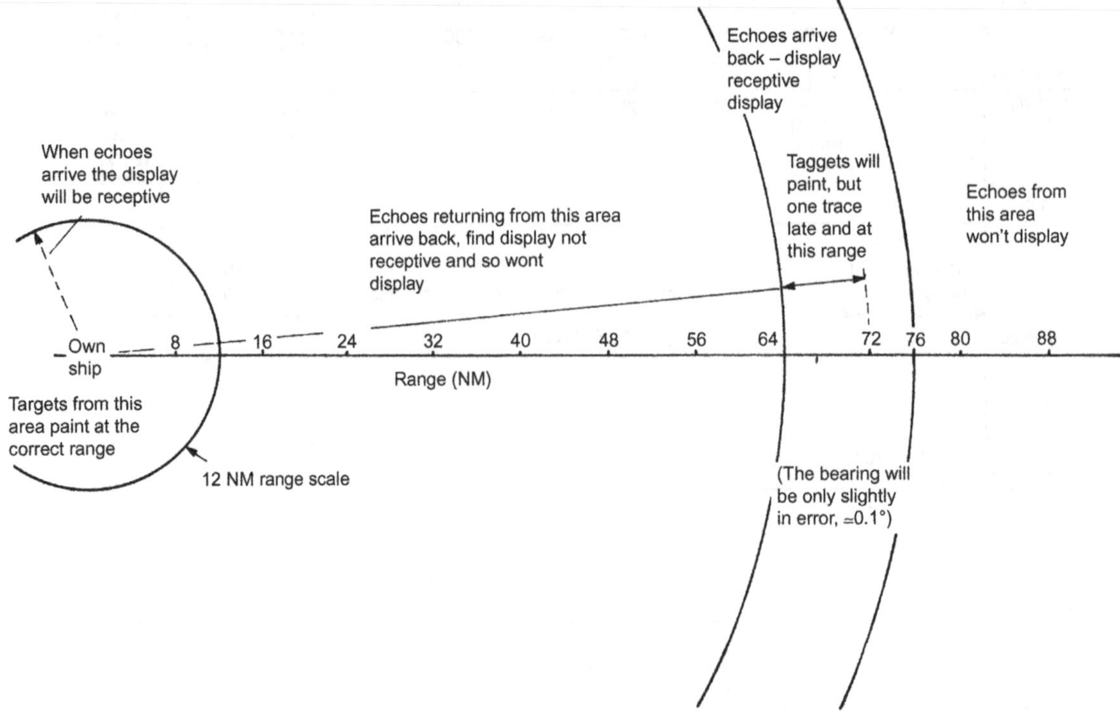

FIGURE 3.62 The standing area coverage pattern.

3.9.6.1 *The Recognition of Second-Trace Echoes*

It should be remembered that second-trace echoes are generally associated with fine weather conditions and so in most cases it will be possible to compare the radar picture with the visual scene. In some areas, notably the Red Sea and Arabian Gulf, some parts of the Mediterranean and off the west coast of Africa, extra super-refraction conditions are accompanied by a dry sand haze which can severely restrict visibility, thus rendering visual comparison with the radar picture impossible.

Where conditions are unfavourable for visual comparison, some indications as to the second-trace nature of the echoes may be recognizable. The radar picture should be compared with echoes from the coastline which might be expected from the vessel's present position. Where the DR position is sufficiently in doubt, other effects should be sought.

Consider a long straight coastline which is returned as a second-trace echo. The coastline would be as depicted in Figure 3.63, with the distortion towards the origin of the display. Also, as the vessel steamed parallel to the coast, the 'point' of land would appear to remain abeam to starboard.

Alternatively, the same coast line can be used as the relative track which an island would follow. When plotted it should be parallel to the heading marker but, as a second-trace echo, would produce a plotted apparent-motion line akin to the curve of the straight coastline. It might also move out of the band, either at long range or short range, and so disappear. The latter experience can be somewhat disconcerting, especially as the 'target' would

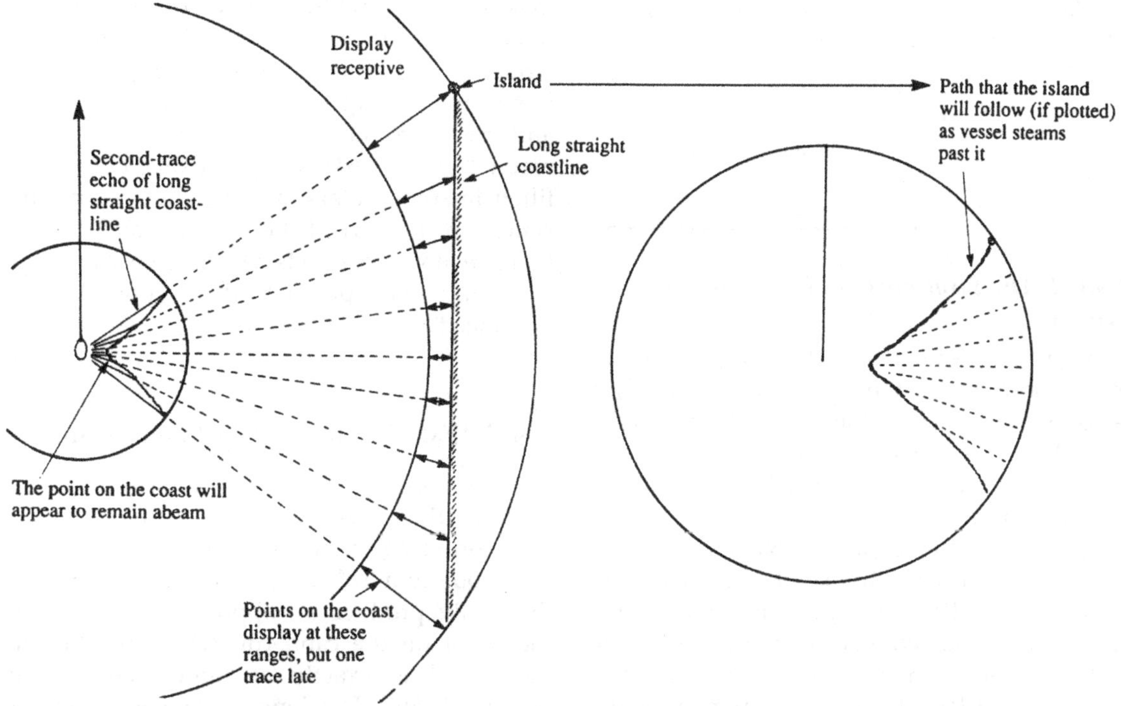

FIGURE 3.63 The distortion of a straight coastline which returns as a second-trace echo.

appear to be on a collision course for much of the time when forward of the beam.

To assess (assuming atmospheric conditions are appropriate) whether it is possible for echoes to be second-trace or multi-trace, it is necessary to study the charted coastline and the accuracy with which the vessel's position is known. This must then be related to the radar's PRF and range scale in use at the time.

EXAMPLE 3.5

An echo is observed at a range of 4 NM but visual observation shows nothing in the vicinity of the vessel. If the PRF is 1000 pps and the 12 NM range scale is in use at the time, what could be the range of the true target? (atmospheric conditions suggest extra super-refraction).
PRF = 1000 pps; therefore, PRP = 1000 μs, which relates to 80 NM.

The bands from which echoes could return are:

80 – 92 NM 160 – 172 NM
240 – 252 NM ···

therefore the target's range could be:
84 NM 164 NM 244 NM···
In general terms, it can be shown that:

$$R_{nt} = \text{PRP in NM} + \text{target's displayed range}$$
$$= \frac{8 \times 10^4 \times (n-1)}{\text{PRF}} + R_d$$

where:

R_{nt} is the range (NM) of an nth-trace echo
R_d is the displayed range of the target
n is the number of the trace on which the echo appears (the trace associated with transmission is counted as 1).

In the example above, if the false echo were third-trace then its range would be:

$$R_{nt} = \left(\frac{80,000 \times 2}{1000}\right) + 4\ \text{NM}$$
$$= 164\ \text{NM (as above)}$$

3.9.6.2 The Elimination of Second-Trace Echoes

By using a knowledge of the PRFs selected by the observer on the various range scales, it is possible to vary the areas from which second-trace echoes are possible.

If, in the example above, it had been possible to change from a PRF of 1000 to one of 500 pps, then the bands at 80 and 240 NM would have been eliminated. By observing what happened to the echo when this change was made, some deduction as to the true range of the target could have been made. If, after the change of PRF, the echo still remained, the true target must have been at a range of 164 NM; if it had disappeared, then its true range could have been either 84 or 244 NM with the former being more likely. The principle here is to change to a lower PRF by selecting a longer range scale. If this is to be done, a knowledge of the parameters of the particular radar is essential.

Some traditional equipment had a false target elimination (FTE) control. The system allowed the observer to deliberately change the PRF to one which is not a sub-multiple of the PRF in use at the time, and so move the second (or multi-) trace bands to totally different areas which eliminate second-trace echoes arising from the original areas.

Most modern radar systems incorporate an automatic elimination circuit which continually varies the PRF at a frequency of, say, 50 Hz on the nominal value. This has the effect of moving the second-trace echo areas 50 times per second, which in turn has the effect of making the individual strikes of any second-trace echoes move in and out. Thus successive strikes would paint at different ranges and so would not line up side by side on the display, and therefore no recognizable paint would result (Figure 3.64). These circuits have been fitted in some radars; because they require no action on the part of the observer and appear to have no side effects which give cause for concern, their presence has gone largely unnoticed.

3.9.7 'False' Echoes from Power Cables

If there are power or other cables crossing a channel, then it is likely they will produce a very confusing radar echo. They act like a very wide but thin reflecting plate. It means that they only produce an appreciable echo from the point on the cable-run where the line to the vessel is exactly at right angles to it (Figure 3.65b). The long cable has a similar effect to the use of a mirror at optical frequencies: you only see the reflection of your eye at the point of the mirror that it is exactly perpendicular to the line joining it and the mirror. The eye is acting like the radar's transmitter, antenna and receiver: the mirror is acting like the cable.

Unfortunately, the rest of the cable does not show up as an echo because all the radar's energy that impinges on the cable away from the perpendicular point is not reflected back to the radar. This means that it can be very difficult to associate the observed echo with the cable and thereby be alerted that it is misleading. Where the cable is at right angles to the channel, the echo from the cable will appear in the channel so that, however the vessel manoeuvres in the channel to avoid it, the 'false' echo will always move into the vessel's path.

Where the cable is angled across the waterway, the false echo may initially appear among

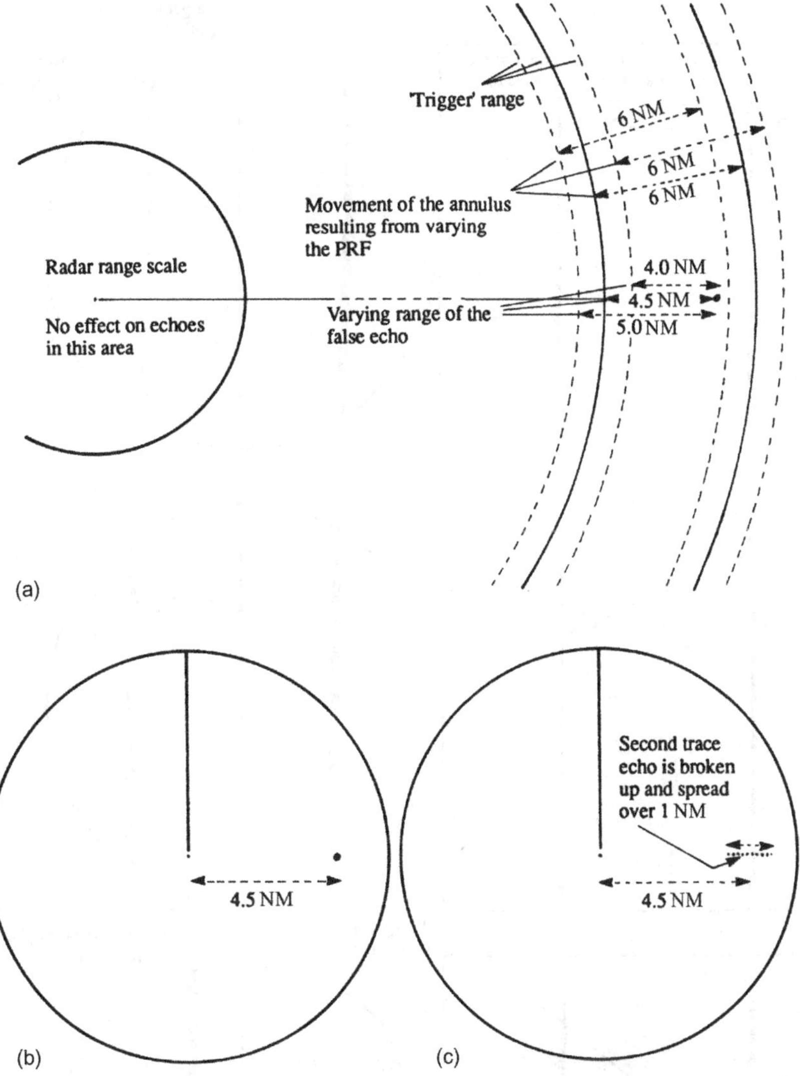

FIGURE 3.64 **Second-trace echoes – the effect of varying the PRF.** (a) Effect on the standing pattern. (b) The target as a second-trace echo with a fixed PRF. (c) The effect of varying the PRF.

the shore echoes and so go unnoticed, but as the vessel approaches the cable, the false echo will appear on the water as if from a vessel on a converging course.

Consider the situations illustrated in Figures 3.65(b) and 3.65(c). As the vessel approaches the cable, a 'vessel' would appear to put out from the starboard bank and proceed on a collision course. Any attempt by the observing vessel to pass under the 'stern' of the false echo would cause the false echo to return towards the bank, i.e. again into the vessel's path. If the observing vessel stopped, the 'target' would also appear to stop.

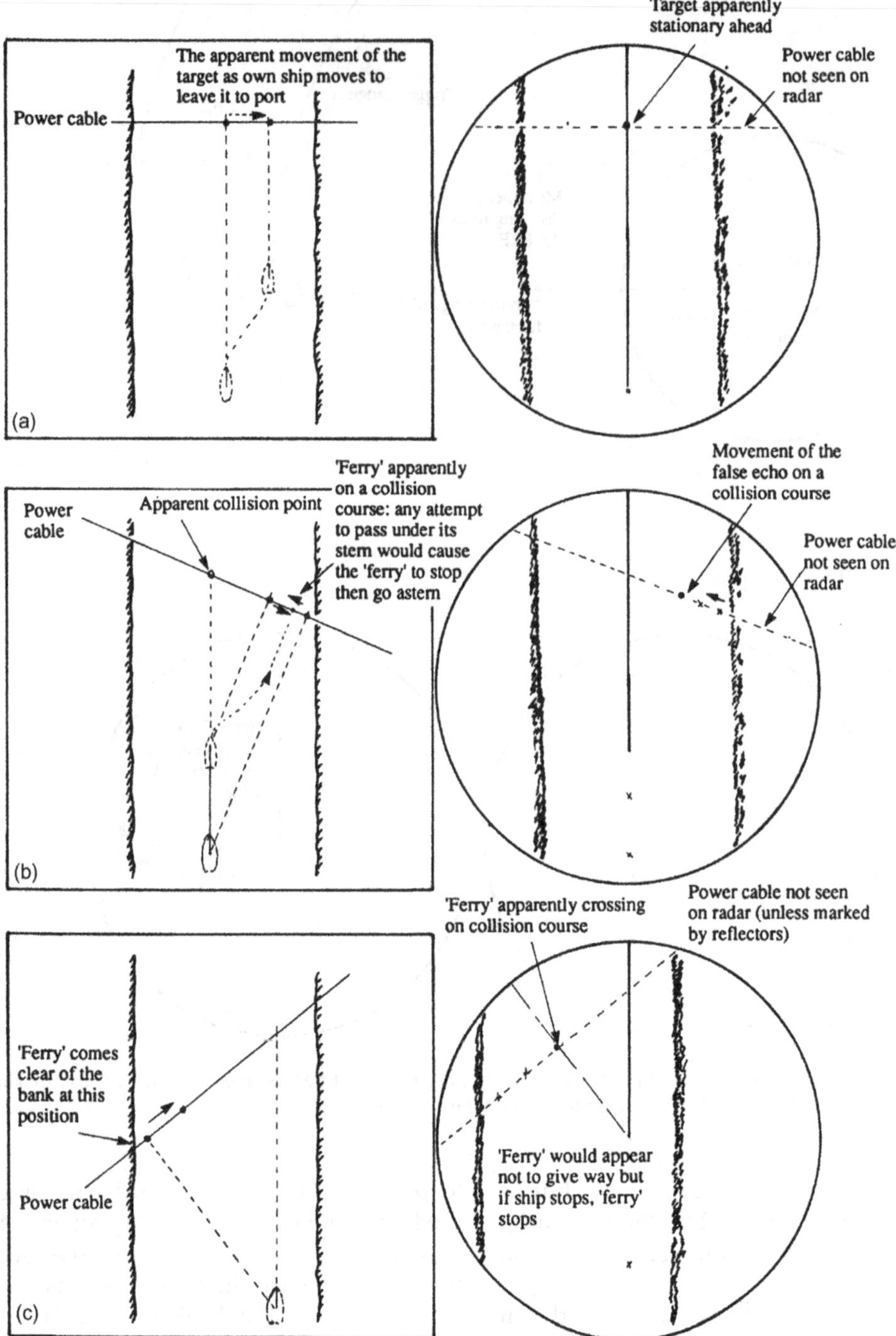

FIGURE 3.65 False echoes due to overhead power cables.

In Figure 3.65(c), the false echo would put out from the port bank, again on a collision course. Here the logical manoeuvre would be for the observing vessel to move farther over to the starboard side of the channel. The false echo would continue to crowd the observing vessel into the starboard bank. The result of stopping or a port manoeuvre would be as described above.

On some waterways, power cables have had radar reflectors fitted to them in order that their line will appear on the radar display. The unusual behaviour of echoes in the vicinity of the cable may thus be associated with the cable and should be treated with due caution.

CHAPTER

4

Automatic Radar Target Tracking, Specified Facilities

4.1 INTRODUCTION

With computers developing at an ever-increasing pace, it was only a matter of time before their capabilities were harnessed to assist the mariner in resolving the continuing problem of tracking targets and analysing movements when faced with heavy traffic.

In the past, many semi-automatic devices were developed to assist in this task. The first device recognized by IMO and other regulatory bodies was called automatic radar plotting aid (ARPA). Performance standards were provided by IMO in 1979 for the ARPA and it became a requirement for there to be a navigator trained on ARPA on the bridge, if one was fitted. It gradually became a requirement that larger ships were fitted with ARPA, although they were often found on other ships.

In IMO Performance Standards for new radar equipment on ships fitted in 1997 or later, a slightly reduced specification was defined for smaller ships or for use as a second plotting device on larger ships. This short-lived device was termed an automatic tracking aid (ATA) which had many (but not quite all) of the features of a full ARPA (see Section 7.10.6). At the same time IMO also specified the standards for an electronic plotting aid (EPA) that aids manual plotting for installation on very small ships and which is covered in Section 7.10.6.

The latest 2004 IMO Performance Standards (see Section 11.2.1) apply to new vessels (and new equipment on existing vessels) from 2008. They have removed the terminology of ARPA and instead refer to automatic 'target tracking' (TT) which also includes an automatic identification system (AIS – see Chapter 5). The ATA and EPA devices defined in the 1997 regulations have both disappeared from the new regulations, as all new SOLAS vessels are now required to have a TT. However, the TT for larger vessels is required to have a trial manoeuvre facility and automatic acquisition (see Table 4.2), while vessels of less than 10,000 grt do not. Thus some of the differentiation intended by the now obsolete ATA remains.

Many parts of industry continue to refer to the TT by its separate components of ARPA and AIS. Indeed, the 2011 UK MCA NAEST training course regulations for electronic navigation use the separate ARPA and AIS terminology.

The 1997 IMO Performance Standards for ARPA state that it should:

> '. . .reduce the workload of observers by enabling them to automatically obtain information about plotted targets, so that they can perform as well with several separate targets as they can by manually plotting a single target'.

Although this statement is not in the latest IMO Performance Standards, it still remains a good description of the function of ARPAs.

4.1.1 Integral and Stand-Alone Displays

When ARPA was first introduced, many of the installations were 'add-ons' to the existing conventional radars. They provided all of the ARPA facilities, but they derived their data from a host radar. The attraction was that the installation expense was reduced and the existing radar retained. Such displays became known as stand-alone ARPAs. It was never an ideal solution and very few, if any, such stand-alone equipment exists today. The concept of a stand-alone display has also disappeared from the latest regulations.

In modern integral equipment all radar, tracked targets and data from other sources (including comparable data from AIS) can be processed in one unit and shown on one display. This has considerable operational and processing advantages.

4.1.2 Carriage Requirements, Standards and Operator Training

In this section the current requirements are considered for new radar equipment installed after 1 July 2008.

Table 4.1 summarizes the carriage requirements for radar and plotting devices on new vessels as stated by IMO-SOLAS Chapter V as amended in 2001 (see Section 11.1). The full ARPA with trial manoeuvre and auto acquisition is only a requirement for one radar on vessels above 10,000 grt, as the second radar can be fitted with a 'reduced' ARPA without trial manoeuvre (see Section 4.4.3) and automatic acquisition (see Section 4.2.3). Ships below 10,000 grt are not required to carry a full ARPA, but only one or two reduced ARPAs depending on their size.

Ever since ARPA was introduced, most maritime administrations have insisted that the operator should be specifically trained to use the ARPA, and not just in radar observer manual plotting techniques. For instance, UK regulations state that:

> 'When a UK ship which is required to be fitted with ARPA is at sea and a radar watch is being kept on the ARPA, the installation shall be under the control of a person qualified in the operational use of ARPA, who may be assisted by unqualified personnel'. (Edited extract from UK SI 1993 No. 69, part IX, see Section 11.4.2.)

TABLE 4.1 Minimum IMO Radar and Plotting Provision for New Ships

Ship Size	Radar	Plotting	Display Diameter
300–499 grt (and all passenger vessels below 300 grt)	1 X-band	Reduced ARPA	180 mm
500–2999 grt	1 X-band	Reduced ARPA	250 mm
3000–9999 grt	1 X-band and 1 S or X-band	Reduced ARPA	250 mm
10,000 + grt	1 X-band and	1 ARPA	320 mm
	1 S or X-band	1 Reduced ARPA	

New SOLAS ships carrying ARPA or reduced ARPA require an integrated input from a gyro and a speed log that measure speed through the water. In the regulations, these are known as the THD (transmitting heading device) and SDME (speed and distance measuring equipment). Without reasonably accurate data from these sources, the ARPA functions have limited value. This is discussed further in Chapter 9.

Table 4.2 gives more detail on display sizes and target capacities and is reproduced from Section 11.2.1. Note that regulations differ slightly for high speed craft, which are defined as those capable of 30 knots or more. The two main functional differences from the previous standards are the general requirement for an automatic target tracker on vessels less than 500 grt and the specific requirement to include information from AIS targets (see Chapter 5 for a full discussion of AIS). Other differences (from previous regulations) are that the minimum number of radar tracked targets has been increased for all ship sizes and the performance standards have split ship sizes into three categories.

The new regulations also make a separate distinction between the operational radar display diameter and the dimensions of the display screen of which the operational radar is a part. There is an implicit assumption that the display is rectangular and therefore that raster display technology is used.

4.1.3 Compliance with IMO Performance Standards

Irrespective of which type of equipment is installed, it is still necessary for it to comply with IMO Performance Standards, as well as with the various requirements of national legislation. Only the facilities which are specified by IMO Performance Standards will be considered in this chapter. Where appropriate, at the beginning of each relevant section, extracts from the current IMO Performance Standards for ARPAs will be quoted (in the form of indented paragraphs) to the extent necessary to support the section.

4.2 THE ACQUISITION OF TARGETS

Acquisition is the term used to describe the process whereby TT is initiated. This may be

TABLE 4.2 Differences in the Performance Requirements for Various Sizes of Ship/Craft to Which SOLAS Applies

Size of Ship/Craft	<500 grt	500 grt to <10,000 grt and HSC <10,000 grt	All Ships/Craft ≥10,000 grt
Minimum operational display area diameter	180 mm	250 mm	320 mm
Minimum display area	195 × 195 mm	270 × 270 mm	340 × 340 mm
Auto acquisition of targets	–	–	Yes
Minimum *acquired* radar target capacity	20	30	40
Minimum *activated* AIS target capacity	20	30	40
Minimum *sleeping* AIS target capacity	100	150	200
Trial manoeuvre	–	–	Yes

'manual', in which case the operator, using the screen marker (see below), indicates to the computer which targets are to be tracked, or it may be 'automatic', when the computer is programmed to acquire targets which enter specified boundaries. When a target is 'acquired', the computer starts collecting data relating to that target.

A graphic symbol known as the *screen marker*, controlled by a joystick or tracker ball, is positioned over the target. When the 'acquire' button is pressed, an area centred on the screen marker is defined within the computer memory. This area is termed the 'tracking gate' or 'tracking window'. The gate is made to appear automatically on some ARPA displays; on others, the operator may display it if desired. Within the gate, the computer will expect to find evidence of a target, that is a binary 1 in the appropriate memory location (see Sections 2.7 and 4.3.6).

4.2.1 The Acquisition Specification

Target acquisition may be manual or automatic for relative speeds up to 100 knots (140 knots for high speed craft radar). However, there should always be a facility to provide for manual acquisition and cancellation. ARPAs with automatic acquisition should have a facility to suppress acquisition in certain areas. On any range scale where acquisition is suppressed over a certain area, the area of acquisition should be indicated on the display.

Although it would seem that anything is possible in today's technological climate, some practical problems still exist and fully automatic acquisition systems do not give quite the results which one might have been led to expect. The main problem with automatic acquisition is that the 'sensitivity' of the detection circuitry, if set too high, will acquire thermal noise and clutter, leading to false alarms, while if its sensitivity is reduced, poor-response targets can evade the plotter.

4.2.2 Manual Acquisition

In this case, the operator specifies the target to be acquired and subsequently tracked. To do this, a joystick and screen marker or tracker ball and screen marker are used. The target is entered into (acquired) or removed from (cancelled) the computer memory when the *acquire* or *cancel* button is pressed (Figure 4.1).

In some ARPAs, tracking can be initiated by touching the position of the target on a special touch-sensitive screen (see Section 2.8.3).

Sections 4.2.3–4.2.6 are not relevant to the reduced ARPA specification (vessels less than 10,000 grt).

4.2.3 Fully Automatic Acquisition

Every echo (up to some 200 maximum) which is detected in the receiver is tested against a set of published criteria. Large (land-sized) targets are rejected and the remainder prioritized according to the set criteria. (*Note*: This may only be on the basis of range, e.g. the

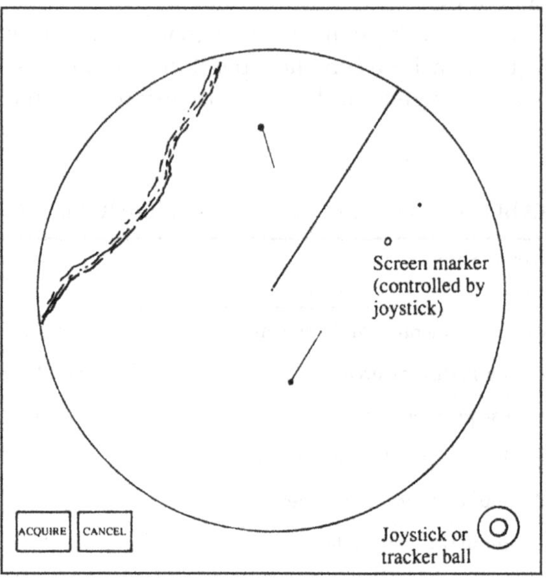

FIGURE 4.1 Manual acquisition.

nearest 20, or, at the other extreme, it may involve tests of CPA, TCPA, range, bearing and other criteria). Tracking is then initiated on the first (say) 20 in the priority order. Targets at the bottom of the list are deleted or acquired as their ranking changes, for example a target moving away may be dropped when there are (say) 20 nearer than it. Some ARPAs have more complicated priority criteria where parameters can be varied by the operator.

4.2.4 Automatic Acquisition by Area

In this type, an area around the ship is specified by the operator and any target that is detected within this area is acquired. Some prioritizing system must operate when the number of targets exceeds the number of tracking channels (Figure 4.2).

4.2.5 Guard Zones

In this system, zones (usually up to two) may be specified by arc and depth. Targets entering the zone will be acquired and an alarm activated. A 'tracks full' warning will be given when all tracking channels are in use and it will then be up to the operator to decide which of the acquired targets can be cancelled (Figure 4.3).

Note: Target 'A' may only be acquired at a late stage in the encounter and target 'B', which appears for the first time inside the inner zone, may not be acquired at all. Careful thought must be given when setting up guard zones and the display must be regularly observed.

4.2.6 Guard Rings and Area Rejection Boundaries

With this method of acquisition, the usual provision is for up to two 'rings' (of predetermined depth) plus up to two area rejection boundaries (ARBs). The rings and ARBs may be positioned by the operator

When a target is automatically acquired in a guard zone/guard area, it is usual for an alarm to be activated to attract the operator's attention (see Section 4.7.1). The target activating the

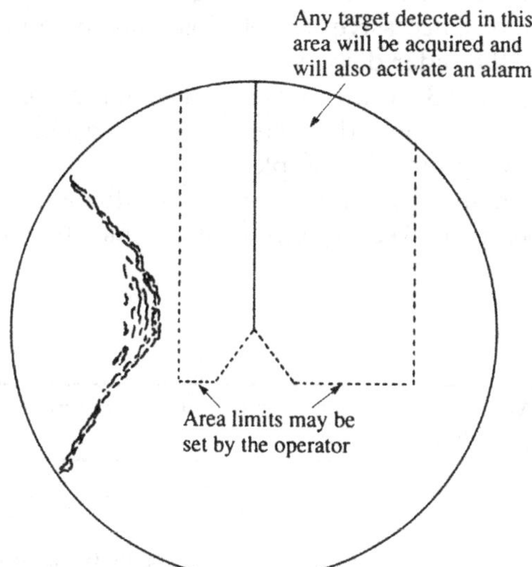

FIGURE 4.2 Automatic acquisition by area.

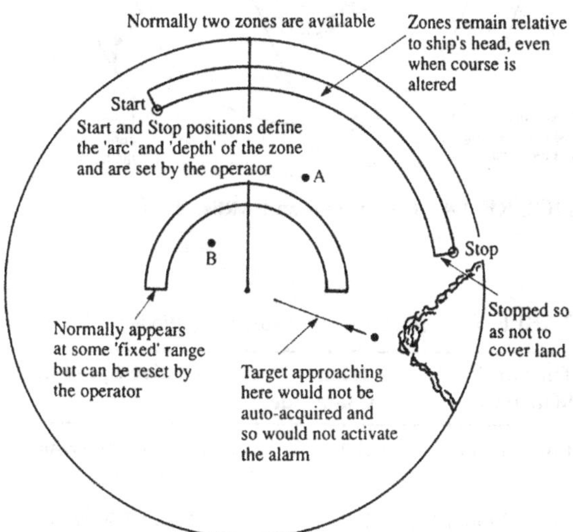

FIGURE 4.3 Guard zones.

alarm will be indicated on the screen by, for example, a flashing symbol (Figure 4.4).

In general, automatic acquisition has traditionally not been as successful as was at first predicted. There is a tendency to acquire sea clutter, rain clutter, noise and interference, while disassociated elements of land echoes will very quickly fill up the available tracking channels. Land echoes can be excluded by careful setting of the zones/areas and ARBs, but spurious targets (e.g. clutter), after having been acquired, are quickly lost and the 'lost target' alarm can sound continually. In future, this situation might be improved with the introduction of coherent radar technology (see Section 2.9.1).

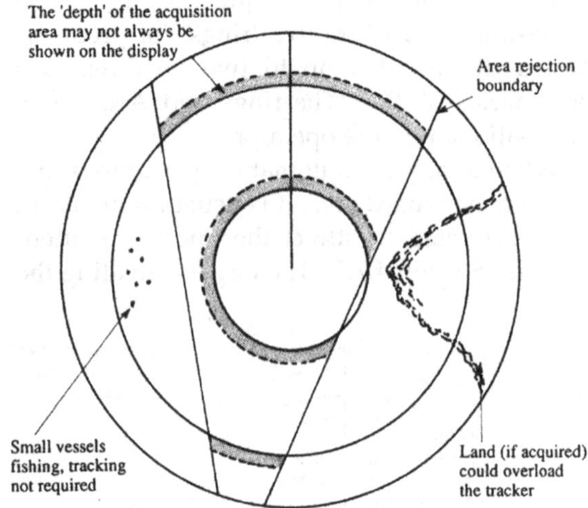
The 'depth' of the acquisition area may not always be shown on the display
Area rejection boundary
Small vessels fishing, tracking not required
Land (if acquired) could overload the tracker

FIGURE 4.4 Guard rings and ARBs.

While it is argued that automatic acquisition will reduce the operator's workload, in practice there is a tendency for it to acquire spurious targets, also to 'over-acquire' and so clutter the screen with unnecessary and unwanted vectors. This has led to auto-acquisition falling out of favour. Enquiries have indicated that it is rarely used in areas of high-density traffic, but can sometimes be useful on long ocean passages where the number of targets is small and there is the danger of loss of concentration by the officer of the watch due to boredom.

Manual acquisition can be very quick and also selective, and hence the perceived need for automatic acquisition has not really materialized.

Guard zones/areas should be regarded as an additional, rather than an alternative, means of keeping a proper lookout.

4.3 THE TRACKING OF TARGETS

4.3.1 The Tracking Specification

In previous performance standards the accuracy of tracking was stated to be not less than that obtainable by using manual recordings of successive target positions obtained from the radar display.

Table 4.3 indicates the tracked target accuracy now required. Accuracy is a compromise with the time taken to plot.

In many cases it may be obvious that a target is being tracked by virtue of the fact that its

TABLE 4.3 Tracked Target Accuracy (95% Probability Figures)

Time of Steady State (min)	Relative Course (deg)	Relative Speed (kn)	CPA (NM)	TCPA (min)	True Course (deg)	True Speed (kn)
1 min: Trend	11	1.5% or 10% (whichever is greater)	1.0	–	–	–
3 min: Motion	3	0.8% or 1% (whichever is greater)	0.3	0.5	5	0.5% or 1% (whichever is greater)

predicted movement will be indicated by a graphic line known as a vector (see Section 4.4). The line originates on the target and its remote end indicates the predicted position of the target after an elapsed time selected by the observer. However, the need for tracked targets to be clearly indicated on the display is important because in the early stages (up to about 1 min) of tracking a fresh target, in most systems the vector is suppressed because the available data is unlikely to be sufficiently accurate or stable. Furthermore, in certain cases, even when the vector is present it may have zero length (e.g. the true vector of a stationary target or the relative vector of a target on the same course and speed as the observing vessel). Some manufacturers overcome this problem by replacing the zero length with the symbol 'Z'.

Once tracking is initiated, by whatever method, the tracker will continue to follow the target until tracking is cancelled (manually, or automatically, because some other criterion has been met, e.g. 'more than 16 miles away and range increasing') or the target is 'lost'.

The precise nature of the algorithms used to ensure that the tracking window will faithfully follow the target varies with manufacturer. The fine details of such methods are beyond the scope of this text, but the general principle can be illustrated by the following description of the technique known as *rate aiding*.

4.3.2 Rate Aiding

When the target is first acquired, a large gate is necessary since there is uncertainty as to the direction in which the target will move.

Figure 4.5 shows how successive positions can be used to improve the forecast of the next position in which the target is expected to appear. The radius of the gate is really a measure of confidence in the tracking and the smaller this value becomes, the more precise the prediction will be. In this way it is possible to establish a feedback loop in the computer which will progressively reduce the size of the tracking gate.

The advantages of a reduced tracking gate are:

a. A lower likelihood of target swap (see Section 4.3.5).
b. An improved ability to track targets through rain and sea clutter.
c. An ability to continue tracking, even when target response is intermittent.

One problem which can arise with reduced gate size is that if a target manoeuvres and, as a result, is not found by the computer in the predicted position, the computer may continue to track and look in the predicted direction and end up by losing the target altogether. To avoid this possibility, as soon as the target is missed, i.e. not found in the predicted position, the gate size is increased (and the tracking duration reduced – see Section 4.3.6). If the target is still detectable and is subsequently found, the tracking will resume and a new track will gradually stabilize.

If, after six fruitless scans, the target is still not found then an alarm is activated and a flashing marker is displayed at the target's last observed position (see Section 4.7.3).

4.3.3 The Number of Targets to Be Tracked

The ARPA should be able to automatically track, process, simultaneously display and continuously update the information on the number of targets specified in Table 4.2, whether manually or automatically acquired. This means that the reduced ARPA requires at least 20–30 tracking channels, depending on ship size. The full ARPA requires a minimum of 40 tracking channels.

It has been suggested that 20 tracking channels might be insufficient in heavy traffic, but

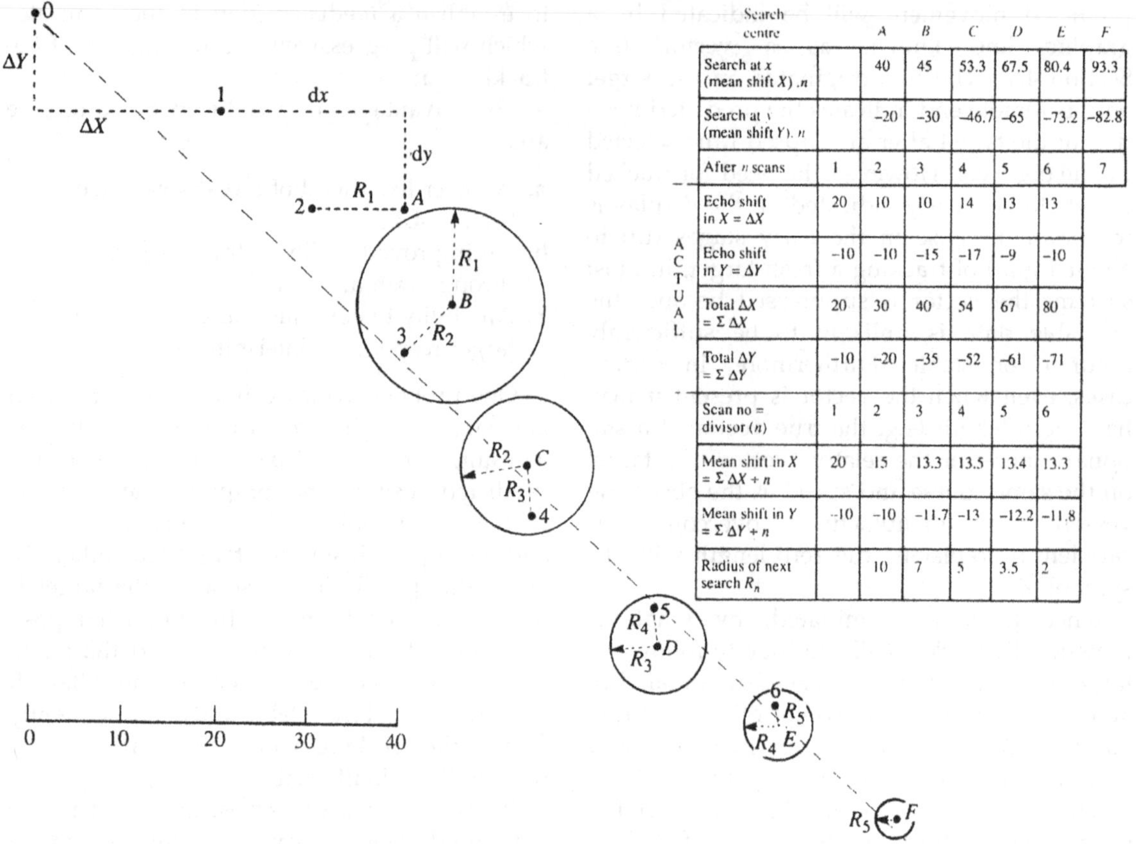

Search centre		A	B	C	D	E	F
Search at x (mean shift X) $.n$		40	45	53.3	67.5	80.4	93.3
Search at y (mean shift Y). n		−20	−30	−46.7	−65	−73.2	−82.8
After n scans	1	2	3	4	5	6	7
Echo shift in $X = \Delta X$	20	10	10	14	13	13	
Echo shift in $Y = \Delta Y$	−10	−10	−15	−17	−9	−10	
Total $\Delta X = \Sigma \Delta X$	20	30	40	54	67	80	
Total $\Delta Y = \Sigma \Delta Y$	−10	−20	−35	−52	−61	−71	
Scan no = divisor (n)	1	2	3	4	5	6	
Mean shift in $X = \Sigma \Delta X + n$	20	15	13.3	13.5	13.4	13.3	
Mean shift in $Y = \Sigma \Delta Y + n$	−10	−10	−11.7	−13	−12.2	−11.8	
Radius of next search R_n		10	7	5	3.5	2	

FIGURE 4.5 Rate aiding.

from practical experience it has been found that ship's officers can quickly identify the targets which need to be tracked and acquire them. (Although at times there will be some 40 plus targets on the screen, not all of them will need to be tracked.) In fact, it has been found that an excess of vectors can produce 'ARPA clutter' (i.e. information overload) and be counter-productive.

It should be noted that a higher tracking capability is required by the performance standard where automatic acquisition is required. This recognizes the tendency of the latter facility to over-acquire (see Section 4.2.6) and compensates for it to some extent.

4.3.4 Target Loss

The system should continue to track an acquired target which is clearly distinguishable on the display for 5 out of 10 consecutive scans or equivalent.

(The term *scan* tends to be used rather loosely in radar terminology. Sometimes it is used to describe one line, as in the term 'interscan period', while on other occasions it refers to one antenna rotation. In the above context it refers to the latter.)

It should be noted here that if, for some reason, a response from a tracked target is not received on a particular scan, the ARPA must not immediately declare the target lost. Also it is implied that some form of 'search' for it must take place, for example by opening the tracking gate rather than merely looking in the limited area in which it was expected but failed to be detected.

4.3.5 Target Swap

The possibility of tracking errors, including target swap, should be minimized by design (Figure 4.6).

Target swap is likely when two targets respond within the tracking gate at the same time. When this happens, the tracker can become confused and the vector(s) may transfer to the wrong target. To minimize this problem, the gate should be made as small as possible, the movement of the target should be predicted and the gate moved on at each scan as described under 'rate aiding' (see Section 4.3.2).

The two requirements that target swap be minimized by the ARPA (and auto-tracking) design and that tracking be continued even if no response is received for a period of time are thus to some extent achieved by the common solution of rate aiding.

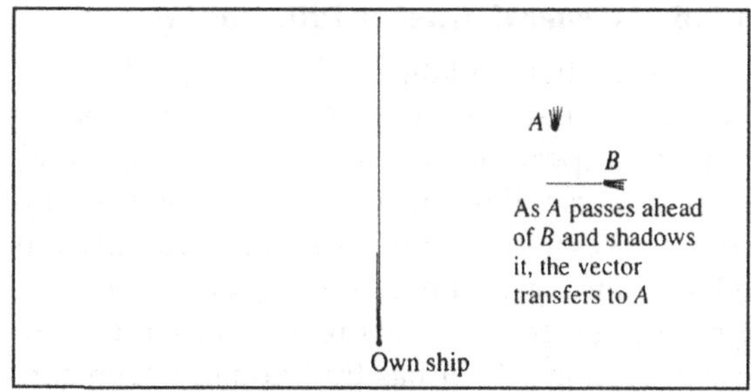

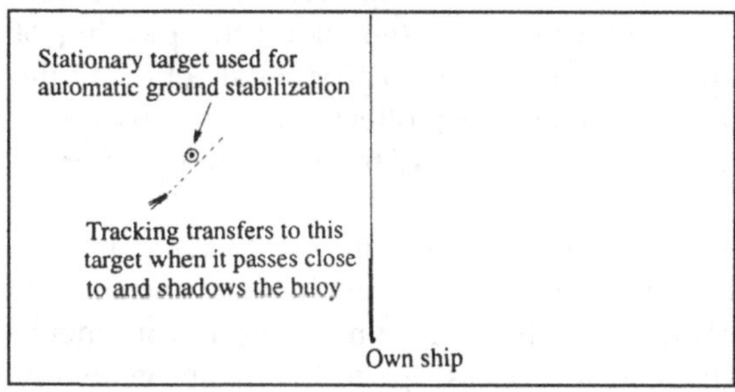

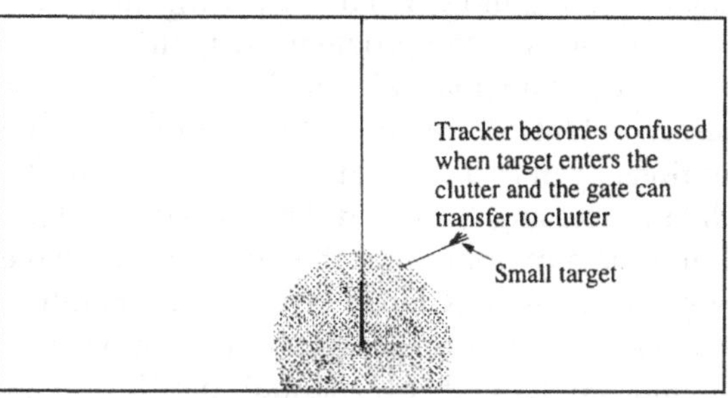

FIGURE 4.6 Target swap.

4.3.6 The Analysis of Tracks and the Display of Data

When a target is acquired the system should present the targets motion within 1 min and the prediction within 3 min.

The performance measures stated in Table 4.3 therefore define two levels of expected accuracy:

1. A lower level (after 1 min) relating to the target's 'trend', which is an early indication of the target's relative track.
2. A higher level (after 3 min) relating to the target's predicted motion; this means the best possible estimate of the target's relative- and true-motion data.

However, for the general discussion of tracking philosophy which follows, it is adequate to appreciate that essentially the performance standard requires the computer to deduce and display relative-motion data to a specified level of accuracy after 1 min of steady-state tracking, and both relative-and true-motion data to a higher specified accuracy after 3 min of steady-state tracking. Steady-state tracking means that neither the target nor the observing vessel has manoeuvred and no significant change has taken place in the errors of the input sensors.

4.3.6.1 General Tracker Philosophy

In order to calculate the data required by the performance standard, the computer must store sequential positions of each tracked target and then analyse the movement represented by such successive positions so as to evaluate the relative-motion and the true-motion of each target. The computer is thus required to automate the operation which has traditionally been carried out manually by the recording of successive target positions (either by the plotting of ranges and bearings on a paper sheet or by the use of a reflection plotter) and the subsequent resolution of the *OAW* triangle (see Chapter 7). Automation of any process normally carried out by an intelligent being is always complex. The injection of pseudo-intelligence and the difficulty of satisfying conflicting requirements give considerable scope to the inventiveness of the designer, and it is not surprising that various solutions to the problem exist, all of which differ in mathematical detail to a greater or lesser extent. A detailed consideration of the options open to the designer in carrying out the data processing is beyond the scope of this text, but it is important for the observer to have some general knowledge of how the operation relates to the way in which the same task would be tackled by manual plotting. This understanding is relevant not only to the above extract from the performance standard, but to the general philosophy that requires the results to be at least as good as would be obtained by using manual recordings, and that auto-tracking should enable observers to perform as well with multiple targets as they can by manually plotting a single target. It should thus be appreciated that auto-tracking is carrying out a task which could be undertaken manually by an observer, given the time and the inclination, and that there is no reason to believe that the results obtained by auto-tracking are any better, or any worse, than those that can be achieved by such an observer.

For acceptance by the ARPA, data must be in digital form, since the necessary calculations must be performed by digital computation. Thus, in the first instance, the analogue radar data must be stored, radial line by radial line, in a digital memory. The presence of a response will be registered as a binary 1 in the appropriate range cell(s) and will be associated with the appropriate bearing word, as was described for the general radar memory in Section 2.7.3. Filtering techniques are used to establish the likelihood of any particular response being a target, and hence whether or not it should be stored for tracking purposes. During each timebase the detected responses are read into what is known as the *prime register*, which is a group of memory elements of the form shown by one line of the memory matrix illustrated in Figure 1.5. During the rest (or interscan) period, the contents of the prime register are transferred into the number one register, which is the first of a group of *N* identical registers. At the same time the contents of number one are transferred into number two, and so on down the group. The contents of the *N*th register (i.e. the oldest data) are discarded. During the transfer period a comparison is made of the contents of the *N* cells in each column and if more than *M* out of *N* cells in the column contain a binary 1, a hit is registered in the correct location in a further section of the tracker memory which holds only filtered responses. The filtering technique helps to reduce the probability of false echoes reaching the tracker and also improves the chances of a real but weak or intermittent response being tracked.

Targets within the filtered area of the memory are selected for tracking when, either manually or automatically, a gate is placed over their responses. As the antenna beam sweeps past a ship-target, it will register a number of strikes (see Sections 2.3.3.3 and 2.5) on successive timebases and it may be that such a target activates more than one successive radial range cell. In the case of picture storage (see Section 2.7.3.2), these digitized responses will aggregate in the memory to generate on the display an echo having the outline of the distinctive echo paint (see Section 2.8.5). Clearly it is neither necessary nor desirable for the computer to track each individual element present in the resolution cell. For this reason the input responses to the tracker are processed to produce a single registration which represents the location of the target to be tracked and about which the gate will in due course attempt to centre itself. The area of memory which is used to store the assembly of such registrations is known as the *hit matrix*.

If the target has been acquired, and is being successfully tracked, a tracking window will be centred on that particular memory location within the hit matrix which corresponds with the target's range and bearing. The co-ordinates of the window can be extracted and stored in a further area of the tracker memory. This area is sometimes referred to as the *track file* and there will have to be a separate track file for each tracked target. Thus, rotation by rotation, as the gate moves in steps following the target's position through the hit matrix, sequential positions of each tracked target can be stored in the appropriate track file.

The processor (which is that part of the computer which manipulates the data and carries out the mathematical operations) must operate on the recorded positions to calculate the most probable track of the target. It is difficult to carry out calculations based on positions which are expressed in terms of range and bearing because the rates at which the bearing and range change are not constant for a target on a straight track. Further, the spatial resolution varies with range (i.e. it is geometrical). For these reasons it is usual to convert the target positions into Cartesian coordinates of northings and eastings, the mathematical operation being the same as that described for scan conversion (see Figure 2.41).

When laying off observed ranges and bearings on a plotting sheet (see Section 7.8), the effect of inherent errors is that, even for a target on a steady track, the plotted positions do not form a perfectly straight line but are scattered about the correct track; the observer has to attempt to draw the line that is the best fit. Exactly the same effect occurs with automatic plotting and it is further exacerbated by quantizing errors introduced by digital storage (see Sections 2.7.3.2 and 9.4). Since the data must eventually be displayed as a stable straight line vector, the processor must calculate a length and direction which represents the best fit to the scattered observations. This operation is known as smoothing and involves the application of quite complex mathematical techniques such as Kalman filtering, the mechanics of which are beyond the scope of this text. However, it is important for the observer to appreciate that such smoothing does take place and to understand the implications.

If the smoothing is carried out over too long a period, the tracker will be insensitive to changes in tracks as a result of which small manoeuvres may go undetected and there will be a long delay before large manoeuvres become apparent. On the other hand, if smoothing is carried out over a short period, the output data will fluctuate, rendering decision-making difficult. Clearly the amount and duration of smoothing is a compromise and the difficulty of reconciling the twin requirements of sensitivity to target manoeuvres and data stability provides the designer with a considerable challenge. As a result, the particular algorithms used for smoothing vary in detail with manufacturer.

When a target is first acquired, the computer will commence storing positions, obtaining updated coordinates each time the antenna sweeps across the target (i.e. about once every 3–5 s). These positions will have an inherent scatter and initially the mean line will be very sensitive to plots which fall some distance from it. However, as the plotting duration increases and more plots are obtained, the mean line will stabilize and accuracy will improve. During the first minute of tracking, the target will normally display only a symbol to indicate that it is being tracked. In most systems the vector (or other graphical indication of target movement) will be suppressed until sufficient observations have been obtained to produce the indication of the target's motion trend to the level of accuracy required by the performance standard. Some systems were designed to display vectors within a few seconds of acquisition. This should not be seen as a sign of instant accuracy. Accuracy demands a number of successive observations and until the 1 min interval has elapsed there is no requirement to meet the performance standard accuracy. Any data derived directly or indirectly from these very early indications could be highly misleading. In general, where such early display takes place, a study of the instability of the vector (or other indication of movement) should convince the user that it is based on insufficient observations. After 1 min the tracker will have smoothed about 12–20 observations and must then produce data to the lower of the two accuracy levels set out in the performance standard. In some systems a graphic symbol on the target is used to indicate that the data is based on more than one but less than 3 min of observation. As long as the target continues to be detected at the location predicted by the rate aiding (see Section 4.3.2), the tracking period is allowed to build up to 3 min, at which stage the processor will be able to smooth some 36–60 observations and must then reach the higher accuracy level. Thereafter as each new plot is added the *oldest* is discarded.

If a target response is not detected in the location forecast by the rate aiding, one possible explanation is that the target has manoeuvred. The tracking gate will be opened out and if the target is detected, tracking will continue. If the departure from the 3 min track is not significant, the processor will conclude that the departure was due to scatter and will continue to smooth the track over a period of 3 min. On the other hand, if the departure is significant, the processor will treat the situation as a target manoeuvre and will reduce the smoothing period to 1 min. This reduction in smoothing period is analogous to the situation in which an observer decides that a target has manoeuvred and therefore discards a previous *OAW* triangle and starts a new plot. If steady-state conditions resume, low-level accuracy must be obtained within 1 min and then the tracking period can again be allowed to build up to 3 min, allowing high level accuracy to be regained.

Most systems have two smoothing periods, a short period of about 1 min and a long period of about 3 min. For compliance with the performance standard the periods must not be more than 1 or 3 min, respectively, but in practice manufacturers quite often use slightly shorter periods and hence reach the required accuracy a little more quickly than the standard requires.

While the mathematical detail of the smoothing is beyond the scope of this work, what is of importance to the observer is the question of whether the smoothing is performed on the relative tracks or on the true tracks. This is a fundamental decision which must be taken by the designer and there are merits in both choices.

In general trackers will *either*:

a. smooth and store the relative track of a target to produce directly the output relative-motion data and hence calculate the true-motion data from the smoothed relative-track data and the instantaneous input course and speed data, which is normally unsmoothed to avoid any loss of sensitivity to manoeuvres by the observing vessel; *or*

b. smooth and store the true track of a target to produce directly the smoothed true-motion data and reconstitute the relative-motion data from the smoothed true-track data and the (normally unsmoothed) input course and speed data.

Note: In order to smooth and store true tracks, the normally unsmoothed course and speed data are applied to the raw relative-motion data (see Section 4.3.6.2).

In the steady-state situation, i.e. where neither tracked target nor the observing vessel manoeuvres and no changes take place in any errors in the input data, both approaches will produce the same result. If a change takes place, the two different approaches will produce differing results over the succeeding smoothing period. To understand the differences, it is necessary to consider in general terms how the calculations are performed.

4.3.6.2 *The Storage of Relative Data*

In this approach the ranges and bearings of the tracked target are scan-converted, smoothed and stored to produce the direction and rate of the target's relative motion. The displayed relative vector (see Section 4.4.1) and the CPA/TCPA (see Section 4.6) data are derived directly from this, as illustrated by Figure 4.7.

The merit of this approach is that, just as when an observer constructs a relative plot, the CPA data is totally independent of the course and speed inputs to the system. This is a feature which many users believe to be of paramount importance because they take the view that the miss-distance ranks in importance above all other data.

If the true-motion data relating to the target is requested, the computer will calculate that information by vector addition of the instantaneous course and speed data to the smoothed relative track. Thus, as might be expected, the accuracy of the true-motion data depends on the accuracy of the input course and speed data (see Section 9.6.3). Where trial manoeuvre data (see Section 4.4.3) is requested, the vector addition will involve the smoothed relative track and the proposed course and speed.

An inherent difficulty which besets relative-track storage is that if the observing vessel manoeuvres, the direction and rate of the relative track of all targets will change. There will thus be a tendency for the relative vectors to become unstable, because the tracker is vainly attempting to smooth a curve to produce a straight line. It follows that, during the time of the manoeuvre and for a short period thereafter, relative and true track data will be suspect. The effect is likely to be particularly dramatic on board vessels such as ferries which can change their heading and speed by a large amount in a fairly short time; the tracking data may give the mistaken impression that all tracked targets are manoeuvring. This is not intended as a criticism of the systems which elect to store relative data, but merely recognition of the basic laws of relative-motion which demand steady-state conditions for a stable and reliable assessment of relative-motion, however it is deduced. In general, the system will attempt to recover stability by smoothing the relative-motion over the shorter period (approximately 1 min) once the computer has decided that a significant track change has taken place.

4.3.6.3 *The Storage of True Data*

In this approach the ranges and bearings are scan-converted to produce Cartesian co-ordinates. Before each position is stored, its co-ordinates are modified by the direction and distance moved by the observing vessel since the first observation (see Figure 4.8).

The effect of this is to produce within the track file a succession of positions whose co-ordinates represent the unsmoothed true track of the selected target. These positions are smoothed to produce the best estimate of the target's true course and speed. The operation

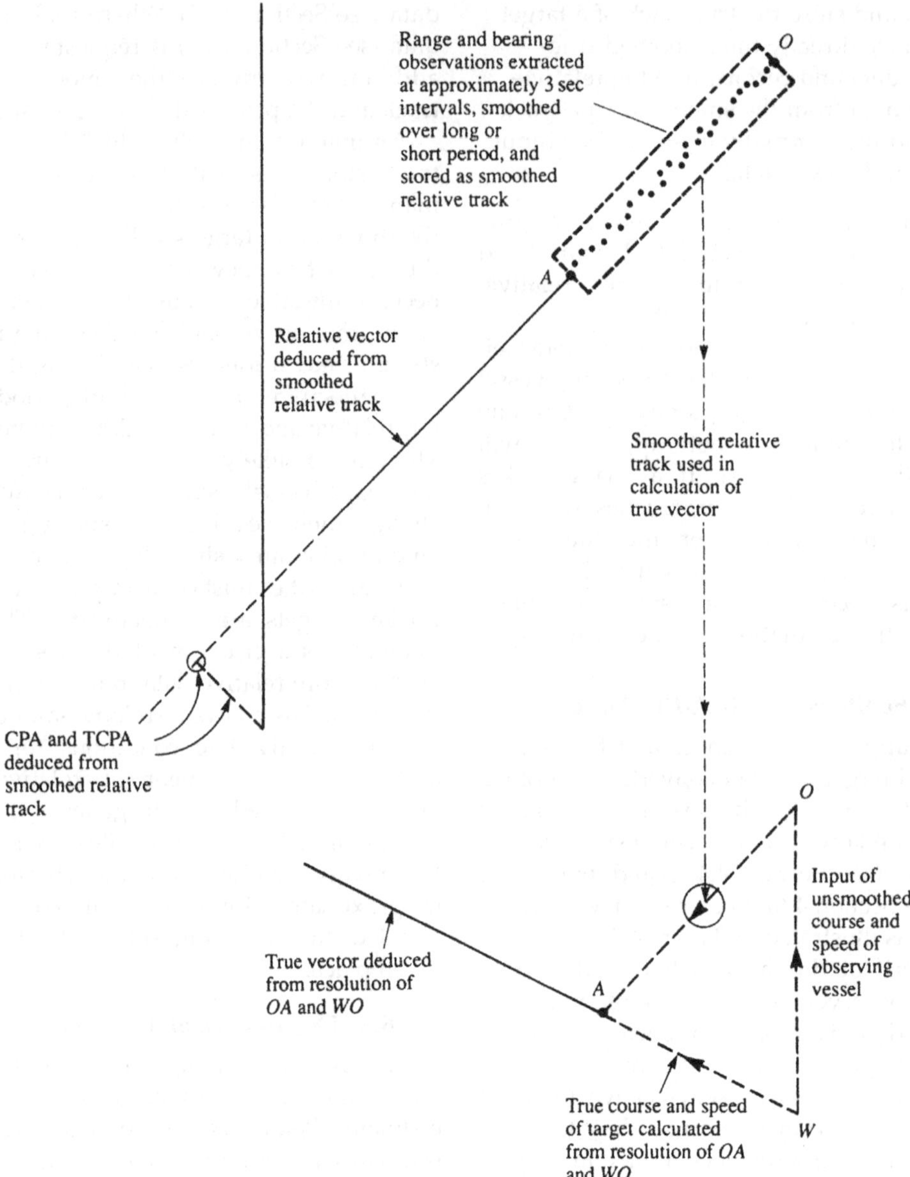

Range and bearing observations extracted at approximately 3 sec intervals, smoothed over long or short period, and stored as smoothed relative track
O
A
Relative vector deduced from smoothed relative track
Smoothed relative track used in calculation of true vector
CPA and TCPA deduced from smoothed relative track
O
Input of unsmoothed course and speed of observing vessel
True vector deduced from resolution of OA and WO
A
True course and speed of target calculated from resolution of OA and WO
W

FIGURE 4.7 The storage of relative data.

is analogous to the construction of a true plot, as described in Section 7.3.

If the relative-motion data relating to the target is requested, the computer will calculate that information by vector addition of the instantaneous input course and speed data to the smoothed true track of the target. Where trial manoeuvre data (see Section 4.4.3) is requested, the vector addition will involve the smoothed true track and the proposed course and speed.

The merit of this approach is that if the observing vessel manoeuvres, the relative vector will remain stable (though changing) because it is updated, rotation by rotation, being calculated from the smoothed true course of the target and the instantaneous course and speed of the observing vessel. On the debit side it would appear at first sight that the calculated relative-motion of the target is dependent on the accuracy of the input course and speed data. This is not the case provided that the input course and speed data are consistent over a full smoothing period. If the input data error is constant for the full smoothing period, the smoothed true track will of course similarly be in error. The computer will then use the wrong input data and the consistently wrong true track and as a result will arrive at the correct relative-motion (see Figures 4.8 and 9.7(b)).

It is thus evident that, provided any error in the input course and speed data is consistent for the full smoothing period, it will not affect the accuracy of the CPA/TCPA data. However, if there is a fluctuating error, for example due to an erratic log input, the relative vector will be inaccurate and unstable. While recognizing the advantage of this approach in ensuring relative data stability during manoeuvres by the observing vessel, many users are concerned about the ability of random input errors to influence the CPA. Most users are prepared to tolerate some degree of instability or inaccuracy of true vectors when the observing vessel is performing a manoeuvre, but most, if not all, believe that the CPA is inviolate.

4.3.6.4 *Storage Philosophies Compared*

The two forms of storage which in general terms are described above represent two different approaches to dealing with the difficulty of maintaining stable tracking data when the observing vessel manoeuvres. This is an inherent difficulty which is experienced whether the plotting is carried out manually or automatically and arises from the fact that the basic information, that is range and bearing, is being measured from a moving vessel. The implications for tracking accuracy are discussed further in Section 9.5.3. However, at this stage it has to be emphasized that the observer must be alert to the likelihood of the data being inaccurate during non-steady-state conditions, and to the way in which the method of track storage will influence the nature of these inaccuracies.

4.3.7 Target Trails and Past Positions

Older ARPA regulations termed past data as 'target history' and the information was required to be provided by four past plots at predefined intervals (2 min was common). At the same time similar information could theoretically be observed from the target trails which at one time were a natural afterglow from the target in the phosphorous of the old CRT display technology then in use. This afterglow was a side effect, but observers found it very useful as an indication of target trend, for reasons explored more fully in the next paragraph. However, as data became digitized, CRT technology moved on and was replaced by LCD and other technologies, these target trails have had to be artificially inserted by the manufacturers. As the trails are now artificial, the time length of these trails can be controlled on the display, which means they provide similar timed historical data as the 'history plots'. The modern regulations (below) allow either or both modes of presentation, the past positions (still often called 'history' plot or dot) and the target trails.

Variable length (time) target trails should be provided, with an indication of trail time and mode. It should be possible to select true or relative trails from a reset condition for all true-motion display modes.

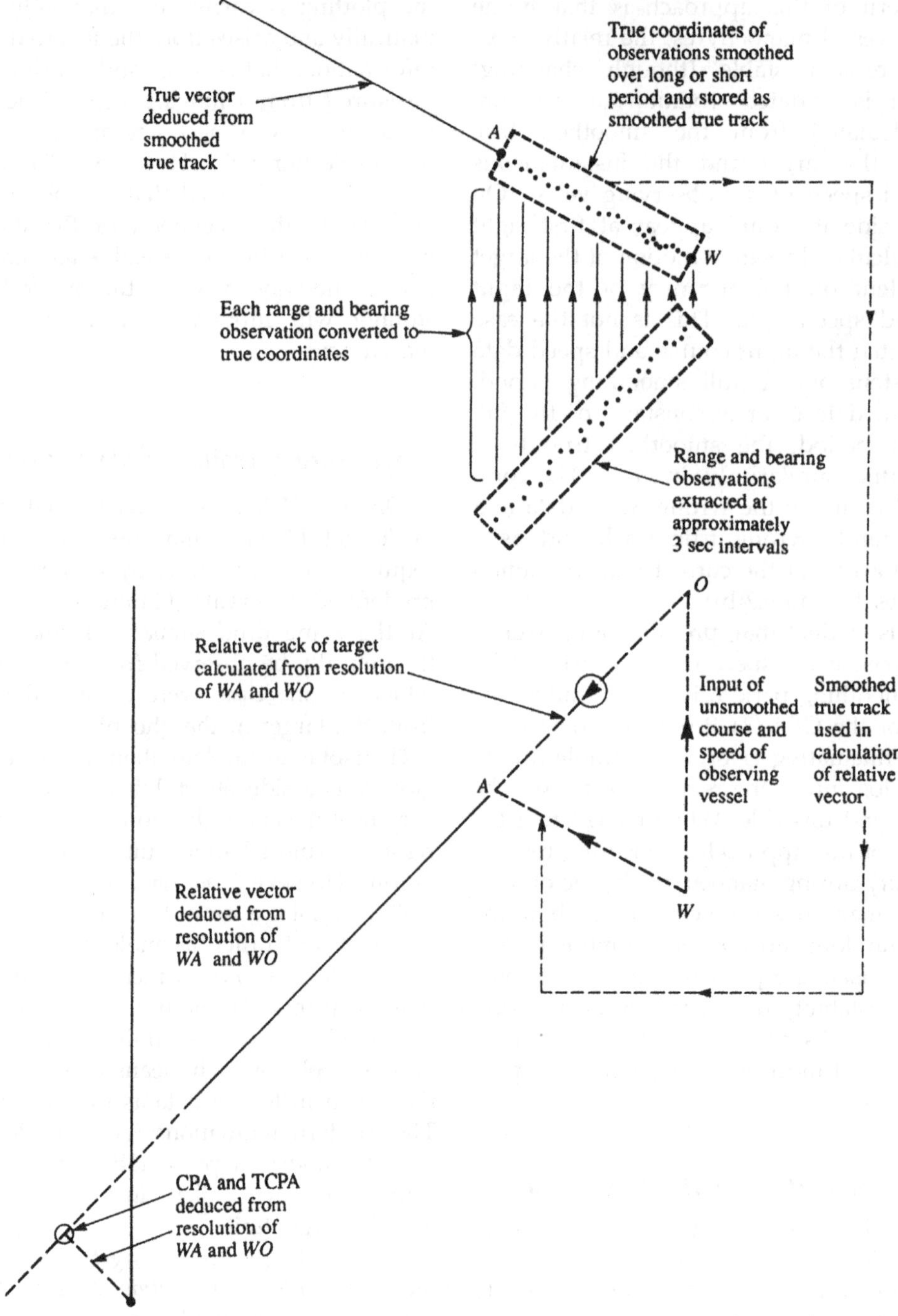
True vector deduced from smoothed true track
True coordinates of observations smoothed over long or short period and stored as smoothed true track
A
W
Each range and bearing observation converted to true coordinates
Range and bearing observations extracted at approximately 3 sec intervals
O
Relative track of target calculated from resolution of WA and WO
Input of unsmoothed course and speed of observing vessel
Smoothed true track used in calculation of relative vector
A
W
Relative vector deduced from resolution of WA and WO
CPA and TCPA deduced from resolution of WA and WO

FIGURE 4.8 The storage of true data.

1. *The trails should be distinguishable from targets.*
2. *Either scaled trails or past positions or both should be maintained and should be available for presentation within two scans or equivalent, following:*
   - *the reduction or increase of one range scale;*
   - *the offset and reset of the radar picture position; and*
   - *a change between true and relative trails.*

The past data enables an observer to check whether a particular target has manoeuvred in the recent past, possibly while the observer was temporarily away from the display on other bridge duties. Not only is this knowledge useful in showing the observer what has happened, but it may well help him to form an opinion of what the target is likely to do in the future. Manoeuvres by targets may be for navigational purposes and it may be possible to gain an indication from the general overview that this is the case. Alternatively, there may be an indication from the nature of the manoeuvre, when considered in relation to the general traffic pattern, that it has been undertaken for collision avoidance and hence the observer can be alert for a resumption of previous conditions.

The past positions or history plots provide slightly better quality information on the history, because although the trails provide good information in changes of direction of target past motion, the only time references are the current position and the end of the trail. The provision of (say) four-time specific history dots would better indicate changes in speed of motion across the display. This is demonstrated by comparing the 'slowing target' in Figure 4.9 and the same target in Figure 4.10a.

However, a problem with the history plots is that they make the screen look more cluttered and they can be very confusing on short ranges and heavy traffic when the plots of different targets start crossing and overlapping each other. These are situations when the observer needs to have no ambiguity. The trails have the advantage of more clearly identifying to which target they belong.

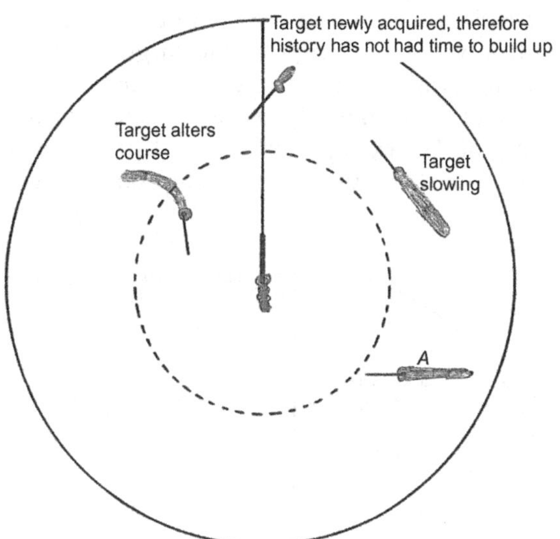

FIGURE 4.9 True Target Trails, with targets maneuvering.

'True history' is without doubt the only meaningful way in which this data can be displayed, since the nature of the manoeuvre is readily apparent. A curve in the trail indicates an alteration of course by the target, whereas a change in the spacing of the plots will occur when there has been a change in speed. Since changes in speed are much slower to take effect, they are consequently more difficult to detect.

'Relative history' is not always provided but where it is, it is important to remember that any change in the direction or spacing of the plots can result from a change of course, a change of speed, or a combination of the two. It is therefore essential that the true manoeuvre is identified. Relative history should be used with great caution (see Section 7.4.2). Relative history will not be considered further.

The history plots can be likened to a vessel dropping buoys (with radar reflectors) at *equal intervals of time*, thus a change in the spacing is indicative of a change in the target's speed, but it is important to appreciate the significance of the distance between the echo and the last plot. For example, in Figure 4.10, target *A* has not just executed a crash stop but has just dropped a buoy and is now steaming away from it.

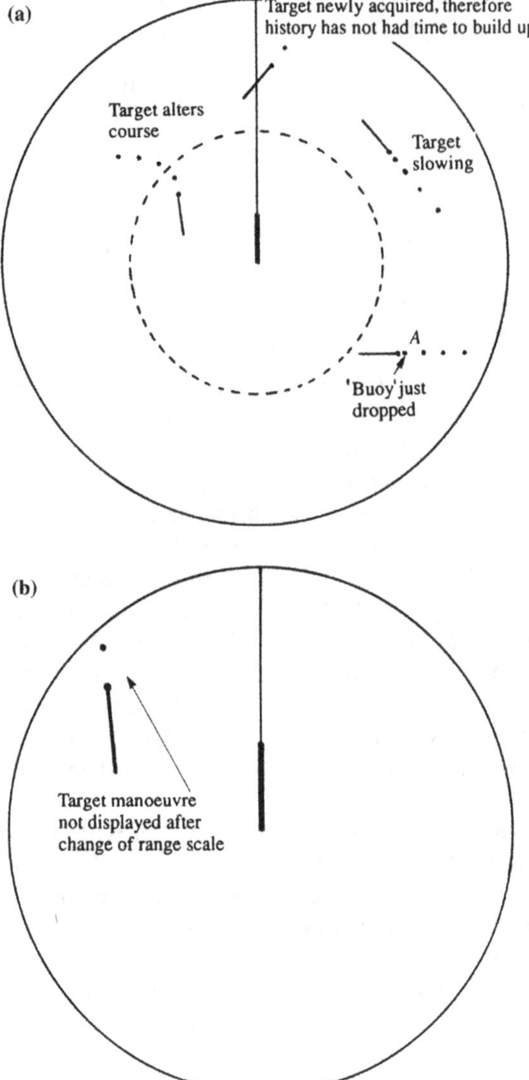

FIGURE 4.10 Tracking history: (a) 12-mile and (b) 6 mile range scale.

When history plots are requested, only those of tracked targets can be displayed, and even then the number of plots which are displayed will depend upon the period for which a particular target has been tracked. A target which has been tracked for 2 min can only display a maximum of 2 min worth of history and so probably only one plot will be displayed, others being added as the tracking period increases.

Some manufacturers cause the plots to flash sequentially, for example 2 min ago, 4 min ago, while others cause the plots to decrease in brightness or size with time. The last plot is always removed as the latest plot is added to the display. Users usually have control of the number of plots and the time spacing between plots.

Uneven tracks of targets or apparent instability of motion may be taken to indicate that tracking of that target is less precise than it might be and the displayed data should be treated with caution.

Because of the variations in the way this facility can operate, great care should be taken when observing history to ensure that one is certain of exactly what is being displayed. In particular, one must establish whether true or relative history is being displayed and also which time spacings are in use.

4.4 VECTORS

The course and speed information generated by ARPA for acquired targets should be displayed in a vector or graphic form (e.g. PADs, predicted areas of danger, see Section 4.12 and Figure 4.10) which clearly indicates the target's predicted motion. In this regard:

1. ARPA (or auto-tracking) presenting predicted (extrapolated) information in

vector form only, should have the option of true and relative vectors. There should be an indication of the vector mode selected and, if true vector mode is selected, the display should show whether it is sea or ground stabilized.

2. An ARPA (or auto-tracking) which is capable of presenting target course and speed information in graphic form should also, on request, provide the target's true and/or relative vector.

   (*Note*: This means that vectors of some kind must be provided, irrespective of the availability of any other method such as PADs, used to display collision-avoidance information.)

3. Vectors displayed should be time adjustable.
4. A positive indication of the timescale of the vector in use should be given.

Vectors must be capable of indicating the rate and direction of the target's relative motion (relative vectors), or indicating the rate and direction of the target's proper motion (true vectors). The direction and rate forecast by a true vector will only be fulfilled if the tracked target maintains its course and speed. In the case of a relative vector, the fulfilment depends additionally on the observing vessel similarly maintaining course and speed.

In all cases, the displayed vector length is time related and is adjustable by using a 'vector length' control. An alternative approach is to have a default physical length which remains the same irrespective of the range scale, for example 3 min on the 6 NM range scale, 6 min on the 12 NM range scale.

Note: Vectors of either of the two types can be displayed on a true- or relative-motion radar picture presentation, that is *true vectors can* be selected to appear on a *relative-motion* presentation and vice versa. It was originally considered in some quarters that this might result in confusion and at least one manufacturer has in the past provided a default condition where relative vectors are displayed when relative-motion is selected and, likewise, true vectors on the true-motion presentation, with the alternative vector mode being temporarily selectable in each case by the user holding down a virtual or real 'spring-loaded' control (see also Section 9.6.1, incorrect interpretation).

4.4.1 Relative Vectors

As described in Section 4.3.6, the ARPA must track the target(s) for a period of time, after which a vector can be displayed as in Figure 4.11(a). Using the vector length control, the vectors can be extended to determine the CPA by observation against the background of the range rings and the TCPA can be read off from the vector length control.

4.4.2 True Vectors

As an alternative, the observer may request that the true vector(s) be displayed (Figure 4.11(b)). In this case, the own ship will also have a vector which will increase in length as the time control is increased. The likelihood of a close-quarters situation developing can be ascertained by running out the true vectors progressively to show the predicted development of the encounter. The dynamic nature of this technique appeals to many users, but it must be borne in mind that any evaluation of CPA/TCPA is a matter of trial and error and thus is better avoided. It is essential to appreciate that the CPA is *not* represented by the point at which the target's true vector intersects the own ship's true vector, except in the case of zero CPA.

True vectors can be 'sea stabilized' or 'ground stabilized' depending on the sensors used or the set-up. A more detailed discussion of these issues is reserved for a later chapter (Section 6.9.6).

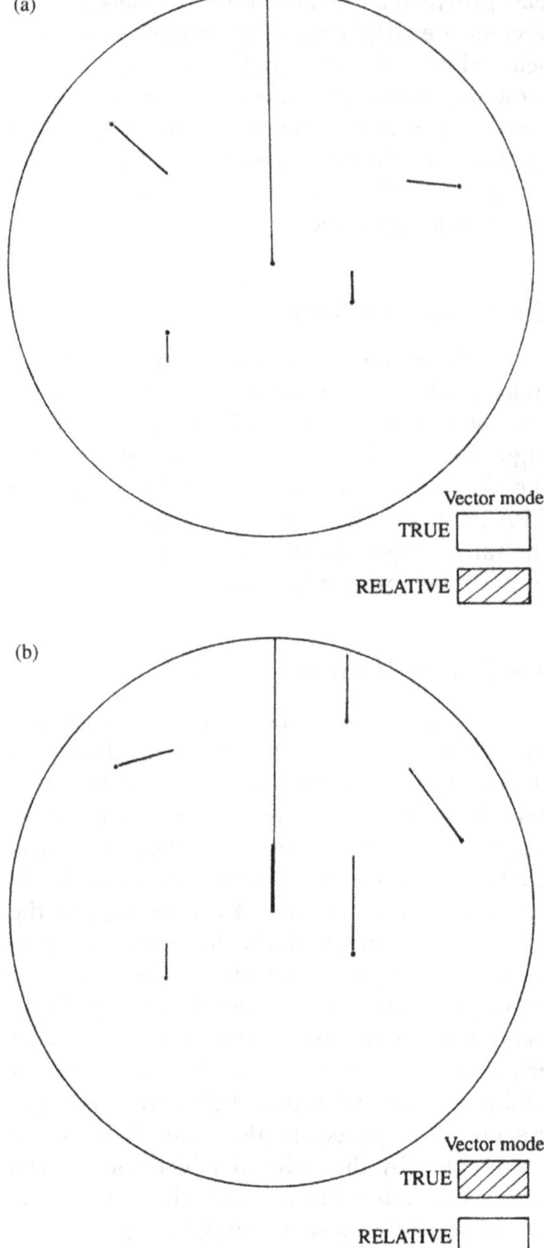

FIGURE 4.11 Vectors. (a) Relative vectors. (b) True vectors for the situation displayed in (a).

4.4.3 Trial Manoeuvre

The system should, where required by Table 4.2, be capable of simulating the predicted effects of the own ship's manoeuvre in a potential threat situation and should include the own ship's dynamic characteristics. A trial manoeuvre simulation should be clearly identified. The requirements are:

- *The simulation of the own ship's course and speed should be variable.*
- *A simulated time to manoeuvre with a countdown should be provided.*
- *During simulation, TT should continue and the actual target data should be indicated.*
- *The trial manoeuvre should be applied to all tracked targets and at least all activated AIS targets.*

The feature is still not a requirement for vessels below 10,000 grt in the latest standards.

With the availability of computer assistance, the problem of predicting the effect of a manoeuvre prior to its implementation by the own ship is much simplified. While it is relatively easy to visualize mentally the outcome of a manoeuvre where two ships are involved, in dense traffic this becomes very difficult. In particular, with large ships and limited sea room, it is necessary to plan and update the whole collision-avoidance strategy as quickly as possible in light of the continually changing radar scene.

While planning, it is important to bear in mind the following points:

a. The own ship may temporarily need to be on a 'collision course' with more distant vessels, that is collisions may require sequential avoidance since it is unlikely that a single manoeuvre will resolve all the problems.

b. Extrapolation of the present situation using the trial manoeuvre facility with current course and speed as inputs can provide valuable information on which of the 'other' vessels in the vicinity may have to manoeuvre in order to avoid collisions between each other. Obvious avoiding manoeuvres may present themselves and should be looked out for.

c. Constraints imposed by navigation may dictate the manoeuvre of 'other' vessels. This should be taken into account when planning strategy and watched for when carrying out the plan and assessing its effectiveness.

d. The ease with which this facility allows the navigator to establish the course to steer for a given passing distance may encourage the choice of a small alteration. This temptation must be avoided at all costs as it loses sight of the need to make a substantial alteration. The rules require the latter in recognition of the fact that other vessels may be using much more rudimentary methods of data extraction and may not be able to detect a small manoeuvre.

Unfortunately, although there is a single requirement for all approved ARPAs to possess a facility for simulating a trial manoeuvre, different methods of providing this have been devised by the various manufacturers.

When information relating to the proposed manoeuvre is fed in, the true or relative vectors which would result from such a manoeuvre are displayed on request. This, combined with the ability to adjust the vector length, can give a clear presentation of potential close-quarters situations between other vessels and the observer's vessel.

It is important to select *relative vectors* when assessing the effect of a manoeuvre, as this will give an indication of how far the target will pass clear (Figure 4.12). It is also possible to vary the inputs while observing this display and note the effect on the CPA.

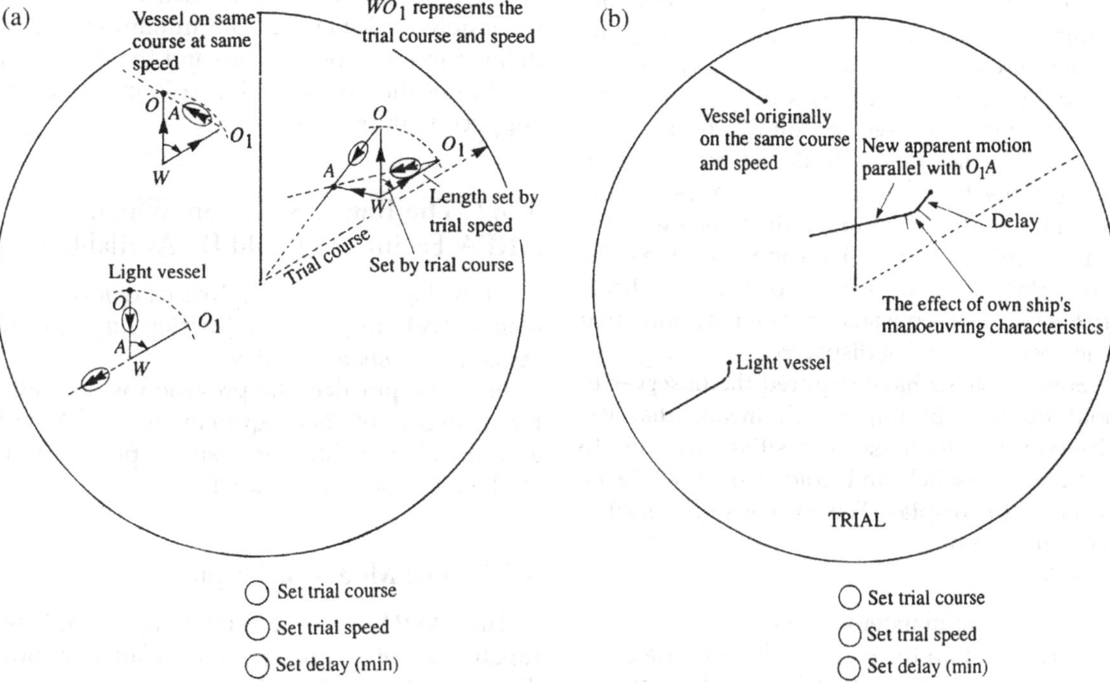

FIGURE 4.12 Trial manoeuvre display in the relative vector mode. (a) Construction of relative vectors. (b) The trial manoeuvre display.

The ARPA has to produce a plot similar to that discussed in Section 7.5. There is a range of facilities available, with an increasing number of factors being taken into account when presenting the trial data. In the simplest form, it is possible to feed in only the intended course, speed and alteration time and observe their effect on the display.

In some ARPAs it is possible for the vessel's handling characteristics to be included in the evaluation, but this will add to the complexity of data required from the user. Given that merchant vessels handling characteristics vary with condition of loading, weather and depth of water, the user must consider whether such data is worth providing or whether an error estimate is allowed in the observed results.

On some equipment, provision is made for two successive manoeuvres to be displayed. This can be extremely helpful when endeavouring to assess the time for which an alteration must be held, as well as modelling complex situations when more than one manoeuvre may be required. Observers should not be tempted into a series of frequent often small manoeuvres (see Section 7.11.6(b)).

In order that there should be no confusion between the 'trial' data and the current situation, when the trial is in operation the screen will display some distinctive indication such as the word SIM or TRIAL. The use of a 'T' to indicate trial is frequently mistaken for an indication that true vectors are being displayed.

Some systems have required the observer to hold down a button, which means that the observer has to make a positive decision to operate the switch and hold it over while he observes the display. Few systems offer such a fail-safe control.

Note:

i. While trial manoeuvres are being presented on the display, the computer continues its normal task of tracking all acquired targets.

ii. Where PADs (see Section 4.12) are provided, all the possible alterations of course should be simultaneously apparent.

iii. A 'trial speed' facility is also provided since any given family of PADs is drawn for a specific speed of the observing vessel.

4.5 THE ARPA DISPLAY

4.5.1 The Size of the Operational Display

The sizes of operational display for different vessel sizes are summarized in Tables 4.1 and 4.2. The ARPA for a vessel above 10,000 grt should be at least 320 mm in diameter, but may be 250 mm in diameter and 180 mm in diameter as vessel size is reduced. This is the size of the circular PPI display and should not be confused with the overall (usually rectangular in practice) screen size, which is defined separately in Table 4.2. The minimum size is defined in terms of a square and the extra area outside of the operational area is intended for supporting information.

4.5.2 The Range Scales on Which ARPA Facilities Should Be Available

TT facilities should be available on at least the 3, 6 and 12 NM range scales. Tracking range should extend to a minimum of 12 NM.

In actual practice, the provision is normally far in excess of this requirement: ARPA and auto-tracking facilities are usually provided on all the range scales provided.

4.5.3 The Modes of Display

The ARPA and auto-tracking should be capable of operating with a relative-motion display with 'north-up' and 'course-up' azimuth stabilization. In addition, the ARPA and auto-tracking may also provide for a true-motion display. If true motion is provided, the operator should be able to select for his display either true or relative motion. There should be a positive indication of the display mode and orientation in use.

The term 'mode' is used here to describe the 'radar' features of picture orientation and picture presentation, which are discussed in detail in Sections 1.4 and 1.5.

4.5.4 The ARPA Data Brilliance Control

Means should be provided to adjust independently the brilliance of the ARPA (or auto-tracking) data, including complete elimination of the ARPA (or auto-tracking) data.

This control allows the observer to remove all ARPA data from the screen in order to see just which is radar and which is ARPA data. It can be particularly useful when searching for a poor-response target (e.g. in clutter) while still wishing to continue tracking the other vessels in the vicinity.

Unfortunately, many a mariner has been caught out by this control and has spent some frustrating minutes trying to find the screen marker, only to realize that the ARPA data brilliance control was turned down.

4.5.5 The Use of the Screen Marker for Range and Bearing Measurement

Provisions should be made to obtain quickly the range and bearing of any object which appears on the ARPA (or auto-tracking) display.

To comply with this requirement, most manufacturers have enabled the range and bearing of the screen-controlled marker to be displayed. In this way, the range and bearing of an object can be obtained by placing the marker over it. Like all range and bearing measuring devices, they should be regularly checked against the range rings and heading marker respectively (see also Section 6.6).

4.5.6 The Effect of Changing Range Scales

After changing range scales on which the ARPA (or auto-tracking) facilities are available or resetting the display, full plotting information should be displayed within a period of time not exceeding one scan.

It should be appreciated that, in order to fulfil this requirement, the ARPA needs to track and plot the acquired targets continually out to some 16 miles, irrespective of the range scale selected by the operator. Because of this, if the shorter range scales are selected and accompanied by a short pulse, targets at a longer range returning a poor response may be lost.

4.6 THE DISPLAY OF ALPHANUMERIC DATA

For each selected tracked radar target, the following data should be presented in alphanumeric form:

1. *Source of data*
2. *Actual range to the target*
3. *Actual bearing of the target*
4. *Predicted target range at the closest point of approach (CPA)*
5. *Predicted time to CPA (TCPA)*
6. *True course of target*
7. *True speed of target.*

The alphanumeric display is usually shared with radar tracked targets and AIS targets. Many ARPAs now designate the targets with a serial number when first plotted or acquired by the AIS. (It should be noted that individual targets will often have two serial numbers, one as a radar tracked target and one as an AIS target.) The alphanumeric display usually makes use of the space to the side of the circular display on the rectangular screen and usually space for the display of one or more sets of target data simultaneously (Figure 4.13).

Some manufacturers also change the colour of the target(s) on the display so it is clear to which target the alphanumerics relate.

When a target is 'marked' or 'designated', the alphanumeric display should change to reflect the data on the target so selected. The system is sometimes programmed by the manufacturer to automatically provide the alphanumeric data when a target becomes dangerous, that is, it first activates the CPA/TCPA alarm (see Section 4.2), although this is not a requirement in performance standards. In some systems the data available also includes the distance at which it is predicted that the target will cross ahead (BCR) and the interval which will elapse until this event occurs (BCT).

Although vectors are suppressed during the first minute of tracking (for the reasons described in Section 4.3.6), the observer can normally select a target during that period and read out the alphanumeric data. This is acceptable as a means of quickly obtaining the range and bearing of the target, but it must be appreciated that other alphanumeric values will at that stage be based on only a few observations and hence can be dangerously misleading.

When *trial manoeuvre* is selected, modern systems continue to provide the real alphanumeric data, while some older ARPA systems produce the trial values. In the case of any given system, it is essential to establish exactly which data are being made available.

4.7 ALARMS AND WARNINGS

It should be possible to activate or deactivate the operational warnings.

4.7.1 Guard Zone Violation

If a user-defined acquisition/activation zone facility is provided, a target not previously acquired/activated entering the zone, or detected within the zone, should be clearly identified with the relevant symbol and an

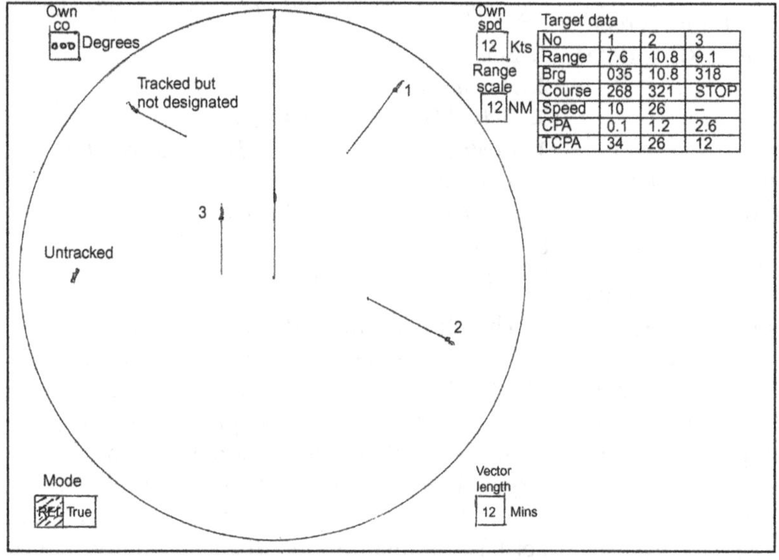

No	1	2	3
Range	7.6	10.8	9.1
Brg	035	10.8	318
Course	268	321	STOP
Speed	10	26	–
CPA	0.1	1.2	2.6
TCPA	34	26	12

FIGURE 4.13 Thedisplay of alphanumeric data.

alarm should be given. It should be possible for the user to set ranges and outlines for the zone.

It is possible to specify an area in the vicinity of the own ship, for example at 10 miles, a zone 5 cables in depth (Figure 4.14), or an area surrounding the own ship, which on entry by a target activates an alarm. It is not unusual to have two zones, one which may be at some preset range and the other at a range which may be varied according to circumstances.

The target which has activated the alarm may be made to 'flash' or alternatively be marked in some other way, such as having a short bearing marker through it. In some systems, the range and bearing of the target entering the zone is displayed in numerical form. It is important to remember that a target which is detected for the first time *at a lesser range than* a guard ring or area will not activate the alarm; this warning system should not be regarded as an alternative to keeping a proper lookout, but rather as an additional means of ensuring the safety of the vessel. Each guard zone may be set for

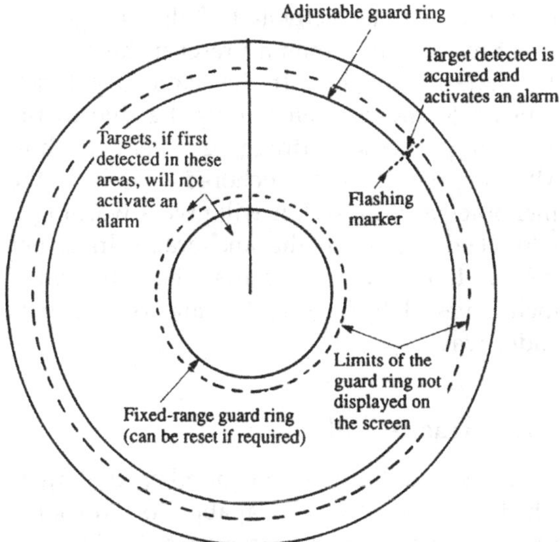

FIGURE 4.14 Guard rings.

360° coverage or for a specific sector of threat. See also Figures 4.2–4.4.

Note: This provision is normally coupled with the 'automatic acquisition' facilities, in which case, in addition to activating the alarm, the target will also be acquired if there is an available tracking channel. Even where automatic acquisition is not provided, the guard zone facility is still required.

4.7.2 Predicted CPA/TCPA Violation

If the calculated CPA and TCPA values of a tracked target or activated AIS target are less than the set limits:

- *A CPA/TCPA alarm should be given.*
- *The target should be clearly indicated.*

As the target has not yet violated BOTH conditions, the alarm has not yet been activated.

It is possible to specify a CPA and TCPA (sometimes referred to as *safe limits*) which will activate an alarm if *both* are violated. For example, if the CPA and TCPA controls are set to 0.5 NM and 30 min, respectively, and a target which is being tracked will come to a CPA of less than 0.5 NM in less than 30 min, then the alarm will be activated (Figure 4.15). This will occur even if the 'relative vector' has not been extended into the specified area. The displayed echo and vector of the target activating the alarm may be made to flash or, alternatively, marked in some other way.

4.7.3 Lost Target

The system should alert the user if a tracked radar target is lost, rather than excluded by a predetermined range or preset parameter. The target's last position should be clearly indicated on the display.

Consider a target which is being tracked but, for one of a number of reasons does not return a

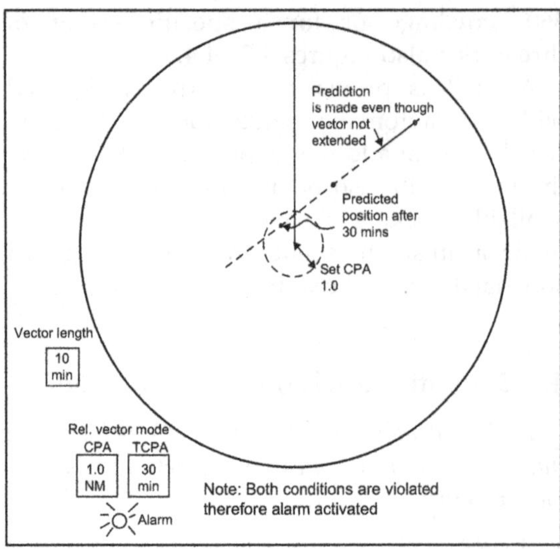

FIGURE 4.15 CPA/TCPA alarm settings.

detectable response on one scan, the tracker will open up the gate and, if it finds a response, will continue to track as described in Section 4.3.2. If it fails to find a response, it is required that the tracker should continue to search for the echo in an area where it might be expected until it has failed to appear for 5 out of 10 consecutive scans. If, after this searching, the target is still not detected, the 'target lost' warning is activated and the last observed position of the echo is marked on the screen. It is also normal to activate an audible alarm.

4.7.4 Time to Manoeuvre

A 'delay' facility is provided with trial manoeuvre (see Section 4.4.3) that an alarm may be provided to warn the officer of the watch that it is (usually) 1 min until 'time to manoeuvre'. Depending on the size of ship and its manoeuvrability, this warning can be used as a reminder to apply helm and/or ring the telegraph, in order that a pre-planned manoeuvre is executed on time.

4.7.5 Track Change

This alarm is associated with an algorithm which quantifies departures from the predicted tracks of targets. The target(s) activating the alarm will be indicated by some graphic symbol. Its operation is likely to be associated with the change from a long to a short smoothing period (see Section 4.3.6.1).

It can be useful, but it does depend on the application of some arbitrary decision as to when the change in track is 'significant'. In some systems, the track change alarm will be activated by large or rapid manoeuvres performed by the observing vessel. This is more likely to arise where relative-track storage is used (see Section 4.3.6.2). In general, this condition can be recognized, as all targets will exhibit the track change symbol.

4.7.6 Anchor Watch

This facility attempts to offer automatic warning of the observing vessel or other vessels dragging in an anchorage. If a known stationary target (e.g. a small isolated navigation mark) is acquired and designated as such, then an alarm will be activated if the designated target moves more than a preset distance from the marked position. If the stationary target appears to move, then it must be due to the observing vessel dragging her anchor. Alternatively, if other anchored vessels in the anchorage are tracked, it will give a warning if a 'tracked' vessel in the anchorage drags her anchor. It will also, of course, be activated if such a vessel heaves up her anchor and gets under way.

4.7.7 Tracks Full

There is a limit to the number of targets which an ARPA is capable of tracking, although this figure has risen considerably over the years. In areas of high traffic density, there may well come a time when all the tracking channels are in use. This is particularly likely when automatic acquisition (see Sections 4.2.4–4.2.6) is in operation. An alarm will warn the operator to inspect the untracked targets for potential dangers and to transfer tracking from less important targets which are being tracked. This is now a compulsory alarm in the current IMO Performance Standards (See Section 11.2.1).

4.7.8 Wrong or Invalid Request

Where an operator feeds in incorrect data or data in an unacceptable form, for example course 370°, an alarm and indicator will be activated and will continue until the invalid data is deleted or overwritten.

4.7.9 Safe Limit Vector Suppression

This uncommon facility, if selected, suppresses the vectors of targets whose predicted motion does not violate the safe limits (see Section 4.7.2) and is an attempt to reduce ARPA 'clutter'. The computer continues to track the targets whose vectors are suppressed. If any of them should manoeuvre in such a way as to violate the set safe limits, the vector of that target will reappear and the safe limit alarm will be activated. If a decision is taken to use this facility, considerable thought must be given to the implications of the selected values of safe limits. In general, it is advisable to switch the facility off before contemplating a manoeuvre.

4.7.10 Trial Alarm

This facility is analogous to the safe limit alarm but operates only when the trial manoeuvre is selected. It is not available on all systems.

4.7.11 Performance Tests and Warnings

All ARPA equipments incorporate some form of self-diagnostic routine which monitors the correct operation of the various circuits. These checks are normally conducted on 'switch on' and repeated at regular intervals (which may range from once per hour to many times per second) or on request from the operator. In the event of a fault, a warning is given to the operator and, in some cases, an indication of the cause or location (e.g. printed circuit board No. 6) of the fault is also given.

However, it must be appreciated that some faults cannot be detected internally, for example a failure of certain elements in a numeric read-out can cause an 8 to appear as a 6 or a 9 to appear as a 3. In such cases, a typical test may provide for the operator to switch all of the elements on so that each indicator should display an 8.

4.8 AUTOMATIC GROUND-STABILIZATION

The difficulties of achieving ground-stabilization on a raw radar display are outlined in Sections 6.2.6.2 and 6.9.6. With ARPA assistance, ground-stabilization can be achieved easily and automatically in many circumstances, but not all. Once selected, this stabilization also applies to all true target vectors displayed, whether or not the operator is showing relative or true-motion display. It is particularly important to appreciate the implications of this, which are discussed later.

An isolated land target with good response is selected as reference. It is acquired and tracked by one of the ARPA tracking channels and then designated as a 'fixed target'. It has also been called 'echo reference' by some manufacturers. This makes it possible for the tracker to calculate the ground track of the observing vessel, and hence to maintain the movement of the electronic origin of the display in sympathy with it. When using this facility the observer should be particularly watchful for other targets which approach the reference target and, in particular, for those which pass between the observing vessel and the reference target. If the intervening target shadows the reference target, the chances of target swap (see Section 4.3.5) may be greatly increased. If target swap does involve the reference target, it may have quite dramatic effects on the presentation of all vectors, particularly if the reference attaches to a fast moving target. If such an eventuality seems likely, it may be expedient to move the reference to another fixed target.

The facility has been developed in such a way that the same stabilization is applied to the radar picture presentation and to the true vectors, i.e. either both are sea-stabilized or both are ground-stabilized. Thus in general, where automatic ground-stabilization is selected, true vectors will indicate the ground tracks of targets and *not* their headings. Failure to appreciate this can render the presentation dangerously misleading if it is mistakenly used in the planning of collision-avoidance strategy. This is illustrated in the right-hand display of Figure 4.16. One might expect the danger of observers being misled in this respect, because (except in case of an along-track tide) there will be angular displacement of the observing vessel's vector from the heading marker. Experience has indicated that this expectation of observers cannot be justified in all cases.

The useful feature is that it makes true-motion parallel indexing a practical proposition, which is scarcely if at all the case with manual ground-stabilization (see Section 8.4.5). It also makes it possible to maintain electronic navigation lines and maps (see Sections 4.9 and 8.4.6) in a fixed position on the screen. Some users find the availability of a fixed map particularly attractive in coastal navigation. However, it must be stressed that the penalty paid is that the presentation may not afford traffic heading information and may therefore in principle be unsuitable for collision avoidance.

Automatic ground-stabilization can also be achieved by using the output from a twin axis Doppler log which is locked to the ground (see Section 6.2.6.2) and by integration with the Global Navigation Satellite System (GNSS) (see Section 10.1).

The ease of integration with GNSS and the accuracy make this a very popular display for modern navigators. However, it is important to remember the potential for misleading anti-collision information in a cross track tide.

The advantage of the ARPA 'fixed target' facility is that it provides a continuous check of ground course and speed, independent of GNSS and other sensors. The disadvantage is that it depends on the availability of a suitable fixed land-based target. Buoys may also be used with caution, particularly in areas like the English Channel, where the large buoys are well monitored and so may be regarded as fixed.

4.9 NAVIGATIONAL LINES AND MAPS (SEE ALSO SECTION 8.4.6.3)

Most modern radar sets (even without ARPA) offer a graphics facility whereby electronic lines can be drawn on the screen. The position, length and orientation of the lines can be adjusted, thus making it possible to produce parallel indexing lines (see Section 8.4) and to delineate navigational limits in channels, traffic separation schemes, poor-response coastlines, etc. It may also be possible to indicate points of interest such as isolated rocks and floating marks with a specific symbol.

The range of facilities available under the general heading of a navigation lines package

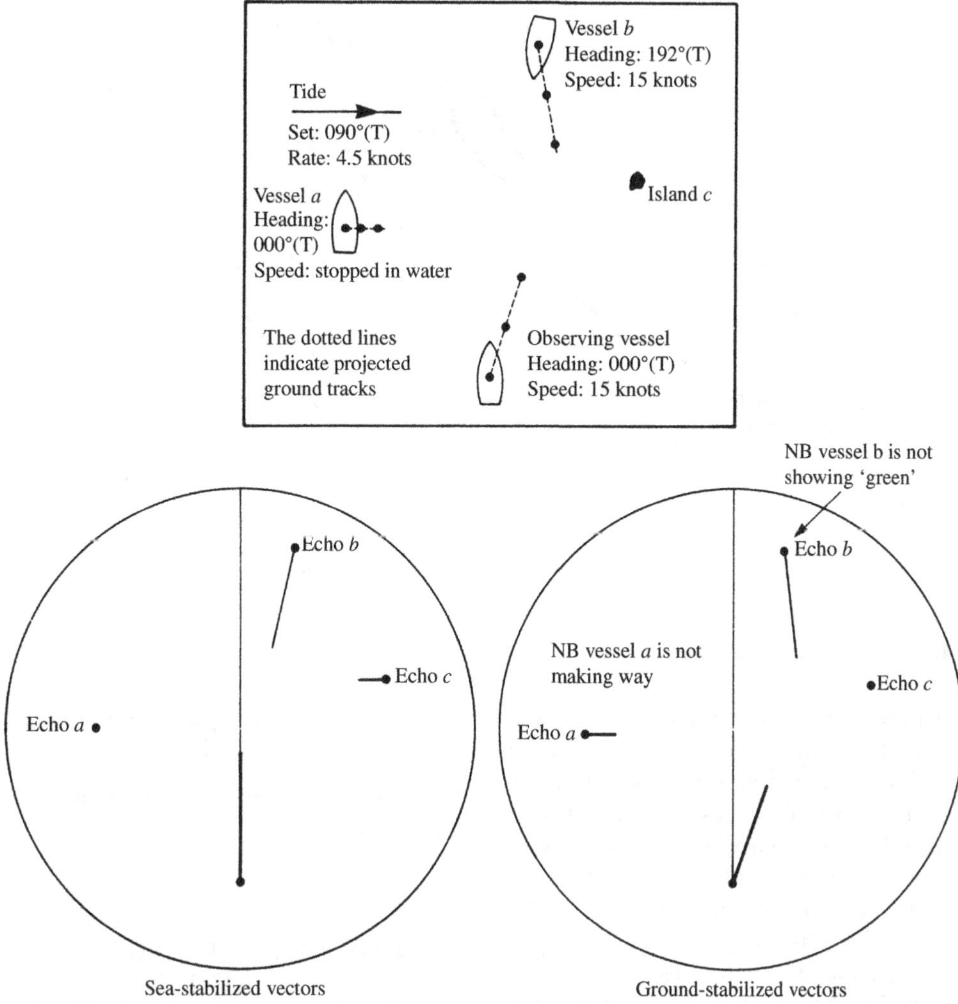

FIGURE 4.16 Sea-stabilized and ground-stabilized displays.

is considerable. With the advent of Electronic Chart Display and Information Systems (ECDIS) integration (see Section 10.2) modern equipments are tending to provide less complex facilities, if at all.

In simple facilities, only some ten to fifteen lines may be available, thus merely allowing the construction of fairly crude patterns or maps (see Figure 4.17). The pattern may be lost if the equipment is switched off, and it may or may not be possible to edit or move the pattern around the screen, once drawn. More advanced packages will allow the observer to prepare and store the pattern at a convenient time when planning the passage and subsequently to recall it when required. It will then be possible to move the pattern around the screen in order to align or realign

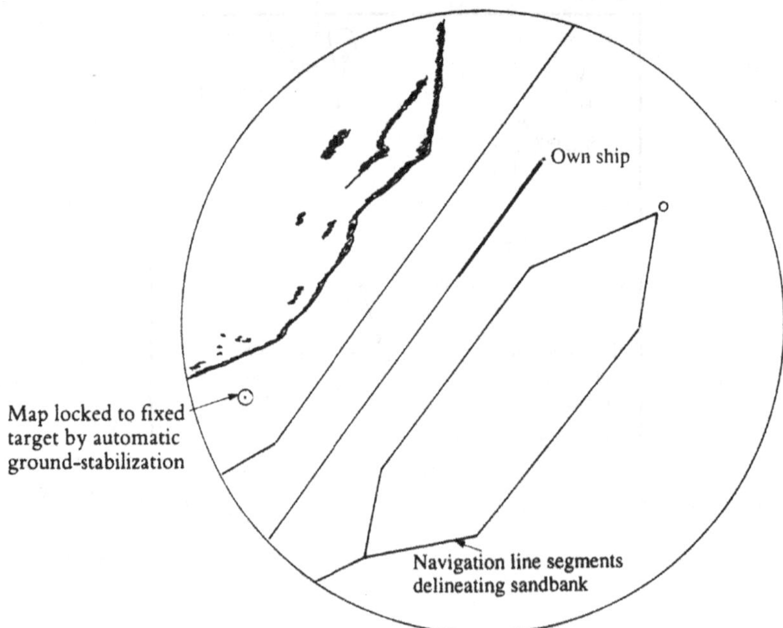

FIGURE 4.17 Map presentation.

it with displayed radar echoes and there will be provision for easy editing. A non-volatile memory allows the storage of several maps even when the equipment is switched off. The most comprehensive mapping facilities will provide for the permanent storage of maps containing as many as a thousand or more elements. While there is no rigid distinction in names between the two extremes, the simpler packages tend to be referred to as navigation lines while the more comprehensive are usually described as mapping facilities. The facility will frequently be used in association with automatic ground-stabilization (see Section 4.8).

The concept of adding maps to the radar display has now extended to the integration of electronic charts (ECDIS) with the radar/ARPA. The advantages are that:

a. The data is not prepared by the navigator which improves the reliability and accuracy as well as being more convenient.

b. It allows a more sophisticated selection of features to be added.

The ECDIS topic is covered in more detail in Section 10.2.

4.10 TARGET SIMULATION FACILITY

A target simulation facility should be provided for training purposes.

Older performance standards required a test programme to check the performance of the radar tracker. Test scenarios were provided with dummy echoes on the display and often these were also found useful for training trainee officers when at sea. The test scenarios are no longer a requirement, but the ability for the system to provide a simulation facility for training purposes is required. When in use, the display needs a clear indication that it is in training mode, in case it is accidently left in this mode and mistaken for a *real* situation.

4.11 THE PREDICTED POINT OF COLLISION

The predicted point of collision (PPC) is that point towards which the observing ship should steer at her present speed (assuming that the target does not manoeuvre) in order for a collision to occur.

Some early ARPA designers recognized that the positions of these points can be quickly calculated and displayed for all tracked targets. The argument made for displaying the PPCs is that they assist in developing a collision-avoidance strategy by showing the navigator, at a glance, the courses which are completely unacceptable because they intersect a collision point. This is the only contribution which can be claimed for PPCs. They do not give any indication *of miss distance* (other than in the zero CPA case) and any attempt to extrapolate the clearing distance either side of the point will be fraught with danger. A safe course is one which, among other things, results in passing at a safe distance, which implies a knowledge of clearing distance. Safe and effective use of PPCs depends upon a thorough understanding of the factors which affect their location and movement. As is evident from the following treatment, this is, in many cases, not a simple matter. Some systems make it possible to display these points as small circles when, but only when, true vectors are selected.

4.11.1 The Concept of Collision Points

When two ships are in the same area of sea, it is always possible for them to collide. The point(s) at which collision can occur may be defined and depends upon:

a. The speed ratio of the two ships.
b. The position of the two ships.

Considering any two ships, usually one is moving faster than the other; the possibility that one is at exactly the same speed as the other and will maintain that ratio for any period of time is remote enough to be disregarded for the moment.

The ship which is the faster of the two will always see displayed one and only one collision point, since it can pursue the target if necessary. This collision point is always on the track of the target, as shown in Figure 4.18.

The ship which is the slower of the two may see displayed two collision points, both of which must be on the target track. One exists where the slow ship heads towards the target and intercepts it, while another exists where the slow ship heads away from the target but is struck by it. The two cases appear in Figure 4.19. Alternatively there may be no way for the slower ship to collide with the faster (even though the faster may collide with the slower) because it is just not fast enough to reach the target (Figure 4.20).

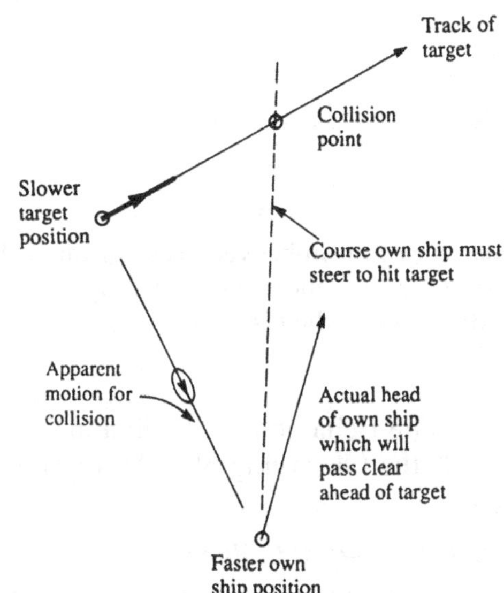

FIGURE 4.18 Single point of collision.

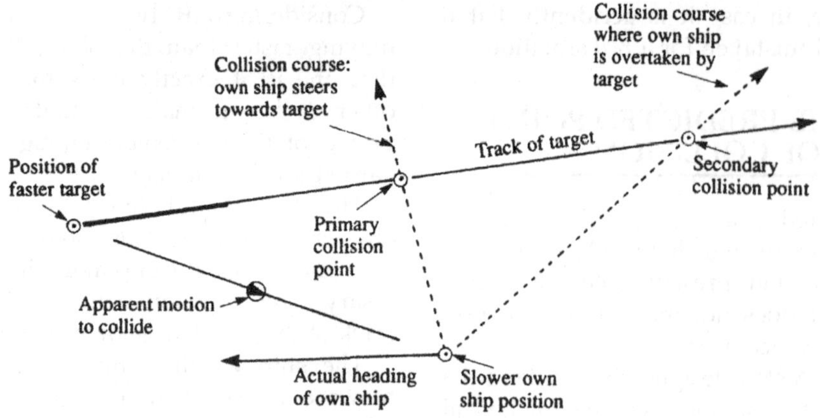

FIGURE 4.19 Dual points of collision.

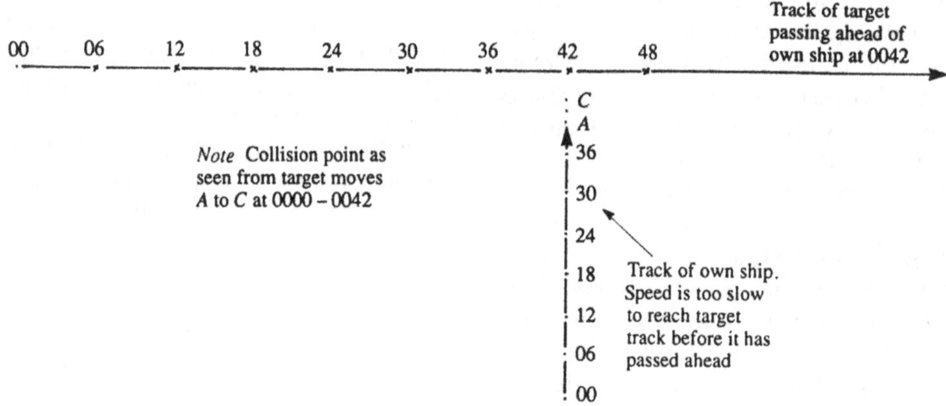

FIGURE 4.20 No points of collision.

Note: A critical in-between case of one collision point exists where the slow ship can just reach the track of the fast ship.

4.11.2 The Behaviour of Collision Points If the Observing Ship Maintains Speed

4.11.2.1 The Initial Collision Case

It is important to realize that collision points exist, whether an actual collision threat exists or not. The only significance is that in the event of an actual collision threat, the collision points are the same for both ships. Figure 4.21 shows how the collision points move in a collision situation and how they will appear to the two ships involved. On the faster ship, the single collision point appears on the heading marker and moves down, decreasing in range as the collision approaches. On the slower ship, one of the two collision points will move down the heading marker, while the other moves down a steady bearing.

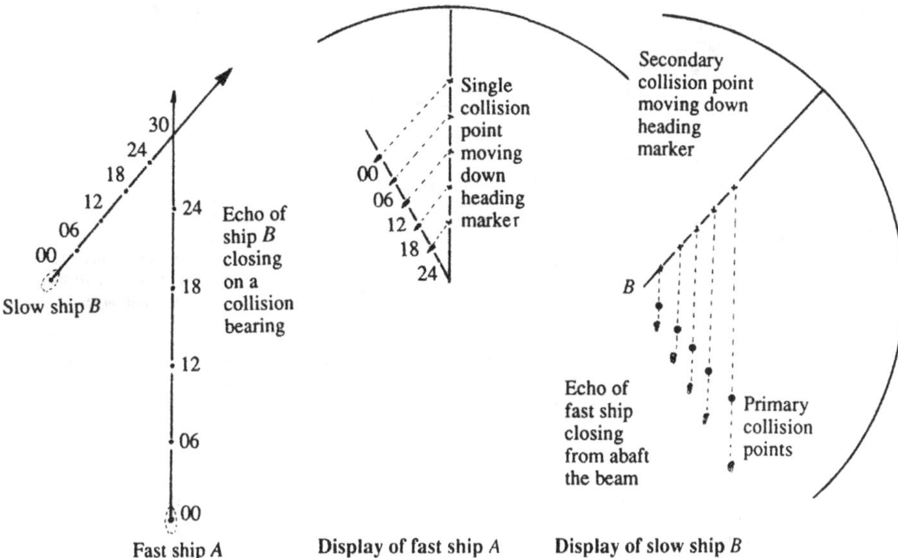

FIGURE 4.21 The movement of the PPC in the collision situation.

4.11.2.2 *The Non-Collision Case*

In a non-collision case, the collision point moves according to well-defined rules, but it will never cross the heading marker. In the case of the faster ship, the movement depends on whether the fast ship will cross ahead of or behind the target ship. Figure 4.22 shows the two possible cases and typical track lines.

The case of the slower ship is more complicated because of the two collision points and the possibility of no collision point existing. If the observing ship's course is to pass between the two collision points, they will pass down either side of the observing ship, generally shortening in range, and then draw together under the stern of the observing ship. As they meet they become one collision point before they finally disappear, as in Figure 4.23.

If the observing ship is steering to pass astern of the fast ship, the collision points will draw together, form one point and then disappear. It is noticeable that the collision point more distant from the target ship, usually termed the secondary point, moves much faster than the point nearer to the target which is termed the primary collision point.

4.11.3 The Behaviour of the Collision Point When the Target Ship's Speed Changes

If the speed ratio is infinitely large, for example when the target is stationary, then obviously the collision point is at the position of the target. If the observing ship maintains speed while the target begins to increase speed, the collision point will begin to move along the target track. When the target speed has increased to that of the observing ship, the secondary collision point will appear at infinity. Further increase of the target speed will move the primary and secondary collision points towards each other (not necessarily by equal amounts); eventually, own speed in comparison to target speed may be so slow that the two points will merge and then

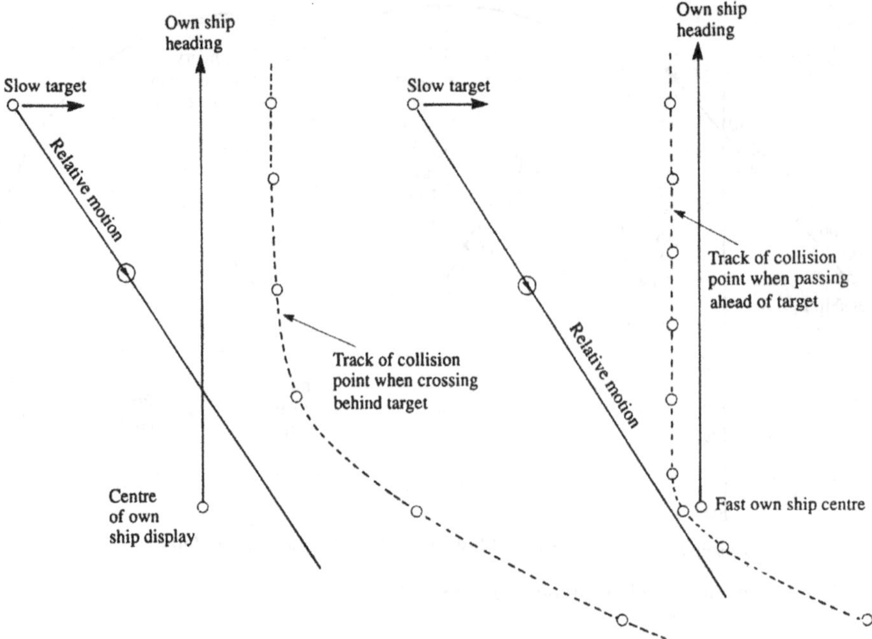

FIGURE 4.22 The track of the collision point.

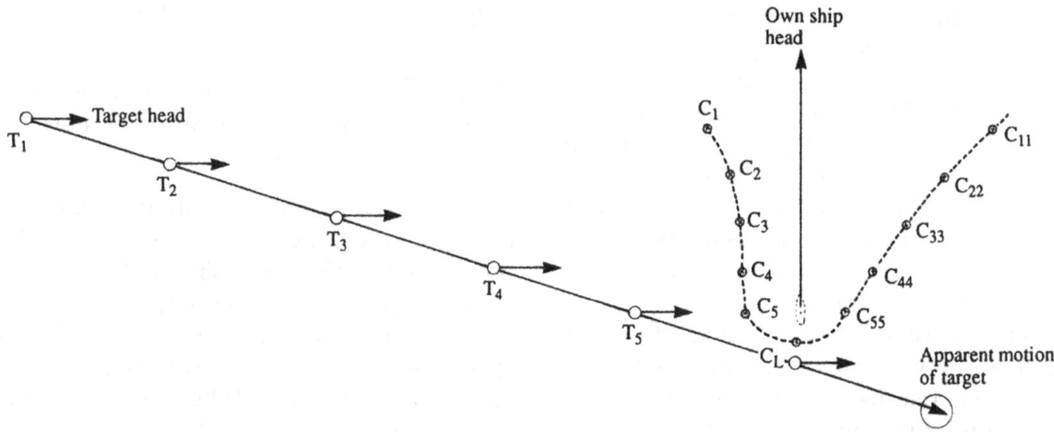

FIGURE 4.23 The movement of dual collision points. Own ship is steering to pass between the collision points at speed 7 kn. Target speed = 24 kn.

$$C_n = \text{primary collision point}$$
$$C_{nn} = \text{secondary collision point when target is at } T_n$$
$$C_{nn}CL = \text{limiting condition where the hazard disappears}$$

disappear. This behaviour is shown in Figure 4.24.

4.11.4 The Behaviour of the Collision Point When the Target Changes Course

If the two ships have the same speed, the collision point moves on a locus which is the perpendicular bisector of the line joining the two ships. The greater the aspect, the farther away the collision point will be (Figure 4.25). Theoretically, the limiting aspect in this case is 90°, but then the collision point would be at infinity and hence an aspect of some 85° plus is considered the practical limit.

4.11.4.1 *For a Slower Observing Ship*

When the observing ship is slower than the target, two collision points exist and they are seen to be on the circumference of a circle

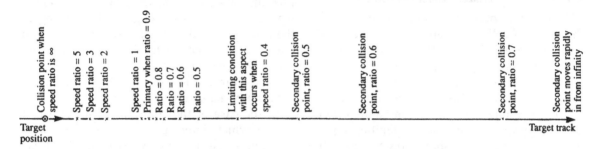

Collision point when speed ratio is ∞
Speed ratio = 5
Speed ratio = 3
Speed ratio = 2
Speed ratio = 1
Primary when ratio = 0.9
Ratio = 0.8
Ratio = 0.7
Ratio = 0.6
Ratio = 0.5
Limiting condition with this aspect occurs when speed ratio = 0.4
Secondary collision point, ratio = 0.5
Secondary collision point, ratio = 0.6
Secondary collision point, ratio = 0.7
Secondary collision point moves rapidly in from infinity
Target position
Target track
Own ship position

FIGURE 4.24 The behaviour of the collision point with the change of speed ratio (i.e. own speed: target speed).

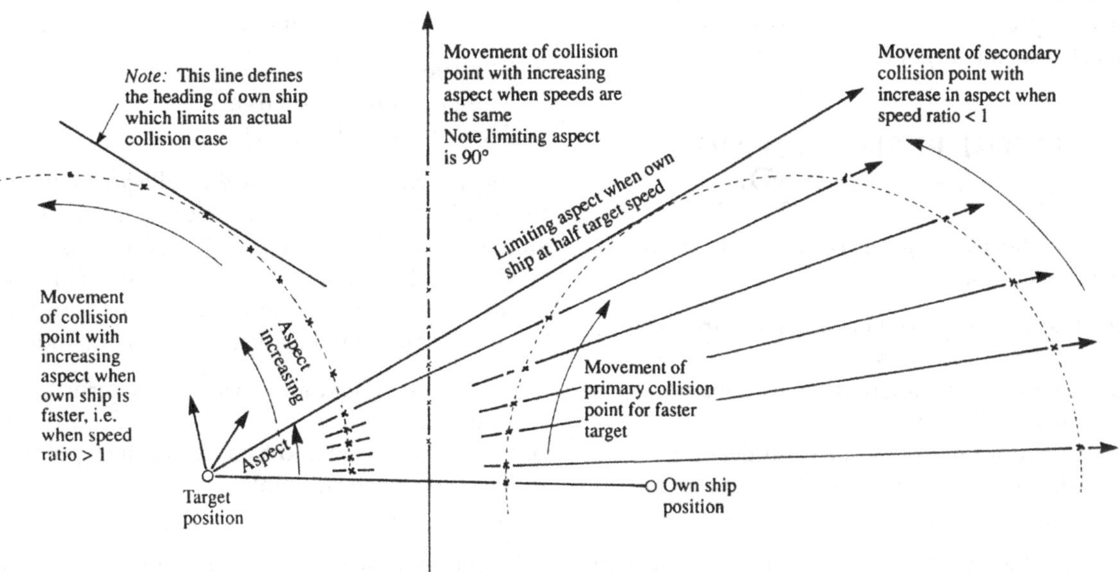

FIGURE 4.25 The movement of the collision point(s) as target aspect changes, drawn for speed ratios of 2:1 and 1:2.

whose centre and radius are dependent on the speed ratio; the circle is always on the observing ship's side of the unity speed ratio locus. A limiting aspect can be defined which is also dependent on the speed ratio. A slower observing ship will mean that a target will have a smaller limiting aspect angle.

Aspects greater than the limit pose no hazard since the observing ship can never catch up with the target.

4.11.4.2 For a Faster Observing Ship

When the observing ship is the faster, the circle of collision points lies on the target side of the equal-speed locus. As the aspect increases, the collision point moves farther away from the observing ship. There is no limiting aspect and collision is always possible.

An interesting point to note in Figure 4.25 is that the inverse of the idea of a limiting aspect to the slow observing ship appears when the observing ship is faster; this is effectively a limiting course for the observing ship. If the actual heading is to the remote side of this line, all collision points appear on the one bow. If the own heading is inside this limiting direction, the collision point will move across the heading marker as the target changes aspect.

4.12 THE PREDICTED AREA OF DANGER (PAD)

The shortcomings of collision points can be listed as follows:

a. Inaccuracies in data acquisition are likely to displace the points.
b. No account is taken of the dimensions of the ships involved.
c. They offer no quantitative indication of miss distance, which is the essential data required for collision avoidance.

The logical development is to construct, around the PPC, a plane figure which is associated with a chosen passing distance and in the calculation of which due margin of safety can be allowed for the effects of data inaccuracies and the physical dimensions of the vessels involved. The area within the figure is to be avoided to achieve at least the chosen passing distance and is referred to as a PAD. The technique was patented by Sperry Marine Systems and is exclusive to the ARPAs produced by that company. The PAD approach is an extremely elegant solution to the problem of how best to present collision-avoidance data. It is essential that the user has a thorough understanding of the principles underlying the presentation with particular reference to the location, movement, shape and change of shape of the PAD. As will be seen from the following explanation, this is not a simple subject.

In the case of the collision point there is a course which intercepts the target's track at the given speed ratio, whereas in the PAD there are generally two intersection points. One of these is where the observing ship will pass ahead of the target and the other where the observing ship will pass astern of the target. The angle subtended by these two limiting courses will depend upon:

a. The speed ratio.
b. The position of the target.
c. The aspect of the target.

As shown in the case of the collision point, a faster observing ship must always generate a single cross-ahead and cross-astern position. A slower observing ship produces much more complex possibilities and, depending on the three variables noted above, these may include:

a. Two cross-ahead and two cross-astern points.
b. One cross-ahead and two cross-astern points.
c. Two cross-astern points.
d. No hazard.

In the case of the single or primary collision point, the position at which the observing ship will cross ahead of the target is always farther from the target than the collision point, while the cross-astern point is always nearer to the target.

In the case of a slower observing ship, where there is a secondary collision point, the second cross-ahead position is nearer to the target and the associated cross-astern position more remote from it (Figure 4.26).

To indicate limits within the 'cross-ahead/cross-astern' arc, it is necessary to draw a bar parallel to the target's track and at the intended miss-distance closer to the observing ship's position.

The limits defined by the arc and the bar are such that, if the observing ship were to cross those limits, then it would be at a less distance than the desired miss-distance from the target. Figure 4.27 shows the generation of the two boundaries in the case of a slower observing ship.

4.12.1 The PAD in Practice

In order to produce an acceptable system for practical operation, these limits are normally encapsulated by a symmetrical figure such as an ellipse or a hexagon.

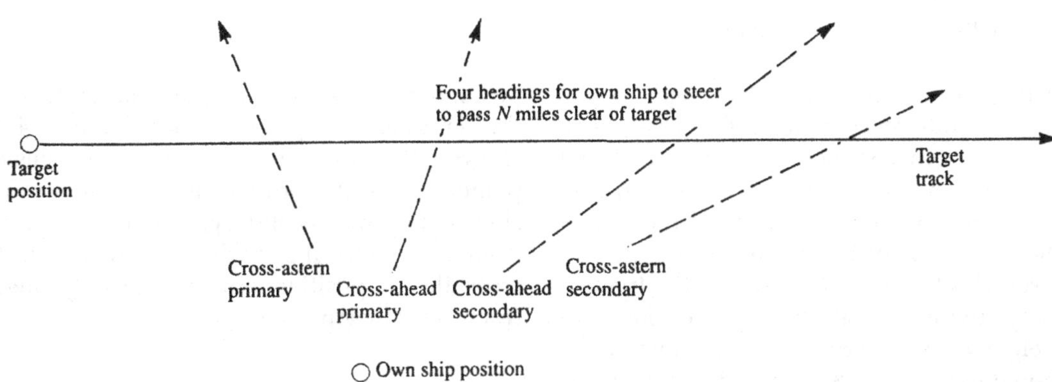

FIGURE 4.26 The sequence of cross-ahead/cross-astern points.

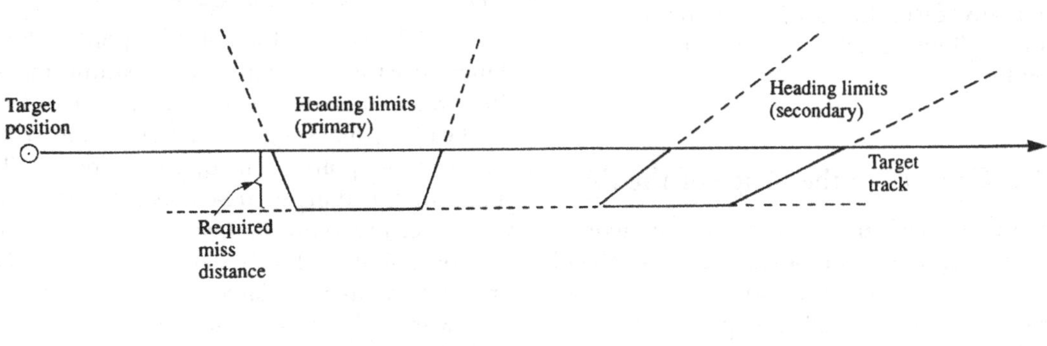

FIGURE 4.27 The development of areas to keep clear: heavy lines indicate boundaries of danger areas.

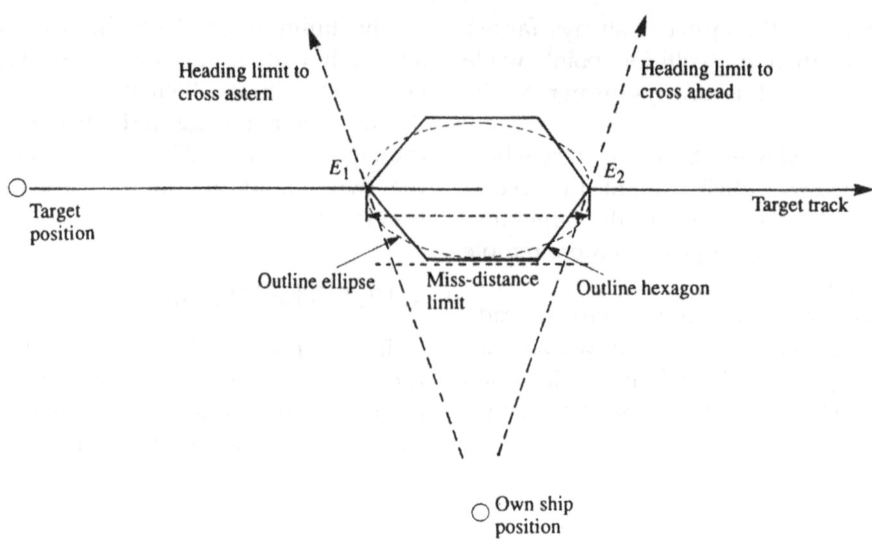

FIGURE 4.28 Acceptable symmetrical figures.

In the case of the ellipse, the major axis is equal to the difference between the cross-ahead and cross-astern distances as measured from the target, while the minor axis is equal to twice the intended miss-distance. In the case of the hexagon, it is drawn from a rectangle and two isosceles triangles. The base of the triangle is always twice the miss-distance and the vertical height is one-quarter of the distance E_1E_2, as shown in Figure 4.28. It should be noted that the collision point is not necessarily at the centre of either of the traditional figures.

In many cases the stylized figures do not follow the limits exactly, but any bias is on the safe side.

4.12.2 Changes in the Shape of the PAD

Due to the lack of symmetry in the geometry which generates the area, the cross-ahead and cross-astern positions do not move symmetrically about the collision point when the miss-distance is changed. The cross-ahead position usually moves more markedly than the cross-astern position (see Figure 4.24, showing the movement of the two collision points, where the primary movement is much slower than the secondary). The overall result is an asymmetrical growth of the area with the cross-ahead position moving rapidly away from the collision point.

4.12.3 The Movement of the PAD

4.12.3.1 The Collision Case

As in the case of the collision point, when a danger area is violated by the heading marker the danger area will continue to move down the heading marker with the cross-ahead and cross-astern points on opposite bows. The shape of the danger area may change, but it will never move off the heading marker. In the case of a slower ship, where either of the two predicted areas is violated, the other will move in towards the target and eventually merge with the one on the heading marker.

In the limiting case where the observing ship's heading marker just touches the limit of either of the PADs, the limit will remain in contact with the heading marker, although the shape of the area may change considerably.

4.12.3.2 The 'Passing Clear' Case

In the non-collision case where the heading marker does not violate one of the danger areas, the areas themselves will move across the screen, changing in shape and position. The movement will be very similar to that described for the collision point in Figure 4.22, depending on whether the observing ship is heading farther ahead than the cross-ahead position or farther astern than the cross-astern position. In the case of the dual areas of danger, although the movement will generally be the same as that shown for the dual collision points, a special case can arise when two danger areas may merge. This special case indicates the possibility of two cross-astern positions existing but no cross-ahead position. It is also possible that cross-astern positions may exist and an area of danger be drawn, which does not embrace an actual collision point.

4.12.3.3 Special Cases

In some cases, for example, an end-on encounter, a cross-ahead and cross-astern position is not valid. In this context it is necessary to consider a pass-to-port and pass-to-starboard as defining the limits of the miss-distance. In the practical case, this results in the generation of a circle about the target's position.

4.12.4 The Future of PADs

PADs are additional to the display of standard relative and true vectors. They have never become recognized by the regulatory authorities and do not have to be part of an ARPA training course. PADs are no longer provided on the present radar models manufactured by the company to which Sperry now belongs, although this does not necessarily mean that PADs will not be introduced on a new model.

PADs require some understanding of the principles and misunderstanding of the data that can occur. There has been some suggestion that PADs take no account of the Collision Avoidance Regulations and encourage less well-trained navigators to alter course for the biggest gap rather than in accordance with the rules. Thus there has been some reluctance to accept PADs at sea, not helped by the fact that availability has been restricted to Sperry models no longer in production.

They remain an elegant solution to the collision-avoidance problem and a study of PADs becomes an interesting test of encounter geometry for the academic. It is not impossible that PADs will again be provided on ARPAs in the future.

CHAPTER

5

Automatic Identification System (AIS)

With the advent of VHF communication systems at sea, the advantages of bridge-to-bridge and bridge-to-shore communications were widely recognized. Unfortunately, outside of pilotage waters and particularly in restricted visibility, there was no certainty as to who was talking to whom. While other vessels might be detected by radar, there was no means of positive identification that would allow meaningful communications or dialogue to be established. Regrettably, attempts to establish a dialogue without this positive identification resulted in the too frequently heard and dangerous practice of calling 'Ship on my starboard bow ...', with all its ramifications, especially where traffic was heavy.

The development of transponders provided the impetus to enable the identification of vessels fitted with the appropriate equipment to be established.

The ability of shore stations, particularly where a vessel traffic service (VTS) is provided, to automatically identify vessels within their surveillance area, has long been an ideal, and the automatic identification system (AIS) provides just such a facility. The AIS also provides useful tracking data for security services and commercial business.

The AIS is based on transponders located on vessels and other locations that transmit and receive information on dedicated VHF frequencies (Figure 5.1). Once set up for a voyage, information is transmitted continuously from each vessel without requiring attention from the mariner, unless something changes. The introduction of AIS provides an alternative way of obtaining information for both collision avoidance and navigation, which has traditionally been the preserve of the radar or automatic radar plotting aid (ARPA). Hence it is particularly important to appreciate the benefits and limitations of this new technology.

The specifications for a marine AIS were adopted by IMO as part of the revisions in 2001 to SOLAS Chapter V. AIS became a requirement for vessels covered by SOLAS in 2004, now termed Class A vessels, this date having been brought forward from 2007 due to international pressure. These are vessels of 300 grt or greater which are approved to operate internationally. Later in 2007, IMO introduced specified minimum standards for non-SOLAS (now termed Class B) vessels, which are generally small vessels, but can be larger vessels that only work in national waters of one state. It should be noted that by definition Class B vessels do not come under IMO jurisdiction, but under national government jurisdiction. IMO specified Class B equipment standards to ensure compatibly with Class A equipment. It

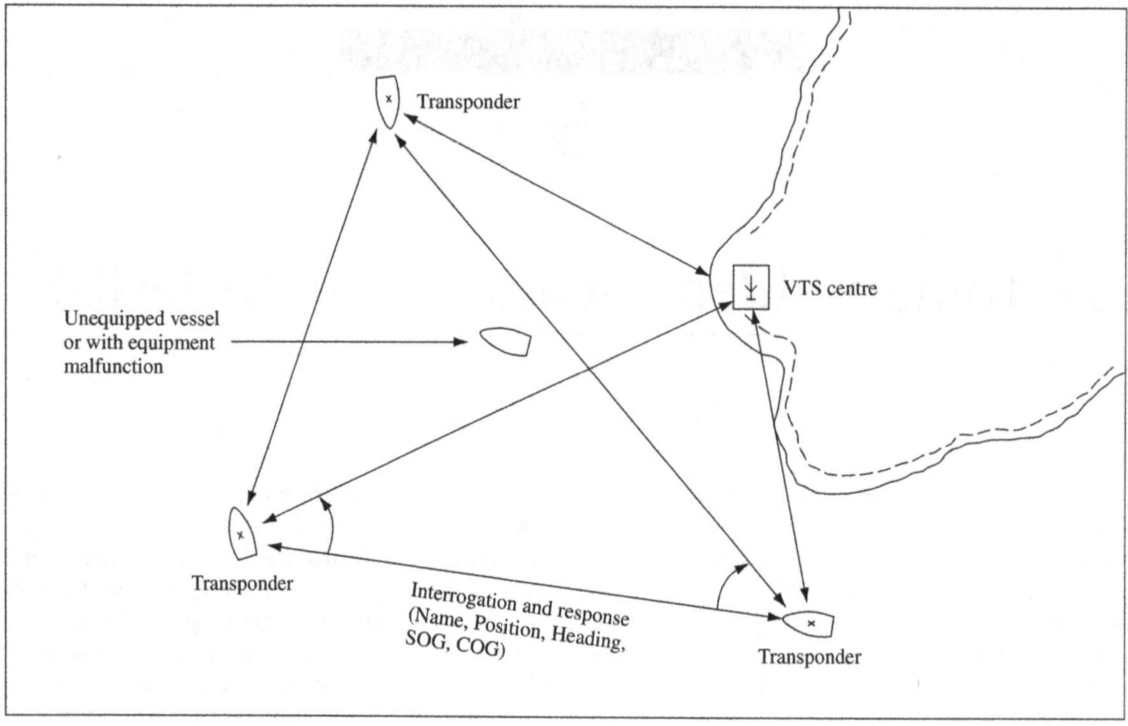

FIGURE 5.1 AIS overview.

also means that the fitting of Class B equipment to these non-SOLAS vessels is optional in most countries unless a country has made it a requirement in their own territorial waters.

Additionally transmissions can also be sent by other types of 'user'. These are SAR (search and rescue) aircraft, aids to navigation (e.g. buoys), AIS base stations and AIS SARTs (search and rescue transponders). More details of these AIS types are provided in this chapter. This chapter also covers AIS displays, and compares the AIS with ARPA data from the perspective of both the shore and sea-based users.

5.1 ORGANIZATION OF AIS TRANSMISSIONS

Two VHF frequencies have been dedicated for AIS transmission. In AIS terms they are designated as AIS channel A (VHF channel 87B or 161.975 MHz) and AIS channel B (VHF channel 88B or 162.025 MHz). The Class A system is designed to be capable of transmission on other frequencies should there be a future demand.

The potential range of Class A AIS on normal power (12.5 W) is therefore similar to VHF voice communications and is usually greater than radar. This is typically 30 or 40 NM for a VHF antenna mounted on a large vessel and 20 miles for a smaller craft. However, if a Class A AIS transponder decides that the AIS is overloaded then it will switch to low power mode (2 W), which is the same power as Class B transmissions and reduces the effective range to under 10 NM.

In order to allow all transponders to share the limited frequencies available, each transponder transmits for very short and precisely controlled time periods. The transmission system is known as TDMA – time division multiple access. The time is regulated by the Global Navigation Satellite System (GNSS) (in practice – GPS) clocks provided by satellite and it is divided into frames, each of 1 min in length. Each frame is divided into 2250 equal slots. Thus there are normally 4500 time slots available for transmission in every minute over the two frequencies (Figure 5.2). Approximately 256 bits, which can transmit the equivalent of 40–50 text characters, are sent in each slot. The allocation of the information into time slots can be made by an individual transponder under three different modes.

5.1.1 Autonomous and Continuous Mode

This is the default Class A mode and its technical name is SOTDMA (self-organizing time division multiple access), where each Class A transponder decides in which slot it is transmitting. The amount of information that is transmitted by each transponder varies, so each transponder may use a variable number of slots on each frame. If a transponder detects that it is transmitting in the same slot as another station, it will change its slot. There is another practical interference problem in that there may be accidental interference with another vessel, as shown by vessels A and C in Figure 5.3, each out of range of the other but both within range of another vessel B. To help reduce the time over which this potential interference occurs, each transponder will periodically change the slots it uses.

5.1.2 Assigned Mode

The second mode of organizing the VHF time slots is the 'assigned mode' in which a suitably equipped shore station will take responsibility for all AIS transmissions within an area and will assign slots to individual AIS transponders. This may be applicable, or indeed become a necessity, in busy waters.

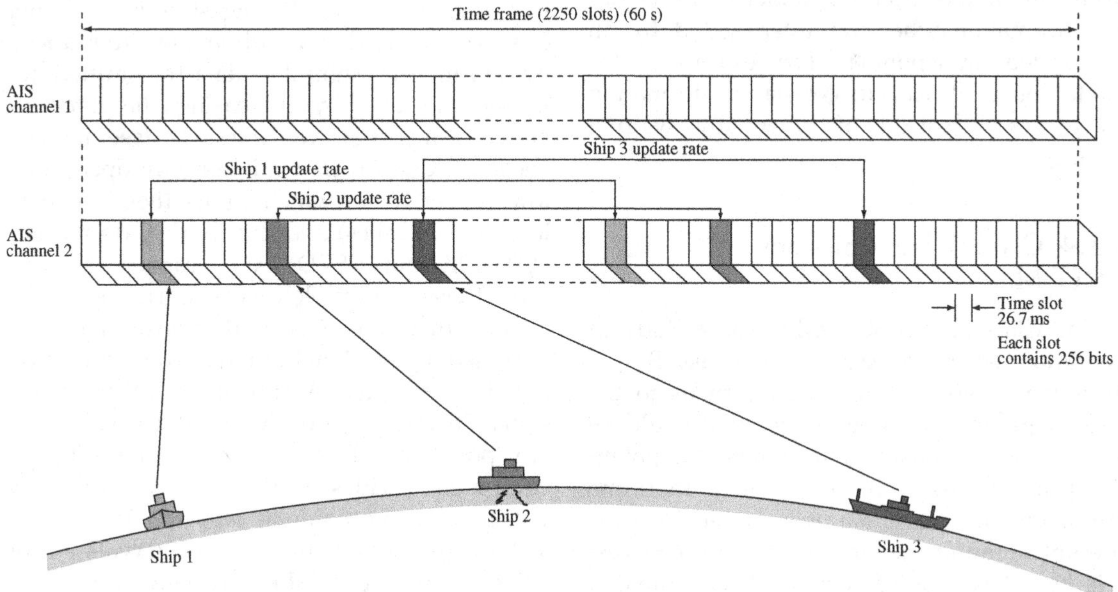

FIGURE 5.2 Principles of SOTDMA.

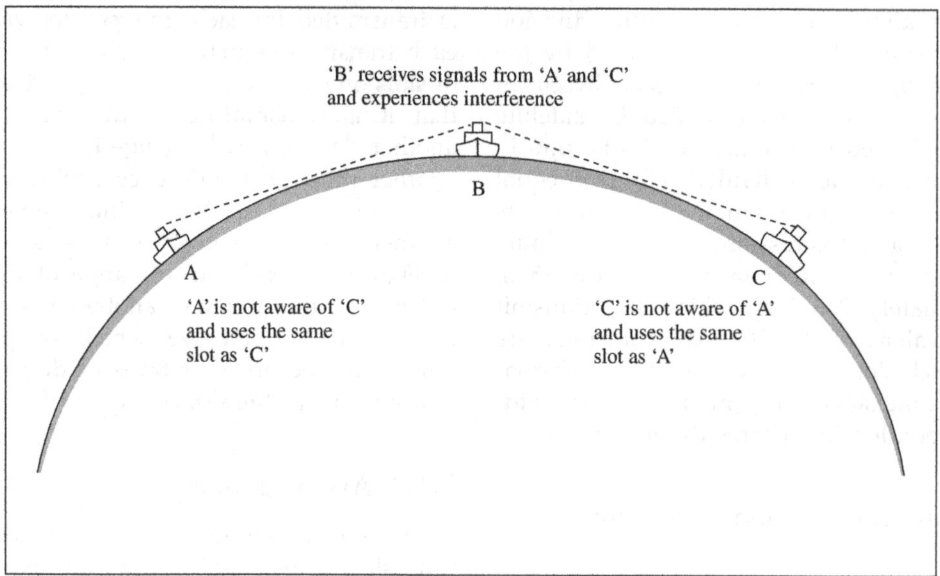
'B' receives signals from 'A' and 'C'
and experiences interference
B
A
C
'A' is not aware of 'C'
and uses the same
slot as 'C'
'C' is not aware of 'A'
and uses the same
slot as 'A'

FIGURE 5.3 Interference to a vessel caused by vessels out of range of each other.

5.1.3 Polling Mode

IMO specification also allows for a third mode of AIS transmission: the 'polling mode'. In this mode the AIS transponder will give out its information whenever interrogated by an authorized transponder. For example, this mode could be used to get static information on a vessel instead of potentially waiting 6 min.

5.1.4 Class B Transmissions – CSTDMA

The Class A AIS SOTDMA was already in operation when the system for Class B was designed. There was much concern as to the AIS congestion and interference that could be caused by the considerable number of potential Class B users. In many busy maritime areas, the theoretical number of leisure craft present outnumbers commercial (SOLAS) vessels by a factor of 10 or more. It is estimated that there are over 10,000,000 potential Class B users in comparison with the potential maximum number of Class A users, which is less than 100,000.

Class B equipment was therefore designed to reduce potential AIS congestion in a number of ways. Firstly data requirements are reduced by limiting the amount of data transmitted per vessel and also by increasing the intervals between message transmissions (see below). Secondly Class B transmitters only operate on low power (2 W) which limits their range to less than 10 miles) and thirdly they do not use SOTDMA but CSTDMA – carrier sensitive time division multiple access. Carrier sensitive means that the Class B transmitters are designed to work within the AIS time slots, but defer to Class A and other AIS transmissions. In other words, Class B transmissions are 'polite' to Class A transmissions. Class B equipment decides on the slots in which they will transmit in a similar way to Class A, but at their prescribed (increased) intervals (30 or 180 or 360 s as per Table 5.1). However, at the start of the time slot, the Class B AIS listens to

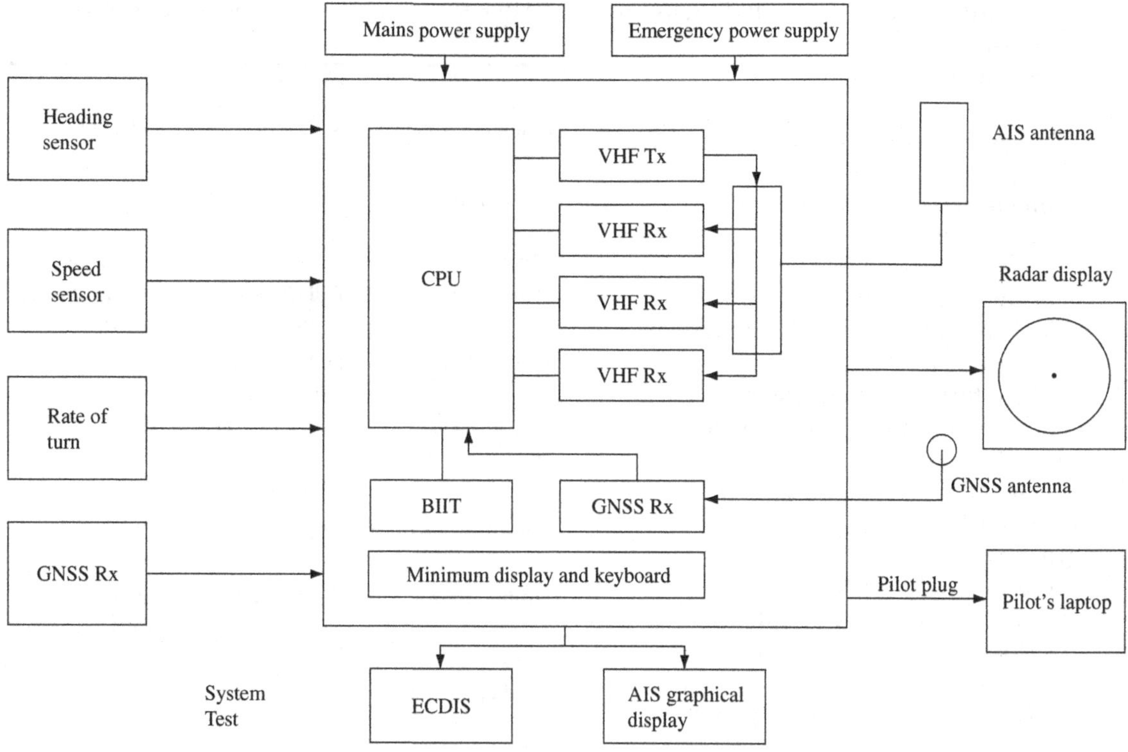

FIGURE 5.4 Block diagram of AIS user equipment.

see if they can detect another station transmitting in that slot and if so, the Class B transmitter does not transmit and defers to the next intended transmission. The main differences between Class B and Class A equipment are summarized in Table 5.2.

5.2 AIS INFORMATION TRANSMITTED BY A CLASS A VESSEL

The standard information transmitted by a SOLAS (Class A) vessel can be divided into four sections, as shown below. Any information sent by a vessel is identifiable as belonging to that vessel by the inclusion of a Maritime Mobile Service Identification (MMSI) number. This is unique to each vessel and enables the receiver to associate messages with a particular ship. For example, it correlates the static data with the dynamic data as these are sent in different messages and time slots. It can also be used by a vessel to communicate to a specific vessel using AIS or Digital Selective Calling (DSC) on VHF, MF, HF or satellite communication frequencies and thus helps avoid communicating with the wrong vessel.

In addition to the MMSI number the following information is sent:

- Static
  - IMO number
  - Callsign
  - Length and beam *(metres)*
  - Type of ship (container, tanker, etc.)
  - Location of position fixing antenna to ship extremities (stern, bow, port and

starboard). Note that overall dimensions are assumed to be the relevant summation)

- Dynamic
  Ship position (with accuracy indication and integrity)
  Time in UTC
  Course over the ground
  Speed over the ground
  Heading
  Navigational status (NUC, underway, at anchor, etc.)
  Rate of turn (where available)
  Angle of heel (optional and where available)
  Pitch and roll (optional and where available)
- Voyage related
  Ship's draught (metres)
  Hazardous cargo type (if any)
  Destination and ETA (at master's discretion)
  Optional – Route plan (waypoints)

TABLE 5.1 Data Reporting Rates

Vessel Condition	Dynamic Data Reporting Interval
Class A Ship at anchor or moored	3 min
Class A Ship 0–14 knots	10 s
Class A Ship 0–14 knots and changing course	3.3 s
Class A Ship 14–23 knots	6 s
Class A Ship 14–23 knots and changing course	2 s
Class A Ship >23 knots	2 s
Class B vessel <2 knots, including at anchor	3 min
Class B vessel >2 knots	30 s
Class A Static and voyage data	6 min
Class B Static data	6 min

TABLE 5.2 Differences Between AIS Class A and Class B Transponders

	Class A	Class B
Communication scheme	SOTDMA	CSTDMA
Dynamic data reporting interval (see Table 5.1)	Dynamic 2–10 s	Dynamic 30 s
Dynamic data	Full Class A required	No rate of turn, navigation status
Voyage data	Full Class A required	No voyage data (draft, ETA, destination)
Static data	Full Class A required	No IMO number
Safety text message	Required	Transmit feature optional
Position source	External GNSS source or AIS internal GNSS	AIS internal GNSS
Heading (compass) input	Required	Optional
Rate of turn input	Required	No
Multiple input/output interfaces	Required	Optional
Transmit power	12.5 W (normal) 2 W (low power mode)	2 W

- Short safety related message
  A short text message can be routinely transmitted.

The information is grouped into different AIS messages (see Section 5.3) and updated at different time intervals, in order to keep the amount of information being transmitted to a minimum. As indicated in Table 5.1, the voyage related data and the static data block are only transmitted every 6 min or if there is an outside request. The safety related message is sent as required. The dynamic data is sent according to the frequency indicated in

Table 5.1, which depends on vessel speed and if changing course. The structure of messages for all ships, shore and aircraft is detailed in the next section.

5.3 AIS MESSAGES AND TYPES

The AIS standards allow up to 63 different messages to be defined, but currently there are only definitions for 27. Different messages are defined for different purposes and different AIS types (Table 5.3).

5.3.1 Class A Transmissions

Messages 1, 2 and 3 are the 'short' messages routinely transmitted by Class A vessels, every few seconds (see Table 5.1). They contain the same positional and dynamic information, but the different message numbers indicate the mode of transmission of the receiver. Message 1 is for use when the AIS transmitter is operating in autonomous mode, Message 2 is sent when the transmitter is in assigned mode as a result of instructions from a competent authority via an AIS base station using Message 16 or 23 (polling mode). Message 3 is a transmitted response to an interrogation (i.e. polling mode) by an authorized user using AIS Message 15.

Message 5 is the 'long' Class A AIS message sent every 6 min. It contains all the expected static and voyage data as indicated in Section 5.2. The exception is the optional route (waypoint) data which is sent using in a subtype of AIS binary message 8 (see Section 5.3.6).

The optional safety text message broadcast to all ships is defined as message 14, but if only sent (addressed) to a specific vessel then Message 12 is used.

5.3.2 Class B Transmissions

The main differences for Class B vessels are summarized in Table 5.2. Class B transponders transmit a shorter Message 18 than Messages 1, 2 and 3 used by Class A. The dynamic information excluded from Class B are the Rate of Turn and Navigational Status fields. There is also no requirement for Class B vessels to provide compass heading, although it is an optional field within Message 18.

Message 24 is equivalent to Class A Message 5 in that it provides the static data. However, the *voyage related* data of Class A is not included. Draft, ETA and destination fields are all not provided by Class B craft. The static data also does not include IMO number, which is logical as these are not IMO ships.

Class B regulations were designed and came into force several years after Class A. On introduction of Class B, some of the older Class A receivers were only partially compatible with Class B messages and they did not show all available Class B information (particularly the Class B static data). These older equipments all should have been (software or hardware) upgraded by now, but no doubt some may still in use.

5.3.3 SAR Aircraft

SAR aircraft can broadcast Message 9 which will provide their positional data on an AIS graphical display. Low flying aircraft can sometimes be picked up in radar vertical lobes, but the marine radar is not designed for this task and the occurrence of aircraft echoes on a radar screen are erratic. The provision of AIS information of SAR aircraft (particularly helicopters) on ship and shore AIS displays could be of potential great benefit in a SAR scenario and could assist in the monitoring and execution of a combined ship/aircraft search.

5.3.4 AIS SART

Although similar in user function to the radar-based SART (see Section 3.5.6), the AIS SART, when activated by the seafarer, instead

TABLE 5.3 AIS Messages

AIS Message ID Number	Name	Description
1	Position report	Scheduled position report (Class A shipborne mobile equipment)
2	Position report	Assigned scheduled position report (Class A shipborne mobile equipment)
3	Position report	Special position report, response to interrogation (Class A shipborne mobile equipment)
4	Base station report	Position, UTC, date and current slot number of base station
5	Static and voyage related data	Scheduled static and voyage related vessel data report (Class A shipborne mobile equipment)
6	Binary addressed message	Binary data for addressed communication
7	Binary acknowledgement	Acknowledgement of received addressed binary data
8	Binary broadcast message	Binary data for broadcast communication
9	Standard SAR aircraft position report	Position report for airborne stations involved in SAR operations only
10	UTC/date inquiry	Request UTC and date
11	UTC/date response	Current UTC and date if available
12	Addressed safety related message	Safety related data for addressed communication
13	Safety related acknowledgement	Acknowledgement of received addressed safety related message
14	Safety related broadcast message	Safety related data for broadcast communication
15	Interrogation	Request for a specific message type (can result in multiple responses from one or several stations)
16	Assignment mode command	Assignment of a specific report behaviour by competent authority using a base station
17	DGNSS broadcast binary message	DGNSS corrections provided by a base station
18	Standard Class B equipment position report	Standard position report for Class B shipborne mobile equipment to be used instead of Messages 1, 2, 3
19	Extended Class B equipment position report	Extended position report for Class B shipborne mobile equipment; contains additional static information
20	Data link management message	Reserve slots for base station(s)
21	Aids-to-Navigation report	Position and status report for aids to navigation
22	Channel management	Management of channels and transceiver modes by a base station
23	Group assignment command	Assignment of a specific report behaviour by competent authority using a base station to a specific group of mobiles
24	Static data report	Additional data assigned to an MMSI Part A: Name, Part B: Static Data
25	Single slot binary message	Short unscheduled binary data transmission (broadcast or addressed)
26	Multiple slot binary message with communications state	Scheduled binary data transmission (broadcast or addressed)
27	Position report for long range applications	Scheduled position report; Class A shipborne mobile equipment outside base station coverage

transmits a distress message on both the AIS channels. IMO now allows SOLAS vessels (which above 500 grt require two such devices) to choose between the traditional radar-based SARTs or the AIS SARTs. Once activated the AIS SART sends eight transmissions a minute, four on each AIS VHF frequency. AIS receivers on other ships in range should alarm and it should appear as a cross surrounded by a red circle on Electronic Chart Display and Information System (ECDIS) and AIS displays. Once activated it should transmit for at least 96 hours. Like radar SARTs they are designed so that they can easily be pole mounted in a life raft or lifeboat. Unlike the radar SART, it does not indicate if it has received AIS signals from other vessels. The transmission consists of a unique number (programmed by the manufacturer before it is fitted to the ship) and the GPS position. Like the radar SART, there is no additional information programmable by the ship's crew at the time of the distress.

5.3.5 Aids to Navigation

5.3.5.1 AIS on Buoys

AIS transponders are fitted to a number of navigationally important buoys. They transmit AIS message 21. This has two main benefits. The buoy will be positively identified to AIS users including its specific name. Furthermore, if the buoy moves out of position, the shore authorities and vessels will be alerted. There are thoughts that these could eventually replace Racons (see Section 3.5.1) as these are not only much cheaper to install, operate and maintain, but additionally AIS self-monitors its position and is not dependent on the user selecting the correct radar frequency for observation.

5.3.5.2 Virtual AIS Transmissions from Shore Stations

Another related area being explored is the use of virtual AIS. In this scenario a shore authority can send AIS messages representing virtual 'buoys' to indicate a safe channel or mark a wreck. These would appear on an AIS display to vessels even though the 'buoys' would not physically exist. This would be of great benefit for marking safe or dangerous areas of a temporary or transient nature, or in advance of real wreck-marking buoys being laid over a new wreck. A further example is in areas where, in winter, real buoys are removed because of floating ice.

5.3.5.3 Synthetic AIS Transmissions from Shore Stations

Another similar proposal is synthetic AIS, where AIS messages for existing marks and buoys are generated by a shore station. This would save the expense of installing and maintaining AIS transmitters on the actual marks and buoys. It could cause confusion if the buoy drifted out of position, but again it could be argued that this is useful information, if an observing vessel with radar and AIS observes and understands the discrepancy.

The standard message used to transmit data on both virtual and synthetic aids to navigation is another example of the message subtype of AIS binary message 8. Other examples of AIS binary messages are covered in the next section.

5.3.6 AIS Binary Messages

Certain AIS messages allowed by IMO performance measures allow competent authorities to define extra messages within the standard IMO message structure. These are often termed ASMs (Application Specific Messages). Thus messages 6, 8, 25 and 26 allow authorized users to transmit ASMs relating to specific areas or additional information. These can be broadcast to all vessels (messages 8, 25, 26) or by using message 6 addressed to specific vessels (or vessel) using MMSI numbers.

Examples of AIS message 8 developed by specific shore authorities include the Saint Lawrence Seaway (which broadcasts data on locks, tidal levels and weather) and the Panama Canal (which broadcasts weather data from specific locations along the canal).

There are also many other more generic binary messages currently defined for authorized users to send on weather reports, tidal information, salinity reports, traffic reports, suggested routes and text messages. Examples of binary messages containing additional information for transmission by vessels include the number of persons onboard, route (waypoint) information, extended static and voyage data.

Individual countries and organizations have often developed these messages and they are then submitted to IMO for approval. These messages are still developing and there is evidence that the development is not always well coordinated. There is overlap and national differences between similar binary messages. The official IMO registry of these binary messages is updated occasionally with some older messages having already been superseded and made obsolete. It emphasizes the need for regular AIS software updates to be provided by equipment manufacturers and updated by individual vessels, if a full AIS service is to be provided.

5.4 AIS UNITS AND BRIDGE DISPLAYS

The Class A AIS unit is fitted on the bridge. When fitted to an existing vessel, much of the installation expense for Class A vessels can be the linking of the unit to other bridge equipment, especially older bridge equipment (e.g. stepper motor type gyro interfaces).

The standard of the provision of AIS data available to the navigator has been very variable and the provision can be divided into three broad types. This is because the initial 2004 standards ensured that vessels transmitted AIS messages, but (at the minimum compliance with IMO standards) did not provide a graphical user-friendly device for the navigator. The compulsory provision of AIS as a graphical interface together with integration of AIS and ARPA data was left until IMO Performance Standards for new radar equipment on ships from 2008.

5.4.1 Stand-Alone Minimum Keyboard Display

The minimum keyboard display (MKD) is a unit with a display that provides a small data screen capable of displaying only text (not graphics) with a limited number of buttons, each performing multiple tasks (similar to mobile phones). Input and extraction of information can be tedious and is prone to keying error. Initially this was the simplest unit and popular because it meets IMO approval at the lowest cost. It is designed for the primary purpose of programming the AIS unit with the vessel's own information. The reception and output of target vessel data to the navigator was clearly of secondary concern to whoever formulated the regulations. There were even reports in the early days of implementation that this philosophy extended on some vessels to the fitting of these units away from the navigational area of the bridge, or not even on the bridge at all.

However, within an MKD, it is possible to get lists of vessels that are in the own ship's geographical area with range and bearing (see Section 6.10.1.2), and this can be used to supplement information on other vessels being observed from traditional sources such as visual scene or radar/ARPA. Alarms can be set up for certain parameters, including collision threats based on GPS position calculations.

5.4.2 Stand-Alone Graphical Display

This unit was more intended for operational use by the navigators of the ship on which the AIS is fitted. Information can be displayed graphically and the display will therefore look similar in function and appearance to a radar screen. However, this graphical display will only display vessels that have AIS transponders.

Additional AIS information on a displayed vessel should therefore be available at the press of a button (See section 6.10.1).

5.4.3 Integration of AIS with ARPA and/or ECDIS

In this system, the AIS unit provides the AIS information on other vessels to an ARPA, radar, ECDIS or integrated bridge system (IBS) display, possibly as part of the network of an IBS.

The advantage of overlaying the AIS information on the ARPA is that radar targets can be compared with the additional AIS information. Vessels that are poor radar targets become obvious and vessels not operating AIS can become identifiable on the radar screen. The navigator can therefore see a summary of both systems on one screen.

The advantages of overlaying the AIS on the ECDIS display is that other vessels' paths can be seen in a navigational context, which is particularly apt as the AIS gives ground track and ground speed. Additionally, the projected routes of target vessels can sometimes also be displayed for consideration.

The obvious disadvantage of AIS information is the danger of information overload on the navigator. In busy situations, navigators have already demonstrated the potential (and often serious) consequences of getting the following information confused:

- ARPA derived relative and true vectors
- ARPA derived ground and water stabilized true vectors.

The provision of additional vectors based on AIS derived information has provoked serious consideration in some quarters to reduce the problem of information overload. There are two results of this research.

1. Standard symbology

   Table 3 in Section 11.2.3 shows IMO approved symbols for AIS targets. These symbols have been selected for their clarity. Note that the triangle points in the direction of the ship's head and should therefore provide aspect. The vector line gives ground course and speed. The curved tail on this vector appears when a vessel is changing course and indicates the direction of the change. The triangle can be replaced by a ship outline of the correct scale dimensions when the display scale is large enough and the static information relating to that vessel is available.
2. Data fusion

   Algorithms have been developed to compare ARPA derived data with AIS derived data. Thus, where the system thinks an ARPA target and AIS source are the same vessel, it will indicate that they are the same on a combined display and will not show them as two separate vessels. The idea here is to reduce the amount of unnecessary information provided to the navigator. These algorithms are manufacturer specific and can vary in operation and in terms of (any) operator adjusted parameters (see Section 6.10.2).

It can be computer intensive for a radar/ARPA/AIS set to track and plot a large number of AIS targets in addition to its normal workload. Radar/ARPA sets that have AIS integration include a maximum number of AIS targets that will be monitored and displayed on the screen. A typical maximum number is 50, but the navigator can usually select a lower number if required. The radar/ARPA/AIS set will plot the nearest AIS targets to the own ship if the maximum number of AIS targets is exceeded.

In the performance standards for new ships built after 2008, the radar installation must be capable of displaying a minimum number of active AIS targets. This number depends on ship size and is, in fact, the same number as the minimum number of radar tracked targets (see Table 4.2). The standards also allow the navigator to designate a certain number as sleeping targets, so the display is not overly cluttered with information. This can be on an individual vessel basis or by designating the area around the own ship in which to show AIS vessels (see Section 6.10.2.1).

5.4.4 Pilot Plug and Display

AIS units are also fitted with a 'pilot plug', which is simply a standard data connector. The pilot brings onboard a display unit (sometimes termed PPU – pilot portable unit) which interfaces directly with the AIS. The advantage is that the pilot can then use a display device with which he is familiar while controlling and using the transponder already installed on the vessel. The pilot is also able to use the information from the ship's transponder in conjunction with other computer-based applications specific to the pilotage, such as local electronic charts and tide data.

Vessels entering US waters are additionally required to provide a convenient 120 V power supply for the use of the pilot display, when using the pilot plug.

5.4.5 Class B Equipment

Class B is (by design) intended to be lower cost and power consumption. The reduced data inputs will also mean less installation costs, as less interfaces are needed. Indeed the external bridge equipment to interface with AIS in Class A is unlikely to be carried by such vessels. In practice they resemble a Class A MKD (Section 5.4.1) or more likely an AIS stand-alone device (Section 5.4.2). Optionally they can be provided for output interfaces for any kind of electronic chart or radar display (full IMO compliant ARPA and ECDIS equipment would be unusual on most Class B vessels). However, pages and menu systems should be simpler due to the lesser data information that Class B vessels transmit. The fact that voyage data and navigational status are not transmitted by such vessels means that the Class B navigator does not normally need to update the AIS on a voyage-by-voyage basis, unlike the Class A counterpart. It can be just switch on and switch off.

5.4.6 AIS Receivers

Also available for small boats are AIS receivers, which do not transmit any AIS messages. They are intended for small craft and are low cost. In particular they have very low power requirements. The advantage of this equipment is that it gives the navigator a graphical picture of larger vessels and any other users in the local area who are transmitting AIS messages. It has the obvious disadvantage that the equipment does not transmit the presence of the 'own ship' to other users, which is effectively missing one half of the true collision avoidance scenario. However, the argument is probably that it is better than no AIS information and at least such vessels are trying to keep out of the way of large vessels. One could also argue that the target market for such equipment represents a large number of vessels and if they adopt this equipment they will not cause AIS congestion. There are no standards for such equipment and no known carriage requirements. These have been unofficially termed Class C AIS by some manufacturers.

5.5 AIS USABILITY

AIS and ARPA information are often compared. In considering the relative capabilities of AIS, a number of issues need to be considered.

5.5.1 Target Swap

A common ARPA issue (onshore or vessel display) is radar target swap (see Section 4.3.5). This is when a target plotted by an ARPA loses its 'tag' to another passing vessel or static radar target. AIS does not have this problem. In fact, on a combined AIS/ARPA display (on vessel or shore VTS), it can provide a warning that radar target swap has occurred.

However, there has been a form of AIS target swap that can occur. It has previously been stressed that AIS quality depends on the information that has been programmed into the AIS equipment. If two vessels are programmed with the same MMSI number and they are both within the observing range of a third vessel or VTS centre, then this will cause a problem on the observing AIS display. The observing AIS equipment will think the messages are all from the same vessel. On the user display, this would result in the vessel jumping from one position to the other as the dynamic data comes in every few seconds. The actual vessel name (and other static data) shown will depend on which vessel last sent its static and voyage message (every 6 min), so the name and static data shown will therefore change probably twice in every 6 min. MMSI numbers are of course supposed to be unique to each vessel, but typographical mistakes can occur and the two vessels may happen to be in the same AIS coverage area. It also does not help that manufacturers usually put the same default information into all their AIS equipment, so it can be an error of omission on commissioning the onboard equipment which results in a common incorrect MMSI being used by several vessels. Furthermore, there has also been a case (hopefully now past) caused by poor AIS equipment design by one AIS manufacturer. The MMSI number on the shipboard unit in question defaulted to the preset manufacturer MMSI number whenever restarted after any power outage. A significant number of vessels therefore were at one time using the same incorrect MMSI number.

In summary, given the very small numbers of vessels likely to be incorrectly programmed with their MMSI number, it is very unlikely that two vessels with the same MMSI number will be in the same ship AIS area and so AIS target swap is unlikely to be a common problem for seafarers. One benefit of the existence of the larger 'global' AIS networks (see Sections 5.8.1 and 5.8.2) is that the existence of such MMSI errors will be known to such networks and they will probably be proactive in advising the culprit vessels of the problem.

5.5.2 The Transmission of AIS Signals Around Large Land Features

AIS VHF transmissions can not only have a longer range than radar, but they also have a limited ability to go round corners and over the top of land masses. This is particularly useful in fjord type regions where radar and visual observer are unable to see far ahead due to blind areas, whereas AIS transmissions can usually be received (Figure 5.5). Such areas can also be blessed with narrow navigation channels with only limited searoom for vessels to pass. The ability to recognize the position and ETA of a vessel approaching around a blind corner, and then communicate a passing strategy in advance can be very useful.

5.5.3 AIS Capacity

It is claimed that the system copes with in excess of 2000 transponders in an area. It can be seen from Table 5.1 that the maximum capacity will depend on whether a large proportion of the vessels are in motion and how fast they are travelling. Thus for such a claim to be possible, it is assuming that the majority of vessels transmitting are stationary and that a significant proportion are Class B, which may well be a reasonable assumption. Class A transmitters can also reduce the power and therefore range of transmissions to reduce congestion, as described in Section 5.1.4.

As the system gets more congested then the occasional transmissions are missed by other craft due to corruption caused interference or because slots are not always available to transmit. Effectively the updates on AIS displays reduce, so whether the AIS is 'coping' depends on the user's perception of what is an acceptable update rate. Certainly most users would not be too concerned if they receive data at intervals significantly larger than IMO specified intervals, and to date there have been no known complaints that the system has failed to work due to congestion.

5.5.4 AIS Dependence on GNSS

An important point about AIS is that the system is completely dependent on an operational GNSS (in practice GPS) for its timing of radio message broadcasts. Indeed AIS units

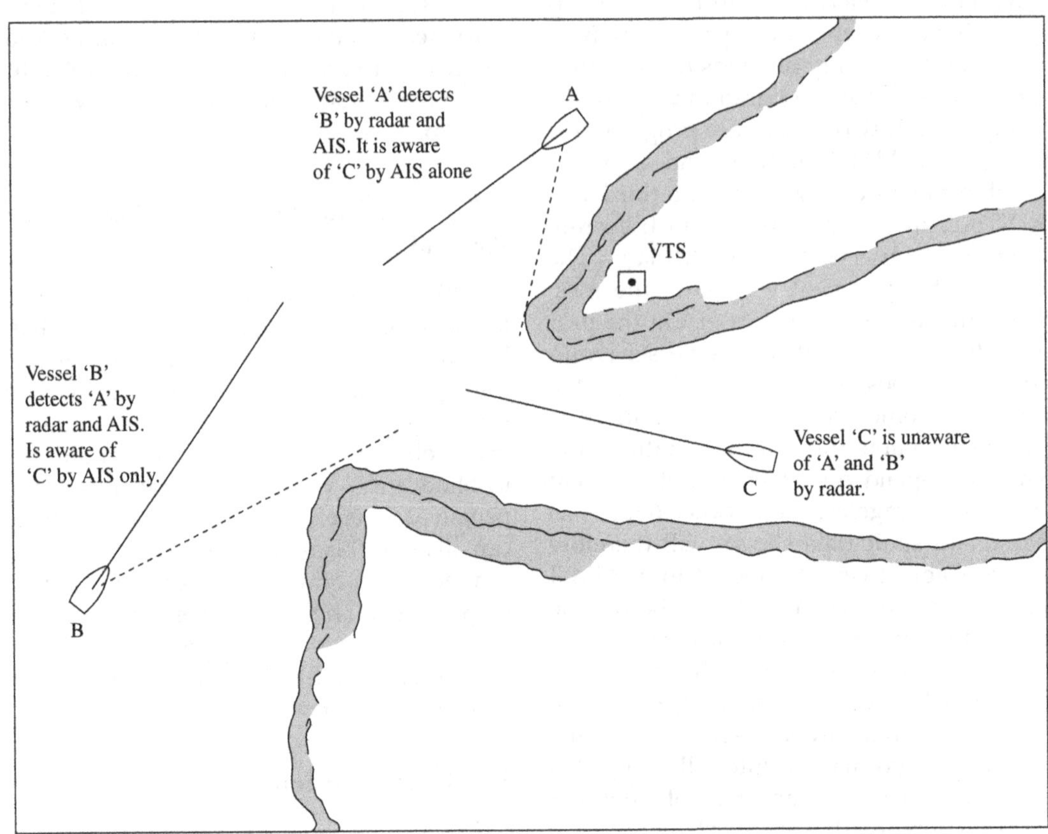

FIGURE 5.5 Fjord effect.

come with an internal GNSS unit incorporated – primarily to provide the timing.

The position, ground speed and ground course will also come from GNSS derived data. However, it should also be noted that these should come from the GNSS unit on the bridge being used by the navigators, which may not be the internal GNSS fitted to the AIS. It is the position of the navigator's GNSS antennae (i.e. not necessarily the antenna of the GNSS internal to the AIS) which should be included in the static message.

The introduction of other GNSS, including GLONASS, Compass and Galileo (see Section 10.1.8) provides the opportunity for the AIS time system not being dependent on GPS alone. This has the political advantage of the AIS not being under the control of one country, but it will not significantly reduce the vulnerability of the system to potential interference to GNSS satellite-based transmissions (see Section 10.1.5.8).

5.5.5 AIS Participation by Vessels

A number of vessels can be exempted by administrations from carrying AIS. These may include warships and security vessels. In practice such vessels have found it beneficial to fit and transmit AIS as (rightly or wrongly) other navigators are getting used to having AIS information as part of their decision-making processes. In should be borne in mind that the transmission of AIS by such vessels is voluntary and that they are fully entitled to switch off AIS without notice. Indeed, as in law enforcement ashore, they may even pretend to be another vessel, when on law enforcement activities.

As already indicated, many Class B vessels are not required to carry or transmit AIS.

It is therefore not certain that every vessel encountered at sea will have AIS even in those waters where the national administration has adopted the Class B AIS. Another potential problem is that there are occasions in many countries when many thousands of leisure craft congregate together in a few square miles. If all these vessels are fitted with AIS, then the system could theoretically become overloaded.

Vessels fitted with AIS do not necessarily have to send all the normal information. As already indicated in Section 5.2, certain parameters, particularly route information, are optional.

The full AIS may be turned off if the master believes that it constitutes a danger to his vessel. This may be the case when transiting known piracy areas or carrying sensitive cargoes such as nuclear waste.

In summary, although some administrations are intending to have AIS fitted and used on all vessels in their waters whenever practical, it is unlikely in the foreseeable future that a situation will be reached where all vessels encountered on a voyage can be depended upon to have AIS. From this perspective it would appear certain that there will be a requirement for vessels to continue to carry and operate radar in order to dependably detect other craft in reduced visibility.

5.5.6 AIS Vulnerability to False Reports, Spoofing and Jamming

Like radar, AIS is susceptible to jamming. The more precise frequency control and low transmission power means that AIS can be jammed far more easily than radar, either by accident or by deliberate act.

As already described in synthetic and virtual AIS (see Sections 5.3.5.2 and 5.3.5.3), it is comparatively easy for one station to send messages referring to another geographic location, and it is therefore very easy for a malignant user to generate false targets on a user's

AIS display. Whilst there has been no reported deliberate occurrence of this type, vessels have had false AIS positions and data reported due to faulty equipment and incorrect data entry by installers and navigators.

5.6 BENEFITS OF AIS TO SHORE MONITORING STATIONS

The speeding up of the AIS implementation by IMO was for the benefit of nations wanting better shore-based surveillance, probably for security and perhaps VTS purposes. Many such VTS centres already operate radars, but having AIS has a number of positive benefits for these centres.

AIS facilitates the positive identification of vessels on a VTS display without recourse to the VHF radio. Many useful details regarding the AIS equipped vessel become available and the information is passed on in computer language. Thus AIS avoids the language difficulties and identification confusions often associated with VHF voice communications.

Many shore VTS stations cover large areas requiring a network of antennae connected to the centre by data communications channels. VHF antennae and transponders are much simpler devices than radar antenna and transceivers, with considerably less power and maintenance requirements. The AIS data communications network should also be of much lower capacity and low cost. Finally, it is much simpler to correlate the data for tracking purposes from multiple AIS receivers than the output from multiple radars. AIS data is digital and it is easy for computer processors to recognize multiple occurrences of the same target from different receivers. Correlation of radar data can be more difficult due the same target being a different size when observed by different radar positions, and the correct range calibration and alignment of each radar in the network is critical.

The net effect of this and two previously mentioned factors, the larger range for Class A vessels (Section 5.5.2) and the 'fjord' effect (Section 5.5.1), means that it normally requires fewer AIS transponders than radar scanners to provide shore coverage of an area, and therefore an AIS shore network is a lot less expensive to operate than a radar-based one. A good example of the benefits of AIS is the Panama Canal VTS. The radar coverage was limited to a very small proportion of the canal before AIS, despite having multiple radar antennae. Once AIS was introduced it was cheap, quick and cost-effective to provide coverage of the whole canal. The Panama Canal has made AIS carriage mandatory for the transit including otherwise exempted vessels (such as warships). AIS sets are available for rental for the transit.

The main disadvantage of the AIS network from a shore perspective is that the vessel messages are dependent upon and under the control of the vessels themselves. If a vessel chooses not to transmit its AIS or its equipment fails, the shore-based VTS would have no indication of the vessel's presence on an AIS network. This indicates that the ideal solution is to have some radar coverage, particularly in the approaches to the VTS area, as well as total coverage by AIS. This would enable the existence of vessels to be confirmed and confirm that their AIS is working. The previously mentioned Panama Canal is a good example of this arrangement. It is also worth mentioning that as well as traffic management, many shore authorities have a security role and the indication of a vessel on the radar which does not have an AIS message output may well make the vessel of interest to security forces.

If the shore authority wishes, AIS can be used to send useful information to vessels autonomously, such as tidal information, lock information, weather and other environmental information (see Section 5.3.6).

The addition of AIS (probably in conjunction with radar) should make remote pilotage more feasible in certain areas, particularly where the pilotage demands are not intense, or where severe weather or other constraints might prevent a pilot from boarding a vessel. The pilot would be able to monitor the course steered and ground track of a vessel very effectively in near real-time and advise the master from the shore. The role of the shore-based radar in emergency pilotage would be to act both as a check that the AIS is operating correctly, and as an emergency fallback should the AIS suffer a malfunction.

Overall the benefits of AIS to a shore authority are thus large and will enable monitoring of vessel traffic in areas not currently feasible by radar due to geographical and/or cost considerations.

5.7 RADAR/ARPA AND AIS COMPARISON FOR COLLISION AVOIDANCE

The AIS information is much faster to update than ARPA. Course changes are indicated as soon as the target ship's heading starts to change instead of waiting for the ARPA to detect the change in apparent motion (it could be 5 more minutes later for large vessels). The AIS information is reliant totally on GPS. The positive side of this is that ground-based information does not depend on other less reliable sensors such as gyro and logs for processing such information.

AIS did have initial teething troubles due to the limited equipment initially installed on the majority of vessel (the stand-alone MKD, see Section 5.4.1), poor bridge installation, incorrect information being programmed, missing information and a complete lack of initial user training. There were some parallels here with the initial early industry introduction of radar in the 1950s which led to the occurrence of several high-profile 'radar assisted collisions'. Fortunately there have been to date very few 'AIS assisted collisions'. However, it has taken time for navigators to see the benefits of having AIS on their displays and only then taking the trouble to ensure that their AIS is transmitting the correct information at the correct stage of the voyage. Vessels with new radar equipment from 2008 have benefited from compulsory integrated AIS and radar display.

In Table 5.4 the comparison of radar and AIS information for anti-collision purposes is summarized. It is true that in the open ocean there is no need for the quality and depth of information that AIS provides in order to easily detect and avoid potential collisions. However, in coastal areas under pilotage type conditions, traffic can be busier, traffic movements more complex and the options to avoid collision become more restricted. In these conditions both the extra information (such as the vessels' destination) and the speed of reception of course/speed changes can be a useful complement to ARPA information for the navigator.

There is also the danger that AIS information could lead to bad practices in potential collision situations, with more vessels exchanging communications and agreeing to actions contrary to good seamanship and the rules of the road, but it is beyond the remit of this book to cover such issues in depth.

Another concern is that the provision of AIS may lead to fewer small craft fitting radar reflectors, even though the ideal would be to have both. A converse point is that there will be practical limitations to implementation of AIS on some small craft due to such problems as available power supply and finding an effective antenna position. From a radar perspective, this is unfortunate as these small vessels are very often poor radar targets and AIS would give confirmation of their positions.

TABLE 5.4 Comparison of ARPA and AIS Data for Anti-Collision Purposes

	ARPA/Radar Derived Data	AIS/VHF Derived Data
Overall accuracy	Similar to AIS at close range but accuracy reduces linearly with range, due mainly to bearing accuracy	Positional errors 10–30 m
Framework for calculations	Relative to ship	Ground based
Derivation of aspect	Derived by calculation and depends on accurate knowledge of the own ship's course and speed through water	Obtained directly from compass of target ship (when available)
Detection of changes in target course and speed	Takes several minutes	Immediate (when compass available), as soon as gyro heading starts to change. Otherwise will still be apparent sooner than radar, when GPS derived ground track changes
Identification of target size	Can be misleading; changes with range and aspect	Good, if static data is transmitted
Reliance on other equipment	All necessary equipment on the own ship; requires compass and log	GPS sensors on other vessels and programming on other vessels
Reliability of detecting other vessels in the vicinity	Dependent on echo strength and weather conditions	Only if fitted with AIS, not significantly weather dependent
Target swap	Possible	Unlikely
Interference and false echoes	A possibility	Less likely
Reduced coverage due to own ship obstructions	Can occur depending on antenna position	Less likely
Reduced coverage due to land mass obstructions (the 'fjord' effect)	Yes, line of sight only	Unlikely
Range	Typically 10–20 miles, maximum depending on antenna heights and environmental factors	Typically 20–40 miles for Class A, 10 miles for Class B or Class A in congested waters. Also depends on antenna heights and environmental factors
Transmission and target response density causing reduction of updates	Unlikely	A possibility

In summary, it is unlikely that AIS will replace radar as an anti-collision aid, because it depends on every other vessel having AIS equipment, for every other vessel to have their equipment in full working order and for every other vessels' AIS to be set up properly. However, there is no denying that when working properly and if provided in a user-friendly display to the navigator, the AIS can provide high-quality and useful information. In poor weather, it may even be superior to radar and visual in detecting small vessels (if fitted with AIS). AIS should therefore be regarded as a complement to radar/ARPA and not a competitor.

5.8 OTHER AIS APPLICATIONS AND APPLICATIONS ASSOCIATED WITH AIS

5.8.1 AIS Internet-Based Networks

The advent of Class A AIS in 2004 also saw the introduction of shore-based AIS networks which were set up for access by Internet users. Technically these are similar to VTS type AIS networks above, but these were set up by coastguards, commercial companies with ambitions to provide a regional/global network, and by local enthusiasts for their local area. It is relatively low cost to set up a VHF antenna and AIS receiver onshore (near a coast) and then transmit the results through an Internet portal to anywhere and everywhere. Some of the commercial companies have very large global networks of antennas/receivers, particularly located in coastal areas next to major sea lanes. The receivers continuously relay the messages to a central server. The server normally does some processing of the messages to remove spurious or corrupted messages and then provides the information on global maps to their clients. Tens of thousands of ships are monitored in this way.

In terms of coverage, although these large networks may have a global nature, they are not and can never be 100% global. Ships are only monitored when within range of a coast station. Most ocean going ships will disappear from coverage for large periods of time, either due to an ocean passage or passing a coastal region with no coverage.

These networks can also provide additional services to their clients, such as time, position and ETA when the ship last reported, and additional correlated non-AIS vessel data such as archived photographs and registry data.

It should also be noted that providers are restricting information in certain areas due to political or other sensibilities. This has included the non-provision of any ship data in areas of high piracy risk.

5.8.2 Satellite AIS

In an attempt to get over the range limitations of terrestrial-based shore AIS networks, several countries and organizations have been using satellite-based technology to receive AIS messages from ships on a truly global basis. They have all used satellites in low earth orbits, to receive messages and then relay them to a shore station. In some cases these are very small satellites, the size of a football, called nano-satellites. Even in these low earth orbits, the footprint of these low earth orbits is such that thousands of AIS transmissions can be received at any one time, something for which AIS was not designed. Furthermore, even if the volume of ships is not high (such as a rarely travelled ocean route), the AIS transmissions will probably overlap and interfere with each other on many occasions, because the AIS messages are being transmitted by ships which are unaware of each others' transmissions. However, it has now proved possible to unscramble some or all the messages being received, even though they are being received simultaneously at the same time and frequency. This can be assisted by the use of directional receivers which can restrict the area being examined and search small area by small area over time. The satellites can only download data when over a suitable shore station, so it is not intended that the service provides instantaneous data, but that individual ship data is only updated over periods of 1–12 hours (Figure 5.6).

These organizations are governmental or quasi-governmental and unlike the shore-based networks (Section 5.8.1) the information is claimed to be strictly controlled. The organizations involved collect the data for safety,

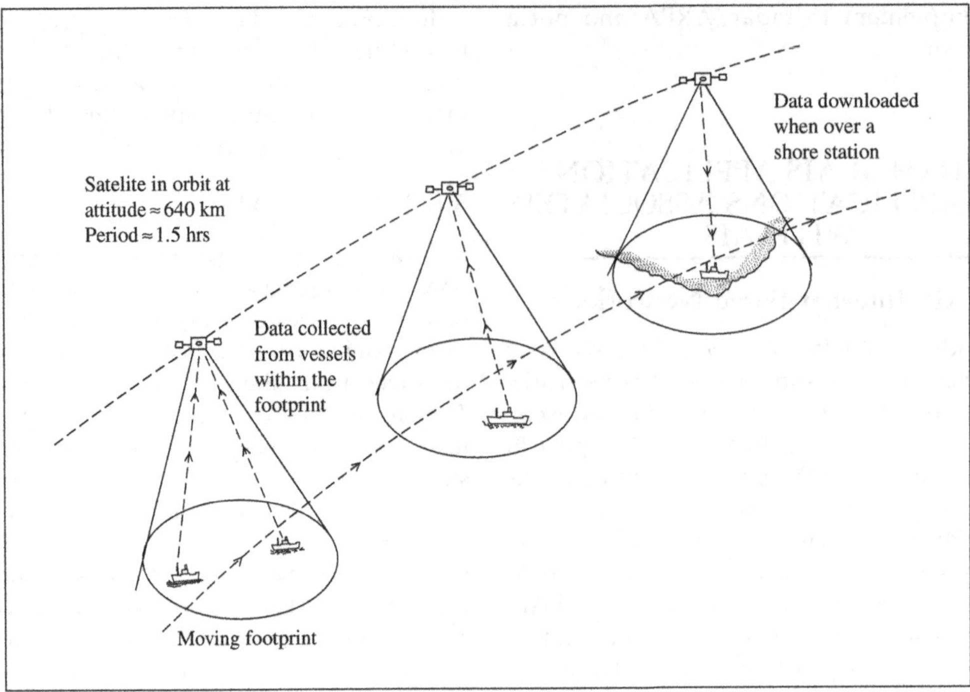

FIGURE 5.6 Satellite AIS data collection and distribution.

search and rescue (SAR), SAR, security and environmental protection. IMO itself does not regulate satellite-based AIS networks.

5.8.3 Long Range Identification and Tracking LRIT

In parallel with the development of satellite-based AIS, IMO itself developed LRIT to correct for the deficiencies of AIS and provide the ability to track vessels on a truly global basis. LRIT is not AIS, it is included here as it is has many objectives which are the same as AIS, and it is useful to make clear the differences.

LRIT applies to all SOLAS vessels, the equipment onboard contains a GNSS device and may contain low-cost text-only dedicated satellite communications equipment (usually INMARSAT MINI-C), or it may be linked to the ships existing satellite equipment. The MINI-C antenna is a static compact antenna without the complications of a gyro stabilized satellite 'dish'. The equipment sends a positional report (at least) every 6 hour via satellite to the LRIT data centre of the flag state. The message contains the vessel ID, date and position. Some flag states have contracted the data collection and supervision out to commercial IT providers, but it still means that the flag states control the data. In the case of the European Union (EU), all European flag data goes straight to an EU LRIT data centre under the control of the European maritime administration (EMSA). Ship data can be requested by approved government parties who have a legitimate interest, through an LRIT international data exchange set up by IMO (Figure 5.7).

Like AIS, the flag state can send an interrogation message to the ship LRIT unit (polling mode), which will generate an immediate positional message by the vessel being polled.

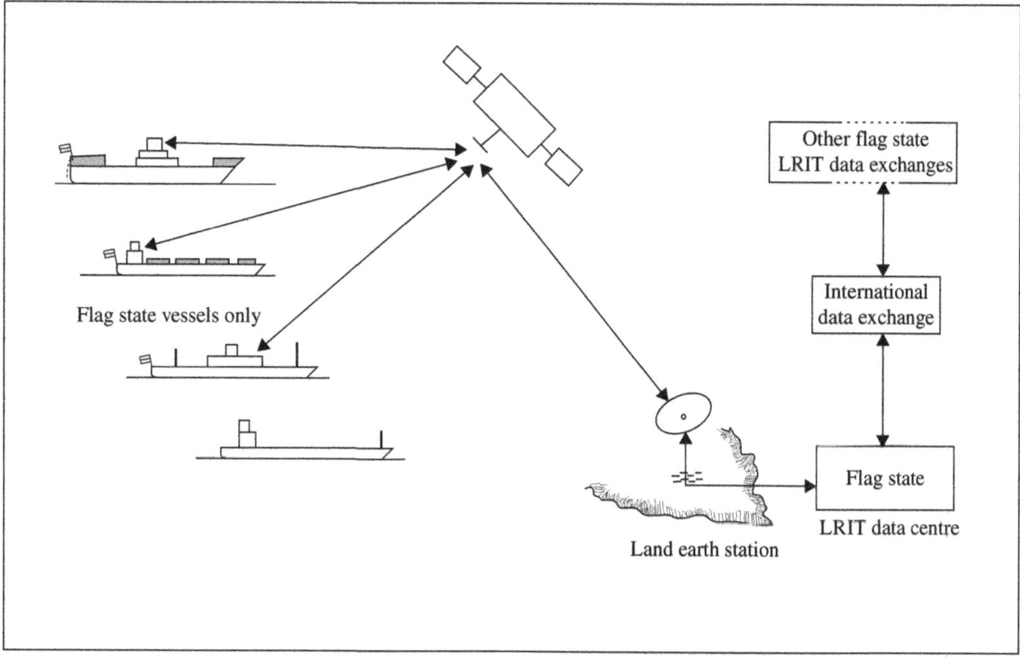

FIGURE 5.7 Block architecture of LRIT.

There are two classes of polling requests, routine and SAR. Requests can be addressed to an individual ship or to ships in a specified area. The flag state can also request, remotely, for the positional message to be sent at much reduced intervals down to a minimum of 15 min.

Unlike Class A AIS units, there are no user inputs on a voyage by voyage basis. However, message transmission intervals can be increased from every 6 hours to every 24 hours if the vessel is in long-term layup or drydock.

CHAPTER

6

Operational Controls

6.1 USE OF CONTROLS AND OPTIMUM PERFORMANCE

This chapter will focus on the use of the main controls for target tracking equipment which essentially can be broken into the radar, the automatic radar plotting aid (ARPA) and the automatic identification system (AIS).

The bulk of the chapter is on radar because unlike the other two components, the performance of traditional radar has depended on the correct setting of several controls. A major aim of the radar observer should be to ensure that all the targets which can be detected *are detected*. This is sometimes described as achieving the maximum probability of detection of targets. For any given radar system, the performance is limited in all circumstances by certain design parameters such as transmitter power, antenna gain and receiver sensitivity (see Section 3.2). On any particular occasion the performance of the system may be further limited by weather, atmospheric conditions and antenna position (see Sections 3.6–3.8). Although design features and environmental conditions place limits on the potential performance of the system, the ability of the observer to set and maintain the operational controls in correct adjustment determines whether or not such potential performance is realized.

Maladjustment of the controls may result in a failure to detect and display targets at a sufficiently early stage or may allow them to escape detection completely. It is therefore of crucial importance that the radar observer knows, understands and implements the correct practical procedures for the setting up of the operational controls and for maintaining them in optimum adjustment according to changing requirements and circumstances.

Further, when basic data is extracted by making measurements from the display, the observer must ensure that the full potential accuracy is realized. Maladjustment of certain controls may impair the accuracy of such measurements and this must be avoided by the use of the proper procedures.

This chapter presents detailed practical procedures which, if followed, will ensure the correct initial setting of the controls at switch-on and the appropriate subsequent adjustments necessary to maintain optimum performance. The procedures also deal with the use of the controls in routine tasks, including the combating of the adverse effects of weather conditions and the extraction of basic radar, ARPA and AIS data from the display. There are different display options of the AIS and ARPA tracking information for a modern observer, and it is important that the observer understands what is being presented.

Traditionally, there existed long established criteria by which the setting of the radar controls for best radar performance was judged. These criteria derived historically from the characteristics of the traditional analogue picture on a cathode ray tube (CRT) display cannot be applied directly to a modern synthetic raster display, and are no longer covered in this book. The recent introduction of new technologies mean that setting up these radars is much simpler in order to obtain an optimum radar picture. The setting up of an ARPA display requires the optimization of the radar controls and the correct adjustment of the ARPA controls to achieve effective and accurate tracking and this is also treated separately (see Section 6.9). In many modern displays, there is the additional complexity of inputs from sensors such as AIS and Global Navigation Satellite System (GNSS). Finally the display options for AIS data are discussed.

6.2 SETTING UP THE RADAR DISPLAY

The operational controls are all located on the display unit and for this reason the procedure is commonly referred to as *setting up the display*. However, it should be borne in mind that some of the controls do not perform display functions. The tuning control, if fitted, operates within the radio frequency section of the transceiver (see Sections 2.7.2 and 2.7.4.1). The gain and sea clutter controls also perform receiver functions (see Section 2.6.4.5), although the section of the receiver in which they operate may well be located within the display unit. The pulse length selector (which may also be affected by the range selector) operates in the transmitter (to control the pulse length and pulse repetition frequency (PRF) – see Section 2.3.3.2) and in the receiver (to control the bandwidth – see Section 2.6.4.2).

6.2.1 Preliminary Procedure

Before switching the equipment on, the following preliminary checks should be carried out:

1. Ensure that the antenna is clear. In particular it is important to check that no personnel are working on or close to the antenna and that there are no loose halyards or other such rigging which may foul it.
2. Check that the power switch that makes the ship's mains available to the installation is on. This switch may not necessarily be located in the wheelhouse.

6.2.2 Switching On

It is convenient to consider that this operation carries out three distinct functions:

1. The supply of power to the antenna.
2. The supply of power to the transceiver and display.
3. The switching of the system from the *stand-by* condition to the *operational* condition (see below).

In some systems an individual switch may be associated with each function. It is perhaps more common to find that a single switch combines functions 1 and 2. In a further traditional variation, a single multi-position switch may service all three functions. It is now common to have a software control to switch between stand-by and operational conditions.

6.2.2.1 *Magnetron Radar*

The application of power to the system initiates a warming-up period. In a magnetron-based radar, very high voltages (EHT) and working temperatures are associated with certain elements of the system, in particular the transmitter (see Section 2.3). Good engineering practice dictates that such elements should be raised gradually to working temperature before the high voltages are applied. Some form of thermal or other delay timer will normally be used to isolate the EHT during the warming-up period, which is likely to exceed one and a half minutes. The maximum duration is set by IMO Performance Standards (see Section 11.2.1) which require that, after switching on from cold, the equipment should become fully operational within 4 minutes. For situations other than that of a cold start the same section of the standard requires the provision of a stand-by condition from which the equipment can be brought to an operational condition within 5 s.

The precise status of the equipment in the stand-by condition may vary with manufacturer, but in all cases the transmitter will be in the quiescent state, that is it will be maintained at or close to working temperature but no pulses will be transmitted. The use of the stand-by condition is discussed in detail in Section 6.5. If the operational condition is selected at initial switch-on (see function 3 above), the system will automatically switch from the stand-by condition to the operational condition when the warming-up period has elapsed.

6.2.2.2 Solid-State Radar

The introduction of this new technology (Section 2.9) means that the radar is no longer high voltage, indeed it may be termed low voltage (under 50 V is typical in the transmitter/receiver). The issues of a significant warming-up period for components are therefore a thing of the past. The only delay is due to booting up time of the computer-based technology which processes the radar signals and provides the display.

6.2.3 Setting the Screen Brilliance

The setting of the brilliance on a raster display is judged by a criterion which differs fundamentally from that traditionally used in radial-scan displays, where the control adjusts the brilliance of the slowly rotating trace, and the criterion for optimum setting can be rigidly specified in terms of the principle on which echoes are displayed. In a modern high-refresh-rate television-type raster display, the screen brilliance controls the brightness of the entire picture frame and the setting is far less critical for good radar performance. The perceived brightness of the picture is largely a function of the ambient light conditions. It is also affected by the screen contrast control which defines the difference between dark and light for both graphics and echoes. In some systems the contrast is preset. The relationship between the brilliance and contrast settings for a green monochrome display is illustrated in Figure 6.1.

Reference to Figure 6.1 shows that the screen contrast adjusts the difference in brilliance level between the darker/darkest areas of the screen and the brighter/brightest areas. In general, the graphics will be single level, i.e. bright on dark, and this will also be the case for target echoes in a simple single-threshold system. Where multi-level video is offered, the number of levels of echo brightness is dependent on the number of levels to which the signal strength has been digitized (see Section 2.6.4.5). The perceived effect of the screen contrast will depend to some extent on the screen brilliance setting. At higher brilliance settings, a greater absolute difference between light and dark will be necessary for an observer to perceive a difference between the two levels. Similarly, at very low levels of brilliance, both light and dark levels may be so dark as to be indistinguishable.

Clearly there is an interrelationship between screen brilliance and screen contrast. Both contribute to the overall brightness of the picture as perceived by the observer, and must be

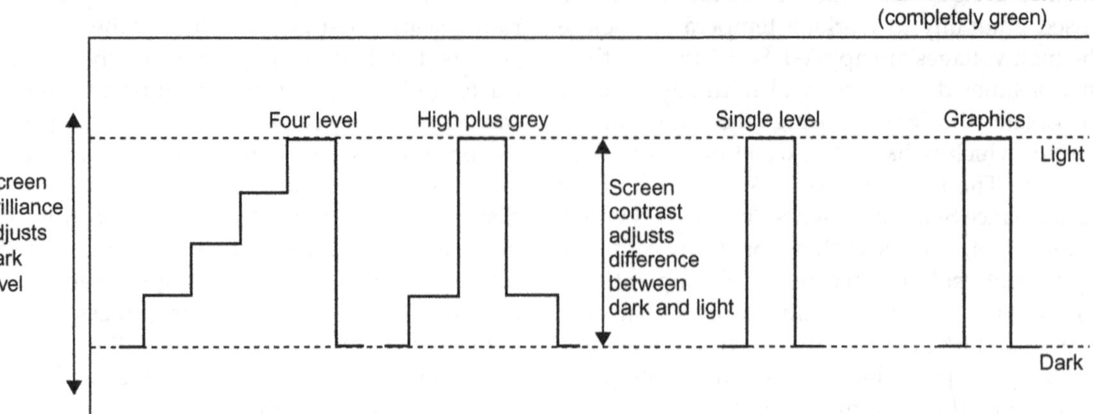

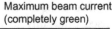

FIGURE 6.1 The relationship between raster-scan screen brilliance and contrast settings.

adjusted alternately until the best joint setting is found. The assessment of 'best' is a matter of subjective judgement for which a simple specific criterion, such as that prescribed for the radial case, cannot be given. However, consideration of the following factors will assist in making the judgement.

Excessive brilliance. If the overall brightness is too high, the picture will be uncomfortable to view in the same way as any source of bright light. Tiredness, stress and headaches may result. At night the high level of illumination associated with the display may impair night vision and distract the watchkeeper by producing reflections on the inside of the wheelhouse windows and other such surfaces. Even at maximum screen contrast setting, it will be difficult to achieve an adequate difference between background and picture content, because both will already be very bright.

Insufficient brilliance. If the overall brightness is too low, it will be difficult to view the display and distinguish picture detail in full daylight, particularly at a distance. Even at maximum screen contrast setting, it will be difficult to achieve an adequate difference between background and picture content, because both will be very dark.

Excessive contrast. If the screen brilliance is set at a level which is comfortable to the eye, excessive screen contrast will make the picture stark and harsh.

Inadequate contrast. Insufficient contrast will result in a picture which is flat and featureless.

In most cases it will be found that if at switch-on the screen brilliance and contrast controls are left in the position in which they were last used, the settings will be sufficiently close to the optimum as to require only a small amount of adjustment to suit the new ambient light conditions. If the controls have been set to zero or have been badly maladjusted, the following procedures should be used:

1. If the system offers alternative daytime and nighttime monochrome or colour combinations, select the option appropriate to the ambient light conditions.
2. Set the screen brilliance and screen contrast controls to the middle of their travel.
   The plan position indicator (PPI) circle,

synthetic bearing scale and other graphics should be visible. (If they are not, advance each control in turn by small amounts until the graphics are visible.)

3. Adjust the screen brilliance to a level consistent with comfortable viewing in the ambient light conditions and such that the light issuing from the display does not interfere with the keeping of a proper lookout.
4. Adjust the screen contrast to achieve an obvious but comfortable difference in brightness between the graphics and the background.
5. It may be necessary to make further repeated small alternate adjustments to the screen brilliance and screen contrast controls to obtain an optimum overall level of brightness.

6.2.4 Default Conditions and Start-Up

In modern systems it is usually found that orientation, presentation, range scale, pulse length and signal processing are selected automatically on switch-on. The conditions automatically selected are known as the *default conditions* and will also probably include status of controls such as range rings and variable range marker (VRM). If the observer does not wish to use any given default setting, appropriate selection can be made after switch-on.

The use of default conditions has been made possible by the availability of modern on-screen controls designed particularly for computer applications. Traditionally, marine radar controls have been analogue in their action, normally taking the form of rotary knobs and multi-position mechanical switches, but most, if not all, of these controls are now replaced by electronic switches which send a digital signal to the display computer when operated by touch or pressure. They may take the form of push buttons, touch-sensitive switches or a tracker ball. The touch-sensitive screen described in Section 2.8.3 is an extension of this technique.

On receipt of the start-up signal, the display computer will carry out some programmed checks on the status of the equipment. The controls can be configured and programmed to operate in a variety of ways. They may operate on the *alternative action* or *toggle* principle, that is one push or touch selects a function and a further push de-selects that function. Alternatively, they may be used to control more than one function, for example the first push selects function 1, the second push selects function 2 and the third push de-selects both functions. They can be used in pairs to perform an incremental action. For example, each push of one control may select the next higher range scale while its partner will reduce, in steps, the range scale selected. The list of variations is almost endless but the common feature of all actions is that, other than the actual push or touch, the action is not mechanical.

The display computer can be programmed to select the default conditions when the power is first applied at switch-on. For example, it might default to true-north-up orientation, relative-motion presentation, 12 mile range scale and long pulse. If the observer wishes to use conditions other than those to which the system defaults, the necessary selection can be made by pushing the appropriate switches immediately after switch-on.

The observer may make the following selections:

Orientation and presentation. This selection is a matter of personal choice and navigational circumstances. This is considered in greater detail in Sections 6.2.5 and 6.2.6. Features which might influence such a selection are considered in Sections 1.4.4, 1.5.3, Chapters 7 and 8.

Range scale. At sea, it is usually best to set up on a medium range scale such as 12 miles. The likelihood of echoes and sea clutter which have not been affected by saturation (see Section 2.7.4) make such a scale best suited for judging the quality of the picture when it is subsequently obtained. In port, when surrounded by large land-based structures and large vessels which prevent the radar pulses travelling long distances, a lower range scale is frequently more applicable.

Pulse length. This should be appropriate to the range scale selected, for example medium range scale would suggest medium pulse length (see Section 2.3.3.1).

6.2.5 Setting the Orientation of the Picture

The picture must be correctly orientated so that the heading marker intersects the correct graduation on the circular bearing scale surrounding the radar picture. In many older systems it was necessary to rotate the picture manually, whereas in modern systems it is an automatic alignment as the azimuth scale is part of the graphics. Nevertheless the display should be observed to ensure that the heading marker has aligned correctly according to the numerical value of the compass course in use, as well as to check that the numerical value indicated on the radar data display is the same as that indicated on the actual compass.

After the heading marker has been checked for alignment, the display should be observed for a short period to ensure that the heading marker is following variations in the ship's heading. The observer must decide between three options, head-up (unstabilized), north-up (stabilized) and course-up (stabilized) as discussed in Section 1.4.1. Modern sets are usually provided with a three button toggle control. The two stabilized options are to be much preferred if the intention is to do manual paper plotting or taking bearings for navigation. They are essential if using an ARPA as stabilized options require the input from a working compass.

The digital read-out of the compass should be clearly displayed on the screen along with the source (e.g. 'gyro compass 2'). There will be a method (usually in the menu system) of changing to different compass sources when available. Older compass repeater systems merely reproduced the changes in headings of the vessel and it was also essential to check that the compass heading on the radar was aligned with the master compass, and to periodically check that this was still the case.

In the particular case of course-up orientation, it will be necessary to reset the reference course each time the vessel makes a sustained alteration of course. In modern systems it is merely necessary to press some form of reset control when the vessel is steady on the new course.

The alignment of the heading marker on the bearing scale should not be confused with that of the alignment of the heading marker with the ship's fore-and-aft line. The latter operation is concerned with ensuring that the heading marker contacts (see Section 2.5.3) close and produce the heading marker signal at the instant the radar beam crosses the ship's fore-and-aft line in the forward direction. This alignment cannot be checked without reference to both visual and radar observations (see Section 6.6.8).

6.2.6 Setting the Presentation of the Radar Picture

The presentation of the radar is matter of observer choice (see Section 1.5.3). In simple terms it is the control of the spot at the centre of the radar picture which represents the own ship (otherwise known as the origin). It also affects the artificial direction of the artificial afterglow or trails of targets and (for true-motion only) the afterglow of the own ship origin. The observer has a choice of relative-motion and true-motion, which is usually provided by two push buttons acting as a toggle switch.

6.2.6.1 Relative-Motion

In relative-motion, the own ship origin has a fixed position on the screen. This can be at the centre so that observer can see an equal range all around the ship, or it can be off-centre – up to 75% of screen radius from the centre of display. The advantage of this latter option is that, for a given range scale, the observer can see further ahead than behind. The theory is that the relative speed of head-on vessels is large and the relative speed of overtaking vessels from behind is small, but care must be taken in case the vessel encounters the odd exceptional fast vessel from astern or quarters. For relatively slow vessels, the offset screen is not recommended.

A push-button control is normally provided to centre the picture and the screen is usually offset by pointing the tracker ball/joystick at the desired position for the own ship origin and clicking a push button.

6.2.6.2 True-Motion

In true-motion presentation, the own ship origin moves across the screen according to the speed and heading obtained from external sensors. The observer has a choice of sea-stabilization or ground-stabilization. This selection of sea-or ground-stabilization also (and more importantly) affects the true vectors on targets (see Section 6.9.6). In the context of presentation mode, in sea-stabilized true-motion, the origin is driven across the display based on compass heading and water speed. While in the ground-stabilized mode, the origin is driven across the display based on heading and speed of vessel with respect to the ground. This ground-based heading and speed may be derived from a number of different types of sensors and sources, as discussed more fully in Section 6.9.6 in the context of effect on displayed vector headings. The ground speed and heading being used should be clearly indicated in a data read-out on the display. In all cases, the inputs of sensor headings and speeds must be verified as correct.

Note that in ground-stabilized true-motion the origin of the own ship moves according to ground speed and track, and the heading line should still be indicating the compass heading of the vessel.

6.2.6.3 True-Motion Reset Controls

As the origin tracks across the screen, the view ahead decreases and from time to time it will be necessary to reset the position of the origin.

In modern systems, with a tracker ball or mouse, the observer can select the new position of the origin at any time, by 'pointing' at the required new position and then giving a 'click'. The origin selected must be a legal position. IMO allows the observer to move the origin anywhere within a fairly large circular limit which must not exceed a maximum. This maximum limit is chosen by the manufacturer, but it must be between 50% and 75% of the radius of the display from the centre of the screen (see IMO Performance Standard in Section 11.2.1). Alternatively, the equivalent action is achieved by holding in a push-button type control and *steering* the origin to the desired reset position by means of a joystick or tracker ball (see Figure 6.11).

Also, often provided is a reset button, which when pressed, resets the origin on the reciprocal of the observing vessel's course at the maximum permitted radius, thus seeking to make the best possible use of the screen area ahead (see Section 6.2.6.3).

If the observer fails to reset the origin and allows it to approach the outer extent of the tracking limit, the display will produce some form of automatic response. The reset position is likely to lie on the reciprocal of the vessel's course at up to 50% of the screen radius from the screen centre, as encouraged by IMO specification (see Section 11.2).

In most systems automatic resetting takes place when a new range scale is selected.

6.2.6.4 True-Motion with Fixed Display Origin

As already mentioned in Section 1.5.2, some manufacturers provide a third presentation option to get over the perceived problem in true-motion of the own ship moving across the radar picture display and the view ahead reducing in range. Effectively it is a true-motion display but instead of the origin moving forward over a fixed radar picture, the origin is in a fixed position on the display and the radar picture is moved backwards as the vessel progresses.

To the observer, this looks like a relative-motion picture, the essential difference is that target trails will be true trails, not relative trails. The different names for this type of display show the confusion as to the complexity of this presentation. For instance, it has been called *true-motion constant display (TMCD)* on one model and *relative-motion true-tracks (RMT)* on another.

6.2.7 Obtaining the Optimum Picture (Magnetron Radar)

The optimum picture is obtained by setting the receiver controls to maximize the detection of weak echoes. These may be automatic controls on some modern radars, but non-coherent magnetron radars always have the option of manual controls that require the adjustment of, firstly, the gain control and, secondly, the tuning control.

The gain, sea clutter and rain clutter controls should be set to zero. In addition, particular attention should be paid to ensuring that signal-processing facilities, such as interference suppression, echo stretch, adaptive gain and rotation-to-rotation correlation, are inoperative (see Sections 3.9.5.1, 3.6.3.3, 3.6.3.4 and 6.8 respectively). In some raster-scan systems these facilities default to the *off* condition at switch-on. The setting of the gain is fundamental to achieving the correct relationship between threshold and noise levels (see Section 2.6.4.4). All of the other foregoing controls and facilities also affect the echo content of the picture and may result in the gain being set incorrectly if not switched to zero effect when the gain level is assessed.

6.2.7.1 Setting the Gain Control

The gain must be set to achieve the optimum relationship between the threshold level and the noise amplitude (see Section 2.6.4.4). In some systems, by operating the gain, the observer controls the threshold level, while in others the receiver gain is affected. The difference is academic as, in both cases, advancing the setting of the gain control brings the threshold closer to the noise.

Ideally one would wish the gain to be set so that the threshold level is sufficiently low to display all detectable echoes, yet sufficiently high to miss the noise (Figure 6.2). Because noise is a very rapidly fluctuating signal, such an ideal cannot be achieved. If the setting of

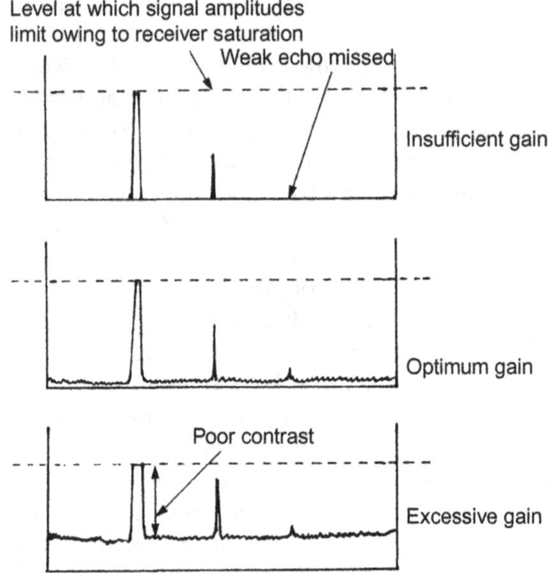

FIGURE 6.2 Setting the gain control.

the gain is sufficiently high for all detectable targets to cross the threshold, many noise peaks will also cross the threshold. Conversely, if the gain is set so that no noise peaks cross the threshold, weak but detectable targets may well be missed. For the case of a single-level digitized picture, the dilemma is illustrated by Figure 6.3. Clearly a compromise setting must be found. The ease with which this setting can be made depends on the number of discrete video levels offered by the system (see Section 2.6.4.5).

In the single-level video case any noise which crosses the threshold will be displayed with the same brightness as a detectable echo. This makes it much more difficult to assess the background noise than in the analogue case, where there is a very obvious amplitude difference between most echoes and the noise background. In some older synthetic displays it is possible to revert to an analogue picture, albeit in some cases on only one range scale. The difficulty can then be avoided by setting the gain while viewing the analogue picture. Later synthetic displays do not offer the analogue option and in such cases it was necessary to live with the difficulty of single-level noise and targets. With practice the observer should be able to identify the noise by its random nature and the radial length of individual noise pulses. The peaks of real noise are sharply spiked, and thus the number of successive range cells (see Section 2.7.3) occupied by a single noise pulse is likely to be noticeably less than that occupied by most genuine targets. This does not mean that every very short digital response is necessarily noise. A weak, but genuine, received echo which barely

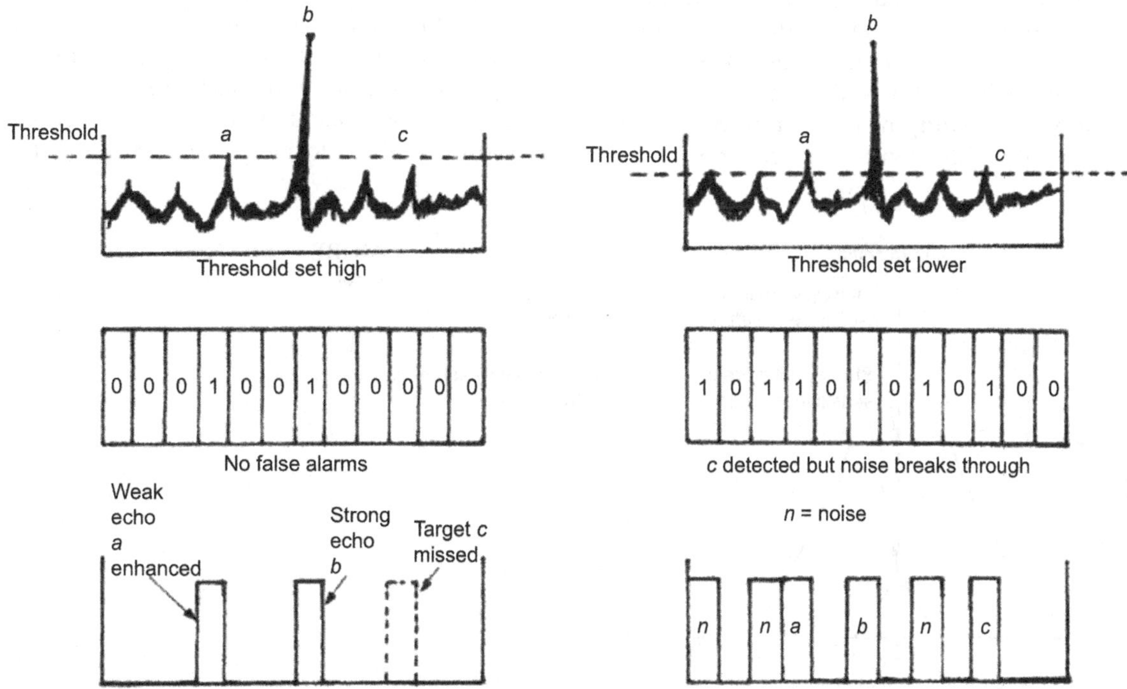

FIGURE 6.3 Target detection and threshold setting.

crosses the threshold may well activate only one range cell, but may appear in the same place on successive scans, whereas the noise will appear randomly.

The difficulty of achieving a suitable noise setting is reduced where more than one video level is offered. Modern equipment with digital signal processing (see Section 2.7.1) allows the signal strength to have multiple digital levels and therefore multiple video levels. Because noise pulses are displayed at a fraction of high-level amplitude, the practised observer will not confuse them with stronger genuine echoes. Further, although low-level echoes will have the same amplitude as the noise, their coherence, persistence and radial length will favour detection when compared with the random noise pattern.

6.2.7.2 *Setting the Tuning Control*

In a magnetron-based radar, the function of the tuning control is to adjust the frequency of the receiver (see Section 2.6.4.2) so that it coincides with that of the transmitter, rather in the same way as one might tune a broadcast receiver to listen for a distant station. Some modern radar equipments offer automatic tuning, while the solid-state coherent radar equipment no longer requires 'tuning' in the traditional sense as the monitoring of the small variations in the received radar signals are part of the 'coherent' principle. This section considers the use of the manual tuning control.

If the receiver is only slightly mistuned, the extent of the bandwidth may still allow stronger echoes to be displayed even to the level of saturation. It is important not to be misled by this because weak echoes will most certainly be lost. It is thus important to establish criteria against which it is possible to make a sensitive assessment of the setting of the tuning control.

One of the most sensitive criteria is the response of a weak echo from land (Figure 6.4). The tuning control should be adjusted while carefully watching such a response. On traditional radar, as the correct frequency was approached, the brightness of the displayed target response increased. It will peak at the correct setting and decay after that setting has been passed. By adjusting the control to achieve the maximum displayed brightness of such a target, the correct setting could be easily and reliably found.

The use of a weak land echo as the sensitive criterion by which to judge the setting of the

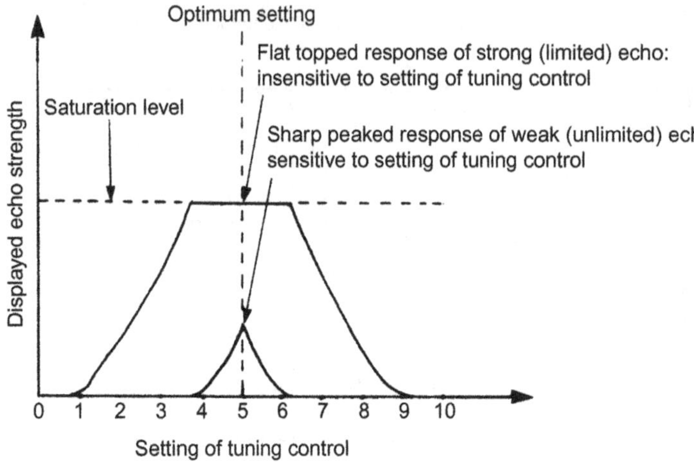

FIGURE 6.4 Setting the tuning control.

tuning control cannot be applied in the case of a modern synthetic display due to the quantizing effect of digital storage. This is illustrated for a single video level system by Figure 6.5.

Reference to Figure 6.5 shows that there may be little or no relation between the amplitude of the synthetic echo and the analogue response which produced it. The setting of the tuning control must not be judged with respect to the amplitude of the synthetic echoes, because their brightness is not sufficiently sensitive to the position of the control. Even if several video levels are offered, the displayed echo strength will change only when the real echo moves into a new inter-threshold band, and even then in steps which may be quite large.

It is therefore necessary to adjust the control to achieve one of the following results:

1. The maximum radial extent of sea clutter echoes: these are described in Section 3.6.
2. The maximum area of precipitation echoes: these are described in Section 3.7.
3. The maximum extent of land echoes (ideally a coastline).
4. The maximum radial extent of the receiver monitor signal: this is discussed in Section 6.5.

It should be noted that all the above criteria are suitable because they are based on the area of response rather than on its brightness.

The tuning may drift as the system warms up and to a lesser extent with the passage of time. The setting of the control should therefore be checked frequently during the first 30 minutes after switching on and periodically thereafter.

If the coarse tuning has been set correctly by service personnel, the correct setting for the display tuning control should lie close to the centre of its range of travel. If the correct position is found to approach either end of the range of travel, the coarse tuning control, usually located in the transmitter and not normally available to the operator, should be readjusted by qualified service personnel.

Most displays offer features such as *magic eyes*, tuning meters and tuning indicators. They should not be used in preference to the

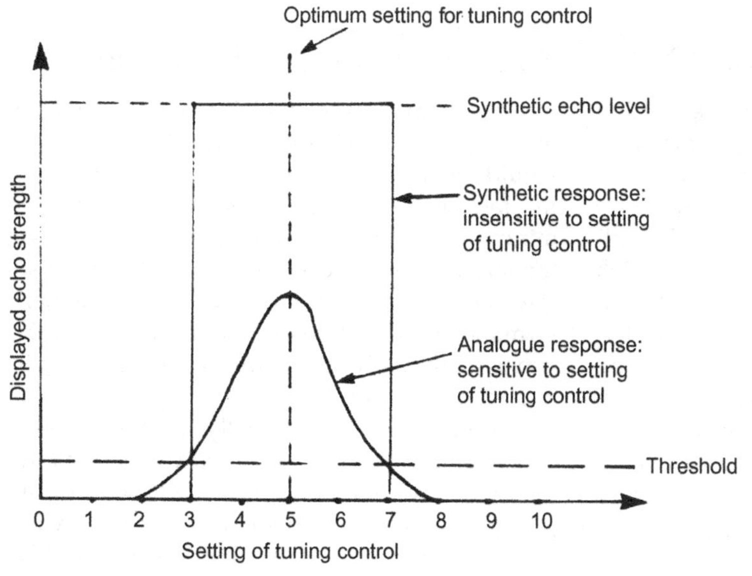

FIGURE 6.5 Setting the tuning control – the effect of quantizing.

above criteria. Such devices are not required by IMO Performance Standards and there is a danger of the observer being misled by such indicators.

Where full automatic tuning is provided, a performance check should be carried out to ensure that optimum tuning is being achieved. The procedure for carrying this out is described in Section 6.3.

6.2.8 Optimum Picture on Coherent Radar Systems

Coherent radars (Section 2.9) provide sophisticated and automated tuning controls, essential to ensure the required coherency of the system. For this reason they are unlikely to have conventional tuning controls. However, they will have manual controls to help the user optimize the picture, which will generally replicate the gain and clutter controls of a conventional magnetron radar. They will also have a number of of automated options for setting the display.

6.3 PERFORMANCE MONITORING

As soon as practical after the initial setting up of the display, the system performance should be monitored. In the case of a shore-based radar system, it may be possible to assess the level of performance from a knowledge of the identity of those targets which were at the limit of detection when the system was installed and the performance was tested and optimized by the manufacturer. However, at sea it is usually not possible to discern whether or not the system is giving optimum performance merely by viewing the displayed picture, because the observer does not usually have exact knowledge of which targets should be seen. Although the picture may look good, the observer cannot be sure that the equipment is operating at the level of performance that was intended by the manufacturer. In the open sea in calm, foggy weather, the absence of echoes, despite the presence of receiver noise, may be due to a loss in performance rather than an empty ocean. Failure to appreciate this fundamental principle could lead to a very dangerous situation. The absence of echoes deprives the observer of the normal criteria by which to judge the setting of the tuning control.

IMO Performance Standard (see Section 11.2.1) requires that means be available, while the equipment is in use operationally, to determine readily a significant drop in performance relative to a calibration standard established at the time of installation, and that the equipment is correctly tuned in the absence of targets. This facility is commonly referred to as a *performance monitor*. The sensitive element which makes monitoring possible is usually a unit known as an echo box and is described in the next section. Some systems use a device known as a transponder (see Section 6.3.4).

6.3.1 The Principle of the Echo Box

Any normal target which was mounted on the ship to reflect signals for test purposes would be unsatisfactory because, apart from the fact that it would probably be inside the minimum detection range (see Section 3.2.4), the strength of the echo would probably be so great as to saturate the receiver even if the performance was well below optimum. The problem is commonly overcome by using what is known as an echo box.

The echo box is essentially a hollow metal container of precisely machined dimensions. In theory, a variety of shapes can be used but, in practice, modern systems normally use a cylinder about the size of a small food can. The theory of the operation of the echo box is quite complex but its behaviour can be described in fairly simple terms. When a pulse of radar energy is transmitted into the box, the energy reverberates within the cavity and if, but only if, the cavity is the correct size in relationship to the wavelength of the radar signal, intense oscillations build up over the duration of the pulse length. At the end of the transmitted pulse the oscillations continue but their amplitude gradually decays over a period which is much longer than the duration of the transmitted pulse. As long as the oscillations persist, radar energy is re-radiated from the box. The energy re-radiated from the echo box can be treated in the same way as other returning radar signals and thus produces a visible, radial response on the display which is known as the *performance monitor signal*. This general principle is illustrated by Figure 6.6, but it should be noted that the angular extent of the monitor signal depends on other factors which are discussed in Section 6.3.2. The echo box and its associated circuitry is referred to as a performance monitor.

It may be found helpful to liken the action of the echo box to that of a bell which rings for a long time after being struck by a comparatively swift blow. In fact, engineers often refer to the re-radiation from the box as 'ringing'. (A full explanation of the action of the echo box is beyond the scope of this work, requiring an understanding of the theory of the resonant cavity: the reader who wishes to pursue this is referred to any good radar engineering text.)

Pursuing the analogy of the bell further, the length of time for which a nearby listener hears the bell ringing depends on the vigour with which it has been struck, and the sensitivity of the listener's ear at the frequency emitted by the bell. Applying this reasoning to the echo box, it should be evident that the radial extent of the monitor signal on the display is a measure of the performance of the radar system since it depends on:

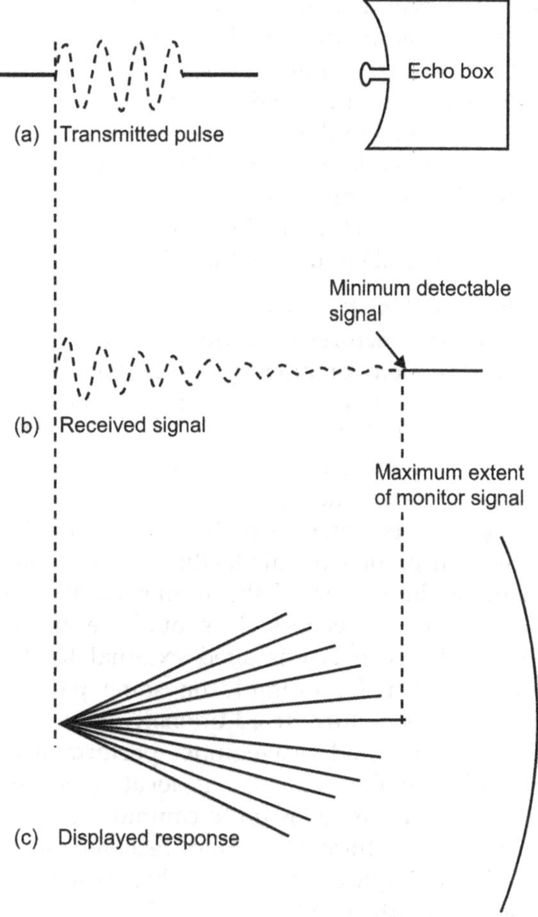

FIGURE 6.6 The echo box response.

1. *The transmitter performance.* The energy entering the echo box is a function of transmitted power and pulse length (see Sections 2.3.3.1 and 2.3.3.2). Thus the greater the amount of transmitted energy, the longer the echo box will ring.
2. *The receiver sensitivity.* The re-radiated signal is capable of detection until its amplitude has decayed to less than that of the receiver noise. Thus the more sensitive the receiver, the greater is the visible extent of the monitor signal (see Figure 6.6).

3. *The receiver tuning*. The frequency re-radiated by the echo box is the same as that of the transmitter. The maximum monitor signal depends on the receiver being tuned to the same frequency.
4. *The display control settings*. Maladjustment of the display controls may prevent the observer detecting the full potential extent of the signal emitted by the echo box.

It is evident that if the system is adjusted by the manufacturer to optimum performance on installation (or after a service visit), and the radial extent of the monitor signal is measured at that time, the observer can be supplied with an objective calibration, in terms of miles and cables, against which to judge the performance of the system in the future. In particular, this facilitates an assessment of the setting of the tuning control in the absence of echoes. It should be noted that if the echo box is sited external to the radar system, for example on some part of the ship's structure, it additionally monitors the antenna and waveguide performance. However, if the echo box is located within the radar system, as is quite common, it only monitors the transmitter and receiver functions. The implications of echo box siting are discussed further in the next section.

6.3.2 Echo Box Siting

In older systems it was common for the echo box to be mounted in a blind sector on a mast, Samson post or other part of the ship's structure. Such boxes were large and rectangular in shape with a parabolic reflector to focus the received energy at the entrance to the cavity. They are now really only of historical interest. One of a number of subsequent variations which developed from this was the incorporation of the echo box within the antenna drive unit and the fitting of a horn with a short length of connecting waveguide below the level of the scanner, as illustrated by Figure 6.7(b).

As the beam sweeps the horn once per revolution, some 10–20 pulses (see Section 1.2.3) will energize the horn and be conveyed down the waveguide to the echo box. This approach, like all external sitings, has the virtue that the length of the monitor signal is a measure of antenna and waveguide performance in addition to that of the transmitter and receiver. Such an echo box and its associated circuitry can justifiably be referred to as an *overall performance monitor* because it monitors the performance of all units of the system. The principal drawback of an external echo box is the general expense of producing such a unit (in particular the cost of making it suitable for above-deck fitting) and the problems which result if water does penetrate the system. The monitor can be switched on and off by means of an attenuator which effectively blocks the short length of waveguide when the signal is not required. Because external echo boxes are only energized as the radar beam sweeps across them, they produce a monitor signal which has a limited angular width the form of which is illustrated in Figure 6.7(c). This form is commonly described as a *plume*.

Many systems site the echo box on the waveguide just on the antenna side of the transmit/receive (T/R) cell (see Section 2.6.3.1), as illustrated in Figure 6.7(a). This dispenses with the need to make the unit suitable for above-deck mounting and allows the unit to be more compact and less expensive. However, it does have the serious shortcoming that it does not monitor antenna or waveguide performance. If the echo box is sited in this position it forms, with the associated circuitry, a *transceiver monitor* and this should be supplemented by some arrangement which monitors the radiation from the antenna. Such an arrangement is normally referred to as a *power* or *output monitor* (see Section 6.3.3).

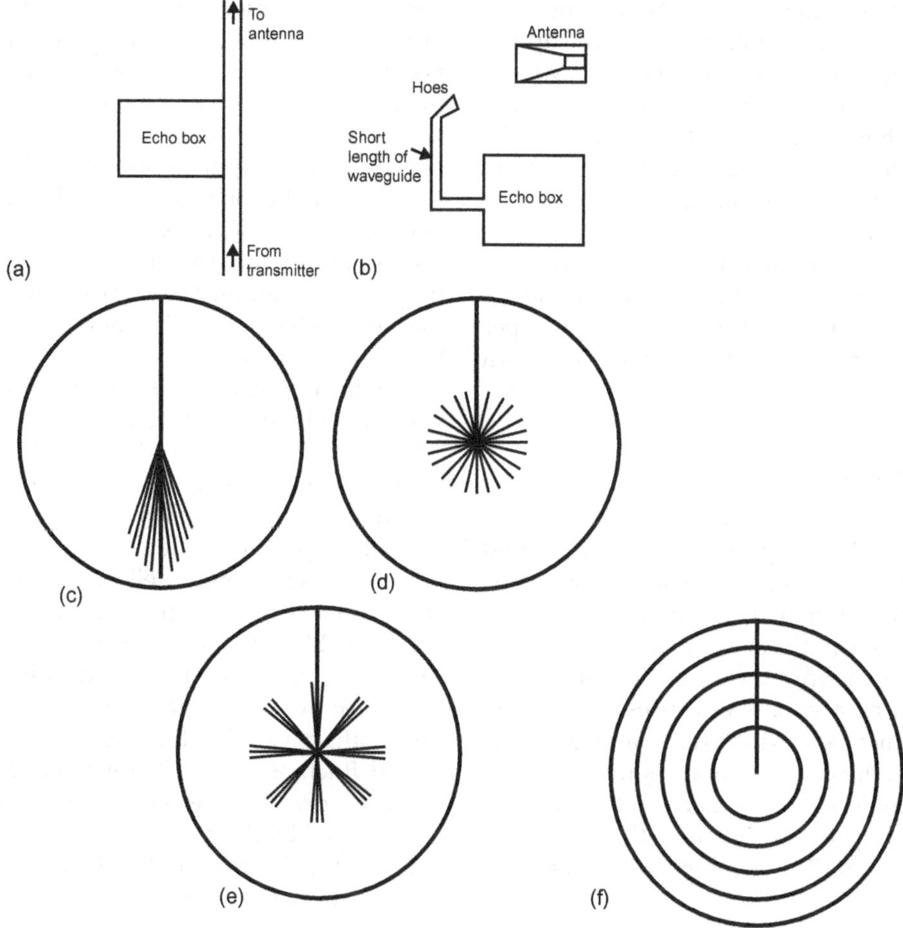

FIGURE 6.7 Echo box siting and the performance monitor signals. (a) Echo box located on waveguide between transmitter and antenna. (b) Echo box located below antenna. (c) 'Plume' from external echo box. (d) 'Sun pattern' from internal fixed-tune echo box. (e) 'Multi-plume' from internal echo box with oscillating plunger. (f) 'Concentric rings'.

In the transceiver monitor the cavity is cylindrical and its size can be varied by adjusting a plunger. This allows service personnel to tune the cavity (i.e. make the cavity the correct size) so that it will ring at the magnetron frequency. The monitor can be switched on or off from the display position by means of some form of electromechanical control which detunes the cavity when the monitor signal is not required. Because the echo box is located on the waveguide, the response is not sensitive to antenna direction and hence the monitor signal extends for the full 360° of azimuth. The resulting pattern is sometimes referred to as a *sun* and is illustrated in Figure 6.7(d). Some observers might have a natural reluctance to switch on a monitor which produces such an extensive pattern. It is also sometimes argued that it is more difficult to discern the extent of a sun pattern than that of a plume, in which the contrast on either side assists in identifying the pattern. Some manufacturers break up the sun pattern as shown in Figure 6.7(e) by driving the tuning plunger in and out of the cavity. The echo box rings each time the plunger passes the tuned position, producing a 'family' of plumes which rotate at a speed governed by the relationship between the periods of the plunger oscillation and that of the antenna rotation. A plunger which makes a single sweep of the cavity and is triggered once per scan by the antenna as it rotates will produce a plume.

6.3.3 Power Monitors

Where a transceiver monitor is fitted, the performance of the system is only monitored as far as the location of the echo box. Thus the presence on the display of a monitor signal of the specified length is no guarantee of optimum radiation from the antenna, which may be reduced by a variety of causes. These may include puncture or distortion damage to the waveguide, the presence of moisture or dirt in the waveguide space, damage to the antenna or the accretion of dirt or ice on the glass reinforced plastic (GRP) envelope which protects the antenna.

The sensitive element commonly used in power monitors is a neon tube – a glass envelope which is filled with the inert gas neon and contains two separate metal electrodes. Neon gas has the particular property that it normally behaves as an insulator but, if it is subjected to an adequate level of radio frequency radiation, the gas is ionized and behaves as a conductor. Thus, within the tube, in the quiescent condition the electrodes are insulated from one another, whereas, when the gas ionizes in the irradiated state, the electrodes are effectively connected by a conducting medium. The neon tube is mounted close to the scanner so that radiated energy from the beam falls on it at each revolution. The tube is connected in series with some circuitry which produces a visual indication at the display position. In the absence of adequate radiation, the de-ionized neon will interrupt the circuitry and give a visual indication of an absence of power. If the output power from the antenna is adequate, the ionized neon will complete the circuitry with a low-resistance path, giving a visual indication of the presence of radiation at the antenna output. (The particular feature of the neon tube is that the ionization level can be fixed very precisely.)

The form of visual indication employed can vary considerably with manufacturer. It may take the form of a simple light which is illuminated when adequate power is present, or the signal from the neon may be used to indicate the output power by the deflection of a meter. The meter scale may well have a red arc or some other feature to give an indication of an unacceptably low-output power level. In both cases it is likely that the indication will be smoothed over a period of one antenna revolution, otherwise the light or meter reading would fluctuate as the beam swept across the neon sensor. Some systems have used the signal detected by the neon to produce on the display a plume whose radial length represents the level of output power.

6.3.4 Transponder Performance Monitors

The sensitive unit of this modern type of performance monitor is an active (as opposed to passive) element known as a transponder, which is located on the antenna drive unit in such a way that it is irradiated by the beam as it sweeps past. The transponder reacts to reception of the radio frequency pulses by transmitting, after a short delay, a low-level coded response at the same frequency. A threshold set within the active element ensures that the response will only be given if the received signal is above a power level determined by the manufacturer.

The power level within the limits of the horizontal beam of the radar system falls off on either side of the axis. Thus, as the beam sweeps past the transponder, the latter responds over an arc bounded by the limits at which the radiated power is above the preset level. The transponder is thus a monitor of transmitted power.

The response returned to the antenna is treated as any other received signal, and thus the ability to receive the response is a measure of receiver response. The response is displayed in the form of a series of arcs (Figure 6.8), the angular limits of which are a measure of performance.

The angular limits are measured by the manufacturer when the performance is known to be optimum and supplied as a calibration level against which performance can be judged.

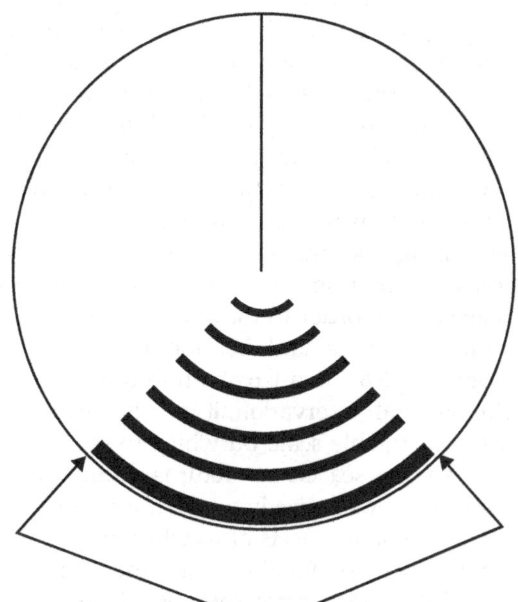
Extent of arc is a measure of overall performance

FIGURE 6.8 Transponder monitor signal.

6.3.5 Calibration Levels

Given the wide variety of purely arbitrary ways in which the monitor information may be displayed, and considering in particular that, for example, one manufacturer may use a plume as an indicator of output power whereas another may use it as a transceiver monitor signal, it should be evident that it is essential to study the operator's manual to establish which signal or indicator to observe and to appreciate precisely what it measures.

The calibration levels should be measured by the manufacturer when the performance of the system is known to be optimum. In the case of a sun or plume pattern, such measurement will be made in miles and cables. The following information should be shown on a tally close to the display:

1. The length of the signal (or status of indicator) consistent with optimum performance.
2. The length of the signal (or status of indicator) corresponding to a significant drop in performance.

If the overall performance is not monitored by a single signal or indicator, the values for both transceiver and power monitors should be shown separately and unambiguously.

6.3.6 Performance Check Procedure

It must be remembered that the length of the monitor signals (or the status of other indicators) is influenced by the position in which the operational controls have been set. When carrying out a performance check it is vital that controls are set as instructed in the maker's manual – otherwise comparison with the calibration level is meaningless.

It cannot be stressed too strongly that the way in which the performance signals are displayed, and the extent to which they monitor the performance, vary considerably from manufacturer to manufacturer and even with different types made by the same company. Thus the important basic rule for carrying out a performance check is: *consult the maker's manual and follow exactly the instructions given therein.*

A performance check should be carried out as soon as practicable after setting up and thereafter at regular intervals. In the United Kingdom, it is a requirement (see Section 11.4.1) that a check should be carried out before sailing and at least every 4 hours when a radar watch is kept. This should be regarded as a minimum.

6.3.7 Modern Trends

The methods described above are still in use today, and even replacements that exist replicate the patterns (above) to the observer.

Manufacturers and suppliers of radars are occasionally seeking ways to avoid the installation of external performance monitors and power monitors. Possibly this is because of economic pressure as there will be an extra exterior unit with cabling to be installed. Certain recent models have been approved without an external monitor, as it is claimed that the internal monitoring of the equipment is sufficient.

Instances have been reported where a significant loss of performance was detected in marine radars (but shore based), by observing that known targets were failing to display in spite of (internal) monitoring indicating that the performance was satisfactory. On visual inspection, the antenna/waveguide section of the system was found to be damaged. Had this happened aboard a vessel in the open sea, in fog where the energy was attenuated/failed to leave the antenna, the existence of vessels in the vicinity would have gone undetected even though the (internal) monitor was indicating that satisfactory performance might be expected.

The most important value of the external performance monitor is the ability to reliably detect a drop in performance. It is a not impossible scenario that the vessel is in poor visibility with little or no sea clutter and there is a defect in radar performance so that weaker targets are no longer seen. The internal monitors may not detect the problem, because some responses are being received. Like the internal performance monitoring, the observer will gain some false confidence from the display of the larger targets, but not be aware that other targets are not being detected. The external monitors are very useful pieces of equipment.

6.4 CHANGE OF RANGE SCALE AND/OR PULSE LENGTH

During normal operation of the radar system it will be necessary for a variety of reasons to change range scale from time to time. For example, having considered factors such as traffic density, speed of the observing vessel and the frequency of observation, a suitable range scale may have been selected on which to carry out radar plotting (see Section 7.2.4), but it may be necessary to change to a longer range scale at intervals in order to fix the vessel's position or to obtain early warning of approaching vessels. When using the radar for position-fixing or progress-monitoring, it may be necessary to change scale in order to maximize the available accuracy of range and bearing measurement (see Sections 6.6.7 and 6.6.8). From time to time during general observation, it may be necessary to select a suitable scale on which to search for targets among sea clutter returns. Many other circumstances will arise in which the range scale selected for some aspects of watchkeeping is not suitable for others. Intelligent use must thus be made of the range scale selector. "Don't keep fiddling with the radar", is a phrase which is frequently heard, when two persons are using the same radar on a bridge. While it might appear to be justified in some cases, it is essential to make intelligent and appropriate use of ALL the radar's facilities and at times, this may appear as fiddling.

A change of range scale may be associated with a change in pulse length, and some pulse lengths may not be available on all range scales (see Section 2.3.3.1). Change of pulse length may be necessary for a variety of reasons such as the need to improve range discrimination, minimum range, detection in sea clutter, long-range detection and detection in and beyond areas of rain (see Sections 2.3.3.1, 3.2.4, 3.6, and 3.7, respectively).

6.5 THE STAND-BY CONDITION

As indicated in Section 6.2.2, IMO Performance Standards require that such a condition be available. This requires the radar to be brought into the full operational condition in 5 seconds (see Section 11.2.1).

The detailed nature of the stand-by condition will vary with manufacturer, but in all cases transmission is inhibited. Thus, although the radar is virtually ready for immediate use, limited-life components in the transmitter are not operational. A further, more general benefit is that the amount of radar-to-radar interference (see Section 3.9.5) experienced by other vessels (which may need to operate their radar systems) will be reduced over a potentially large geographical area. The life of components and hence the reliability of the radar system will be far less affected by continuous running than by frequent switching on and off, so that in periods of uncertain visibility, it is better to leave the radar in full operation or in the stand-by condition.

Rule 7(b) of the Collision Regulations requires that 'proper use shall be made of radar equipment if fitted and operational'. There are many occasions on which *proper use* means continuous running. However, there are other occasions, for example in the open sea on a clear day with fog forecast, when the stand-by condition constitutes proper use. Thus, compliance with the letter and spirit of the rule is ensured as there is no excessive delay should the equipment be required quickly, yet there is no unnecessary transmission in the meantime.

If the equipment is to be left on stand-by, the controls may be left in the optimum position, but the settings, particularly that of the tuning control, should be checked immediately on returning to the operational condition.

If the equipment is likely to be on stand-by for a long period, it is probably better on a magnetron radar for the gain, sea clutter and rain clutter controls to be set to minimum effect and all signal-processing facilities should be switched off, all being subsequently reset on returning to the operational condition. On coherent radar there are fewer controls to be concerned about. The concepts of gain and tuning no longer apply. There will be different filter settings which can help reduce different types of clutter.

6.6 CONTROLS FOR RANGE AND BEARING MEASUREMENT

A wide range of facilities are provided to enable the observer to measure the range and bearing of targets. Some arrangements measure range and bearing while others measure one quantity or the other. Modern equipments employ a joystick/tracker ball and a graphic read-out to measure range and bearing. There is considerable variation in the precise detail of facilities provided by different manufacturers, but the principle is essentially the same in each case.

The new IMO Performance Standards (see Section 11.2.1) also require that all radar displays are fitted with compensation for an offset antenna position so that all positional data is relative to a consistent common reference point (normally the conning position). This could be very useful for multiple antenna installations.

6.6.1 Fixed Range Rings

IMO Performance Standard requires that fixed rings should be provided. The number of rings required depends on the range scale selected; six rings are typical for range scales above 1 NM, but this can be reduced to two rings for range scales below 1 NM. There is also a requirement that the range scale displayed and the interval between the range rings should be clearly indicated at all times.

The fixed range rings take the form of a pattern of equally spaced circles concentric with the electronic origin of the picture.

In a raster-scan display, a slight 'staircase' effect may be noticed due to the quantizing effect of scan conversion (see Section 2.7.1). The effect will be most noticeable close to bearings of 0° and 180°, because in that area the circumference of the circle runs parallel or nearly parallel to the raster lines and will be exacerbated at lower line standards. The effect may offend the eye of the observer, but it should be appreciated that it does not in itself represent a ranging error. This can be understood from consideration of the fact that a circular stretch of coastline at the range represented by the ring would have the same steps due to quantizing.

A facility will be provided to switch the rings on/off and often to adjust their brilliance. This may take the form of a single control or two separate controls. Usually only two or three different levels of brilliance are provided. Brighter brilliance will usually make the rings thicker and fuzzy, thus making accurate range measurement more difficult. There also exists the risk that a weak target will be obscured by the brightness of the ring.

IMO Performance Standards require that the fixed range rings should enable the range of an object to be measured with an error not exceeding 1% of the maximum range of the scale in use or 30 m, whichever is greater. Thus, when using the range rings to measure the range of a target, the most open range scale appropriate should be selected so as to maximize the inherent accuracy of the measurement.

The measurement of range should be taken to the nearer edge of the echo, as this represents the leading edge of the returned echo. Time measurement commences at the leading edge of the transmitted pulse, and thus elapsed time must be measured to the same reference on the received pulse (see Figure 2.4). It must be appreciated that the leading edge of the echo represents the range of the nearest *detectable* facet of the target and not necessarily the nearest part of the target (e.g. possibly the bridge rather than the bow of a ship approaching head-on).

The only situation in which an observer can conveniently fix the vessel's position with a sufficient degree of precision to be able to check the accuracy of the rings is when the vessel is alongside a berth. It is possible for an observer to check the accuracy of the rings by observing the echo of a known small, isolated and steep-sided target when secure in a known charted position.

The range rings are highly suited to the measurement of the range of a target which lies on a ring and there may well be situations in which it is convenient to measure the range of a target as it crosses a ring. In general it will be necessary to interpolate between the rings and this is best done with the aid of the VRM (see Section 6.6.6) or the joystick/tracker ball control.

If the decision is taken to interpolate by eye, a visual check should be made to ensure that the rings are equally spaced.

The factors affecting the inherent accuracy of range measurement and the procedure for ensuring that this accuracy is realized are summarized in Section 6.6.7.

6.6.2 Variable Range Marker

The new IMO Performance Standards (see Section 11.2.1) require that two electronic range markers with a numerical read-out of range be provided. It also requires that arrangements are made so that it is possible to vary the brilliance of each marker and to remove it completely from the screen. Such arrangements will in general be similar to those described for the fixed rings in Section 6.6.1. The accuracy requirement, that is 1% of the maximum range of the scale in use or 30 metres (whichever is greater), is also the same as that required for the fixed range rings.

The VRM will be similarly liable to exhibit the staircase effect described for the fixed range rings in Section 6.6.1.

It is good practice to check the accuracy of the VRM regularly by placing it over a ring and comparing the read-out value with that of the ring. To maximize the inherent accuracy of any range measurement made with the VRM, the most open range scale appropriate should be selected. The VRM brilliance should be adjusted to obtain the finest possible line and the range should be taken to the nearer edge of the echo (see also Section 6.6.1). Errors in VRM of this kind are not very likely, but it is prudent to check the VRM anyway.

The factors affecting the inherent accuracy of range measurement and the practical procedure for ensuring that this accuracy is realized are summarized in Section 6.6.8.

6.6.3 Parallel Index Lines

These lines are available electronically for display on the screen. For a full discussion of their use, see Chapter 8.

6.6.4 The Electronic Bearing Line

This may also be referred to as the electronic bearing indicator (EBI) or electronic bearing marker (EBM). It takes the form of a continuous or dashed line which is generated electronically (Figure 6.9(a)). It can also be available as an electronic bearing and range line (EBRL), where the length as well as direction is controlled by the observer (see Figure 6.9(b)). Because it emanates from the electronic origin it can be used even if the origin is not centred. In raster-scan displays the brilliance levels may be restricted to on/off or possibly to off/low/high.

The electronic bearing line (EBL) or ERBL may be controlled by incremental touch controls, a joystick or a tracker ball (see Section 6.6.6).

Although errors are unlikely in modern equipment, the EBL can be checked by superimposing it on the heading marker and checking that the bearing read-out is in agreement. Similarly, the ERBL should be checked for range measurement against the rings.

In general, the type of bearing (i.e. relative or true) which is read off using the EBL will depend on the orientation selected, as described in Section 6.2.5. In some systems arrangements are made for the EBL to read off true bearings even when a ship's head-up orientation is selected. This is only possible where there is a gyro repeater input, as the ship's instantaneous heading must be added automatically to the relative bearing to produce a

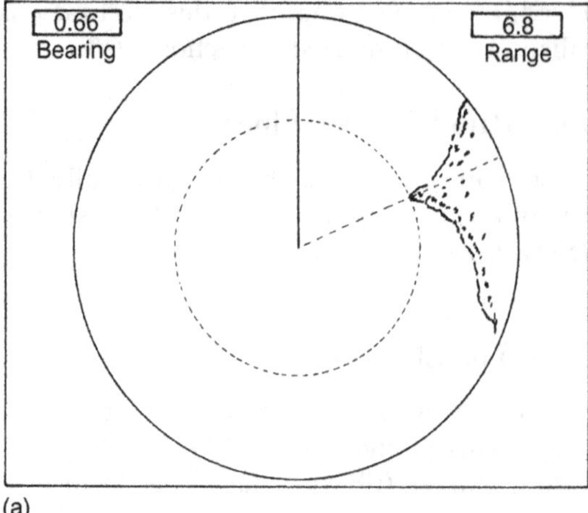

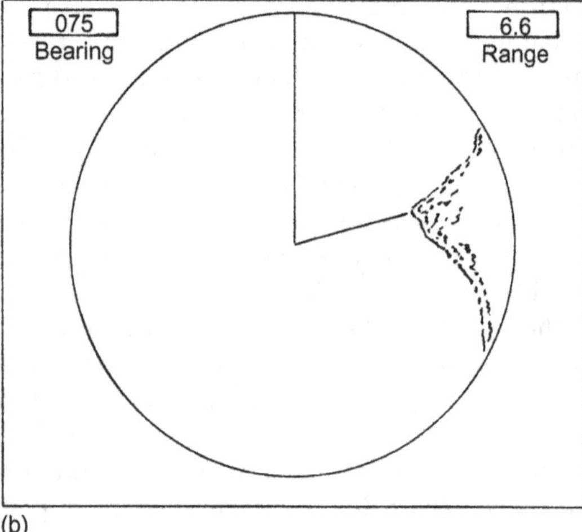

FIGURE 6.9 Electronic bearing line, VRM and EBRL.

true read-out. For any given system, care must be taken to establish the type of bearings available with each orientation.

When taking a bearing the most open range scale appropriate should be selected, and the bearing should be taken through the angular centre of the echo. The inherent accuracy of radar bearings and the procedures for ensuring that the potential accuracy is realized are discussed in detail in Section 6.6.8.

6.6.5 Free Electronic Range and Bearing Line

This facility takes the form of an ERBL, the origin of which can be positioned at any point on the screen. It allows the observer to measure the range and bearing between two displayed targets, as illustrated by Figure 6.10.

The following three examples serve to indicate that there is a range of circumstances in which this facility might be found useful.

1. Approaching port may be useful to ascertain the distance of other approaching vessels from the pilot station.
2. Having heard another vessel broadcast its position in terms of bearing from a well-defined charted object, it may be possible to identify positively the radar echo of that vessel. (*Note:* the incorrect use of VHF in collision avoidance can contribute to the development of a dangerous situation; see also AIS, Chapter 5.)
3. Having positively identified one land echo, it may be possible to positively identify others by measuring their range and bearing from the known echo.

Other applications which are particularly appropriate to a given trade will no doubt suggest themselves to readers. The facility can also be used to draw *navigational lines* in systems which offer automatic tracking (see Section 8.4.6.3).

6.6.6 Joystick/Tracker Ball and Screen Marker

This facility makes use of the fact that in modern synthetic displays, range and bearing data are stored in digital form (see also Sections 2.7 and 4.5.5).

A joystick is a short spindle which is mounted on the display adjacent and normally to the right of the screen (Figure 6.11). In its neutral position it lies perpendicular to the

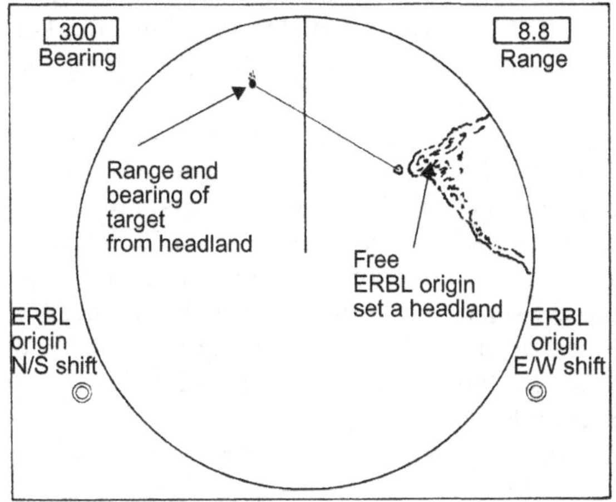

FIGURE 6.10 Free electronic range and bearing line: (a) EBL and VRM and (b) ERBL.

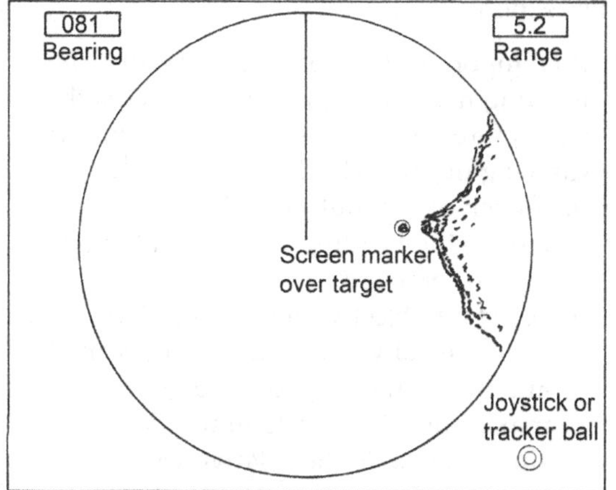

FIGURE 6.11 Joystick/tracker ball and screen marker.

plane of the screen. A small sphere or other suitable hand-grip is fitted on the outer end of the spindle, while the inner end is pivoted in such a way that it can be moved fore-and-aft, athwartships or in any direction which lies between these two. The movement of the joystick controls two independent voltages which are analogues of the displacement of the joystick from its neutral position. If these analogue signals are converted into digital numbers they can be used to control the X and Y screen coordinates of a graphics marker and to generate the corresponding digital angle and range in the form of a read-out of bearing and range. The engineering details of the technique by which the displacement of the joystick is translated into marker position (or bearing and range) may vary in detail with manufacturer.

The graphics marker is commonly a small circle or cross which is located at the origin of the picture when the joystick is in its neutral position. By steering the marker with the joystick, it is placed over the chosen target and the bearing and range are read out on a computer-controlled digital scale. This facility offers a simple and rapid method of range and bearing measurement. As with any form of range and bearing measurement, it should be monitored for errors by checking it against the rings and heading marker.

In some systems a tracker ball is used as an alternative to a joystick. This comprises a sphere, half of which projects above the top panel of the display. It is pivoted in such a way that it can be rotated in any direction, thus allowing control of the graphics marker in a similar fashion to that of the joystick. It effectively operates as an inverted computer mouse (see Section 2.8.6).

6.6.7 Range Accuracy

IMO Performance Standards (see Section 11.2.1) require that the fixed range rings and the VRM enable the range of a target to be measured with an error not exceeding 1% of the maximum range of the scale in use or 30 m, whichever is greater. Where appropriate in the preceding descriptions of the various facilities available for range measurement, mention was made of procedures for ensuring that the potential accuracy is realized. For ease of reference it is appropriate to summarize these procedures in this section.

1. If the ability exists, adjust the brilliance of the VRM, or ERBL, to obtain the finest possible line.
2. Measure the range to the nearer edge of the displayed echo.
3. Use the rings when the target is on or close to a ring.
4. Use the VRM, ERBL or joystick marker to interpolate between the rings.
5. Regularly check the VRM, ERBL or joystick marker against the rings.
6. Use the most open scale appropriate.

6.6.8 Bearing Accuracy

IMO Performance Standards require that the means provided for measuring bearings should enable the bearing of a target whose echo appears at the edge of the display to be measured with an accuracy of ±1° or better.

It is essential to appreciate that the permitted error of ±1° refers only to the measurement of the angle (whose accuracy will depend on the correctness of the alignment of scanner and trace/digital bearing word and the graduation accuracy of the measuring device). It does not take into account the accuracy of the heading marker from which the angle is measured, or of the accuracy of the gyro signal used where azimuth stabilization is provided. The performance standards specify an accuracy of ±1° for the heading marker and ±0.5° for the gyro input. If these error sources are aggregated, a bearing measured from the display could be in error by 2.5° without the provisions of the standards being contravened. However, it must be said that each error will have a component which is substantially constant throughout the period of taking a series of bearings. For example, only in the case of mechanical fault would one expect the error from the heading marker source to change from +1° to −1° during the taking of a series of bearings and there are no mechanical components in modern equipment. The significance of the aggregate error will depend on whether the bearings are being used for radar plotting or for position-fixing. The implications of bearing errors are discussed separately in terms of their significance in the accuracy of radar plotting in Section 7.8 and of the accuracy of position-fixing in Section 8.3.

Where appropriate in the preceding descriptions of the various facilities available for bearing measurement, mention was made of procedures for ensuring that the potential accuracy is realized. For ease of reference it is appropriate to summarize these procedures in this section:

1. It is important to check regularly that the heading marker accurately represents the ship's fore-and-aft line. The procedure for carrying out this check is as follows:
   a. Switch off azimuth stabilization and ensure the heading marker is aligned with 0° on bearing scale.
   b. Select an object which is conspicuous but small visually and whose echo is small and lies as near as possible to the maximum range scale in use. Measure simultaneously the relative visual bearing of this object and the bearing on the PPI relative to the bearing scale. It is important that the visual bearing is taken from a position near the radar antenna in plan. Repeat at least twice and calculate the mean difference between bearings obtained visually and by radar.
   c. If an error exists, adjust the heading marker contact adjustment to remove the error. Traditionally, the adjustment was in the antenna unit, but modern systems allow adjustment to be done in the

processing unit and so could be accessible to an operator.

d. Repeat (b) above to check the accuracy of the adjustment.

2. Ensure that the picture is correctly orientated.

3. Use an appropriate range scale with the target as near to the edge of the screen as possible.

4. If using a ship's head-up unstabilized orientation, the ship's head must be read at the same instant as taking the bearing.

5. Check the EBL, ERBL or joystick marker by superimposing them on the heading marker, at which time the bearing and heading should agree. Any error should be noted, applied to bearings and the cause investigated.

6. For a small isolated target align the cursor, EBL, ERBL or marker with the centre of the target.

7. Temporarily reduce the gain if it will give a more clearly defined echo.

In general it should be borne in mind that positions derived from radar ranges have a much higher inherent accuracy than is obtainable from position lines derived from radar bearings. Further, radar bearings are not as accurate as visual bearings.

6.7 CONTROLS FOR THE SUPPRESSION OF UNWANTED RESPONSES

Controls for the suppression of unwanted responses naturally group themselves under the headings of sea clutter, rain clutter and interference. Each of these controls will be dealt with in turn, but it should be borne in mind that the usefulness of their effects is not necessarily limited to the heading under which they are considered. For example, the sea clutter control may assist in the detection of a target masked by proximate rain clutter or strong sidelobe response (see Sections 3.7.4 and 3.9.4) and the rain clutter control may be efficacious in improving range discrimination (see Section 3.7.4.4).

6.7.1 Sea Clutter Suppression

The theory of the sea clutter response and of the various techniques available for its suppression are treated in detail in Section 3.6 for non-coherent radar and it is intended in this chapter merely to deal with the appropriate practical procedures.

It should be appreciated that many targets will be visible *before* they enter the clutter area and the probability of detecting them within the clutter will be improved if their position is regularly monitored as they approach. As the range of any target decreases its response will improve (see Section 3.6), but once within the clutter it must compete for attention with clutter signals which will be weak at the point of entry but the response of which improves with decreasing range. If the clutter response overtakes that of the target the latter will be undetectable. Thus, where weak targets are approaching, it is important to attempt to detect them before this occurs. Very weak targets may of course be within the clutter area before they return a detectable response.

The above principle is well illustrated by the situation in which a vessel engaged in a search and rescue mission is attempting to locate the echoes of survival craft close to her in clutter conditions. Under such circumstances it may be that the rescue vessel cannot detect the echoes of the survival craft because of the clutter. Another vessel stationed sufficiently far away that the first vessel and the survival craft responses are clear of the clutter response may well be able to locate all the craft involved, because it does not have to view them against a clutter background.

6.7.1.1 *The Manual Sea Clutter Control*

It is evident from the theory (see Section 3.6) that there is no single correct setting for the sea clutter control and that the correct use of the control is to perform regular *searching* operations. The practical procedure for carrying out searches can be set out as follows:

1. Turn the manual sea clutter control to maximum effect. This will immediately suppress the responses within the receiver but the artificial afterglow of previously displayed echoes will persist on the screen for a short period. To maximize the possibility of detecting targets it is necessary to pause and allow the afterglow to fade.
2. Reduce the suppression by a small step and study the screen for the appearance of echoes.
3. Continue to reduce the suppression in small steps, pausing at each step to examine the screen for the appearance of echoes. In due course the clutter peaks will begin to appear and, as the suppression continues to be removed, it will become progressively difficult to detect targets against the clutter background. In this connection it should be remembered that sea clutter paints will tend to be random while those of targets will tend to be relatively steady.
4. Repeat the search procedure at frequent intervals. The frequency with which this is done must be matched to the prevailing conditions. In particular, consideration should be given to the speed of the observing vessel, the visibility and the type of floating object which is likely to be encountered.
5. Between searches it is prudent to set the control so that just a few sea clutter echoes are painting.
6. The effectiveness of the searching operation may be assisted by the overall reduction in response which can be achieved by the use of a short pulse length, a logarithmic receiver and the selection of S-band transmission. In very heavy clutter, it may be found that even the full suppression offered by the sea clutter control is inadequate, in which case its effect will have to be supplemented by that of the gain control.

Where two transmitters of different powers are available, it may be helpful to select the one having the lower power.

Effective sea clutter searching requires considerable skill which can only be developed by practice. It is recommended that, wherever possible in clear weather, the observer should make use of opportunities to carry out such practice. It must always be remembered that with non-coherent radar, targets weaker than the clutter may not be detected although a weak target in clutter may be detected due to its persistence.

6.7.1.2 *Adaptive Gain*

When adaptive gain is selected, the manual sea clutter and rain clutter controls are rendered inoperative, a fixed amount of differentiation is applied, and the gain is varied instantaneously and automatically according to the dictates of a suppression waveform derived from the signals present on the current or previous radial timebase (see Section 3.6.3.3).

The name 'adaptive gain' is a general engineering term but the control which selects the facility may be variously named by different manufacturers. Examples include video processor, automatic clutter, auto sea clutter and CFAR *(constant false alarm rate* – a statistical term used in the theory of threshold detection). In this, unwanted targets, such as noise and clutter, which cross the threshold are referred to *as false alarms*. In the case of noise, which is a random phenomenon as described in Section 2.6.4.4, the false alarm rate, that is the number of false alarms per unit time, should be the same whatever the screen location. This makes it comparatively easy to make a judgement on threshold setting. In the case of sea clutter, the false alarm rate is likely to have a high value at the centre of the screen falling to a lower value where only noise is present, making the choice of threshold setting a problem. Adaptive gain seeks to simplify the problem of threshold setting by producing a *constant false alarm rate*, that is reducing the clutter to a noise-like response (see Section 3.6.3.3).

The particular virtue of adaptive gain is that, if correctly set up, it will provide near optimum instantaneous clutter suppression, line by radial line, hour after hour, without tiring, losing concentration or becoming distracted. In general, it is not possible to set up the circuitry so as to guarantee optimum performance over the entire range of sea conditions that may be experienced, and exclusive reliance must not be placed on its ability to remove *all* clutter responses and display *all* targets. In some conditions, particularly heavy clutter, it may well be found that skilled use of the manual sea clutter control in a 'searching' action (see Section 3.6.3.1) will provide a higher probability of detection. The adaptive gain facility should be regarded as having the ability to display *most* of the targets masked by clutter for *most* of the time. However, the observer must from time to time switch off the adaptive gain and use the manual clutter control to check for targets close to clutter level. Commonly, the control for the selection of adaptive gain is integral with the manual clutter control.

The manual gain control remains operative and thus the observer maintains control of the gain level which obtains in the absence of automatic suppression. At initial setting up the manual gain control should be set with the manual and adaptive clutter controls at zero effect. If the adaptive gain circuitry is functioning correctly, further adjustment of the gain should not be necessary. However, at any time that the observer suspects that excessive suppression is being applied, the gain may be adjusted to produce a higher level of residual speckle in order to ensure maximum probability of detection of targets.

For the reasons set out in Section 3.6.3.3, it may be necessary to switch off the adaptive gain facility in the following circumstances:

1. When attempting to detect radar beacons. In some systems, separate automatic sea and rain adaptive gain controls are provided and it may be possible to detect a distant radar beacon while maintaining the ability to use the limited range of the automatic sea clutter adaptive gain facility.
2. When carrying out a performance check.
3. When attempting to identify coastlines.
4. In rivers, narrow channels and enclosed dock areas.

6.7.1.3 Rotation-to-Rotation Correlation

This facility seeks to remove clutter responses by identifying their random character (see Section 3.6.3.4). The signal produced as a result of the processing is sometimes referred to as *integrated video* and the control may be similarly labelled.

Only two states, *on* and *off*, are associated with the facility. In some systems the facility defaults to *off* at switch-on and the observer is required to make a positive selection if the correlation is required. In other systems the facility defaults to the *on* condition and the observer must switch it off if not desired. In some cases the switch is spring-loaded. In cases where the facility does default to the *on* state, it is important to switch it off when setting the gain control because it will produce some reduction in the displayed noise.

The use of the facility in combination with adaptive gain will greatly reduce the displayed clutter response. It will also make manual searching a simpler operation. However, it must be borne in mind that echoes close to clutter level, such as those from small boats, buoys and small icebergs, may also be random in their character. Small fast-moving targets may also be displayed at low level if the facility is selected.

6.7.2 Rain Clutter Suppression

The theory of the rain clutter response and of the various techniques available for its suppression is treated in detail in Section 3.7 and it is intended in this chapter merely to deal with the appropriate practical procedures.

As in the case of sea clutter, targets may well be visible before they enter areas of precipitation and it is self-evident that the task of detecting any target is made easier by knowing roughly where to look. Thus every effort should be made to track such targets so as to improve the probability of detecting them once they are within the clutter area. One must of course always be alert to the danger of a target manoeuvring within the clutter region. Rain clutter differs from sea clutter in that the echoes may appear in any part of the screen and may move quite rapidly. The probability of problems arising with targets being lost in precipitation can be anticipated if efforts are made to track the direction in which particular areas of precipitation are moving.

6.7.2.1 The Rain Clutter Control

As described in Sections 2.7.4.2 and 3.7.4.4, this control seeks to deal with the saturation problem by displaying only the leading edges of echoes. Because it is so called, there is a tendency for observers to assume that it is the most effective way of dealing with rain echoes. On the basis of practical experience, many observers hold the view that the use of the gain control in a searching fashion is much more effective. It can of course be argued that this view is to some extent subjective. The important point to appreciate is that suppression and differentiation are two quite different attempts to solve the same problem. Given the subjective element that exists, an observer should take every opportunity to establish which particular technique he or she finds most helpful, and whether the preference applies to particular circumstances.

6.7.2.2 Manual Searching for Targets Within Rain

The masking effect of rain can be dealt with as follows:

1. If the rain is close to the ship, searching with the sea clutter control may be effective.
2. In general the most effective technique is to use the gain control in a searching fashion.
3. If preferred, the rain clutter control may be used in a searching fashion, either alone or in combination with some suppression of the gain.
4. The effectiveness of the searching operation may be assisted by the overall reduction in response which can be achieved by the use of a short pulse length, a logarithmic receiver, the selection of S-band transmission, the selection of a low-power transmitter or the selection of circular polarization if available.

6.7.2.3 Adaptive Gain

This is an effective way of providing automatic rain clutter suppression. Attention however is drawn to the practical limitations of the technique set out in Sections 3.7.4 and 6.7.1.2.

6.7.2.4 Searching for Targets Beyond Precipitation

Because of the attenuating effect of precipitation (see Section 3.7.4), it may be difficult to detect weak targets which lie beyond areas of precipitation. It may be possible to overcome this difficulty by using S-band transmission, a higher power transmitter (if fitted) or selecting a longer pulse length. While searching for specific targets it may be useful temporarily to turn the gain above the normal setting, or use an echo-stretching facility if available (see below).

6.7.3 Interference Suppression

Section 2.6.5.1 explains how interference suppression works. It is based on pulse-to-pulse (also known as line-to-line) correlation. The facility should be switched off when initially setting the gain control because it also reduces the noise content of the picture. Its use will reduce the visibility of weak targets and may also affect the display of certain types of racon (see Section 3.9.5).

6.7.3.1 Second-Trace Echo Elimination

Some systems provide a facility whereby second-trace echoes can be eliminated. The facility can normally be switched on or off. The theory of the techniques used to eliminate the second-trace echoes is described in Section 3.9.6.

6.8 ECHO STRETCH

In a synthetic display, the radial length of the displayed echo is normally fixed by the number of elements of memory activated by the detected response. Some systems offer an additional facility whereby the number of radial elements representing the target can be increased, as a result of which the observer can cause the displayed echoes to be stretched radially away from the origin in order to make them more obvious, while preserving the leading edge at the correct range. The rules which govern the stretching vary somewhat with manufacturer, but in general the availability of the facility is limited to:

1. The longer range scales.
2. Targets beyond a preset minimum range.
3. Targets whose received echoes exceed a preset duration.

Such a facility may be particularly useful when trying to detect distant land echoes. Once detection has been achieved and identification becomes a priority, the stretching effect will become counter-productive (see Section 8.2.1).

6.9 USING AN AUTOMATIC RADAR PLOTTING DISPLAY

Most ARPAs offer a host of facilities, some of which are required by IMO Performance Standards and others which are not (see Chapter 4). The precise way in which the facilities are controlled varies considerably with manufacturer. This general treatment seeks to deal with the setting up of the display for basic tracking of radar targets in terms of the important guiding principles which are common to all systems.

6.9.1 The Input of Radar Data

It is essential to ensure that the radar system supplying the raw data to the ARPA tracker is correctly set up (see Section 6.2). Particular attention on a magnetron radar must be paid to pulse length selection, tuning and *any* control that affects receiver gain (e.g. the sea clutter control – see Sections 3.6.3.1 and 6.7.1). Failure to use the correct pulse length or to tune the receiver will reduce the probability of targets being detected and tracked by ARPA although, in some cases, once tracking has been initiated, the tracker will continue to track the target even though the target may not appear on the display (see below). The effect of the setting of the receiver gain controls on the video fed to the tracker varies from one manufacturer to another. In most modern systems the signal level fed to the tracker (see Section 4.3.6) is independent of the operator's gain controls, being continuously set by some form of adaptive gain or other form of automatic signal processing. In other systems, the manual gain control determines the signal level fed to both the tracker and the CRT. When first using an ARPA with which the observer is not familiar it is prudent to establish which gain control settings affect the input to the tracker. This can be done quite simply by acquiring a target, adjusting each relevant control in turn, and observing any consequent loss of tracking (see Section 4.7.3).

6.9.2 Switching on the Computer

In most systems the ARPA computer is switched on when the radar is switched to the 'operate' condition, while there may still be some older models in which a separate 'ARPA' switch is provided. After the computer is switched on, it may be necessary to wait for a short time, normally less than a minute, for the computer to carry out a self-checking programme.

When the computer becomes active, many ARPAs have default conditions by which all but the basic facilities are switched off. This ensures that the observer obtains the additional facilities only if they are positively requested and is thus not confused by uncertainty as to the status of the controls, nor plagued by the need to find the control necessary to switch off unwanted facilities.

6.9.3 Heading and Speed Input Data

The observer should check that the correct heading and *water speed* information is being fed to the computer for the target tracking; if not, the necessary adjustments should be made. It is vital that this check is made, otherwise no reliance can be placed on the tracking. The importance of heading and speed inputs cannot be overstressed and the observer must always be watchful for errors in these fundamental data sources. The use of ground-stabilized heading and speed is discussed in Section 6.9.6.

6.9.3.1 Compass Heading Input

The checking and setting up of the compass input has been covered in Section 6.2.5

6.9.3.2 Water Speed Input

As in the case of the course input, it is vital that the correct value of speed is fed in, otherwise all the displayed tracks will be in error and may dangerously mislead the observer (see Section 7.9.3).

There were will be an input speed selector that enables the observer to choose the input either as a signal from the ship's log or an artificial log that is usually referred to as *manual speed*. In the latter case the observer must input the value of estimated speed, perhaps based on engine revolutions with an allowance for propeller slip which will vary according to weather and the loaded condition of the vessel.

Many shipboard logs measure speed through the water and will thus, if selected, produce the correct water track true vector. The Doppler log is an exception to this rule. Where a Doppler log is used to provide the speed input, particular care must be taken to establish the answers to the following questions:

1. Is the log measuring speed through the water or speed over the ground?
2. Is the log a single-axis or a dual-axis (sometimes referred to as a Janus Array) type?

Most Doppler logs can be set to operate either on signals returned from the water mass or on signals returned from the sea bed. These two options are commonly referred to as *water lock* and *ground lock*, respectively. In very shallow water it will only be possible to obtain ground lock, while in very deep water only water lock will be possible.

A single-axis Doppler log uses one transducer which measures fore-and-aft movement only. If the log is ground locked and the vessel is experiencing a tide having an athwartships component, the reading will be neither the speed through the water nor the speed over the ground (see paragraph 3 below).

A dual-axis Doppler log has a second transducer which additionally measures athwartships movement. The output from the dual-axis Doppler log can be in the form of the individual velocity components or their resultant. Some radar displays are designed to accept both a fore-and-aft input and an athwartships input and to cause the origin to track in sympathy with the resultant from both, whereas others will only accept a single input. The possible effects on the true-motion tracking of these various options are illustrated by Figure 6.12 (see also Section 7.9.3) and can be analysed by considering the following three possible cases:

1. If the Doppler log is correctly water locked, the input speed to the true-motion computer will represent the speed through the water whether it is a single-axis or a dual-axis log and irrespective of whether the radar display is designed to accept single-axis or dual-axis inputs (assuming that leeway can be neglected). The effect of this input will be to produce a correct water track true vector on targets and (when selected by the observer) a correct sea-stabilized presentation.
2. If the Doppler log is correctly ground locked, and the radar display is designed to accept a dual-axis input, then this will represent the vessel's speed over the ground in two dimensions and can be used to set up a ground-stabilized presentation and ground track true vectors of target vessels will be displayed, as covered in Section 6.9.6. Attention is drawn to the danger of using this presentation when planning a collision avoidance strategy (see Section 1.5.2.2).
3. If the Doppler log is ground locked, and it is of the single-axis type and/or the radar display is designed to accept a single-axis input only, then it will be neither the vessel's speed over the ground nor the vessel's speed through the water: It will be the vessel's speed through the water plus that component of the tide which lies in the direction of the vessel's course (see Figure 6.12). The effect will be to produce an erroneous calculation of target vessels true course and effectively a presentation which is neither sea-stabilized nor ground-stabilized. Such a display will be highly misleading and could lead to the development of a dangerous situation. The effect of errors on the true vectors and true-motion presentation is discussed in Section 7.9.

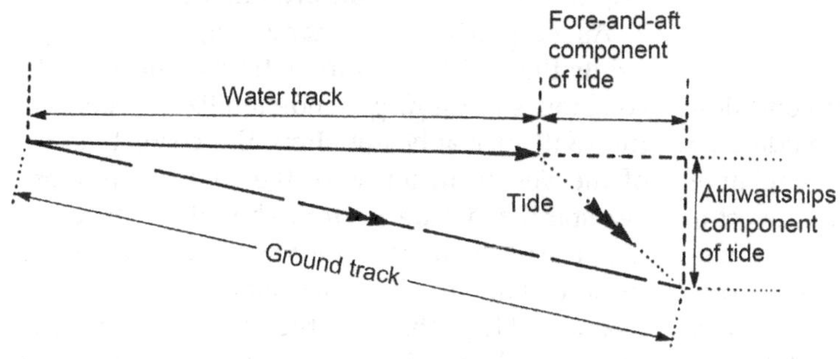

FIGURE 6.12 The use of a Doppler log as speed input.

It is evident that, where a Doppler log is used, great care must be taken to establish the nature of the input to the ARPA calculations of true vector. Failure to do so may result in inadvertent use of the ground-stabilized true vectors or the use of true vectors with an inherent error in the speed input.

See also Section 6.9.6 for a further discussion of heading and speed data in the context of ground-stabilized true vectors.

6.9.4 Setting the Vector Time Control

In many systems the vector time control will default to some non-zero setting, probably in the range of 6–12 minutes. This ensures that, when vectors are initially displayed after acquisition has commenced, they will in general have a length which is sufficient to be obvious but not excessive. In due course the length can be set to suit the traffic conditions and will of course have to be adjusted from time to time as the vectors are used to extract data from the display. Where there is no default condition the observer should, when setting up, adjust the vector time control to a suitable non-zero value, usually in the range 6–12 minutes. Some observers have recommended using a value equivalent to range scale in use, thus 6 minutes vectors for 6 NM scale as their 'default' as it represents a good compromise for many conditions.

6.9.5 Selecting the Relative or True Vector Mode

Most systems will have a default condition which will select a particular vector mode (see Section 4.4) at switch-on, either true or relative vectors. Some systems default to relative vector mode, some to true vector mode, while in others the default condition depends on the radar picture presentation that has been selected. The observer clearly must be able to select the alternate mode to the default condition, but in some systems the vector mode is biased to the default condition, so that in order to select the other mode the observer must activate a control which returns to the default condition on release. Such are the variations found in practice that it is essential to consult the operating manual and discover what options are available.

There should be a clear indication on the display of the vector mode. Even so, more than one collision has been partially attributed to confusion between which vectors were being observed by the navigators.

6.9.6 Selecting Ground- or Sea-Stabilized True Vectors

Ground- and sea-stabilization is normally selected by a software button operating as a toggle switch. The main importance to the observer (assuming that ground/sea speeds and headings are not identical) is that choosing ground- or sea-stabilization affects the displayed course, speed and aspect of targets, and therefore the direction and the length of true vectors. It does not affect the direction and size of the relative vector (and therefore does not affect CPA and TCPA readings).

As already covered in Section 6.2.6, the selection of ground- or sea-stabilization of vectors will also affect the movement of the own ship origin across the display and the direction of artificial afterglow of targets and own ship origin, if in true-motion presentation.

As emphasized in many other sections, if using the ARPA or target tracker in an anti-collision situation, ground-stabilized true vectors will probably not show the correct aspect of the vessel, in terms of the rule of the road. Sea-stabilized true vectors should be used for this purpose. In this respect, it is interesting to note that in IMO performance specification (Section 11.2) that in the event of ground course and speed sensor failure, the observer should select sea-stabilized display, while in the event of failure of the water track speed sensor, the user should use a manual estimate of speed.

As with all displayed information, the displayed information is only as good as the input data from which it is derived. The input and issues in getting the correct values of compass and water speed for sea-stabilized true vectors is covered in Section 6.9.3. The rest of this section will focus on the possible inputs to obtain the ground heading and speed of the own vessel, in order to obtain ground-stabilized true vectors.

6.9.6.1 GNSS or Other Navigational Input

It is relatively straightforward for the manufacturers to input the positional signal from a GNSS (currently most commonly GPS) into the radar (see Sections 10.1 and 10.3), and also to get the ground speed and heading which the GNSS has obtained by integration of the positional data over time. The GNSS equipment itself must be switched on and checked that it is operational. Then the GNSS must be selected in the ARPA/target tracking display (usually in the menu system). There may be a choice of GNSS equipment in some vessels. The read-out of position, ground heading and ground speed displayed on the ARPA should be checked against the original equipment.

The same can apply to other non-satellite navigational equipment such as LORAN C, if available in the vessels geographical region and with a LORAN C receiver onboard. However, this is not commonly fitted to vessels in practice.

6.9.6.2 Manual Estimate of Tide/Current

It is clear from Figure 6.12 that given that all ARPAs should have compass and water speed inputs, it is also possible for the ARPA to estimate ground speed and heading if the observer can provide a value of tide or current.

The manual tidal inputs can simply take the form of two data entry fields. One is the estimated direction of the tidal stream or current between 0° and 360°, while the other is the estimated rate of the tidal or current stream in knots. The ground speed and heading used for true vector calculations will be the vector addition of these values of current and the compass heading and water speed (see Figure 6.12). In theory the estimate of current can be obtained from current information from charts and tables or by radar plotting of buoys and fixed objects. In the latter case, consideration should be given to use the fixed target reference mode of ground-stabilization. However, if persevering with this method of manual tide/current tide settings:

a. Set a sea-stabilized display with zero current/tide settings. Any radar plots of known stationary targets will indicate the set and rate of the current. However, the actual direction of current will be the reciprocal of the 'true course' indicated on the fixed target.
b. Enter this value into tide current settings and select ground-stabilized vectors using this manual entry.
c. The effectiveness of the value of current entered can be monitored by continuing to plot the fixed targets in ground-stabilized vector mode. If the speed of the fixed targets is not zero or near zero, then the value of current entered into the ARPA must be adjusted until it does become zero or near zero.

In summary, the availability of easier to operate and more accurate alternates do not make this option attractive to use.

6.9.6.3 Dual-Axis Doppler Log in Ground Track Mode

As already discussed in Section 6.9.3 the output of the dual-axis Doppler log when used in ground track mode will make a valid estimate of ground track. The ARPA must add together by vector addition, the fore-and-aft component of ground track with the athwartship component of ground track to obtain a value of ground heading and speed to be indicated on the ARPA display (see Figure 6.12). The dual-axis log option (if available) is simply selected as an alternative to GNSS or manual entry of current data in the menu system.

The concerns of a Doppler log probably not remaining ground locked throughout the passage are well covered in Section 6.9.3 and equally apply to this section. If used, the status of the dual-axis ground lock must be checked frequently.

6.9.6.4 *Ground-Stabilization Using an ARPA Plotted Fixed Target as Reference*

This technique is well covered in Sections 4.8 and 8.4.6. When possible, this is a very effective way of monitoring the instantaneous vessel's navigational position and obtaining ground speed and heading independently from GNSS and other navigation equipment.

1. A fixed target must be selected for use as a reference. Some systems allow a number of such targets to be selected. Ideally it is a rock or platform or tower fixed to the sea bed, small enough to be trackable by the ARPA. In extremis, a buoy can be used with caution, particularly if more than one can be used or they are the large channel buoys fitted with racons and monitored by the shore radar, such as in the Dover Strait.
2. The selected fixed target(s) must be selected as ARPA targets and tracked for the requisite 3 minutes.
3. The selected fixed target(s) must then be 'designated' in the ARPA menu.
4. Finally, the option to use the designated target(s) as a reference for ground-stabilization can be selected. The designated targets will be indicated on the display by a special symbol. 'R' (Reference) is in the latest IMO approved symbol list, but 'F' (Fixed) is quite common. It will, of course, have no true vector, when used as such.

Even if limited to only one designated target at a time, it is good practice to have several suitable targets plotted at the same time, in order to change as the vessel proceeds. Whilst ground-stabilized true vectors are selected, they should be showing as a 'stopped' target or nearly stopped target, which is a good check that the observer has selected a stationary target. It also allows the observer to quickly swap the designated target to another one, should the tracking of the existing designated target be lost. Indeed the prudent observer will change the designated target in advance of the designated target being lost. His passage plan should indicate which target(s) to designate next.

6.9.7 Safe Limits

The observer should set suitable safe limits in terms of CPA and TCPA (see Section 4.7.2). The judgement of what is suitable is to some extent subjective. Excessive alerts can be distracting and counter-productive whereas, by contrast, if very small limits are used the dangerous target alarm will not be activated by a vessel which is standing in a close-quarters situation.

6.9.8 Preparation for Tracking

If the observer is unfamiliar with the system, time should be spent in locating the joystick (or tracker ball), the *acquire* button and the *cancel* button and practicing their use. This can be done using live targets if available or, failing that, use can be made of the synthetic targets provided by the training programme supplied with some ARPAs.

6.9.8.1 *Manual Tracking*

Once the radar is set up, tracking of radar targets can commence. This requires putting the tracker ball/joystick control into 'tracking' or 'acquire' mode (by pressing an appropriate software button) and then using the pointing device to hover over a target and then 'click' a push button to manually acquire. Then one can move to another target and repeat to acquire this target (if required). Most equipments do not have 'acquire' mode as the default setting, so after a short delay the joystick/tracker ball pointer will default to another mode. There will also be another push button which will allow the observer to select 'delete target' mode for the tracker ball/joystick pointer and ARPA targets with which target vectors can be deleted by pointing and clicking. Tracking information (containing minimum of CPA, TCPA, course, speed, range and bearing) on a small number of selected targets can be shown textually on the display. These targets are selected and clicked in the default mode. It should be noted that these same functions are also used for AIS targets in AIS-enabled equipment.

6.9.8.2 *Selecting Automatic Acquisition*

Larger ships are required to have automatic acquisition facilities. There will be a simple software touch button to press to activate this. However, before doing this, it is important to take the trouble to make sure that an observer understands how to control the area or region of the screen that is being automatically searched for targets to track. It may seem to be a good thing to be searching a large area, but each time a new target is found, an alarm will go off and if these alarms are continually sounding for targets which are of absolutely no interest to the observer, then the observer's responsiveness to alarms will be reduced. Also the maximum target capacity of the tracker may be reached although many of the targets are of no interest.

As discussed in Section 4.2.3, there are as many different ways of defining the shapes of automatic acquisition areas (sometimes called guard zones) as there are manufacturers. Figures 4.2–4.4 give some indications of the shape and controls of some types of area, but there are others. It is important that the observers (who may be intending to use automatic acquisition) are familiar with the system on their own particular equipment. The outline of the area can be shown graphically (and hidden) on the graphical display at the touch of a software button (often within the display menu system), but there will also be controls to alter the boundaries of the area to suit user requirements.

It should be pointed out that automatic acquisition should not be used as an excuse for an observer to not examine the ARPA for new targets. The manual operator can often find weak targets missed by the automatic tracking. Also, the automatic acquisition may be subject to many false alarms in heavy clutter conditions.

6.10 AIS OPERATIONAL CONTROLS

As already discussed in Chapter 5, AIS data can be provided from a stand-alone device or it can be fed to an electronic chart display and information system (ECDIS) or radar type screen. There is a multitude of models and types of AIS enabled equipment, including all SOLAS radars installed since 2008 (see Sections 4.1 and 4.2). This section will discuss the basic operations required for both types of AIS-enabled target tracking equipment with a particular focus on that likely to be installed on Class A vessels.

6.10.1 Stand-Alone AIS Equipment

Class A vessels could either have an MKD (minimum keyboard display; see Section 5.4.1) or an AIS with a graphical display as the main stand-alone device onboard (see Section 5.4.2). The keyboards on such devices are typically a minimum set of multi-function keys similar to an early model mobile phone with a few extra keys for navigating the menu and display (text or graphical display). Much of the data broadcast is derived automatically from its own GPS and other external sensors, all of which do not need any attention from the observer, except to monitor that it is working. There are some manual fields which will need occasional attention.

The AIS devices have a series of pages containing a few lines of data or fields for the observer to enter data. These multiple pages are arranged in a tree-like hierarchy and they can be cumbersome to operate and use. The observer has to find the appropriate screen for the task required, be it to update or input data, or to get data on other vessels or other information from the AIS.

6.10.1.1 Input of Data

Certain important data must be programmed into the unit at the appropriate time, by the user. These can be divided into permanent data, voyage-related data and safety messages:

a. *Permanent data*. This data is normally programmed in by the installer and will only need changing should this information be changed during the life of the ship. It includes the AIS static data, MMSI number, name of ship, IMO number, type of ship and dimensions of ship in relation to the navigational GPS antenna (port, starboard, forward and aft). It may also include equipment settings such as the data rates to communicate with other shipboard devices.
b. *Voyage data*. This is data which normally needs updating at the start of each voyage and which may occasionally need updating while on the voyage. This includes draft, destination, ETA and navigational status of the vessel (underway, restricted in ability to manoeuvre, etc.). There are also optional extras such as the route waypoints and number of persons onboard.
c. *Safety messages*. The equipment also has to be programmed for the ship to broadcast safety messages to all users or a specific address or addresses. This is by the MMSI number(s), but the navigator need not type the target MMSI direct, but can select the ship or other AIS user to address from the target list (see below).

6.10.1.2 Output of Data

In the menu system of the AIS, the stand-alone units can display lists of available targets, which are usually ordered in terms of range from the own ship. If the data list cannot be fitted into one page then the data is spread over multiple pages, accessed by the arrow up/down keys (Figure 6.13(a)). These simple lists usually only display ship name, range, true bearing and class of user (usually class A or B, but other transmissions from SAR aircraft, buoys, etc. can be received as discussed in Section 5.3). The class B ships can be hidden from the list, but all class A vessels must be shown. As the user goes up and down the list, selecting an AIS vessel, further details of that vessel can obtained by pressing the appropriate button, which will take the observer to one or more pages of information giving the full details of its AIS message. Pressing a 'back' button will normally take the navigator back to the target list (Figure 6.13(a) and (b)).

In the case of stand-alone graphical devices, the input and output features are the same, but additionally a radar-like situation display that only displays the AIS targets is included (Figure 6.14). Typically each ship is shown as an isosceles triangle on the situation display with the sharp angle indicating the heading of the vessel. The full range of symbols that can be used is detailed in Section 11.3. The heading of the own ship is displayed as a single line, and typically a single range ring is provided to

(a)
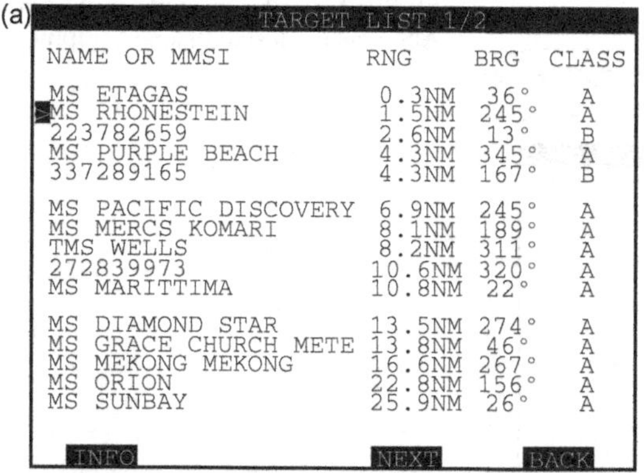
TARGET LIST 1/2

NAME OR MMSI	RNG	BRG	CLASS
MS ETAGAS	0.3NM	36°	A
MS RHONESTEIN	1.5NM	245°	A
223782659	2.6NM	13°	B
MS PURPLE BEACH	4.3NM	345°	A
337289165	4.3NM	167°	B
MS PACIFIC DISCOVERY	6.9NM	245°	A
MS MERCS KOMARI	8.1NM	189°	A
TMS WELLS	8.2NM	311°	A
272839973	10.6NM	320°	A
MS MARITTIMA	10.8NM	22°	A
MS DIAMOND STAR	13.5NM	274°	A
MS GRACE CHURCH METE	13.8NM	46°	A
MS MEKONG MEKONG	16.6NM	267°	A
MS ORION	22.8NM	156°	A
MS SUNBAY	25.9NM	26°	A

INFO NEXT BACK

(b)
TARGET INFORMATION

NAME: MECKLENBURG VORPOMME
CALL SIGN: DQLV
IMO: 211245200
MMSI: 009131797 CLASS: A
SHIP TYPE: 063/Passenger ship
CARGO: Hazard category C
LENGTH: 200m
BREADTH: 32m DRAUGHT: 5.7m
NAV.STATE: under way using engine
DESTINATION: Trelleborg
ETA: Nov 17 / 21:00 UTC
BRG: 010.3° LAT: 54:24.45N
RNG: 14.5NM LON: 012:09.31E
HDG: 016° COG: 024.7°
ROT: >123°/min SOG: 17.0kt
SEND MSG BACK

FIGURE 6.13 Examples of textual AIS data: (a) textual data display from vessels transmitting in the AIS area and (b) textual data display relating to one vessel in the AIS text list.

give the observer an appropriate reference for scale. The observer can change the range being displayed. The keyboard arrow will allow the user to select one of the targets and then go to the appropriate AIS individual target's data pages (Figure 6.13(b)).

6.10.2 AIS Integrated with ARPA

The facility to have integrated ARPA and AIS data on one target tracking display is a prominent feature of IMO Performance Standard in Section 11.2. The AIS data is fed from the stand-alone device and can be superimposed on top of the ARPA data. The observer can choose to display or not display the additional AIS vector data. The navigator can also select a specific AIS target and can:

a. Obtain the equivalent information to an ARPA target (range, bearding, course, speed, CPA, TCPA), plus the actual target name. This is usually displayed alongside the comparable radar target text displays of data.

b. Request a further display of detailed AIS information associated with that target (IMO number, navigational status, AIS class, ETA, destination, number of persons onboard, etc.). This is usually accessed through a menu page.

Additionally, in this combined ARPA/AIS/target tracking display menu system, there is usually a copy of many of the AIS functions found on the stand-alone AIS equipment indicated in Section 6.10.1. Thus, AIS information displayed by the ownship can be checked on the ARPA/AIS/target tracking device and also some of the updating of the own ship AIS information can be done within the ARPA/AIS/target tracking display which is then sent to the AIS device which transmits the AIS messages.

6.10.2.1 Sleeping and Active AIS Targets

As well as having the ability to totally remove all AIS targets from the display, it is also possible to filter the targets being shown. These targets are not removed from the display completely, but are all designated as 'sleeping' targets and are typically shown as very small isosceles triangles which display no vector or target identification detail. However, the acute angle of the triangle should be indicating the heading of the AIS target. The full IMO approved list of symbols is given in Section 11.2.3.

The criteria for targets to become sleeping targets may vary from manufacturer to

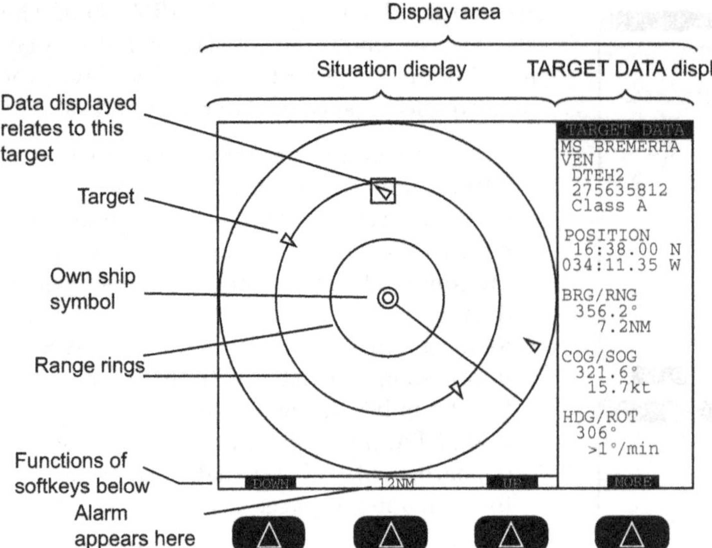

FIGURE 6.14 Example of graphical AIS display on stand-alone equipment.

manufacturer. Some allow individual selection, but the manufacturers can provide the ability to filter the AIS targets on the basis of range, CPA/TCPA and target class (Class A, B, etc.). Most also have some kind of area control, so that targets outside a designated area are automatically designated as sleeping, while those inside the area are designated as active AIS targets and are shown on the display with name and appropriate vector. The filter status of AIS targets should be clear to users.

The controls to move the boundaries of the area again vary between manufacturers, but IMO have specified that it is the same area shape that defines the limits of both 'active' AIS data and automatic acquisition of radar targets (see Sections 6.9.8.2 and 4.2.3). It can by a point, click and drag on the display itself or more usually within the menu system of the display. In this latter option, the observer selects digital values of the distance ahead, port, starboard and astern of vessel and/or relative bearings defining radial lines which define the area. This sharing of the same area works quite well in practice, partly because the same areas of the target display are of interest to the observer, and partly because automatic acquisition is not actually used by many observers.

Finally, at low-range scales, the isosceles triangle symbol can be replaced by an outline (to scale) of a target vessel, based on the dimensions of the target being broadcast in the AIS message.

6.10.2.2 Data Fusion

IMO (Section 11.2.1) also requires that where a tracked radar target and AIS information are clearly the same physical target, they should appear as one target and the AIS symbols and alphanumeric text should be displayed by default. This is called *data fusion* and is intended to make the display less cluttered for the observer. However, the observer should have the ability to change the display of the target symbol and alphanumeric data to that provided by the ARPA or the AIS.

The algorithms to make this decision as to whether an AIS and ARPA target is the same target are specific to the manufacturers. Within the AIS section of the menu system, there are normally criteria which the observer can adjust which are used for making this decision. For example, it could specify the maximum differences in:

a. Distance between the AIS and radar positions (probably in metres).
b. Heading (in degrees).
c. Speed (in knots).

In this case, if the AIS and ARPA targets are less than all three limits, then the targets would appear as one.

Again, it is important that the observer understands how the display is making decisions on what information is being displayed.

CHAPTER

7

Radar Plotting Including Collision Avoidance

7.1 INTRODUCTION

It may be argued that the prime function of the radar system is avoidance of collision, either with other vessels, with navigation marks or land in restricted visibility. After the initial euphoria of being able to detect other vessels in fog, the plethora of accidents involving vessels equipped with radar left no doubt as to the need for systematic observation of the radar display and for proper techniques for the extraction of data relating to other vessels. The method by which this information has been displayed to the navigator has seen a steady progression from paper plotting sheets via true-motion displays to the present-day automatic radar plotting aids (ARPAs). All have sought to provide the mariner with the most meaningful portrayal of the situation in the proximity of his ship and with the least amount of distraction to the other bridge routines. In order to understand the more complex operations performed by ARPAs and for the benefit of those who are not so fortunate as to have that aid or where it has broken down, it is still necessary to have a fair degree of skill and dexterity in practical plotting. To maintain this dexterity, it is essential to engage in regular practice, preferably in clear weather where visual confirmation of results will build confidence in the information obtained when using the radar in restricted visibility.

This chapter covers the basics of plotting from a manual plotting perspective. It also covers electronic aids to plotting. This chapter concludes with a section on the use of radar and ARPA for collision avoidance.

7.2 THE RELATIVE PLOT

Since the radar is carried along by the vessel as it proceeds, the direct measurements obtainable are always *relative to the observing vessel* (see Section 1.5.1). Thus, in order to determine the true courses and speeds of other vessels, it is necessary to resolve this observed relative or apparent track into its components by using a knowledge of the own ship's true course and speed. The means by which this may be achieved ranges from pencil-on-paper plotting sheets via reflection plotters and true-motion displays to computer-based collision-avoidance systems. Irrespective of the mechanism used, if a complete appreciation of the dynamic situation in the vicinity of the own ship is to be obtained, the relative track of targets must be systematically recorded and the true track derived by means of a plot (manual or ARPA). The techniques employed to extract the relevant data by manual methods and the use to which it is put will form the major part of this chapter. Failure to carry out systematic observation has resulted in many dangerous misconceptions (Figure 7.1). An older generation of navigators reliant on manual methods often used the excuse that the method was laborious and slow. However, such lapses still occasionally occur with modern navigators with ARPA equipment.

If systematic observations are made of the bearing and range of a displayed echo and the positions are plotted on a traditional plotting sheet, the line joining the plotted positions will depict the target's apparent track *OA* (Figure 7.2), which if extended will enable a measurement of the target's closest point of approach, CPA, to be obtained. The time to closest point of approach, TCPA, can be obtained by stepping off the rate from *O* to *A* along *OA* extended to CPA. This does of course assume that both vessels maintain their present courses and speeds. If this condition is not met, the apparent track will not be uniform, that is the echo will not move across the screen in a constant direction at a constant rate.

It is important to realize that the measurements obtained thus far are only subject to those errors inherent in the radar system itself, the azimuth stabilization (see Sections 7.9.1 and 7.9.2) and in the observations. They are

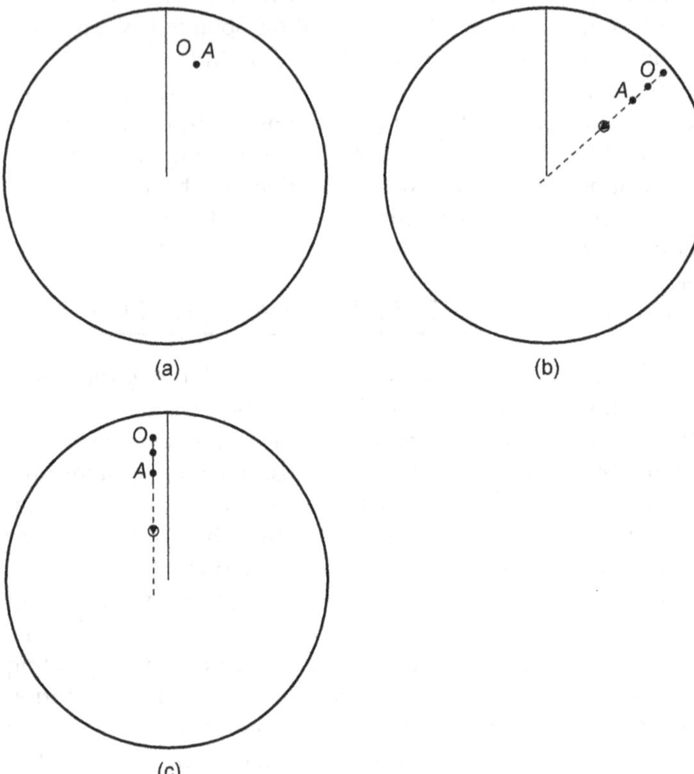

FIGURE 7.1 Common misinterpretations of relative-motion displays. (a) The target, believed to be stopped, is in fact on the same course and proceeding at the same speed as the own ship. (b) A vessel which might be expected, if observed visually at night, to be showing both sidelights, i.e. to be head-on and to present no threat, is in fact crossing broad starboard-to-port and on a collision course. (c) The target appears to be proceeding slowly on a reciprocal course, showing a red light; it is actually being overtaken and showing a stern light.

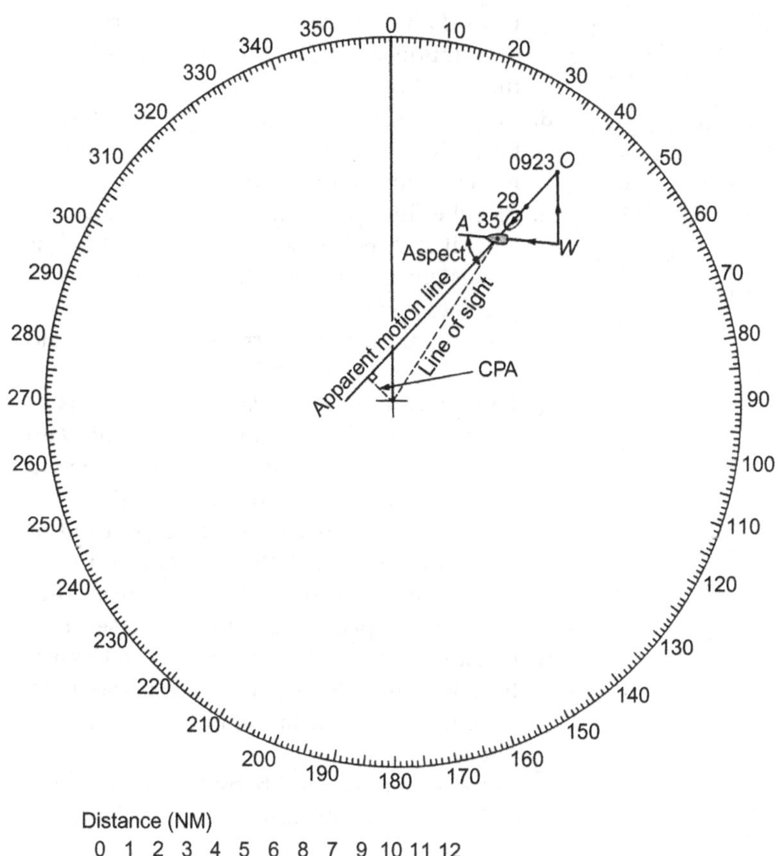

FIGURE 7.2 The relative-motion plot.

therefore uncontaminated by errors (or blunders) to which other inputs – notably the own ship's speed – might be subjected. For this reason, relative track (or the relative plot) is probably the most reliable, although not the only data on which to base decisions.

While CPA and TCPA are two important pieces of information, it is essential that, prior to manoeuvring the own ship, the true course and speed of the target(s) are derived and this can only be done by resolving the vector triangle.

7.2.1 The Vector Triangle

By plotting the apparent track of a target and with a knowledge of the own ship's true course and speed, it is possible to determine the true course and speed of the target (Figure 7.3).

7.2.2 The Plotting Triangle

EXAMPLE 7.1 (USED IN FIGURES 7.2 AND 7.4)

While steering 090° (T) at 12 knots, the echo of a vessel is observed as follows:

0923 echo bearing 127° (T) at 9.5 NM
0929 echo bearing 126° (T) at 8.0 NM
0935 echo bearing 124° (T) at 6.5 NM.

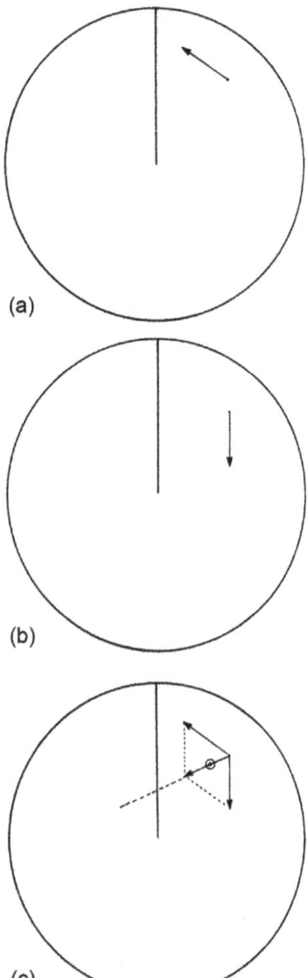

FIGURE 7.3 The vector triangle. (a) With the own ship stopped, target moves with true course and speed. (b) With target stopped, the own ship movement causes the target to appear to move. (c) The resultant track of the echo when both vessels are moving.

At 0935, determine the target's true course and speed; CPA and TCPA; and aspect.

The data relating to the target in Figure 7.4 as extracted from the plot at 0935 is as follows:

Course: 008° (T); CPA: 1.0 NM; Aspect: Red 65° Speed: 11 kn TCPA: in 26 minutes.

7.2.3 The Construction of the Plot

a. Draw in the own ship's heading line on the plotting sheet.
b. Plot the first position of the target. Label it *O*. Insert the time.
c. Lay off a line to represent the apparent track of a stationary target, from *O*, opposite to the direction of the heading line.
d. Plot at least one intermediate position of the target to ascertain that its apparent track is not changing. Insert the time.
e. Plot the final position, with the time, and label it *A*. Ensure that the apparent track is consistent in direction and rate. Join *OA* and label it thus.
f. Produce the apparent-track line *OA* to find the CPA and the TCPA.
g. The 'plotting interval' is the time between the readings of range and bearing for *O* and *A*. Calculate the distance that the own ship has steamed in this time and plot the position in which you would expect to find the stationary target *W* at the end of the plotting interval. (*OW* is derived from the own ship's speed and course reversed.)
h. If *A* and *W* coincide, the target is stationary. If they do not, the line *W* to *A* represents the proper track of the target in the plotting interval.
i. 'Aspect' is measured between the 'line of sight' and the *WA* direction as in Figure 7.4.

It can be seen that, whichever orientation is used for the plot, the answers obtained will be the same, but some consideration must be given to the actual practicalities of plotting and the way in which the results are to be used. Where it is not possible to compass-stabilize the radar display (see Section 1.4.1), the ship's head-up display will be the obvious choice. Where the radar display is capable of being compass-stabilized, it makes good sense to orientate the plot in the same way as the radar display, that is true-north-up when the radar display is true-north-up, especially when the vessel is on a southerly heading. There can be considerable confusion and even potential danger in trying to relate a ship's head-up plot to a displayed radar picture which is 'upside down', for example when heading south. Course-up radar

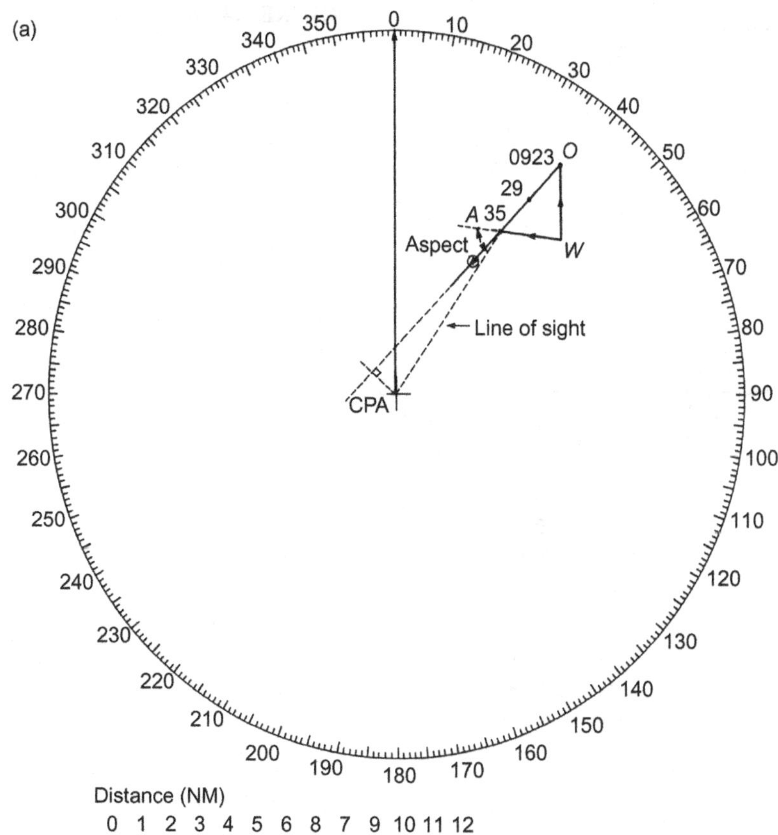

FIGURE 7.4 (a) The relative plot for Example 7.1 ship's head-up. (b) True-north-up.

display and paper plot orientation is also a possible option, but is best avoided as it is a common outcome of radar target plotting that the own ship alters course and therefore loses the benefit of the 'new' course being at the top of the display. On the other hand, when conning the vessel and therefore relating the plot to the visual scene ahead of the vessel, it is sensible to orientate the display and to plot the ship's head-up or course-up accordingly.

7.2.4 The Practicalities of Plotting

The way in which the plot is constructed in practice is very different from the way in which one theoretically tackles a plot, such as the example in Section 7.2.2. In the first place, ranges and bearings are obtained only at intervals of some 3–6 minute (not all at the same time), and it is unlikely that the plotter will be able to enjoy the luxury of radar plotting as a dedicated task, having rather to dovetail this activity into the many other bridge duties. If plotting is not to become all-consuming, or if it results in peaks and troughs of activity, some wrinkles will have to be adopted to assist in spreading the load:

a. Always draw in the heading line before starting to plot.

b. When a target is first plotted, draw in *OW* and graduate it in 3 minute steps up to

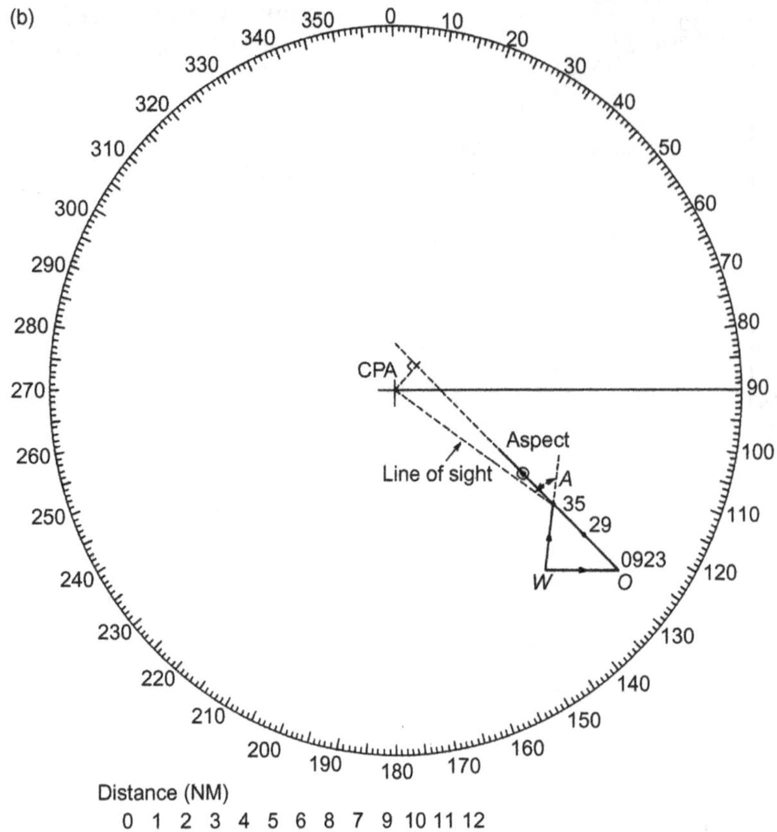

FIGURE 7.4 (Continued)

about 12 minutes. This is not to suggest that one should plot slavishly at 3 minute intervals but, rather, that one should plot when the opportunity arises and interpolate visually between the 3 minute graduations on the *OW* line.

The frequency with which the target should be plotted is dependent on a number of factors, namely, the range, the approach rate and the CPA. A target which is closing fast will require more frequent plots than, say, one which the observing vessel is slowly overhauling.

c. Make a quick, early plot after, say, 3 minutes. Although the triangle will be small and not particularly accurate, it will give early warning of the potential of the encounter which is arising and allow time for some pre-planning. *Do not wait 12 minutes* before drawing the first triangle.

d. Obtain at least three consistently spaced positions in a reasonably straight line before being prepared to make a decision based on the plot.

e. Plotted positions should be in a reasonably straight line and spaced at distance intervals which are related to the time intervals between the plots. Where a plotted position appears significantly different from what is anticipated, it should be investigated immediately for an error either in reading the range and bearing from the radar or in plotting the position. Where a change in apparent track is found to have occurred, a new triangle should be started (see Section 7.4).

f. Only essential lines should be put on the plot and then kept to a minimum in length. Extending *WA* beyond *A* should be avoided as it can be dangerously misleading and mistaken for an indication of CPA. On the other hand, *OA* should *not* be stopped short but extended at least beyond CPA.

g. It can be extremely helpful, especially when other officers are likely to observe the plot, to adopt some standard form of labelling (the labelling used in all figures in this text conforms with the United Kingdom Department of Transport's recommendations).

h. Times should be placed alongside each plotted position. This is essential when the plotting interval varies as it can justify the unequal spacing of the positions. 'Minutes' are quite sufficient, with the occasional indication of the hour.

i. TCPA should be given as the 'time to elapse', that is 'CPA in 16 minutes' rather than 'CPA at 1743' which is far less meaningful to someone who has then to check the time and subtract mentally. Also, time to elapse will in itself convey the degree of urgency in the situation.

j. After some 12 minutes there should be no real need to continue plotting bigger and bigger triangles, but it is essential to continue to plot positions and ensure that the target's apparent track is being maintained. As soon as the apparent track is observed to change, a new plot should be commenced (see Section 7.4).

k. Neatness is essential at all times. An untidy plot – lines everywhere, no labels, with times not related closely to plots, crossings out, etc. – is likely to be more of a hindrance than a help. There is no merit in a plot which even the plotter has difficulty in understanding.

l. The range scale on which to plot requires some consideration. As a general rule, plots should be initiated on the 12 NM range scale, but two differing requirements can arise. Where targets are likely to be closing fast, the earlier the plot is commenced the better, in which case the 12 NM range scale should be adequate; when targets are close, a shorter (3 or 6 NM) scale should be in use, possibly off-centred. The advantages of the shorter range scale are that there is better intrinsic accuracy, and that changes in the target's movement are quickly and easily identifiable. As a target closes, a shorter range scale should be selected, but the plotter should, at intervals, temporarily return to the longer range scale(s) to search for newly arrived targets which could be a potential threat.

7.2.5 The Need to Extract Numerical Data

Plotting theory provides for the extraction of data in numerical form but this takes time and is needed only rarely. Also, the precision with which it is displayed often belies its accuracy and therefore the reliance which can be placed upon it.

Of the information normally available on a clearly constructed plot, CPA and the target's aspect/course should be directly observable with sufficient accuracy and without precise measurement. The target's speed should be mentally assessable by visual comparison with the own ship's speed vector, that is by comparing *WA* with *WO*, which is known. By this means, the target's aspect/course, speed and CPA should be obtainable with sufficient accuracy for the practical assessment of the situation, and only TCPA will require some form of measurement.

7.2.6 The Plot in Special Cases Where No Triangle 'Appears'

As can be seen in the cases shown in Figure 7.5, all of the vector values are represented but in some of the cases their value is zero, while in others, the vectors are superimposed, all of which requires a little more care in the drawing of the plot and its subsequent analysis.

7.3 THE TRUE PLOT

The data given in Example 7.1 in Section 7.2.2 can alternatively be plotted on graph paper or on the chart as a 'true plot' (Figure 7.6), although it has never been in common use.

In this case, the own ship's course line is laid off on the plotting sheet and the own position marked. From this position the bearing and range of the target are laid off. The own ship's position at, say, 3 minute intervals is then marked off along the course line. What will be obtained directly is the target's course and the distance travelled in the plotting interval but, as already stated, perhaps the most important single piece of information available from a plot is the target's CPA, which is not immediately available and must be determined by the following geometrical construction:

a. Lay off a line *WO*, parallel with the course line and of a length equal to the distance that the own ship has travelled in the plotting interval.

b. Join *O* to the current position of the target *A* and extend *OA* to pass O_2, that is

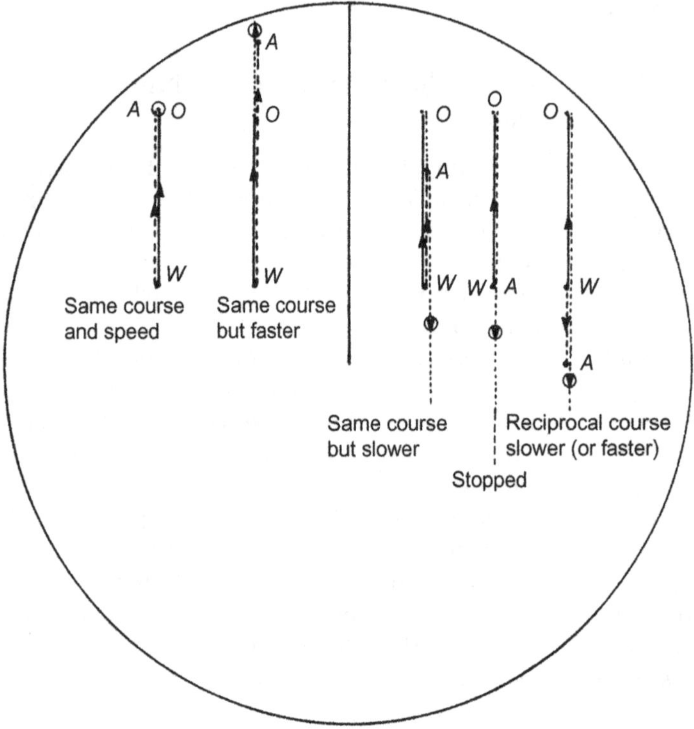

FIGURE 7.5 Plots where no triangle appears.

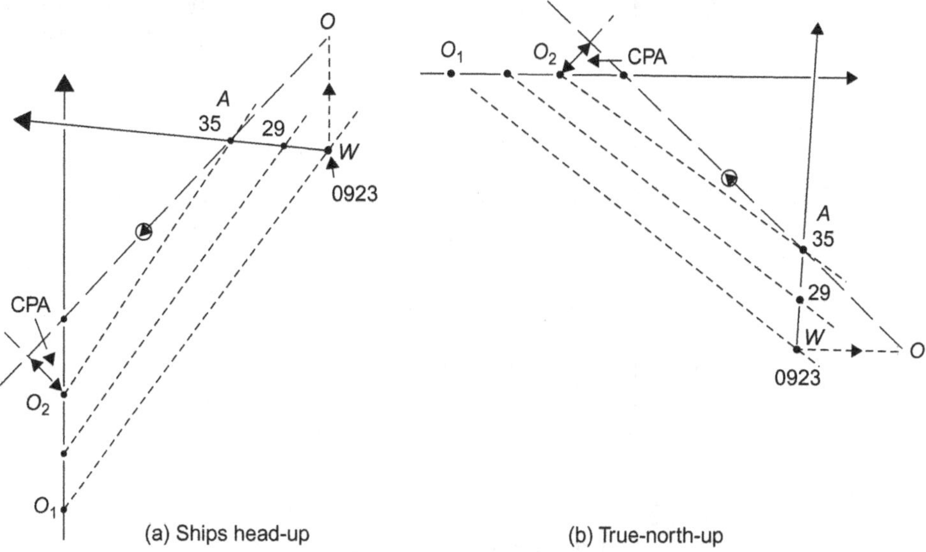

FIGURE 7.6 The true plot for Example 7.1.

the own ship's current position, thus giving CPA. The static situation is identical with that depicted in Figure 7.4 and the required information is extracted in the same way. To this extent, relative and true plots are equally valid, but some practical points are worth noting:

i. The convenience of a bearing scale and range circles on a relative plotting sheet are not possible with a true plot where the own position is continually moving. In plotting the target's position, protractor (or equivalent) and dividers will have to be used and the operation will be somewhat slower.

ii. Each target position will have to be laid off from the own ship's current position, which can be very time-consuming if a number of targets are being plotted.

iii. Any error in plotting the own ship's position (i.e. an error in the course and/or speed input) will result in an error in the directly obtained answer, that is the course and speed of the target. *Note.* If this error is repeated when laying off *WO* the CPA of the target will in fact be correct.

iv. Perhaps the most disconcerting feature of this plotting technique is the temporary nature of the apparent-track line (Figure 7.7). It is essential to appreciate that the 'displayed' CPA relates to *only one* the own ship's position, after which the apparent-track line can be at worst misleading. Where a number of targets are being plotted on the same sheet, apparent-track lines can relate to *different* own ship positions.

In general, where manual plotting has to be carried out, time is of the essence and in this respect the relative plot, is better. For this reason the subsequent treatment will assume the use of the relative plot, except in the now obsolescent case of reflection plotting on a true-motion display. In the latter case, a true plot was easily and quickly produced.

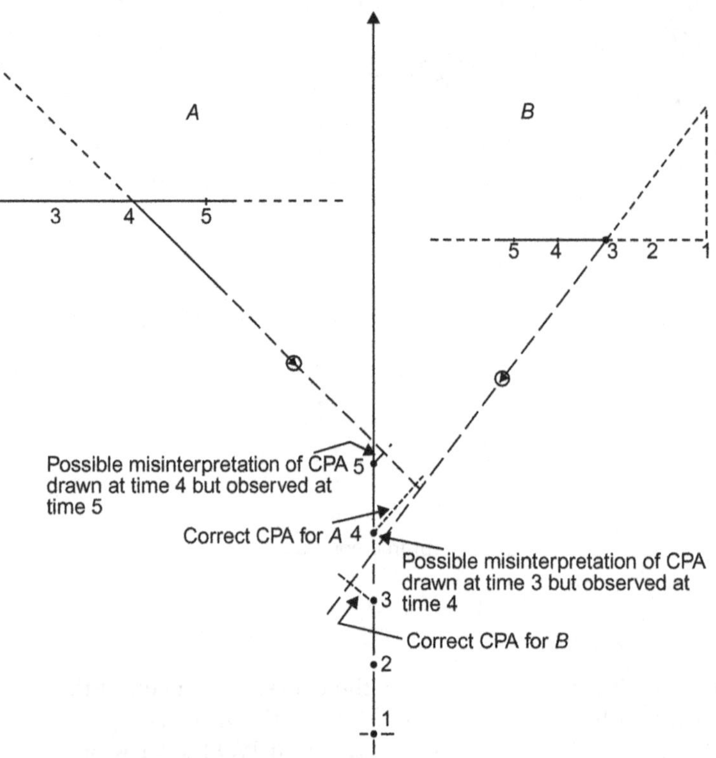

FIGURE 7.7 The temporary nature of the apparent-track line on a true plot.

7.4 THE PLOT WHEN ONLY THE TARGET MANOEUVRES

EXAMPLE 7.2 (USED IN FIGURE 7.8)

Your own ship is steering 310° (T) at a speed of 12 knots. Echoes are observed as follows:

Time	Echo A	Echo B
0923	brg 270° (T) at 9.0 NM	brg 347° (T) at 9.5 NM
0929	brg 270° (T) at 7.5 NM	brg 346° (T) at 8.0 NM
0935	brg 270° (T) at 6.0 NM	brg 344° (T) at 6.5 NM

(a) Determine the CPA, TCPA, course and speed of each target.

Plotting is continued as follows:

0941	brg 270° (T) at 4.5 NM	brg 341° (T) at 5.0 NM
0947	brg 253° (T) at 2.8 NM	brg 350° (T) at 4.0 NM
0953	brg 206° (T) at 1.7 NM	brg 004° (T) at 3.1 NM

(b) Say what action each target has taken. (Your vessel maintains course and speed throughout.)

The apparent-track line *OA* which is produced when determining CPA and TCPA depends upon four factors: the own ship's course and speed, and the target's course and speed. If any of these factors change then the apparent-track line will also change. Changes in the own ship's proper track are known (but in this example remain unchanged) and the resulting apparent track can be predicted (see Section 7.5), but where a change in the apparent track of the target is observed, the procedure set out in Section 7.4.1 should be followed to determine the change in the target's proper track.

Answer

		Target A	*Target B*
(a)	CPA:	Collision	1.0 NM
	TCPA:	24 minutes	25 minutes
	Course:	040° (T)	226° (T)
	Speed:	10 kn	10.5 kn
(b)	Target has altered course only: 50° to starboard		Target has stopped

Note that target *B* is similar to that used in the previous example and although, in this case, the bearings are re-orientated, the plots CPA and TCPA are identical.

7.4.1 The Construction of the Plot (Figure 7.8)

a. The own ship's heading line should have been drawn in.

b. The *OAW* triangle should have been constructed and the basic information extracted.

c. Continue to plot the apparent track of the target, making sure the direction and rate are as predicted in the initial plot.

d. *As soon as a change in the target's apparent track is noticed*, plot more frequently and, when a new steady apparent track is established, choose a position, label it O_1 and insert the time.

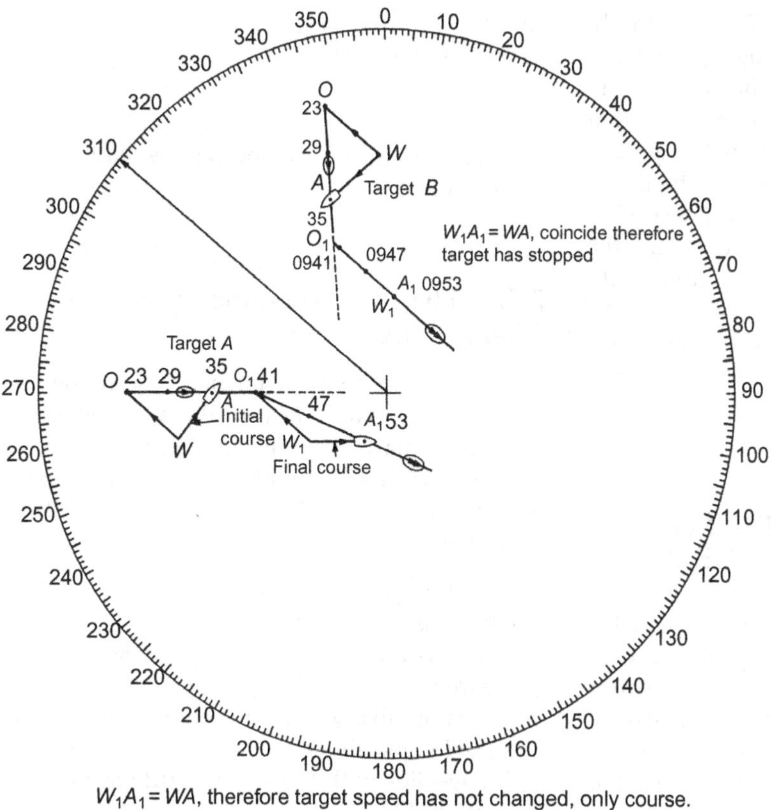

FIGURE 7.8 The plot when only the target manoeuvres (Example 7.2).

e. Plot the new apparent track as if it were a new target. Predict the new CPA and TCPA, also course and speed, from which any changes from the previous proper track may be noted.

7.4.2 The Danger in Attempting to Guess the Action Taken by a Target

Since the change in the target's apparent track can result from virtually any combination of changes in its course and speed, one should always construct a new plotting triangle and never attempt to assume a particular manoeuvre has taken place even though it fits in with what is both logical and desirable under the collision-avoidance rules. In the typical crossing situation shown in Figure 7.9, one might expect the target to give way by altering her course to starboard. While the change in apparent track would seem to indicate that this is what has happened, when the plot is completed it is clearly indicated that the target has in fact reduced her speed. If the observing vessel is subsequently presented with the development of a close-quarters situation with another vessel, the assessment of the appropriate manoeuvre will be dangerously flawed if it assumes the wrong heading for the pre-existing target.

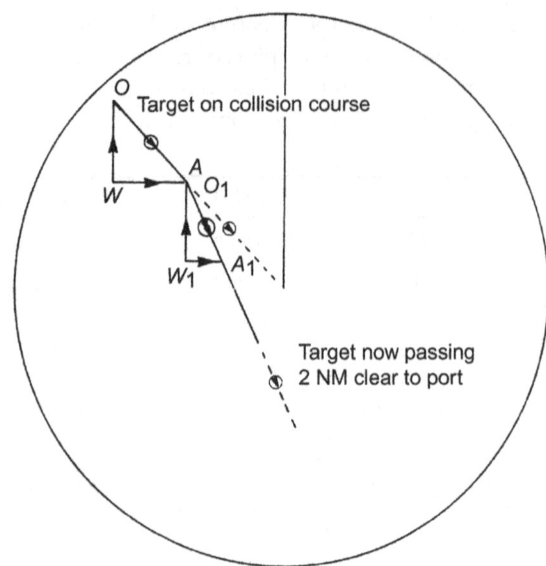

FIGURE 7.9 The danger in attempting to guess the action taken by a target.

7.5 THE PLOT WHEN THE OWN SHIP MANOEUVRES

When it is decided (after assessment of the initial plot) that it is necessary for the own ship to manoeuvre, it is essential to determine the effect of that manoeuvre *prior to* its execution and to ensure that it will result in a safe passing distance. After the manoeuvre has been completed, plotting must be continued to ensure that the manoeuvre is having the desired effect.

7.5.1 The Plot When the Own Ship Alters Course Only

Because of the time taken for a change in speed to have any effect on the apparent-track line, the mariner will frequently select a change in course if it will achieve a satisfactory passing distance. This has some distinct advantages:

a. It is quick to take effect.
b. The vessel retains steerage way.
c. The encounter may be more quickly cleared.
d. It is more likely to be detected if the other vessel is plotting.
e. It complies with the spirit of rule 8c (see Section 7.11.6).

EXAMPLE 7.3 (USED IN FIGURE 7.10)

With the own ship steering 000° (T) at a speed of 12 knots, an echo is observed as follows:

0923	Echo bears	037° (T)	at	9.5 NM
0929	Echo bears	036° (T)	at	8.0 NM
0935	Echo bears	034° (T)	at	6.5 NM

At 0935 it is intended to alter course 60° to starboard (assume this to be instantaneous).

a. Predict the new CPA and TCPA.
b. Predict the new CPA and TCPA if the manoeuvre is delayed until 0941.
c. Predict the range and bearing of the echo at 0953, if the (instantaneous) manoeuvre is made at 0941.

Answer

a. CPA is 4.4 NM, 13 minutes after the alteration, that is at 0948.
b. CPA is 3.6 NM, 10 minutes after the alteration, that is at 0951.
c. Predicted range is 3.8 NM on a bearing of 334° (T).

7.5.2 The Construction of the Plot (Figure 7.10)

a. The original *OAW* triangle should have been drawn.
b. With compasses at *W* and radius *WO*, draw an arc.
c. Draw from *W* a line at an angle to port or starboard of *WO*, equal to the proposed alteration of course.
d. Label as O_1 the point at which this line cuts the arc.
e. Join O_1 to *A* – this now represents the new apparent track in direction and rate.
f. Draw in the ship's *new heading line* and expunge the old heading line.
g. Until the manoeuvre takes place, the target will continue to move down the original apparent-track line. Predict the position of the target at the time at which it is proposed to alter the own ship's course and label it O_2.
h. Draw O_2A_2 parallel with and equal to O_1A and produce if necessary to find the new CPA.

Note:

a. The prediction is based on the assumption that the target will maintain its course and speed. If the target also manoeuvres then the target will not follow the predicted apparent track in both direction and rate, and a new plot will have to be started.
b. It is assumed that the alteration of course is instantaneous. For a more practical prediction, when manoeuvring at sea, to obtain a more realistic indication of the new CPA, one could predict the effect of the manoeuvre and check that it is acceptable. If so, bring the ship round to the new course and, when steadied on the new course, plot the position of the target and transfer the new apparent track, O_2A_2, through this position.

7.5.3 The Plot When the Own Ship Alters Speed Only

EXAMPLE 7.4 (USED IN FIGURE 7.11)

With the own ship steering 000° (T) at a speed of 12 knots, an echo is observed as follows:

0923	Echo bears	037° (T)	at	9.5 NM
0929	Echo bears	036° (T)	at	8.0 NM
0935	Echo bears	034° (T)	at	6.5 NM

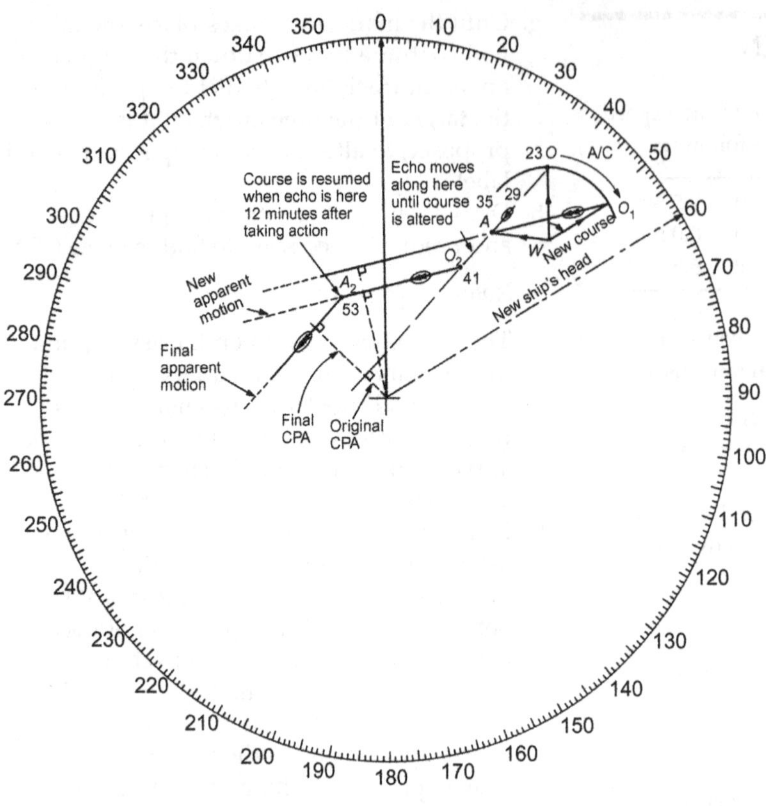

FIGURE 7.10 The plot when the own ship alters course only (Example 7.3).

At 0935 it is intended to reduce speed to 3 knots (assume this to be instantaneous).

a. Predict the new CPA and TCPA.

b. Predict the new CPA and TCPA if the manoeuvre is delayed until 0941 (assume an instantaneous reduction of speed).

c. Predict the bearing and range of the echo at 0953, if the manoeuvre is made at 0941.

Answer

a. New CPA is 4.8 NM, 24.5 minutes after the alteration of speed, that is at 0959.5.

b. New CPA is 3.9 NM, 19 minutes after the alteration of speed, that is at 1000.

c. At 0953 the bearing should be 007° (T), range 4.0 NM.

7.5.4 The Construction of the Plot (See Figure 7.11)

a. The original *OAW* triangle should have been drawn.

b. From *W*, measure off in the direction *WO*, the distance at the new speed that the own ship will steam in the plotting interval and label the point O_1. Join O_1 to *A*. This represents, in direction and rate, the new apparent track.

c. Until the proposed alteration takes place, the target should continue to move along the old apparent-track line. Predict the position of the target at the time it is proposed to alter the own ship's speed and label it O_2.

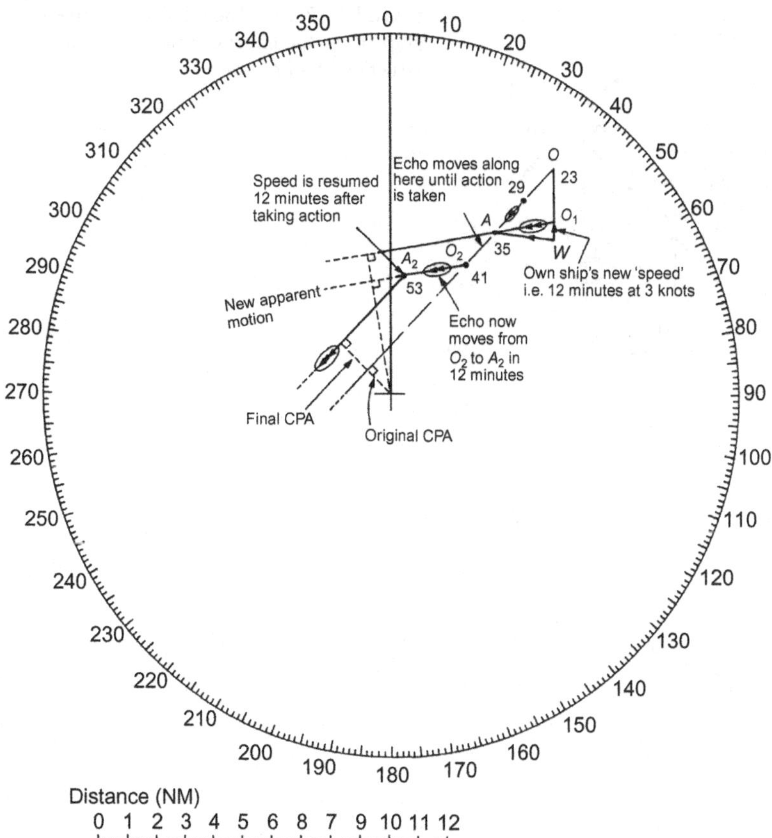

FIGURE 7.11 The plot when the own ship alters speed only (Example 7.4).

d. Lay off O_2A_2 parallel with and equal to O_1A and extend it if necessary to find the new CPA.

Note:

a. This prediction is based on the assumption that the target will maintain its course and speed and that the own change in speed is instantaneous.

b. It should be appreciated that when changing speed, especially in large vessels, it can take a considerable time to achieve the required new speed. Because of this, the new apparent-track line, O_2A_2, will be much nearer the own ship than predicted in the example above.

This is demonstrated in Figure 7.12 where a ship's engines are stopped after a 12 minute plotting triangle has been constructed. The own ship's head reach (stopping distance) is 2.3 NM. The limit line constructed on Figure 7.12 will represent a line the target will not cross, but will eventually meet as the own ship comes to a complete stop.

7.5.5 The Use of 'Stopping Distance' Data in the Form of Tables, Graphs and Formulae

7.5.5.1 Stopping Distance Tables

In the radar period before the arrival of ARPA, several organizations produced stopping

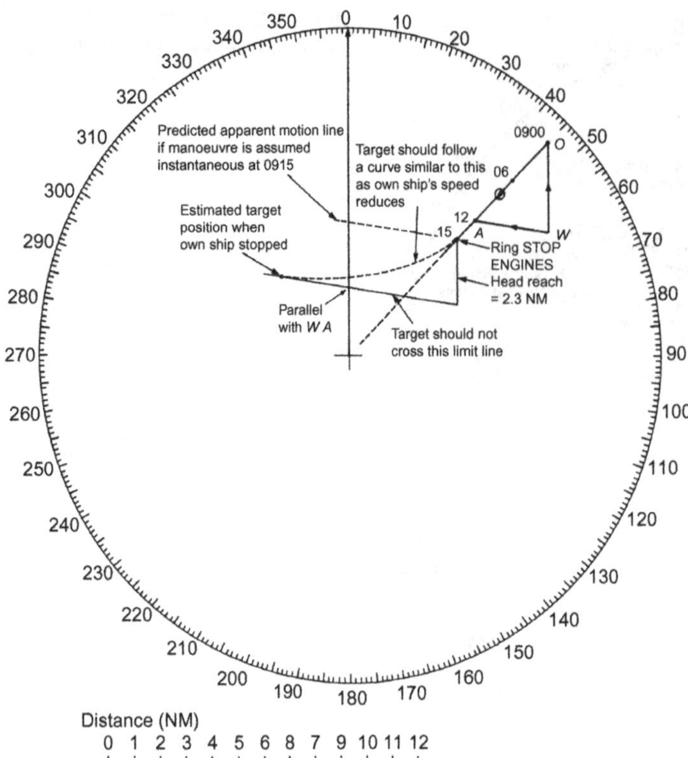

FIGURE 7.12 The plot when the own ship stops: the effect of head reach and construction of a limit line.

distance tables to enable the manual plotter to estimate the stopping distances. The most notable of these was produced by the Honourable Company of Master Mariners in London, which included tables of stopping distances for a variety of ship types and tonnages.

7.5.5.2 Manoeuvring Data Which Should Be Available on Board

The International Maritime Organization (IMO) recommends in Resolution A.601 (15) that booklets containing manoeuvring information should be on board and available to the masters of large ships and also on board smaller ships, especially those carrying dangerous chemicals and gas carriers.

The booklet should contain manoeuvring data and/or diagrams (preferably the latter), typical of which are the turning circle and stopping curve diagrams (Figures 7.13 and 7.14).

The form of the booklet should be such that additional information can be added from 'trials' and during the normal course of the voyage. Shipmasters are encouraged to add to this basic information as they gain experience in the handling of the vessel in conditions not covered by the original data.

The data should be presented in the form of pilot cards, wheelhouse posters and manoeuvring booklet.

7.5.5.3 The Use of Stopping Curves Drawn Up from Data Recorded During Trials

Consider a vessel proceeding at 16 knots. If the engines are put to *stop*, it will take some

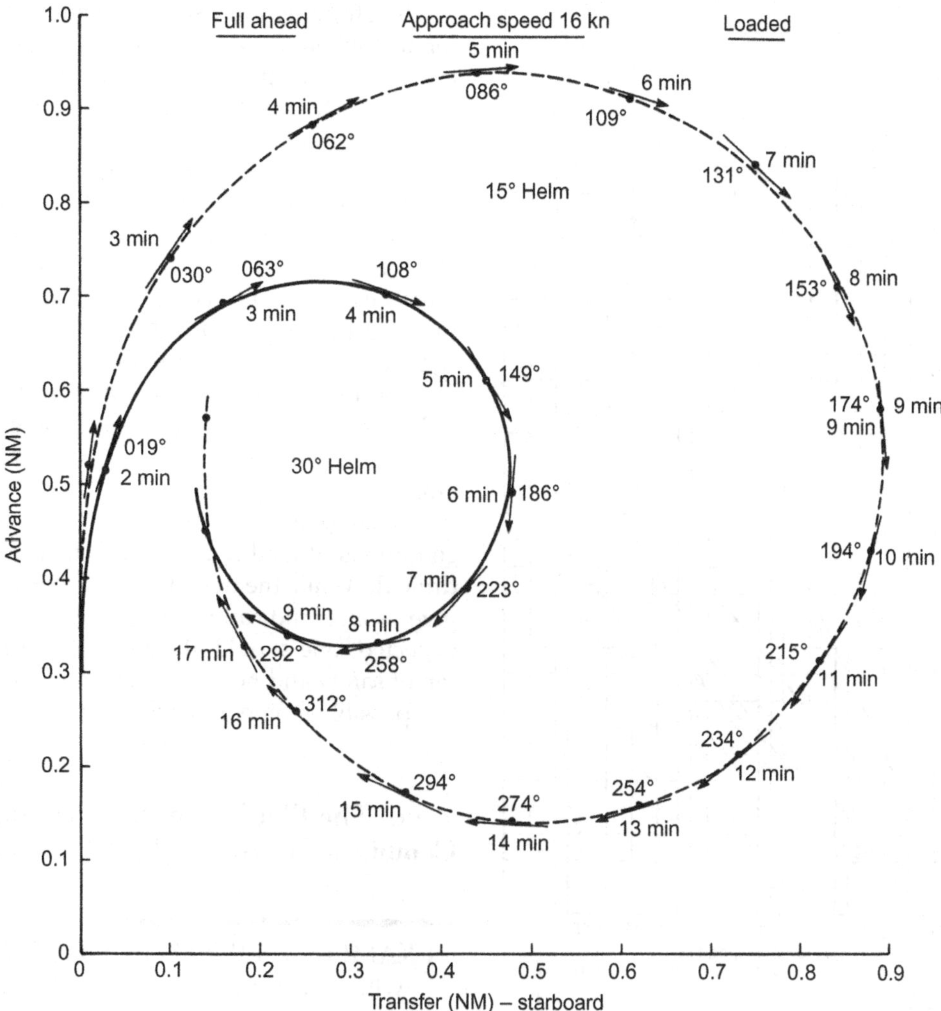

FIGURE 7.13 **Turning circles.** The response actually experienced may differ from that shown if any of the following conditions is not met:

Calm weather	Displacement 97,440 tonnes
No current or tidal stream	Draught forward 9.14 m
Water depth at least 2 × vessel's draught	Draught aft 9.14 m
Clean hull	

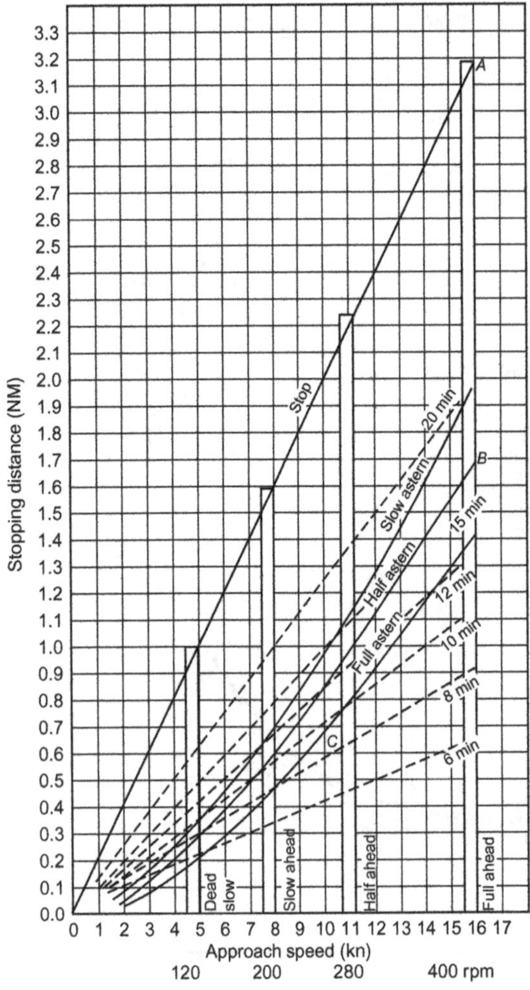

FIGURE 7.14 Stopping curves.

3.2 NM to run off the way, that is the readings associated with *A*, Figure 7.14.

If the engines are put to *half astern*, the vessel will be stopped in the water in some 17 minutes having covered some 1.7 NM in running the way off, that is the readings associated with *B*.

If proceeding at 10 knots and full astern is demanded, the vessel will be stopped in the water in 10 minutes having covered some seven cables while running the way off, that is the readings associated with C.

The stopping curves are intended primarily for use when the engines are to be put astern in order to stop the vessel in the water, for example when picking up a pilot, anchoring or in an emergency.

Note: The curves assume that the environmental and loaded conditions, etc. are the same as when the trials data was obtained. They should only be used as a rough guide as to what might be expected. It should also be borne in mind that as speed reduces, a critical speed will be reached when the vessel will no longer answer to the helm. When it is necessary to reduce speed even more quickly, other techniques such as rudder cycling may have to be employed. The prediction of how the speed may change or the vessel may turn under any conditions of loading, weather, etc. can be very difficult. While the documents (see Section 7.5.5) give some indication as to what may be expected, one should always work with a margin of safety and not leave manoeuvres until the last possible moment which theory suggests.

7.5.6 The Plot When the Own Ship Combines Course and Speed Alterations

EXAMPLE 7.5 (USED IN FIGURE 7.15)

With the own ship steering 000° (T) at a speed of 12 knots, an echo is observed as follows:

0923	Echo bears 037° (T) at 9.5 NM
0929	Echo bears 036° (T) at 8.0 NM
0935	Echo bears 034° (T) at 6.5 NM

At 0935 it is intended to alter course 40° to starboard and reduce speed to 6 knots (assume this to be instantaneous).

a. Predict the new CPA and TCPA.

b. Predict the new CPA and TCPA if the manoeuvre is delayed until 0941.

c. Predict the range and bearing of the echo at 0953 if the manoeuvre was made at 0941.

Answer (Figure 7.15)

a. CPA is 4.4 NM, 20 minutes after the manoeuvre. *Note*: The new apparent-track direction is the same as that when course only was altered (see Figure 7.10) but the new rate, O_1A, is much slower.
b. CPA is 3.6 NM, 14.5 minutes after the manoeuvre at 0941, that is at 0955.5.
c. Predicted range is 3.6 NM on a bearing of 355° (T).

7.5.7 The Plot When the Own Ship Resumes Course and/or Speed

When taking action to avoid a close-quarters situation, the action should be *substantial enough* and *held for long enough* to show up in the plotting of radar observers on other vessels.

In order to determine the effect of resuming the own ship's course and/or speed:

a. The original *OAW* triangle should have been drawn.
b. The apparent track after manoeuvring should have been predicted and, provided that the target has maintained its course

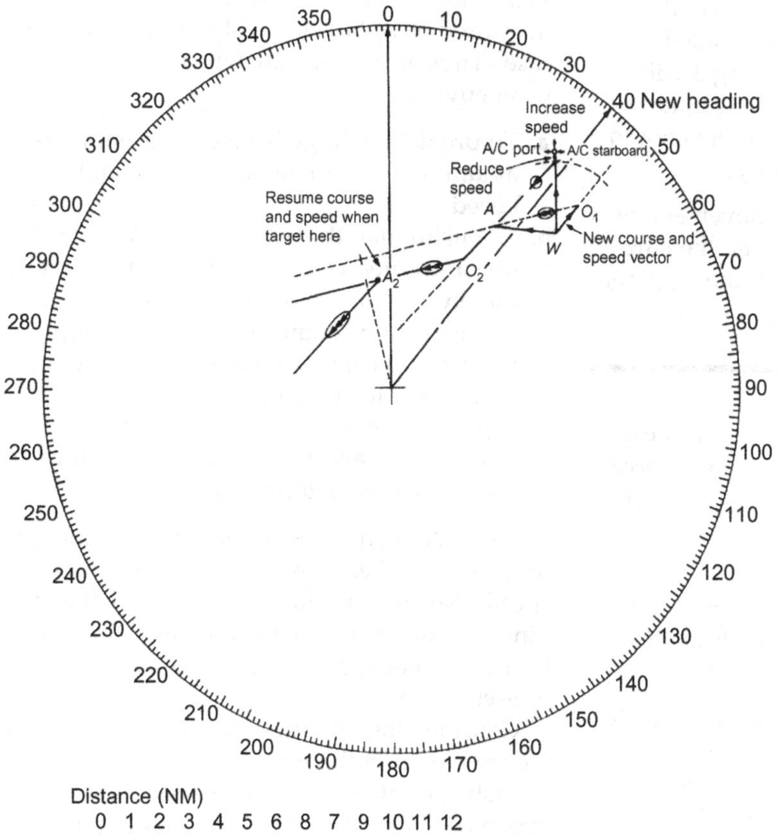

FIGURE 7.15 The plot when the own ship combines course and speed alterations (Example 7.5).

and speed, the target will follow the new apparent-track line, O_2A_2.

c. If the target still maintains its course and speed and the own ship resumes her original course and/or speed, the direction and rate of the final apparent track will be the same as in the original *OAW* triangle. Thus, from the target's predicted position at the time of resumption, draw O_3A_3 parallel and equal to *OA* and produce if necessary to find the new CPA after resumption.

d. If it is required to find the time to resume course/speed so that the CPA will not be less than, for example, 4 NM, lay off a line parallel to *OA* and tangential to the 4 NM range circle. The point at which this final apparent-track line crosses the second apparent-track line should be labelled O_3. Predict the time at which the target will reach O_3. This will be the earliest time to resume course/speed if the target is not to come within the required CPA.

Note: When predicting the movement of a target along an apparent-track line, be sure to use the appropriate rate along the apparent track.

EXAMPLE 7.6

Using Examples 7.3 and 7.4, predict in each case the CPA if the original course and speed respectively are resumed at 0953 (see Figures 7.10 and 7.11).

Answer

a. If course is resumed at 0953, CPA = 3.5 NM.

b. If speed is resumed instantaneously at 0953, CPA = 2.4 NM.

Note: If the order for an increase in speed is given at 0953, then by the time the vessel has worked up to the required speed the target will have passed in excess of the 2.4 NM CPA.

7.5.8 The Plot When Both Vessels Manoeuvre Simultaneously

It has been assumed until now that only one of the two vessels in the encounter manoeuvres. In such cases, the resulting apparent track can be predicted when the own ship manoeuvres, or resolved when only the target manoeuvres. When ships of about the same size meet, it is not uncommon for them to commence plotting at about the same time and subsequently manoeuvre simultaneously or nearly so. If each vessel has plotted initially and completed the construction showing the effect of an own ship manoeuvre, then the target in each case will not follow the predicted apparent track. In some cases this will result in each 'target' clearing by a greater distance than predicted, but this will not always be the case. Thus it is essential when the own ship manoeuvres:

a. To predict the target's new apparent track, assuming that it maintains course and speed.

b. To ensure that the manoeuvre is having the (desired) predicted effect. It is not sufficient merely to assume that, because the target 'appears' to be going clear, it is moving in the direction *and rate* predicted. Where the predicted direction and rate are not as expected, a new plot must be started immediately and the action taken by the target must be determined.

The latter can have important consequences in deciding when it is safe to resume course/speed. No longer can it be assumed that the 'final' apparent track will be parallel and equal to the original apparent track, as was assumed in Section 7.5.7.

Consider the situation in fog, as set out in Figure 7.16. If both vessels manoeuvre simultaneously or at nearly the same time, *B* as observed from *A* will move along the new

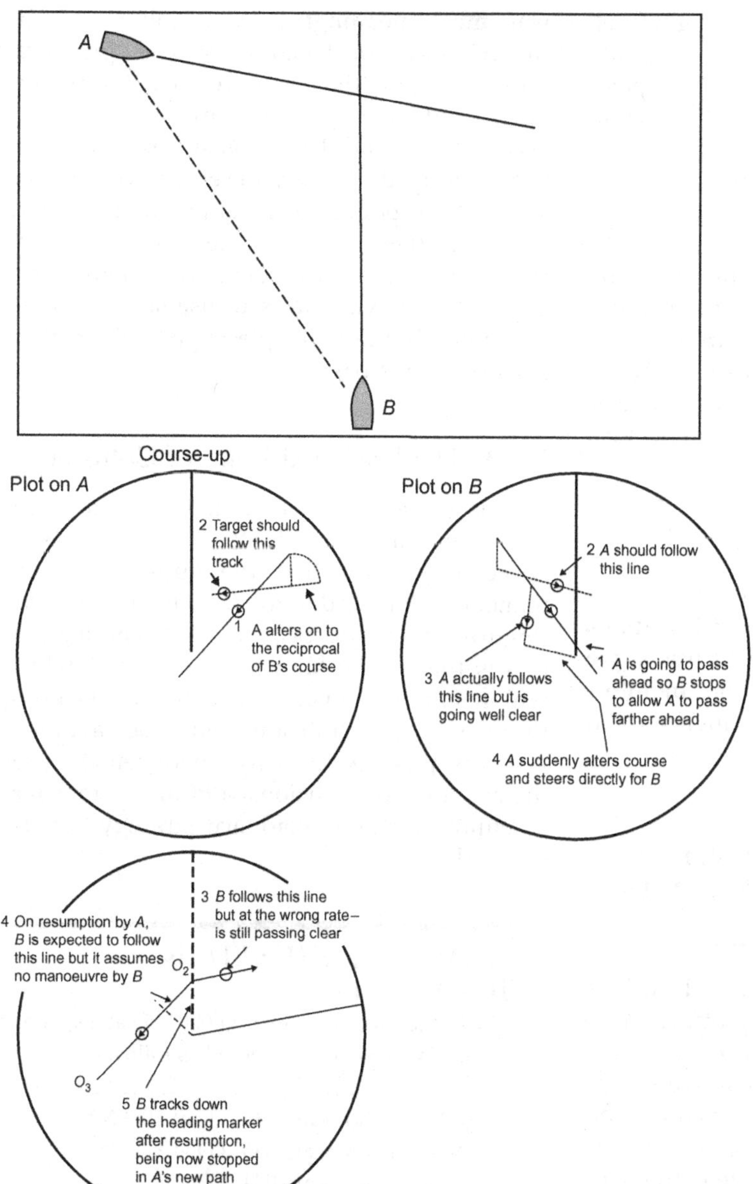

FIGURE 7.16 The danger of not determining the target's actual manoeuvre when apparently passing clear.

apparent-track line (but at the wrong rate, which is now much slower) and so appear to be going clear. No longer will *B* move along O_2O_3 when *A* resumes; it will move down parallel to the heading marker, since *B* is now stopped in the water.

If the apparent track of *B* had not been predicted, monitored and reappraised after the manoeuvre was detected, *A* might well have resumed course when *B* was at O_2, only to find a vessel stopped directly in its path. This would have manifested itself by *B* coming down *A*'s heading marker and thus requiring sudden emergency action, possibly by both vessels at a very late stage in the encounter.

If systematic observation and interpretation is not carried out on *B*, as the way is run off, the relative-motion plot would start to relate to the true situation, in which case *B* would have observed a vessel steering to pass clear down the port side but then appearing to turn to port and steer directly towards it. Having very little way on at this stage, there is very little that *B can* do to avoid the close-quarters situation.

The encounter considered above is frequently carried out with the best of intentions but it is clear that, where manoeuvres are executed without proper prediction, monitoring and reappraisal, the consequences can be dire.

7.6 THE THEORY AND CONSTRUCTION OF PPCs, PADs, SODs AND SOPs

The harnessing of computers and ARPAs to assist in the resolution of the plotting triangle led to the question of whether some assistance could be given which would be of a higher order than that provided by the *trial manoeuvre* facility on the ARPA (see Section 4.4.3). Without going to the extreme of an *expert system* (see Section 7.12), some attempts were made to give a graphical representation of the encounter geometry which would assist the mariner in selecting an acceptable manoeuvre. The major problem was to devise a way in which this 'advice' should be presented to the mariner so that it was most meaningful. What follows is the theory and construction of the graphical representations that were devised; the philosophy and shortcomings are discussed in Sections 4.11 and 4.12. It is stressed that it is not envisaged that an observer would construct these points or areas manually. Where available, they would be generated by computer graphics. However, it is essential that any observer who elects to use such facilities understands the principles upon which their construction is based.

7.6.1 The Predicted Point of Collision

This is the point or points towards which the own ship should steer at her present speed (assuming that the target does not manoeuvre) in order for a collision to occur (Figure 7.17(a) and Section 4.11). The logic of this is that if one knows the course(s) which will result in collision then, by not steering those courses, collisions will be avoided. Also, by picking one's way between the predicted point of collisions (PPCs), a more far-reaching collision-avoidance strategy can be evolved.

EXAMPLE 7.7 (USED IN FIGURE 7.17)

With the own ship steering 000° (T) at a speed of 10 knots, an echo is observed as follows:

0923	Echo bears 037° (T) at 10.3 NM
0929	Echo bears 036° (T) at 8.5 NM
0935	Echo bears 034° (T) at 6.7 NM

Determine the bearing and range of the PPC(s).

Answer (Figure 7.17(b))

a. P_1 bearing 337° at a range of 4.4 NM.
b. P_2 bearing 270° at a range of 18.0 NM.

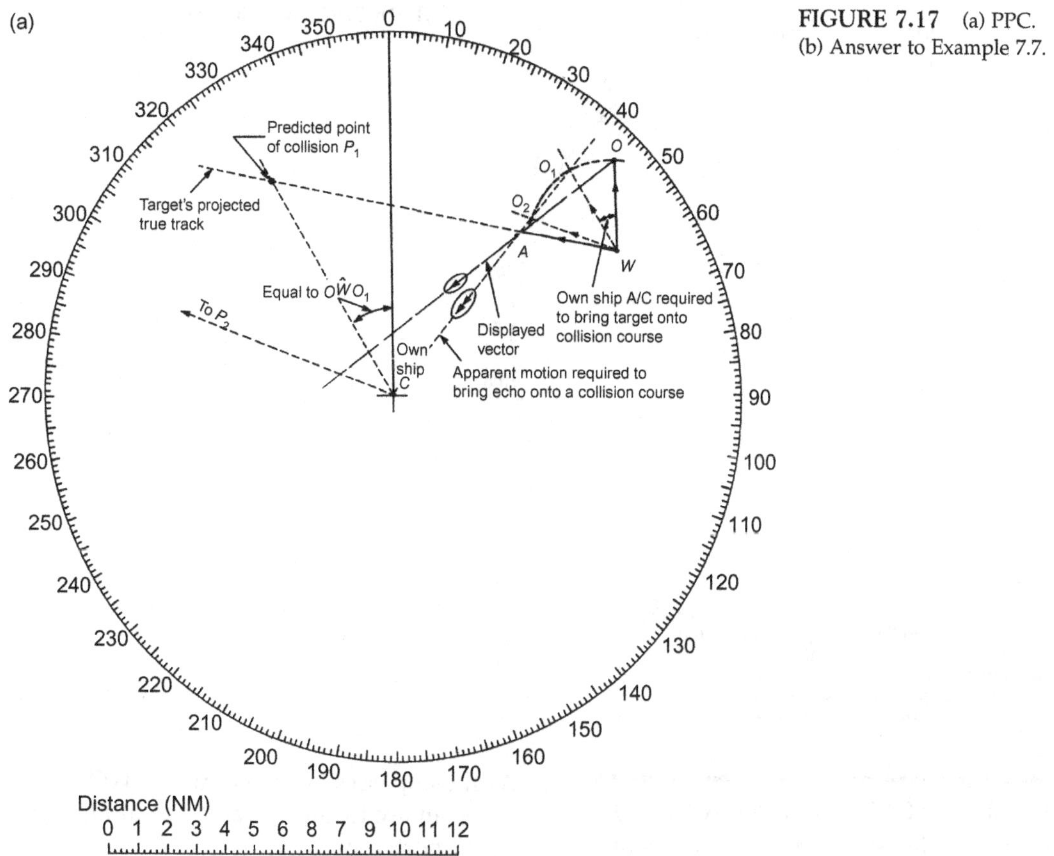

FIGURE 7.17 (a) PPC. (b) Answer to Example 7.7.

7.6.2 The Construction to Find the PPC

For a full explanation of PPCs, refer to Section 4.11. The plot is constructed as follows:

a. Plot the target and produce the basic triangle.
b. Join the own ship position '*C*' to the target's position '*A*' and extend beyond '*A*'.
c. With compasses at *W* and radius *WO*, scribe an arc to cut *CA* produced at O_1 or, if the own ship is the slower (i.e. $WO < WA$), at O_1 and O_2.
d. Join WO_1 (and WO_2).
e. Draw CP_1 parallel to WO_1 to cut *WA* produced at P_1 (and CP_2 parallel to WO_2 to cut *WA* produced at P_2).
f. P_1 (and P_2) is the PPC.

For a clearer appreciation of the determination of the PPC, Example 7.7 should be drawn out full size and to scale on a plotting sheet.

7.6.3 The Predicted Area of Danger

The PPC gives no indication of the course which needs to be steered to clear the PPC by some specific distance; to this end, the predicted area of danger (PAD) was devised (Figure 7.18(a)). A further improvement is that the target is connected to its PAD on the display, which was not the case with the PPC.

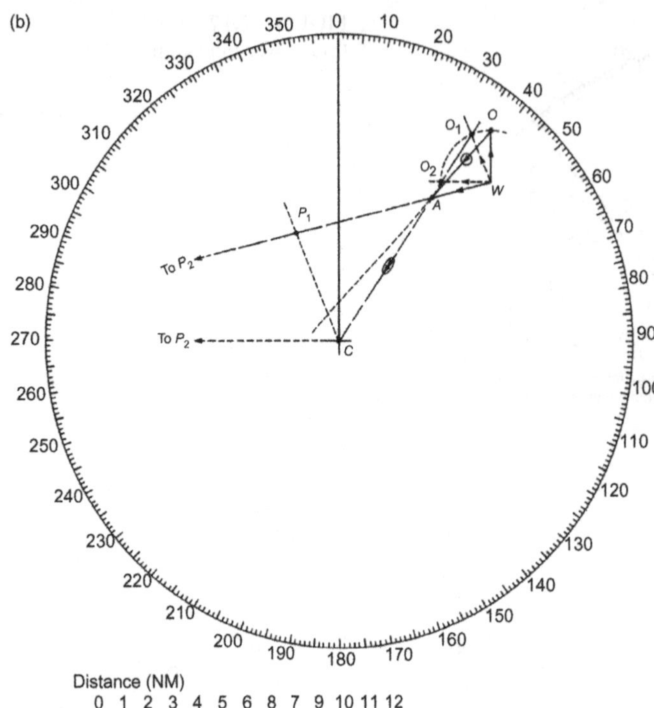

FIGURE 7.17 (Continued)

EXAMPLE 7.8 (USED IN FIGURE 7.18)

With the own ship steering 000° (T) at a speed of 12 knots, an echo is observed as follows:

1000	Echo bears 045° (T) at 10.0 NM
1006	Echo bears 045° (T) at 8.5 NM
1012	Echo bears 045° (T) at 7.0 NM

Plot the target and draw in the PAD for a 2.0 NM clearing.

Answer See Figure 7.18(b).

7.6.4 The Construction of the PAD

a. Plot the target and produce the basic triangle.

b. Draw lines AT_1 and AT_2 from the target's position A, tangential to a circle of radius equal to the required CPA. Extend beyond A.

c. With compasses at W and radius WO, scribe an arc to cut T_1A and T_2A at O_1 and O_2, respectively.

d. Join WO_1 and WO_2. These represent the limiting courses to steer to clear the target by the required CPA.

e. Draw CE_1 and CE_2 parallel to WO_1 and WO_2, respectively, to cut WA produced at E_1 and E_2, respectively.

f. At the mid-point of E_1E_2, draw the perpendicular to E_1E_2 and extend it in both directions. In each direction, mark off the 'required CPA' and label the points E_3 and E_4.

g. Draw in the ellipse (Figure 7.18(a)) passing through the points E_1, E_2, E_3 and E_4 (or the hexagon as indicated in Figure 7.18(b)). In the interest of simplicity, the ellipse was replaced by the hexagonal PAD in later equipment.

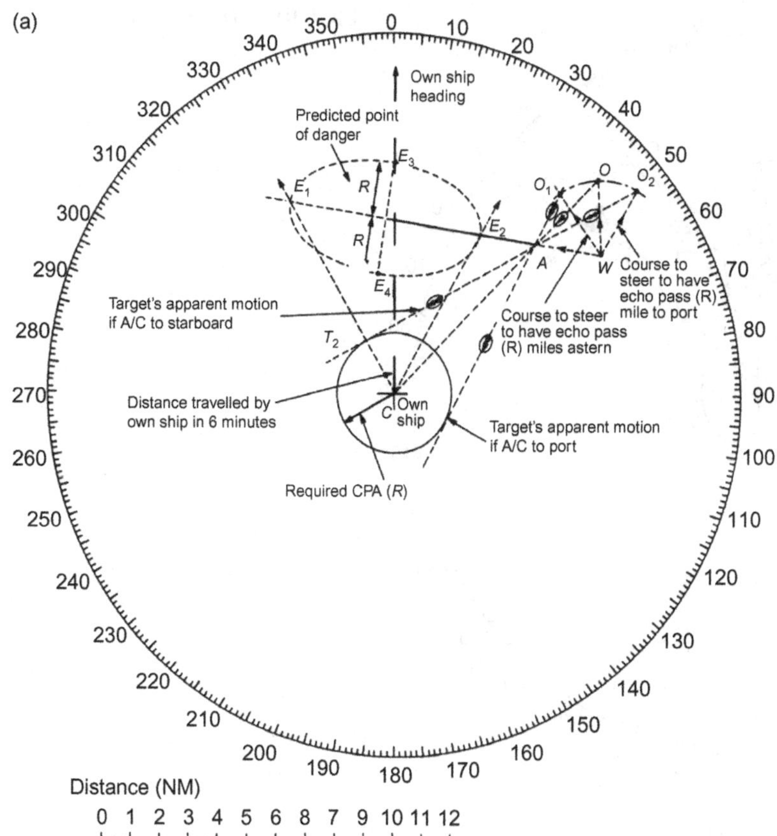

FIGURE 7.18 (a) The PAD (b) Answer to Example 7.8.

For a clearer appreciation of the PAD, Example 7.8 should be drawn out full size and to scale on a plotting sheet.

7.6.5 The Sector of Danger

In this approach, a line *CH*, equal in length and direction to *OW*, is drawn from the centre of the plotting sheet. This line is referred to as the *own ship vector.* The sector of danger is an area constructed to provide a chosen passing distance. If the remote end of the own ship vector lies outside the sector of danger (SOD), a passing distance greater than the chosen value will be achieved, provided that the target does not manoeuvre.

EXAMPLE 7.9 (USED IN FIGURE 7.19)

With the own ship steering 000° (T) at a speed of 12 knots, an echo is observed as follows:

1000	Echo bears 045° (T) at 11.0 NM
1010	Echo bears 045° (T) at 9.0 NM
1020	Echo bears 045° (T) at 7.0 NM

Plot the SOD and suggest the minimum alteration of course to starboard at 1020 (instantaneous) which will achieve a 2 NM CPA if the own ship's speed is maintained throughout.

Answer. If the own ship alters course to 027° (T) at 1020, then the target's new apparent track will be along AT_1 (Figure 7.19).

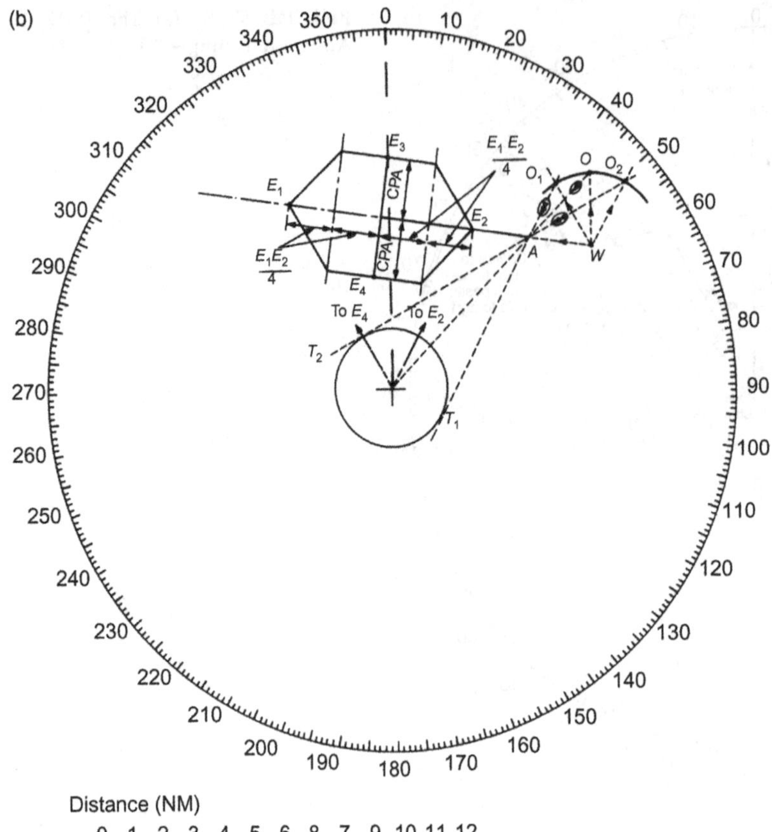

FIGURE 7.18 (Continued)

7.6.6 The Construction of a Sector of Danger

a. Plot the target for a set period.

b. Draw the own ship's vector for the same period – *CH*, equal to the distance that the own ship will travel in the plotting interval, that is equal to *OW*.

c. Draw *HS* parallel and equal to *OA*.

d. Draw AT_1 and AT_2 tangential to the 'required CPA' range circle. (*Note*: If it were intended to delay the manoeuvre, then the tangents would be drawn from the predicted position of the target at the manoeuvre time.)

e. Draw parallel lines to T_1A and T_2A through *S*.

f. Any manoeuvre which takes *H* outside the shaded sector will achieve a CPA in excess of the required CPA.

g. Swing *CH* until it just touches the sector boundary at H_1. This is the minimum alteration of course to starboard which the own ship can make at the present speed to keep the target to a 2 NM CPA.

Alternatively, 'speed' could have been reduced to CH_2 which would also have taken the own ship's vector just outside the sector.

While this will take care of the geometry, it is essential to ensure that the manoeuvre

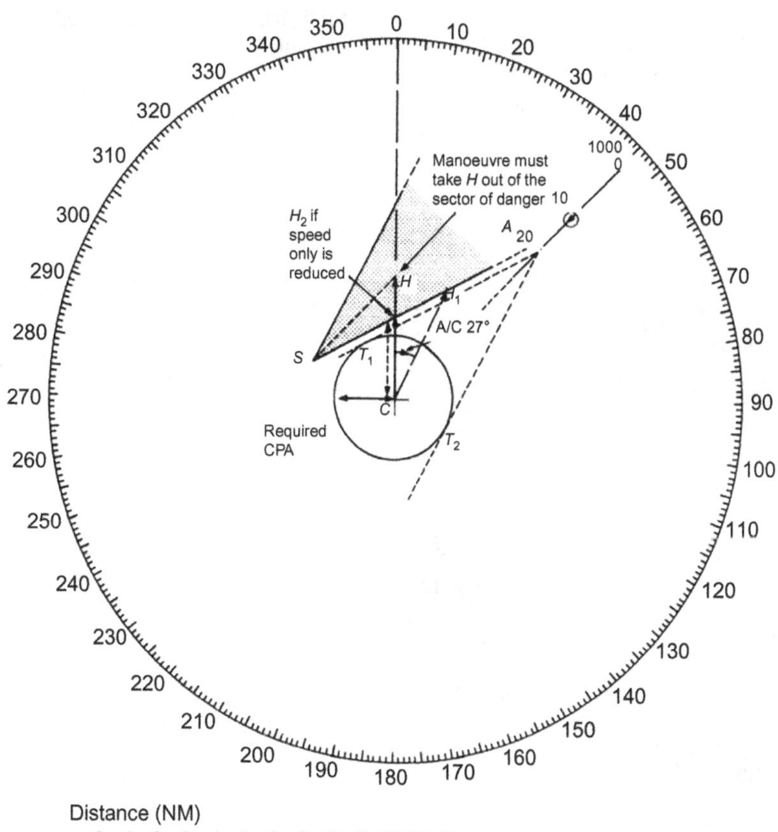

FIGURE 7.19 The SOD – Example 7.9.

selected is still in accordance with the collision-avoidance rules and the practice of good seamanship (see Section 7.11).

7.6.7 The Sector of Preference

This relates to a stationary target or navigation mark and is a sector into which one must steer to achieve a passing distance within prescribed limits (Figure 7.20).

EXAMPLE 7.10 (USED IN FIGURE 7.20)

With the own ship steering 310° (T) at a speed of 15 knots, a light vessel, identified by its racon, is observed as follows:

0900	Echo bears 296° (T) at 10.0 NM
0906	Echo bears 290° (T) at 8.6 NM
0912	Echo bears 282° (T) at 7.4 NM

It is intended to manoeuvre at 0924.

a. Construct the sector of preference (SOP) if the light vessel is required to pass at not less than 2 NM nor more than 4 NM to port.
b. Determine the course alteration which would be needed to achieve a CPA of 2 NM to port (assume the alteration to be instantaneous).

Answer. (a) See Figure 7.20. Alter course to 258° (T), that is 52° to port.

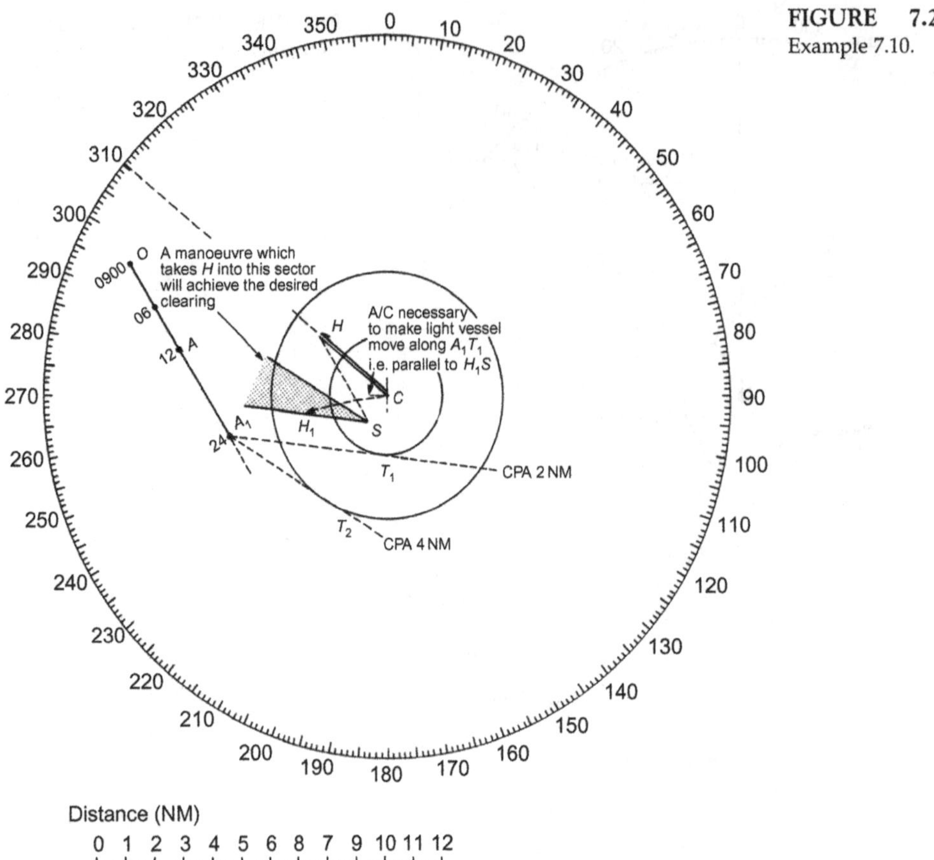

FIGURE 7.20 The SOP – Example 7.10.

7.6.8 The Construction of an SOP

a. Plot the stationary target for a set period of time.

b. Draw the own ship's vector for the same period, *CH*, equal to the distance that the own ship will travel in the plotting interval, that is equal to *OW*.

c. Draw *HS* parallel and equal to *OA*.

d. Plot the position of the target at the time at which it is intended to manoeuvre, A_1.

e. From A_1, draw tangents to range limits for clearing the target, A_1T_1 and A_1T_2.

f. Draw parallel lines to T_1A_1 and T_2A_1 through *S*.

g. Any manoeuvre which takes *H* into the shaded sector will result in the target clearing within the limits for which the sector was drawn (provided that any tidal effect remains constant). This could be achieved by an alteration of course or speed, or both. If *CH* is swung until it intersects either of the limits of the shaded sector then that will be the course required to achieve the specific CPA.

7.7 THE PLOT IN TIDE

EXAMPLE 7.11 (USED IN FIGURE 7.21)

With the own ship steering 310° (T) at a speed of 15 knots, the echo from a light vessel fitted with a racon is observed as follows:

0900	Echo bears 346° (rel) at 10.0 NM
0906	Echo bears 340° (rel) at 8.6 NM
0912	Echo bears 332° (rel) at 7.4 NM

a. Find the set and rate of the tide.

b. It is required that your ship should pass 2 NM off, leaving the light vessel to port.

i. If action is taken instantaneously at 0912, find the alteration of course which would be necessary.

ii. If instead, the action is to be taken at 0924 instead of 0912, find the alteration of course which would then be necessary.

Answer. See Figure 7.21.

a. Set = 032° (T) at 5 knots

b. **i.** Course to steer at 0912 is 278° (T)

ii. Course to steer at 0924 is 260° (T)

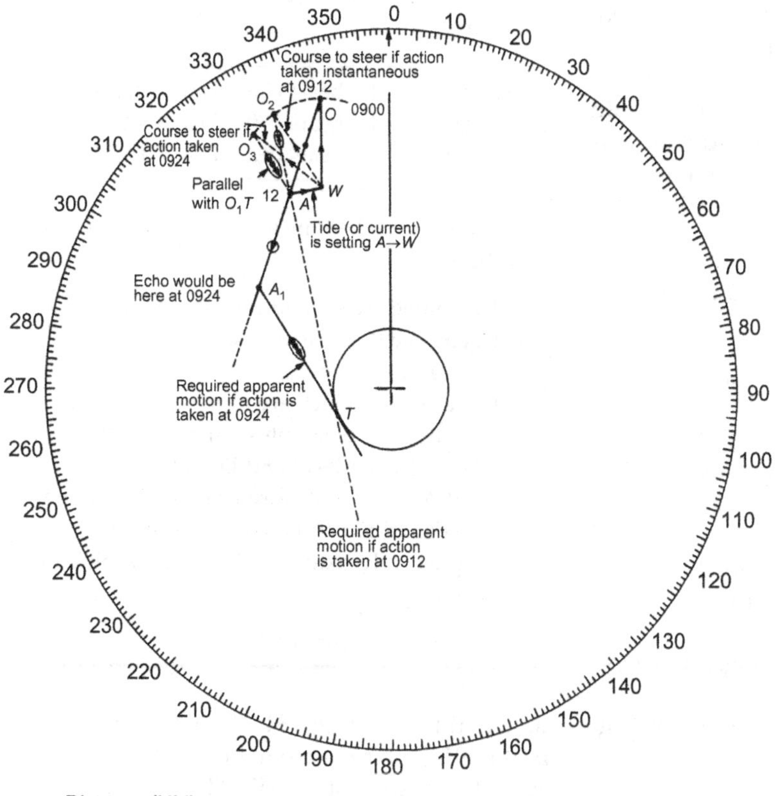

FIGURE 7.21 The plot in tide – Example 7.11.

7.7.1 The Construction of the Plot

a. Establish the identity of the target beyond all doubt (see Section 8.2). Plot the target and construct the basic triangle.

b. *W* and *A* should coincide if there is no tide. If they do not, then *A* to *W* is the direction in which the tide is setting. Note that this is the reverse of what is normally deduced from a plot but since the target cannot be moving, it is the ship which must be experiencing set. The distance *WA* is the drift of the tide in the plotting interval and it is from this that the tidal rate can be calculated.

Note: When plotting other (moving) targets while experiencing tide, it is their movement through the water which is evaluated. It is precisely this which needs to be known when applying the collision-avoidance rules, and so it is essential that *no* attempt should be made to apply the effect of tide or current to their courses as determined from the plot.

7.7.2 The Course to Steer to Counteract the Tide

a. Plot the target and determine the tide or current.

b. Predict the position of the target at the time at which it is intended to manoeuvre.

c. Draw in the apparent track which is required, for example A_1T.

d. Draw in a line in the reverse direction through *A*, i.e. AO_2 (and AO_3 for the manoeuvre at 0924).

e. Swing *WO* to cut the line through *A* at AO_2 (and AO_3).

f. WO_2 is the new course to steer (and WO_3 if the manoeuvre is delayed).

Note: All of the above assume instantaneous manoeuvres. When handling large ships in tide, due allowance will need to be made for their handling characteristics, especially when a speed change is demanded.

7.7.3 The Change of Course Needed to Maintain Track When Changing Speed in Tide

When proceeding while experiencing tide, any change in the own ship's course and/or speed and/or tidal set and/or rate will affect the vessel's ground track.

EXAMPLE 7.12 (USED IN FIGURE 7.22)

While steering 090° (T) at a speed of 16 knots, a beacon marking the anchorage is observed as follows:

0900	Echo bears 076° (T) at 10.0 NM
0906	Echo bears 070° (T) at 8.6 NM
0912	Echo bears 062° (T) at 7.4 NM

It is intended to anchor 2 NM due south of the beacon.

a. Determine the set and rate of the tide.

b. Determine the course to steer at 0912 to counteract the tide.

c. If the speeds which are represented by the engine room telegraph settings of Half Ahead, Slow Ahead and Dead Slow Ahead are 9, 6 and 3 knots, respectively, determine the courses to be steered as the speed is progressively reduced when approaching the anchorage.

Answer. See Figure 7.22.

a. The tide is setting 184° (T) at 5 knots.

b. At 0912, alter course to 060° (T).

c. Half Ahead – steer 052° (T).

Slow Ahead – steer 035° (T).

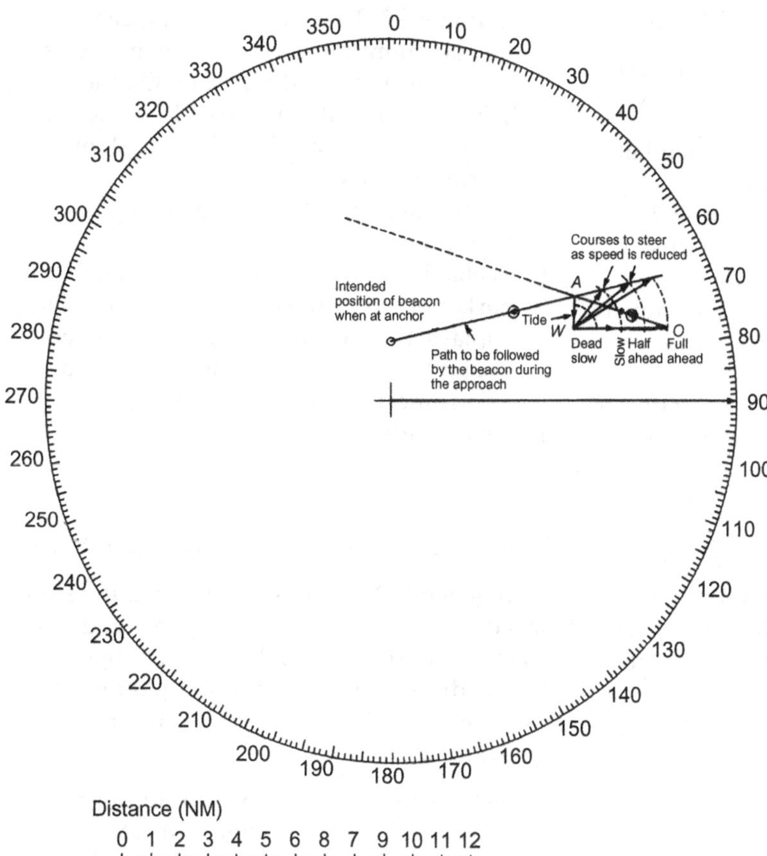

FIGURE 7.22 The course to steer in tide as speed is reduced – Example 7.12.

Dead Slow Ahead: it is necessary to steer 004° (T) and stem the tide, but note that the tide is 1 knot greater than the ahead speed.

7.8 MANUAL PLOTTING – ACCURACY AND ERRORS

The error in the result from any computation depends upon the accuracy of the data used. The intrinsic sources of error in the radar system relate to the measurement of ranges and bearings. Other data needed to complete the radar plot and subject to error are the own ship's course and speed, and the plotting interval. Also, there are of course personal errors and blunders. While a *constant* (or systematic) error in input will result in a constant error in the answer obtained, it is the errors which are of a random nature which govern the size of the 'circle of uncertainty' around the plotted position which, if not actually drawn, should be borne in mind when plotting a position.

Note: Precision of measurement and accuracy of measurement should not be confused. For example, it was no good being able to measure bearings to 0.1° when the 'free play' in traditional mechanical gearing was ±2°.

7.8.1 Accuracy of Bearings as Plotted

Errors in bearings may arise from any of the following causes:

1. The existence of inherent errors which fall within the limits allowed by IMO Performance Standard (see Section 11.2.1). The individual error sources are discussed in detail in Section 6.6.8 where it is shown that they may aggregate to as much as ±2.5°. It is also indicated that some components of the total error can be expected to remain constant over a series of bearings and these will have the effect of slewing all plotted positions by a fixed amount. There will, of course, be a random component which is most likely to arise from an instantaneous misalignment of the antenna and trace, and should not exceed ±1°. This error will have the effect of scattering the observed positions about the correct apparent track.
2. Parallax when bearings are taken with a Perspex cursor (no longer seen at sea).
3. Failure to centre the origin correctly when use is to be made of a Perspex cursor, (again not a modern problem).
4. Errors of alignment of the electronic bearing line. This error is likely to be constant over a series of bearings (sec Section 6.6.8).
5. Failure to check the heading at the time a bearing is taken when a ship's head-up unstabilized orientation is selected. This error is likely to be random.
6. Personal errors and blunders.

7.8.2 Accuracy of Ranges as Plotted

Sources of error in the measurement of range may include any or all of the following:

1. Range errors inherent in radar systems which comply with IMO Performance Standards. These are described in Section 6.6.7. They should not exceed 1% of the maximum range of the scale in use or 30 m, whichever is the greater. On the 12 NM range scale this gives 1.2 cables or 222 m. These errors should be constant over a series of ranges.
2. Inaccurate interpolation. This will give rise to random errors.
3. Mechanical errors in the variable range marker of old radial cathode ray tube (CRT) displays. Such errors should normally be constant over a series of ranges unless a serious mechanical fault exists.
4. Personal errors or blunders.

7.8.3 Accuracy of the Own Ship's Speed

In general, the means of obtaining the ship's speed can be flawed in the extreme and is the quantity most susceptible to error. Speed (or rather, distance travelled in the plotting interval) can be derived from a variety of sources, for example:

1. *Distance (towed) log.* A very traditional mechanical system in which it was not possible to know one's speed quickly, particularly when altering speed (e.g. in poor visibility). Also, it was common practice to hand the log when the engines are put on stand-by, so as to avoid fouling the propeller when the engines were put astern. Some towed logs did have an additional unit which provide a read-out of speed as well as distance.
2. *Speed (pitot, impeller, electromagnetic) log.* Although the speed may be read at any instant, the sensor is frequently withdrawn when the vessel is in shallow water, for example in port approaches.
3. *Engine revolutions.* This is only accurate in so far as how well the 'slip' is accurately known and this is rarely the case when

changing speed such as when manoeuvring in fog.

4. *Doppler log.* It should be borne in mind that if ground locked, this indicates speed over the ground which in tide can lead to misinterpretation of the aspects of other ships (see Section 6.9.3). Also, there is some uncertainty as to just what 'speed' is being measured if using a single-axis sensor which is 'ground locked'.
5. *Speed derived from positions plotted on the chart or Global Navigation Satellite System (GNSS) (see Section 10.1).* There is a common misconception that this is the ship's 'correct' speed, and on this basis it is used for plotting and as the manual input to the true-motion unit and ARPA (see Section 6.9.3). It must be remembered that the speed derived is measured over the ground, whereas it is the speed through the water which is required for plotting. Thus, if there is any tide involved, its effect *must* be allowed for in order to deduce the water speed.

Note: The slower the own ship's speed, the greater will be the proportionate effects of errors in the knowledge of the own ship's speed. Unfortunately, the plot can be at its most inaccurate when both vessels are moving slowly, as they might be when proceeding in fog.

7.8.4 Accuracy of the Own Ship's Course

Compass error should be small and relatively constant so, although it will produce errors in target course and speed, they too should be small and constant. Where the ship is off course for minutes at a time and this is not taken into account in the plot, errors in the target's course and speed will result (Figure 7.23).

7.8.5 Accuracy of the Plotting Interval

The times of each plot are normally recorded to the nearest minute and so an error of up to half a minute can be quite common when plotting a position. This can mean an error of up to 1 minute in a plotting interval of, say, 5 minutes, i.e. 20%. As the plotting interval is used to calculate the distance run (*OW*), an error here will have the same effect as an error in speed.

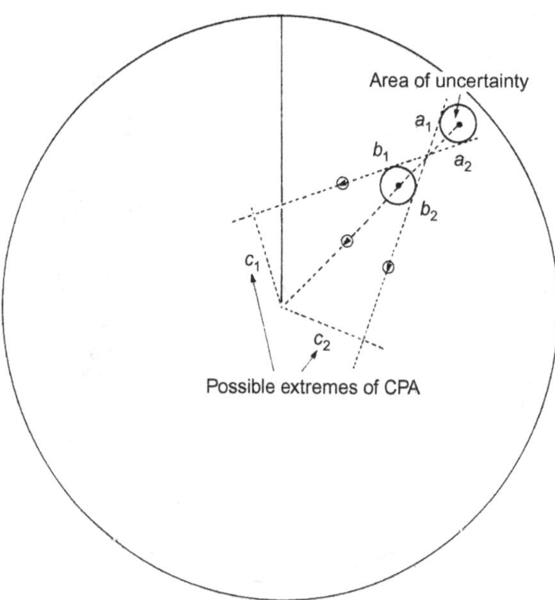

FIGURE 7.23 Threat assessment markers on a true-motion presentation.

7.8.6 The Accuracy with Which CPA Can Be Determined

Random errors in obtaining the range and bearing of a target will mean that when a position is plotted, it should be surrounded by an *area of uncertainty* based on the errors referred to in Sections 7.8.1 and 7.8.2. In Figure 7.24, the apparent-track line through the plotted positions indicates a vessel on a collision course. However, if 'worst accuracy' is considered, that is that the target's position is at a_1 and some time later at b_2 or, alternatively, is at

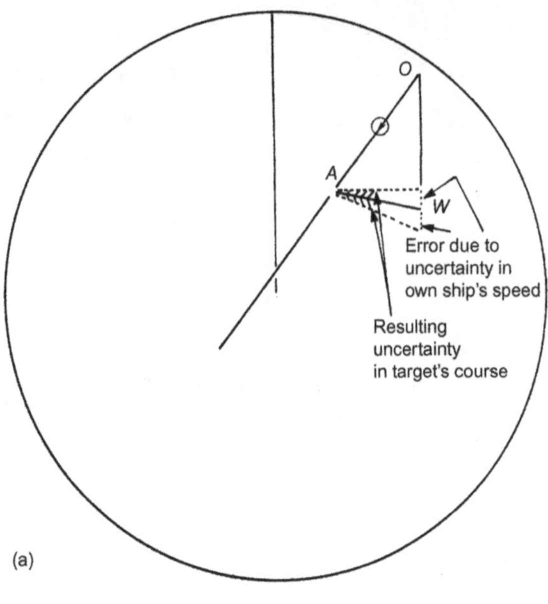

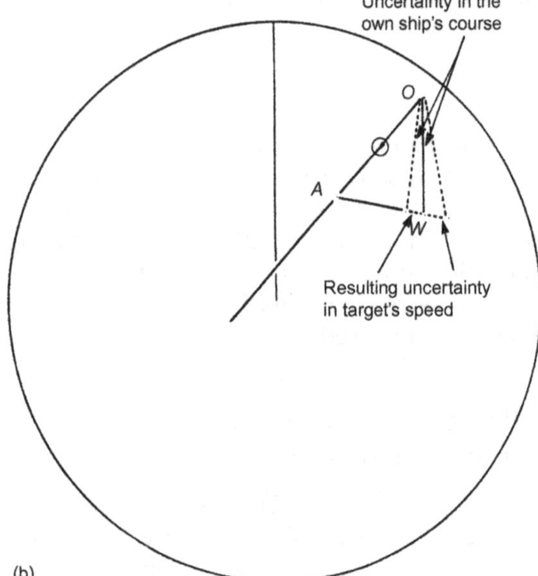

FIGURE 7.24 The accuracy with which CPA can be determined.

a_2 and later at b_1, then the CPA of the target will lie in the range between c_1 to port and c_2 to starboard.

The means by which accuracy can be improved are:

a. Improve the inherent accuracy of the system, that is decrease the size of the circles of uncertainty.
b. Plot frequently. This has the advantage of being a quick check for blunders while also allowing for a better 'mean' line to be drawn through the plotted positions. It has been suggested that the ideal time period for which plotting should be continued before accepting that the result is of sufficient accuracy should be calculated by allowing the target to traverse some $0.2 \times$ radius of screen, for example *OA* should be some 2.4 NM when using the 12 NM range scale.

7.8.7 The Consequences of Random Errors in the Own Ship's Course and Speed

We assume in the following that positions *O* and *A* are without error.

In the *OAW* triangle (Figure 7.25), the position of *W* is determined from a knowledge of the own ship's course and speed. It is important to note here, that, when dealing with a slow-moving target, if *A* falls within the error circle around *W* then it is impossible to determine the course and/or speed of the target (Figure 7.26).

7.8.8 Summary

1. It has been estimated that the full maximum error occurs only in some 1% of encounters.
2. In deciding what is an acceptable CPA, the consequences of inaccuracies should be borne in mind and one should err on the safe side.
3. Small alterations of course and speed by plotted targets can be completely swamped by errors, especially when speeds are low.

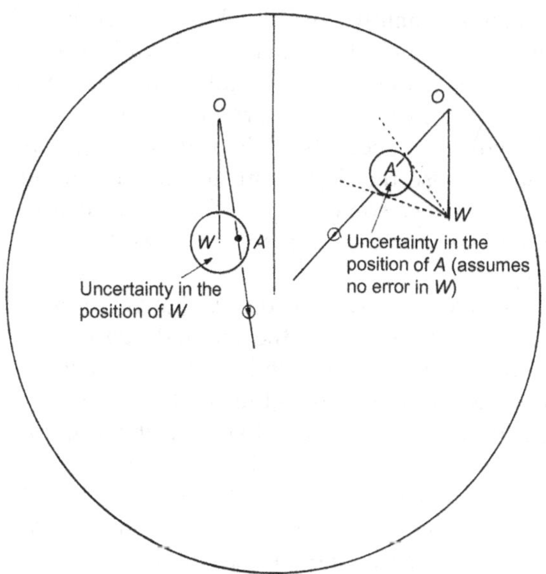

FIGURE 7.25 **Errors resulting from inaccuracy (a) in the own ship's speed and (b) in the own ship's course.** *Note*: Assume no error in *O* or *A*.

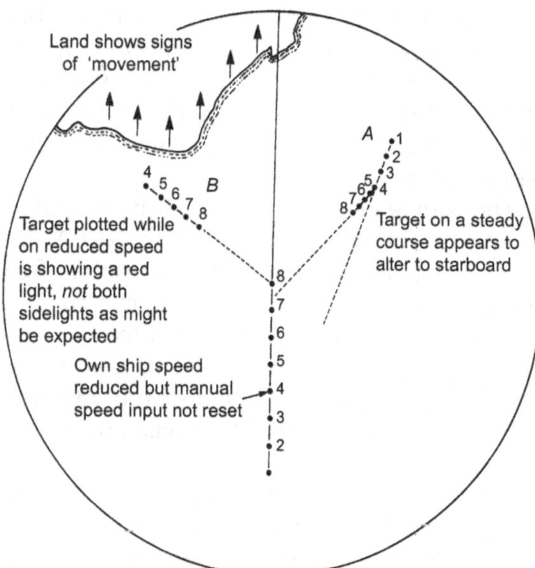

FIGURE 7.26 Slow-moving targets.

In an attempt to evaluate the errors, the following have been suggested:

$$\text{Error in CPA} = \frac{\pm 0.03 \times \text{mean range}}{\text{range change}}$$

$$\text{Error in vessel's estimated course (degrees)} = \frac{103 \times \text{mean range}}{\left(\begin{matrix}\text{estimated speed}\\ \text{of target}\end{matrix}\right) \times \left(\begin{matrix}\text{plot interval}\\ \text{in minutes}\end{matrix}\right)}$$

$$\text{Error in target's speed (\%)} = \frac{180 \times \text{mean range}}{\left(\begin{matrix}\text{estimated speed}\\ \text{of target}\end{matrix}\right) \times \left(\begin{matrix}\text{plot interval}\\ \text{in minutes}\end{matrix}\right)}$$

7.9 ERRORS ASSOCIATED WITH THE TRUE-MOTION PRESENTATION

The true-motion display (see Section 2.6.4) is, in fact, a contrived mode of display in that the 'received data' is processed by applying the ship's course and speed in order to achieve the desired method of presentation. Because of this, the accuracy of this form of presentation is subject, in addition to the errors discussed in Sections 7.8.7 and 7.8.8, to errors which may arise in the accuracy with which the origin tracks across the screen.

The use of a true-motion display for manual plotting is associated historically with the reflection plotter (see Section 7.10.3). The true plot is not as easy to use as the relative plot when using paper-based manual plotting (see Section 7.3). This discussion of errors is still relevant to the modern ARPA display.

7.9.1 Incorrect Setting of the True-Motion Inputs

When setting up the true-motion presentation, the method of course and speed input will have to be selected by the operator. The course input will invariably be from a repeating compass and will be automatic, but the repeater in the radar will have to be aligned with the master compass. Provided that this has been done correctly, errors from this source should be minimal (see Section 6.2.6.2).

The source of speed input, can be provided by log, by manual input, or with ARPA, by means of automatic ground-stabilization (see Section 4.8). The points discussed in Section 7.8.7 apply here and will affect the displayed movement of the target(s) (Figures 7.27 and 7.28), but it is the setting of the 'manual' input of speed that warrants special attention.

While steaming at a steady speed it is possible to set the manual speed input and virtually forget about it, but in fog, heavy traffic and in port approaches it may be necessary to change speed frequently. Amid all the other bridge activity, it must be remembered that the manual speed control on the radar will also need to be reset each time there is a change in the demanded speed. It is logical at times like these to use the 'log' input to the true-motion unit, but on some vessels these are also the times when the log is least likely to be deployed.

It is very easy to order a change in speed and forget to change the manual speed input to the radar and so have it portray a most misleading presentation of the situation. Such forgetfulness will be less likely if the observer maintains a plot of the origin.

Consider the own ship steaming at 16 knots, using manual speed input to the radar which is correct during the initial plotting period. Having passed ahead of vessel *A*, the own ship now reduces speed (the other vessels maintain course and speed throughout) but forgets to alter the manual speed input to the radar.

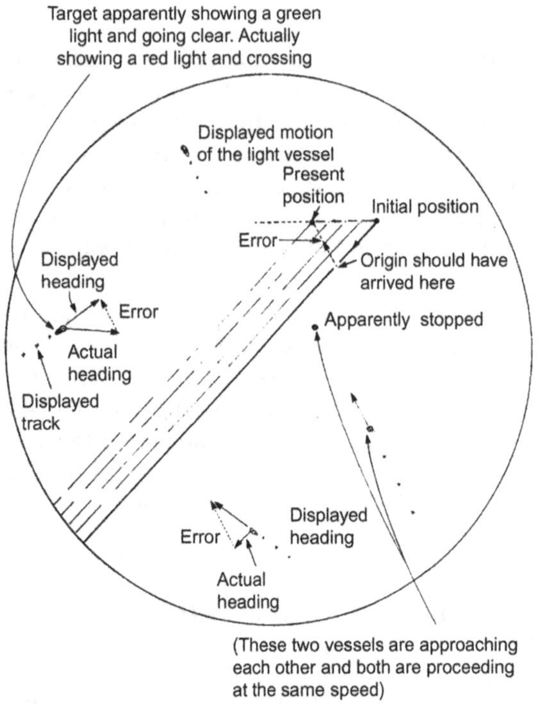

FIGURE 7.27 Tracking course errors on a true-motion presentation. True course 225°; Speed 10 kn. Origin tracking 270°; Speed 10 kn.

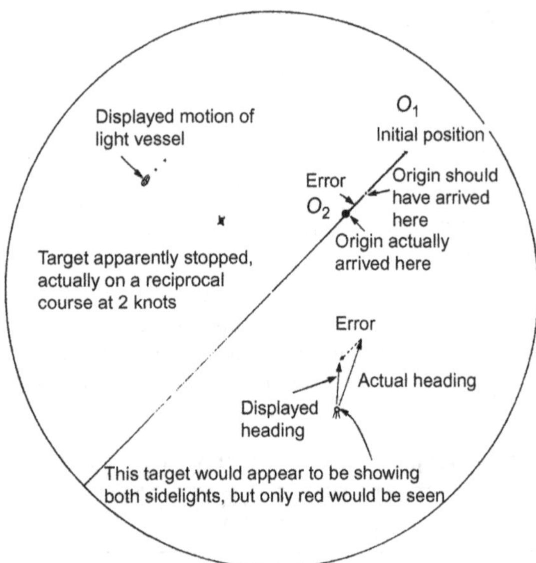

FIGURE 7.28 Tracking speed errors on a true-motion presentation. True course 225°; Speed 5 kn. Origin tracking 225°; Speed 7 knots.

The target *A* would *appear* to alter course to starboard as indicated in Figure 7.29, but on 'visual' observation would still be showing a green light (her starboard side) instead of the expected red light. Careful observation of the land would show that it too is now 'moving' – but at times like this, one rarely has time to analyse the more subtle discrepancies on the display.

One must also be conscious of the fact that an error in the log input will produce a similar effect.

Other inputs which can inadvertently cause problems are the tide input controls (see Sections 1.5.2 and 6.9.6.2). When setting up the true-motion display, it is essential to check that these controls are set as required (see Section 6.9.6) or are completely inoperative. If allowed to remain as when previously in use, they can cause the origin to track in completely the wrong direction and thus give a totally incorrect impression of the direction in which the targets are tracking. Again, the importance of plotting the origin must be stressed.

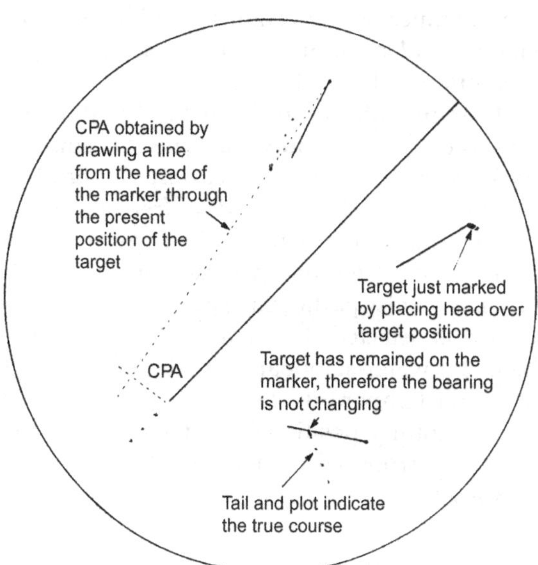

FIGURE 7.29 The misleading effect caused by failing to reset the manual speed input to the true-motion unit after a manoeuvre – a reduction of speed in this case.

7.9.2 Tracking Course Errors

In achieving true-motion, the origin must track in the direction in which the vessel is travelling. If it does not, then the target's courses and/or speeds will be in error. The magnitude and direction of these errors will vary from target to target and will be dependent upon the encounter geometry, and the magnitude and direction of the tracking error (see Figure 7.27).

Causes of tracking course error can include a fault in the true-motion unit, incorrect alignment of the compass repeater (heading marker) and incorrect setting of the tidal correctors (see Section 6.9.2).

7.9.3 Tracking Speed Errors

As with tracking course errors, these will result from errors in the movement of the origin across the display (see Figure 7.28). Causes may include a fault in the true-motion unit, an error in the transmitting log, an error in the estimated speed, failure to reset the manual speed control after ordering a change in speed and failure to reset the tidal correctors.

It is essential not only to plot the origin of the display when using the true-motion presentation, but also to ensure that the movement is correct in both direction and scale speed. This may mean that a future position of the origin must be predicted and marked on the reflection plotter. If the prediction is not achieved then the cause should be investigated.

7.10 RADAR PLOTTING AIDS

Many pieces of equipment have been developed over the years which were intended to assist the mariner in plotting and thereby interpreting the display. Most are now only of historical interest, having been superseded by ARPAs, but one or two are worthy of individual mention.

7.10.1 The Radar Plotting Board

This took a number of forms as various manufacturers attempted to provide something more permanent and durable than the paper plotting sheet. Effectively, this was little more than a plotting sheet, overlaid with Perspex or heavy transparent film, which was free to rotate and so assist in predicting the effect of an alteration of course by the own ship.

It is not intended here to discuss its construction and use since it has been superseded, but this was a fruitful area for personal innovation and some boards are still in use in the pleasure boat market.

7.10.2 Threat Assessment Markers ('Matchsticks' or 'Pins')

It has long been the practice of mariners to observe the compass bearing of an approaching vessel and, where there is no appreciable change, to assume that risk of collision exists. Unwary mariners have been seriously misled when using an unstabilized display as the source of bearing measurement.

With the increased use of compass-stabilized displays, the principle of observing the compass bearing of a target again manifested itself by mariners adopting the practice of placing the cursor on the target and observing how its bearing changed in relation to this datum.

Where it is necessary to monitor more than one target, the technique is difficult to implement. Recognizing this problem, Decca Radar in 1966 made provision for the observer to use up to five 'threat assessment markers'. These are, in effect, sections of electronic cursor which can be positioned in range and azimuth so that, when placed on a target as shown in Figure 7.23, they will quickly show whether the bearing of the target is changing. The remote end of the marker, which represents a constant range and bearing from the observing vessel, is brightened and thus the marker resembles a match or pin.

Correct use of the markers provides much more information than merely an indication of change of bearing. It enabled the observer to deduce the relative- and true-motion data for each marked target. To exploit this facility, the true-motion sea-stabilized presentation must be selected. Used in this way, the true plot of the target will indicate its true motion, but the position of the target in relation to the bright end of the marker will indicate its potential threat. By joining the 'head' of the marker to the current position of the target and extending the line, the CPA of the target can be assessed.

Note: When used with a true-motion display, the markers track across the screen in the same manner as the origin but with their orientation and position in relation to the origin remaining unchanged.

Other manufacturers have produced variations on this basic idea. In one system, the markers are bright spots (i.e. only the head of the 'match' is shown) and a free electronic bearing line, having time divisions on it, is used to lay off the relative track of the chosen target. In this particular system, if relative-motion is selected, the marker traces out the track of a water-stationary target and there is a provision temporarily to move the marker to the O_1 plotting position in order to assess the effect of a proposed manoeuvre by the observing vessel.

7.10.3 The Reflection Plotter

The work involved in the transferring of data from the radar display to a plotting sheet is both time-consuming and a potential source of errors and so is a discouragement to practical plotting. For those having to plot manually, the advantage of being able to plot directly on the screen surface was recognized at a very early stage in the development of radar. As a result, the anti-parallax reflection plotter was developed.

It has rapidly disappeared from general use due to the introduction of the raster-scan display. The UK regulations (see Section 11.4.1) still require that plotting facilities shall be at least as effective as a reflection plotter.

7.10.3.1 The Construction of the Reflection Plotter

As can be seen from Figure 7.30, the plotting surface has the same curvature as the CRT. By inverting it and placing a flat partial reflecting surface midway between the two curved surfaces, it is possible to put a mark on the concave plotting surface such that its *reflection* will be aligned with the target's response on the CRT surface, irrespective of where the observer's eye is positioned. This overcomes the problems of parallax which would arise if one were to try to plot on the plastic cursor placed some inches above the CRT surface.

7.10.3.2 The Practical Use of Reflection Plotters, Including the Use of the Perspex Cursor and Parallel Index

Plotting is carried out on the plotting surface as if working on a plotting sheet, but it is the 'reflections' which must be continually observed. Two techniques are peculiar to reflection plotting (Figure 7.30):

a. It was necessary to make a 'scale' rule for measuring distance on the plotting surface. This was done by brightening the range rings, placing a mark on each ring and then, using stiff card, marking the position of each ring on the card. Sub-divisions may be put in by eye or more precisely, using the variable range marker. See Figure 7.31.

b. It was necessary to draw parallel lines on the plotter surface, for example *OW* parallel to the heading marker. This was done by lining up the Perspex cursor with the heading marker. The parallel index lines on the cursor were then parallel with the heading marker. The edge of the scale rule was then aligned with the nearest parallel index line below it – this may mean that one had to move the position of one's head slightly, or slide the card a short distance.

Note: When drawing a line from a point, tangential to a range ring, for example when determining the time to resume, it was best first to mark the position of the range ring on the plotter surface.

7.10.3.3 Changing Range Scale

The reflection plotter had two distinct advantages: namely, it eliminated the need for reading and transferring of ranges and bearings to a plotting sheet and it maintained the immediate contact between display and plot. However, there was often a certain reluctance on the part of the observer to change range scale because of the perceived potential loss of the plot. Nonetheless, it is important that when targets are close, the most appropriate range scale was used. This meant adapting the plot to the new range scale. As can be seen in Figure 7.32, once the target is within the inner portion of the screen, the range scale was changed and the predicted apparent-track line A_2A_3 drawn parallel through the new position of the target.

It can often prove useful to retain the original plot and, provided that the lines do not become confusing, they can be left on. If available, different colour wax pencils could be used for each range scale to aid clarity.

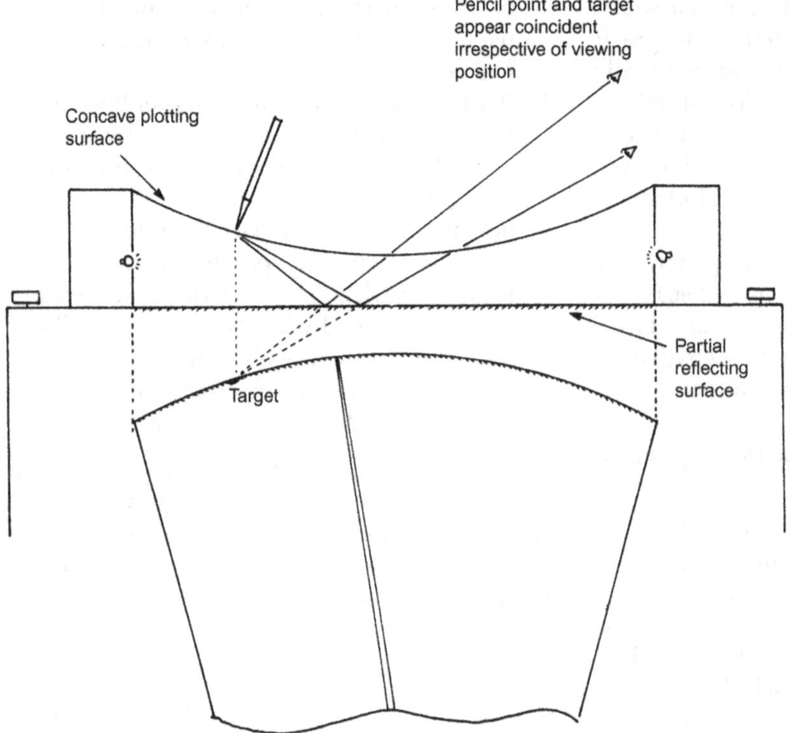

FIGURE 7.30 The reflection plotter.

7.10.3.4 The Use of the 'Free' EBL to Draw Parallel Lines (See Section 6.6.5)

The 'free' EBL was often extremely useful for transferring parallel lines in conjunction with a reflection plotter in preference to the use of the parallel lines on the Perspex cursor.

The EBL origin is positioned on the reflection of the line to be transferred and is rotated so that it is aligned with the line to be transferred. The EBL maintained its orientation as it was moved about the screen using the joystick (Figure 7.33).

7.10.3.5 Fixed and Rotatable Surfaces – Use with a Ship's Head-Up Unstabilized Display

While using a reflection plotter on a ship's head-up unstabilized display, if it became necessary to alter the own ship's course, the target and plot became out of alignment as the own ship changes course. For example, in Figure 7.34, after an initial plot, the own ship alters course 55° to starboard; at the completion of the manoeuvre, the plot and target will be as depicted. To realign the plot to the target, it was necessary to rotate the plotter surface anticlockwise (i.e. in the opposite direction to the alteration of course) by the amount which the own ship has altered course.

When a rotatable surface has not been provided, there are means by which the plot can be continued. However, they can be complicated and consequently confusing, and are best avoided. When a fixed-surface plotter was provided, it was virtually essential that the true-north-up stabilized presentation was selected (see Section 1.4.1.2).

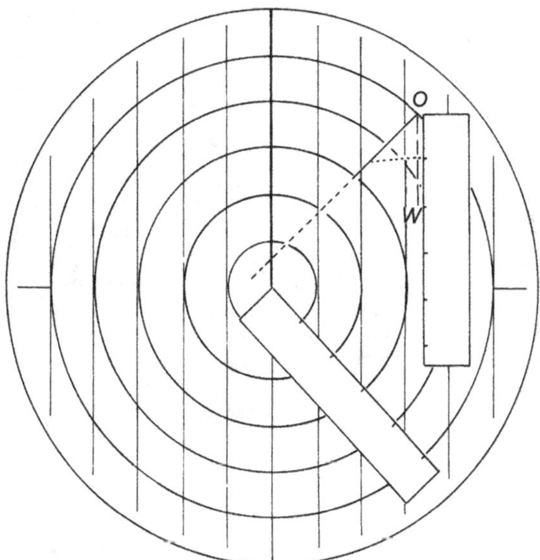

FIGURE 7.31 Constructing and using a scale rule.

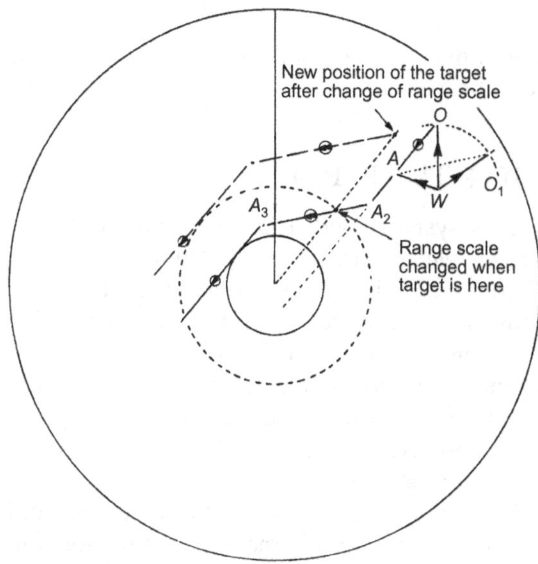

FIGURE 7.32 Changing range scale on a reflection plotter.

7.10.3.6 *Flat and Concave Surfaces*

The concave surface can be rather awkward to work on and so more complex plotters were

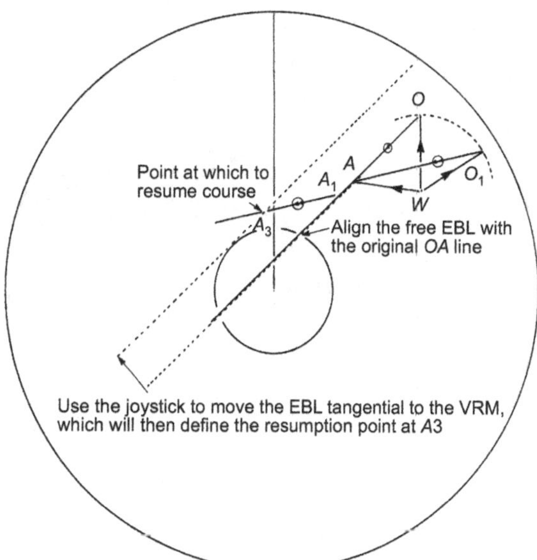

FIGURE 7.33 **The use of the 'free' EBL to draw parallel lines.** *Note*: Transfer the line to the plotting surface if the EBL is required for another use; if not, leave it in place.

provided with flat plotting surfaces. This was made possible by using a curved partial reflecting surface. However, apart from that, its use was exactly the same as for a plotter with a concave plotting surface.

7.10.3.7 *Use in Conjunction with Parallel Indexing*

The provision of a reflection plotter made it possible for a prepared navigation plan to be marked out on the plotter and the movement of particular navigation marks observed in relation to their predicted movement. Any deviation from the pre-planned track will be readily apparent and compensation can be made (see Section 8.4).

7.10.3.8 *Reflection Plotters and Raster-Scan Displays*

No manufacturer has produced an optical reflection plotter which can be used with a television-type rectangular CRT or liquid

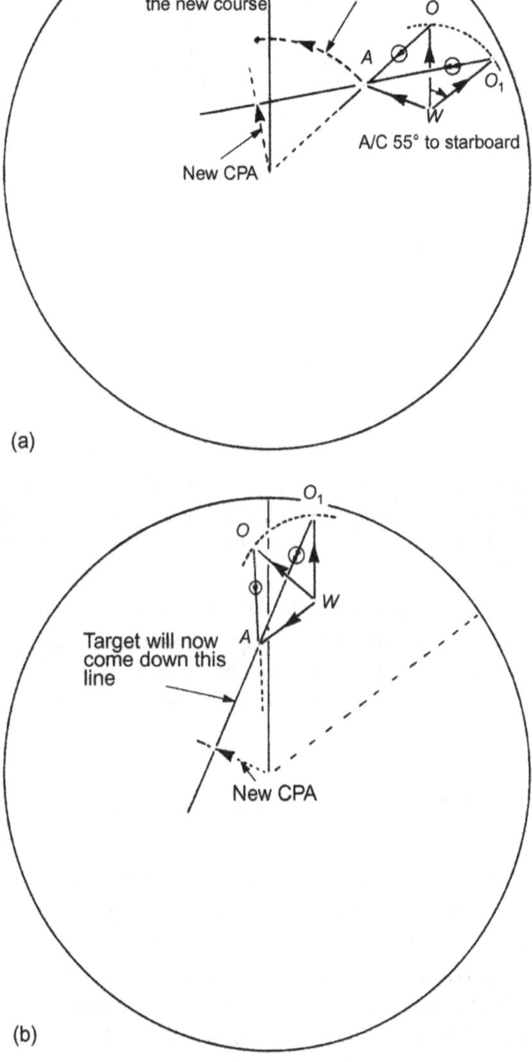

FIGURE 7.34 The use of a rotatable plotting surface. (a) Before or during alteration of course. (b) After rotation of the plotting surface.

crystal display (LCD). There is an inherent assumption that in the unlikely event that the ARPA is not working but the radar is, then any manual plotting will have to be done on paper.

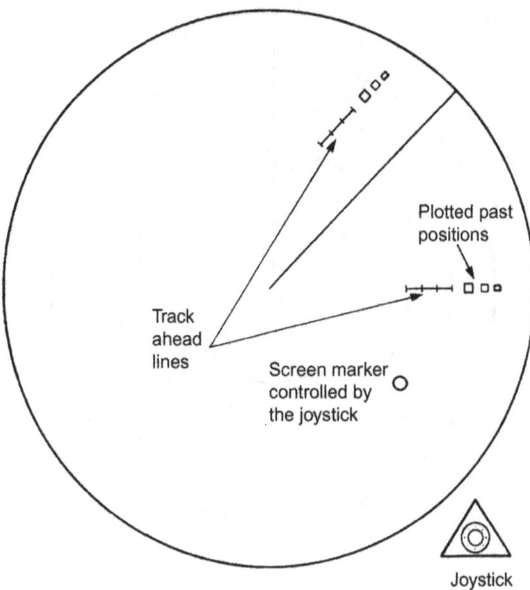

FIGURE 7.35 The 'E' plot.

7.10.4 The 'E' Plot

This system used a joystick and marker to record 'electronically' the position of a target (Figure 7.35). The operator recorded the positions of targets at regular intervals, after which the data was processed and extrapolated to provide the information which would be available from a normal plot, that is course, speed, CPA and TCPA. Up to 10 targets could be manually tracked.

Manual recording as practiced here should not be confused with manual acquisition followed by automatic tracking which takes place in the simplest of ARPAs.

The system was unique to Kelvin Hughes. However, other types of plotting aids using graphics were common. Development of such facilities is a natural result of the need to provide plotting facilities on a raster-scan display which are at least equivalent to a reflection plotter (see Section 7.10.3), and has been assisted by the ease with which lines and points can be drawn electronically on such a display.

The 'E' plot has now been superseded by the EPA.

7.10.5 Electronic Plotting Aid (EPA)

In 1997, IMO produced performance standards for an electronic aid to manual plotting. It was a refinement of earlier graphical devices such as the E plot covered in the previous section. IMO called the device an electronic plotting aid (EPA) and for a short period of time it was the minimum requirement for the smallest vessels (under 500 grt) that come under SOLAS (see Section 4.1.3), although not for high-speed craft. This regulation and the device have since disappeared.

The performance standards required that the radar/EPA has azimuth stabilization, which could be provided by either a gyro or transmitting magnetic compass. There is no requirement for an electronic speed input, thus the speed may have to be input manually.

The observer simply plots a target initially by using the joystick or tracker ball controls to position a marker symbol over the target to be plotted. A second plot is added any time between 30 seconds or 15 minutes later. A target symbol and vector then appear on the display. The symbol will move in an extrapolation of the two plots across the screen. The observer should enter more plots in time, particularly if the symbol's relative motion does not match the underlying relative track of the target across the screen. Individual plots can be corrected if not plotted accurately.

As soon as a second position has been plotted, the observer can obtain the traditional alphanumeric target information associated with ARPA equipment, i.e. present range and bearing, CPA, TCPA, true course and speed. The observer is also able to switch between true and relative vectors, as well as adjust the time length of vectors. As the observer added more plots to the target, the information and vectors were updated based on *calculations from the last two plots only*. It obviously follows that best results were obtained by plotting at reasonably regular intervals. Plotting at too small an interval caused minor plotting errors to have a large effect on the accuracy of the result. On the other hand, if the time interval was stretched too much then any target manoeuvres are less accurately detected and followed. The best interval to use will vary according to the target relative speed and direction and the range scale in use at the time (see also Section 7.9).

The standard called for a minimum capability of 10 targets to be tracked in this way, so each target symbol must be assigned a target number, which (when appropriate) is included on the alphanumeric display as well as on the radar picture. The observer selects an appropriate target first, before reading the alphanumeric display or adjusting its plot.

7.10.5.1 Alarms

The traditional target violation of preset CPA and TCPA limits found on the ARPA (see Section 4.7.2) was provided on an EPA.

In addition there is a warning if a target plot has not been updated by the observer for more than 10 minutes. The plot is dropped completely if this update time exceeds 15 minutes.

7.10.5.2 Comparison Between EPA and the Reflection Plotter

For a short time the EPA effectively superseded the requirement for a reflection plotter on non-ARPA radars. The modern raster CRT and LCD displays do not lend themselves to a reflection plotter and the EPA eliminated the need for calculations by the observer that are associated with hand plotting.

The plotting errors due to working with a thick grease pencil were reduced and a major advantage is that it is no longer a requirement for the navigator to record the precise time of the plot as it is done by the EPA. Indeed, while it may improve accuracy of EPA to plot at a suitable regular interval (as discussed above), the EPA was more likely to obtain accurate results than a reflection plotter if the plotting periods were a little irregular.

The EPA also maintained plots at the correct range and predicted speed when the range scale was changed, unlike the reflection plotter.

The main disadvantage of the EPA over the reflection plotter was that only two plots are used for the calculations and it is up to the observer to monitor the target track to ensure that the target achieved the second plot in a straight line and at constant speed. The results would not be valid if (unbeknown to the observer) the target adjusted its course or speed and ran a dogleg to the second plot.

It is of course possible for an observer using a reflection plotter to make this error, but in this case it is less likely as the observer was encouraged to take three or more plots and the change in relative track would be more obvious. The human observer may also obtain a more accurate indication of target relative track by averaging it over three or more target plots. This is likely to be particularly true if, for example, the azimuth input was from a transmitting magnetic compass or the vessel is yawing badly.

As in all manual plotting, the results are still only as good as the diligence of the observer.

7.10.6 Auto-Tracking Aid

IMO Performance Standards of 1997 also introduced the auto-tracking aid or ATA. Essentially it was an ARPA, but with certain features removed (Table 7.1). It was introduced as the minimum standard for vessels between 500 and 10,000 grt and even as the second radar tracker on ships greater than 10,000 grt. Like the EPA, it has been short-lived, existing in the regulations for a period. It had limited takeup by the industry and it is no longer part of current regulations.

As noted in Section 4.2.6, automatic acquisition has not been found as useful as was first predicted (particularly for a vessel underway) when ARPA standards were first introduced. This omission from ATA standards perhaps reflected this low usage and usefulness. The logic for the lower minimum number of tracked targets follows mainly by removing this automatic acquisition option. There is also a school of thought that too many tracked targets simply clutter up the screen. It follows that 10 tracked targets are more than adequate, provided the navigator selects the 10 targets having the greatest potential threat. Similarly, the history dot option (see Section 4.3.7) is one

TABLE 7.1 Differences Between Minimum ARPA and ATA Standards (IMO 1997 and 2001)

	ATA	*ARPA*
Acquisition (see Section 4.2)	Manual	Manual and (optionally) automatic
Minimum tracked target capability (see Section 4.2)	10	20
History dots (see Section 4.3.7)	None required	Four equally spaced past positions over time period suitable for range scale
Trial manoeuvre (see Section 4.4.3)	None	Yes, including with or without a time delay

not often used in practice, especially as manufacturers provide a traditional smear 'snail' trail (true or relative track) to show target progress. The biggest loss of function between the ATA and the ARPA is probably the lack of a trial manoeuvre facility (see Section 4.4.3).

All told, the manufacturing differences between ARPA and ATA were actually quite small and one would speculate that the difference in costs of production between a minimum specification ARPA and a minimum specification ATA is actually quite small (mainly software and computer memory). A significant cost (at the time of writing the ATA regulations) would probably have been a smaller minimum screen size, but display screens subsequently went down dramatically in price.

A separate point is that ever since ARPA was first introduced, most maritime administrations have insisted that the operator should be specifically trained to use the ARPA and not just in radar observer manual plotting techniques. For instance UK regulations state that:

> 'When a UK ship which is required to be fitted with ARPA is at sea and a radar watch is being kept on the ARPA, the installation shall be under the control of a person qualified in the operational use of ARPA, who may be assisted by unqualified personnel'. (Edited extract from UK SI 1993 No. 69, part IX, see Section 11.4.2.)

The ATA had no such provision and would therefore come under the regulation for radar, not ARPA. For instance in UK ships, it would come under the requirement:

> 'While a UK ship which is required to be fitted with a radar installation is at sea and a radar watch is being kept, the radar installation shall be under the control of a qualified radar observer who may be assisted by unqualified personnel'. (Extract from UK SI 1993 No. 69, part IV, see Section 11.4.1.)

Whilst it is hard to justify the difference in minimum training between an ATA and an ARPA on the basis of the difference in features, it would appear that lesser training was required. The only basis for this difference in training is that personnel who are not ARPA trained are effectively restricted to smaller vessels, and traditionally smaller vessels have always had lesser qualifications. However, it is a tradition that does not stand up to close scrutiny as smaller vessels often spend more time in high-density traffic and navigate in waters with little sea room.

Although the ATA has disappeared from the performance standards, the philosophy (rightly or wrongly) that small ships need a reduced specification for ARPA and this need still exists in the current IMO performance, as covered in Sections 4.1 and 4.2.

7.11 THE REGULATIONS FOR PREVENTING COLLISIONS AT SEA AS APPLIED TO RADAR AND ARPA

7.11.1 Introduction

The specific function of radar plotting (whether carried out on a paper plotting sheet, with the aid of an ARPA, or with any of the intermediate facilities or techniques) is to provide the data on which a collision-avoidance strategy can be based. It is not intended that this section should represent a treatise on collision-avoidance strategy in restricted visibility. This section aims to provide the radar observer with an appreciation of the need to carry out radar plotting (or equivalent systematic observation) in order to comply with the rules for preventing collisions at sea and with an understanding of the relationship between the data extracted and the provision of the regulations to the extent necessary to provide informed support for the master or other officer responsible for the collision-avoidance strategy.

A notable feature of the 1972 Collision Regulations was that many specific references to the use of radar were made in the body of the rules. Subsequent amendments have not changed the role of radar as an anti-collision aid.

When considering the application of radar and ARPA to collision avoidance, it is particularly pertinent that the term *radar plotting or equivalent systematic observation of detected objects* (rule 7b) appears for the first time. Although the rules do not define this term, consideration of the various instructions and cautions given in the specific references to radar make it possible to deduce a procedure which would enable competent personnel to comply with both the letter and spirit of the rules. In this respect, ARPA should merely be seen as readily providing data which would otherwise have to be obtained from the radar by lengthy and tedious manual extraction.

7.11.2 Lookout – Rule 5

Every vessel shall at all times maintain a *proper lookout* by sight and hearing, as well as by *all available means appropriate* in the prevailing circumstances and conditions so as to make a full appraisal of the situation and of the risk of collision.

Although this rule does not specifically mention radar, there seems little doubt that radar is embraced by the term 'all available means'; its ability to detect targets and its role as a source of information allow an observer to make a more complete appraisal of the situation. It would also appear that there is an implied requirement to use the equipment in clear weather where it can augment or clarify the visual scene, for example in dense traffic, especially at night.

7.11.3 Safe Speed – Rule 6

Every vessel shall at all times proceed at a safe speed so that she can take proper and effective action to avoid collision and be stopped within a distance appropriate to the prevailing circumstances and conditions. In determining a *safe speed*, the following factors shall be among those taken into account:

a. By all vessels:
  - **i.** the state of visibility;
  - **ii.** the traffic density including concentrations of fishing vessels or any other vessels;
  - **iii.** the manoeuvrability of the vessel with special reference to stopping distance and turning ability in the prevailing conditions;
  - **iv.** at night the presence of background light such as from shore lights or from backscatter of her own lights;
  - **v.** the state of wind, sea and current, and the proximity of navigational hazards;
  - **vi.** the draught in relation to the available depth of water.

b. Additionally, by vessels *with operational radar*:
  - **i.** the characteristics, efficiency and limitations of the radar equipment;
  - **ii.** any constraints imposed by the radar range scale in use;
  - **iii.** the effect on radar detection of the sea state, weather and other sources of interference;
  - **iv.** the possibility that small vessels, ice and other floating objects may not be detected by radar at an adequate range;
  - **v.** the number, location and movement of vessels detected by radar;
  - **vi.** the more *exact assessment of the visibility* that may be possible when radar is used to determine the range of vessels or other objects in the vicinity.

In listing the factors to be considered when determining a safe speed, this rule devotes a complete section to those factors which can be determined by the use of radar. It is important to realize that the factors listed extend beyond the context of basic radar into that of collision-avoidance systems such as ARPA.

7.11.4 Risk of Collision – Rule 7

a. Every vessel shall *use all available means* appropriate to the prevailing circumstances and conditions to determine if risk of collision exists. If there is any doubt, such risk shall be deemed to exist.

b. *Proper use shall be made of radar equipment* if fitted and operational, including long range scanning to obtain early warning of risk of collision and *radar plotting* or equivalent systematic observation of detected objects.

c. Assumptions shall not be made on the basis of scanty information, especially *scanty radar information.*

d. In determining if risk of collision exists, the following considerations shall be among those taken into account:

  i. such risk shall be deemed to exist if the compass bearing of an approaching vessel does not appreciably change;

  ii. such risk may sometimes exist even when an appreciable bearing change is evident, particularly when approaching a very large vessel or a tow, or when approaching a vessel at close range.

Any discretion as to whether to make use of radar if fitted and operational, which may have existed under the previous rules, appears to have been removed by rule 7b. This section specifies that proper use shall be made of such equipment and the well-established warning about 'scanty information' is now embodied in this rule.

7.11.5 Conduct of Vessels in Restricted Visibility – Rule 19

a. This rule applies to vessels not in sight of one another when navigating in or near an area of restricted visibility.

b. Every vessel shall proceed at a safe speed adapted to the prevailing circumstances and conditions of restricted visibility. A power-driven vessel shall have her engines ready for immediate manoeuvre.

c. Every vessel shall have due regard to the prevailing circumstances and conditions of restricted visibility when complying with the rules of section i of this part.

d. A vessel which *detects by radar alone* the presence of another vessel *shall determine* if a close-quarters situation is developing and/or *risk of collision exists*. If so, *she shall take avoiding action* in ample time, provided that when such action consists of an alteration of course, so far as possible *the following shall be avoided*:

  i. an alteration of course to port for a vessel forward of the beam, other than for a vessel being overtaken;

  ii. an alteration of course towards a vessel abeam or abaft the beam.

e. *Except where* it has been determined that a *risk of collision does not exist, every vessel* which hears apparently *forward of her beam the fog signal* of another vessel, or which cannot avoid a close-quarters situation with another vessel forward of her beam, *shall reduce her speed to the minimum* at which she can be kept on her course. She shall if necessary take all her way off and in any event navigate with extreme caution until danger of collision is over.

7.11.5.1 The Development of a Close-Quarters Situation

In cases where a target is detected by radar alone, rule 19d places a specific obligation on the observer to determine whether a close-quarters situation is developing. To comply with this requirement, the target should be plotted and its CPA and TCPA determined. With ARPA, this information can be made available in alphanumeric form if the target is designated. (The TCPA will assume particular importance if a close-quarters situation is developing.)

At this juncture, it is the duty of the officer of the watch to decide if the CPA constitutes a close-quarters situation; the CPA which would suggest that the encounter be deemed 'close-quarters' will depend upon:

a. the geographical position of the vessels,
b. the handling capability of the ship,
c. the density of the traffic.

7.11.5.2 *Manoeuvres to Avoid Collision*

If it is decided that a close-quarters situation is developing, the observer must take action to resolve the situation in ample time, subject to the recommendations laid down in rule 19d (i) and (ii). Any avoiding manoeuvre not based on a knowledge of the true vector of all relevant targets would be unseamanlike and could certainly attract criticism as an assumption based on scanty radar information.

When ARPA is being used, the true vector data may be obtained in alphanumeric form and cross-checked by interpretation of the graphical presentation but, since these are often generated from the same database, a good comparison should not be taken as an assumption of accuracy of the information. It is important to remember that the success of the action by the own ship may be influenced by recent changes in the target's true vector. It is thus useful to check, by using the *history* presentation on the ARPA, whether such changes have occurred.

7.11.6 Action to Avoid Collision – Rule 8

a. Any action taken to avoid collision shall be taken in accordance with the rules of this part and shall, if the circumstances of the case admit, *be positive*, made *in ample time* and with due regard to the observance of *good seamanship*.
b. Any alteration of course and/or speed to avoid collision shall, if the circumstances of the case admit, *be large enough* to be readily apparent to another vessel *observing* visually or *by radar*; a *succession of small alterations* of course and/or speed should be avoided.
c. If there is sufficient sea room, alteration of course alone may be the most effective action to avoid a close-quarters situation provided that it is made in good time, is substantial and does not result in another close-quarters situation.
d. Action taken to avoid collision with another vessel shall be such as to result in passing at a safe distance. *The effectiveness of the action shall be carefully checked* until the other vessel is finally past and clear.
e. If necessary to avoid collision or *allow more time to assess* the situation, a vessel shall *slacken her speed* or take all way off by stopping or reversing her means of propulsion.

7.11.6.1 *The Forecast*

Rule 8d requires that action taken to avoid collision with another vessel shall be such as to result in a safe passing distance. This rule applies to all states of visibility but, in restricted visibility, the safe passing distance may have to be greater than that which would be tolerated in clear weather.

When employing radar and ARPA, passing distances may and should be determined by the construction of a 'predicted OA line' on the plot or the use of the 'trial manoeuvre' facility which all ARPAs are required to have. Such a forecast will also enable the observer to check that the manoeuvre will not result in a close-quarters situation with other vessels (rule 8c) and can often indicate where targets may manoeuvre to avoid each other.

It is essential to remember that a further criterion for an acceptable manoeuvre is that it must be readily apparent to another vessel observing by radar (rule 8b). In making this judgement, one must consider the own ship's speed and the speeds of target vessels.

Further, it must be appreciated that, even if the target vessel is equipped with radar, the plotting facilities may be very basic. Hence the rate at which the other observer can extract data, and thus become aware of changes, may be slow and the ability to identify small changes may be particularly limited (for this reason, rule 8b gives a warning about a succession of small alterations).

Finally, the possibility that a target vessel may not be plotting, or may not even have an operational radar, must be continually kept in mind.

7.11.6.2 *The Effectiveness of a Manoeuvre*

Rule 8d requires that the effectiveness of the avoiding action be checked until the other vessel is finally past and clear. This requirement can be satisfied by monitoring:

a. *The true track* of all relevant targets (to ensure early detection and identification of target manoeuvres).
b. *The relative track* of all relevant targets to check for the fulfilment of the forecast nearest approach.

7.11.6.3 *Resumption*

When the target is finally past and clear, the decision must be made to resume course and/ or speed.

As in the case of the avoiding manoeuvre, prior to resuming, the apparent track of the target after resumption (see Section 7.5.7) or the trial manoeuvre on the ARPA should be employed to verify that a safe passing distance will be achieved with respect to all relevant targets. It should be remembered that the resumption will be most obvious to other observing vessels if it is performed as a single manoeuvre as opposed to a series of small alterations. The common practice of resuming course in steps by 'following the target round' will make it difficult for other observing ships to identify the manoeuvre positively and could be considered to be in contravention of the spirit of rule 8b.

7.11.7 The Cumulative Turn

This set of related manoeuvres based on wrongly interpreted radar information (in restricted visibility) has probably done more than anything else to give radar a bad name and to indicate the need for a proper appreciation and use of the data displayed on the screen.

A number of elements have been identified which are common in the majority of the cases which have come before the courts and which are of interest when endeavouring to understand just how, in spite of repeated warnings, the same fatal scenario has been played out on so many occasions.

The common elements are:

a. The encounter is end-on or nearly end-on.
b. In most cases, only the two ships are involved, that is there are no constraints by other vessels or by the land.
c. There was no proper assessment of the situation based on acceptable plotting techniques, but rather a subjective judgement based on casual observation of the display. Because of the small change in bearing, each assumes (on scanty information) that the other vessel is on a reciprocal course.
d. With only one *other ship* on the screen, each considers that there is no need to go to the trouble of a formal plot.
e. Speeds have invariably been excessive.
f. Small alterations of course have been seen as adequate and usually the justification has been that it was done 'only to give the other vessel a wider berth'.
g. Each vessel alters into the apparently 'clear water', that is *A* alters to starboard while *B* alters to port (Figure 7.36). In other words, each vessel alters course away from the target.

h. The misunderstanding of the true situation can be aggravated by the use of a ship's head-up unstabilized relative-motion display where, after the completion of the manoeuvre, the impression obtained from the casual observation of the display is that the target will now pass farther off (Figure 7.37).

i. In the agony of the final moments any manoeuvre might be ordered but, in general, the wheel will be put hard over and the engines put astern. The outcome or the intention of the manoeuvre is extremely difficult to predict and rarely has time to take effect.

With the aid of ARPA, the true situation can be more readily understood at an early stage in the encounter so that if early and substantial action is taken to avoid the collision followed by rigorous subsequent monitoring, the irretrievable situation should not be allowed to develop. Where there is no ARPA assistance, it is essential that both the letter and spirit of the collision-avoidance rules are adhered to in terms of plotting, predicting ahead, monitoring target manoeuvres, while at the same time making an early and substantial avoiding manoeuvre. The importance of 'making time to plot' by reducing speed at an early stage in the encounter cannot be too strongly stressed.

Classical cases which have followed the above scenario with fatal consequences are the *Stockholm–Andrea Dona; Crystal Jewel–British Aviator; Dalhanna–Staxton Wyke*; as well as the *Canopic–Hudson Firth* and many more which followed the same basic pattern, but in which no lives were lost.

7.11.8 Conclusion

A summary of the regulations for preventing collision as applied to radar and ARPA is presented in Table 7.2.

In any potential collision situation, particularly in restricted visibility, the interpretation of displayed radar information facilitates the determination and execution of action to avoid close-quarters situations. Traditionally this is

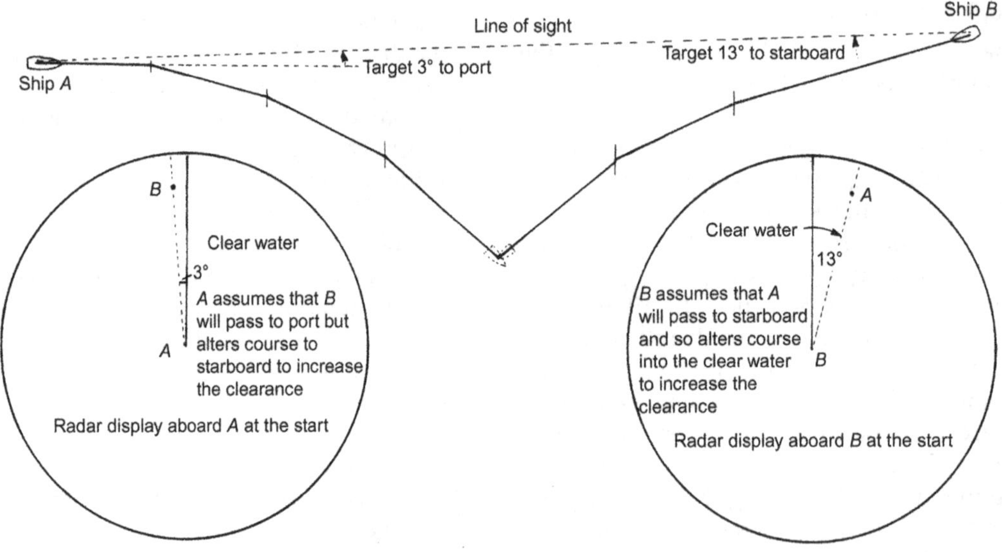

FIGURE 7.36 The cumulative turn.

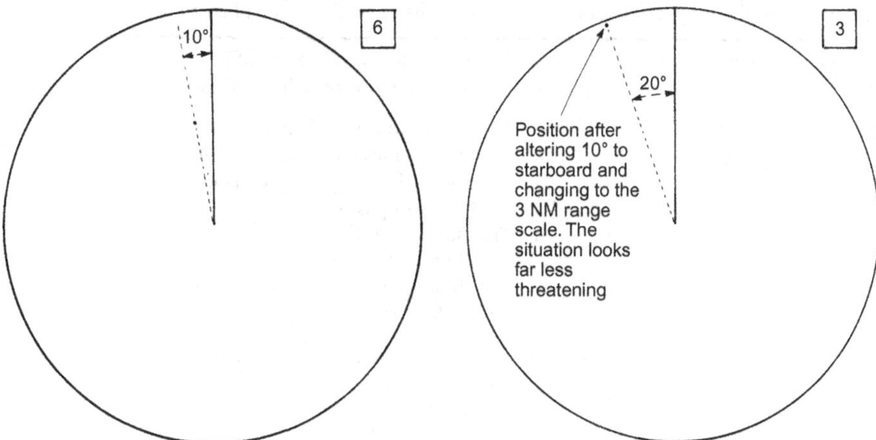

FIGURE 7.37 The effect of altering course and changing range scale on a ship's head-up unstabilized radar presentation.

achieved by a decision based on data extracted by manual plotting. ARPA should be seen as a device that extracts and presents such data. It thus reduces the workload on the observer by carrying out routine tasks and allows him more time to carry out decision-making on the basis of the data supplied.

The ability of the equipment to carry out routine tracking and computation in no way relieves the observer of the need to understand fully the principles of radar plotting, or of being capable of applying such principles to a practical encounter, such as might be the case in the event of an equipment failure. The traditional Merchant Navy training in manual radar plotting is useful both as an insight to potential issues with ARPA derived results, as well as providing a backup strategy in case of failure of the ARPA tracker.

It is imperative that the observer is capable of interpreting and evaluating the data presented by the ARPA system. Equally essential is the ability to detect any circumstances in which the equipment is producing data which is inconsistent with the manner in which a situation is developing as observed, say, from the raw radar.

In general, it is vital that the observer understands the limitations of the system in use and hence is aware of the dangers of exclusive reliance on the data produced by an ARPA. In particular, the implicit reliance on the validity of the prediction of small non-zero passing distances should be avoided. The CPA errors tabulated in sections of the Performance Standards for ARPA (see Section 9.3) clearly indicate that predicted passing distances of less than one mile should be treated with the utmost caution.

The importance of regular comparison of the visual situation with that presented by the radar/ARPA cannot be stressed too strongly.

7.12 INTELLIGENT KNOWLEDGE-BASED SYSTEMS AS APPLIED TO COLLISION AVOIDANCE

Systems have been developed to supply the navigator with advice as to the manoeuvre (or manoeuvres) that will be most effective in the existing circumstances.

The International Regulations for Preventing Collisions at Sea provide a framework within which all manoeuvres must conform.

TABLE 7.2 Summary of the Regulations for Preventing Collision at Sea as Applied to Radar and ARPA

1. Use ARPA and radar if they can be of the slightest assistance	**a.** Clear weather with heavy traffic **b.** Proper lookout (especially at night)
2. Does risk of collision exist, and/or is a close-quarters situation developing?	**a.** Analyse displayed information • Relative tracks, CPA and TCPA • True tracks – course and speed • History of past position
3. Determine best manoeuvre	**a.** Trial facility (ARPA only) • Will targets pass safely? • Will targets have to manoeuvre? **b.** Avoid A/C to port, or turning towards converging overtaking ships
4. Manoeuvre	**a.** Bold – avoid succession of small A/C **b.** Large – obvious to others **c.** In good time – time to correct **d.** Good seamanship
5. Ensure its effectiveness	**a.** Continuous check of relative track and history **b.** Watch for changes in true track of targets
6. Resume	**a.** Are targets clear? **b.** Trial – check other vessels for likely manoeuvres (ARPA only) **c.** Make one manoeuvre

Unfortunately, the regulations only work well for two-ship encounters in open water. The presence of other vessels and restrictions in navigable water may cause apparent conflicts with the rules. Furthermore, terms such as 'early', 'substantial' and 'close-quarters' are quantitatively imprecise and so application and encoding of the rules is far from simple.

Computational techniques can facilitate the encoding of experience. If experienced mariners are presented with complex encounters that are known to cause difficulty, their solutions can be analysed. The principles behind the solutions are learned by the computer which then applies them to a future similar (but not necessarily identical) encounter.

These systems at present are mainly experimental and do not have any validation from international agencies. Systems based solely on ARPA information have the following features:

1. The navigator must select whether the restricted visibility or unrestricted visibility rules apply.
2. The user must designate manually the targets believed to have extra rights under the rules (e.g. vessels engaged in fishing, restricted in ability to manoeuvre and NUC).
3. The navigator must designate manually if there are areas not navigable by the own ship.
4. The systems also should present the navigator with the rationale behind the recommended manoeuvre strategy.
5. The system can also predict the likely manoeuvres of other vessels based on the rules, which also aids the decision-making process, although it should be remembered that the other vessel might be constrained by vessels not seen on the own ship's radar.

These systems will also benefit from the increased use of other navigation technologies.

The integration with ECDIS and GNSS (see Sections 10.1 and 10.2) means that geographical information can be input into the computer and overcomes point 3 above. The universal adoption of the Automatic Identification System (AIS) (see Section 5.4.3), if it happens, will overcome many of the above limitations. Targets will automatically be transmitting their status under the rules (e.g. engaged in fishing and NUC). They also can transmit their intended route so that changes in course for navigational reasons can be included in the analysis. The detection range of AIS is potentially further than radar and is less affected by blind areas (see Section 5.5.2) overcoming the problem identified in point 5.

The recommendation in point 5 above of a particular set of manoeuvres is not intended to take over the navigational function of the navigator, but rather to support, in quantitative terms, the overall strategy that should be followed.

CHAPTER

8

Navigation Techniques Using Radar and ARPA

8.1 INTRODUCTION

The availability of information from radar and Automatic Radar Plotting Aid (ARPA) forms the basis of a number of techniques which may assist in the safe navigation of vessels. Successful and safe use of these require an ability to relate the echoes displayed by the radar to the information shown on the chart and an understanding of the levels of performance and accuracy which can be achieved under given circumstances. Where radar information alone is used in making a landfall, the ship's position may be in considerable doubt and it may be difficult to positively identify specific echoes, particularly if the observer is unfamiliar with the locality. In routine coastal navigation there may be more general certainty as to the vessel's position, but effective use of these techniques will require organization, skill, practice and a thorough awareness of the capability of the radar system. They will also be found to be of great assistance in certain pilotage situations, but it has to be said that current civil marine radar equipment has a very limited ability to contribute to the docking of vessels.

Before Global Navigation Satellite System (GNSS), these actions were of prime importance in vessel navigation, as there were many areas of the world without an alternate positioning technology until visual sightings were possible. With the advent of GNSS these activities can be considered as firstly to confirm the position indicated by the GNSS and secondly to provide an alternate means of navigation in case of GNSS failure. The ability of a navigator to make best use of the radar in these circumstances requires a certain amount of skill and experience, so the prudent navigators should be performing these tasks to confirm their position even when GNSS is thought to be fully operational. Radar has the advantage that all the equipment is onboard the vessel, while GNSS requires reliance on signals from an external source.

The role of Automatic Identification System (AIS) in this chapter is very limited. It is true that (when provided by shore authorities) the AIS identification of buoys and prominent radar targets can aid identification of these important marks in the same way as racons currently do. However, AIS on buoys depends on GNSS for timing control of its transmission signals as well as the position of these marks, so the AIS information only works if the GNSS as a whole works. The assumption in this chapter is either that GNSS is not available or, at least, the navigator is preparing for a situation when GNSS is not available or is inaccurate.

This chapter discusses the difficulties of making comparisons between the radar picture and charted features before passing on to describe the navigation techniques in terms of the underlying theory covered elsewhere in this text.

It is often stressed that radar is *only an aid* to navigation. This does not mean that radar information is necessarily of any less value than that obtained from other sources. What it does mean is that radar data should not be used in isolation and to the exclusion of that available from other sources. The radar system should be seen as one element in a variety of data sources which must be taken into account in arriving at decisions related to directing the safe navigation of the vessel. The exercise of command decision-making based on an evaluation of navigation information derived from all sources is referred to as *navigation control.*

8.2 IDENTIFICATION OF TARGETS AND CHART COMPARISON

It is sometimes suggested that the radar picture offers a *bird's-eye view* of the area surrounding the observing vessel. This analogy is imperfect on a number of counts, but two are particularly evident in the use of radar for navigation. The radar antenna does not look down on the terrain from a great height and thus its *view can* be obstructed. Further, it does not offer the optical resolution (see Section 2.2.2) which the use of the word 'eye' may imply. As a consequence, the radar picture may be an incomplete and fairly coarse version of the chart's finely detailed plan view of the terrain. This may limit the observer's ability to identify positively elements of the terrain echoes and relate them to the charted representation.

Interpretation of the displayed picture involves consideration of a number of factors each of which will be discussed in turn.

8.2.1 Long Range Target Identification

When making a landfall, the radar must be carefully observed in order to obtain an early indication of the presence of the terrain. In the absence of clutter (see Sections 3.6 and 3.7) the first echoes will have to be found against the background of receiver noise (see Section 2.6.4.4). If the observer knows the approximate bearing and range at which to expect the first echoes, early detection may be assisted by, from time to time, temporarily setting the gain control a little higher than the normal optimum level (see Section 6.2.7.1). A slight loss of contrast is traded for an increase in received signal amplification and this may be beneficial when looking in a specific area as opposed to scanning the entire screen area. Where an echo stretch facility (see Section 6.8) is provided, it may be found similarly helpful. A knowledge of where to look for the expected echoes on the screen pre-supposes other sources of information concerning the vessel's likely position and the probable detection range of specific terrain features. Such information may stem from dead reckoning techniques, knowledge of leeway and tidal streams, other position-fixing systems (GNSS or LORAN C), and an assessment of specific target detection ranges in the light of radar, target and environmental characteristics (see Chapter 3). This emphasizes the complementary nature of the various data sources available for the safe navigation of the vessel.

Initially the presence of the land may be indicated by only a few responses and these will be considerably distorted by the angular width of the resolution cell (see Sections 2.8.5 and 8.2.2) which will be large at long range. Under these circumstances it will be extremely difficult, if not impossible, to identify positively specific terrain features from the few distorted echoes which are being observed. It must be appreciated that the use of the echo stretch technique is likely to exacerbate the problem by adding radial distortion and it should be switched off after it has fulfilled its role of assisting initial detection. Identification may be assisted if the observer knows which parts of the terrain are likely to show first and also the approximate range and bearing at which they should appear. As indicated in the previous paragraph this presupposes other sources of information related to probable position and target detection considerations.

To exploit fully the use of radar in making a landfall, adequate preparation should be made in terms of collating the information from other sources. An up-to-date and best available estimate of the ship's position should be maintained using information from all available sources. Prior to making the landfall, the chart and the Admiralty Sailing Directions should be consulted in order to assess the ranges at which the radar should detect specific terrain features which are likely to be easy to identify. This assessment should take into account the characteristics of the radar system, the characteristics of the target, the atmospheric conditions and any limits to detection which may be imposed by clutter or attenuation, all of which are dealt with in detail in Chapter 3. In particular it should be remembered that frequently coastlines are backed by higher terrain and at long ranges it must be borne in mind that the first land to show is not necessarily the *coastline* (see Section 3.8.1). Failure to appreciate this can be dangerously misleading because the measured range will suggest that the vessel is farther to seaward than is in fact the case. Also, as the land is closed, responses will be obtained from land which is lower and closer than that originally detected and as it comes above the horizon it may give an exaggerated impression of the speed at which the vessel is approaching the land. When a particular landfall has been made on more than one occasion, it may be possible to establish a list of good landfall targets which may supplement those tabulated in the Sailing Directions.

It has to be recognized that, while civil marine radar offers early warning of the presence of most land formations, long range target identification is not a function at which it excels. A coded racon (see Section 3.5.2.1) is probably the only sure source of early positive target identification and its range may well be limited by the height of its antenna. Positive identification of just one target is a major step forward because it may then be possible to identify other terrain echoes by virtue of their known range and bearing from the identified target. *A free* electronic bearing line *(EBL)* (see Section 6.6.4) is ideally suited to making the necessary measurements.

Care must be taken not to jump to conclusions when a radar echo appears in the general area in which a particular point of land is expected by dead reckoning (DR). There appears to be a great temptation to ascribe immediately and unquestioningly the hoped-for identity to the target. This temptation must be resisted until cross-checks have established that any fix so obtained is consistent with that obtained from all other available sources of information.

8.2.2 The Effect of Discrimination

In the previous section attention was drawn to the difficulty of identifying targets in the landfall situation because of the effect of the size of the resolution cell on the few echoes which may be detected at extreme ranges. Even at the more moderate ranges likely to be employed when using the radar for routine coastal navigation, the effect of resolution (see

Sections 2.7.3 and 2.8.5) may still make it difficult to identify specific coastal features, despite the fact that there will be a larger aggregation of land responses forming some sort of *chart-like* coastline on the plan position indicator (PPI).

The angular width of the resolution cell produces an angular distortion on both sides of all targets by an amount, each equal to half the horizontal beamwidth plus the spot size effect. Neglecting for the moment the spot size effect, it is evident that if two headlands forming the entrance to a bay are separated by less than the horizontal beamwidth, their echoes will overlap. Thus the effect of the limited bearing discrimination may be to mask coastal features such as bays, river entrances, sea lochs and other similar inlets. The echoes of an island close to the mainland may appear as a peninsula and small islands close together may appear as one large island. Thus a charted coastline having many indentations may translate to a featureless coastline on the PPI. It must be borne in mind that, while the width of the beam is a fixed angle, the linear measurement of the arc it cuts off increases in direct proportion to the range at which it is considered. Thus features which are masked at a distance may become identifiable as their range decreases.

The pulse length/spot size dimension of the resolution cell will produce radial distortion of all responses. Except at very short ranges this will have an effect which is much less significant than that of the angular distortion. However, where two features lie one behind the other, for example an island or a buoy located close to the shore, they may appear as one feature due to the limited range discrimination.

The observer's ability to minimize the effects of the angular and radial distortion is fairly limited. In both cases the spot size effect will be reduced by using the shortest appropriate range scale. Where a dual or interswitched system (see Section 2.10.8) is available, selection of the antenna having the narrower beamwidth will assist, as would a modern raster-scan display offering beam processing. A temporary reduction in the setting of the gain control may help to locate features which have been masked by angular distortion, but this requires some practice and is not invariably successful. Use of differentiation (see Section 3.7.4.4) may improve the picture by combating the radial distortion. However, this technique must be used with care, as if two echoes overlap and the farther is the weaker of the two, the effect of differentiation may be to remove the remote echo.

Thus, due to the inherent difficulty of relating the radar picture to the chart, care must be taken to ensure positive identification of any target before selecting it for use in position fixing (see also Section 8.3).

8.2.3 Shadow Areas

Because the line of sight of the radar system is substantially horizontal, some features of the terrain may be shadowed by others. These shadows are a further factor which may make it difficult to relate the radar picture to that of the chart, the problem being compounded by the fact that the shadow pattern will vary with the position of the observing vessel (Figure 8.1).

8.2.3.1 Vertical Shadowing

Figure 8.2 shows an example of vertical shadowing. To a vessel which is close inshore, the lower range of coastal hills may shadow higher mountains lying inland, whereas from another vessel farther offshore, the radar may be able to 'see' the mountains over the top of the coastal range.

Bearing in mind that, as described in Section 8.2.1, a vessel making landfall on such a coastline may well detect the higher inland mountains first, it is evident that if a ship

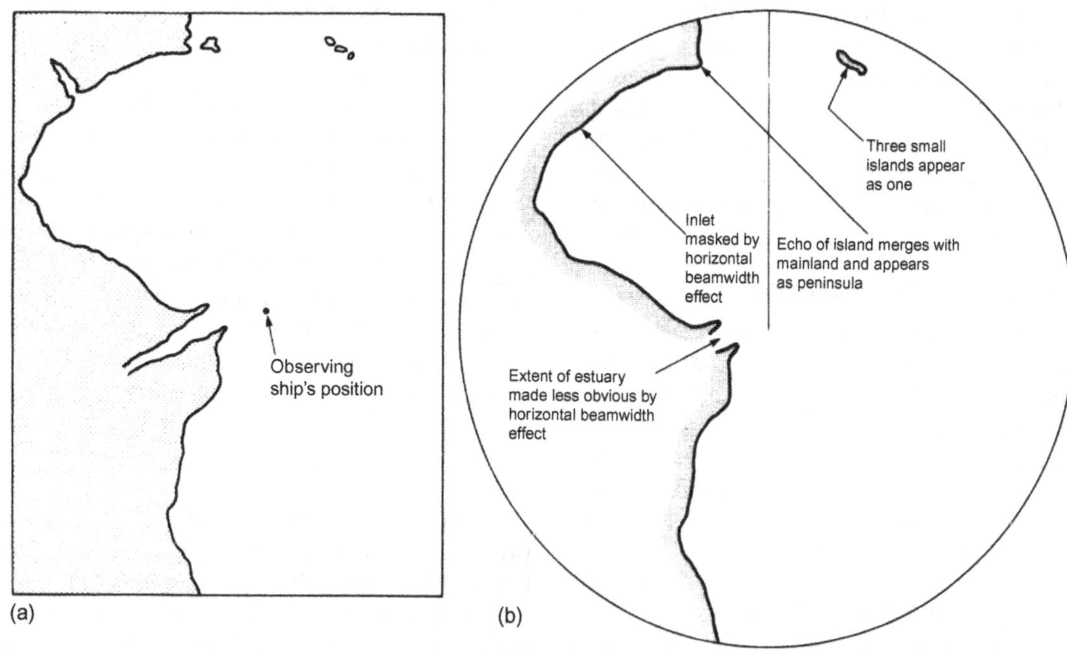

FIGURE 8.1 The radar display of charted features (a) Chart. (b) Radar.

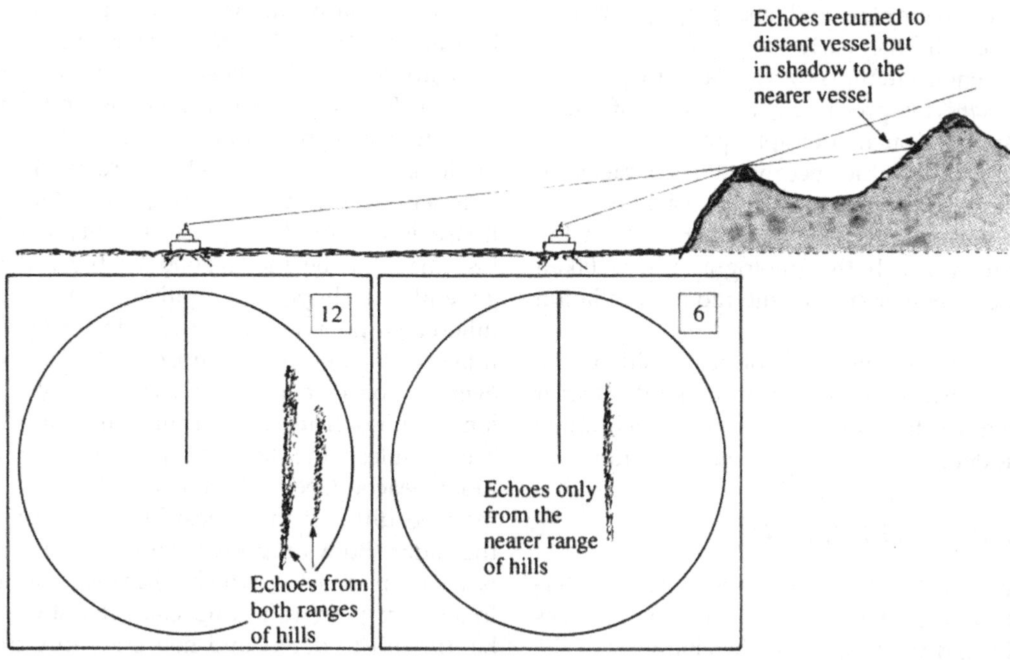

FIGURE 8.2 Vertical shadowing.

closes the coastline from seaward the character of the radar picture will change through three distinct phases. Initially the vessel will observe the higher mountains only, then both coastal and inland mountains and finally only the coastal range. If there are gaps in the coastal range, horizontal shadowing (see below) will also take place, further complicating the pattern.

8.2.3.2 *Horizontal Shadowing*

Figure 8.3 shows an example of horizontal shadowing from which it is clear that the radar picture obtained by a vessel at location *A* looks quite different from that which would be obtained at location *B*. It follows that a ship whose course takes it from *A* to *B* will notice a progressive change in the character of the radar picture.

8.2.3.3 *Composite Shadowing*

In many cases both vertical and horizontal shadowing occur and the interaction between the two as a ship steams past the terrain may produce a complex and changing pattern of responses. In the early days of radar some experimental attempts were made to produce radar *maps* by producing a collage of radar photographs. Such attempts proved to be of little practical value, because the character of the radar picture portrayed by each element in the collage was heavily dependent on the location at which the photograph was taken and the characteristics of the radar installation used.

If a ship is uncertain of its position, the complex shadow pattern may make it difficult to recognize the particular stretch of coastline or to identify specific targets on the terrain.

8.2.4 Rise and Fall of Tide

The radar picture may appear quite different at various states of the tide. The changes will depend very much on the character of the area. For example, at low water an area with off-lying, drying sandbanks may produce large areas of no response (where the smooth sloping surfaces reflect the radar energy away from the antenna) surrounded by fringes of clutter (where the waves break on the shore) which depend on the strength and direction of the wind. By contrast, at high water, in calm conditions there may be no response at all, whereas in strong winds there may be extensive clutter in the shallow water over the banks. The ability to understand and recognize such effects may be of value to the radar observer in identifying specific targets.

8.2.5 Radar-Conspicuous Targets

A radar-conspicuous target is one which produces a good response that can be positively identified. While the ability of radar waves to penetrate fog is of great assistance to vessels navigating in restricted visibility, all targets obscured visually by the fog are not necessarily conspicuous when viewed by the radar.

A particularly important example is a lighthouse, which traditionally represents a key navigational mark whose charted position is known. It is visually conspicuous and can be identified beyond doubt. Unless other adequate sources of position lines are available, it is a serious loss to the navigator if the lighthouse is obscured by poor visibility and that loss may not be made good by the radar. In general, the shape of a lighthouse makes it an inherently poor target (see Section 3.3.4) and, unless it is situated on an isolated rock, whatever response there is tends to be lost in the land echoes from the terrain surrounding it. Thus, unless the lighthouse is fitted with a radar beacon (see Section 3.5), in many cases it may be difficult or impossible to identify on the radar. Buoys represent another example of navigation marks which, because of their inherently poor reflecting characteristics, may be difficult to detect and hence identify unless fitted with a reflector or a racon (see Sections

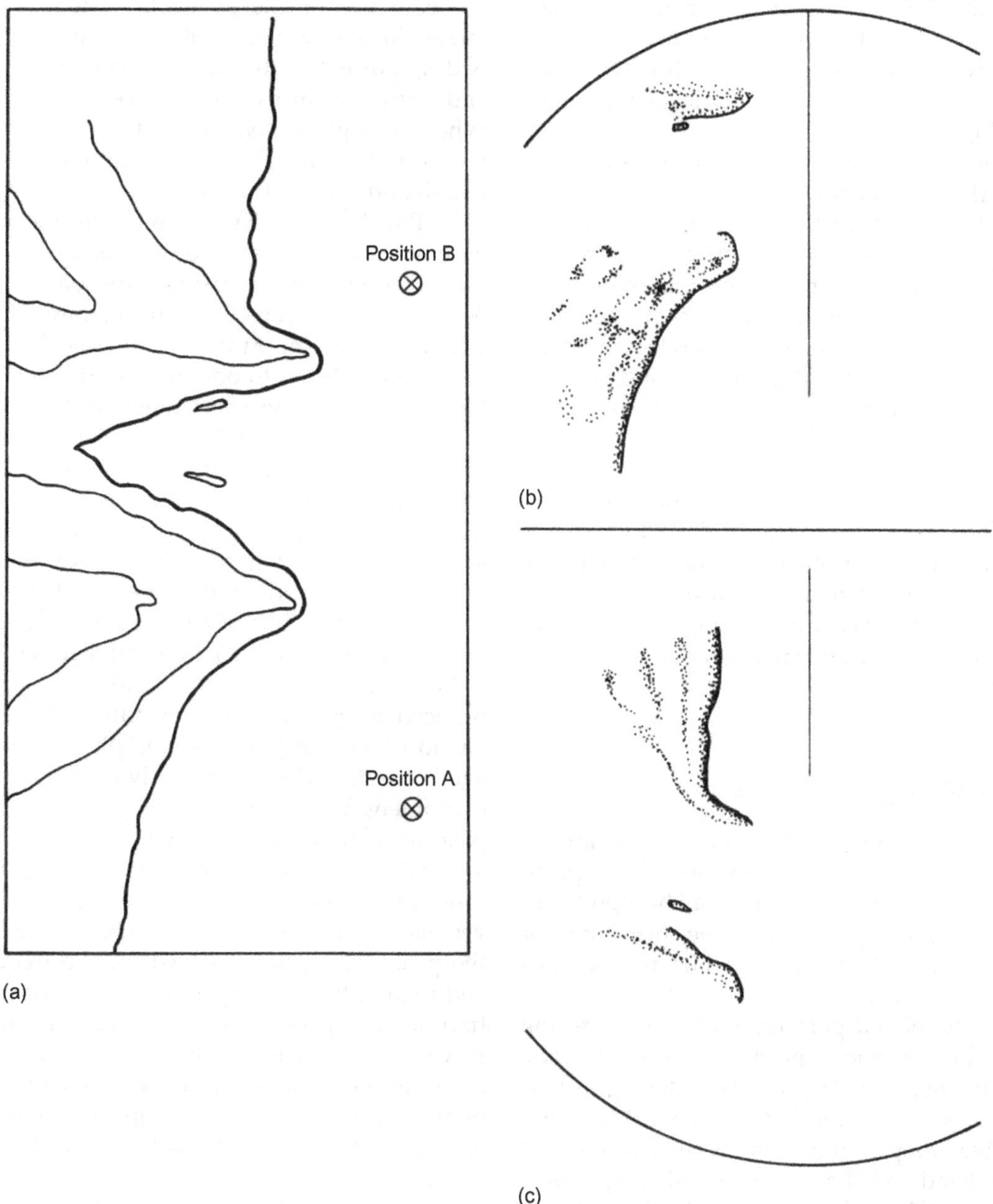

FIGURE 8.3 Horizontal shadowing. (a) The chart. (b) The radar picture from position A: shadowing takes effect particularly in the south, west and northwest areas of the bay. (c) The radar picture from position B: shadowing takes effect particularly in the north, west and southwest areas of the bay.

3.4 and 3.5) or an AIS transmission (see Section 5.3.5.1). This may be particularly so in cases where vessels anchor or fish in the vicinity of a floating navigation mark such as a fairway buoy.

However, some man-made features are particularly conspicuous when observed on radar. Tank farms and small built-up areas are particularly good examples. Their response is among the strongest likely to be encountered and, if isolated from other areas of response, they may be very conspicuous. Naturally occurring features such as small isolated islands offer a further example of radar-conspicuous and potentially identifiable targets.

Sloping sandy coastlines may be quite conspicuous to the eye even when fairly low-lying, but their radar response is likely to be poor because most of the energy is not reflected back towards the antenna. The response may be improved if waves are breaking on the beach, but this will only be experienced at fairly short range.

8.2.6 Pilotage Situations

In most pilotage situations targets are so close that detection scarcely presents a problem. However, identification may be a problem, and an urgent one given the proximity of hazards and the speed with which manoeuvres must be effected.

In estuary and port approach situations, the need for continuous position monitoring may well be met by using parallel-indexing techniques (see Section 8.4), provided that suitable preparation has been carried out beforehand and the necessary indexing targets can be easily and positively identified. Some high-definition radars are designed with very short range scales, low minimum range and good discrimination that suit them for use in pilotage situations. In the absence of such equipment the observer must attempt to make the best use of the available system. In an interswitched system, antenna siting, beamwidth, pulse length, and the setting of gain and clutter controls should receive attention. Where a single system is fitted, fine detail can be improved by use of the shortest pulse length and differentiation.

In the docking phase of pilotage, the presence at very close range of targets which produce strong and spurious responses (see Section 3.9) will conspire with the limited discrimination of the radar system (see Sections 2.8.5.2 and 2.8.5.3) to obscure the fine detail of the picture. On the short range scales in use under such circumstances, only relative-motion presentation will be available and the consequent movement of the land echoes on the screen, together with the changes in shadow areas and indirectly reflected echoes, will greatly exacerbate this effect and make it very difficult to identify the extent of berths, lock entrances and other essential features.

While the problem can be to some extent reduced by judicious use of differentiation, it has to be said that the docking phase of pilotage is one for which current civil marine radar equipment is not particularly suited. In the past the docking of very large vessels has been assisted by the siting of Doppler equipment ashore to measure the rate of approach of the vessels. There is considerable debate as to how the problem might be solved for the frequent and regular berthing of ferries in fog. The more traditional suggestion is to use radar having a wavelength of a few millimetres, while more recent applications use differential global positioning system (DGPS) in conjunction with an electronic chart (see Sections 10.1.6 and 10.2).

8.3 POSITION FIXING

Essentially two types of position lines are available directly from the radar, namely radar range circles and radar bearings. These can of course be used in association with position lines from other sources. In making a decision as to which particular position lines should be used to obtain a fix in any given circumstances, consideration must be given to the accuracy that can be obtained. This will depend on the targets chosen and the type of position lines selected.

8.3.1 Selection of Targets

In selecting targets for position fixing, attention must be given to the certainty with which they can be identified and the suitability of their angular disposition.

8.3.1.1 Target Identification

The importance of positive target identification and the difficulties of achieving this have been discussed at length in the proceeding sections of this chapter. Mistaken identification can seriously mislead the observer as to the vessel's most probable position. Even where the feature has been correctly identified, unless it is small, it is essential to identify which part of the feature has reflected the radar energy; otherwise, accuracy may suffer. For example, when measuring a range from a sloping surface, the high level of ranging accuracy can be lost if the range is not laid off from the correct contour on the chart.

Similarly, if taking a bearing from the edge of the radar echo of a point, considerable thought must be given to the effect of any slope of the terrain and the beamwidth effect of the radar before deciding from which charted position to lay off the observed bearing. In general, if radar bearings are used it is better to attempt to avoid the effect of the half beamwidth distortion by using the centre of small isolated targets or, if it is necessary to use a point of land, to take a bearing which runs along the axis of the headland rather than one which is tangential to it.

8.3.1.2 Angle of Cut

In selecting targets for position fixing, the considerations given to angles of cut are not confined to radar but are based on the general principles of position lines. Where two position lines are involved, the angle of cut should be as close to 90° as possible as this minimizes the displacement of the fix due to any errors in the position lines used. To provide cross-checking, good practice dictates that where possible at least three position lines should be used. Under these circumstances, angles of cut less than 30° or greater than 150° should be avoided as small errors in the position lines can produce relatively large errors in the fix.

8.3.2 Types of Position Line

The position lines available from radar and other sources can be used in a variety of combinations. The considerations of radar accuracy affecting these combinations are discussed in turn.

8.3.2.1 The Use of Radar Ranges

The inherent accuracy of radar range is very high. IMO Performance Standards (see Section 11.2.1) require that the fixed range rings, the variable range marker and cursor should all enable the range of a target to be measured with an error not exceeding 1.0% of the range scale in use or 30 m, whichever is the greater. On the 12 mile range scale the error should not exceed 222 m and this is adequate for most navigation at this range from the coast. Care must be taken in using the facilities provided for range measurement to ensure that the full potential accuracy is realized. The practical procedures for doing this are summarized in Section 6.6.7. Additionally, as indicated in Section 8.3.1.1, it is essential to establish with certainty the charted location of the reflecting surface. There is little point in measuring with a high degree of accuracy the range of a target which has been incorrectly identified. Assuming that the foregoing procedures are followed, radar ranges have the potential to produce a highly accurate fix.

8.3.2.2 The Use of Radar Bearings

When compared with that of radar ranges, the inherent accuracy of radar bearings is very much lower. In Section 6.9 the various error sources set out in IMO Performance Standards are discussed and it is shown that the inherent accuracy of radar bearings is such that a bearing measured from the display can be in error by as much as 2.5° without the system being in breach of the Standard. The arc subtended by an angle of 2.5° at 12 miles is approximately 870 m, which does not compare favourably with the inherent accuracy offered by radar ranges. Suitable practical procedures must be followed to ensure that the potential accuracy is realized; these are summarized in Section 6.6.7. As already stressed, care must be exercised to ensure that the target has been correctly identified (see Section 8.3.1.1).

In general, given the relative inherent accuracy levels, wherever possible radar ranges should be used in preference to radar bearings.

8.3.2.3 The Combination of Radar Ranges and Bearings

Inevitably, circumstances will arise in which a combination of ranges and bearings will have to be used. Where only one feature is available for fixing, a single range and bearing fix does have the virtue that the angle of cut is 90°. However, in such circumstances the respective accuracies of the two position lines must be borne in mind.

8.3.2.4 The Use of Single Position Lines

A single radar range or bearing can be used in the same way as any other single position line to exploit the various general navigation techniques, such as the running fix, which maximizes the use of such observations.

8.3.2.5 The Combination of Radar Position Lines with Those from Other Sources

Radar ranges and bearings should be seen as one of several sources of position lines, all of which may be combined to arrive at a decision as to the vessel's most probable position.

A particular example is the situation which arises in clear weather when only one feature can be positively identified. Under such conditions the accuracy of a visual bearing (which is higher than that of a radar bearing) can be combined with that of a radar range circle to produce a fix having a potentially high degree of accuracy. In conditions of poor visibility of course the lower accuracy of a radar bearing would have to be substituted for the visual.

In general, the radar should be seen as one source of position-fixing information which should be compared and combined with the others that are available. Any disparity between the information available from different sources such as is manifest by a 'cocked hat' should alert the observer to consider carefully why the disparity exists. Serious thought should precede any decision to discount a particular data source. In particular, a large 'cocked hat' should be a warning to check all position lines and not an excuse for discarding the one which puts one furthest from one's preconceived position.

8.4 PARALLEL INDEXING

8.4.1 Introduction

While navigating from one port to another, it is inevitable that for part of the time the ship will be in confined waters, be it the approaches to the port and berth, or in a busy waterway such as the Dover Strait or Malacca Strait. Leaving aside for the moment any consideration of avoiding collision with other vessels, restrictions on the available sea room require the navigator to monitor the vessel's position, not just with an increased accuracy commensurate with the reduced safety margins and clearing distances imposed upon him, but also with an increased frequency to ensure that environmental and other forces that take the vessel off her desired track are recognized in sufficient time for corrective action to be taken and the vessel to be maintained on a safe track.

Traditionally the navigator would identify the vessel's proximity to the track or danger by putting a fix on the chart, the data for this fix having been obtained from a variety of navigation sources. Depending upon which navigation system is being used, the time needed to establish the fix will probably be, at best, about 2 minutes. However, a single fix does not reveal the whole story, merely the ship's position some 2 minutes ago. Before the navigator can take corrective action, it is necessary to know the trend of the movement, that is a series of fixes is required – probably three or more, bearing in mind the inherent imprecision of most fixing systems. Consequently, there may be a time delay of the order of 6–10 minutes between the vessel beginning to deviate from its desired track to the time when proper considered action is taken to return the vessel to safety.

If the reason for the deviation is a five knot crosscurrent and the shoal water is only a few cables away, a reaction time in excess of 6 minutes is too great.

One might expect that, under conditions such as these, where the shoal water is so close, there ought to be sufficient visual navigation marks nearby to enable the person conning the vessel, be it the master, pilot or officer of the watch, to react almost instantly to any deviation from the planned track. This will probably be the situation in a port approach with moderate to good visibility. In poor visibility, however, when all the visual marks disappear, conning becomes extremely difficult even with the radar giving the relative positions of some of these marks. Consider also the very large vessel fully laden. Its deep draft means that it may be 'confined' to waters well away from port approaches in an area where the navigation marks necessary for the visual conning are sparse.

The usefulness of parallel indexing in the above circumstances is indisputable. There have been numerous incidents of grounding which have resulted from the navigator using a position-monitoring method that had too long a reaction time for the conditions in which the vessel was operating (for instance, the *Metaxa*, and also the *Sundancer* casualties), or where the navigator failed to recognize that the data he was appraising was insufficient on which to base remedial action.

It is in situations like these that parallel indexing shows its true worth by enabling the navigator to monitor the vessel's progress moment by moment and by providing enough data to allow corrective manoeuvres to be made in a timescale which is very similar to that of visual conning, that is about 2–3 minutes.

8.4.1.1 Parallel Index Facilities

The facilities for parallel indexing have evolved from the reflection plotter (see Section 7.10.3) to electronic index lines or electronic maps drawn by the user (see Section 4.9). Effectively, parallel indexing has further evolved into the integration of ARPA with Electronic Chart Display and Information Systems (ECDIS) and GNSS, where all fixed radar features can be compared with the chart for accuracy, simultaneously checking the GNSS system against the radar. This is covered fully in Sections 10.1 and 10.2. These advances in other technologies may have led some manufacturers to drop maps (with permanent storage) as many sets only have the four electronic index lines (IMO minimum requirement).

Unlike map lines, the user enters direction and offset for the index line, so there is no control of the length of these lines only, they are automatically drawn from the edge of the display to the edge of the display.

The simplest parallel index facilities possible are an azimuth-stabilized radar with a reflection plotter (traditional radar) or the electronic index lines (modern radar). With this simple system it is usual to use a relative parallel index plot on a relative-motion display. It is difficult to use a true parallel index plot on a ground-stabilized true-motion display for reasons covered in Section 8.4.5. The relative-motion parallel index plot is covered in Sections 8.4.2–8.4.4.

The ability for the user to draw electronic lines on the display has replaced the Perspex reflection plotter. In fact the lines provided can be relative lines (which remain relative to the own ship) or true lines (which are fixed to where the positional information that the system sensors tell the radar). The facilities associated with drawing these lines are covered in Section 8.4.6.3.

The use of electronic lines also means that it is now more straightforward for an azimuth-stabilized radar to use a true parallel index plot, where the user moves the map manually over the radar plot keeping the radar marks in line with marks on the electronic map as covered in Section 8.4.5.

The advantage of the true parallel index technique is that the index lines have the same orientation as the chart and take less skill to interpret. However, this technique becomes far more powerful when the display is additionally automatically ground-stabilized. The automatic ground-stabilization can be provided by:

1. Reference to a known fixed radar target or targets plotted by an ARPA.
2. Input from a ground track dual-axis Doppler log, although this would be unusual.
3. Input from a position-fixing system such as a GNSS.

These automatic ground-stabilized displays are covered in Section 8.4.6.

In the following sections the techniques specifically for using the reflection plotter are mentioned when appropriate, as some may still be in use. The relative-motion parallel index plot still has a useful future using electronic lines, particularly for use on simple low-cost radar displays without speed input but also on more complicated systems for simple parallel index plotting including the 'unplanned' techniques covered in Section 8.4.7.

8.4.2 Preparations and Precautions

8.4.2.1 Pre-Planning

Navigators who conduct their vessels in confined waters using blind pilotage techniques such as parallel indexing must never lose sight of the fact that safety margins are often minimal and on no account must operator errors be allowed to creep in to the parallel-indexing data. By the time an error becomes apparent it can very likely be too late to recover the situation. Considerable care must be taken therefore in deriving the parallel-indexing data from the navigation chart and transferring it onto the radar display, to ensure that it is as accurate as possible. All data should be cross-checked and, indeed, the whole process of acquiring the parallel-indexing data is best carried out when there is a minimum of pressure and distraction. It would be inviting trouble to attempt to derive and use parallel-indexing data in quantity when the vessel is already proceeding through the confined area. Most 'off the cuff' work suffers from inadequate checking and hence is susceptible to error. On occasions, parallel indexing may be employed directly from the radar screen without any pre-planning (see Section 8.4.7), but generally the need for parallel indexing should be assessed during the passage-planning stage and all the necessary data to carry out the task, extracted from the chart and stored in suitable form (see Section 8.4.3.5), hours before it is intended to make use of the data. Only by doing this can one be sure of reliable data of sufficient quality to realize the full benefits of parallel indexing.

8.4.2.2 *Preparing the On-Board Equipment*

The radar set is an integral part of the parallel-indexing process and must be in proper working order, able to display the *indexing* target, be operated correctly, orientated and stabilized and with proper discrimination and accuracy. Particular care must be exercised with the tuning, gyro compass error and heading marker accuracy (see practical setting-up procedures in Chapter 6). The electronic index lines will be used, so the readout accuracy must be confirmed. Similarly, if the bearing lines and variable range markers are also to be used, then any errors must be removed or known so that due allowance can be made.

8.4.2.3 *The Radar Presentation*

In theory, parallel indexing can be employed either on a relative-motion north-up azimuth-stabilized display or on a ground-stabilized true-motion display.

There are practical difficulties in achieving a reliable ground-stabilized display with a simple radar without ARPA or a GNSS position input (see also Sections 6.2.6.2 and 8.4.5). For instance:

1. Obtaining the present tidal rate and set can be time-consuming and not necessarily very accurate. Even if good values are obtained, changing conditions may mean that the values need to be frequently reassessed.
2. An accurate knowledge of our own ship's course and speed through the water is not always easily available, especially when using unreliable speed logs.
3. Given good information regarding the vessel's movement over the ground, it still remains for the radar equipment to track accurately in response to this information. This cannot always be taken for granted.

Bearing in mind that parallel indexing is intended to improve navigational safety, the operational difficulties mentioned above that could influence the reliability of this work means that true-motion parallel indexing must be considered second best to relative-motion parallel indexing for a simple radar with azimuth stabilization.

8.4.2.4 *Selecting the Indexing Target and its Effect on Accuracy*

The *indexing target* is a chosen radar target which appears on the PPI and whose movements relative to the observing vessel as it transits a particular confined area will be closely monitored. If everything goes according to plan, the ship will be manoeuvred to make this target track along the lines that have been drawn on the reflection plotter, or at least within certain limits from these index lines, and in so doing the navigator can ensure that the vessel follows the chosen track line.

For a target to be suitable, there are certain conditions which must apply to the indexing target:

1. It must be a good radar target, that is clearly visible on the PPI at the ranges at which it is intended to be used.
2. It must be identifiable among all the other land targets in the area and also there must be some recognizable feature on the target to which all the measurements can be related. This latter requirement can be difficult to resolve in the case of non-point

targets, since most change their radar aspect and therefore their appearance as the relative position of the ship changes (see Section 8.2).

The navigator, in preparing for parallel indexing, must study the chart of the area with particular care to find the most suitable indexing targets, such as the end of a breakwater, the tip of a headland, small islands, isolated rocks, isolated lighthouses, etc. Lighted navigation marks are of no consequence in this context of course and it is usually advisable to not rely on a floating navigation mark as an indexing target (except as provided in Section 8.4.6.1), not so much because of their slight variations in position but because of the possibility of their not being there at all when the vessel arrives. It is impossible to change all the pre-worked parallel-indexing data to another target at short notice.

Note: Whenever a vessel is operating in a confined area where parallel indexing can be useful, the navigator is advised to watch the radar picture carefully with a view to identifying suitable indexing targets for use on a future occasion. This method is far more certain than studying the chart. This information should be stored for future use, for example by marking the chart.

Another important consideration is the choice of the working range scale for the radar. Inevitably, the longer the range scale, the lower the accuracy with which one can monitor the position. The navigational requirements will determine the accuracy required and also the safety margins which will apply. The navigator must therefore be sure that the working range scale of the radar meets this accuracy requirement and that the selected indexing target will be 'in range'. This should not be taken to imply that a complete parallel index transit should be completed on one range scale. As accuracy requirements change, so should the working range scale. Hand-drawn index lines need to be redrawn at different range scales. It is a great benefit that modern electronic index lines change their position automatically with changes in radar range scale. Similarly, as the vessel progresses through the confined area, so the indexing target will also need to be changed.

8.4.2.5 *Preparing the Navigation Data*

Parallel indexing is part of the navigation of the ship and therefore any guidance lines that appear on the display should be directly related to actual data of navigational significance on the chart, for example the safety margins and course lines (Figure 8.4).

In planning a passage through a confined area, the adjacent dangers need to be assessed in detail. Adjacent dangers in addition to shoal water can include areas of strong tidal sets and the boundaries of traffic separation schemes. The navigator should also assess the state of preparedness of the vessel for the passage, that is for the time of the intended transit, speed and engine readiness, method of steering and turning circles, preparedness for anchoring and manning levels, etc. and, with these factors in mind, decide on a safe distance at which to clear the adjacent dangers.

This data is put onto the chart as a safety or 'margin' line spaced away from the danger by the estimated safe distance.

Common sense and experience must be relied upon here when estimating where to position the margin lines, as it is impracticable to draw a 'contour' line around every adjacent danger. Furthermore, it is extremely difficult to reproduce curved lines from the chart on a reflection plotter with any degree of accuracy and impossible on modern radars equipped with electronic lines. The purpose of this part of the passage-planning process is to identify clearly all the clear water that can be used safely by the vessel during a transit of the area. Naturally the intention is to maintain the course line, but unfortunately the seaways have to be shared with many other vessels and

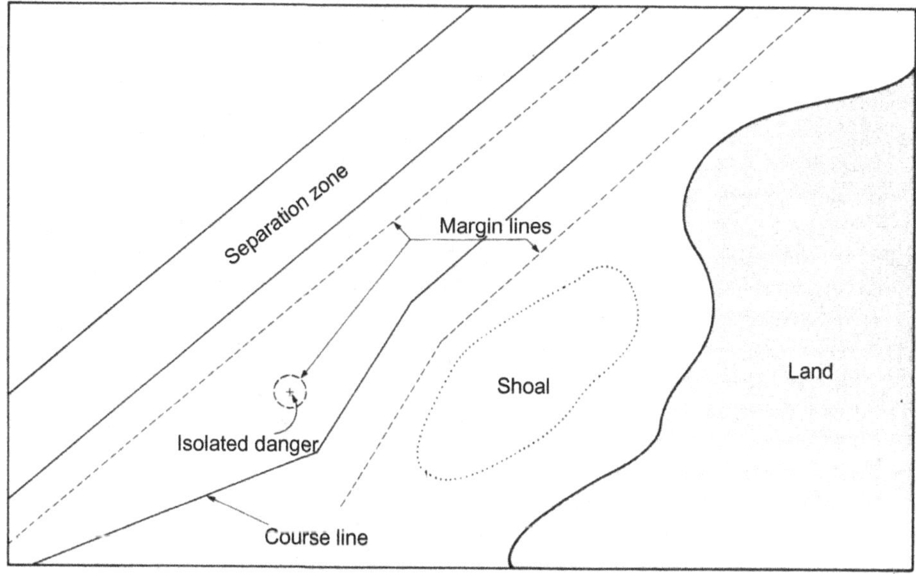

FIGURE 8.4 Adjacent dangers, margin lines and course lines.

therefore one must be prepared to leave the course line should an anti-collision manoeuvre be required at any time.

The final stage of this navigation process is to draw in the course lines on the chart, taking care at all times to keep within the defined safe/clear water areas.

8.4.3 Relative Parallel Indexing: The Technique

8.4.3.1 The Relative Parallel Indexing Principle

On a relative-motion compass-stabilized radar display, a land target will always move across the screen in a direction which is the reciprocal of the observing vessel's course made good over the ground and at a rate directly related to the vessel's speed over the ground.

In the example shown in Figure 8.5(a), as the vessel steams from position *A* to position *B* (3.6 miles off island *C*), the radar echo of the island on the PPI will move from a position on the starboard bow until it is 3.6 miles off on the starboard side bearing 100° (*T*), having moved in a direction 190° (*T*), that is reciprocal to the ship's course made good over the ground.

With an understanding of this principle, it is possible for the navigator to predict the movements of an echo on the PPI, of a 'fixed' object while the ship follows a series of course lines near this object and to draw lines (index lines) on the display to represent these movements prior to arriving in the area (see Figure 8.5(b)).

8.4.3.2 Positioning the Index Line on the Radar Display

In the example illustrated in Figure 8.6(a) it is intended to make good a course of 010° (*T*) and to pass 3.6 miles off the island *C* (the indexing target).

The electronic index line is usually positioned on the display by entering the desired angle and offset required (see Figure 8.7(c)). The latest standards require four of these lines (see Section 11.2.1).

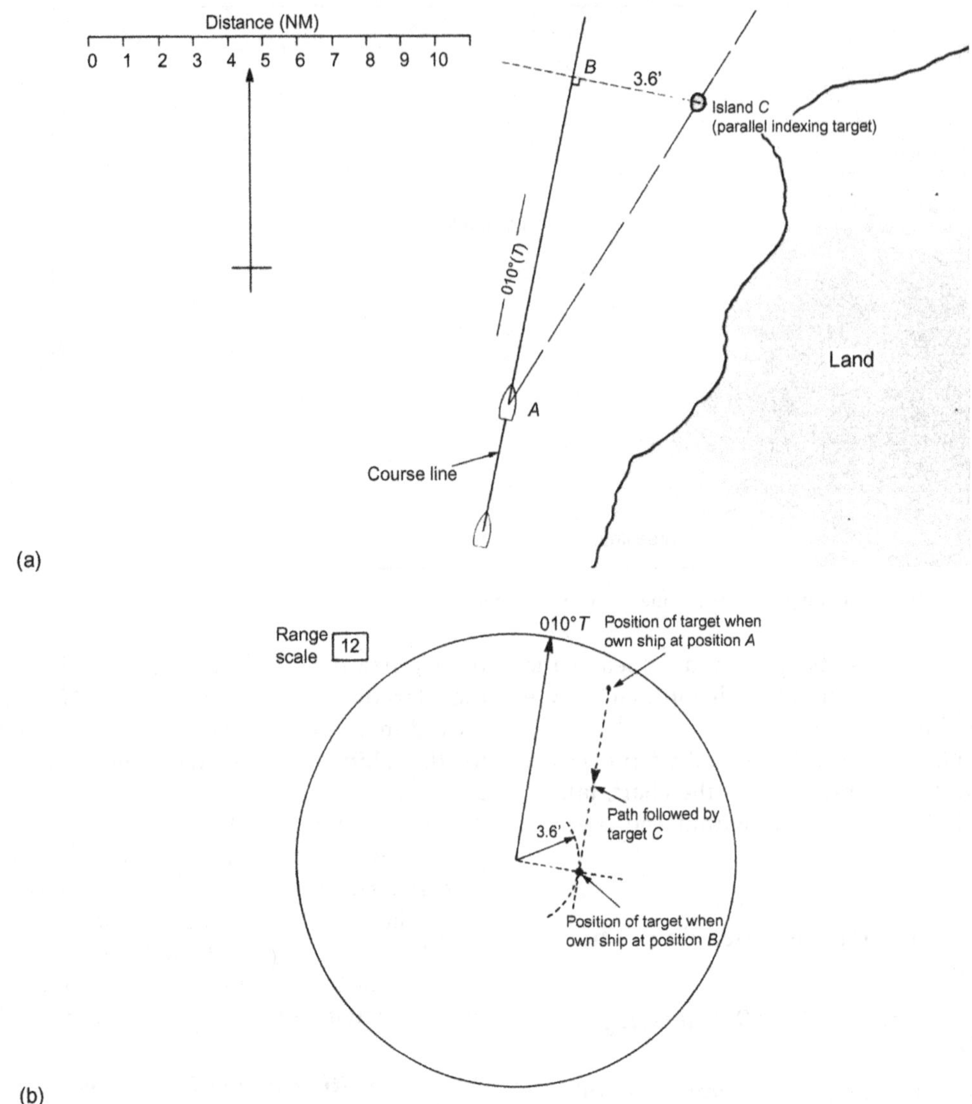

FIGURE 8.5 A simple index line. (a) A charted plan. (b) The relative movement of a fixed target on the PPI as prepared in (a).

On equipment with maps or manual lines drawn on the reflection plotter it was often necessary to set up the lines in one of two ways, either:

a. Set 3.6 miles on the variable range marker and then turn the cursor to 010°/190° (*T*). Now, lay off the index line at a tangent to the variable range marker and parallel to the cursor as shown in Figure 8.5(b); or

b. Measure the range and bearing of the indexing target from a chosen position on the course line (any position will do, but

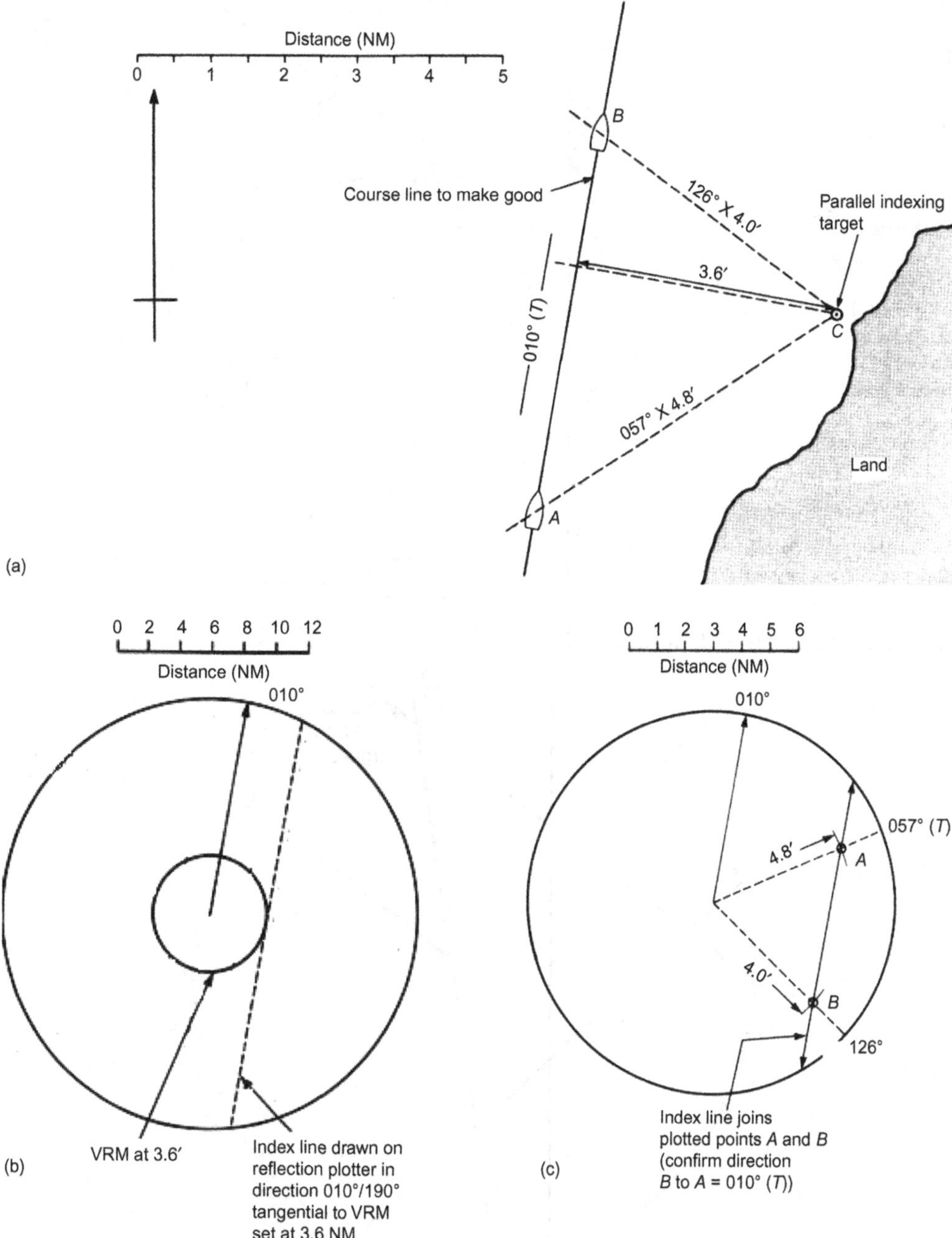

FIGURE 8.6 Positioning the relative index lines. (a) A charted navigation plan. (b) Positioning the index line by beam distance using a mechanical cursor. (c) Positioning the index line by ranges and bearings.

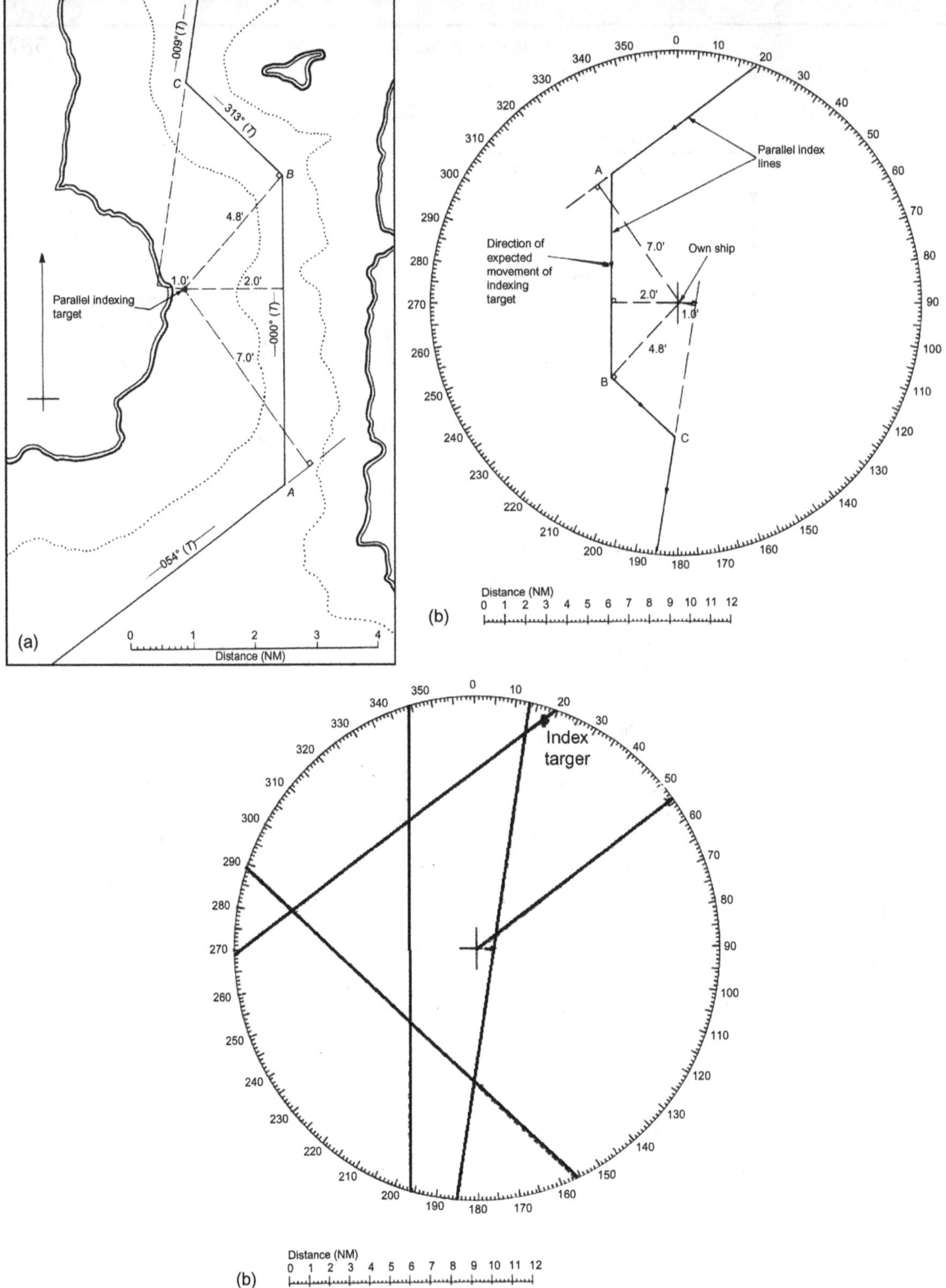

FIGURE 8.7 More complex indexing. (a) A complex charted plan. (b) A series of parallel index lines. (c) Undesirable effect of using all the four minimum IMO specified index lines together.

preferably a beam position, an alter-course position or another position where the highest accuracy is needed). Plot the range and bearing, turn the cursor to 010°/190° (*T*) and lay off the index line, parallel to the cursor and passing through the plotted position.

As a variation on method (b), if two points on the charted course line are fixed and plotted on the reflection plotter then they can be simply joined by a straight line. This method is useful in that it allows a simple cross-check, that is fix the start and end of a particular index line and then confirm from the chart, the actual direction and distance between the two (see Figure 8.6(c)). Example 8.1 and Figure 8.7 (a) shows a more complex series of intended tracks within a parallel-indexing area and Figure 8.7(b) illustrates how this data translates onto the display.

EXAMPLE 8.1

Figure 8.7(a) has been drawn to scale. Extract the appropriate data and then, using either (a) the beam distance method or (b) the range and bearing method, construct the parallel-indexing lines as they would be drawn out on the display. The accuracy of the construction should be verified by measuring the directions and distances *A* to *B* and *B* to *C* or the offset distances, as appropriate. The plot should appear as in Figure 8.7(b).

Figure 8.8(c) shows how this would appear using all four electronic lines, at once, as on a modern IMO specification display. Clearly the extra portions of line do not lend themselves to clarity in this example. It is strongly advised in this type of scenario, that navigators have all four electronic index lines entered and ready, but when on passage, only display the current index line and the next one (i.e. have two turned off).

8.4.3.3 Transferring the Margin Lines to the Radar Display

a. *Normal lines.* Adding the margin line data to the parallel-indexing plot is simply an extension of what has already been done, that is draw another series of lines on the display representing the acceptable limits for deviation of the indexing target. When possible, these are usually drawn in such a way as to distinguish them from the normal course keeping index lines. See the straight margin lines on Figure 8.8(a) and (b).
b. *The danger circle.* A position circle could be constructed on the now obsolete reflection plotter. It is not usual to find this option to plot a circle electronically on modern raster display equipment. If the option is there, Figure 8.8(a) and (b) illustrates the special case of an isolated danger with a circular margin of safety or 'danger circle'. In this case, its position on the display is determined by first measuring the position of the indexing target from the isolated danger and then transferring that to the display. Then the margin of safety is drawn as a circle of appropriate radius centred on the plotted point *X*. Some modern radar sets allow the addition of symbols, which could indicate the centre of such a circle.

   Note: Although position *X* has been plotted on the display, this should not be misinterpreted as the actual position of the isolated danger relative to the observing vessel. As the ship transits the area, the indexing target moves along the index line. When the index target reaches position *Z* (the alter-course position), the vessel is 2.8 NM away from the danger, that is by the distance *XZ*.
c. *An alternative method of defining margin lines.* On traditional displays and in confined waters where the margin lines are relatively close to the course line, it may be found more convenient to fix the margin lines on

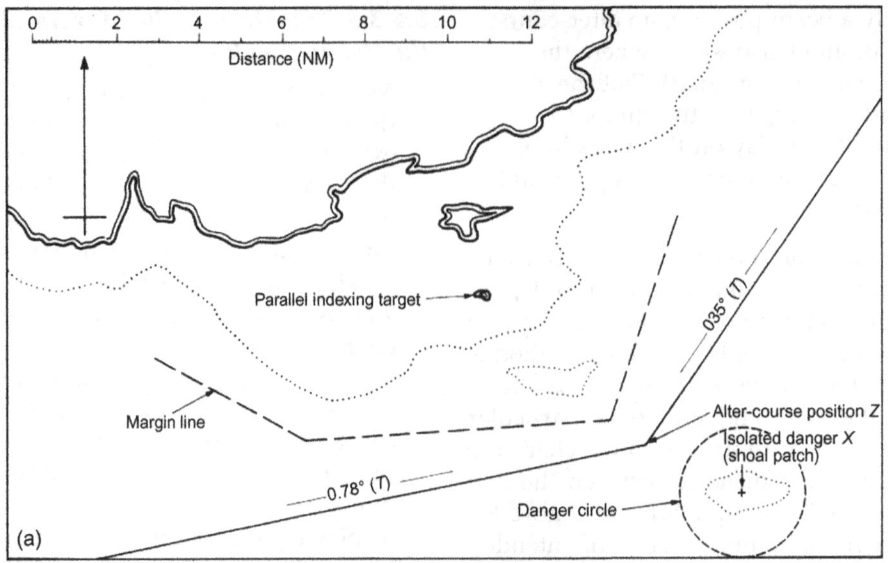

FIGURE 8.8 Margin lines and danger circles. (a) A navigation plan including margin lines and a danger circle. (b) A parallel-indexing plot of margin lines and a danger circle.

the plotter using the index lines as the datum rather than the indexing target. This technique is depicted in Figure 8.9(a) and (b).

Note that the distances being measured from the chart are the beam distances between the alter-course positions 1 and 2 and the adjacent margin line for both headings. The margin lines on the final 310° course are additionally referenced by the directions and distances.

As confirmation that the resultant margin lines on the plotter are in the correct position, check that their individual directions and lengths agree with those laid off on the chart. Provided that the distances *A*, *B*, *C*, etc. are relatively small, it will not be found necessary to measure the beam angle but to lay off the direction by eye and measure the appropriate distance.

8.4.3.4 The 'Wheel-Over' Position

When transiting a very confined channel and/or conning a large vessel, it is advisable to plan the wheel-over positions using a knowledge of the ship's turning characteristics including any interaction effects. Use of this data should make it possible to keep on the track line required. This information will appear on the chart as a point on the course line spaced a calculated distance prior to the alter-course position. This distance can be used on the display to indicate the point on the index line where the ship should begin to turn. See Figure 8.10(a) and (b).

Note: It is advisable to make these turning predictions for a moderate value of helm (10°–15°) on the understanding that, if the conditions are other than calm (and assuming deep water), the actual turn as executed may not correspond with the predictions. Consequently the person conning the ship should be prepared to increase or reduce the amount of helm applied in order to meet the precise requirements of the turn.

The track line may be regained using the wheel-over position when the vessel is not exactly on the course line and is approaching

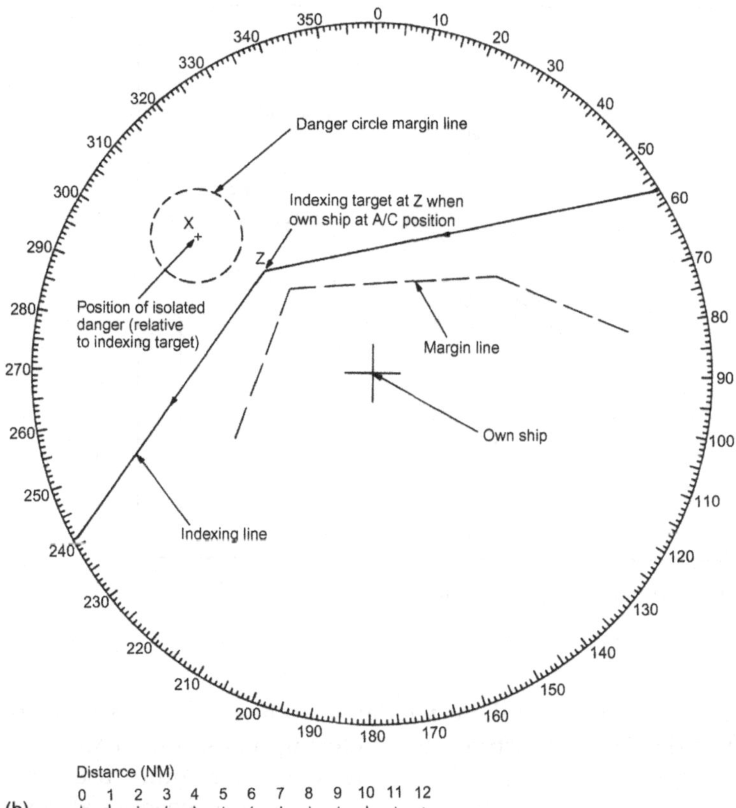

FIGURE 8.8 (Continued)

an alteration point. Provided circumstances allow, it is common practice to take the new course either early or late, as necessary, in order to recover the course line (see Figure 8.11(a)).

To identify the position at which to begin the alteration of course, a construction line is drawn through the wheel-over position and parallel to the new course line. On the parallel index plot (see Figure 8.11(b)), when the indexing target reaches the construction line through the wheel-over position, the turn begins.

8.4.3.5 Recording the Parallel Indexing Data

Having spent the necessary time and effort in planning a coastal passage or port approach and having defined the areas where parallel indexing will be used, the data needed for the plot must be extracted and recorded in a suitable form, that is either in a notebook or in a diagrammatic form so that, when the time comes to use the data, it can be used with an absolute minimum of delay. It is essential to remember that no mistakes can be tolerated at this stage.

The diagrammatic method of data storage gives the operator the additional advantage of being able to cross-check by direct comparison of the indexing lines drawn on the plot with the previously prepared data.

An extension of the diagrammatic method when used with the traditional reflection plotter involves drawing the parallel-indexing data

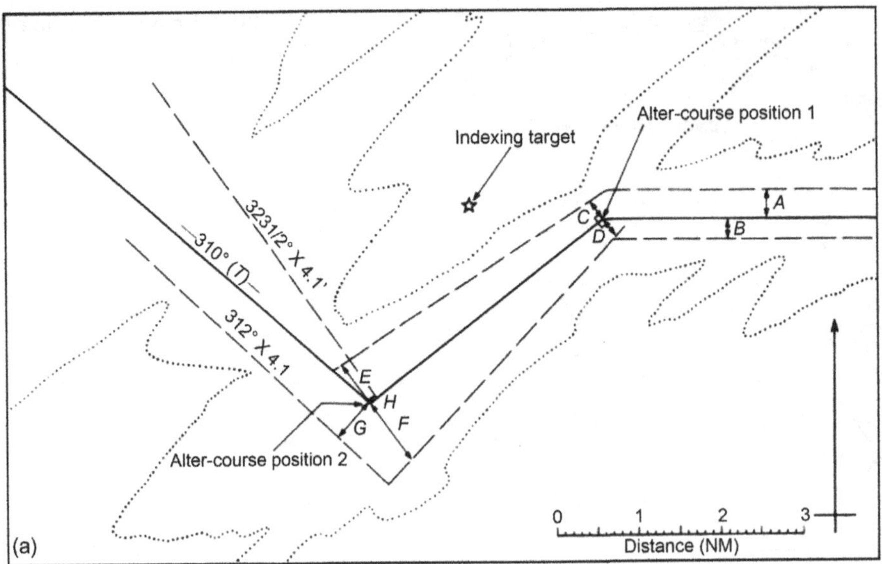

FIGURE 8.9 Narrow margins. (a) A navigation plan with narrow margins.

accurately on one or more acetate sheets which have been cut to fit the face of the display. Apart from the normal parallel indexing data, these sheets would need careful marking with reference points for centre and north.

The recording of this navigational data is part of the normal passage-planning process and the operator should not look on the data as permanent and able to be used repeatedly on future occasions without amendment. It should be remembered that, in deciding on the margin-of-safety lines, the tide and present vessel status, for example, are important factors. Hence the parallel indexing data contained on the acetate overlays should not be used subsequently without first checking that all the data is still valid.

Included with the parallel indexing lines will be the operational instructions, such as the point at which to begin using the data, the range scale to use, the appropriate point/time to change range scale or to change the indexing target.

Where a radar 'mapping' facility is used to construct the indexing lines, the map may instead be simply stored for future use. The fact that an index map has been used and tested is useful as the possibility of an error in its construction has been removed. However, the map should always be rechecked before use in case of changes in the navigational regime which may have occurred in the interim.

8.4.4 Progress Monitoring

As described in Section 8.4.2.4, the intention while navigating using parallel indexing is to keep the indexing target on the relative indexing line that has been drawn for it, or at least within the margin lines that may apply. To do this the navigator must continuously monitor the movement of the indexing target and take particular note of (a) the actual position of the indexing target relative to the index line and

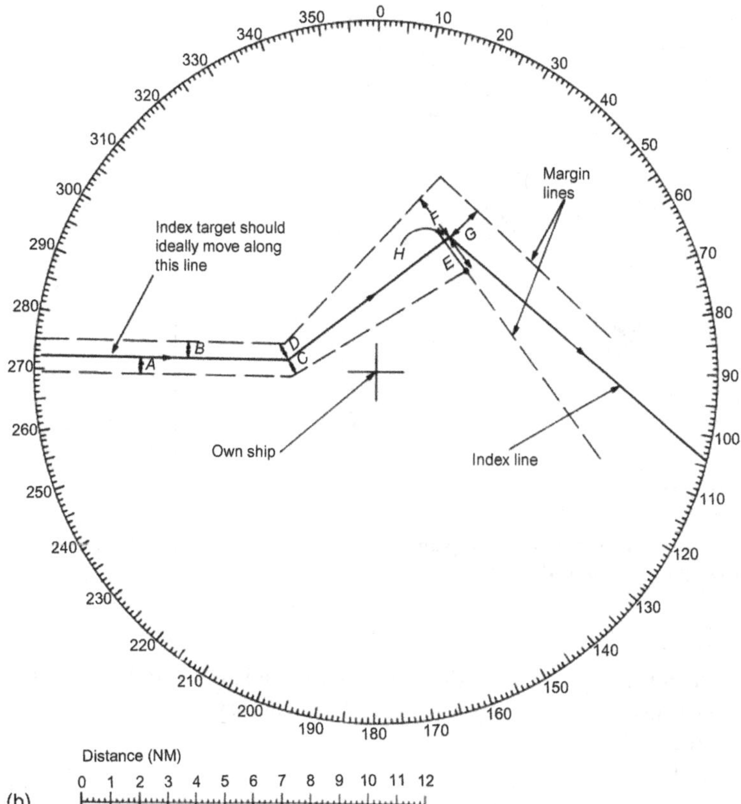

FIGURE 8.9 (Continued)

safety margin lines and (b) the present trend of movement relative to the desired direction of movement.

Simple observation of the echo and its afterglow can provide this information, but on the traditional reflection plotter it was possible to plot the indexing target with grease pencil to make the data more obvious. The actual position of the indexing target relative to the margin lines confirms that the vessel is in safe water and how much sea room is currently available. The trend of movement as provided by the plotting confirms whether or not the vessel is at that time making good the required track. The navigator can now decide, on the basis of these two pieces of information, whether corrective action is needed, either to regain the planned course line or to make good the required track direction.

For example, in Figure 8.12, monitoring the indexing target as the vessel transits the narrow channel shows that over the past few minutes the indexing target has left the indexing line and is approaching the margin line. This indicates that the ship is being set to the west and that the required corrective action is to turn to starboard by a few degrees so that the indexing target moves back onto the index line, or at least does not move any farther away from it.

Notice that no reference is made to the ship's head in this process because what in fact is relevant here is the course which is being made good rather than the direction in

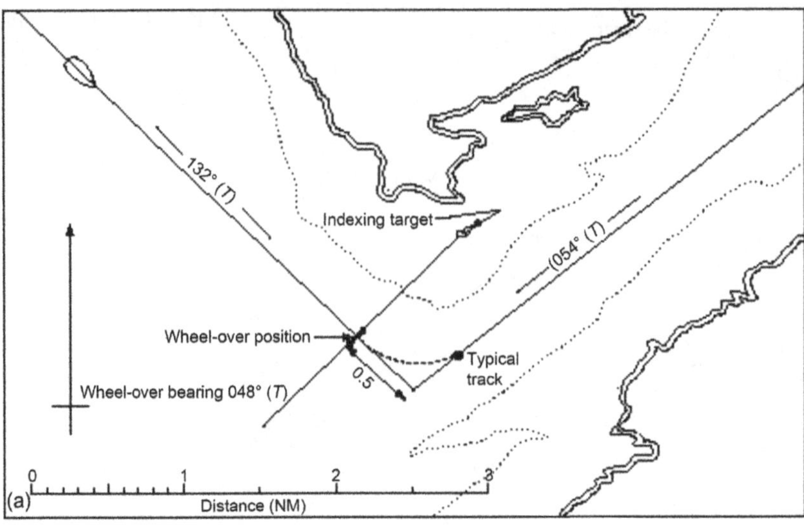

FIGURE 8.10 Incorporating the wheel-over position. (a) A navigation plan with alter-course and wheel-over positions. (b) A parallel-indexing plot with the wheel-over position marked.

which the ship is pointing. The difference between the two can be large in a cross tideway and when the ship's speed is low. When visibility is restricted, these differences can be very disorientating. It is not being suggested here that the ship's head should be totally disregarded; obviously there should be some logical relationship between the two directions provided by knowledge of the tidal set and rate.

As the indexing target moves along the indexing line there is no need to keep the earlier information and plots behind it and these may be progressively cleared from the plotter. While the vessel is being navigated using parallel indexing it is very important to fix the ship's position on the chart by normal methods at frequent intervals as confirmation that the ship *is* in fact in safe water and *is* in fact moving in agreement with the parallel indexing information.

On modern equipment with electronic lines, track monitoring is limited (but still very effective) to the target trails or history dots. Care should be taken that relative target trails or history dots are being observed for this purpose (not true trails or history dots).

8.4.5 Parallel Indexing on a True-Motion Display

As explained in Section 8.4.2.3, the use of a true-motion radar presentation for parallel indexing requires an ability to ground-stabilize the display reliably. Assuming that this condition can be met, the PPI may be used as an extension of the chart. The navigational data additional to the radar targets already on the PPI, such as course lines and margin-of-safety lines, can be marked on the display.

The data needed to be transferred to the plotter is identified during passage planning in exactly the same way as when working with relative indexing, except that the lines will be located relative to any fixed identifiable target on the PPI rather than to a single indexing target. This implies that, if the data is to be maintained at a high level of accuracy, careful

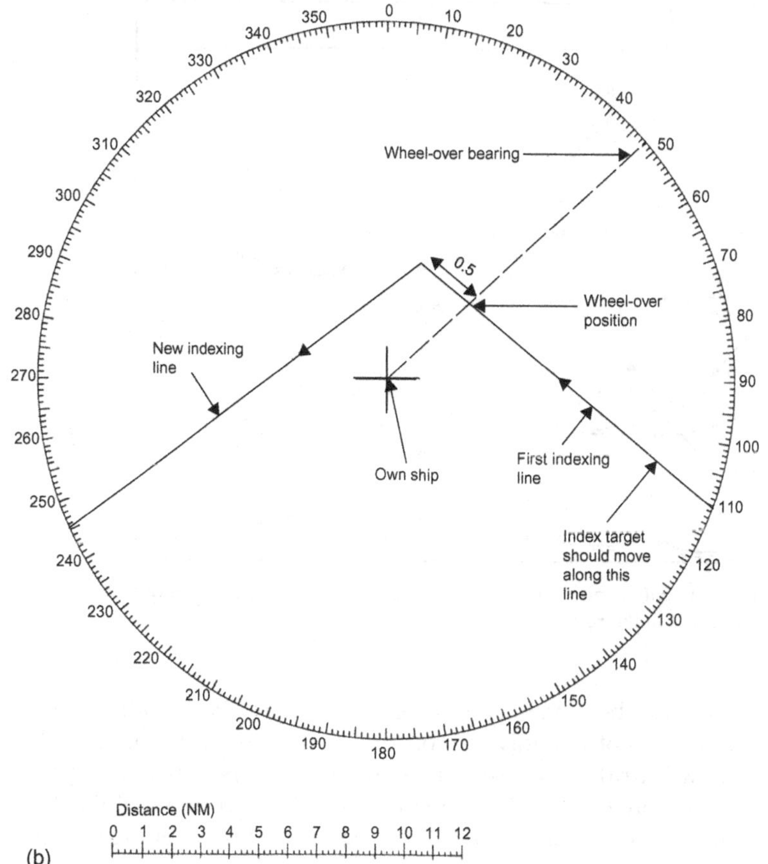

FIGURE 8.10 (Continued)

consideration must be given to the identification of these fixed targets. In addition, the overall accuracy of this process will be enhanced by using the shortest range scale which is capable of showing the required data and by using reference navigation marks that are the nearest available. Each individual line should also be cross-checked by confirming from the chart the direction and length of each leg. Figure 8.13(a) shows the navigational situation with the own ship approaching an alter-course position north of island *Y*, to be followed by the transit of a narrow passage. Suitable radar navigation marks are identified at point *X*, island *Y* and point *Z*. The alter-course position and margin lines are referenced relative to these points. As our vessel approaches the area (see Figure 8.13(b)), the radar echoes of the selected navigation marks are identified, marked on radar and the navigation lines are drawn on the display using the previously recorded ranges and bearings. Each individual line should also be cross-checked by confirming, from the chart, its length and direction. As the ship transits the area in question, the navigator must be prepared to reset the origin and/or change the range scale at the appropriate time. Where the navigation data is on a reflection plotter, this necessitates redrawing, therefore the point at which to do this needs to be pre-planned and the data stored in a form suitable for rapid

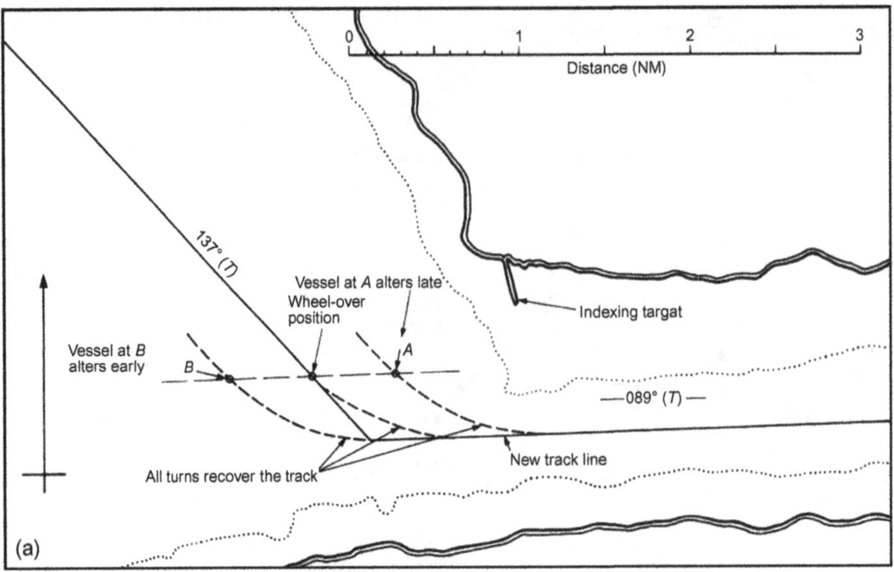

FIGURE 8.11 Recovering the track. (a) A navigation plan to recover the track line using the wheel-over line. (b) A parallel-indexing plot with wheel-over line to assist in the recovery of the track line.

application to the plotter. It has to be recognized that if ground-stabilization is not perfectly maintained, the fixed targets will drift away from their marks on the reflection plotter. The picture reset controls are normally fairly coarse and it may be difficult or even impossible to recover from this situation without redrawing the lines. Many users believe that this limitation of true parallel indexing rendered the technique unsuitable for practical use on older radars equipped with azimuth stabilization and older reflection plotter.

However, if electronic lines or maps are employed it becomes more feasible for the observer to move the map as it drifts due to tide and other effects.

The process is further improved by the use of automatic ground-stabilization. This technique requires the use of more advanced facilities of either ARPA, dual-axis ground track speed log, GNSS or other positioning system. However as an ECDIS display would also probably be present on such a vessel and the use of the ECDIS display would almost certainly be of preference, with perhaps radar overlay if available. A possible exception might be using the ARPA ground-stabilization method (described next) in preference to the ECDIS.

8.4.6 Modern Radar Navigation Facilities

8.4.6.1 Ground-Stabilization Using ARPA (See Also Section 4.8)

The target-tracking capability of the ARPA can provide the navigator with an important and potentially very useful piece of data to help with the navigation problem, namely a continuously updated rate and set of the current being experienced. This is a significant improvement on the previous situation where relatively long periods of plotting or position fixing were needed to be able to deduce the rate and set experienced over that period. The

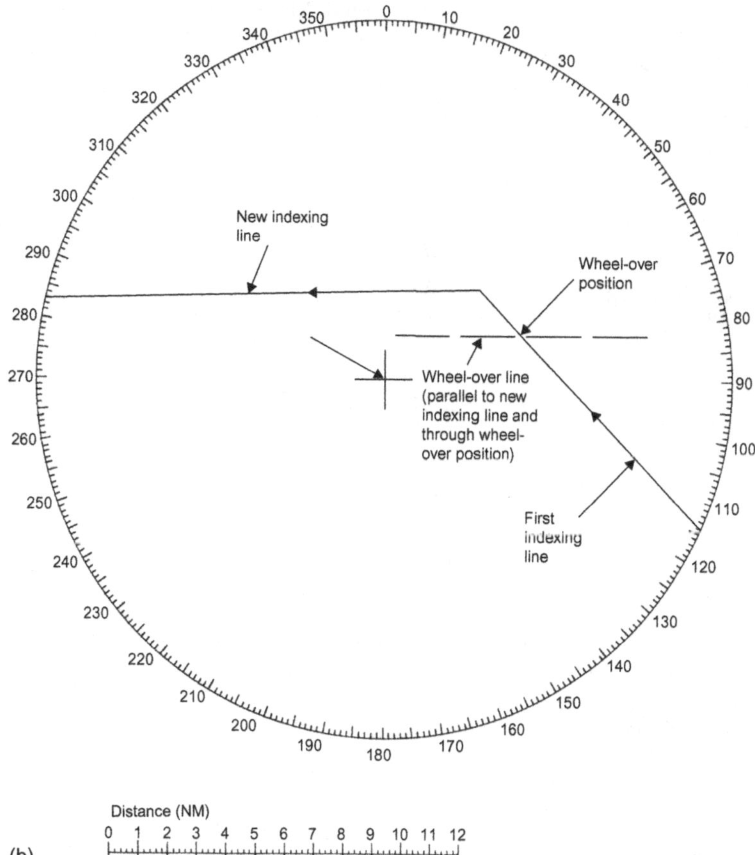

FIGURE 8.11 (Continued)

navigator then had to assume that this data was constant and use it to plan future action, naturally with varying degrees of success as this data can be very time-variable.

There are two prerequisites for the ARPA to provide the navigator with accurate tidal data:

1. The computer must be provided with accurate values of the own ship's motion *through the water*, that is true heading and speed (see Section 6.9.3).
2. There must be a radar echo from a known fixed reference target showing on the PPI and capable of being tracked by the ARPA. In most cases this means that it must be small and isolated, otherwise the point being tracked will tend to wander randomly within the echo area and so provide random spurious data. In some geographical areas, this type of fixed target can be hard to find. However, the rigorous selection requirements associated with parallel-indexing targets (see Section 8.4.2.4) do not apply in this case and, provided that the target is small and isolated and does not move its geographical location during the period of time that it is being used for this purpose, it will suffice. Hence floating marks are acceptable with due caution in this regard.

The accuracy of the tidal data calculated is dependent not only on the accuracy with which the reference target is tracked (yielding

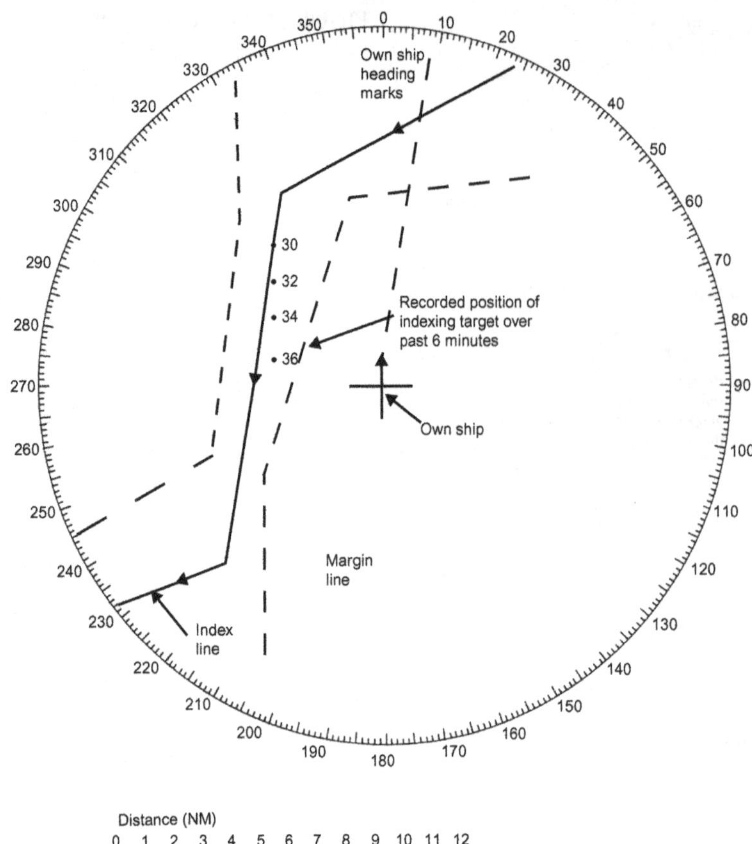

FIGURE 8.12 A parallel-indexing plot with the target movement recorded.

the ground track), but also on the accuracy of the input of the observing vessel's water track since the tidal values are deduced from the vector difference of these two tracks. For practical purposes, the accuracy of the set and drift so obtained should be at least as good as would be obtained by using more traditional methods, i.e. better than ±0.5 knot for rate and ±20° for direction. The fact that it has been obtained very rapidly and with only token effort by the navigator makes it an ideal source of information with which to ground-stabilize the radar presentation. This may be done by setting in the relevant values on separate controls, or the radar manufacturer may provide a direct internal route such that, at the touch of the appropriate control, the tidal data is automatically applied to the tracking of the own ship (see Sections 4.8 and 6.9.6). Since the tidal data is continually updated, the result is an almost perfect degree of ground-stabilization where the own ship and all target motions are now indicated with respect to the ground.

Note: When the display is being automatically ground-stabilized by a reference target, the accuracy of the own ship's perceived movement through the water does not affect the indicated ground course and speed on the screen. If, however, a read-out is taken of tidal rate and set, then the computation involves the own ship's input course and speed data, and this factor must therefore be correct.

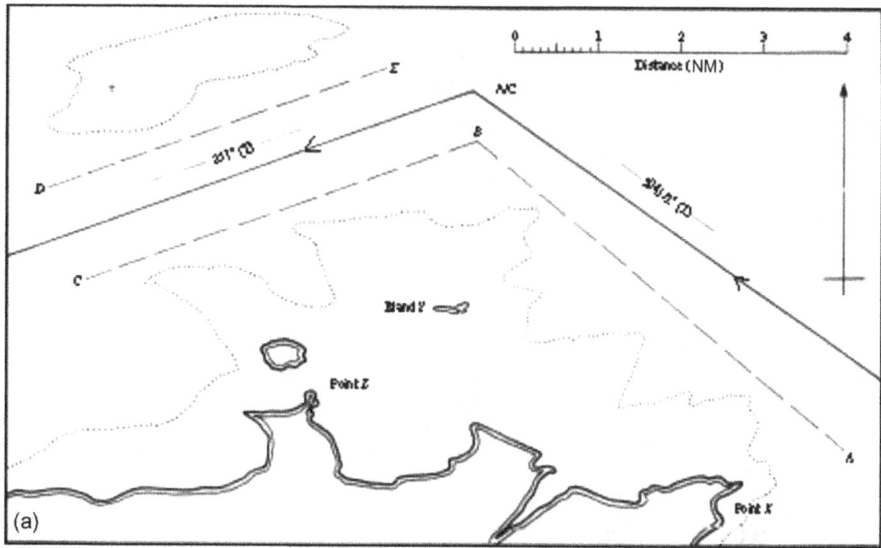

FIGURE 8.13 Parallel indexing on a true-motion ground-stabilized presentation. (a) A navigation plan to transit a narrow passage. (b) A parallel-indexing plot on a true-motion ground-stabilized presentation.

The reliability with which the ARPA continues as a ground-stabilized display depends on how well it continues to track the *reference* fixed target. Tracking can be lost under certain circumstances (see Sections 4.8 and 6.9.6) and, remembering the purpose for which ground-stabilization is selected for the radar presentation, it is advisable to pay close attention to the performance of the ARPA in this respect. In particular, always try to keep the reference target visible on the PPI and under observation. Other fixed targets suitable for the purpose of ground-stabilizing should already be being tracked and available to take over the role of reference target at short notice. Some ARPA manufacturers allow the simultaneous tracking of several fixed targets from which the mean value of tidal data is derived and applied. This makes the system less vulnerable to the loss of tracking of a single reference target.

Once the radar presentation has been stabilized, the additional navigational data can be applied to the reflection plotter as described in Section 8.4.5.

8.4.6.2 Other Automatic Ground-Stabilization Methods

The main disadvantage of the reference target method of stabilization is the availability of fixed small radar targets. Other methods of automatic ground-stabilization exist that do not require ARPA plotting facilities.

The inputs from a dual-axis ground track Doppler log can be used, but this is not common because of accuracy issues with logs in general and the limited depth range in which the equipment will operate.

The most popular method of ground-stabilization tends to be the input from GNSS (see Sections 10.1 and 10.3) as they have a high degree of accuracy and availability and are easily connected. It is also possible to use other position-fixing systems such as LORAN C, but the much superior accuracy, reliability and availability of GNSS mean that other systems are not popular for this purpose.

When using these sources for fixing position, the observer has no need to worry about the availability of reference targets and the

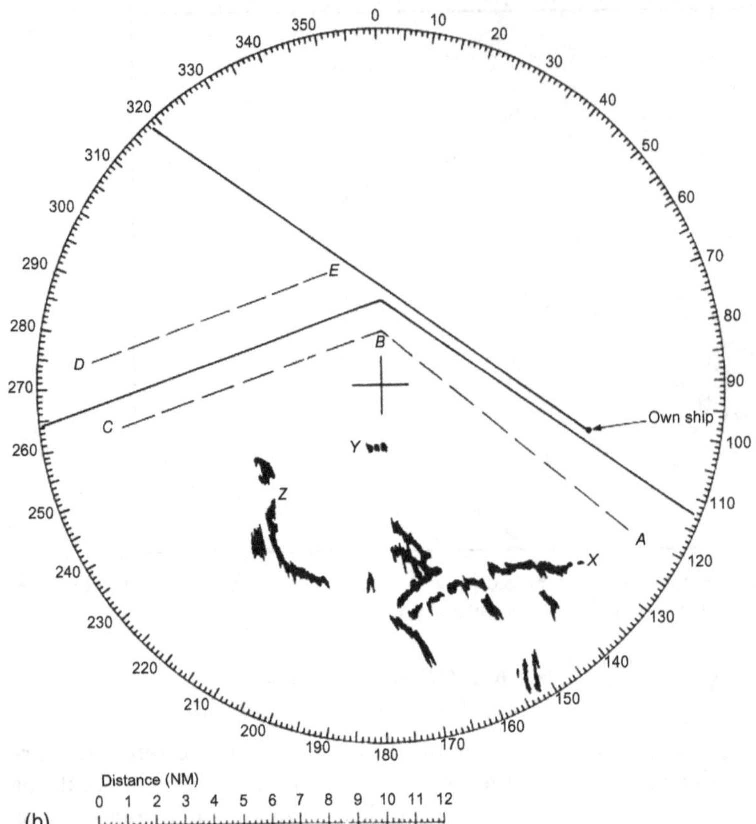

FIGURE 8.13 (Continued)

system will run until switched off. However, in using GNSS data the observer should remember that the operation of the GNSS depends on equipment outside the vessel over which he has no control and may not be informed of system problems. GNSS position fixing should therefore be checked regularly. This can be done externally to the radar set by comparing radar, GNSS and other data positions on the chart. However, if the facilities are available, it can be done on the radar display by comparing the radar image with an electronic true map (see Section 8.4.6.3 below) or integration with ECDIS (see Section 10.2). If not available, conventional parallel indexing may be used.

The comments on the accuracy of derived tidal data in the previous sections also apply to GNSS automatic ground-stabilization.

8.4.6.3 Electronic Navigation Lines (See Also Section 4.9)

Many radars provide some form of optional navigation line package. This facility allows the navigator to use graphics to draw electronic lines directly on the screen. A major advantage of such a facility is that the observer has complete freedom to reset the origin of the picture and change range scale, because the lines are automatically scaled and referenced to suit the selected range scale and presentation.

In theory, the pattern of lines may be referenced to the observing vessel, or to the ground. There is considerable variety in the way manufacturers have approached the provision of this facility, and both modes are not necessarily available in any given package.

Where the lines can be referenced to the observing vessel they will maintain constant position on the screen when the relative-motion presentation (see Section 1.5.1) is selected, thus facilitating relative-motion parallel indexing as described previously.

The ready availability of automatic ground-stabilization in radar and ARPA systems makes it simple to generate a line pattern which remains stationary on the screen when the ground-stabilized true-motion presentation (see Section 1.5.2.2) is selected. This facilitates true-motion parallel indexing as described previously in Section 8.4.5 and, additionally, allows the navigator to arrange the lines in such a way as to form simple maps giving a *chart-like* presentation. It is essential to appreciate that in the case of many systems this will be accompanied by ground-stabilization of the true vectors, as a result of which they will *not* represent the headings of other vessels. If the radar is also being used for collision avoidance, the observer may be dangerously misled. This important limitation of ground-stabilization is discussed in Sections 8.4.8.2, 4.8 and 6.9.6. In some ARPA systems it is possible to display ground-referenced lines when the relative-motion presentation is selected. In such a situation the map or other pattern should maintain its registration with a ground-stationary target.

The format in which navigation line data is read into the computer is not standardized and many different systems exist while yet others are being developed. In general, the method will involve one of the following:

a. A series of ranges and bearings
b. A Cartesian coordinate system
c. A series of latitudes and longitudes.

In each case the navigator must have some datum (or reference) from which to measure so as to establish the relationship between the chart and the radar picture (e.g. if a navigation line is referenced on the chart relative to a prominent navigation feature such as the end of a breakwater, then, when the vessel arrives in the area in question and the breakwater echo appears on the radar screen, the navigation line can be positioned in the correct relationship to the radar echo of the breakwater and will thus take up its proper geographical location).

The navigator must be very aware of the accuracy with which the datum can be defined and, similarly, of the accuracy with which the lines can be positioned on the screen. It is also a condition that the datum used must be an identifiable radar target.

Many radar 'navigation' packages provide a storage facility such that the map can be drawn at some convenient time prior to use and put into the computer's memory. Depending on the sophistication of the particular package, several maps of varying sizes may be stored simultaneously and called up when required. When recalled to the display, the map needs precise positioning; to assist in this function, a *map align* control is usually provided. It is worth remembering that it may be impossible to turn the map in azimuth once drawn and therefore, before constructing any map, the navigator must check that the radar presentation is accurately north-referenced and stabilized.

A fundamental precaution that must be taken by all navigators designing maps which are to be used to assist with navigation in confined waters is the inclusion of check marks within the map such that, at any time while the maps are being used, it is possible for the navigator to confirm by simple inspection of the PPI screen that the lines drawn are in correct registration with geographical features. Happily, the lines drawn on the paper chart do not move without warning, but this may happen with electronic lines. For any radar there is likely to be one or more sets of circumstances that would allow the maps to shift their position. Usually the drift is very slow but therein lies the danger, for the gradual movement easily escapes detection unless there are obvious marks within the map to highlight the discrepancy. Generally, when the radar presentation has been ground-stabilized by *fixed target* reference, the map stabilization is reliable provided that the ground-stabilization is accurately maintained. The commonest form of check is to include points or lines in the map which correspond to recognizable fixed radar targets. Only when these markers lie on the appropriate radar echoes is the map known to be properly positioned. These markers, if they include floating reference marks, have the complementary function of attracting the attention of the navigator to buoys which are out of position or missing.

8.4.7 Unplanned Parallel Indexing

Occasionally, the navigator may realize a need for parallel-indexing data while actually conning the vessel in an area where the passage-planning appraisal had not foreseen the need. In such a case it is necessary to devise a strategy which, with the minimum of measurements, will allow a simple and rapid application of data to the reflection plotter or a rapid input of data to the facility which generates the navigation lines.

For example, suppose that the vessel is experiencing difficulty, due to tide or wind, in making an alter-course position off a buoy or other radar target:

a. With a relative-motion radar presentation, set the variable range marker (VRM) at the required passing distance (CPA) from the radar target and draw a line on the display from the radar target at a tangent to the VRM. Thereafter, *steer the vessel* so that the echo remains on the index line.

b. With a true-motion ground-stabilized radar presentation, set the VRM at the required passing distance from the radar target, then draw a line on the display through the radar echo at a tangent to the VRM. Thereafter, *steer the vessel* to maintain the tangent relationship between the VRM and the index line.

It is essential to appreciate that '*steer the vessel*' means just that, and that no 'course to steer' is directly available.

The advantage of these methods is their instant response to a particular need. Navigators should not, however, lose sight of the dangers inherent in failing to pre-plan thoroughly, such as the absence of cross-checking and the inability to detect the proximity of danger. The navigator also needs to be able to pick out, at short notice, a reliable indexing target.

8.4.8 Anti-Collision Manoeuvring While Parallel Indexing

Parallel indexing is purely a radar navigational technique which is unlikely to be employed unless navigational constraints are present and which requires the continuous attention of the navigator. However, non-navigational obstructions in the form of other vessels may appear at any time requiring anti-collision action. This action must not take the vessel into unsafe water and it is in integrating this anti-collision action with the navigational constraints that the *margin-of-safety lines* included in the parallel indexing data show their true value.

8.4.8.1 Using a Relative-Motion Presentation

Unless ARPA data is available, the relative-motion display with parallel indexing data on the display is most unsuitable as a source of anti-collision data, mainly due to the disruption it causes to the navigation process while the plotting is taking place. To determine the motion of a target, it is preferable to use one of the following techniques:

a. Use a second radar set solely for anti-collision data acquisition. In a confined water situation with no ARPA, true-motion would probably be of assistance as it allows the tactical situation to be more apparent and, consequently, possible action by targets to be more predictable and detectable.

b. Plot the target's position on the navigation chart. A period of such plotting puts the anti-collision problem into its proper context and the navigator can then plan his actions to avoid the traffic while still remaining in safe water.

When working on the relative-motion plot, it can be difficult to assess rapidly the importance of any particular moving target and much time and effort can be used in plotting each one as it appears. A 'rule of thumb' method to resolve this problem is to take an approximate range and bearing of the moving target from the own ship, transfer the line defined by this vector to the indexing target (the end of the line which intersected the target to be placed on the indexing target). The other end of the line now indicates the position of the moving target relative to the fairway. If it is in or close to the fairway margin lines then it should be given further consideration immediately. If, however, it falls completely away from the parallel-indexing fairway then it can be discounted for the time being as it is not in or close to the intended route. Obviously it could subsequently enter the fairway so it cannot be ignored completely.

In Figure 8.14, while indexing on the present heading of 110° (*T*), target *a* has appeared on the starboard bow. By transferring line *Oa* so that *a* lies on the indexing target at *A*, it is established that the target is well outside the fairway at O_1 and the observing vessel expects to turn to 050° (*T*) before reaching the target.

Where ARPA facilities are provided, it should be possible to extract collision-avoidance data in the usual way.

8.4.8.2 Using a True-Motion Ground-Stabilized Presentation

a. *Without automatic ground-stabilization.* In this case stabilization is achieved by the navigator observing a known fixed mark and making adjustments based on these observations (see Sections 4.8 and 6.9.6). Moving targets within the vicinity are soon recognized after a small amount of plotting. Their positions and approximate movements with respect to the navigation problem can be deduced. However, it is essential to bear in mind that all targets will now exhibit their *ground* tracks.

This limitation is resolved by turning the tidal speed control to zero for long enough to assess the course and speed of the target through the water, that is 2–3 minutes, depending on the range scale. (*Note*: The parallel-indexing lines on the plotter are now out of position and the origin will need to be reset to restore the parallel-indexing plot.)

b. *With automatic ground-stabilization.* In this case the ground stabilization is provided by a fixed target reference, dual-axis log or position-fixing system such as GNSS. Echoes of moving targets can be tracked in the normal manner and their true vectors displayed. The relative vectors or data will provide CPA and TCPA information as required.

However, the true vectors may well represent the movement of the targets *over the ground* and if the tidal component is significant there could be a large discrepancy between the

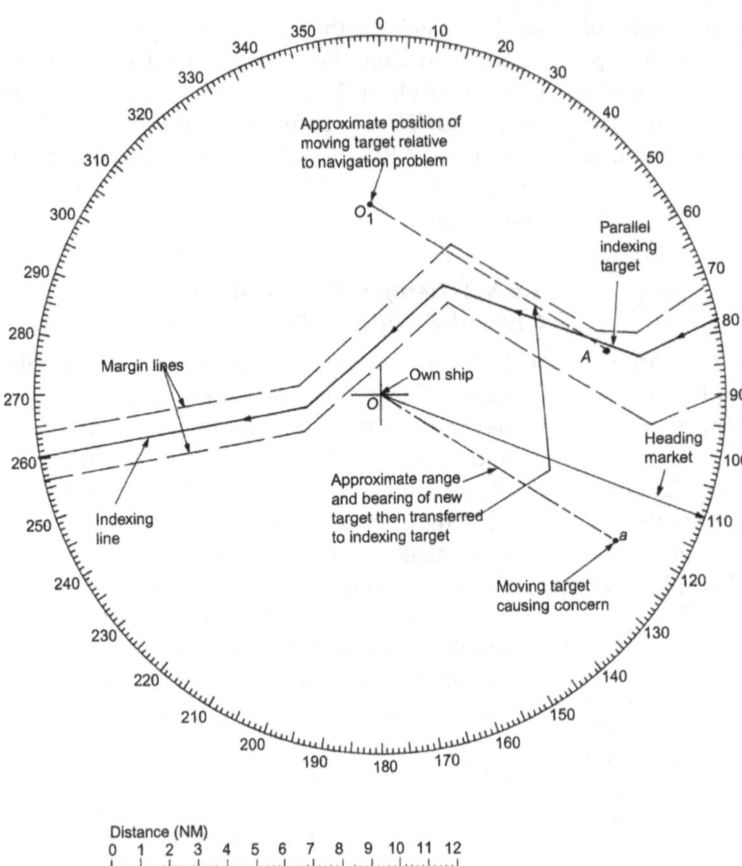

FIGURE 8.14 Approximate identification of the relative position of a target on a parallel-indexing plot.

ARPA ground tracks and the target's course and speed *through the water* (see Sections 4.8 and 6.9.6). Since it is the *course through the water* which provides the *aspect*, the distortion of this piece of information can result in a complete misunderstanding of the situation which is facing the navigator. For example, in clear visibility at night, as a vessel approaches a port and is presented with an array of lights from moving targets against a backdrop of shore lights, unless there is a fairly close correlation between observed aspects and those illustrated on the PPI, the navigator may become disorientated or, alternatively, begin to lose faith in the equipment, all at a critical point in the passage. In reduced visibility, the fact that the data is distorted may go completely unnoticed and result in the navigator having a completely erroneous understanding of the traffic situation. This could lead to the wrong anti-collision action being taken.

The solution to this particular problem is that the navigator should not use a radar presentation that is primarily intended for navigation to deal with an anti-collision problem. The correct information on which to base collision-avoidance action comes from a sea-stabilized display and should always be made use of prior to making an anti-collision decision. The observer should remove the fixed target reference or GNSS input and return to the display of sea-stabilized data. Most of these changes of presentation are simple and quick with modern radars, and there is no excuse for attempting to guess the effect that removing the tide would have. Having assessed the relevant information, the navigator can then return to his navigation problem by reselecting the reference and realigning the navigation lines which will have drifted.

CHAPTER

9

ARPA – Accuracy and Errors

9.1 INTRODUCTION

Errors present in the data displayed on the automatic radar plotting aid (ARPA) screen or by the alphanumeric read-out will affect decision-making. The observer must therefore have a knowledge of the level of accuracy that can be expected and the errors which will affect it. It is not a simple matter to specify the accuracy because it depends on, among other things, the geometry of the plotting triangle. For this reason, current IMO Performance Standards specify accuracy in terms of what must be achieved by the ARPA in Section 9.2. Pre-2008 performance standards used four carefully chosen test scenarios, but this has now been simplified to a generic level of tracking performance.

9.2 THE ACCURACY OF DISPLAYED DATA REQUIRED BY THE PERFORMANCE STANDARD

The ARPA should provide accuracies not less than those given in Table 9.1. Two different categories of equipment are specified. For craft with a speed of less than 30 knots, the standards should be met for targets moving at a relative speed of 100 knots to the own ship. For craft whose speed is greater than 30 knots, Table 9.1 should be met with targets moving at a relative speed of 130 knots.

An ARPA should present within 1 minute of steady-state tracking, the relative-motion trend of a target with the accuracy values shown on top line of Table 9.1 (95% probability values).

An ARPA should present within 3 minutes of steady-state tracking, the motion of a target with the accuracy values shown in the bottom line of Table 9.1 (95% probability values).

When a tracked target, or the own ship, has completed a manoeuvre, the system should present in a period of not more than 1 minute, an indication of the target's motion trend and display within 3 minutes the target's predicted motion, in accordance with Table 9.1.

The ARPA should be designed in such a manner that under the most favourable conditions of the own ship's motion, the error contribution from the ARPA should remain insignificant compared to the errors associated with the input sensors.

It should be noted that the table gives 95% probability values. This is necessary because any such error treatment is essentially statistical in its nature and means that the results must be within the tolerance values on 19 out of 20 occasions.

It is important to appreciate that even though the data is analysed by a computer, this does

TABLE 9.1 Tracked Target Accuracy (95% Probability Figures) for Installations on Post-2008 New Vessels

Time of Steady-State (minutes)	Relative Course (degrees)	Relative Speed (knots)	CPA (NM)	TCPA (minutes)	True Course (degrees)	True Speed (knots)
1 minute: trend	11	1.5% or 10% (whichever is greater)	1.0	–	–	–
3 minutes: motion	3	0.8% or 1% (whichever is greater)	0.3	0.5	5	0.5% or 1% (whichever is greater)

not mean that the results are perfect. Further, the errors considered in arriving at these tabulated values *do not* include 'blunders' or errors that result from the input of incorrect data.

Data obtained from the radar, log and gyro compass will be subject to random variations which all contribute to the uncertainty of the results as predicted by the computer. It should be appreciated that, because of the very short time period over which the computer plots (usually 1 or 3 minutes, see Section 4.3.6), and the consequent positional precision required, a whole new appreciation of radar errors is required. Errors which were not even considered with conventional radars in the past now become significant. In any case, the tables included in the specification should be taken as a guide as to what can be expected, especially with regard to the closest point of approach (CPA). The performance may also be reduced below the stated values by factors such as rolling by the vessel on which the ARPA is located. The performance standards are slightly vague on how much degradation is permitted, but do state that performance should not be substantially impaired for rolls up to 10°, which indicates that degraded results are to be expected when the vessel is rolling.

9.3 THE CLASSIFICATION OF ARPA ERROR SOURCES

The accuracy levels discussed in the previous section are dependent on the chosen level of sensor errors, but can also be affected by other errors. All errors which can affect the accuracy of displayed data can be conveniently arranged in three groups as follows:

1. Errors which are generated in the radar installation itself, the behaviour of the signals at the chosen frequency and the limitations of peripheral equipment such as logs, gyro compasses and dedicated trackers.
2. Errors which may be due to inaccuracies during processing of the radar data, inadequacies of the algorithms chosen and the limits of accuracy accepted.
3. Errors in interpretation of the displayed data.

9.4 ERRORS THAT ARE GENERATED IN THE RADAR INSTALLATION

Errors in the radar, gyro compass and log which feed data to the ARPA system will result in errors in the output data. Range and bearing errors which remain constant or nearly so during the encounter, for example a steady gyro compass error of a few degrees, will introduce an error into the predicted vectors of other ships, but are unlikely to cause danger since all data will be similarly affected, including the own ship. The effect of errors on the predicted data depends on the kind of error, the situation and the duration of the plot for which the data is stored for processing and prediction. This time is typically in the range of 1–3 minutes and in this respect it must be appreciated that errors which in the past could be considered to be negligible may have a significant effect on derived data. In the following examples, the situation is assumed to be a near miss or a collision.

9.4.1 Glint

As a target ship rolls, pitches and yaws, the apparent centre of its radar echo moves over the full ship's length; this is termed glint. Its distance from amidships is random with a standard deviation of one-sixth length, that is for a 200 m ship it is probable that the error does not exceed ±33 m. Since the beam of a ship is usually small by comparison with its length, transverse glint is negligible. If the target ship's aspect is beam-on, glint introduces random bearing errors.

9.4.2 Errors in Bearing Measurement

These cause false positions to be recorded on each side of the relative track of the other ship, leading to errors in the observed relative track and therefore in the predicted CPA, and also in the displayed aspect of the other ship. Unfortunately, the greatest errors in displayed aspect occur in those cases where the real aspect is near end-on. Bearing errors may result from the following causes.

9.4.2.1 Backlash in Gearing

Backlash can occur between the rotating antenna and its azimuth transmitter. Air resistance on the rotating antenna will tend to maintain geartooth contact, but bounce and reverse torque due to aerodynamic forces will break the contact and allow some backlash to occur. This problem has been to a large extent overcome by the use of more modern forms of bearing transmission.

9.4.2.2 Unstable Platform or Antenna Tilt

Ship motion causes the axis of rotation of the radar antenna to tilt. When the ship is heeled to an angle of B radians, a bearing error of $-(\frac{1}{2}B^2 \sin\theta \cos\theta)$ radians is produced, where θ is the bearing of the target off the own ship's bow. This error is quadrantal, that is zero ahead, astern and abeam, rising to alternate plus and minus maxima at 45° and 135°, etc. It will not be reversed by the opposite roll since B is squared.

When the ship is rolling, the tilt has two components: a random variation between zero and a maximum, according to the value of B (i.e. the actual roll angle which happens to be present when the antenna is directly on the bearing of the target) and a rise and fall of the maximum over periods of about 1 or 2 minutes with wave height variation. For a relative bearing of 45° and a roll of 7.5° towards or away from the other ship, the error is −0.25° maximum.

9.4.2.3 Parallax Due to Roll of the Own Ship (Figure 9.1)

If the radar antenna is mounted at a height H above the roll axis of the ship and the ship rolls to an angle B, the antenna moves transversely by $H \sin B$. The measured bearing of a target at a bearing of θ from the ship's head and at a range R will be in error by an angle e which is given by:

$$\tan e = \frac{H \sin B}{R} \times \cos\theta$$

(*Note*: H and R must be in the same units and θ is the relative bearing.)

Since e is small:

$$e = \frac{180\, H \sin B \cos\theta}{\pi R} \text{degrees}$$

This error will vary sinusoidally with time and has a period equal to the roll period.

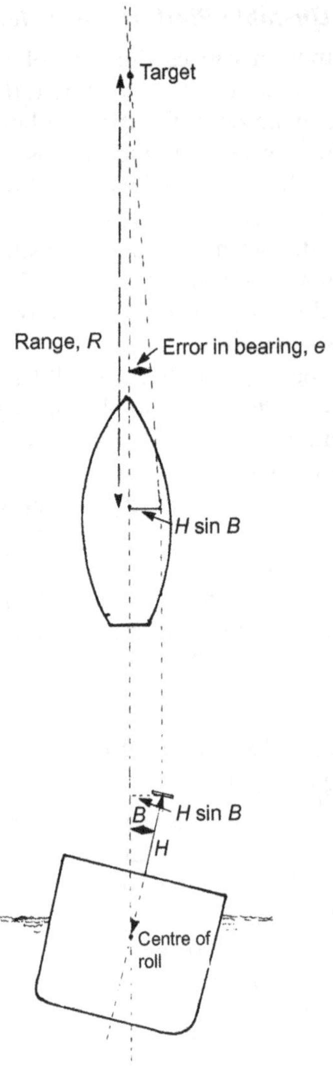

FIGURE 9.1 Parallax due to roll of the own ship.

The error is maximum when $\theta = 0°$ and 180° minimum when $\theta = 090°$ and 270°, i.e. it varies with $\cos\theta$.

9.4.2.4 *Asymmetrical Antenna Beam*

The ARPA should take the bearing of the target as that of the centre of the echo. If the antenna beam is asymmetrical, the apparent position of the echo may change with the echo strength. Errors due to this cause can become very large in some systems if the echo strength is sufficient for the close-in sidelobe pattern of the antenna to become apparent. At least one system employs special techniques to eliminate this problem.

9.4.2.5 *Azimuth Quantization Error (Figure 9.2)*

The antenna position must be converted to digital form before it can be used by the computer (see Section 2.7.3.2). An azimuth defined by a 12-bit computer word has a least significant bit (LSB) equivalent to 0.09° (i.e. 360°/4096) so that the restriction to 12 bits introduces a quantization error of 0.045°. The same error will arise if the computer truncates the input azimuth information to 12 bits. Antenna azimuth is often taken to a resolution of either 12 or 13 bits.

Note: Since gyro compass error is of the order 0.1–0.5° at best, there is no real point in making the antenna encoder bit size very much smaller.

9.4.3 Errors in Range Measurement

9.4.3.1 *Range Change Due to Roll of the Own Ship*

If the radar antenna is mounted at a height H above the roll axis of the ship and the ship rolls to an angle B, the antenna moves transversely by $H \sin B$. For a target bearing θ from the ship's head, the measured range will be in error by a distance d which is given by:

$$d = H \sin B \sin \theta$$

Pitch error is much less significant, but if roll and pitch occur together, the effects add non-linearly and must be worked out separately.

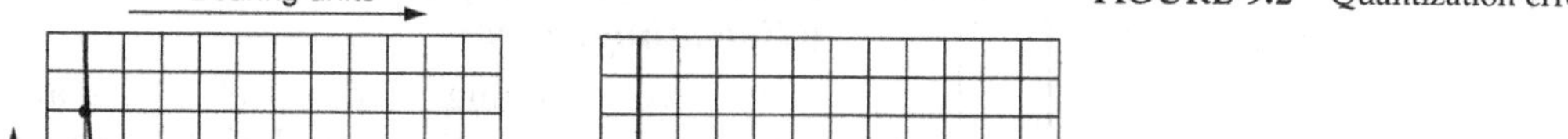

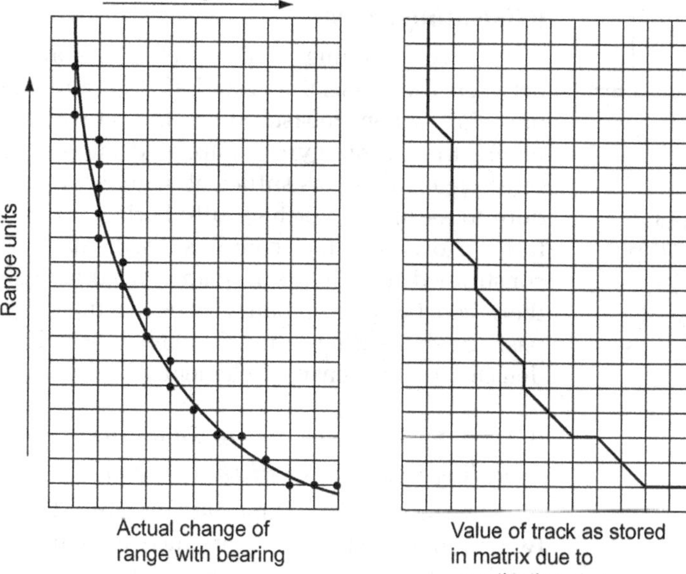

FIGURE 9.2 Quantization errors.

9.4.3.2 Range Quantization Error (Figure 9.2)

The range of a target must be converted into a digital number for the computer to use and it is likely that this will be done by measuring the range using counting techniques. A convenient clock rate is that corresponding to 0.01 NM steps on the 12-mile range scale (see Section 2.7.3.1).

Typical step functions due to range and bearing quantizing are shown in Figure 9.2.

Note: During some periods, it will appear that the target is on a collision course, that is steady bearing, although this is never the case in fact.

9.4.3.3 Pulse Amplitude Variation (Figure 9.3)

The equipment will typically measure the range of an echo at the point at which the echo strength rises above a preset threshold. Because of the finite bandwidth of the radar receiver, the echo pulse will have a sloping leading edge and the measured range will vary with the pulse amplitude (see Section 2.3.3.4). For the pulse lengths commonly used on anti-collision range scales, the receiver bandwidths are chosen for long range performance rather than for discrimination, so that it is likely that the leading edge slope will be nearly as long as the transmitted pulse.

The resulting apparent range variation will depend upon some assumptions about echo amplitude variations but is likely to be about 40 m as a maximum.

9.4.4 The Effect of Random Gyro Compass Errors

A gyro compass master unit, mounted at an arbitrary height above the roll axis of the ship, is subject to transverse acceleration at each extremity of the roll. This includes a false vertical as the pendulous unit tilts in its gimbals.

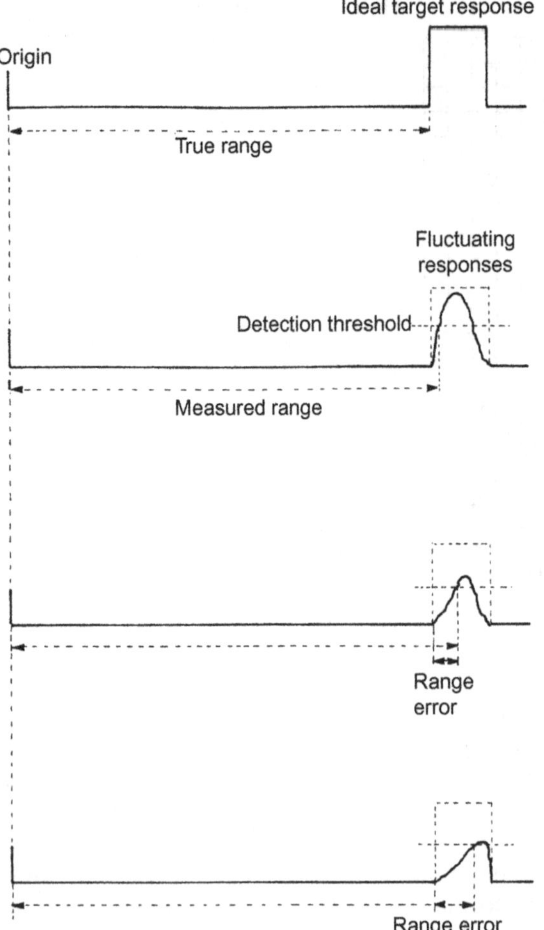

FIGURE 9.3 The error due to pulse amplitude variation.

This puts an error into the gyro compass output affecting all bearings; it has random and slowly varying components in just the same way as the radar tilt error. Observation at sea indicates that 0.25° is the error in many typical installations.

The gyro compass also has other errors. Long-term errors (e.g. settling point ±0.75°) are unimportant if they remain sensibly constant, as they normally do, but short-term random errors (e.g. settling point difference ±0.2°) are significant.

9.4.4.1 *Gyro Compass Deck-Plane (Gimballing) Errors*

The true heading of a vessel is the angle between the vessel's fore-and-aft line and the meridian when measured in the horizontal plane. In several gyro compass designs, the sensitive element has sufficient degrees of freedom to assume a north–south, horizontal attitude. However, the compass card may be constrained to the deck plane. In this case there can be a discrepancy between the compass card reading and the ship's heading detected by the sensitive element.

9.4.4.2 *Yaw Motion Produced by the Coupling of Roll and Pitch Motions*

When a ship is rolling and pitching, these two motions interact to produce a resultant yawing motion. The motion can be resolved into horizontal and vertical components. The horizontal component is the yaw motion and is detected by the gyro compass sensing element.

9.4.5 The Effect of Random Log Errors

An error in the own ship's log will produce a vector error in every tracked ship's true speed and/or course. This will also result in an error in the displayed aspect of other ships; however this aspect error is minimum in all cases where the real aspect is end-on. A further effect will be to produce non-zero speed indications on all stationary targets being tracked. This cause may also produce large errors in the aspects of very slow-moving targets. If this error is assumed not to exceed 0.4 knots, it will give rise to a positional error of some 15.5 m in a plot time of 75 s.

Where true tracks are stored by the tracker (see Section 4.3.6.2), a fluctuating log error can also affect the relative vectors. However, the errors being considered here are the small random variations in the log output and not the large fluctuations which can occur if the log's performance becomes erratic due to technical malfunction or problems caused by outside influences such as fouling or aeration. Such larger fluctuations are more appropriately considered under errors in input data (see Section 9.5.3).

9.5 ERRORS IN DISPLAYED DATA

9.5.1 Target Swap

When two targets are close to each other, it is possible for the association of past and present echoes to be confused so that the processor is loaded with erroneous data.

The result is that the historical data on one target may be transferred to another target and the indicated relative (and true) track of that ship will be composed of part of the tracks of two different target motions. Target swap can occur with any type of tracker, but is least likely in those which use a diminishing gate size as confidence in the track increases and those which adopt rate aiding. It is most likely to occur when two targets are close together for a comparatively long time and one target echo is much stronger than the other, see Figure 9.4. It is particularly likely to occur if one target shadows the other (see also Section 4.3.5).

9.5.2 Track Errors

The motion of a target is rarely completely steady and even steady motion will return positions which are randomly scattered about the actual track, due to basic radar limitations. Quantization errors in range and bearing which are introduced by the translation of the basic radar information to the processor database further exacerbate the effects of these system errors. The only way in which the tracker can deal with these is to use some form of smoothing over a period of time by applying more or less complicated filtering techniques (see also Section 4.3.6). The aim of the filtering is to give the best possible indication of the steady track and at the same time detect real changes quickly.

Given the need to satisfy these two conflicting requirements, it is inevitable that the tracker will be limited in its ability to predict precisely the relative- and true-motion of a tracked target at any instant and thus tracking errors will result. The effect of these errors should of course fall within the limits set out in Section 9.2, but the prudent observer will use suitable clear weather opportunities to gain some evaluation of the practical performance of the tracker which is producing the ARPA data. Such performance can be usefully judged against two important criteria. These are the stability of the track shown on the display for a vessel which is observed to be standing-on and the rapidity with which the track responds to a target which has been observed to manoeuvre. It has to be recognized that alterations of the target's course are easy to detect by visual observation, whereas speed changes are not. However, in clear weather, the former are very much more common than the latter, except in very confined waters.

Accuracy is most difficult to achieve with targets whose tracked movement is slow. In the case of relative track storage (see Section 4.3.6.2), this will affect a target whose course and speed are close to those of the observing vessel. The length of the relative track will be small and thus the system errors are a much more significant proportion of track length than would be the case with a target having a rapid relative-motion. Thus the inherent accuracy of CPA data will be low. Conversely, in the case of true track storage (see Section 4.3.6.3), a near-stationary target will suffer the same low accuracy in the prediction of its true-motion.

In general the tracker is likely to offer the best indication of both the relative-motion and

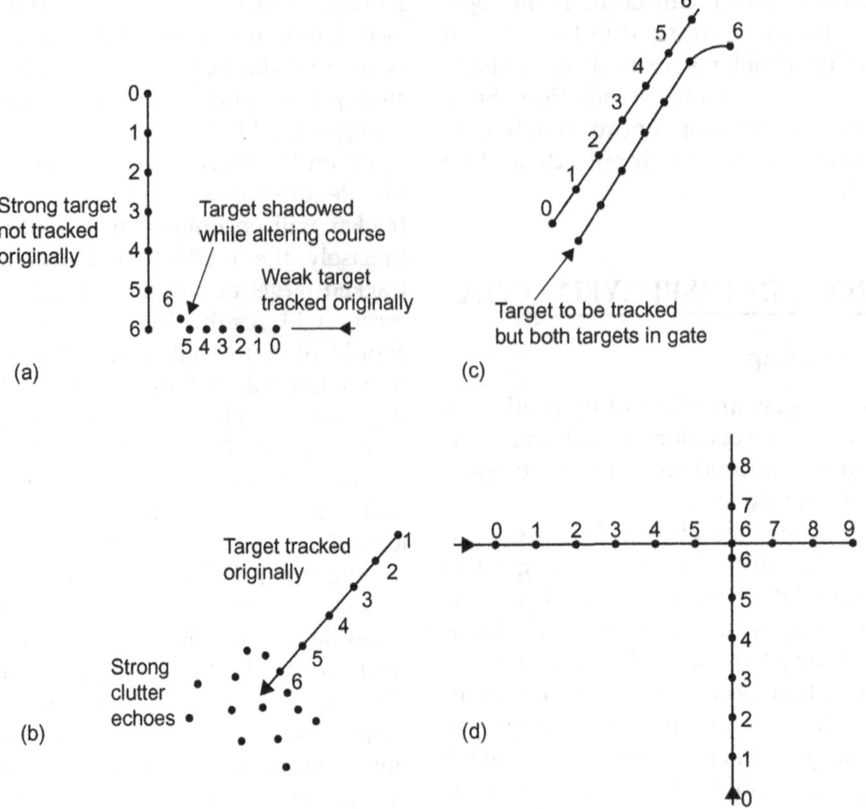

FIGURE 9.4 **Examples of radar target swap. (a) At position 6, tracker transfers to stronger target. (b) At position 6, and later, the profusion of echoes confounds tracker accuracy. The tracker may easily pick up random clutter instead of the target ship. (c) Targets travel close together for a period, then separate. The tracker may not follow the diverging target ship.** *Note*: It is often difficult to acquire a target that is close to others already being tracked. **(d) Two tracked targets pass close to each other, so that both are in the tracking gate at one time.** *Note*: In this case, longer rate aiding may be an advantage.

the true-motion when both the target and the observing vessel maintain their course and speed for a full smoothing period. In the changing situation the track errors will depend on the nature of the changes and the form of track storage adopted (see Sections 4.3.6.2 and 4.3.6.3).

Where only the target manoeuvres there will be a finite response time in which the displayed vector will seek to follow the change and to stabilize on the new track. Under such circumstances, irrespective of whether the tracker smooths relative or true tracks, there should in theory be no difference in the tracker performance when measured in terms of the accuracy with which it provides output of both relative and true vectors. In both cases the vectors may be erratic when the processor reverts to smoothing over the short period.

Where only the observing vessel manoeuvres, the method of storage is significant because the relative tracks of all targets will be curves for the duration of the manoeuvre. If the smoothing is applied to relative tracks, the tracker will be faced with the task of trying to produce a straight line from a curve and will hence obtain a mean track. Errors in the relative track will result and the relative vectors of all targets may be erratic in the short term. True-motion data derived from this will also be in error, just as where a manual plotter constructs an *OAW* triangle on the basis of an apparent motion which is not consistent (see Section 7.2). The effect may be exacerbated by the fact that, during the vessel's manoeuvre, the path traced out by the mass of the vessel, and hence the antenna, may differ significantly from that indicated by the gyro compass and log. Systems which smooth true tracks should derive a more accurate indication of the target's true track during the observing vessel's manoeuvre, as the true track is in theory rendered independent of changes in the observing vessel's course and speed. This independence will be reduced by any difference between the velocity of the ship's mass during the manoeuvre and the direction and speed fed in by the gyro and log. Again, all errors must fall within the limits of the performance standards, but the prudent observer can assess the effect of manoeuvres on the performance of the tracker by observing a known stationary target during a manoeuvre. In this connection it must be remembered that, even in steady-state conditions, a land-stationary target may display some component of motion due to the effect of tide, and water-stationary targets may have small non-zero vectors due to system errors.

Where both observing vessel and target manoeuvre at the same time, it is unlikely that any system will provide a reliable indication of any target data until either the observing vessel or the target ceases to manoeuvre.

If targets are tracked down to very close ranges, the relative-motion will give rise to very rapid bearing changes and this may make it impossible for the tracker to follow the target; thus the 'target lost' condition may arise, not because the echo is weak, but because the gate cannot be moved fast enough or opened up sufficiently to find it. It is also worth remembering that the use of true vectors as an indication of target heading is based on the assumption that the target is moving through the water in the direction in which it is heading. Leeway is the prime example of a case where this may not be correct. Unless one can see the target, it is impossible even to begin to make an estimate of leeway. In poor visibility and high winds, the observer must be alert to the possibility and use the displayed data with additional caution.

In summary, it must be remembered that whenever the steady-state conditions are disrupted, there will be a period in which the data will be particularly liable to the track errors described above, in the same way as is the case when a target is first acquired. When the steady-state is regained, accuracy and stability will improve, first over the short smoothing period and then over the long period. Any track data extracted during periods of non-steady-state conditions must be viewed with suspicion.

9.5.3 The Effect on Vectors of Incorrect Course and Speed Input

From the theory of manual radar plotting (see Section 7.2), it is evident that it is possible to deduce the relative-motion of a target without using a knowledge of the true motion of the observing vessel (other than to produce stabilized bearings). Deduction of the true-motion of the target requires a knowledge of the true-motion of the observing vessel to allow resolution of the *OAW* triangle, and the accuracy of the result depends largely on the accuracy of the course and speed data used. Extrapolation of this reasoning suggests that the accuracy of relative vectors and the associated CPA data are independent of the accuracy of the course and speed input, whereas the accuracy of true vectors and the associated data are dependent on the accuracy of the observing vessel's input of course and speed. In the case of systems which smooth relative tracks, this is invariably correct; in the case of systems which store true tracks, it is correct subject to the qualification that the input errors are constant. In the case of a fluctuating error input, the two storage approaches will produce different results. For this reason the effect of steady input errors on relative and true vectors is discussed in Sections 9.5.3.1 and 9.5.3.2, respectively, while the case of a fluctuating error is treated separately in Section 9.5.3.3.

Whatever approach the tracker uses, it is essential for the observer to ensure that the correct course and speed inputs are fed in when setting up (see Section 6.9.3), and that regular and frequent checks are made to ensure that the values remain correct. Failure to do this will in general result in the erroneous display of true data which may seriously mislead the navigator when choosing a suitable avoiding manoeuvre strategy.

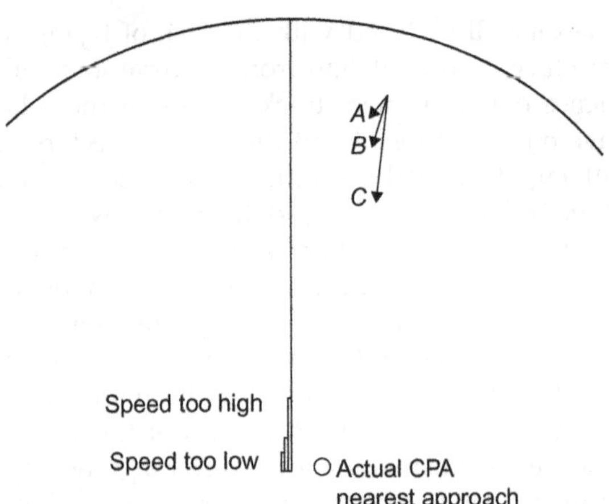

FIGURE 9.5 The effect of incorrect speed input on a true-vector presentation. Vector *B* is correct.
Speed input 'too high' shows vector *A*, a crossing slow ship.
Speed input 'too low' shows vector *C*, a fast target on near parallel crossing.

9.5.3.1 *Relative Vectors*

The relative vectors and the associated CPA/TCPA data should be unaffected if the observer allows a fixed erroneous input of course and speed data to be applied. In the case of relative storage the information is not used in the calculation. In the case of true storage it is used twice, and one could say that this illustrates a classic case of 'two wrongs making a right', as illustrated by Figure 4.8.

The presence of a fixed gyro compass error at the point at which the bearings are digitized would result in the picture being slewed, but would not affect the CPA/TCPA data.

9.5.3.2 *True Vectors*

The true vectors will be displayed incorrectly if the observer allows an erroneous input of course and speed data to be applied, irrespective of the storage format. This may give the observer a seriously misleading impression of the other vessel's heading and speed and may prompt an unsafe manoeuvre. Such a situation is illustrated by Figures 9.5 and 9.6.

Figure 9.5 shows a target which is moving at a similar speed to the observing vessel, is showing a red light (vector *B*) and which will pass clear down the starboard side. Incorrect speed input could make this appear to be a slower vessel in a broad crossing situation (vector *A*) or a faster ship passing (vector *C*) green to green.

The effect of an incorrect heading input will have a similar capacity to produce misleading results, as illustrated in Figure 9.6.

9.5.3.3 *The Effect of Fluctuating Input Error*

The most probable source of a fluctuating input error is the log. There are a number of circumstances in which this might arise. An example is the case of some Doppler logs

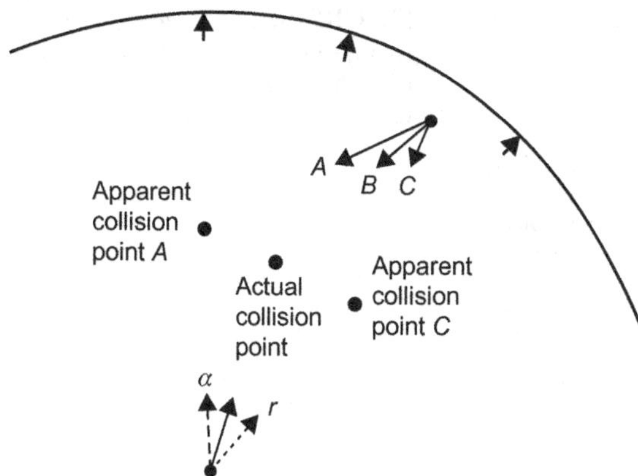

FIGURE 9.6 **The effect of incorrect compass input on a true-vector presentation.** Vector *B* is correct. Incorrect compass input to the left at α shows a target broad crossing to port and faster than the own ship. Incorrect compass input to the right at r shows a target fine crossing to starboard and slower than the own ship.

which tend to give erratic output in bad weather when the transducer has to operate through an aerated layer. The fluctuating effect will affect the display of the true vectors whatever the form of storage, but the behaviour of the relative vectors will depend on the mode of storage used. Where relative tracks are smoothed, the fluctuating error will have no effect on the relative vectors and the associated CPA/TCPA data. Where true tracks are smoothed, the relative vector will tend to change erratically in sympathy with the input fluctuations, since the relative vector is derived from the smoothed true track and the instantaneous input course and speed data.

The difference in effect can be considered by an example shown in Figure 9.7. Consider the case where the observing vessel has been on a steady course and speed for a full smoothing period, and the correct course and speed data has been consistently fed in. Both methods of smoothing will have settled to produce the correct relative and true vectors.

Suppose that the log develops an instantaneous fault and reads half the correct speed. The relative track system, shown in Figure 9.7(a), will show no change in the relative vector (because the relative track is smoothed), but the true vector will immediately go to the erroneous value because it is derived from the smoothed relative track and the instantaneous course and speed input.

The true track system will show no immediate change in the true vector, because it is insulated by the smoothing, whereas the relative vector will immediately go to an erroneous value since it is derived from the smoothed true track and the instantaneous course and speed input. If no further change takes place in the error and the vectors are observed over a full smoothing period, the true vector will gradually change to the erroneous value while the relative vector will gradually come back to the correct value as the previously smoothed track is progressively discarded. Thus, in this type of storage, if the speed input fluctuates the relative vector will also fluctuate. This disadvantage has to be set against the advantage gained by being able to maintain a stable relative vector when the observing vessel is manoeuvring (see Section 9.5.2).

An erratic course input would have a similarly disruptive effect. However, such a condition would also affect the digitization of bearings. This would tend to cause targets to jump and it would be fairly obvious to even a casual observer that something was wrong.

9.5.3.4 *Comparison of Relative and True Vectors*

Given that one or other of the vectors can be affected by input errors in a way which may be dependent on the tracker philosophy, it is important to stress the need for the observer continuously to compare one data source with another to ensure that, in all cases, indications given from relative vectors and true vectors sensibly agree.

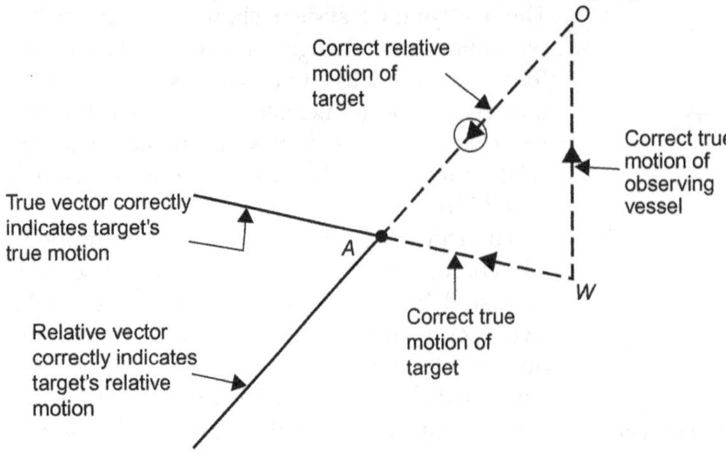

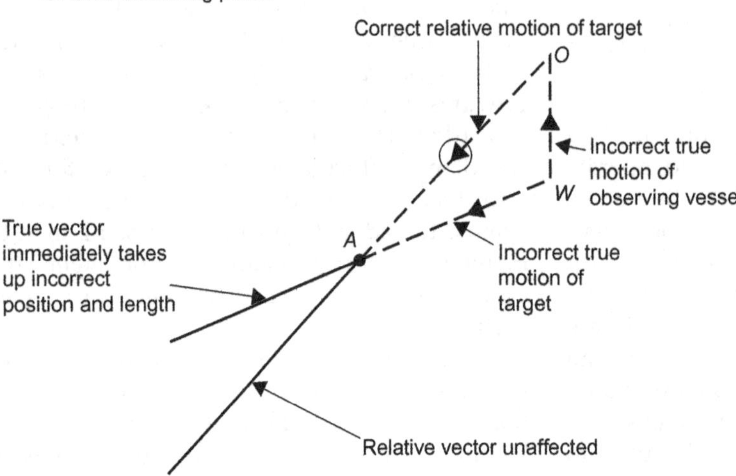

FIGURE 9.7 The effect of a step change in the error of speed input: (a) where relative track storage is used.

9.5.4 The Effect on the PPC of Incorrect Data Input

9.5.4.1 Errors in Speed Input

If incorrect speed is input to a collision situation, the collision point (see Section 4.11) will still appear correctly on the heading marker, but at an incorrect range, and will move down the heading marker at an incorrect speed.

In the case where there is, in fact, a miss distance, the collision point will appear in the wrong position, which may give rise to a misjudgement of the danger or urgency of the situation. Figure 9.8 shows how the collision point may be displaced due to a speed error in the two cases where the target is crossing ahead and crossing astern.

9.5.4.2 Errors in Course Input

The behaviour of the collision point when an error in the course is input is too complex

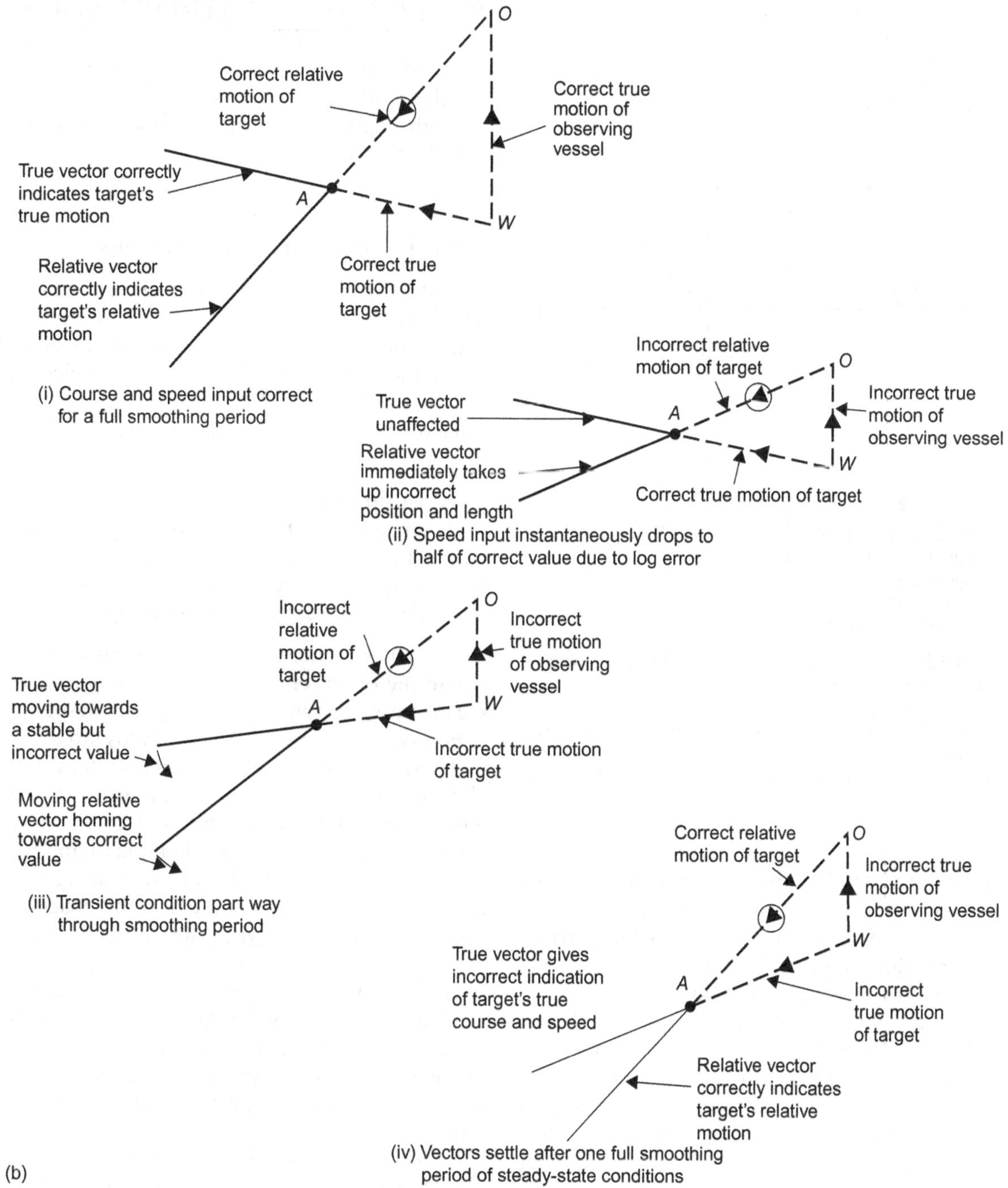

FIGURE 9.7 (b) where true track storage is used.

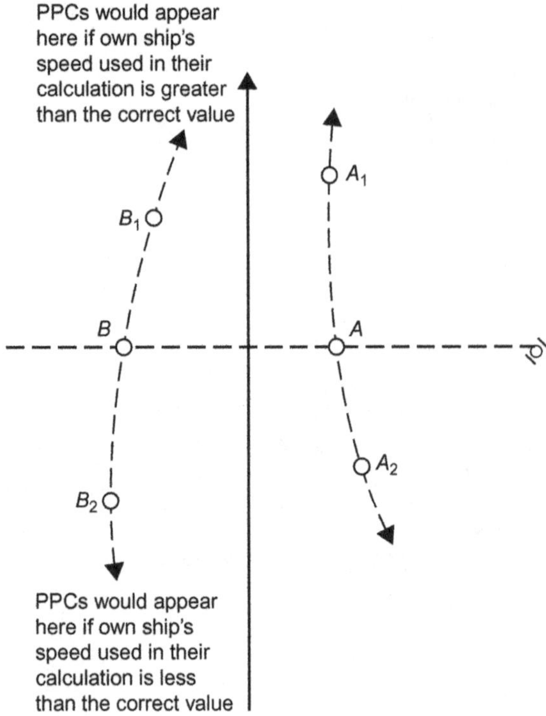
PPCs would appear here if own ship's speed used in their calculation is greater than the correct value
A_1
B_1
B
A
hdg
A_2
B_2
PPCs would appear here if own ship's speed used in their calculation is less than the correct value

FIGURE 9.8 The effect on the PPC of a speed error. *A*, target passing astern, correct speed used.
B, target passing ahead, correct speed used.
A_1, B_1, PPC appears here if a speed greater than the correct value is used.
A_2, B_2, PPC appears here if a speed less than the correct value is used.

to allow definition of a pattern. If the error occurs only in the calculation and does not appear in the position of the heading marker, the collision point could appear on the heading marker in a miss situation. More dangerously, a collision point could appear off the heading marker in a collision situation. When the same error appears in both heading marker and calculation, as might occur due to a gyro compass error, the collision case will always show the collision point on the heading marker.

Similarly, if a miss distance exists, the collision point will not be on the heading marker.

9.6 ERRORS OF INTERPRETATION

These errors are not within the system but are those likely to be made by the operator through misunderstanding, inexperience or casual observation.

9.6.1 Errors with Vector Systems

In the case of vector systems, the most common mistakes arise because the observer, either from lack of concentration due to stress of the moment or through lack of knowledge, confuses relative and true vectors (as in the collision between *Norwegian Dream* and *Ever Decent*). Typical blunders are:

a. measuring the CPA as the tangential distance at which the true vector passes the origin;
b. mistaking the direction of the relative vector for the target's true heading.

A further source of error sometimes occurs where the observer runs out the true vector to see the possible dynamic development of the situation (which in itself is a useful ploy) to assist in determining collision avoidance strategy, but deduces the point of closest approach as being where the vectors cross. This is of course only correct in the collision case. Attempts to find the passing distance by trial and error using this technique also frequently mislead the observer and are not necessary when the CPA is so easily available from other sources.

Some manufacturers fit spring-loaded switches or hold-down buttons to ensure that the equipment always reverts to either true or relative vector mode in an attempt to reduce the chance of misinterpretation of the data in this way.

Other common errors include the confusion of real and trial values of CPA and omitting to set in the correct trial speed. Where the displayed vectors and history of a different type

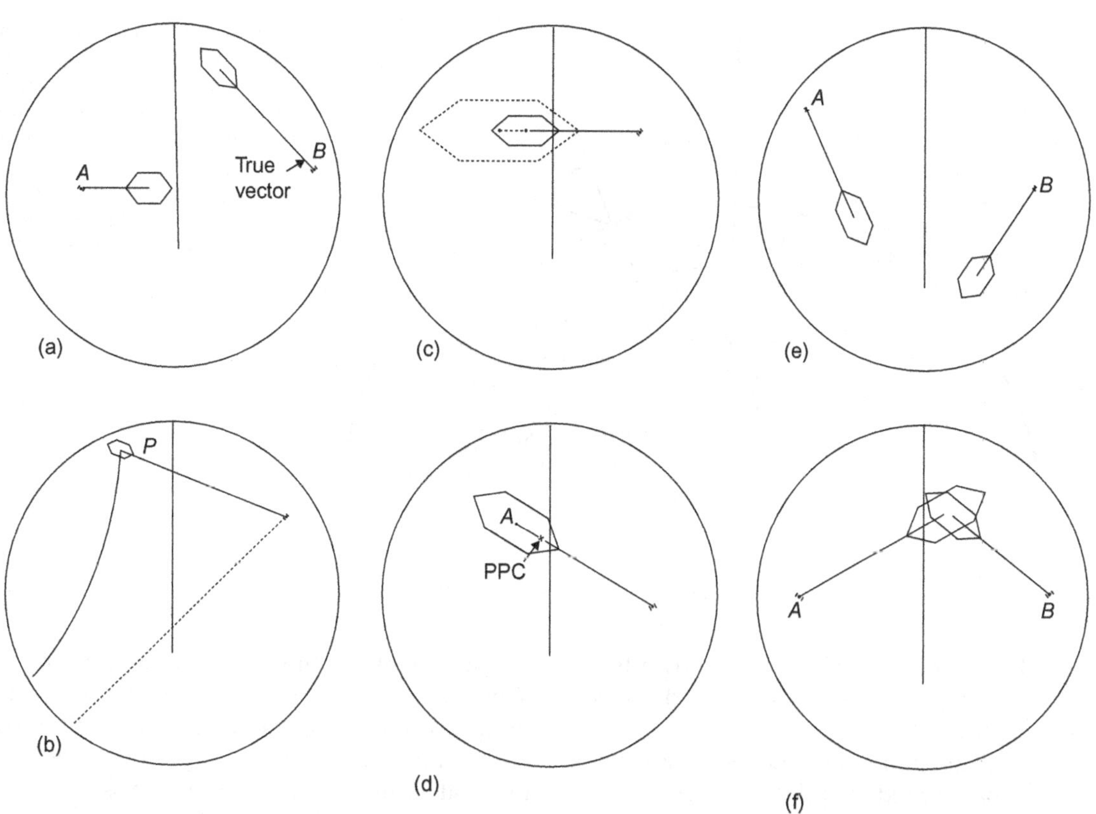

FIGURE 9.9 Errors of interpretation. (a) Target *A* is faster and target *B* is slower than the own ship, despite appearances. *Note*: Vectors will show this. (b) The solid line shows the track of the PPC from *P*. The apparent track of the echo is shown by the broken line. (c) The broken line shows the PAD for 2 miles. The solid line shows the PAD for 1 mile. (d) The PPC is not at *A*, the centre of the hexagon. (e) Targets *A* and *B* will collide, although this is not apparent from the display. (f) Targets *A* and *B* will not collide, although they may pass within the miss distance.

are simultaneously presented, the difference between the two may be mistaken for a manoeuvre by the target.

During the second and third minutes of tracking, the vectors will be stabilizing and care must be taken not to be misled into assuming that this is an alteration by the target or that it is yawing.

9.6.2 Errors with PPC and PAD Systems

In the case of predicted points of collision (PPCs) and predicted areas of danger (PADs), the commonest mistakes arise when attempting to interpolate or extrapolate data from the display. As mentioned in Section 4.12.4, PADS are no longer found on ships, but this discussion of PPCs and PADs is included for completeness. Typical errors arise because of failure to appreciate the following:

1. The line joining the target to the collision point is not a time-related vector and does not indicate speed (Figure 9.9(a)).
2. The collision point gives no indication of miss distance.
3. Changes in collision point positions do not necessarily indicate a change in the target's true course or speed.

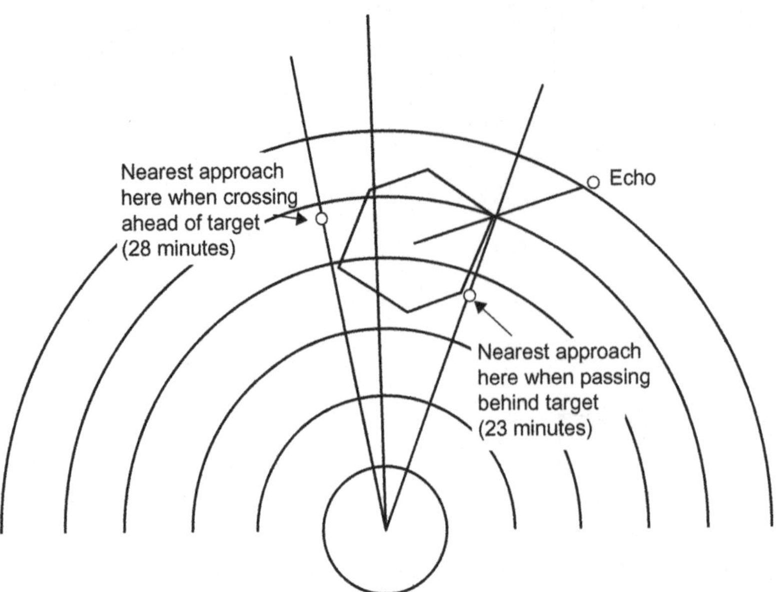

FIGURE 9.10 The misleading effect of using the bearing cursor to determine the time to resume course. *Note*: There are four intervals to the point of crossing ahead and five intervals to the point of crossing astern, but nearest approaches occur, as indicated, at 4.9 and 3.6 intervals, respectively.

4. The PAD does not change symmetrically with a change in the selected miss distance (see Figure 9.9(c)).
5. The collision point is not necessarily at the centre of the predicted danger area (see Figure 9.9(d)).

It is always important to realize that the areas of danger generated on the screen apply only to the own ship and the target, and that they do not always give warning of a mutual threat between two targets. If two areas of danger overlap, it is reasonable to suppose that the two targets involved will also pass each other within the stated miss distance (see Figure 9.9 (f)), but separated danger areas do not imply safe passing between the targets. Two targets may eventually have a close passing although their danger areas, as applied to the own ship, appear to be well separated (see Figure 9.9(e)).

9.6.2.1 Resumption of Course

Where a 'chained' bearing cursor is available and the chain divisions are an indication of time, care must be exercised in the measurement of time to resume course. As shown in Figure 9.10, the marker correctly indicates the time the own ship will cross ahead and astern on the target track, but the time at which the required miss distance occurs cannot be determined.

9.6.3 The Misleading Effect of Afterglow

Because the vector mode (i.e. relative or true) is not necessarily the same as the radar presentation which has been selected, vectors and afterglow trails may not match. When true vectors are selected on a relative-motion presentation, the vectors and the afterglow will not correlate. When relative vectors are selected while using a true-motion presentation, the true afterglow will not match the relative vector.

9.6.4 Accuracy of the Presented Data

Over-reliance on, and failure to appreciate inaccuracies in, presented data which has been derived from imperfect inputs should be avoided at all costs. It must always be borne in mind that a vector/PAD/alphanumeric read-out is not absolutely accurate, just because it has been produced by a computer, no matter how many microprocessors it may boast. An indication by the ARPA that a target will pass one cable clear of the own ship should not be regarded as justification for standing-on into such a situation.

The errors given in Table 9.1 are quite typical and should always be allowed for.

9.6.5 Missed Targets

An automatic acquisition system may totally fail to detect and acquire a target of vital importance, for one of a number of reasons. Similarly, it may also drop or cancel a fading target. In the latter case, the target may subsequently be re-acquired and present a course and speed which may indicate that the target has manoeuvred when, in fact, the track is new and has not yet established its long-term accuracy.

CHAPTER

10

Ancillary Equipment

Whereas, in the past, the radar and target tracking equipment was essentially a stand-alone module and could, apart from cross-checking, be considered in relative isolation, this is no longer the case. Now that virtually all data is digitized prior to display, it has become possible and desirable for the display of data to be integrated (see Section 10.3). For this reason, some insight into the way other major data sources interact with the radar and target tracking system is included here.

10.1 GLOBAL NAVIGATION SATELLITE SYSTEMS

With the availability of artificial earth satellites and highly accurate clocks, it was soon recognized that there was the potential for a satellite-based global navigation system. At first, the 'Transit' system was developed in the United States. The ship's position was derived from a series of hyperbolic position lines based on measurements obtained from a single satellite in a low earth orbit. With the advance of technology and the US requirement for a 24 hour high precision position fixing system for military purposes, the global positioning system (GPS) was developed and the Transit system has been discontinued. It is not intended to discuss global navigation satellite systems (GNSS) to any great depth here, but rather to consider the manner in which they interact with the radar and target tracking equipment. Where a deeper understanding of a particular GNSS is required, readers should consult a more specialized text. Several GNSS's now exist or are in an advanced stage of implementation. The American GPS will be discussed first (with a little more detail), as it was for many years the only globally available system and is typical of a satellite system from which positional data is derived.

10.1.1 Global Positioning System

The first experimental GPS satellite was launched in 1978, but the full system was only declared operational with 18 satellites in 1994. The system comprises three segments, namely the tracking and control segment, the satellite or space segment, and the user segment.

10.1.1.1 The Tracking and Control Segment

In general, the overall control and operation of the system is in the hands of the United States Air Force who monitor the performance of the satellites via a series of ground tracking stations around the world. They also update the information transmitted by the satellites and from time

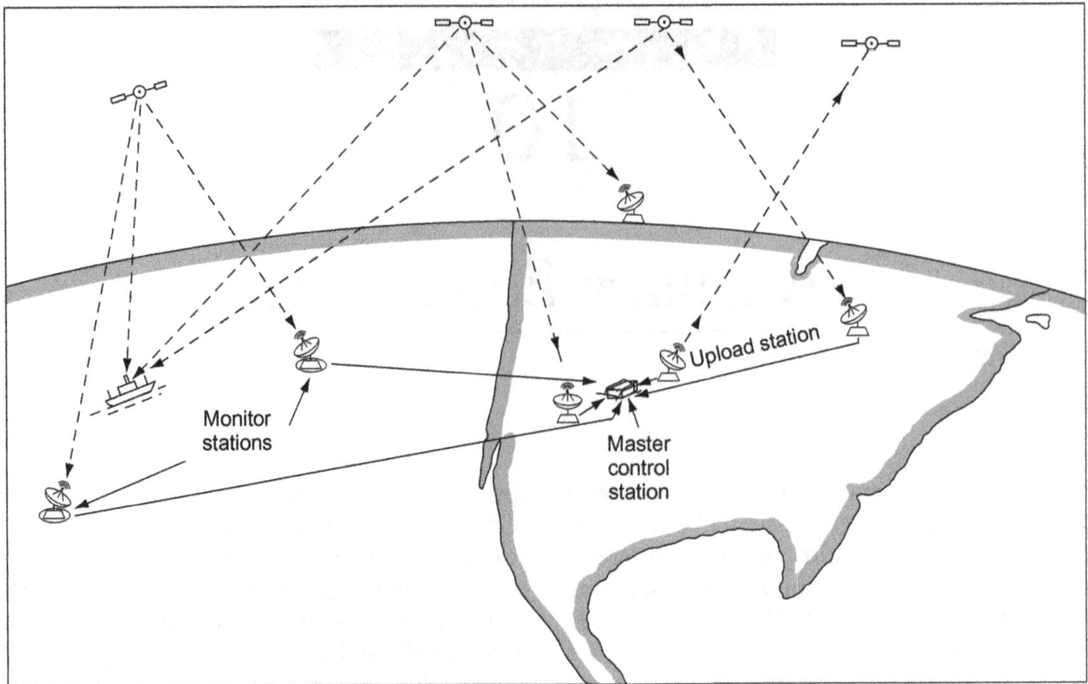

FIGURE 10.1 Overall supervision of the system.

to time correct the satellite positions to ensure that their orbits remain within the desired tolerances. The whole of the system is supervised from a master control station (MCS) in the United States. They also launch new satellites to replace those that have come to the end of their useful life or have malfunctioned (Figure 10.1).

10.1.1.2 The Space Segment

This is now based on a theoretical constellation of some 24 satellites in six near-circular orbits, each orbit being inclined at an angle of 55° to the equator and separated by some 60° in longitude. Within each orbit, the four satellites are spaced at 90°. The relationship between satellites in consecutive orbits is that as one satellite is crossing the equator heading northwards, the one in the orbit 'ahead' will have crossed the equator 60° to the east and will now be some 15° to the east along its track and north of the equator. The maximum declination of each orbit is 55° north and south (Figures 10.2 and 10.3). The orbits have a mean altitude of 20,200 km, which is designed to result in the satellite completing two orbits in precisely one sidereal day. While the satellites move along the orbits, the earth will rotate within this 'bird cage'. There is a tendency for the orbits to be perturbed, but this is corrected by the MCS as necessary.

For practical reasons, the constellation has been developed over a period of years and in a climate of advancing technology. Satellites have been replaced and in most orbits, there is at least one spare satellite. Consequently, the actual constellation is not precisely as described above, but the basic principle, that is to have at least four visible and usable satellites available worldwide at all times, is achieved and surpassed (Figure 10.5).

Each satellite carries a number of high precision 'clocks' on board with all clocks in the

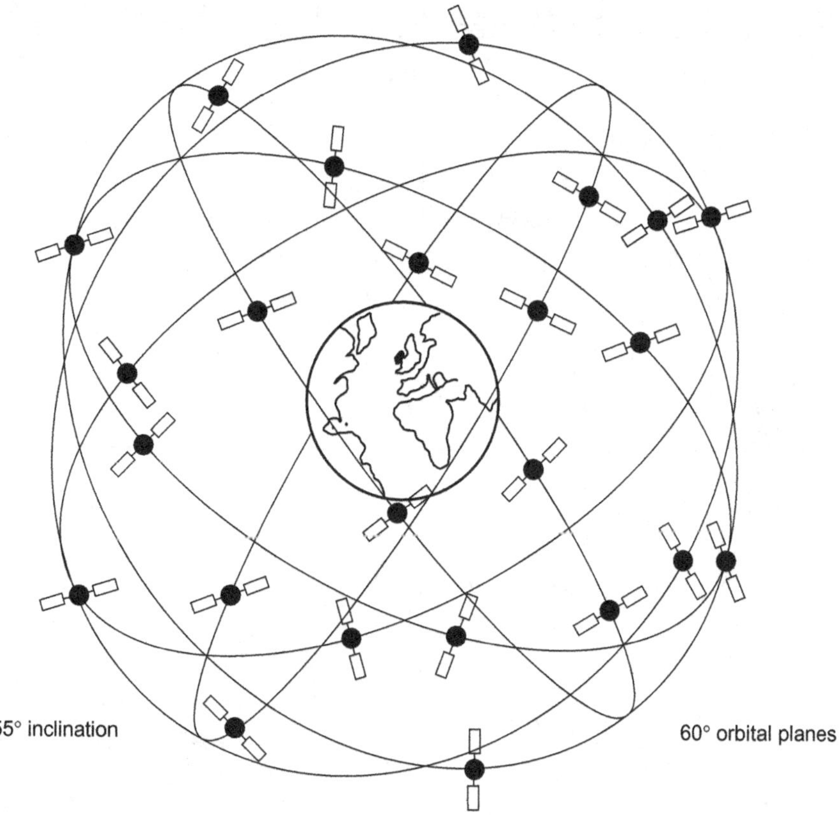

FIGURE 10.2 The satellite constellation.

system synchronized to a common standard. In the initial specification, all satellites transmit signals on two frequencies in the 'L'-band – L1 = 1575.42 and L2 = 1227.6 MHz – derived from a basic clock frequency of 10.23 MHz, both of which are uniquely coded to provide satellite identification and orbital data and also a signal which provides the means for signal transit time to be measured (see Figure 10.4). Currently only the L1 frequency is available for civilian users (but see Section 10.1.7).

10.1.1.3 The User Segment (See Performance Specifications Section 11.3.1)

As each satellite proceeds along its orbit, it transmits information regarding its identity, position and also 'timing pulses' which allow a receiver to determine the time taken for a particular pulse to travel from the satellite to the receiver. With knowledge of the speed of the pulse, the range from the satellite can be determined. Then, using the transmitted data, the satellite's position can be fixed at that instant and using the distance from the observer, a position sphere around the satellite can be defined. If similar measurements are made from a second satellite, the intersection of the two position spheres gives a circular ring or position circle in space. The position sphere derived from a third satellite cuts this position circle in two places. Given that one of these positions will be in the

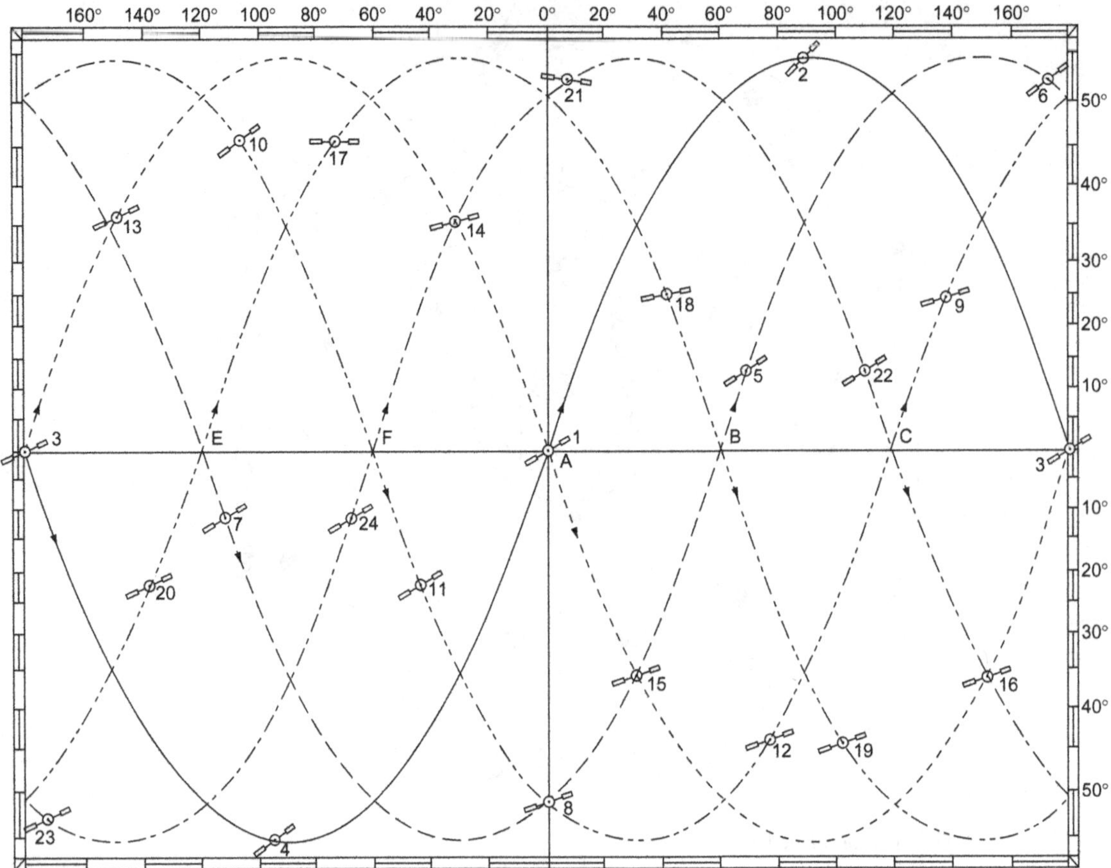

FIGURE 10.3 Mercator diagram of the satellite orbits.

reaches of outer space and can be discarded, the receiver's position can therefore effectively be calculated in three dimensions from three satellites. However, this also assumes that there is a very accurate clock in the receiver see Figures 10.5 and 10.8.

The timing signals have two levels of potential precision. The higher level, or Precision code (P-code), is primarily intended for military and other specialized users and is available on both L1 and L2 frequencies, while the Coarse/Acquisition code (C/A code) is currently freely available to all users on the L1 frequency only.

10.1.2 The Measurement of Range and Time from the Satellite

The timing signal transmitted by the satellite can be thought of as a continuous train of pulses having a unique format. Within the user's receiver, having identified the satellite from its navigation message, an identical train of pulses relating to that satellite is generated which is then time adjusted until it matches the incoming signal. The amount of this time adjustment is a measure of the signal travel time, and hence the range from the satellite is calculated (Figure 10.6). The detailed process

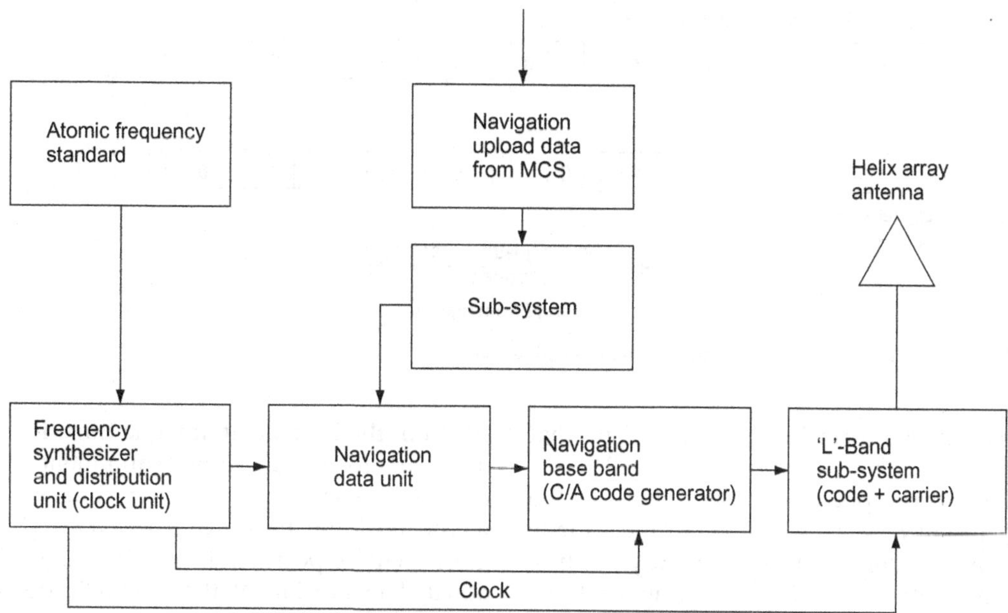

FIGURE 10.4 Satellite block diagram.

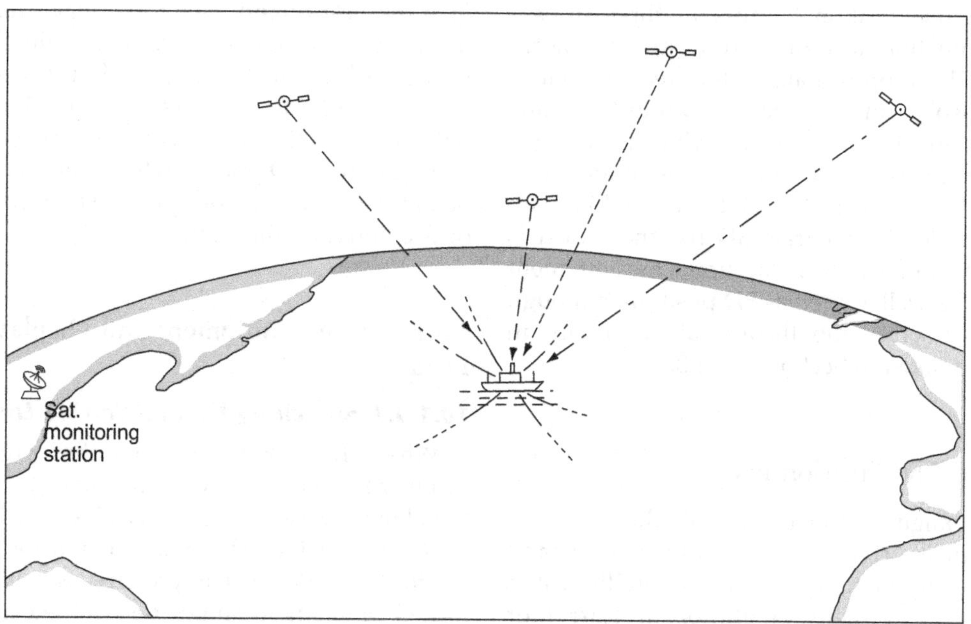

FIGURE 10.5 The ranging principle.

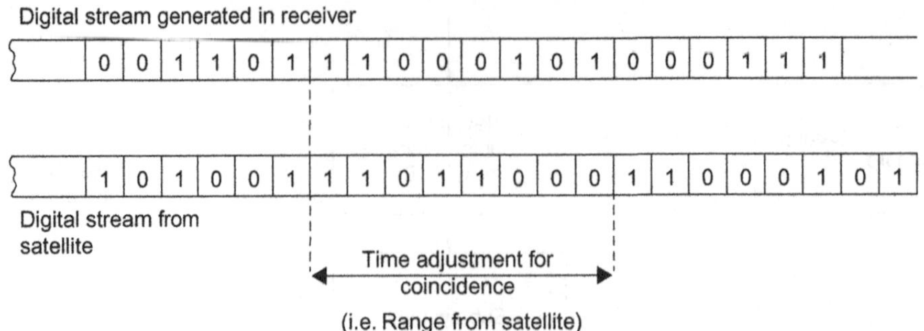

FIGURE 10.6 Time signal digital matching using satellite code.

uses correlation principles, similarly to that described for coherent radar in Section 2.9.

All this presupposes that the clock in the user's receiver is as accurate as the atomic clocks in the satellites. This is patently not true as the majority of receivers are designed to be low cost, portable and robust. However, the error in ranges due to any receiver clock error will be the same for all observations. Also, the system is specified to be able to receive four satellites worldwide (not the theoretical three) so that the positional errors in three dimensions and time can all be resolved very accurately. This provision of the receiver clock error also means that very accurate GPS time is maintained at the receiver, which is also useful for many (very often non-navigational) users. In simplistic terms, three satellites are required for an accurate 3D fix; the fourth is required to allow very accurate measurement of time as well as well as 3D position, although it also resolves the theoretical ambiguity in position noted in Section 10.1.1.3.

10.1.3 The Position Fix

The angular movement of the satellites along their orbits is comparatively slow and with a constellation of some 24 satellites, it is intended that there should be at least four usable satellites above the preferred altitude of 10° from the horizon at any one time. It is then necessary for the receiving equipment to analyse the geometry and select four satellites whose position lines are expected to give the most accurate position. Included in the navigational data portion of the satellite's transmission are details relating to the complete constellation which makes it possible to analyse the geometry of the available satellites and select the four which will give the best angle of cut for the position spheres and which give the best quality signals. In particular, the use of low altitude satellites is avoided whenever possible, due to the probability of significant atmospheric errors (see Section 10.1.5). If only three satellites are available, many receivers go into 2D mode where the receiver's height is estimated from previous observations or is entered by the user.

10.1.4 User Equipment and Display of Data

10.1.4.1 Switching On and Setting Up

When first switched on, the pre-tuned receiver locks on to any satellite and its navigational message is used to determine which satellites are available for use in that area and which are to be preferred. The preferred satellites are locked onto, their ranges determined and the position of the observing equipment fixed.

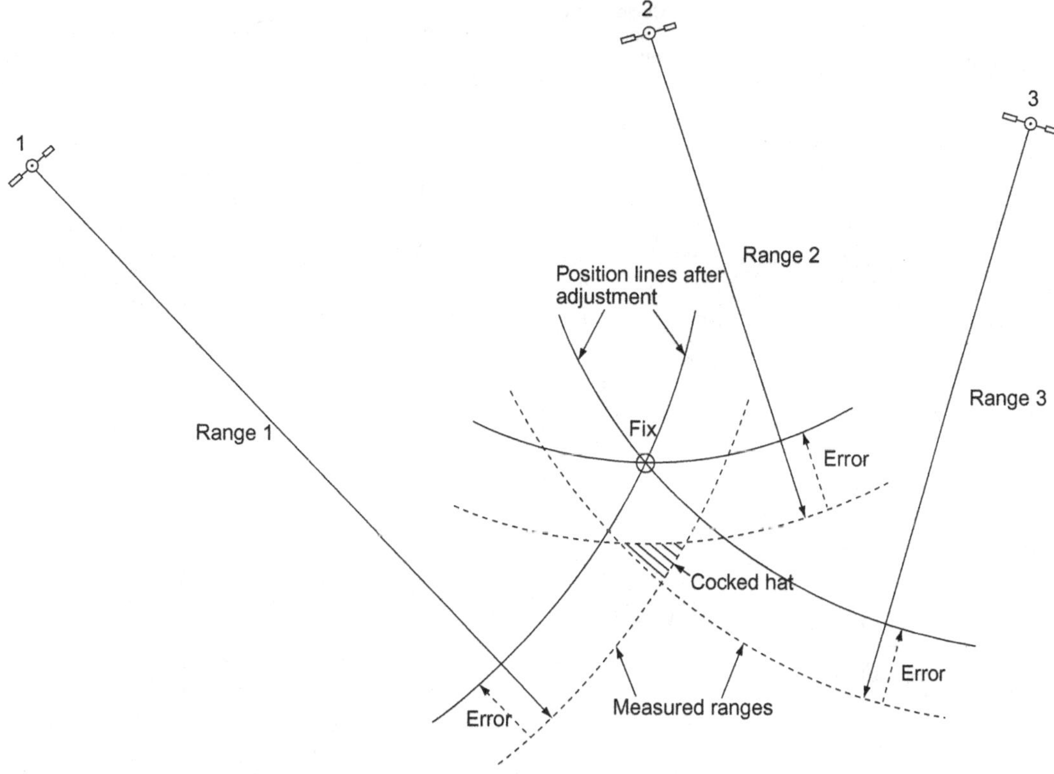

FIGURE 10.7 Effect of systematic errors (including time) on measured ranges.

10.1.4.2 Continuity – Transfer to Other Satellites with Time

With the passing of time, the user equipment will change the four satellites being observed, as the positions in the sky change. Thus the observed satellites are changed to those whose position is navigationally more favourable, either through altitudes being better or improved angle of cut of position line or higher signal quality.

10.1.4.3 Frequency of Update of Position

This is to all intents and purposes continuous, automatic and seamless. The whole operation after switch-on, that is satellite acquisition, lock-on, analysis of data, timing to determine ranges and the derivation of position, as well as continual assessment of the changing circumstances, continues without the operator's intervention.

10.1.4.4 Output of Position Coordinates

As a first choice, positions are given in latitude/longitude coordinates which relate to the earth as defined by the World Geodetic System 1984 (WGS 84), and for relative navigation this is sufficient. Where it is necessary to relate the output position to a navigational chart which is based on some other datum, there is usually a facility for the user to enter the name of the datum into the receiver which will then make the necessary adjustments to the output readings. Where positions are being automatically

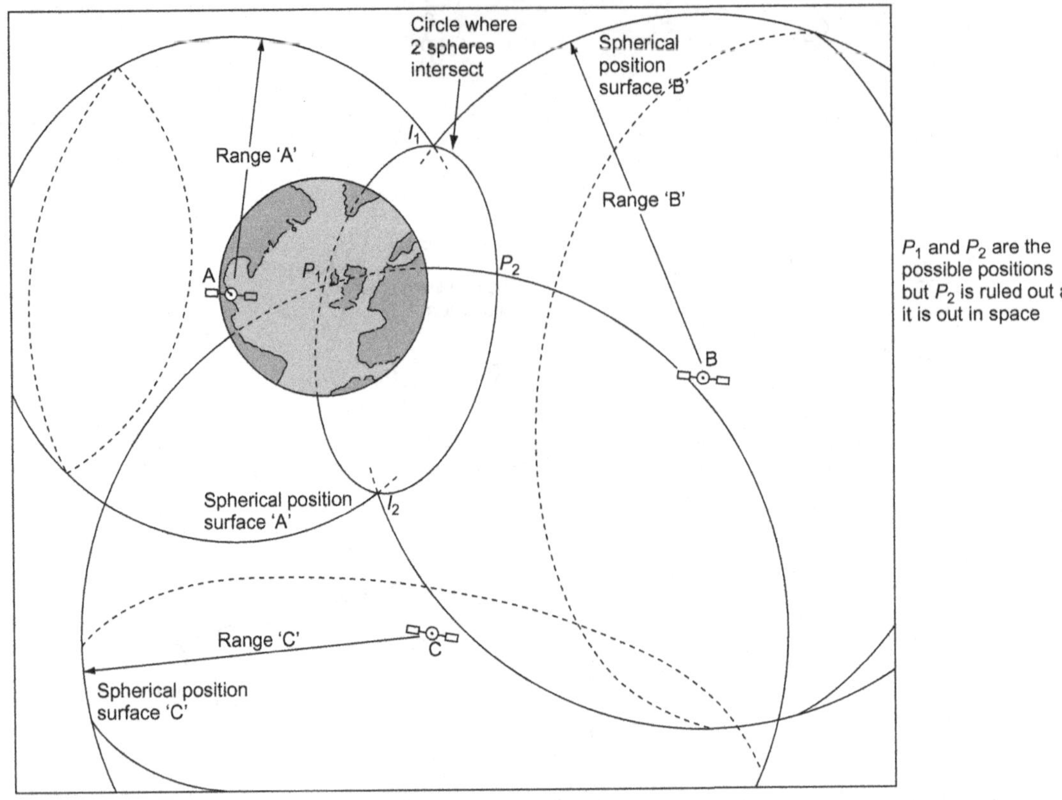

FIGURE 10.8 Three-dimensional position fix.

transferred to an ECDIS (see Section 10.2.3), it is essential that the correct datum has been fed in. It is also necessary to check regularly that the correct datum is being used. It is required that '... the display should indicate that coordinate conversion is being performed, and should identify the coordinate system in which the position is expressed' (see Section 11.3.2).

10.1.5 Accuracy and Errors

The positional accuracy that is being achieved by civil marine users is exceptionally high compared with earlier position fixing methods. Nonetheless, it is important to appreciate why the quoted accuracy figures might be compromised.

10.1.5.1 Sources of Error

Atmospheric conditions. As the signals travel through the various layers of the atmosphere, the path followed and the medium through which they pass will affect their speed and will therefore affect the calculated range from the satellite.

Signal path. The time taken to pass through the various layers and therefore the delaying effect will be different, depending on the altitude of the satellite (Figure 10.10).

It is necessary to make some estimates based on models of the atmosphere through which the satellite signal has passed. In equipment which is capable of receiving at more than one frequency, it is possible to determine a more accurate estimate of the value of error due to atmospheric conditions by using a more

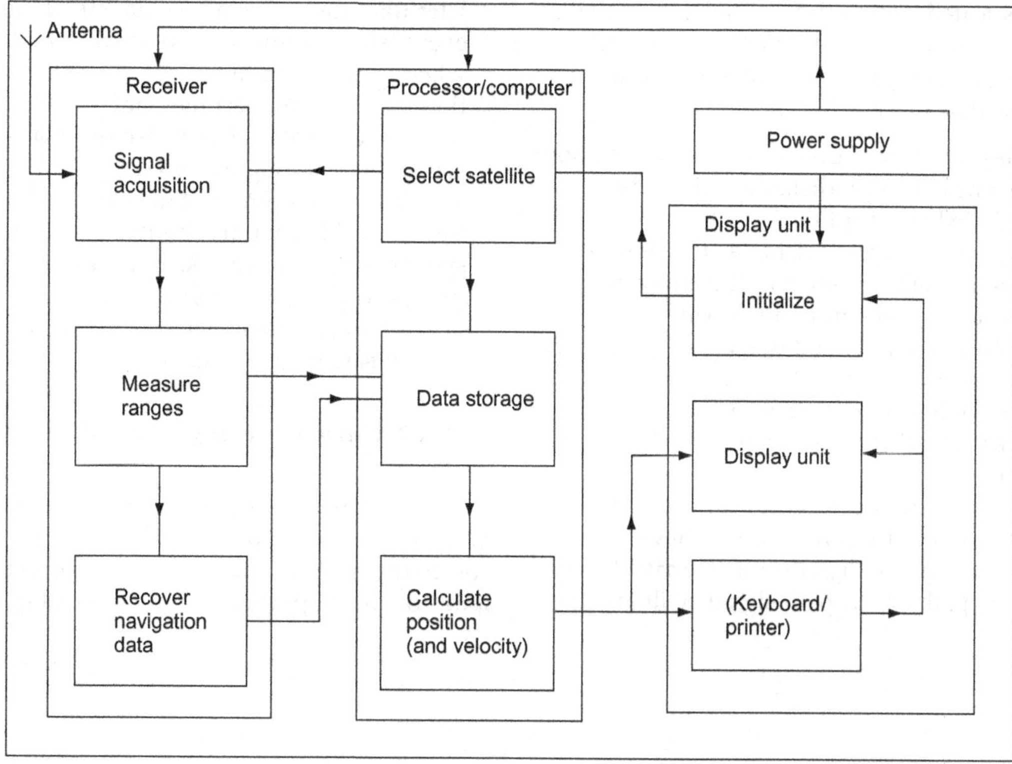

FIGURE 10.9 Block diagram of user equipment.

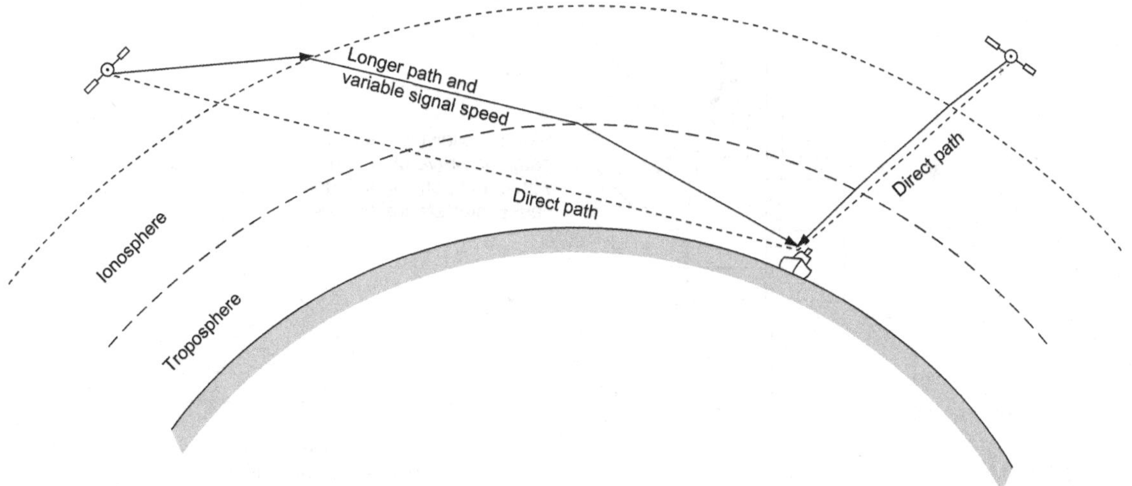

FIGURE 10.10 Effect of signal path.

sophisticated error model. Effectively, civilian users have access only to the single L1 frequency, although this situation is changing (see Section 10.1.7).

- *Ionosphere*. Errors due to this source are now to some extent predictable and can be modelled and applied.
- *Troposphere*. Errors arising as the signals pass through this layer of the atmosphere can vary widely and will depend on the existing weather conditions – in particular, water vapour. Very little can be done in practice to eliminate this source of error.
- *Multi-path*. With range being measured to such a high precision, signals arriving as a result of reflections from nearby surfaces on the vessel or from the sea will have an effect on the required signals which arrive by the direct path. Multi-path signals will arrive later than those that arrive directly. In any event, all the signals are very weak and where signals arrive out of phase, the timing signal can become undetectable. Note that signals suffer phase reversal at each reflection (Figure 10.11).
- *Relativity*. According to Einstein, as a clock travels in different gravity fields and at speeds that are a significant proportion of the speed of light, it will tend to slow down. Corrections for this effect are taken care of in each satellite's transmission.

10.1.5.2 Cross-Checking with Other Systems

It cannot be stressed too strongly that total reliance should not be placed on a single position fixing system and GPS is no different. In spite of its reputation, cross-checking GPS

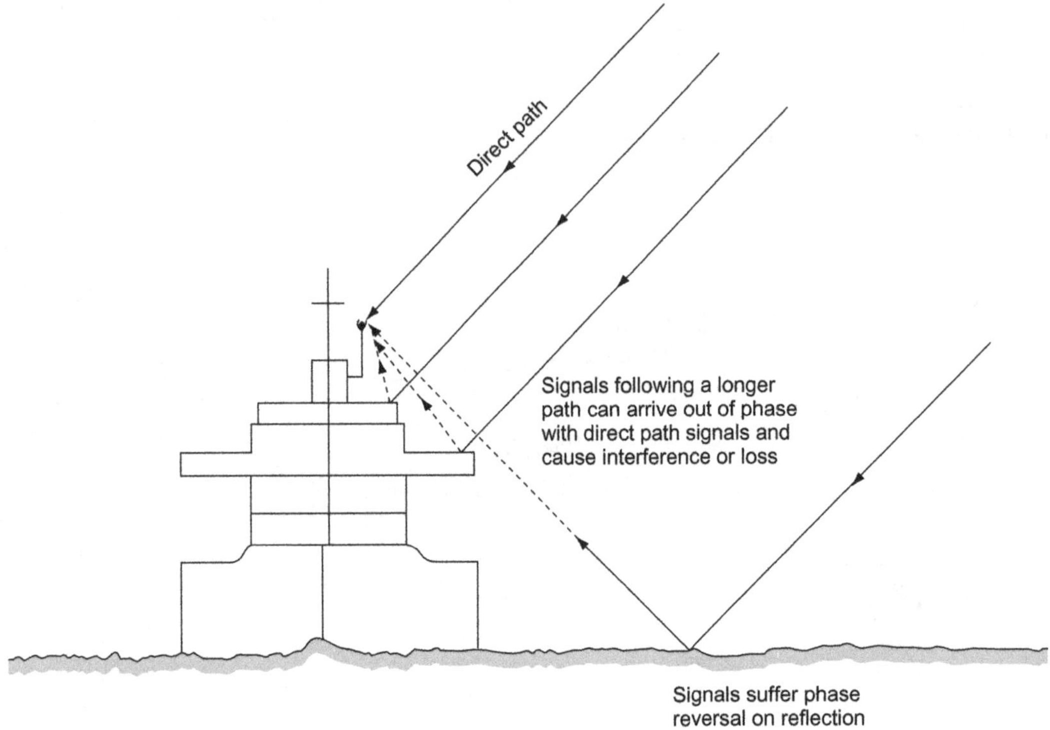

FIGURE 10.11 Multi-path signals.

derived positions against those derived by other position fixing methods, for example visual or radar means, is essential. Serious examples have already occurred where signal loss or system malfunction has resulted in the receiver reverting to dead reckoning mode and although this was indicated on the display, its significance was missed as the displayed position continued to be updated on the equipment that was in use.

10.1.5.3 Loss of Signal

There are a number of possibilities why this might occur. Typical is that as time progresses or after an alteration of the ship's course, a satellite signal might be temporarily obstructed by the ship's funnel or by some other part of the ship's structure. When this occurs, the receiver will endeavour to select another satellite and use its signal to replace the one that has become unreliable. Not all receivers indicate that lock has been lost or that transfer has taken place.

10.1.5.4 System Malfunction

The satellite constellation's 'health' is being continually monitored by the ground tracking stations, and at any sign of trouble this is communicated to the MCS and action is taken to resolve and rectify the problem. However, satellites are not always in sight of the monitoring stations and they can only be corrected when they are in sight of the MCS. Both detection and correction processes may take some time (see also Section 10.1.5.7) and there have been documented instances when significant outages have occurred.

10.1.5.5 Expected Accuracies

When originally set up, the signals available to civil users (using the single L1 frequency and C/A code) were deliberately downgraded. This downgrade was referred to as selective availability (SA) and ensured that accuracies of the order of 100 m (95% accuracy) were the norm. In 2000, SA was discontinued and long-term average accuracies of around 35 m (95% accuracy) should therefore have been achieved by a 24 satellite constellation. However, the system currently has more satellites (usually around 30) than the minimum constellation and accuracies of 10–15 m (95% accuracy) are now considered typical under standard conditions. This satellite over-population is not guaranteed to continue but has been the norm for many years. It provides redundancy for satellites that are near or past their designed life span.

10.1.5.6 Quality of Fix or Dilution of Precision

This is a concept that is essentially intended to give the user a confidence figure that will indicate the reliance that he can place on the position that is being displayed. The figure is based on the geometry of the position fix (Figure 10.12). The minimum attainable DOP which indicates a perfect fix is 1. IMO Performance Standards require that the horizontal dilution of precision (HDOP) $\leq$ 4 (see Section 11.3.2).

Note: DOP may be given for the horizontal plane – HDOP – or the vertical plane – VDOP – or in three dimensions – position dilution of position (PDOP).

10.1.5.7 Warnings (See Also Section 11.3.2 – Warnings and Status Indications)

The receiver is required to give a warning within 5 s if the HDOP has been exceeded, i.e. $>$4, or if a new position has not been calculated for more than 1 s, in which case, the time and position of the last valid fix must be displayed, with a clear indication of what has occurred and a warning of loss of position.

When DGPS (see Section 10.1.6) is in operation, there must be an indication that DGPS signals are being received and that DGPS

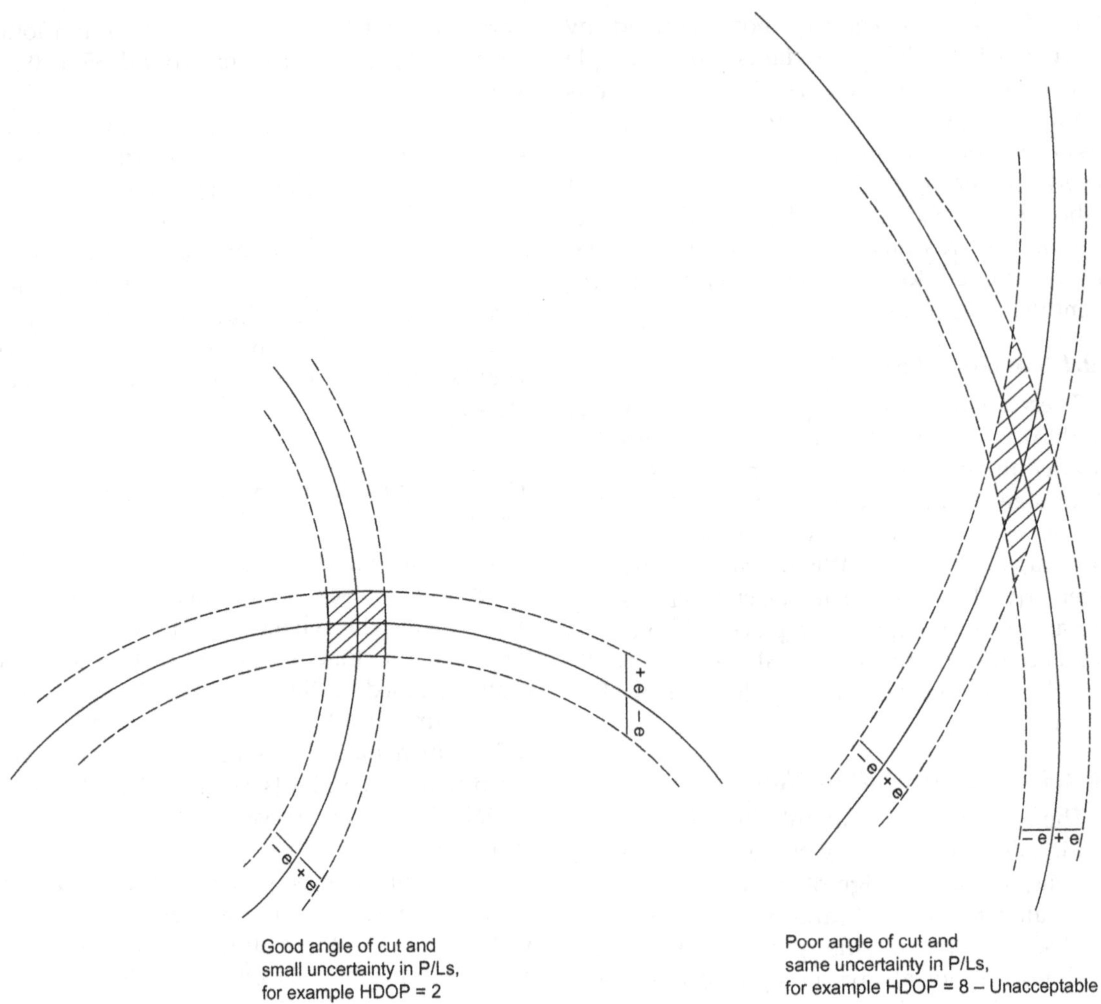

FIGURE 10.12 Dilution of precision.

corrections are being applied to the displayed position. There must also be a display of DGPS integrity status and alarms (if any) plus any DGPS text messages.

10.1.5.8 Vulnerability of GPS Signals

GPS signals are very low power when received on the earth's surface and the received signal strength is about the level of background noise. The code system allows receivers to pick out an individual satellite from the transmission of other satellites and background noise all operating at the same frequency. However, if the interfering signals are powerful enough, they will swamp the GPS signals and render the receiver ineffective.

The main natural cause of concern is the potential for interference from solar flares from the sun. Solar flares are associated with sunspot activity on the sun. The flares are a large energy release from the sun and they are accompanied by extensive magnetic radiation across a wide spectrum of radio frequencies. If the earth is hit by such radiation, it is likely that GPS (and any other radio equipment) will be affected. The frequency of occurrence of solar flares follows an 11-year cycle, but precise prediction is very problematic.

GPS jammers are widely available to the public, as well as for military applications. There have also been a number of occurrences of unintentional interferences that cause GPS 'blackspots'. Many of the causes of unintentional interference have been traced to faulty or illegal radio equipment and they usually occur in inland regions. However, there has been more than one occasion when harbours and harbour entrances were affected. There have also been a number of GPS blackspots that have not been satisfactorily explained.

It is also understood that GPS 'spoofers' exist that give false satellite signals and would enable the receiver to give a false but believable position on a user's display.

In summary, although GPS has proved to be a very reliable and secure device, it is theoretically very vulnerable to attack and unintentional interference. The user is advised not to rely on it uniquely, but heed the advice given earlier and cross-check between position fixing systems.

10.1.6 Differential GPS

10.1.6.1 System Principle

The position of a shore-based antenna is accurately surveyed and its geo-coordinates calculated. The differences given by satellite fixes and the surveyed position are transmitted as corrections to be applied to positions obtained by vessels using GPS in that area. In order to receive these signals, a DGPS receiver is required. The application of these corrections is normally automatic (see Figure 10.12).

The UK government has set up a public system of DGPS stations using the sites and frequencies previously used for marine RDF beacons (see Figure 10.14). The US Coastguard has done much the same on the US coasts and commercial organizations can provide corrections in many areas for a fee. Also available are wide area DGPS networks which provide continuous DGPS corrections over continental areas using geostationary communication satellites. They include the European EGNOS, the US WAAS and the Japanese MSAS. The generic term for such systems is satellite based augmentation systems (SBAS). Sometimes conventional ground based DGPS is called ground based augmentation systems (GBAS).

10.1.6.2 Potential Accuracies Using DGPS

Using DGPS equipment, accuracies in the order of 5 m are expected and even better results have been reported. With such high precision, it is important to appreciate that it is the position of the receiving antenna that is being displayed. Updating of the direct signal is typically within 15 s. The improvement in accuracy of DGPS over GPS is irrelevant for most marine navigators now that SA (see Section 10.1.5.5) has been removed. However, DGPS has a large advantage over GPS in that there is immediate notification of system malfunction, unlike stand-alone GPS (see Section 10.1.5.4). Thus many users are using DGPS because of the gain in reliability due to external monitoring and not necessarily because of its increased accuracy.

10.1.7 Improvements to GPS

The US military is currently implementing three main improvements in the GPS service for civilian users, as well as a new military service (M code) not for civilian use. The process is a slow one because the improvements are only being fitted to satellites that have been either launched relatively recently or are still in reserve on the ground. Each improved

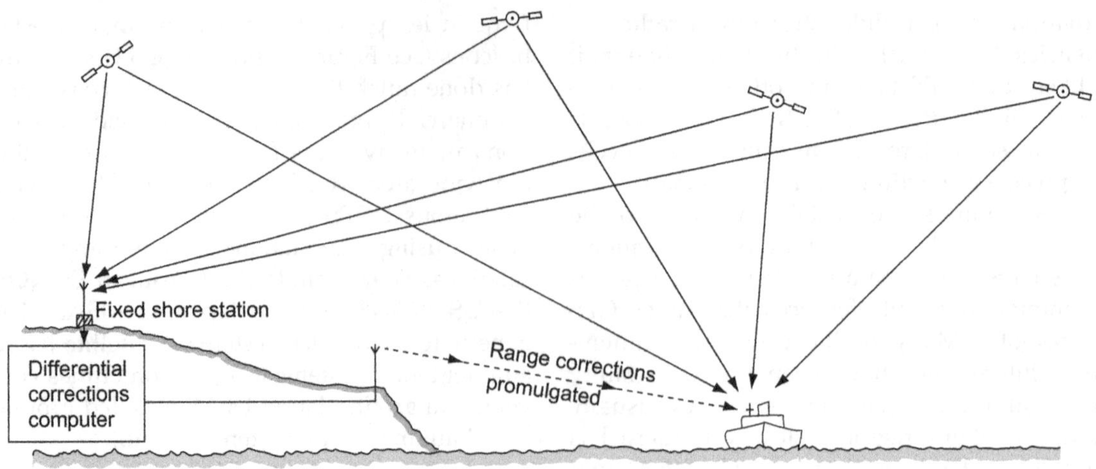

FIGURE 10.13 Principle of DGPS.

service will therefore only become properly operational when a minimum of about 18 improved satellites is in orbit. It is an ironic fact that the impressive reliability and longevity of the older satellites is actually holding up further progress.

The first civilian improvement is to open up L2 to civilian users. They are not allowed to use the P-code, so the modification is to transmit the equivalent of the C/A code as well as the P-code on the L2 frequency. This is called the L2C, it currently transmits from a limited number of satellites and it is expected to be fully operational in 2018. The addition of observations on a second frequency will enable better modelling of the atmospheric errors and hence a more accurate fix will be achieved (see Section 10.1.5.1). Moreover, the L2C code will be inherently more accurate and more reliable in weak reception conditions than the C/A code used on the L1 frequency.

The second civilian improvement is the addition of a third frequency, 1176.45 MHz, termed L5. As well as potentially allowing more sophisticated atmospheric error modelling from using up to three frequencies simultaneously, the civilian code available on L5 will be a more sophisticated signal allowing higher accuracy and reliability than both the L2C and L1 C/A codes. Satellites with the L5 frequency are just beginning to be launched and a full constellation is expected by 2021.

The third civilian improvement is an additional civilian code on the L1 frequency termed the L1C. This code will again offer improved detection and accuracy in weak reception areas. However, its major importance is that it was developed with GALILEO GNSS as a common code and it is also now intended to be implemented on navigation satellites of other countries (see next section).

10.1.8 Other Navigation Satellite System Developments

The GPS is entirely under the control of the US military. It is fully operational, is freely available and is expected to remain so for the foreseeable future. However, other GNSS and regional systems are available or about to become available. It is foreseen that for most users, the use of these additional systems will not be as an alternative to GPS. Instead receivers will be become GNSS receivers capable of receiving satellites' signals operated by many

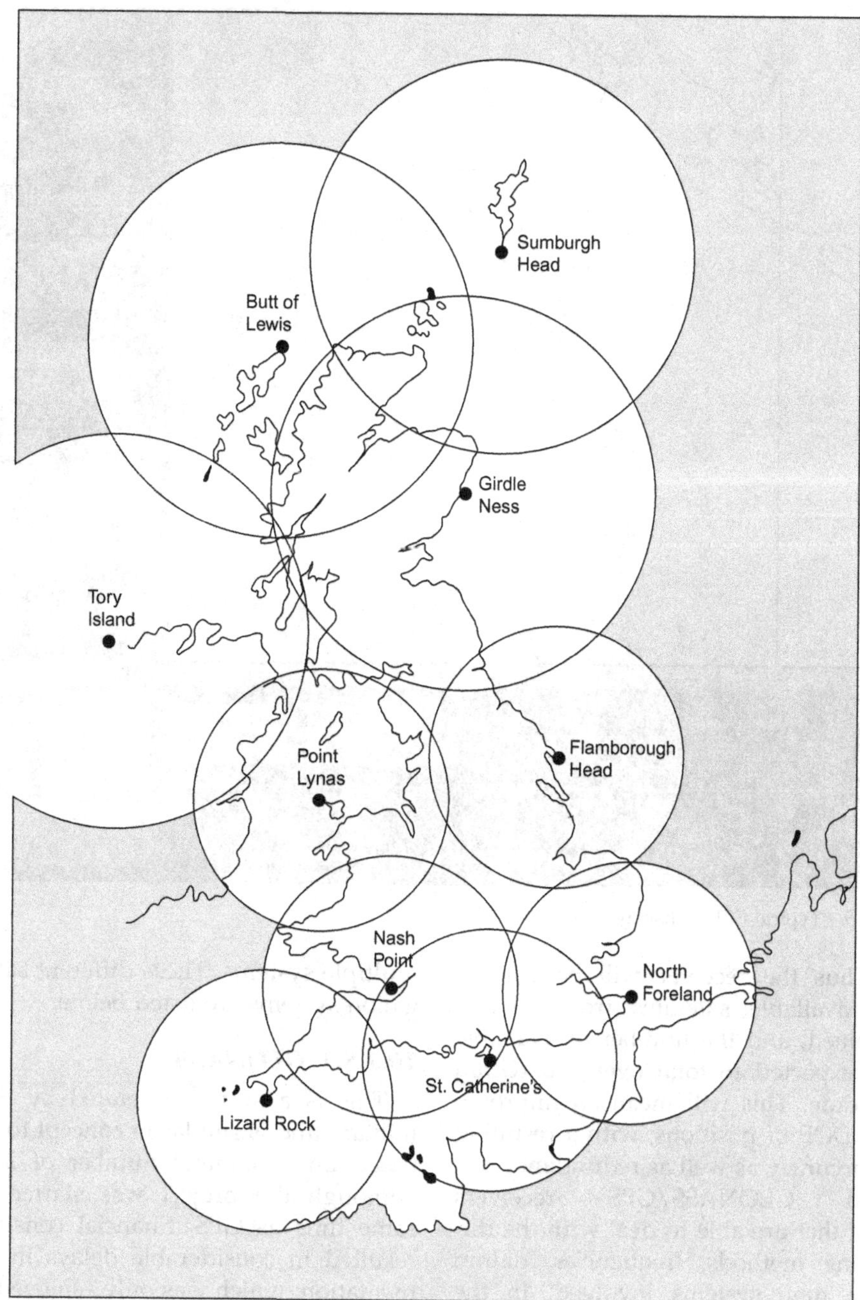

FIGURE 10.14 DGPS coverage in UK waters.

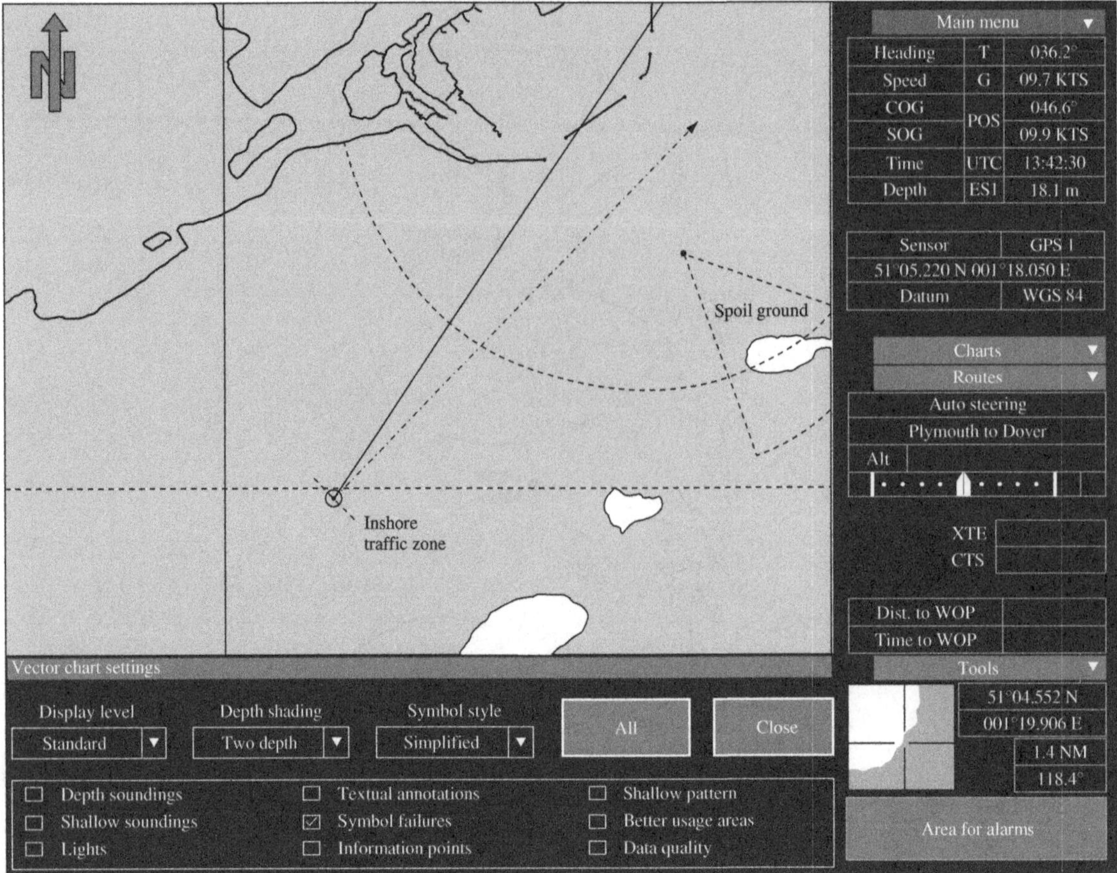

FIGURE 10.15 Typical ECDIS display.

countries. Thus the receiver will be able to select best available satellites from all the GNSS combined, and the number of available satellites is expected to total over 100 within the next decade. This will mean an improvement in the DOP of positions, with a resulting increase in accuracy as well as redundancy.

Combined GLONASS/GPS receivers already exist that are able to deal with the different ranging methods, frequencies, datum and satellite time systems involved. In the future, this will be even easier for the other systems discussed below, as these are intended to have the new L1C code as part of their specification, which will facilitate the integration of multiple systems. These different satellite navigation systems are listed below.

10.1.8.1 GLONASS

This is a system operated by the Russian military and is similar in concept to GPS, being based on a nominal number of 24 satellites. Although the project was started about the same time as GPS, financial constraints have resulted in considerable delays in full implementation, which was only achieved in 2011.

The orbit configuration is three equally spaced orbits, inclined at 64.8° to the equator. Each orbit is designed to have eight equally spaced satellites. The orbit altitude is slightly less than GPS at 19,130 km, which gives an orbit period of 11.25 hours.

The GLONASS satellites currently transmit on different frequencies, unlike the GPS, which shares a single frequency with different satellite codes. Note that GLONASS satellites that are 180° apart can use the same frequency, which makes slightly more efficient use of the frequency spectrum. GLONASS achieves similar accuracies to GPS. One advantage of the system is that GLONASS satellites can communicate with each other and that satellite errors can therefore be corrected even when the satellite is not visible to the MCS.

The Russians have also developed a code-based signal for broadcast from GLONASS satellites. This will make building combined GNSS receivers that include GLONASS even easier.

10.1.8.2 Galileo

This is a satellite system for navigation and communications, planned and led by the European Union (EU) with an ever-advancing completion date. Unlike GPS and GLONASS, it is civilian controlled and the project has partnerships with a number of countries outside the EU. It also plans to charge some users for some of its high-quality and specialist services. Although primarily a GNSS for navigation, it is expected to have some communications facilities including use by emergency and distress services. It has had considerable funding difficulties and delays over the project's lifetime.

The basic design has a 30-satellite system (27 satellites plus three active spares). There are three separate orbits (10 satellites in each) inclined at 56° to the equator with an altitude of 23,616 km and an orbit period of about 14 hours. Every satellite will transmit on several frequencies with E5A (1176.45 MHz), E5B (1207.14 MHz) and E2 (1575.42 MHz) being planned for public use. Another feature of the Galileo system is the large number of ground stations to control and monitor the satellites, that is more than provided by GLONASS and GPS. At the time of writing, funding is committed to launch two-thirds of the full constellation, with a small number already launched.

The system plans to offer a number of services including a 'free to user' service of similar accuracy to GPS. A single frequency receiver should achieve 25 m (95% accuracy), while a dual frequency receiver should achieve 5 m (95%). Higher accuracy 'fee-paying' services will also be available.

Galileo operates like GPS using satellite codes and is thus able to use very similar frequencies to GPS. Galileo is to adopt different datum and time systems to GPS, but the differences in datum are expected to be minimal and measured in centimetres. It has been intended from the outset that Galileo and GPS will have a degree of compatibility and receivers capable of receiving both systems will be relatively easy to manufacture. One of the frequencies to be used is the internationally agreed L1C (see Section 10.1.7).

10.1.8.3 BeiDou Navigation Satellite System (Aka COMPASS)

This is a Chinese GNSS which already offers a regional service and is intended to offer a full global service by 2020. It will consist of 30 medium earth orbit satellites (that is similar to GPS characteristics) as well 5 geostationary satellites which serve regional areas. Like Galileo and now GPS, it is planned to offer code-based signals on several frequencies. This also includes the L1C signal (see Section 10.1.7).

10.1.8.4 Regional Satellite Navigation Systems

Although not global, it should be noted that a number of countries are developing regional satellite navigation systems. These include the

Japanese QZSS and the Indian IRNSS, both of whom intend to use the L1C signal.

10.1.9 Inter-Relationship of GNSS with Radar

Where radar equipment is fitted with mapping facilities, it may be possible to feed co-ordinates to the radar in latitude/longitude. Where this can be done, it is usual to be able to automatically feed the GNSS derived position of the ship to the radar and thereby fix the latitude/longitude of radar derived features on the display, thus aiding identification (see also cross-checking, Section 10.1.5.2).

The availability of high-quality GNSS derived positions has led to the decline in using radar as the prime navigational tool. Obviously positions derived from radar are not as accurate as GNSS, but the radar remains a good source of secondary navigational information that is independent of any external transmission and external control. The good navigator will use the radar to check the accuracy of GNSS positions, but the concern is that the less professional navigators who over-rely on the 'easiest' positional source will ignore the radar as a navigational tool and thus leave themselves exposed to increased risk in the event of a GNSS failure.

10.1.10 Inter-Relationship of GNSS with ECDIS (See Sections 10.2 and 11.3.3)

GNSS equipment is able to supply position and time, as well as (optionally) Course Over the Ground (COG) and Speed Over the Ground (SOG) to a digital interface which can then provide an input to an ECDIS.

Radar images and ARPA information may also be added to the ECDIS but should not degrade the chart information and should be clearly distinguishable from it. Where this information is added, the data should match in scale and orientation. It should be possible to remove this radar data from the ECDIS by a single operator action.

10.2 ELECTRONIC CHARTS (ECDIS)

Electronic charts are an example of a geographical information system (GIS). A GIS has a database of geographical information that can be filtered and arranged in a display for the convenience of the user.

For many hundreds of years, the navigator has used paper-based charts. It is only relatively recently that a computer-based alternative has become practical due to the improvements in and the affordability of two computing related technologies. The first is the ability to store and process large amounts of data and the second is the display technology that is able to show data to an adequate resolution and sufficiently large view of an area of the chart.

The standards for an approved electronic chart system were being developed from the 1980s by IMO and IHO (International Hydrographic Organization).

The new system was called ECDIS or electronic chart display and information system. The standards for ECDIS are very high. The opportunity was taken to:

- set electronic and hydrographic data storage standards to allow easy interchange of information between hydrographic services and also update ships at sea;
- set minimum hydrographic standards of accuracy and reliability;
- set a common chart datum (basically the one used by GPS).

IMO ECDIS standards came into force in 1996 and it was then theoretically possible for a vessel operating under SOLAS regulations to replace its approved paper chart system with an ECDIS. The reality was that although manufacturers rushed out ECDIS software display systems for ships, there was no approved ECDIS chart data. ECDIS data was slow to emerge due firstly to the time it takes to convert existing chart data into ECDIS format, but more particularly due to the need to resurvey hydrographically many areas to the new ECDIS standards. More recently the hydrographic standards required have been relaxed, making a worldwide portfolio of ECDIS charts a practical reality. New merchant ships are now required to have ECDIS fitted.

10.2.1 Vector Charts

The data format of ECDIS is vector storage. In this system, the ECDIS data is stored in data tables of separate information. Thus there are data tables of buoys, navigation aids, digitized coastlines, sounding lines, text labels, etc. Each item is stored with details of geographical location, its characteristics and the range of chart scales at which it should be shown. When data on a specific area, at a specific scale, is requested for display, the system selects the data in the requested area, of the right type for display. Important points to note are:

- The chart data shown on the chart display can be manipulated to remove data not required by the navigator and which may be cluttering up his screen. There is no need to show all data available in a specific area. Thus information on lights could be removed from the display if the vessel is conducting a daylight transit. IMO standards place careful limits on the information that can be removed by the navigator and also what should be displayed. For instance, all information on depths need not be shown, but there must be a minimum safety contour, which shows safe/unsafe water for the own ship.
- The vector chart can easily change scale or zoom. As the navigator zooms in or out, the data displayed will change according to scale parameters stored with each item of data. Thus a buoy off the Liverpool Bar will not be shown when viewing the whole Atlantic Ocean, but as the navigator zooms into the Liverpool region, the buoy will appear. All depths or soundings have minimum and maximum scale parameters so that as the navigator zooms into an area, more soundings appear so that an appropriate density of soundings is maintained. It will be appreciated that the attachment of these scale factors is a highly skilled and laborious job for a marine cartographer and this is one reason why it has taken such a long time to get approved ECDIS data published, even when the hydrographic data was up to ECDIS standard.
- The responsibility for obtaining and maintaining approved ECDIS data is divided up between different nations and all data is stored centrally by the IHO. Participating nations can then obtain approved ECDIS data from other areas through the IHO.
- Vessels that are required by SOLAS to carry an approved chart navigation system can replace their paper chart system with an approved ECDIS which uses approved ECDIS vector data. However, it should be noted that there must be a backup system as well as regulations on installation, training and the provision of regular chart corrections.
- Vector ECDIS displays are required to show chart datum soundings in a similar way to paper and raster charts. The predicted tidal height is shown as a separate figure, when available.

10.2.2 Raster Charts

An alternative method of chart data storage is raster storage. Raster data is stored in picture format. In marine terms, the paper chart is scanned into a digital file using the same techniques as a photocopier or fax machine. The picture to be stored is divided into lines of very small dots or pixels. The scanner decides what the colour number of the original document was for that specific pixel, and it is the colour number that is stored for every pixel on the document. Principal points to note are:

- Raster data on display cannot be manipulated by the user. Effectively we have a photocopy of a paper chart on screen. It is not possible to suppress chart data.
- A raster chart should be displayed at the correct scale on screen as the original paper chart from which it was derived. Some chart displays do allow a limited zoom function, but the navigator should note that the system simply magnifies or shrinks the chart. At extreme zoom levels the chart becomes unreadable and/or the amount of data shown is inappropriate for the scale. Thus systems which allow zoom display warnings that the chart is being shown underscale or overscale, as appropriate. The navigator can of course select another chart of different scale if one is available. This process is similar to using paper charts, but much faster as courses and positions will normally be transferred automatically.
- In the western world, only two organizations maintain a full worldwide paper chart system; the chart authorities of the United Kingdom and the United States. Both these countries have produced raster charts based on their existing paper charts and effectively these are also the only two raster systems that provide comprehensive and approved raster electronic chart systems. These systems were available before ECDIS received IMO approval in 1996.
- Approved raster data can be used in ECDIS where ECDIS vector data is not available. This is sometimes termed a 'dual-fuel' system. However, IMO has determined that raster chart data is not an acceptable replacement for paper charts. The official position is that the paper chart is the approved navigation system and the raster electronic chart is an aid to navigation. Therefore, when the system switches to raster data, because of lack of vector data, the prime means of navigation switches to the paper chart. However, some administrations are allowing a significant reduction in the number of paper charts carried, when the raster chart system is used in conjunction with the appropriate training, backup, installation and correction system.

10.2.3 Comparison Between Paper and Electronic Charts

The biggest single advantage of electronic charts over paper systems to the professional navigator is the ease of correction. Both raster and vector systems allow for automatic electronic correction of the charts onboard, with no real work on the part of the navigator. This means that the corrections are inserted exactly as intended by the marine cartographer. The potential labour reductions for a ship with a worldwide portfolio of 2000–3000 charts are also significant. It is estimated that there are 10,000–15,000 manual corrections which need to be made by the navigator in a typical year for a worldwide paper portfolio of this size. The electronic corrections also lend themselves to electronic transmission so that corrections can be passed immediately to the ship, even while at sea. This is a vast improvement on the traditional paper-based system where corrections often took months to reach a ship by post, although paper chart small corrections can now be transmitted electronically and printed out onboard. The electronic systems also allow the electronic transmission of whole new charts. This could be either because of unexpected passage changes at sea or the issue of a new chart edition when it would usually be impractical to get the chart to the ship at sea in a paper-based system.

Another significant advantage is the ability to easily add extra information, overlaid on top of the electronic chart. This can include routes, notes, links to extra material (pictures, videos and text documents), tidal currents and lines of safety. Moreover this information can be saved as both a permanent record and also brought back from storage for reusing when a vessel undertakes a second transit of the same area in the same direction. The paper chart requires much laborious preparation and rubbing out on every occasion it is used. It will wear out in time and need replacement.

Another pencil saving advantage is that the electronic chart lends itself to electronic fixing methods. GNSS or LORAN C systems can feed the position electronically to the chart (see Section 10.1). The improvement over the paper system means that the ship's position is continuously plotted rather than sampled at intervals convenient to the navigator and the possibility of human plotting error is eliminated. There is usually an option to stabilize the chart display on the vessel's position so the chart always shows the area around the current position, that is the chart moves past the ship. In these cases, the ship does not have to be at the centre of the display, but can be offset to show a feature of land or to see further ahead than astern. There is also a provision to put in the vessel's length and breadth so that a scaled ship shape can be displayed when the chart scale in use is appropriate.

There can be an issue with the electronic chart datum being different to the datum of the electronically derived position, but provided the datum adjustments are known to the navigator, they can easily be adjusted automatically in the electronic plotting (see also Section 10.2.4).

Currently, the main disadvantage of the electronic chart is screen size, although large displays of more than 40 inches are becoming affordable. The minimum size of display for an approved ECDIS is 21 inches, which provides about a sixth of the coverage area of the traditional paper chart. This is why zooming and panning facilities are useful, together with the ability to have different windows on the chart area open and available for easy viewing.

The other main disadvantage is the reliance on electronic equipment and the need for a power supply. Hence, there is a requirement for emergency power supplies and a backup chart system that may be a duplicate system and/or a minimal number of paper charts intended to get a vessel to a pilot station.

10.2.4 Comparison Between Vector and Raster Charts

One major advantage of the ECDIS vector system is that the navigator can be assured that all the data is surveyed to an approved standard. The raster (and paper) charts traditionally represent the best available data for an area, which is not quite the same thing.

The main strength of the raster chart system has been that approved worldwide portfolios are available from both the United Kingdom and the United States. Now that equivalent vector ECDIS data is available, the use of the raster charts is expected to decline.

The ECDIS vector charts are all based on a model of the earth that is effectively the same as the WGS 84 datum, so no corrections are needed when using the American GPS. Raster charts are copied from paper charts that traditionally tended to be based on a local chart datum developed for a particular region. As an example, the British Admiralty paper chart system uses about 300 different chart datums over its worldwide portfolio. However, in their raster electronic chart version (known as ARCS), the corrections are included electronically within approved raster chart data and applied automatically, so the navigator can still use GPS positional data directly without datum conversion. Moreover, as new paper chart editions are produced, the WGS 84 chart datum is being adopted for paper charts, so the theoretical problem is further diminished. For example, all paper and raster charts of the UK region produced by the British Admiralty are now based on WGS 84.

The raster charts have a chart edge, just like the paper chart. The navigator has to change chart or the system will itself decide to change to the next chart. However, most mariners are already well used to dealing with chart edges and changing charts in the paper-based system. In contrast, ECDIS vector datasets or 'charts' can be integrated seamlessly together, so no join shows on the user display.

Both raster and vector systems are easily correctable by electronic means and while each electronic chart system may have different data storage and computer processing power requirements to cover a particular area in raster or vector format, the actual requirements of either are not expected to be significant or expensive by modern computer standards.

A major benefit of the vector system is the potential for an electronic chart to have 'intelligence'. In a vector system, the computer knows that it is a buoy, wreck or coastline ahead and the depth expected. Thus it is possible for the chart to warn the navigator if he is approaching shallow water or a danger, the times at which lights should be picked up or if he has inadvertently laid a course line over a shoal or dangerous wreck. The data stored in a raster system is simply the colours of different dots and the computer would not know if a specific black pixel was part of a buoy, coastline or text.

10.2.5 Unapproved Electronic Chart Data

A number of commercial companies offer unapproved vector data for electronic chart systems. The data first appeared due to the long wait for approved ECDIS data. The standard of this unapproved data is very variable. The companies rarely use original hydrographic data but usually digitize existing paper charts. They also do not provide a continuous system of chart corrections. Nevertheless there is nothing to stop them being used on small vessels not required to carry any approved charts or as an 'aid' on larger vessels that still have the paper system as the official navigation system. Electronic charts of this type can sometimes be associated with specialized tasks such as fishing or other underwater activities. The term Electronic Chart System (ECS) is used by IMO to denote a system that does not meet the standards for ECDIS.

10.2.6 Publications Associated with Charts

The same regulations that require charts to be carried on ships also require that associated publications be carried. These include tide tables, tidal stream atlases, light lists, radio signal lists and sailing directions. These publications are slowly moving over to an electronic format, which would also simplify usage and improve the quality of corrections. The ideal is an integration of the combined chart and publication information into one large GIS so that the relevant textual information can be derived on a geographical basis from a visual display and be selected as required.

10.2.7 Relationship of ECDIS with Radar and Target Tracking

The ECDIS Performance Standards (see Section 11.3.3) allow for the display of radar and target tracking information on the ECDIS display. In practice, there are two levels of integration on the display.

The first is the display of radar and AIS tracked targets on the ECDIS. These look similar to the symbols/vectors displayed on a standard radar display and this can be achieved through a standard IEC/NMEA interface (see Section 10.3.4). The advantage of this transfer of data is that the traffic situation can be analysed in the context of the chart's geographical information. This is particularly useful in coastal areas as the flow of traffic around geographical features may not be obvious on the pure radar screen. These may include shoals, underwater hazards, traffic separation schemes and political boundaries. Charts can cover a larger geographical area than the radar and this can be useful for planning manoeuvres in advance.

The second level of radar integration is the superimposition of the radar image onto the ECDIS display. This will allow the easy identification of some radar targets (e.g., which target is the buoy and which target is a possible fishing vessel). The radar overlay will also check that the land features or buoy patterns on the ECDIS line up with the radar overlay, which is the quickest and easiest way to check that GNSS, ECDIS, radar and compass are all working correctly.

The navigator must be able to suppress or remove all radar and AIS tracked target information from the ECDIS in one operation (see Section 11.3.3).

There is also the reverse possibility of superimposing selected ECDIS data on the radar display (see Section 11.2.1). Care must be taken that the radar information has priority and is not masked by ECDIS data.

10.3 INTEGRATED SYSTEMS

Traditionally bridge (and other) instruments have tended to be developed in isolation, with each having an individual (possibly mechanical) method of measuring and displaying data. Ergonomics was a discipline for the future and each instrument had to fight for its position on the bridge. As each new instrument was added to existing bridges, they were invariably placed wherever a vacant spot existed. In time, the grouping and placement of instrument displays was given some consideration in vessels under construction. It is only relatively recently, with the widespread digitization of data, that the ability to integrate and display the data has become a possibility. With the development of an international standard for data format and transmission protocols (see Section 10.3.4), integration is now becoming more commonplace. Whilst initially some manufacturers did integrate instruments of their own manufacture, it is now possible to integrate instruments from different manufacturers who adhere to these common standards.

10.3.1 Integrated Bridge Systems

An integrated bridge system (IBS) was defined as '... a combination of systems which are interconnected in order to allow centralized access to sensor information...'. 'The IBS should support systems performing two or more of the following operations: passage execution; communications; machinery control; and safety and security'.

While IMO Performance Standard for IBS dealt with all aspects of vessel operation and safety, it is intended here to deal only with the integration of those systems which have a navigational function and are in some way related to the radar/ARPA as a sub-system of the integrated system. It should also be noted that IMO no longer recognizes the term IBS in the current standards.

10.3.2 Integrated Navigation Systems

An integrated navigation system (INS) aims to '... enhance safe and expeditious navigation and to complement the mariner's capabilities ...'. Three levels of performance standard are specified by IMO (see Section 11.3.4).

10.3.3 Typical Systems That May Be Integrated

Given that a sensor has the ability to export data in a universally recognized format, virtually any combination of instruments may be integrated into a single display system, but the method of display and data grouping does require careful consideration.

- *Integration on smaller craft*. For some time now, sailing boats, and in particular those intended for racing, have tended to concentrate on those instrument displays, crucial for strategic decision-making in a single console, readily visible (in all weathers) by the tactician.

  Although integration implies that everything is centralized in one unit, this is not necessarily the case, nor is it particularly desirable. The final navigation package may well incorporate a variety of configurations.

  While the integrated system console is intended to centralize the display of data for command and control of the craft, each of the instruments is required to be independently operable should the link to the main console or the main console itself fail or malfunction. It is usual for any 'permanent' setting-up of the sensor to be done at the sensor with data and status being transmitted to the central position. Where operating controls are in regular use (such as the radar range change control), these are usually duplicated at the display. There may also be remote keypads to facilitate operations such as steering and autopilot inputs which may need to be done at some distance from the console, for example during berthing. The system may also include facilities for data logging, recording of VHF communications and the printout of text transmissions and navigational warnings (Figure 10.16).
- *Integration on larger pleasure craft and commercial vessels*. This follows all the same basic principles as above with all the same sensors being integrated but expanded to embrace additional sensors and to take account of different navigational requirements. The additional sensor inputs may include radar target tracking; AIS, rate of turn indicator; bow/stern thruster demand and status; engine and steering control demand and status, with the extra data being displayed on an additional monitor.

  The three monitors would be identical and what is displayed would be interchangeable in order to provide redundancy should one display malfunction.

  While virtually anything is possible, for example the inclusion of AIS (see Section 10.3) and GMDSS, the desirability of incorporating additional functions in a single facility, and the need to display data in a readily assimilated format, needs to be carefully considered. If one bears in mind the integration of other shipboard systems on the bridge (see Sections 10.3.1 and 11.3.4), especially with the additional alarms and warnings, overkill and information overload become a distinct possibility (see Figures 10.17 and 10.18).

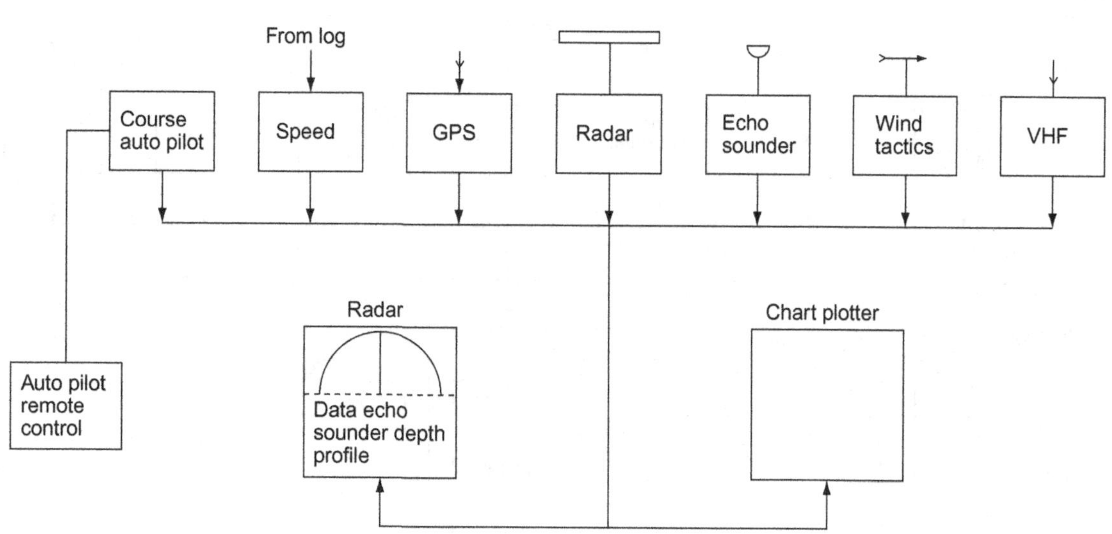

FIGURE 10.16 Instruments typically integrated on smaller craft.

10.3.4 Connectivity and Interfacing

Initially, the interfacing between different items of navigational equipment was individual to certain manufacturers. However, more open general standards have been developed. The NMEA 0183 (National Marine Electronics Association) data standard was developed in the early 1980s and is still extensively used where the data rate required is low. Most radar devices have some NMEA connections so the simple NMEA 0183 interface standard is covered here. The international standard based on NMEA-type messages is known as IEC 61162-1, which is mandated for use on SOLAS vessels. Other standards issued by the IEC and NMEA are typically also used for faster and more flexible interconnection.

10.3.4.1 *Binary and the ASCII Standard*

Computers use the binary language for all their communications and calculations. For example, in terms of electric wires, the '0's and '1's are represented by high and low voltages, whereas in magnetic storage locations (as found on traditional hard disks) use the polarity of the magnetic field at the location.

Standard sequences of '0's and '1's are used to store text characters and the most common one is the ASCII standard which is used by NMEA. The ASCII standards were developed before computers for telex systems and ASCII refers to the American Standard Code for Information Interchange. The code has 7-bit and 8-bit versions. The 7-bit version contains 128 characters, the additional bit in the 8-bit version allows for an additional 128 characters, 256 characters in all. The first 128 characters (numbered 0 to 127) are the same in both versions, the extra 128 bits are used for symbols for simple graphics. Some of the original ASCII characters have strange names, originally developed as a function in the telex system, and do not necessarily represent printable characters. These include the code for line feed or 'LF' which is ASCII code 12 (i.e. 00001100 in 8-bit binary, see Section 2.7.1) and carriage return or 'CR' ASCII code 13 (i.e. 00001101). The capital letters start at the binary for 'A' at 65 (i.e. 01000001); 'B' is 66 (i.e. 01000010), etc. The small letters start for 'a' at

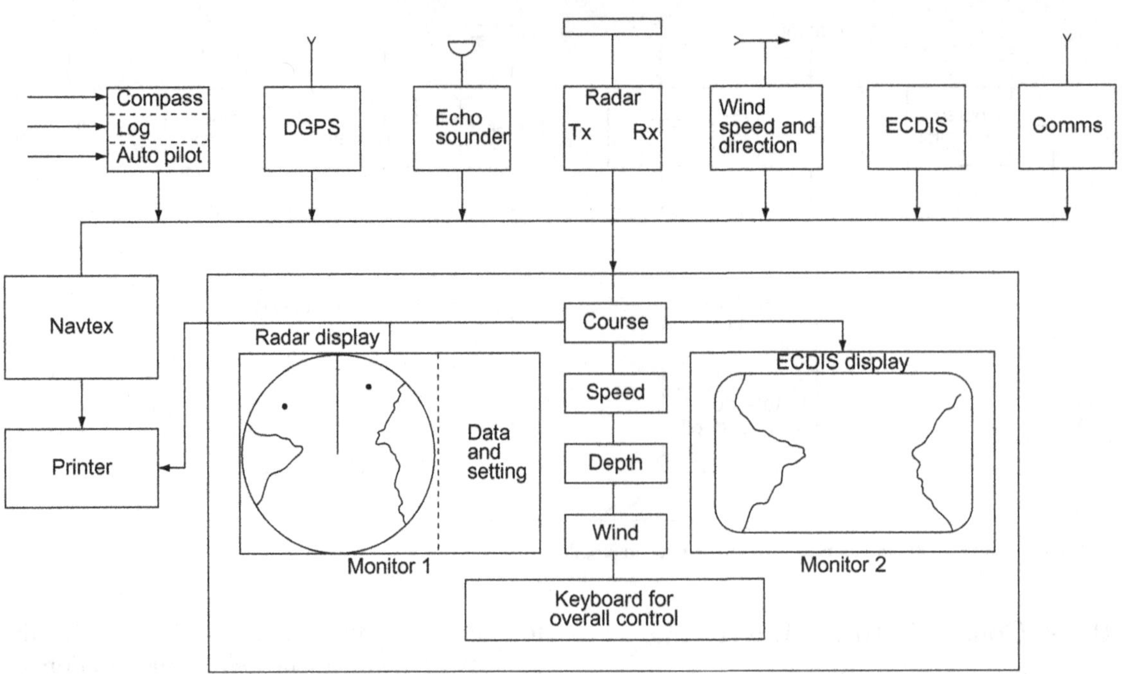

FIGURE 10.17 Integrated system console.

97 (i.e. 01100001); 'b' is 98 (i.e. 01100010), etc. Number characters are also included: '0' is 48 (i.e. 00110000); '1' is 49 (i.e. 001100001), etc. and most symbols commonly found on keyboards are also included, for example '$' is 36 (i.e. 00100100).

10.3.4.2 NMEA 0183 Message Standard

The current most popular interface standard is NMEA 0183 (IEC 61162-1). It is the simplest way of connecting one device to another. Other more complex standards are available, such as NMEA 2000 (IEC 61162-3), that allow multiple device two-way communication as well as much faster data rates. These are not compatible with NMEA 0183.

The adoption of new open standards is a very time consuming and difficult procedure and manufacturers have to make difficult decisions on what interfaces to include such that they can be easily and effectively connected to other manufacturers' products. For this reason, modern equipment can either offer multiple interfaces (often capable of operating in parallel), or else just a simple NMEA 0183 interface, which is compatible with most equipment but is relatively unsophisticated.

Each line (or sentence) of an NMEA 0183 message is made up of a number of ASCII characters. Each sentence starts with a '$'. Five characters immediately follow the '$'. The first two characters identify the device from which the message originated, for example 'HE' for a gyro compass, 'GP' for a GPS device and 'RA' for a radar or ARPA. The last three characters identify the NMEA standard sentence. After the identification group, the data fields follow that are associated with that sentence. Each data field is separated by a comma. Many of the sentences include the option of a checksum at the end. Sometimes it is compulsory. The checksum is the numerical binary addition

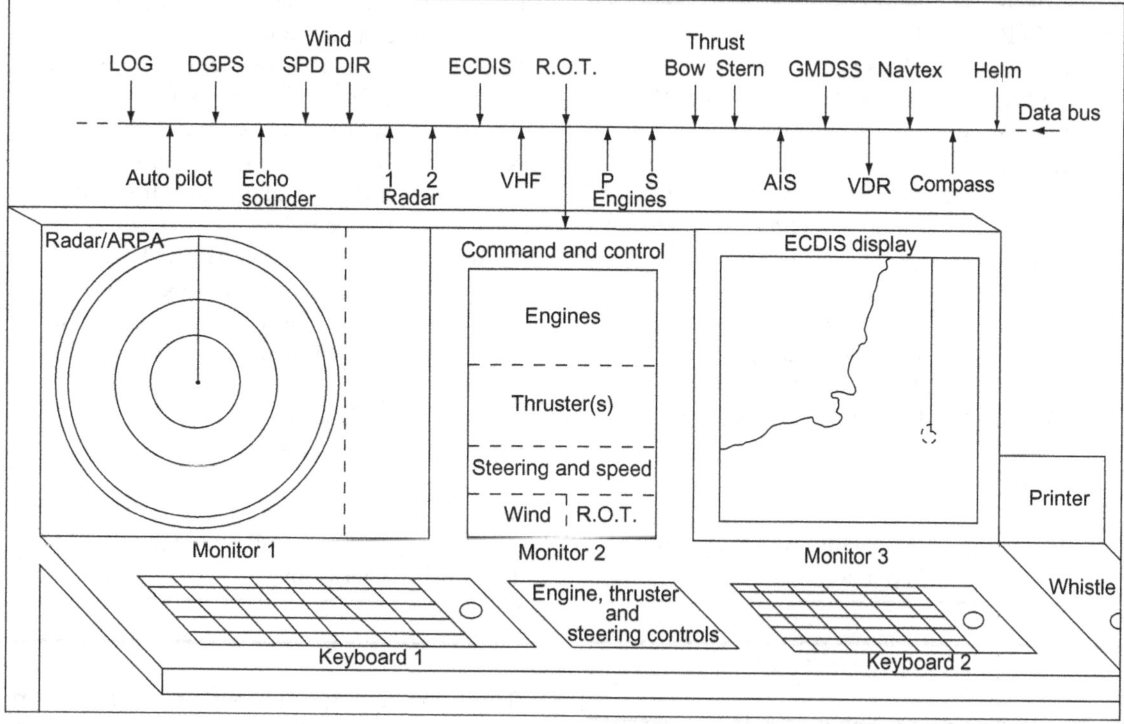

FIGURE 10.18 Integrated system console as it may appear on a larger vessel.

(using exclusive OR logic) of all the binary code sent in a sentence between the '$' and the '*' included as the second character of the checksum (for format see Example 10.1). The checksum is a communication error detection device. If the message received does not add up to the checksum, then there has been a corruption of the sentence between transmission and reception. Each sentence is terminated with carriage return 'CR' and line feed 'LF' ASCII codes.

The range of information and instructions that can be sent via NMEA 0183 is large and ever increasing. For example, it can send information as diverse as water temperature, navigation waypoints, GPS almanac, location and type of trawler gear, autopilot and engine instructions.

Some example sentences with their potential radar related applications are given below.

EXAMPLE 10.1

HDT: Heading – true.

Used to send heading information from gyro compass to other devices including radar.

$HEHDT,035.9,T,3*56	
HE	Sender ID (north seeking gyro compass)
HDT	Sentence ID (true heading)
035.9,T	Heading (035.9 true)
3*56	Checksum

EXAMPLE 10.2

GLL: Geographical position.

Used to send position from GPS, LORAN C, etc. to other devices such as radar.

$GPGLL,2325.53,N,03423.67,W,143341,1	
GP	Sender ID (GPS)
GLL	Sentence ID (geographical position, latitude and longitude)
2325.53,N	Latitude (23 deg, 25.53 min N)
03423.67,W	Longitude (034 deg, 23.67 min W)
143341	Fix time UTC (14:33:41 UTC)
1	Fix validity (valid), 0 if fix is doubtful

EXAMPLE 10.3

TTM: Tracked target status.

Used to send tracked target information from radar to ECDIS. A separate sentence is sent for each target currently plotted.

$RATTM,04,5.2,034.3,T,10.5,135.7,T,0.5,15.6,N, TAR01,T,095643.67,M,R,4*43	
RA	Sender ID (radar)
TTM	Sentence ID (tracked target status)
04	ARPA assigned target number (04), can be 00 to 99
5.2	Target distance (5.2 NM)
034.3	Target bearing from own ship (034.3°)
T	Target bearing orientation (true), R is relative
10.5	Target speed (10.5 knots)
135.7	Target course (135.7°)
T	Course orientation (true), R is relative
0.5	Closest point of approach (0.5 distance units)
15.6	Time to closest point of approach (15.6 minutes), a negative number indicates the target is past its CPA
N	Distance and speed units (NM and knots), k is kilometres, S is statute miles
TAR01	Target label on ARPA (TAR01)
T	Target status (tracking), L is lost target, Q is target being acquired
R	Reference target (i.e. designated fixed target), left blank if not designated
095643.67	Time of data in UTC (09 h 56 minutes 43.67 seconds)
M	Type of target acquisition (manual), A is automatic
4*43	Checksum

There are a large number of standard sentences available and each manufacturer selects and incorporates the sentences that a specific piece of equipment can use, either for receiving or transmission. In many devices, it is also possible for the navigator to select from the manufacturer's list the sentences for actual transmission or reception.

NMEA 0183 is continually evolving and different versions of sentences sometimes exist. Usually later versions are longer, so some earlier equipment may only read the first part of the sentence. Another feature of NMEA 0183 is that many manufacturers add extra proprietary sentences designed to interface only with other equipment made by the same manufacturer. The format of the identifier data field is different for proprietary codes. These sentences are all prefixed by $P, the next three characters are the approved NMEA code for that manufacturer (e.g. 'GRM' is Garmin) and the last character is the sentence ID specified by the manufacturer.

10.3.4.3 Other Connectivity Standards

NMEA 0183 is not fast enough to send picture data so other faster methods are used (for example) to transfer the radar picture to ECDIS. Such standards can be proprietary and confidential so that such a transfer of information is only possible between equipment provided by a single manufacturer or between two manufacturers who have a working relationship.

10.3.5 Advantages of Integration

There are distinct advantages to having all data upon which decisions are to be based readily available at one command and control position. However, not all of the data needs to be shown all of the time. There are unique phases in a voyage – as diverse as docking and open-sea collision avoidance – where only some of the data is required and it is important during these operations that irrelevant information is not cluttering up the displays. Hence, great care is necessary when deciding on what data is to be displayed and in what format.

Having all the data outputs centrally located makes it easy for cross-checking similar data derived from different sensors and sources, for example the vessel's position may be derived from visual observations, ECDIS/GNSS and radar. A quick check of compass heading and COG as derived from the GNSS will give a sense of how the vessel is being set.

While all this data is available, it is essential that great care is exercised when making decisions and that no unjustifiable weight is given to some of the data while other data is overlooked.

In earlier sections in this chapter, we considered the advantages of integrating ECDIS and GPS/GNSS (Section 10.1.10); radar target tracking with ECDIS (Section 10.2.7); ARPA, AIS and ECDIS (Section 5.4.3). Clearly a full integration of bridge information does not stop at integrating these suggested combinations of equipment alone.

10.3.6 Potential Dangers of Data Overload

There is an increasing awareness of the potential problem of information overload on the user. As has already been stated, not all the data is needed all of the time, especially when one includes other data displayed on the bridge (see Section 10.3.1). It is essential, when designing the display of data, that only data essential to the operation taking place be directly displayed, while additional data is available on demand.

10.3.7 System Cross-Checking, Warnings and Alarms

In order for the early detection of malfunction in a system, cross-checking between systems (either electronically or manually) is important. It is also important to check progress against a predetermined plan. Warnings and alarms can be built into the system at various levels of urgency, but the navigator can be overwhelmed and distracted by buzzers, whistles and flashing lights. Again care in design is essential.

10.3.8 Sensor Errors and Accuracy of Integration

One would hope that similar data derived from different sources would have a high correlation factor, but this is not necessarily the case due to system and other errors, which can be random. Where there is unacceptable divergence an alarm should be activated and the navigator will need to exercise his judgement as to the possible cause of the discrepancy or if it is due to the malfunction of a specific sensor.

10.3.9 Data Monitoring and Loggings

In addition to voyage data recorders (VDRs) (see Section 10.4), data derived from a multitude of shipboard systems (see Section 10.3.1) is in digital form and may be accumulated and stored on a variety of media to facilitate record keeping (e.g. log book) and for voyage performance analysis. In many organizations, there is still a preference for a hard copy of information required for permanent record and legal purposes. A printer is therefore still part of most systems.

10.4 VOYAGE DATA RECORDERS (SEE ALSO SECTION 11.3.5)

The purpose of a VDR is to record operational information of a vessel so that it can assist investigators should the vessel become involved in an incident. The idea has long been used in the aviation industry, where it has often been called the 'black box'. Commercial aircraft actually record different information in two separate boxes that are coloured bright orange.

The VDR helps to overcome two distinct problems in ascertaining causes of marine incidents. Firstly, there is occasionally a situation when key personnel do not survive the incident. Secondly, even when personnel survive an incident, the truth is not always forthcoming. Survivors may not be aware of certain facts. They could have a mistaken impression of what happened due to the trauma of the event and they may not tell the truth in order to protect themselves or colleagues. Marine incidents are not unique in that one party has a different version of an accident to a second party. The lack of objective evidence has prevented firm conclusions in many enquiries.

VDR regulations were first introduced in the revisions to SOLAS Chapter V in 2001. All new vessels of more than 3000 grt and all new passenger vessels are required to have them fitted. There is also a phased requirement for fitting them to some existing ships, especially passenger ships. All other existing vessels above 3000 grt are required to have a simplified VDR (SVDR), which is less onerous on the requirements of the data capsule and the number of interfaces connected.

10.4.1 Equipment

The VDR equipment has two main items of hardware. There is a single data storage capsule located externally on top of the vessel and a data collection unit probably located inside the bridge (Figure 10.19). Replay facilities are normally located ashore.

The data storage capsule is designed to survive a total loss of the vessel. It is a capsule designed to withstand an intense fire as well as immersion in over 4000 m of water. Its construction is therefore very thick steel. The capsule would normally be located towards the top of the vessel in an open space. Its securing arrangements are designed to be easily released by a remotely operated vehicle (ROV) or a deep-sea diver encumbered in a heavy suit. It will not actually contain much electronic equipment, the main component being enough solid-state memory to record at least the last 12 hours of the vessel's operations. Solid-state memory does not actually require a battery to maintain the data recorded. The capsule is fully sealed to meet its demanding specifications with only a data wire penetrating the casing. To aid detection, it is brightly coloured, covered in retro-reflective material and an underwater locator beacon is attached.

The data collection unit is usually a standard personal computer, although the usual

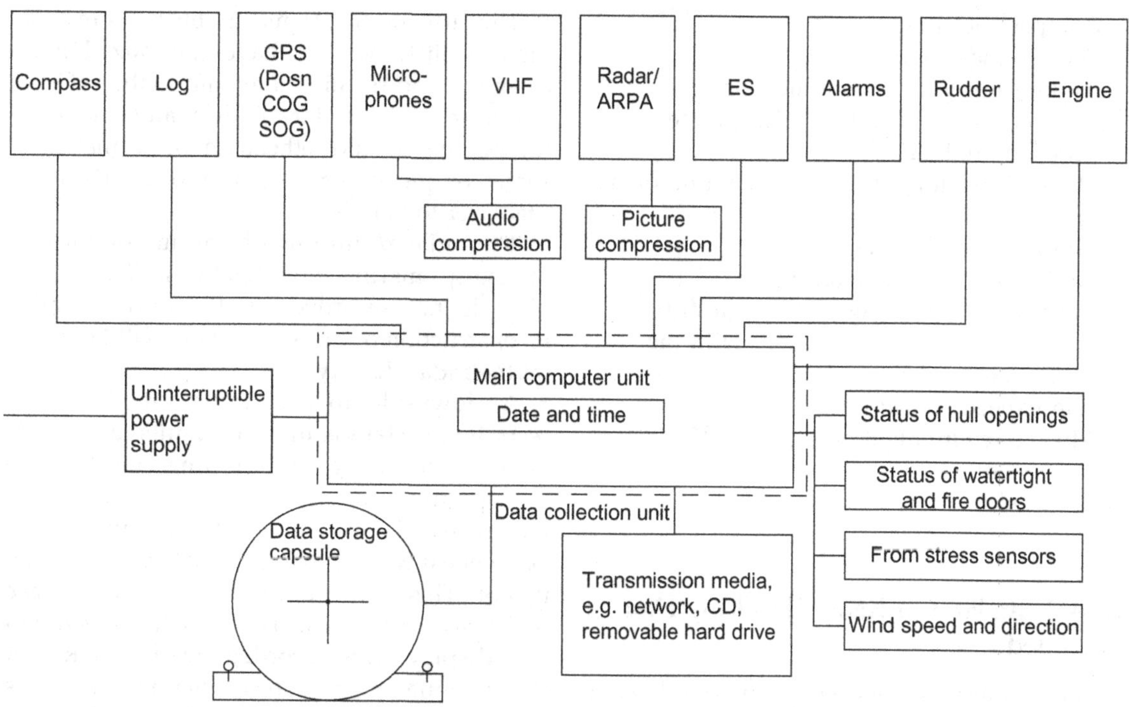

FIGURE 10.19 Block diagram of VDR system.

controls and displays need not normally be provided for ships' staff. There is little for the navigator to operate except to check that the system is on and functioning, indicated by some manufacturers as a simple flashing light. Connections from all the required bridge equipment are made direct to the data collection unit. The data is stored internally as well as sent to the data storage capsule. Data compression techniques are employed to enable efficient use of the storage systems.

The data storage capsule only needs to contain the last 12 hours of operation. It is also the single most expensive item in the VDR equipment and once opened cannot be easily reused (if at all). It would be an unnecessary expense to have to cut open the capsule to investigate the vast majority of marine incidents when the bridge and ship are still essentially intact. Following an incident, the stored data in the data collection unit will be maintained over a much longer period. (more than a week is typical) and standard electronic means are provided to extract the data much more cheaply than by opening up the data storage capsule. The method of extraction is at the choice of the supplier and buyer but can be via an electronic network or by standard computer storage media such as compact disk or removable hard drive.

10.4.2 Non-Radar Data Recorded

The data recorded includes the following from bridge equipment:

Date and time
Position from electronic position fixing systems
Speed

Compass heading
Echo sounder depth
Audio from VHF communications
Audio from a network of microphones installed on the bridge
Main (mandatory) alarms to be found on a bridge
Rudder order and response
Engine order and response
Status of hull openings, watertight/fire doors, hull stresses (on vessels with the equipment)
Wind speed and direction
AIS is a requirement only on SVDR equipment where radar display is not recorded.

10.4.3 Radar and Radar Tracking Data Recorded

There is also a requirement to record radar and tracking data. The system adopted makes use of the fact that all modern radar displays are raster. A software technique called 'frame grabbing' to store an instantaneous frame (the raster picture) is well established in the computer industry and the picture data can be compressed for efficient storage. The frame grab of the full raster display is completed at least every 15 seconds from one of the ship's radars.

The raster display on a modern raster-scan radar does not just show the picture, it also shows the main radar settings (e.g. range scale in use, true or relative vectors in use, vector time and amount of sea clutter control applied). This trend has increased as 'hard' analogue controls have largely disappeared and been replaced by software driven menu controls. Therefore, the frame grabber will record much information of value to investigators as to the way the equipment was being used, as well as the operational state of the radar equipment and the outside scenario in which the vessel was participating.

This use of the frame grabber means that there is little need to have specialized interfaces to capture this radar information. Given the number of radar models and manufacturers, this would otherwise be a matter of some complexity and expense for a VDR manufacturer to handle.

The radar picture will be an important part of many marine investigations. The visual view is not recorded, so the only dynamic reconstruction of physical events will be found on the radar display.

A downside to the frame grabber radar recording system is that the results are dependent on the navigator adjusting the radar to the optimum settings. If the radar is set up incorrectly, the true situation (e.g. missing target echoes) will not be retrievable by the investigator. This would not necessarily be the case if the raw radar data (i.e. prior to digitization and display) was recorded by the VDR, but the potential gain in recording raw radar is small compared with the potential complexity of recording the raw radar data on a large variety of radar models.

An advantage of the frame grabber system is that it will record when navigators have not used the radar display correctly, and will probably be able to detect the occasions when human error on the part of the navigator in the use of the radar was a factor in a collision.

10.4.4 Playback Equipment

It is not intended that the vessel has playback equipment. The playback software is usually installable on a standard multimedia personal computer and would normally be available in the ship manager's office. The software and the retrieved data would be made available to investigators after an accident. Most playback software allows rewinding, fast-forwarding and freeze framing of data in a similar fashion to a standard video player. A selection of different data streams (radar, helm-controls position, audio channel) can be played back simultaneously and the graphical data (radar picture, helm orders, etc.) are shown in different windows so that an overall view of the incident is obtained. The positional data can be fed to a separate electronic chart to further aid the replay analysis.

10.4.5 Future of VDRs

Operational experience with VDRs is still very limited. Whilst a lot of vessels now carry them, they have only been of use in a limited number of investigations. Like the ship's lifeboat, very few of the installations will actually be used for their intended purpose. Nevertheless it is anticipated that, the few times that they are used, they will provide vital and objective information as to the causes of the incident. IMO have also introduced a simplified VDR (SVDR) which has been installed on existing ships over 3000 grt. This does not require data to be recorded if it's not economically viable to provide, such as from old analogue equipment. If a radar picture is not recorded then AIS data must be provided in its place. The regulations for SVDR also allow for a float free data capsule if preferred to the VDR data capsule designed to sink with the vessel.

The data collected by VDR also has potential for research and analysis of vessel operation when not concerned with an incident. There are already schemes where navigators report (usually anonymously) examples of a near miss that otherwise would not have been analysed. If it were possible to treat data records of these near misses in the same way, future safety would be significantly improved. In fact, it would be useful for VDRs to be examined for examples of good (and successful) navigational practice and for navigational data. For instance, this could include radar pictures of port approaches so that parallel index targets can be identified. This VDR information could be usefully fed back to navigators. However, it would need a large culture shift by navigators, regulators and shipowners to make this data freely available.

CHAPTER

11

Extracts from Official Publications

This chapter contains extracts from official publications relating to the topics covered by this volume. They have been edited and the reader is referred to the full original text if more details are required.

Within the extracts the paragraph, table and appendix numbering of the original publications have been retained

11.1 EXTRACTS FROM REGULATION 19, CHAPTER V, SAFETY OF NAVIGATION, OF IMO-SOLAS CONVENTION

Edited to show application of radar and AIS to new vessels.

1.1 Ships constructed on or after 1 July 2002 shall be fitted with navigational systems and equipment which will fulfil the requirements prescribed in paragraphs 2.1 to 2.9.

2 SHIPBORNE NAVIGATIONAL EQUIPMENT AND SYSTEMS

2.1 All ships irrespective of size shall have:

2.1.7 if less than 150 gross tonnage and if practicable, a radar reflector, or other means, to enable detection by ships navigating by radar at both 9 and 3 GHz.

2.3 All ships of 300 gross tonnage and upwards and passenger ships irrespective of size shall be fitted with:

2.3.2 a 9 GHz radar, or other means to determine and display the range and bearing of radar transponders and of other surface craft, obstructions, buoys, shorelines and navigational marks to assist in navigation and in collision avoidance;

2.3.3 an electronic plotting aid, or other means, to plot electronically the range and bearing of targets to determine collision risk;

2.3.4 a speed and distance measuring device, or other means, to indicate speed and distance through the water;

2.3.5 a properly adjusted transmitting heading device, or other means to transmit heading information for input to the equipment referred to in paragraphs 2.3.2, 2.3.3 and 2.4.

2.4 All ships of 300 gross tonnage and upwards engaged on international voyages and cargo ships of 500 gross tonnage and upwards not engaged on international voyages, and passenger ships irrespective of size shall be fitted with an automatic identification system (AIS), as follows:

2.4.1 ships constructed on or after 1 July 2002.

2.4.5 AIS shall:

.1 provide automatically to appropriately equipped shore stations, other ships and aircraft, information including the ship's identity, type, position, course, speed, navigational status and other safety-related information;

.2 receive automatically such information from similarly fitted ships;

.3 monitor and track ships; and

.4 exchange data with shore-based facilities;

2.4.6 The requirements of paragraph 2.4.5 shall not be applied to cases where international agreements, rules or standards provide for the protection of navigational information; and

2.4.7 AIS shall be operated taking into account the guidelines adopted by the Organization. Ships fitted with AIS shall maintain AIS in operation at all times except where international agreements, rules or standards provide for the protection of navigational information.

2.5 All ships of 500 gross tonnage and upwards shall, in addition to meeting the requirements of paragraph 2.3 with the exception of paragraphs 2.3.3 and 2.3.5, and the requirements of paragraph 2.4, have:

2.5.1 a gyro compass, or other means, to determine and display their heading by shipborne non-magnetic means, being clearly readable by the helmsman at the main steering position. These means shall also transmit heading information for input to the equipment referred to in paragraphs 2.3.2, 2.4 and 2.5.5;

2.5.5 an automatic tracking aid, or other means, to plot automatically the range and bearing of other targets to determine collision risk.

2.7 All ships of 3000 gross tonnage and upwards shall, in addition to meeting the requirements of paragraph 2.5, have:

2.7.1 a 3 GHz radar or where considered appropriate by the Administration a second 9 GHz radar, or other means to determine and display the range and bearing of other surface craft, obstructions, buoys, shorelines and navigational marks to assist in navigation and in collision avoidance, which are functionally independent of those referred to in paragraph 2.3.2; and

2.7.2 a second automatic tracking aid, or other means, to plot automatically the range and bearing of other targets to determine collision risk which are functionally independent of those referred to in paragraph 2.5.5.

2.8 All ships of 10,000 gross tonnage and upwards shall, in addition to meeting the requirements of paragraph 2.7 with the exception of paragraph 2.7.2, have:

2.8.1 an automatic radar plotting aid, or other means, to plot automatically the range and bearing of at least 20 other targets, connected to a device to indicate speed and distance through the water, to determine collision risks and simulate a trial manoeuvre.

11.2 IMO PERFORMANCE STANDARDS FOR RADAR EQUIPMENT

11.2.1 Extracts from IMO Resolution MSC.192(79), Performance Standards for Radar Equipment for New Ships Constructed After 1 July 2008

INDEX

1. SCOPE OF EQUIPMENT
2. APPLICATION OF THESE STANDARDS
3. REFERENCES
4. DEFINITIONS
5. OPERATIONAL REQUIREMENTS FOR THE RADAR SYSTEM
6. ERGONOMIC CRITERIA
7. DESIGN AND INSTALLATION
8. INTERFACING
9. BACKUP AND FALLBACK ARRANGEMENTS

1 SCOPE OF EQUIPMENT

The radar equipment should assist in safe navigation and in avoiding collision by providing an indication, in relation to the own ship, of the position of other surface craft, obstructions and hazards, navigation objects and shorelines.

For this purpose, radar should provide the integration and display of radar video, target tracking information, positional data derived from the own ship's position (EPFS) and geo referenced data. The integration and display of AIS information should be provided to complement radar. The capability of displaying selected parts of Electronic Navigation Charts and other vector chart information may be provided to aid navigation and for position monitoring.

The radar, combined with other sensor or reported information (e.g. AIS), should improve the safety of navigation by assisting in the efficient navigation of ships and protection of the environment by satisfying the following functional requirements:

- in coastal navigation and harbour approaches, by giving a clear indication of land and other fixed hazards;
- as a means to provide an enhanced traffic image and improved situation awareness;
- in a ship-to-ship mode for aiding collision avoidance of both detected and reported hazards;

TABLE 1 Differences in the performance requirements for various sizes/categories of ship/craft to which SOLAS applies

Size of ship/craft	<500 gt	500 gt to <10,000 gt and HSC < 10,000 gt	All ships/craft ≥ 10,000 gt
Minimum operational display area diameter	180 mm	250 mm	320 mm
Minimum display area	195 × 195 mm	270 × 270 mm	340 × 340 mm
Auto acquisition of targets	–	–	Yes
Minimum ***acquired*** radar target capacity	20	30	40
Minimum ***activated*** AIS target capacity	20	30	40
Minimum ***sleeping*** AIS target capacity	100	150	200
Trial manoeuvre	–	–	Yes

- in the detection of small floating and fixed hazards, for collision avoidance and the safety of the own ship; and
- in the detection of floating and fixed aids to navigation (see Table 2, note 3).

2 APPLICATION OF THESE STANDARDS

These Performance Standards should apply to all shipborne radar installations, used in any configuration, mandated by the SOLAS Convention 1974, as amended, independent of the:

- type of ship;
- frequency band in use; and
- type of display;

providing that no special requirements are specified in Table 1 and that additional requirements for specific classes of ships (in accordance with SOLAS chapters V and X) are met.

The radar installation, in addition to meeting the general requirements as set out in resolution A.694(17)*, should comply with the following performance standards.

Close interaction between different navigational equipment and systems makes it essential to consider these standards in association with other relevant IMO standards.

3 REFERENCES

References are in Appendix 1.

4 DEFINITIONS

Definitions are in Appendix 2.

5 OPERATIONAL REQUIREMENTS FOR THE RADAR SYSTEM

The design and performance of the radar should be based on user requirements and up-to-date navigational technology. It should provide effective target detection within the safety-relevant environment surrounding the own ship and should permit fast and easy situation evaluation.**

5.1 Frequency

5.1.1 Frequency Spectrum

The radar should transmit within the confines of the ITU allocated bands for maritime radar and meet the requirements of the radio regulations and applicable ITU-recommendations.

5.1.2 Radar Sensor Requirements

Radar systems of both X and S bands are covered in these performance standards:

- X band (9.2–9.5 GHz) for high discrimination, good sensitivity and tracking performance; and
- S band (2.9–3.1 GHz) to ensure that target detection and tracking capabilities are maintained in varying and adverse conditions of fog, rain and sea clutter.

The frequency band in use should be indicated.

5.1.3 Interference Susceptibility

The radar should be capable of operating satisfactorily in typical interference conditions.

*IEC Publication 60945.

**Refer to MSC/Circ 878 - MEPC/Circ.346 on the application of the Human Element Analysing Process (HEAP).

5.2 Radar Range and Bearing Accuracy

The radar system range and bearing accuracy requirements should be:

- *Range* – within 30 m or 1% of the range scale in use, whichever is greater;
- *Bearing* – within 1°.

5.3 Detection Performance and Anti-clutter Functions

All available means for the detection of targets should be used.

5.3.1 Detection

5.3.1.1 Detection in Clear Conditions

In the absence of clutter, for long range target and shoreline detection, the requirement for the radar system is based on normal propagation conditions, in the absence of sea clutter, precipitation and evaporation duct, with an antenna height of 15 m above sea level.

Based on:

- an indication of the target in at least 8 out of 10 scans or equivalent; and
- a probability of a radar detection false alarm of 10^{-4};

the requirement contained in Table 2 should be met as specified for X- and S-band equipment.

The detection performance should be achieved using the smallest antenna that is supplied with the radar system.

Recognizing the high relative speeds possible between the own ship and target, the equipment should be specified and approved as being suitable for classes of ship having normal (<30 kn) or high (>30 kn) own ship speeds (100 kn and 140 kn relative speeds, respectively).

5.3.1.2 Detection at Close Range

The short-range detection of the targets under the conditions specified in Table 2 should be compatible with the requirement in paragraph 5.4.

TABLE 2 Minimum detection ranges in clutter-free conditions

Target description	Target feature	Detection range in NM[6]	
Target description[5]	Height above sea level in metres	X-Band NM	S-Band NM
Shorelines	Rising to 60	20	20
Shorelines	Rising to 6	8	8
Shorelines	Rising to 3	6	6
SOLAS ships (>5000 gross tonnage)	10	11	11
SOLAS ships (>500 gross tonnage)	5.0	8	8
Small vessel with Radar Reflector meeting IMO Performance Standards[1]	4.0	5.0	3.7
Navigation buoy with corner reflector[2]	3.5	4.9	3.6
Typical Navigation buoy[3]	3.5	4.6	3.0
Small vessel of length 10 m with no radar reflector[4]	2.0	3.4	3.0

[1]*IMO performance standards for radar reflectors – Radar Cross Section (RCS) 7.5 m^2 for X band, 0.5 m^2 for S band.*

[2]*The corner reflector (used for measurement), is taken as 10 m^2 for X-Band and 1.0 m^2 for S-Band.*

[3]*The typical navigation buoy is taken as 5.0 m^2 for X-Band and 0.5 m^2 for S-Band.*

[4]*RCS for 10 m small vessel taken as 2.5 m for X-Band and 1.4 m for S-Band.*

[5]*Reflectors are taken as point targets, vessels as complex targets and shorelines as distributed targets (typical values for a rocky shoreline, but are dependent on profile).*

[6]*Detection ranges experienced in practice will be affected by various factors, including atmospheric conditions (e.g. evaporation duct), target speed and aspect, target material and target structure. These and other factors may either enhance or degrade the detection ranges stated. At ranges between the first detection and own ship, the radar return may be reduced or enhanced by signal multi-path, which depend on factors such as antenna/target centroid height, target structure, sea state and radar frequency band.*

5.3.1.3 Detection in Clutter Conditions

Performance limitations caused by typical precipitation and sea clutter conditions will result in a reduction of target detection capabilities relative to those defined in 5.3.1.1 and Table. 2.

5.3.1.3.1 The radar equipment should be designed to provide the optimum and most consistent detection performance, restricted only by the physical limits of propagation.

5.3.1.3.2 The radar system should provide the means to enhance the visibility of targets in adverse clutter conditions at close range.

5.3.1.3.3 Degradation of detection performance (related to the figures in Table 2) at various ranges and target speeds under the following conditions should be clearly stated in the user manual:

- light rain (4 mm per hour) and heavy rain (16 mm per hour);
- sea state 2 and sea state 5; and
- a combination of these.

5.3.1.3.4 The determination of performance in clutter and specifically, range of first detection, as defined in the clutter environment in 5.3.1.3.3, should be tested and assessed against a benchmark target, as specified in the Test Standard.

5.3.1.3.5 Degradation in performance due to a long transmission line, antenna height or any other factors should be clearly stated in the user manual.

5.3.2 Gain and Anti-Clutter Functions

5.3.2.1 Means should be provided, as far as is possible, for the adequate reduction of unwanted echoes, including sea clutter, rain and other forms of precipitation, clouds, sandstorms and interference from other radars.

5.3.2.2 A gain control function should be provided to set the system gain or signal threshold level.

5.3.2.3 Effective manual and automatic anti-clutter functions should be provided.

5.3.2.4 A combination of automatic and manual anti-clutter functions is permitted.

5.3.2.5 There should be a clear and permanent indication of the status and level for gain and all anti-clutter control functions.

5.3.3 Processing

5.3.3.1 Means should be available to enhance target presentation on the display.

5.3.3.2 The effective picture update period should be adequate, with minimum latency to ensure that the target detection requirements are met.

5.3.3.3 The picture should be updated in a smooth and continuous manner.

5.3.3.4 The equipment manual should explain the basic concept, features and limitations of any signal processing.

5.3.4 Operation with SARTs and Radar Beacons

5.3.4.1 The X-band radar system should be capable of detecting radar beacons in the relevant frequency band.

5.3.4.2 The X-band radar system should be capable of detecting SARTs and radar target enhancers.

5.3.4.3 It should be possible to switch off those signal processing functions, including polarization modes, which might prevent an X-band radar beacon or SARTs from being detected and displayed. The status should be indicated.

5.4 Minimum Range

5.4.1 With the own ship at zero speed, an antenna height of 15 m above the sea level and in calm conditions, the navigational buoy in Table 2 should be detected at a minimum horizontal range of 40 m from the antenna position and up to a range of 1 NM, without changing the setting of control functions other than the range scale selector.

5.4.2 Compensation for any range error should be automatically applied for each selected antenna, where multiple antennas are installed.

5.5 Discrimination

Range and bearing discrimination should be measured in calm conditions, on a range scale of 1.5 NM or less and at between 50% and 100% of the range scale selected:

5.5.1 Range

The radar system should be capable of displaying two point targets on the same bearing, separated by 40 m in range, as two distinct objects.

5.5.2 Bearing

The radar system should be capable of displaying two point targets at the same range, separated by 2.5° in bearing, as two distinct objects.

5.6 Roll and Pitch

The target detection performance of the equipment should not be substantially impaired when own ship is rolling or pitching up to ±10°.

5.7 Radar Performance Optimization and Tuning

5.7.1 Means should be available to ensure that the radar system is operating at the best performance. Where applicable to the radar technology, manual tuning should be provided and additionally, automatic tuning may be provided.

5.7.2 An indication should be provided, in the absence of targets, to ensure that the system is operating at the optimum performance.

5.7.3 Means should be available (automatically or by manual operation) and while the equipment is operational, to determine a significant drop in system performance relative to a calibrated standard established at the time of installation.

5.8 Radar Availability

The radar equipment should be fully operational (RUN status) within 4 minutes after switch ON from cold. A STANDBY condition should be provided, in which there is no operational radar transmission. The radar should be fully operational within 5 sec from the standby condition.

5.9 Radar Measurements – Consistent Common Reference Point (CCRP)

5.9.1 Measurements from the own ship (e.g. range rings, target range and bearing, cursor, tracking data) should be made with respect to the consistent common reference point (e.g. conning position). Facilities should be provided to compensate for the offset between antenna position and the consistent common reference point on installation. Where multiple antennas are installed, there should be provision for applying different position offsets for each antenna in the radar system. The offsets should be applied automatically when any radar sensor is selected.

5.9.2 The own ship's scaled outline should be available on appropriate range scales. The consistent common reference point and the position of the selected radar antenna should be indicated on this graphic.

5.9.3 When the picture is centred, the position of the Consistent Common Reference Point should be at the centre of the bearing scale. The off-centre limits should apply to the position of the selected antenna.

5.9.4 Range measurements should be in nautical miles (NM). In addition, facilities for metric measurements may be provided on lower range scales. All indicated values for range measurement should be unambiguous.

5.9.5 Radar targets should be displayed on a linear range scale and without a range index delay.

5.10 Display Range Scales

5.10.1 Range scales of 0.25, 0.5, 0.75, 1.5, 3, 6, 12 and 24 NM should be provided. Additional range scales are permitted outside the mandatory set. Low metric range scales may be offered in addition to the mandatory set.

5.10.2 The range scale selected should be permanently indicated.

5.11 Fixed Range Rings

5.11.1 An appropriate number of equally spaced range rings should be provided for the range scale selected. When displayed, the range ring scale should be indicated.

5.11.2 The system accuracy of fixed range rings should be within 1% of the maximum range of the range scale in use or 30 m, whichever is the greater distance.

5.12 Variable Range Markers (VRM)

5.12.1 At least two variable range markers (VRMs) should be provided. Each active VRM should have a numerical readout and have a resolution compatible with the range scale in use.

5.12.2 The VRMs should enable the user to measure the range of an object within the operational display area with a maximum system error of 1% of the range scale in use or 30 m, whichever is the greater distance.

5.13 Bearing Scale

5.13.1 A bearing scale around the periphery of the operational display area should be provided. The bearing scale should indicate the bearing as seen from the consistent common reference point.

5.13.2 The bearing scale should be outside of the operational display area. It should be numbered at least every 30° division and have division marks of at least 5°. The 5° and 10° division marks should be clearly distinguishable from each other. 1° division marks may be presented where they are clearly distinguishable from each other.

5.14 Heading Line (HL)

5.14.1 A graphic line from the consistent common reference point to the bearing scale should indicate the heading of the ship.

5.14.2 Electronic means should be provided to align the heading line to within 0.1°. If there is more than one radar antenna (see 5.35) the heading skew (bearing offset) should be retained and automatically applied when each radar antenna is selected.

5.14.3 Provision should be made to temporarily suppress the heading line. This function may be combined with the suppression of other graphics.

5.15 Electronic Bearing Lines (EBLs)

5.15.1 At least two electronic bearing lines (EBLs) should be provided to measure the bearing of any point object within the operational display area, with a maximum system error of 1° at the periphery of the display.

5.15.2 The EBLs should be capable of measurement relative to the ship's heading and relative to true north. There should be a clear indication of the bearing reference (i.e. true or relative).

5.15.3 It should be possible to move the EBL origin from the consistent common reference point to any point within the operational display area and to reset the EBL to the consistent common reference point by a fast and simple action.

5.15.4 It should be possible to fix the EBL origin or to move the EBL origin at the velocity of the own ship.

5.15.5 Means should be provided to ensure that the user is able to position the EBL smoothly in either direction, with an incremental adjustment adequate to maintain the system measurement accuracy requirements.

5.15.6 Each active EBL should have a numerical readout with a resolution adequate to maintain the system measurement accuracy requirements.

5.16 Parallel Index Lines (PI)

5.16.1 A minimum of four independent parallel index lines, with a means to truncate and switch off individual lines, should be provided.

5.16.2 Simple and quick means of setting the bearing and beam range of a parallel index line should be provided. The bearing and beam range of any selected index line should be available on demand.

5.17 Offset Measurement of Range and Bearing

There should be a means to measure the range and bearing of one position on the display relative to any other position within the operational display area.

5.18 User Cursor

5.18.1 A user cursor should be provided to enable a fast and concise means to designate any position on the operational display area.

5.18.2 The cursor position should have a continuous readout to provide the range and bearing, measured from the consistent common reference point, and/or the latitude and longitude of the cursor position presented either alternatively or simultaneously.

5.18.3 The cursor should provide the means to select and de-select targets, graphics or objects within the operational display area. In addition, the cursor may be used to select modes, functions, vary parameters and control menus outside of the operational display area.

5.18.4 Means should be provided to easily locate the cursor position on the display.

5.18.5 The accuracy of the range and bearing measurements provided by the cursor should meet the relevant requirements for VRM and EBL.

5.19 Azimuth Stabilization

5.19.1 The heading information should be provided by a gyrocompass or by an equivalent sensor with a performance not inferior to the relevant standards adopted by the Organization.

5.19.2 Excluding the limitations of the stabilizing sensor and type of transmission system, the accuracy of azimuth alignment of the radar presentation should be within 0.5° with a rate of turn likely to be experienced with the class of ship.

5.19.3 The heading information should be displayed with a numerical resolution to permit accurate alignment with the ship gyro system.

5.19.4 The heading information should be referenced to the consistent common reference point (CCRP).

5.20 Display Mode of the Radar Picture

5.20.1 A true-motion display mode should be provided. The automatic reset of the own ship may be initiated by its position on the display, or time related, or both. Where the reset is selected to occur at least on every scan or equivalent, this should be equivalent to true-motion with a fixed origin (in practice equivalent to the previous relative-motion mode).

5.20.2 North-up and course-up orientation modes should be provided. Head-up may be provided when the display mode is equivalent to true-motion with a fixed origin (in practice equivalent to the previous relative-motion head-up mode).

5.20.3 An indication of the motion and orientation mode should be provided.

5.21 Off-Centring

5.21.1 Manual off-centring should be provided to locate the selected antenna position at any point within at least 50% of the radius from the centre of the operational display area.

5.21.2 On selection of off-centred display, the selected antenna position should be capable of being located to any point on the display up to at least 50% and not more than 75% of the radius from the centre of the operational display area. A facility for automatically positioning the own ship for the maximum view ahead may be provided.

5.21.3 In true-motion, the selected antenna position should automatically reset up to a 50% radius to a location giving the maximum view along the own ship's course. Provision for an early reset of selected antenna position should be provided.

5.22 Ground and Sea Stabilization Modes

5.22.1 Ground- and sea-stabilization modes should be provided.

5.22.2 The stabilization mode and stabilization source should be clearly indicated.

5.22.3 The source of the own ships' speed should be indicated and provided by a sensor approved in accordance with the requirements of the Organization for the relevant stabilization mode.

5.23 Target Trails and Past Positions

5.23.1 Variable length (time) target trails should be provided, with an indication of trail time and mode. It should be possible to select true or relative trails from a reset condition for all true-motion display modes.

5.23.2 The trails should be distinguishable from targets.

5.23.3 Either scaled trails or past positions or both, should be maintained and should be available for presentation within 2 scans or equivalent, following:

- the reduction or increase of one range scale;
- the offset and reset of the radar picture position; and
- a change between true and relative trails.

5.24 Presentation of Target Information

5.24.1 Targets should be presented in accordance with the performance standards for the Presentation of Navigation-related Information on Shipborne Navigational Displays adopted by the Organization and with their relevant symbols according to SN/Circ. 243.

5.24.2 The target information may be provided by the radar target tracking function and by the reported target information from the Automatic Identification System (AIS).

5.24.3 The operation of the radar tracking function and the processing of reported AIS information is defined in these standards.

5.24.4 The number of targets presented, related to display size, is defined in Table 1. An indication should be given when the target capacity of radar tracking or AIS reported target processing/display capability is about to be exceeded.

5.24.5 As far as practical, the user interface and data format for operating, displaying and indicating AIS and radar tracking information should be consistent.

5.25 Target Tracking (TT) and Acquisition

5.25.1 General

Radar targets are provided by the radar sensor (transceiver). The signals may be filtered (reduced) with the aid of the associated clutter controls. Radar targets may be manually or automatically acquired and tracked using an automatic Target Tracking (TT) facility.

5.25.1.1 The automatic target tracking calculations should be based on the measurement of radar target relative position and the own ship motion.

5.25.1.2 Any other sources of information, when available, may be used to support the optimum tracking performance.

5.25.1.3 TT facilities should be available on at least the 3, 6 and 12 NM range scales. Tracking range should extend to a minimum of 12 NM.

5.25.1.4 The radar system should be capable of tracking targets having the maximum relative speed relevant to its classification for normal or high own ship speeds (see 5.3).

5.25.2 Tracked Target Capacity

5.25.2.1 In addition to the requirements for processing of targets reported by AIS, it should be possible to track and provide full presentation functionality for a minimum number of tracked radar targets according to Table 1.

5.25.2.2 There should be an indication when the target tracking capacity is about to be exceeded. Target overflow should not degrade the radar system performance.

5.25.3 Acquisition

5.25.3.1 Manual acquisition of radar targets should be provided with provision for acquiring at least the number of targets specified in Table 1.

5.25.3.2 Automatic acquisition should be provided where specified in Table 1. In this case, there should be means for the user to define the boundaries of the auto-acquisition area.

5.25.4 Tracking

5.25.4.1 When a target is acquired, the system should present the trend of the target's motion within one minute and the prediction of the target's motion within 3 minutes.

5.25.4.2 TT should be capable of tracking and updating the information of all acquired targets automatically.

5.25.4.3 The system should continue to track radar targets that are clearly distinguishable on the display for 5 out of 10 consecutive scans or equivalent.

5.25.4.4 The TT design should be such that target vector and data smoothing is effective, while target manoeuvres should be detected as early as possible.

5.25.4.5 The possibility of tracking errors, including target swap, should be minimized by design.

5.25.4.6 Separate facilities for cancelling the tracking of any one and of all target(s) should be provided.

5.25.4.7 Automatic tracking accuracy should be achieved when the tracked target has achieved a steady-state, assuming the sensor errors allowed by the relevant performance standards of the Organization:

5.25.4.7.1 For ships capable of up to 30 kn true speed, the tracking facility should present, within 1 minute of steady-state tracking, the relative-motion trend and after 3 minutes, the predicted motion of a target, within the following accuracy values (95% probability), (see Table 3).

TABLE 3 Tracked target accuracy (95% probability figures)

Time of steady-state (minutes)	Relative course (degrees)	Relative speed (kn)	CPA (NM)	TCPA (minutes)	True course (degrees)	True speed (kn)
1 minute: Trend	11	1.5 or 10% (whichever is greater)	1.0	—	—	—
3 minutes: Motion	3	0.8 or 1% (whichever is greater)	0.3	0.5	5	0.5 or 1% (whichever is greater)

Accuracy may be significantly reduced during or shortly after acquisition, own ship manoeuvre, a manoeuvre of the target, or any tracking disturbance and is also dependent on the own ship's motion and sensor accuracy.

Measured target range and bearing should be within 50 m (or ±1 per cent of target range) and 2°.

The testing standard should have detailed target simulation tests as a means to confirm the accuracy of targets with relative speeds of up to 100 kn. Individual accuracy values shown in the table above may be adapted to account for the relative aspects of target motion with respect to that of the own ship in the testing scenarios used.

5.25.4.7.2 For ships capable of speeds in excess of 30 kn (typically high-speed craft (HSC)) and with speeds of up to 70 kn, there should be additional steady-state measurements made to ensure that the motion accuracy, after 3 minutes of steady-state tracking, is maintained with target relative speeds of up to 140 kn.

5.25.4.8 A ground referencing function, based on a stationary tracked target, should be provided. Targets used for this function should be marked with the relevant symbol defined in SN/Circ. 243.

5.26 Automatic Identification System (AIS) Reported Targets

5.26.1 General

Reported targets provided by the AIS may be filtered according to user-defined parameters. Targets may be sleeping or may be activated. Activated targets are treated in a similar way to radar tracked targets.

5.26.2 AIS Target Capacity

In addition to the requirements for radar tracking, it should be possible to display and provide full presentation functionality for a minimum number of sleeping and activated AIS targets according to Table 1. There should be an indication when the capacity of processing/display of AIS targets is about to be exceeded.

5.26.3 Filtering of AIS Sleeping Targets

To reduce display clutter, a means to filter the presentation of sleeping AIS targets should be provided, together with an indication of the filter status (e.g. by target range, CPA/TCPA or AIS target class A/B, etc.). It should not be possible to remove individual AIS targets from the display.

5.26.4 Activation of AIS Targets

A means to activate a sleeping AIS target and to deactivate an activated AIS target should be provided. If zones for the automatic activation of AIS targets are provided, they should be the same as for automatic radar target acquisition. In addition, sleeping AIS targets may be automatically activated when meeting user-defined parameters (e.g. target range, CPA/TCPA or AIS target class A/B).

5.26.5 AIS Presentation Status (see Table 4).

5.27 AIS Graphical Presentation

Targets should be presented with their relevant symbols according to the performance standards for the Presentation of Navigation-related Information on Shipborne Navigational Displays adopted by the Organization and SN/Circ. 243.

5.27.1 AIS targets that are displayed should be presented as sleeping targets by default.

5.27.2 The course and speed of a tracked radar target or reported AIS target should be indicated by a predicted motion vector. The vector time should be adjustable and valid for presentation of any target regardless of its source.

5.27.3 A permanent indication of vector mode, time and stabilization should be provided.

5.27.4 The consistent common reference point should be used for the alignment of tracked radar and AIS symbols with other information on the same display.

TABLE 4 The AIS presentation status should be indicated as follows:

Function	Cases to be presented		Presentation
AIS ON/OFF	AIS processing switched ON/graphical presentation switched OFF	AIS processing switched ON/graphical presentation switched ON	Alphanumeric or graphical
Filtering of sleeping AIS targets	Filter status	Filter status	Alphanumeric or graphical
Activation of targets		Activation criteria	Graphical
CPA/TCPA alarm	Function ON/OFF Sleeping targets included	Function ON/OFF Sleeping targets included	Alphanumeric and graphical
Lost target alarm	Function ON/OFF Lost target filter criteria	Function ON/OFF Lost target filter criteria	Alphanumeric and graphical
Target association	Function ON/OFF Association criteria Default target priority	Function ON/OFF Association criteria Default target priority	Alphanumeric

5.27.5 On large scale/low range displays, a means to present the true scale outline of an activated AIS target should be provided. It should be possible to display the past track of activated targets.

5.28 AIS and Radar Target Data

5.28.1 It should be possible to select any tracked radar or AIS target for the alphanumeric display of its data. A target selected for the display of its alphanumeric information should be identified by the relevant symbol. If more than one target is selected for data display, the relevant symbols and the corresponding data should be clearly identified. There should be a clear indication to show that the target data is derived from radar or from AIS.

5.28.2 For each selected tracked radar target, the following data should be presented in alphanumeric form: source(s) of data, actual range of target, actual bearing of target, predicted target range at the closest point of approach (CPA), predicted time to CPA (TCPA), true course of target, true speed of target.

5.28.3 For each selected AIS target, the following data should be presented in alphanumeric form: source of data; ship's identification; navigational status; position where available and its quality, range, bearing, COG, SOG, CPA, and TCPA. Target heading and reported rate of turn should also be made available. Additional target information should be provided on request.

5.28.4 If the received AIS information is incomplete, the absent information should be clearly indicated as 'missing' within the target data field.

5.28.5 The data should be displayed and continually updated, until another target is selected for data display or until the window is closed.

5.28.6 Means should be provided to present the own ship AIS data on request.

5.29 Operational Alarms

A clear indication of the cause for all alarm criteria should be given.

5.29.1 If the calculated CPA and TCPA values of a tracked target or activated AIS target are less than the set limits:

- A CPA/TCPA alarm should be given.
- The target should be clearly indicated.

5.29.2 The preset CPA/TCPA limits applied to targets from different radar and AIS should be identical. As a default state, the CPA/TCPA alarm functionality should be applied to all activated AIS targets. On user request the CPA/TCPA alarm functionality may also be applied to sleeping targets.

5.29.3 If a user-defined acquisition/activation zone facility is provided, a target not previously acquired/activated entering the zone, or detected within the zone, should be clearly identified with the relevant symbol and an alarm should be given.

It should be possible for the user to set ranges and outlines for the zone.

5.29.4 The system should alert the user if a tracked radar target is lost, rather than excluded by a predetermined range or preset parameter. The target's last position should be clearly indicated on the display.

5.29.5 It should be possible to enable or disable the lost target alarm function for AIS targets. A clear indication should be given if the lost target alarm is disabled.

If the following conditions are met for a lost AIS target:

- The AIS lost target alarm function is enabled.
- The target is of interest, according to lost target filter criteria.
- A message is not received for a set time, depending on the nominal reporting rate of the AIS target.

Then:

- The last known position should be clearly indicated as a lost target and an alarm be given.
- The indication of the lost target should disappear if the signal is received again, or after the alarm has been acknowledged.
- A means of recovering limited historical data from previous reports should be provided.

5.30 AIS and Radar Target Association

An automatic target association function based on harmonized criteria avoids the presentation of two target symbols for the same physical target.

5.30.1 If the target data from AIS and radar tracking are both available and if the association criteria (e.g. position, motion) are fulfilled such that the AIS and radar information are considered as one physical target, then as a default condition, the activated AIS target symbol and the alphanumeric AIS target data should be automatically selected and displayed.

5.30.2 The user should have the option to change the default condition to the display of tracked radar targets and should be permitted to select either radar tracking or AIS alphanumeric data.

5.30.3 For an associated target, if the AIS and radar information become sufficiently different, the AIS and radar information should be considered as two distinct targets and one activated AIS target and one tracked radar target should be displayed. No alarm should be raised.

5.31 Trial Manoeuvre

The system should, where required by Table 1, be capable of simulating the predicted effects of the own ship's manoeuvre in a potential threat situation and should include the own ship's dynamic characteristics. A trial manoeuvre simulation should be clearly identified. The requirements are:

- The simulation of the own ship course and speed should be variable.
- A simulated time to manoeuvre with a countdown should be provided.
  - During simulation, target tracking should continue and the actual target data should be indicated.
- Trial manoeuvre should be applied to all tracked targets and at least all activated AIS targets.

5.32 The Display of Maps, Navigation Lines and Routes

5.32.1 It should be possible for the user to manually create and change, save, load and display simple maps/navigation lines/routes referenced to the own ship or a geographical position. It should be possible to remove the display of this data by a simple operator action.

5.32.2 The maps/navigation lines/routes may consist of lines, symbols and reference points.

5.32.3 The appearance of lines, colours and symbols are as defined in SN/Circ. 243.

5.32.4 The maps/navigation lines/route graphics should not significantly degrade the radar information.

5.32.5 The maps/navigation lines/routes should be retained when the equipment is switched OFF.

5.32.6 The maps/navigation lines/route data should be transferable whenever a relevant equipment module is replaced.

5.33 The Display of Charts

5.33.1 The radar system may provide the means to display ENC and other vector chart information within the operational display area to provide continuous and real-time position monitoring. It should be possible to remove the display of chart data by a single operator action.

5.33.2 The ENC information should be the primary source of information and should comply with IHO relevant standards. Status of other information should be identified with a permanent indication. Source and update information should be made available.

5.33.3 As a minimum, the elements of the ECDIS Standard Display should be made available for individual selection by category or layer, but not as individual objects.

5.33.4 The chart information should use the same reference and coordinate criteria as the radar/AIS, including datum, scale, orientation, CCRP and stabilization mode.

5.33.5 The display of radar information should have priority. Chart information should be displayed such that radar information is not substantially masked, obscured or degraded. Chart information should be clearly perceptible as such.

5.33.6 A malfunction of the source of chart data should not affect the operation of the radar/AIS system.

5.33.7 Symbols and colours should comply with the performance standards for the Presentation of Navigation-related Information on Shipborne Navigational Displays adopted by the Organization (SN/Circ. 243).

5.34 Alarms and Indications

Alarms and indications should comply with the performance standards for the Presentation of Navigation-related Information on Shipborne Navigational Displays adopted by the Organization.

5.34.1 A means should be provided to alert the user of 'picture freeze'.

5.34.2 Failure of any signal or sensor in use, including; gyro, log, azimuth, video, sync and heading marker, should be alarmed. System functionality should be limited to a fallback mode or in some cases, the display presentation should be inhibited (see fallback modes, section 9).

5.35 Integrating Multiple Radars

5.35.1 The system should safeguard against single point system failure. Fail-safe condition should be applied in the event of an integration failure.

5.35.2 The source and any processing or combination of radar signals should be indicated.

5.35.3 The system status for each display position should be available.

6 ERGONOMIC CRITERIA

6.1 Operational Controls

6.1.1 The design should ensure that the radar system is simple to operate. Operational controls should have a harmonized user interface and be easy to identify and simple to use.

6.1.2 The radar system should be capable of being switched ON or OFF at the main system radar display or at a control position.

6.1.3 The control functions may be dedicated hardware, screen accessed or a combination of these; however, the primary control functions should be dedicated hardware controls or soft keys, with an associated status indication in a consistent and intuitive position.

6.1.4 The following are defined as primary radar control functions and should be easily and immediately accessible:

Radar Standby/RUN, range scale selection, gain, tuning function (if applicable), anti-clutter rain, anti-clutter sea, AIS function on/off, alarm acknowledge, cursor, a means to set EBL/VRM, display brightness and acquisition of radar targets.

6.1.5 The primary functions may also be operated from a remote operating position in addition to the main controls.

6.2 Display Presentation

6.2.1 The display presentation should comply with the performance standards for the Presentation of Navigation-related Information on Shipborne Navigational Displays adopted by the Organization.

6.2.2 The colours, symbols and graphics presented should comply with SN/Circ. 243.

6.2.3 The display sizes should conform to those defined in Table 1.

6.3 Instructions and Documentation

6.3.1 Documentation Language

The operating instructions and manufacturer's documentation should be written in a clear and comprehensible manner and should be available at least in the English language.

6.3.2 Operating Instructions

The operating instructions should contain a qualified explanation and/or description of information required by the user to operate the radar system correctly, including:

- appropriate settings for different weather conditions;
- monitoring the radar system's performance;
- operating in a failure or fallback situation;
- limitations of the display and tracking process and accuracy, including any delays;
- using heading and SOG/COG information for collision avoidance;
- limitations and conditions of target association;
- criteria of selection for automatic activation and cancellation of targets;
- methods applied to display AIS targets and any limitations;
- principles underlying the trial manoeuvre technology, including simulation of the own ship's manoeuvring characteristics, if provided;
- alarms and indications;
- installation requirements as listed under section 7.5;
- radar range and bearing accuracies; and
- any special operation (e.g. tuning) for the detection of SARTs; and
- the role of the CCRP for radar measurements and its specific value.

6.3.3 Manufacturer's Documentation

6.3.3.1 The manufacturer's documentation should contain a description of the radar system and factors which may affect detection performance, including any latency in signal processing.

6.3.3.2 Documentation should describe the basis of AIS filter criteria and AIS/radar target association criteria.

6.3.3.3 The equipment documentation should include full details of installation information, including additional recommendations on unit location and factors that may degrade performance or reliability.

7 DESIGN AND INSTALLATION

7.1 Design for Servicing

7.1.1 As far as is practical, the radar system should be of a design to facilitate simple fault diagnosis and maximum availability.

7.1.2 The radar system should include a means to record the total operational hours for any components with a limited life.

7.1.3 The documentation should describe any routine servicing requirements and should include details of any restricted life components.

7.2 Display

The display device physical requirements should meet those specified in the performance standards for the Presentation of Navigation-related Information on Shipborne Navigational Displays adopted by the Organization (SN/Circ. 243) and those specified in Table 1.

7.3 Transmitter Mute

The equipment should provide a mute facility to inhibit the transmission of radar energy over a preset sector. The mute sector should be set up on installation. An indication of sector mute status should be available.

7.4 Antenna

7.4.1 The antenna should be designed to start operating and to continue to operate in relative wind speeds likely to be encountered on the class of ship on which it is installed.

7.4.2 The combined radar system should be capable of providing an appropriate information update rate for the class of ship on which it is installed.

7.4.3 The antenna sidelobes should be consistent with satisfying the system performance as defined in this standard.

7.4.4 There should be a means to prevent antenna rotation and radiation during servicing, or while personnel are in the vicinity of up-mast units.

7.5 Radar System Installation

Requirements and guidelines for the radar system installation should be included in the manufacturers' documentation. The following subjects should be covered:

7.5.1 The Antenna

Blind sectors should be kept to a minimum and should not be placed in an arc of the horizon from the right ahead direction to 22.5° abaft the beam and especially should avoid the right ahead direction (relative bearing 000°). The installation of the antenna should be in such a manner that the performance of the radar system is not substantially degraded. The antenna should be mounted clear of any structure that may cause signal reflections, including other antenna and deck structure or cargo. In addition, the height of the antenna should take account of target detection performance relating to range of first detection and target visibility in sea clutter.

7.5.2 The Display

The orientation of the display unit should be such that the user is looking ahead, the lookout view is not obscured and there is minimum ambient light on the display.

7.6 Operation and Training

7.6.1 The design should ensure that the radar system is simple to operate by trained users.

7.6.2 A target simulation facility should be provided for training purposes.

8 INTERFACING

8.1 Data

The radar system should be capable of receiving the required input information from:

- a gyro-compass or transmitting heading device (THD);
- speed and distance measuring equipment (SDME);
- an electronic position fixing system (EPFS);
- an Automatic Identification System (AIS); or
- other sensors or networks providing equivalent information acceptable to the Organization.

The radar should be interfaced to relevant sensors required by these performance standards in accordance with recognized international standards*.

8.2 Input Data Integrity and Latency

8.2.1 The radar system should not use data indicated as invalid. If input data is known to be of poor quality, this should be clearly indicated.

8.2.2 As far as is practical, the integrity of data should be checked, prior to its use, by comparison

*Refer to IEC publication 61162.

with other connected sensors or by testing to valid and plausible data limits.

8.2.3 The latency of processing input data should be minimized.

8.3 Output Data

8.3.1 Information provided by any radar output interface to other systems should be in accordance with international standards*.

8.3.2 The radar system should provide an output of the display data for the voyage data recorder (VDR).

8.3.3 At least one normally closed contact (isolated) should be provided for indicating failure of the radar.

8.3.4 The radar should have a bi-directional interface to facilitate communication so that alarms from the radar can be transferred to external systems and so that audible alarms from the radar can be muted from external systems, the interface should comply with relevant international standards.

9 BACKUP AND FALLBACK ARRANGEMENTS

In the event of partial failures and to maintain minimum basic operation, the fallback arrangements listed below should be provided. There should be a permanent indication of the failed input information.

9.1 Failure of Heading Information (Azimuth Stabilization)

9.1.1 The equipment should operate satisfactorily in an unstabilized head-up mode.

9.1.2 The equipment should switch automatically to the unstabilized head-up mode within 1 minute after the azimuth stabilization has become ineffective.

9.1.3 If automatic anti-clutter processing could prevent the detection of targets in the absence of appropriate stabilization, the processing should switch off automatically within 1 minute after the azimuth stabilization has become ineffective.

9.1.4 An indication should be given that only relative bearing measurements can be used.

9.2 Failure of Speed through the Water Information

A means of manual speed input should be provided and its use clearly indicated.

9.3 Failure of Course and Speed Over Ground Information

The equipment may be operated with course and speed through the water information.

9.4 Failure of Position Input Information

The overlay of chart data and geographically referenced maps should be disabled if only a single Reference Target is defined and used, or the position is manually entered.

9.5 Failure of Radar Video Input Information

In the absence of radar signals, the equipment should display target information based on AIS data. A frozen radar picture should not be displayed.

9.6 Failure of AIS Input Information

In the absence of AIS signals, the equipment should display the radar video and target database.

9.7 Failure of an Integrated or Networked System

The equipment should be capable of operating equivalent to a stand-alone system.

APPENDIX 1 – REFERENCES

IMO SOLAS chapters IV, V and X	Carriage rules.
IMO resolution A.278 (VII)	Supplement to the recommendation on PS for navigational radar equipment.
IMO resolution A.424 (XI)	Performance standards for gyrocompasses.
IMO resolution A.477 (XII)	Performance standards for radar equipment.
IMO resolution A.694 (17)	General requirements for shipborne radio equipment forming part of the global maritime distress and safety system and for electronically navigational aids.
IMO resolution A.817 (19) as amended	Performance standards for ECDIS.

IMO resolution A.821 (19)	Performance standards for gyrocompasses for high-speed craft.
IMO resolution A.824 (19)	Performance standards for devices to indicate speed and distance.
IMO resolution MSC.86 (70)	Performance standards for INS.
IMO resolution MSC.64 (67)	Recommendations on new and amended performance standards (Annex 2 revised by MSC.114 (73)).
IMO resolution MSC.112 (73)	Revised performance standards for shipborne global positioning (GPS) receiver equipment.
IMO resolution MSC.114 (73)	Revised performance standards for shipborne DGPS and DGLONASS maritime radio beacon receiver equipment.
IMO resolution MSC.116 (73)	Performance standards for marine transmitting heading devices (THD).
IMO MSC Circ. 982	Guidelines on ergonomic criteria for bridge equipment and layout.
IHO S-52	Colour and symbol specification for ECDIS.
IEC 62388	Radar Test Standard (replacing 60872 and 60936 series of test standards).
IEC 60945	Maritime navigation and radio communication equipment and systems – General requirements – Methods of testing and required test results.
IEC 61162	Maritime navigation and radio communication equipment and systems – Digital interfaces.
IEC 61174	Maritime navigation and radio communication equipment and systems – Electronic chart display and information system (ECDIS) – Operational and performance requirements, methods of testing and required test results.
IEC 62288	Presentation and display of navigation information.
ISO 9000 (all parts)	Quality management/assurance standards.

APPENDIX 2 – DEFINITIONS

Activated AIS target — A target representing the automatic or manual activation of a sleeping target for the display of additional graphically presented information. The target is displayed by an 'activated target' symbol including:

- a vector (COG/ SOG);
- the heading; and
- ROT or direction of turn indication (if available) to indicate initiated course changes.

Acquisition of a radar target — Process of acquiring a target and initiating its tracking.

Activation of an AIS target — Activation of a sleeping AIS target for the display of additional graphical and alphanumeric information.

Acquired radar target — Automatic or manual acquisition initiates radar tracking. Vectors and past positions are displayed when data has achieved a steady-state condition.

(*Continued*)

APPENDIX 2 – DEFINITIONS (Continued)

AIS	Automatic Identification System.
AIS target	A target generated from an AIS message. *See* activated target, lost target, selected target and sleeping target.
Associated target	If an acquired radar target and an AIS reported target have similar parameters (e.g. position, course, speed) complying with an association algorithm, they are considered to be the same target and become an associated target.
Acquisition/activation zone	A zone set up by the operator in which the system should automatically acquire radar targets and activate reported AIS targets when entering the zone.
CCRP	Consistent Common Reference Point: A location on the own ship, to which all horizontal measurements such as target range, bearing, relative course, relative speed, closest point of approach (CPA) or time to closest point of approach (TCPA) are referenced, typically the conning position of the bridge.
CPA/TCPA	Closest Point of Approach/ Time to the Closest Point of Approach: Distance to the closest point of approach (CPA) and time to the closest point of approach (TCPA). Limits are set by the operator related to the own ship.
Course Over Ground (COG)	Direction of the ship's movement relative to the earth, measured on board the ship, expressed in angular units from true-north.
Course Through Water (CTW)	Direction of the ship's movement through the water, defined by the angle between the meridian through its position and the direction of the ship's movement through the water, expressed in angular units from true-north.
Dangerous target	A target whose predicted CPA and TCPA are violating the values as preset by the operator. The respective target is marked by a 'dangerous target' symbol.
Display modes	**Relative-motion:** means a display on which the position of the own ship remains fixed, and all targets move relative to the own ship. **True-motion:** a display across which the own ship moves with its own true motion.
Display orientation	**North-up display:** an azimuth stabilized presentation which uses the gyro input (or equivalent) and north is uppermost on the presentation. **Course-up display:** an azimuth stabilized presentation which uses the gyro input or equivalent and the ship's course is uppermost on the presentation at the time of selection. **Head-up display:** an unstabilized presentation in which the own ship's heading is uppermost on the presentation.
ECDIS	Electronic Chart Display and Information System.
ECDIS Display Base	The level of information which cannot be removed from the ECDIS display, consisting of information which is required at all times in all geographic areas and all circumstances. It is not intended to be sufficient for safe navigation.
ECDIS Standard Display	The level of information that should be shown when a chart is first displayed on ECDIS. The level of the information it provides for route planning or route monitoring may be modified by the mariner according to the mariner's needs.
ENC	Electronic Navigational Chart. The database standardized as to content, structure and format according to relevant IHO standards and issued by, or on the authority of, a government.

EPFS	Electronic Position Fixing System.
ERBL	Electronic bearing line carrying a marker, which is combined with the range marker, used to measure range and bearing from the own ship or between two objects.
Evaporation duct	A low lying duct (a change in air density) that traps the radar energy so that it propagates close to the sea surface. Ducting may enhance or reduce radar target detection ranges.
Heading	Direction in which the bow of a ship is pointing expressed as an angular displacement from north.
HSC	High-speed craft (HSC) are vessels which comply with the definition in SOLAS for high speed craft.
Latency	The delay between actual and presented data.
Lost AIS target	A target representing the last valid position of an AIS target before the reception of its data was lost. The target is displayed by a 'lost AIS target' symbol.
Lost tracked target	Target information is no longer available due to poor, lost or obscured signals. The target is displayed by a 'lost tracked radar target' symbol.
Maps/Nav lines	Operator defined or created lines to indicate channels, Traffic Separation Schemes or borders of any area important for navigation.
Operational display area	Area of the display used to graphically present chart and radar information, excluding the user dialogue area. On the chart display, this is the area of the chart presentation. On the radar display, this is the area encompassing the radar image.
Past positions	Equally time-spaced past position marks of a tracked or reported target and the own ship. The past positions' track may be either relative or true.
Radar	Radio direction and ranging. A radio system that allows the determination of distance and direction of reflecting objects and of transmitting devices.
Radar beacon	A navigation aid which responds to the radar transmission by generating a radar signal to identify its position and identity.
Radar detection false alarm	The probability of a radar false alarm represents the probability that noise will cross the detection threshold and be called a target when only noise is present.
Radar target	Any object fixed or moving whose position and motion is determined by successive radar measurements of range and bearing.
Radar target enhancer	An electronic radar reflector, the output of which is an amplified version of the received radar pulse without any form of processing except limiting.
Reference target	Symbol indicating that the associated tracked stationary target (e.g. a navigational mark) is used as a speed reference for the ground-stabilization.
Relative bearing	Direction of a target's position from own ship's reference location expressed as an angular displacement from own ship's heading.
Relative course	Direction of motion of a target relative to own ship's direction. (Bearing.)
Relative-motion	Combination of relative course and relative speed.
Relative speed	Speed of a target relative to the own ship's speed data.
Rate of turn	Change of heading per time unit.

(Continued)

APPENDIX 2 – DEFINITIONS (Continued)

SART	Search And Rescue Transponder.
SDME	Speed and Distance Measurement Equipment.
Selected target	A manually selected target for the display of detailed alphanumeric information in a separate data display area. The target is displayed by a 'selected target' symbol.
Sleeping AIS target	A target indicating the presence and orientation of a vessel equipped with AIS in a certain location. The target is displayed by a 'sleeping target' symbol. No additional information is presented until activated.
Stabilization modes	**Ground-stabilization:** Display mode in which speed and course information are referred to the ground, using ground track input data, or EPFS as reference. **Sea-stabilization:** Display mode in which speed and course information are referred to the sea, using gyro or equivalent and water speed log input as reference.
Standard display	The level of information that should be shown when a chart is first displayed on ECDIS. The level of the information it provides for route planning or route monitoring may be modified by the mariner according to the mariner's needs.
Standard radar reflector	Reference reflector mounted 3.5 m above sea level with 10 m^2 effective reflecting area at X-band.
Steady-state tracking	Tracking a target, proceeding at steady motion: – after completion of the acquisition process, or – without a manoeuvre of a target or own ship, or – without target swap or any disturbance.
Speed Over Ground (SOG)	Speed of the ship relative to the earth, measured on board the ship.
Speed Through Water	Speed of the ship relative to the water surface.
SOLAS	International Convention for the Safety of Life at Sea.
Suppressed area	An area set up by the operator within which targets are not acquired.
Target swap	Situation in which the incoming radar data for a tracked target becomes incorrectly associated with another tracked target or a non-tracked radar echo.
Target's predicted motion	Prediction of a target's future course and speed based on linear extrapolation from its present motion as determined by past measurements of its range and bearing on the radar.
Target tracking (TT)	Computer process of observing the sequential changes in the position of a radar target in order to establish its motion. Such a target is a Tracked Target.
Trails	Tracks displayed by the radar echoes of targets in the form of an afterglow. Trails may be true or relative.
Trial manoeuvre	Graphical simulation facility used to assist the operator to perform a proposed manoeuvre for navigation and collision avoidance purposes, by displaying the predicted future status of at least all acquired or activated targets as a result of the own ship's simulated manoeuvres.
True bearing	Direction of a target from the own ship's reference location or from another target's position expressed as an angular displacement from true-north.

True course	Direction of motion relative to ground or to sea, of a target expressed as an angular displacement from north.
True-motion	Combination of true course and true speed.
True speed	Speed of a target relative to ground, or to sea.
Vector modes	**True vector:** Vector representing the predicted true motion of a target, showing course and speed with reference to the ground.
	Relative vector: Predicted movement of a target relative to the own ship's motion.
User Configured Presentation	A display presentation configured by the user for a specific task at hand. The presentation may include radar and/or chart information, in combination with other navigation or ship-related data.
User Dialogue Area	Is an area of the display consisting of data fields and/or menus that is allocated to the interactive presentation and entry or selection of operational parameters, data and commands, mainly in alphanumeric form.

11.2.2 Extracts from IMO Resolution MSC.191(79) Performance Standards for the Presentation of Navigation-Related Information on Shipborne Navigational Displays for New Ships Constructed After 2008

1 PURPOSE

These performance standards harmonize the requirements for the presentation of navigation-related information on the bridge of a ship to ensure that all navigational displays adopt a consistent human – machine interface philosophy and implementation.

These performance standards supplement and, in case of a conflict, take priority over presentation requirements of the individual performance standards adopted by the Organization for relevant navigational systems and equipment, and cover the presentation of navigation-related information by equipment for which performance standards have not been adopted.

2 SCOPE

These performance standards specify the presentation of navigational information on the bridge of a ship, including the consistent use of navigational terms, abbreviations, colours and symbols, as well as other presentation characteristics.

These performance standards also address the presentation of navigation information related to specific navigational tasks by recognizing the use of user selected presentations in addition to presentations required by the individual performance standards adopted by the Organization.

3 APPLICATION

The general principles of these standards are applicable for all displays on the bridge of a ship.*

These performance standards are applicable to any display equipment associated with the navigation systems and equipment for which individual performance standards have been adopted by the Organization. They also address display equipment associated with navigation systems and equipment for which individual performance standards have not been adopted.

In addition to the general requirements set out in resolution A.694 (17),** display equipment should meet the requirements of these performance standards, as applicable.

4 DEFINITIONS

Definitions are given in the Appendix.

5 GENERAL REQUIREMENTS FOR THE PRESENTATION OF INFORMATION

5.1 Arrangement of information

5.1.1 The presentation of information should be consistent with respect to screen layout and arrangement of information. Data and control functions should be logically grouped. Priority of information should be identified for each application,

*The general principles are addressed in paragraphs 5 and 8.

**IEC Publication 60945 (see Appendix 1).

permanently displayed and presented to the user in a prominent manner by, for example, use of position, size and colour.

5.1.2 The presentation of information should be consistent with respect to values, units, meaning, sources, validity and, if available, integrity.

5.1.3 The presentation of information should be clearly separated into an operational display area (e.g. radar, chart) and one or more user dialogue areas (e.g. menus, data, control functions).

5.2 Readability

5.2.1 The presentation of alphanumeric data, text, symbols and other graphical information (e.g. radar image) should support readability from typical user positions under all ambient light conditions likely to be experienced on the bridge of a ship, and with due consideration to the night vision of the officer of the watch.

5.2.2 Alphanumeric data and text should be presented using a clearly legible non-italic, sans-serif font. The font size should be appropriate for the viewing distance from user positions likely to be experienced on the bridge of a ship.

5.2.3 Text should be presented using simple unambiguous language that is easy to understand. Navigation terms and abbreviations should be presented using the nomenclature defined in SN/Circ. 243.

5.2.4 When icons are used, their purpose should be intuitively recognized by appearance, placement and grouping.

5.3 Colours and intensity

5.3.1 The colours used for the presentation of alphanumeric data, text, symbols and other graphical information should provide sufficient contrast against the background under all lighting conditions likely to be experienced on the bridge of a ship.

5.3.2 The colours and brightness should take into account the light conditions of daylight, dusk and night. The presentation should support night viewing by showing light foreground information on a dark non-reflecting background at night.

5.3.3 The background colour and contrast should be chosen to allow presented information to be easily discriminated without degrading the colour coding aspects of the presentation.

5.4 Symbols

5.4.1 Symbols used for the presentation of operational information are defined in SN/Circ. 243.

5.4.2 Symbols used for the display of charted information should comply with relevant IHO standards.

5.5 Coding of information

5.5.1 When colour coding is used for discrimination or conspicuousness of alphanumeric text, symbols and other graphical information, all colours in the set should clearly differ from one another.

5.5.2 When colour coding is used, the colour red should be used for coding of alarm related information.

5.5.3 When colour coding is used, it should be used in combination with other symbol attributes, such as size, shape, and orientation.

5.5.4 Flashing of information should be reserved for unacknowledged alarms.

5.6 Integrity marking

5.6.1 The source, validity, and where possible, the integrity of information should be indicated. Invalid information or information with low integrity should be clearly marked, qualitatively and/or quantitatively. Invalid information or information with low integrity may be quantitatively indicated by displaying absolute or percentage values.

5.6.2 When colour coding is used, information with low integrity should be qualitatively marked by using yellow, and invalid information should be qualitatively marked by using red.

5.6.3 In order to show that the screen is being refreshed, means should be provided to immediately make the user aware of a presentation failure on an operational display (e.g. 'picture freeze').

5.7 Alarms and indications

5.7.1 The operational status of information should be indicated as follows:

Status	Visual indication	Audible signal
Alarm, not acknowledged	Red, flashing	Accompanied by an audible signal
Alarm, acknowledged Invalid Information	Red	Suppression of audible signal
Important Indications (Warnings) (e.g. low integrity)	Yellow	Silence unless otherwise specified by the Organization
Normal state	None required, optionally green	Silence

5.7.2 A list of alarms should be provided based on the sequence of occurrence. Additional indication of priority, as set by the user, should be provided on displays showing alarms from multiple sources. Alarms that have been acknowledged and are no longer relevant should be deleted from the list of alarms, but may be retained in an alarm history list.

5.7.3 When a single display is used to present information from multiple navigation systems and equipment, the presentation of alarms and indications should be consistent for the display of the time of alarm occurrence, the cause of the alarm, the source of the alarm and the status of the alarm (e.g. acknowledged, not acknowledged).

5.8 Presentation modes

If displays are capable of presenting information in different mode(s), there should be a clear indication of the mode in use, for example orientation, stabilization, motion and chart projection.

5.9 User manuals

The user manual and operator instructions should be available in the English language at least. The user manual or reference guide should include a list of all terms, abbreviations and symbols, and their explanations presented by the equipment.

6 PRESENTATION OF OPERATIONAL INFORMATION

6.1 Presentation of own ship information

6.1.1 When a graphical representation of the own ship is provided, it should be possible for the user to select either a scaled ship's outline or a simplified symbol as specified in SN/Circ. 243. The size of the ship's outline or the simplified symbol in the graphical presentation should be the true scale size of the ship or 6 mm, whichever is greater.

6.1.2 A heading line, and where appropriate a velocity vector, should be associated with the own ship symbol and should originate at the position of the consistent common reference point (CCRP).

6.2 Presentation of charted information

6.2.1 The presentation of charted information that is issued by, or on the authority of a government authorized hydrographic office, or other relevant government institution should comply with the relevant IHO standards.

6.2.2 The presentation of proprietary charted information should comply with relevant IHO standards, as far as practical. There should be a clear indication when the presentation is not in accordance with IHO standards.

6.2.3 The presentation of user-added charted information should comply with the relevant IHO standards, as far as practical.

6.2.4 If chart data derived from different scales appear on the display, the scale boundary should be clearly indicated.

6.3 Presentation of radar information

6.3.1 Radar images should be displayed by using a basic colour that provides optimum contrast. Radar echoes should be clearly visible when presented on top of a chart background. The relative strength of echoes may be differentiated by tones of the same basic colour. The basic colour may be different for operation under different ambient light conditions.

6.3.2 Target trails should be distinguishable from targets and clearly visible under all ambient light conditions.

6.4 Presentation of target information

6.4.1 General

6.4.1.1 Target information may be provided by radar target tracking and/or by reported target information from the Automatic Identification System (AIS).

6.4.1.2 The operation of the radar target tracking function and the processing of reported AIS information, including the number of targets presented, related to screen size, is defined within the Performance Standards for Radar Equipment as adopted by the Organization. The presentation of radar target tracking and AIS information is defined within these performance standards.

6.4.1.3 As far as practical, the user interface and data format for operating, displaying and indicating radar tracking and AIS information should be consistent.

6.4.2 Target capacity

6.4.2.1 There should be an indication when the target tracking and/or reported target processing/display capacity is about to be exceeded.

6.4.2.2 There should be an indication when the target tracking and/or reported target processing/display capacity has been exceeded.

6.4.3 Filtering of AIS sleeping targets

6.4.3.1 To ensure that the clarity of the total presentation is not substantially impaired, it should be possible to filter the presentation of sleeping AIS targets (e.g. by target range, CPA/TCPA or AIS target class A/B, etc.).

6.4.3.2 If a filter is applied, there should be a clear and permanent indication. The filter criteria in use should be readily available.

6.4.3.3 It should not be possible to remove individual AIS targets from the display.

6.4.4 Activation of AIS targets

6.4.4.1 If zones for the automatic activation of AIS targets are provided, they should be the same as for automatic radar target acquisition, if available. Any user-defined zones (e.g. acquisition/activation zones) in use should be presented in graphical form.

6.4.4.2 In addition, sleeping AIS targets should be automatically activated when meeting user-defined parameters (e.g. target range, CPA/TCPA or AIS target class A/B).

6.4.5 Graphical presentation

6.4.5.1 Targets should be presented with their relevant symbols according to SN/Circ. 243.

6.4.5.2 AIS information should be graphically presented either as sleeping or activated targets.

6.4.5.3 The course and speed of a tracked radar target or reported AIS target should be indicated by a vector that clearly shows the predicted motion. The vector time (length) should be consistent for presentation of any target regardless of its source.

6.4.5.4 The presentation of vector symbols should be consistent irrespective of the source of information. The presentation mode should be clearly and permanently indicated, including for example: true/relative vector; vector time and vector stabilization.

6.4.5.5 The orientation of the AIS target symbol should indicate its heading. If the heading information is not received, the orientation of the AIS symbol should be aligned to the COG. When available, the turn or rate of turn (ROT) indicator and/or the path prediction should indicate the manoeuvre of an activated AIS target.

6.4.5.6 A consistent common reference point should be used for the alignment of tracked target symbols and AIS target symbols with other information on the same display.

6.4.5.7 On large scale/low range displays, a means to present a true scale outline of an activated AIS target should be provided.

6.4.5.8 It should be possible to display the past positions of activated targets.

6.4.6 Target data

6.4.6.1 A target selected for the display of its alphanumeric information should be identified by the relevant symbol. If more than one target is selected for data display, the symbols and the corresponding data should be clearly identified.

6.4.6.2 There should be a clear indication to show that the target data is derived from radar or AIS or from a combination of these.

6.4.6.3 For each selected tracked radar target, the following data should be presented in alphanumeric form: source(s) of data; measured range of target; measured bearing of target; predicted target range at the closest point of approach (CPA); predicted time to CPA (TCPA); true course of target; true speed of target. Additional target information should be provided on request.

6.4.6.4 For each selected AIS target, the following data should be presented in alphanumeric form: source of data; ship's identification; position and its quality; calculated range of target; calculated bearing of target; CPA; TCPA; COG; SOG; navigational status. Ship's heading and rate of turn should also be made available. Additional target information should be provided on request.

6.4.6.5 If the received AIS information is incomplete, the absent information should be clearly indicated in the target data field as missing.

6.4.6.6 The data should be displayed and continually updated, until another target is selected for data display or until the window is closed.

6.4.6.7 Means should be provided to present the own ship AIS data on request.

6.4.6.8 The alphanumeric displayed data should not obscure graphically presented operational information.

6.4.7 Operational alarms

6.4.7.1 A clear indication of the status of the alarms and of the alarm criteria should be given.

6.4.7.2 A CPA/TCPA alarm of a tracked radar or activated AIS target should be clearly indicated and the target should be clearly marked by a dangerous target symbol.

6.4.7.3 If a user-defined acquisition/activation zone facility is provided, a target entering the zone should be clearly identified with the relevant symbol and for tracked radar targets an alarm should be given. The zone should be identified with the relevant symbology, and should be applicable to tracked radar and AIS targets.

6.4.7.4 The last position of a lost target should be clearly marked by a lost target symbol on the display, and the lost target alarm should be given. The lost target symbol should disappear if the signal is received again, or after the alarm has been acknowledged. There should be a clear indication whether the lost target alarm function for AIS targets is enabled or disabled.

6.4.8 AIS and radar target association

6.4.8.1 An automatic target association function serves to avoid the presentation of two target symbols for the same physical target. If target data from AIS and radar tracking are both available and if the AIS and radar information are considered as one target, then as a default condition, the activated AIS target symbol and the alphanumeric AIS target data should be automatically selected and displayed. The user should have the option to change the default condition to the display of tracked radar targets and should be permitted to select either radar tracking or AIS alphanumeric data.

6.4.8.2 If the AIS and radar information are considered as two distinct targets, one activated AIS target and one tracked radar target should be displayed. No alarm should be raised.

6.4.9 AIS presentation status

The AIS presentation status should be indicated as follows:

Function	Cases to be presented		Presentation
AIS ON/OFF	AIS processing switched ON/ graphical presentation switched OFF	AIS processing switched ON/ graphical presentation switched ON	Alphanumeric or graphical
Filtering of sleeping AIS targets (6.4.3)	Filter status	Filter status	Alphanumeric or graphical
Activation of targets (6.4.4)		Activation criteria	Graphical
CPA/TCPA alarm (6.4.7)	Function ON/OFF CPA/TCPA criteria Sleeping targets included	Function ON/OFF CPA/TCPA criteria Sleeping targets included	Alphanumeric and graphical
Lost target alarm (6.4.7)	Function ON/OFF Lost target filter criteria	Function ON/OFF Lost target filter criteria	Alphanumeric and graphical
Target association (6.4.8)	Function ON/OFF Association criteria Default target priority	Function ON/OFF Association criteria Default target priority	Alphanumeric

6.4.10 Trial manoeuvre

A trial manoeuvre simulation should be clearly identified by the relevant symbol positioned astern of the own ship within the operational display area of the screen.

7 OPERATIONAL DISPLAYS

7.1 General

7.1.1 If the display equipment is capable of supporting the presentation of multiple functions, then there should be a clear indication of the primary function supported by the presentation (e.g. radar, ECDIS). It should be possible to select the radar presentation (see 7.2) or the ECDIS presentation (see 7.3) by a single operator action.

7.1.2 If a radar image and an electronic chart are displayed together, the chart and the radar image should use a consistent common reference point and match in scale, projection and orientation. Any offset should be indicated.

7.1.3 Range scales of 0.25, 0.5, 0.75, 1.5, 3, 6, 12 and 24 NM should be provided. Additional range scales are permitted. These range scales do not apply when presenting raster chart data. The range scale should be permanently indicated.

7.1.4 When range rings are displayed, the range ring scale should be indicated.

7.1.5 No part of the operational display area should be permanently used for presentation of information that is not part of the navigation presentation (e.g. pop-up displays, drop-down menus and information windows). Temporary, limited and relevant alphanumeric data may be displayed adjacent to a selected symbol, graphic or target within the operational display area.

7.2 Radar display

7.2.1 General

7.2.1.1 Radar video, tracked radar targets and AIS targets should not be substantially degraded, masked or obscured by other presented information.

7.2.1.2 It should be possible to temporarily suppress all graphical information from the display, retaining only radar video and trails.

7.2.1.3 The brightness of radar echoes and associated graphic symbols for tracked radar targets should be variable. It should be possible to control the brightness of all displayed information. There should be independent means to adjust the brightness of groups of displayed graphics and alphanumeric data. The brilliance of the heading line should not be variable to extinction.

7.2.2 Display of chart information on radar

7.2.2.1 Vector chart information may be displayed on a radar presentation. This should be accomplished using layers selected from the chart database. As a minimum, the elements of the ECDIS Standard Display should be available for individual selection by category or layer, but not as individual objects. As far as practical, chart information should be presented in accordance with the ECDIS performance standards and with these presentation standards.

7.2.2.2 If chart information is displayed within the operational display area, the display of radar information should have priority. The chart information should be clearly perceptible as such. The chart information should not substantially degrade, mask or obscure the radar video, tracked radar targets and AIS targets.

7.2.2.3 When chart information is displayed, there should be a permanent indication of its status. Source and update information should also be made available.

7.2.3 Display of maps on radar

Map graphics may be displayed, but should not substantially degrade, mask or obscure the radar video, tracked radar targets and AIS targets.

7.3 ECDIS display

7.3.1 General

7.3.1.1 The ENC and all updates to it should be displayed without any degradation of their information content.

7.3.1.2 Chart information should not be substantially degraded, masked or obscured by other presented information.

7.3.1.3 It should be possible to temporarily suppress all supplemental information from the display, retaining only chart related information contained in the Display Base.

7.3.1.4 It should be possible to add or remove information from the ECDIS display. It should not be possible to remove information contained in the Display Base from the ECDIS display.

7.3.1.5 It should be possible to select a safety contour from the depth contours provided by the ENC. The safety contour should be emphasized over other contours on the display.

7.3.1.6 It should be possible to select a safety depth. Soundings equal to or less than the safety depth should be emphasized whenever spot soundings are selected for display.

7.3.1.7 An indication should be provided if the information is displayed at a larger scale than that contained in the ENC, or if the own ship's position is covered by an ENC at a larger scale than that provided by the display.

7.3.1.8 Overscaled areas shown on the ECDIS display should be identified.

7.3.2 Display of radar information on ECDIS

7.3.2.1 Radar and target information may be displayed on ECDIS but should not substantially degrade, mask or obscure the chart information. As far as practical, radar and target information should be presented in accordance with the radar performance standard and with these presentation standards.

7.3.2.2 Radar and target information should be clearly distinguishable from the chart information. It should be possible to remove this information by a single operator action.

7.3.3 Display of additional information on ECDIS

7.3.3.1 Information from additional sources may be displayed on ECDIS but should not substantially degrade, mask or obscure the chart information.

7.3.3.2 Additional information should be clearly distinguishable from the chart information. It should be possible to remove this information by a single operator action.

7.4 User selected (task orientated) presentation

7.4.1 The user may configure a presentation for a specific task at hand. The presentation may include radar and/or chart information, in combination with other navigation or ship related data. When not fully compliant with the Radar or ECDIS performance standards, such a presentation should be identified as an auxiliary presentation.

7.4.2 As far as practical, the presentation of any radar and/or ECDIS related functions should be compliant with the requirements of the relevant performance standards and of these presentation standards, with the exception of size requirements for the operational area. Chartlets or windows of radar information may be presented along with other information associated with the task at hand.

8 PHYSICAL REQUIREMENTS

8.1 Display adjustment

8.1.1 It should be possible to adjust the contrast and brightness of the display provided, as applicable to the display technology. It should be possible to dim the display. The range of control should permit the display to be legible under all ambient light conditions.

8.1.2 It should be possible for the navigator to reset the values of contrast and/or brightness to a preset or default condition.

8.1.3 Where magnetic fields degrade the presentation of navigation information, a means to neutralize the effect of magnetic fields should be provided.

8.2 Screen size

8.2.1 Display equipment should be of sufficient size to support the requirements of the relevant performance standards adopted by the Organization.

8.2.2 The operational display area of the chart presentation for route monitoring should be at least 270 × 270 mm.

8.2.3 The operational display area of the radar presentation should be at least a circle of diameter:

- 180 mm for ships smaller than 500 gross tonnage;
- 250 mm for ships larger than 500 gross tonnage and high-speed craft (HSC) less than 10,000 gross tonnage;
- 320 mm for ships larger than 10,000 gross tonnage.

8.3 Colours

8.3.1 Multicoloured display equipment should be used except where monochrome displays are permitted within individual performance standards adopted by the Organization.

8.3.2 Multicoloured operational displays including multifunction displays (e.g. conning displays) should provide a minimum of 64 colours except where permitted or not required by the Organization, or when used for a single specific purpose (e.g. speed log, echosounder).

8.4 Screen resolution

Operational display equipment including multifunction displays (e.g. conning displays) should provide a minimum screen resolution of 1280 × 1024, or equivalent for a different aspect ratio, except where permitted or not required by the Organization, or when used for a single specific purpose (e.g. speed log, echosounder).

8.5 Screen viewing angle

The display should support the reading of information under all ambient light conditions, simultaneously, by at least two users, from standing and sitting operator positions likely to be found on the bridge of a ship.

APPENDIX – DEFINITIONS

Activated AIS target	A target representing the automatic or manual activation of a sleeping target for the display of additional graphically presented information.
AIS target	A target generated from an AIS message.
Associated target	A target simultaneously representing a tracked radar target and AIS target having similar parameters (e.g. position, course, speed) and which comply with an association algorithm
CCRP	The Consistent Common Reference Point is a location on the own ship, to which all horizontal measurements such as target range, bearing, relative course, relative speed, closest point of approach (CPA) or time to closest point of approach (TCPA) are referenced, typically the conning position of the bridge.
Dangerous target	A target with a predicted CPA and TCPA that violates values preset by the operator. The respective target is marked by a 'dangerous target' symbol.
Display base	The level of information which cannot be removed from the ECDIS display, consisting of information which is required at all times in all geographic areas and all circumstances. It is not intended to be sufficient for safe navigation.
ENC	Electronic Navigational Chart. The database standardized as to content structure and format according to relevant IHO standards and issued by, or on the authority of, a government.
Heading	Direction in which the bow of a ship is pointing expressed as an angular displacement from north.
Important indication	A marking of an operational status of displayed information which needs special attention, e.g. information with low integrity or invalid information.
Lost target	A target representing the last valid position of a target before its data was lost. The target is displayed by a 'lost target' symbol.

(Continued)

APPENDIX – DEFINITIONS (Continued)

Operational display area	Area of the display used to graphically present chart and radar information, excluding the user dialogue area. On the chart display this is the area of the chart presentation. On the radar display this is the area encompassing the radar image.
Past positions	Equally time-spaced past position marks of a tracked or reported target and own ship. The coordinates used to display past positions may be either relative or true.
Sleeping AIS target	A target indicating the presence and orientation of a vessel equipped with AIS in a certain location. The target is displayed by a 'sleeping target' symbol. No additional information is presented until activated.
Selected target	A target selected manually for the display of detailed alphanumeric information in a separate data display area. The target is displayed by a 'selected target' symbol.
Standard display	The level of information that should be shown when a chart is first displayed on ECDIS. The level of the information it provides for route planning or route monitoring may be modified by the mariner according to the mariner's needs.
Trial manoeuvre	Facility used to assist the operator to perform a proposed manoeuvre for navigation and collision avoidance purposes, by displaying the predicted future status of all tracked and AIS targets as a result of the own ship's simulated manoeuvres.
User dialogue area	An area of the display consisting of data fields and/or menus that is allocated to the interactive presentation and entry or selection of operational parameters, data and commands, mainly in alphanumeric form.
User-selected presentation	An auxiliary presentation configured by the user for a specific task at hand. The presentation may include radar and/or chart information, in combination with other navigation or ship related data.

11.2.3 Extracts from IMO SN/Circ. 243 Guidelines for the Presentation of Navigation-Related Symbols, Terms and Abbreviations (As Amended)

ANNEX 1

Guidelines for the Presentation of Navigation-Related Symbols

1 PURPOSE

The purpose of these annexed guidelines is to provide guidance on the appropriate use of navigation-related symbols to achieve a harmonized and consistent presentation.

2 SCOPE

The use of these guidelines will ensure that the symbols used for the display of navigation-related information on all shipborne navigational systems and equipment are presented in a consistent and uniform manner.

3 APPLICATION

These guidelines apply to all shipborne navigational systems and equipment. The symbols listed in the Appendix should be used for the display of navigation-related information to promote consistency in the symbol presentation on navigational equipment.

The symbols listed in the Appendix should replace symbols which are currently contained in existing performance standards. Where a standard symbol is not available, another symbol may be used, but this symbol should not conflict with the symbols listed in the Appendix.

Appendix Navigation-related symbols

TABLE 1 Own ship symbols

Topic	Symbol	Description
Own ship		Double circle, located at the own ship's reference position. Use of this symbol is optional, if the own ship position is shown by the combination of Heading Line and Beam Line.
Own ship true scale outline		True scale outline located relative to the own ship's reference position, oriented along the own ship's heading. Used on small ranges/large scales.
Own ship radar antenna position	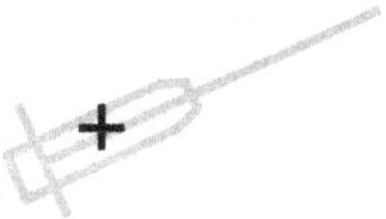	Cross, located on a true scale outline of the ship at the physical location of the radar antenna that is the current source of displayed radar video.
Own ship heading line		Solid line thinner than the speed vector line style, drawn to the bearing ring or of fixed length, if the bearing ring is not displayed. Origin is at the own ship's reference point.
Own ship beam line		Solid line of fixed length; optionally length variable by operator. Midpoint at the own ship's reference point.
Own ship speed vector	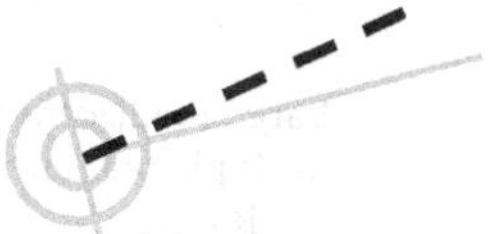 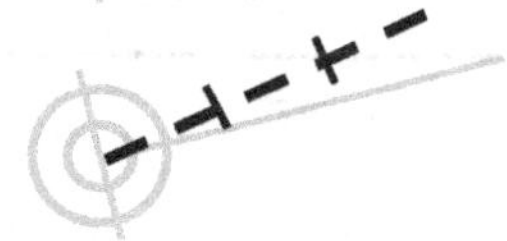	Dashed line – short dashes with spaces approximately twice the line width of heading line. Time increments between the origin and endpoint may optionally be marked along the vector using short intersecting lines. To indicate water/ground-stabilization optionally one arrowhead for water-stabilization and two arrowheads for ground-stabilization may be added.
Own ship path prediction	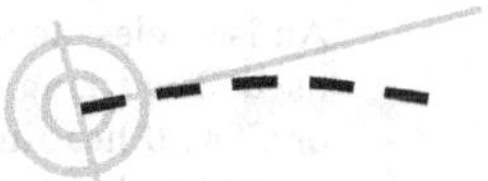	A curved vector may be provided as a path predictor.
Own ship past track	 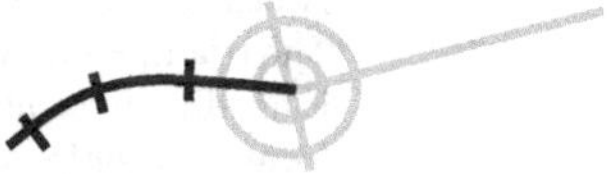	Thick line for primary source. Thin line for secondary source. Optional time marks are allowed.

TABLE 2 Tracked radar target symbols

Topic	Symbol	Description
Tracked target including dangerous target		Solid filled or unfilled circle located at target position. The course and speed vector should be displayed as a dashed line, with short dashes with spaces approximately twice the line width. Optionally, time increments, may be marked along the vector. For a '**Dangerous Target**', bold, red (on colour display) solid circle with course and speed vector, flashing until acknowledged.
Target in acquisition state		Circle segments in the acquired target state. For automatic acquisition, bold circle segments, flashing and red (on colour display) until acknowledged.
Lost target		Bold lines across the circle, flashing until acknowledged.
Selected target		A square indicated by its corners centred around the target symbol.
Target past positions		Dots, equally spaced by time.
Tracked reference target	R	Large R adjacent to designated tracked target. Multiple reference targets should be marked as R1, R2, R3, etc.

TABLE 3 AIS Target Symbols

AIS target (sleeping)		An isosceles, acute-angled triangle should be used. The triangle should be oriented by heading, or COG if heading missing. The reported position should be located at centre and half the height of the triangle. The symbol of the sleeping target should be smaller than that of the activated target.
Activated AIS target including dangerous target		An isosceles, acute-angled triangle should be used. The triangle should be oriented by heading, or COG if heading missing. The reported position should be located at centre and half the height of the triangle.

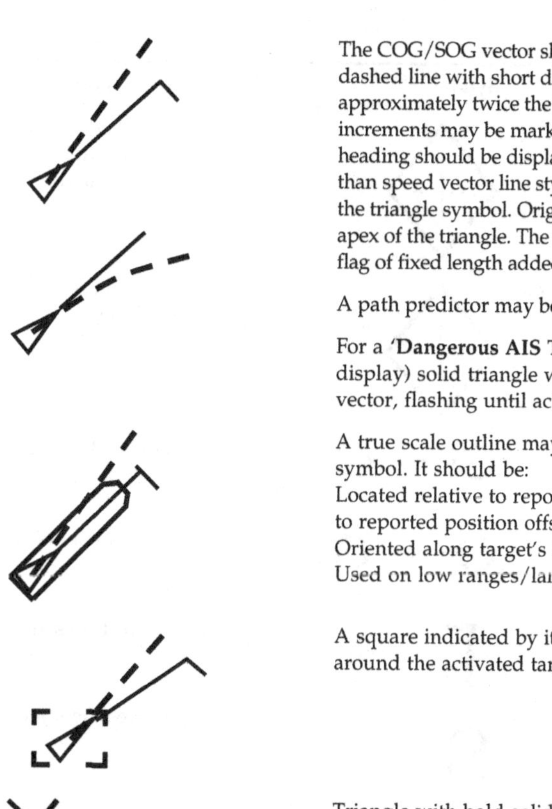

The COG/SOG vector should be displayed as a dashed line with short dashes with spaces approximately twice the line width. Optionally, time increments may be marked along the vector. The heading should be displayed as a solid line thinner than speed vector line style, length twice the length of the triangle symbol. Origin of the heading line is the apex of the triangle. The turn should be indicated by a flag of fixed length added to the heading line.

A path predictor may be provided as curved vector.

For a **'Dangerous AIS Target'**, bold, red (on colour display) solid triangle with course and speed vector, flashing until acknowledged.

AIS target – true scale outline

A true scale outline may be added to the triangle symbol. It should be:
Located relative to reported position and according to reported position offsets, beam and length.
Oriented along target's heading.
Used on low ranges/large scales.

Selected target

A square indicated by its corners should be drawn around the activated target symbol.

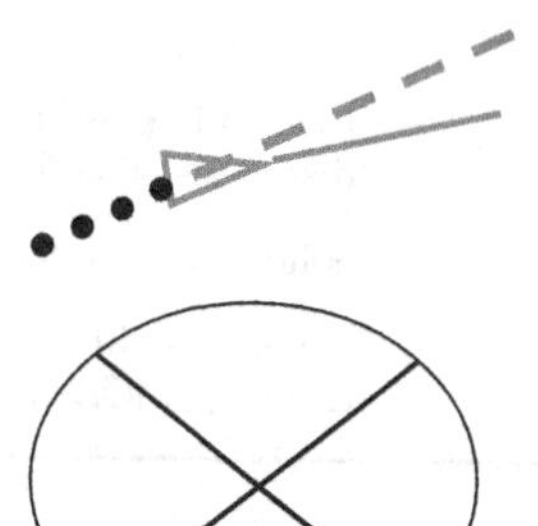

Lost target

Triangle with bold solid cross. The triangle should be oriented per last known value. The cross should have a fixed orientation. The symbol should flash until acknowledged.
The target should be displayed without vector, heading and rate of turn indication.

Target past positions

Dots, equally spaced by time.

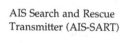

AIS Search and Rescue Transmitter (AIS-SART)

A circle containing a cross drawn with solid lines

TABLE 4 Other symbols

Topic	Symbol	Description
AIS based AtoN Real position of charted object		Diamond with crosshair centred at reported position. (Shown with chart symbol. Chart symbol not required for radar.)
AIS based AtoN Virtual position		Diamond with crosshair centred at reported position.
Monitored route		Dashed bold line, waypoints (WPT) as circles.
Planned or alternate route		Dotted line, WPT as circles.
Trial manoeuvre	T	Large T on screen.
Simulation mode	S	Large S on screen.
Cursor		Crosshair (two alternatives, one with open centre).
Range rings		Solid circles.
Variable Range Markers (VRM)		Circle. Additional VRM should be distinguishable from the primary VRM.
Electronic Bearing Lines (EBL)		Dashed line. Additional EBL should be distinguishable from the primary EBL.
Acquisition/activation area		Solid line boundary for an area.
Event mark		Rectangle with diagonal line, clarified by added text (e.g. 'MOB' for man overboard cases).

ANNEX 2

Guidelines for the Presentation of Navigation-Related Terms and Abbreviations

1 PURPOSE

The purpose of these guidelines is to provide guidance on the use of appropriate navigation-related terminology and abbreviations intended for presentation on shipborne navigational displays. These are based on terms and abbreviations used in existing navigation references.

2 SCOPE

These guidelines are issued to ensure that the terms and abbreviations used for the display of navigation-related information on all shipborne navigation equipment and systems are consistent and uniform.

3 APPLICATION

These guidelines apply to all shipborne navigational systems and equipment including radar, ECDIS, AIS, INS and IBS. When navigation-related information is displayed as text, the standard terms or abbreviations listed in the Appendix should be used, instead of using terms and abbreviations which are currently contained in existing performance standards.

Where a standard term and abbreviation is not available, another term or abbreviation may be used. This term or abbreviation should not conflict with the standard terms or abbreviations listed in the Appendix and should provide a clear meaning. Standard marine terminology should be used for this purpose. When the meaning is not clear from its context, the term should not be abbreviated.

Unless otherwise specified, standard terms should be shown in lower case, while abbreviations should be presented using upper case.

APPENDIX – LIST OF STANDARD TERMS AND ABBREVIATIONS

Term	Abbreviation	Abbreviation	Term
Acknowledge	ACK	ACK	Acknowledge
Acquire, Acquisition	ACQ	ACQ	Acquire, Acquisition
Acquisition Zone	AZ	ADJ	Adjust, Adjustment
Adjust, Adjustment	ADJ	AFC	Automatic Frequency Control
Aft	AFT	AFT	Aft
Alarm	ALARM	AGC	Automatic Gain Control
Altitude	ALT	AIS	Automatic Identification System
Amplitude Modulation	AM	ALARM	Alarm
Anchor Watch	ANCH	ALT	Altitude
Antenna	ANT	AM	Amplitude Modulation
Anti-clutter Rain	RAIN	ANCH	Anchor Watch
Anti-clutter Sea	SEA	ANCH	Vessel at Anchor (applies to AIS)
April	APR	ANT	Antenna
Audible	AUD	APR	April
August	AUG	AUD	Audible
Automatic	AUTO	AUG	August
Automatic Frequency Control	AFC	AUTO	Automatic
Automatic Gain Control	AGC	AUX	Auxiliary System/Function
Automatic Identification System	AIS	AVAIL	Available
Auxiliary System/Function	AUX	AZ	Acquisition Zone
Available	AVAIL	BITE	Built in Test Equipment
Background	BKGND	BKGND	Background
Bearing	BRG	BRG	Bearing

(*Continued*)

APPENDIX – LIST OF STANDARD TERMS AND ABBREVIATIONS (Continued)

Term	Abbreviation	Abbreviation	Term
Bearing Waypoint-to-Waypoint	BWW	BRILL	Brilliance
Brilliance	BRILL	BWW	Bearing Waypoint To Waypoint
Built in Test Equipment	BITE	C	Carried (e.g. carried EBL origin)
Calibrate	CAL	**C UP** (See note 2)	Course Up
Cancel	CNCL	CAL	Calibrate
Carried (e.g. carried EBL origin)	C	CCRP	Consistent Common Reference Point
Centre	CENT		
Change	CHG	CENT	Centre
Circular Polarized	CP	CHG	Change
Clear	CLR	CLR	Clear
Closest Point of Approach	CPA	CNCL	Cancel
Consistent Common Reference Point	CCRP	COG	Course Over the Ground
Consistent Common Reference System	CCRS	CONT	Contrast
Contrast	CONT	CORR	Correction
Correction	CORR	CP	Circular Polarized
Course	CRS	CPA	Closest Point of Approach
Course Over the Ground	COG	CRS	Course
Course Through the Water	CTW	CTS	Course To Steer
Course To Steer	CTS	CTW	Course Through the Water
Course-up	C UP (See note 2)	CURS	Cursor
Cross Track Distance	XTD	D	Dropped (e.g. dropped EBL origin)
Cursor	CURS	DATE	Date
Dangerous Goods	DG	DAY/NT	Day/Night
Date	DATE	DEC	December
Day/Night	DAY/NT	DECR	Decrease
Dead Reckoning, Dead Reckoned Position	DR	DEL	Delete
December	DEC	DELAY	Delay
Decrease	DECR	DEP	Departure
Delay	DELAY	DEST	Destination
Delete	DEL	DEV	Deviation

Departure	DEP	DG	Dangerous Goods
Depth	DPTH	DGAL (See note 2)	Differential Galilleo
Destination	DEST	DGLONASS(See note 2)	Differential GLONASS
Deviation	DEV	DGNSS(See note 2)	Differential GNSS
Differential Galilleo	DGLA (See note 2)	DGPS (See note 2)	Differential GPS
Differential GLONASS	DGLONASS(See note 2)	DISP	Display
Differential GNSS	DGNSS (See note 2)	DIST	Distance
Differential GPS	DGPS(See note 2)	DIVE	Vessel Engaged in Diving Operations (applies to AIS)
Digital Selective Calling	DSC	DPTH	Depth
Display	DISP	DR	Dead Reckoning, Dead Reckoned Position
Distance	DIST	DRG	Vessel Engaged in Dredging or Underwater Operations (applies to AIS)
Distance Root Mean Square	DRMS(See note 2)	DRIFT	Drift
Distance To Go	DTG	DRMS(See note 2)	Distance Root Mean Square
Drift	DRIFT	DSC	Digital Selective Calling
Dropped (e.g. dropped EBL origin)	D	DTG	Distance To Go
East	E	E	East
Electronic Bearing Line	EBL	EBL	Electronic Bearing Line
Electronic Chart Display and Information System	ECDIS	ECDIS	Electronic Chart Display and Information System
Electronic Navigational Chart	ENC	ENC	Electronic Navigational Chart
Electronic Position Fixing System	EPFS	ENH	Enhance
Electronic Range and Bearing Line	ERBL	ENT	Enter
Enhance	ENH	EP	Estimated Position
Enter	ENT	EPFS	Electronic Position Fixing System
Equipment	EQUIP	EQUIP	Equipment
Error	ERR	ERBL	Electronic Range and Bearing Line
Estimated Position	EP	ERR	Error
Estimated Time of Arrival	ETA	ETA	Estimated Time of Arrival
Estimated Time of Departure	ETD	ETD	Estimated Time of Departure

(Continued)

APPENDIX – LIST OF STANDARD TERMS AND ABBREVIATIONS (Continued)

Term	Abbreviation	Abbreviation	Term
Event	EVENT	EVENT	Event
Exclusion Zone	EZ	EXT	External
External	EXT	EZ	Exclusion Zone
February	FEB	FEB	February
Fishing Vessel	FISH	FISH	Fishing Vessel
Fix	FIX	FIX	Fix
Forward	FWD	FM	Frequency Modulation
Frequency	FREQ	FREQ	Frequency
Frequency Modulation	FM	FULL	Full
Full	FULL	FWD	Forward
Gain	GAIN	GAIN	Gain
Galilleo	GAL	GAL	Galilleo
Geometric Dilution Of Precision	GDOP	GC	Great Circle
Global Maritime Distress and Safety System	GMDSS	GDOP	Geometric Dilution Of Precision
Global Navigation Satellite System	GNSS	GLONASS	Global Orbiting Navigation Satellite System
Global Orbiting Navigation Satellite System	GLONASS	GMDSS	Global Maritime Distress and Safety System
Global Positioning System	GPS	GND	Ground
Great Circle	GC	GNSS	Global Navigation Satellite System
Grid	GRID	GPS	Global Positioning System
Ground	GND	GRI	Group Repetition Interval
Group Repetition Interval	GRI	GRID	Grid
Guard Zone	GZ	GRND	Vessel Aground (applies to AIS)
Gyro	GYRO	GYRO	Gyro
Harmful Substances (applies to AIS)	HS	GZ	Guard Zone
Head-up	H UP(See note 2)	H UP(See note 2)	Head-up
Heading	HDG	HCS	Heading Control System
Heading Control System	HCS	HDG	Heading
Heading Line	HL	HDOP	Horizontal Dilution Of Precision
High Frequency	HF	HF	High Frequency

High Speed Craft (applies to AIS)	HSC	HL	Heading Line
Horizontal Dilution Of Precision	HDOP	HS	Harmful Substances (applies to AIS)
Identification	ID	HSC	High Speed Craft (applies to AIS)
In	IN	I/O	Input/Output
Increase	INCR	ID	Identification
Indication	IND	IN	In
Information	INFO	INCR	Increase
Infrared	INFRED	IND	Indication
Initialization	INIT	INFRED	Infrared
Input	INP	INFO	Information
Input/Output	I/O	INIT	Initialization
Integrated Radio Communication System	IRCS	INP	Input
Interference Rejection	IR	INT	Interval
Interswitch	ISW	IR	Interference Rejection
Interval	INT	IRCS	Integrated Radio Communication System
January	JAN	ISW	Interswitch
July	JUL	JAN	January
June	JUN	JUL	July
Latitude	LAT	JUN	June
Limit	LIM	LAT	Latitude
Line Of Position	LOP	LF	Low Frequency
Log	LOG	LIM	Limit
Long Pulse	LP	LOG	Log
Long Range	LR	LON	Longitude
Longitude	LON	LOP	Line Of Position
Loran	LORAN	LORAN	Loran
Lost Target	LOST TGT	LOST TGT	Lost Target
Low Frequency	LF	LP	Long Pulse
Magnetic	MAG	LR	Long Range
Manoeuvre	MVR	MAG	Magnetic
Manual	MAN	MAN	Manual

(Continued)

APPENDIX – LIST OF STANDARD TERMS AND ABBREVIATIONS (Continued)

Term	Abbreviation	Abbreviation	Term
Map(s)	MAP	MAP	Map(s)
March	MAR	MAR	March
Maritime Mobile Services Identity number	MMSI	MAX	Maximum
Maritime Pollutant (applies to AIS)	MP	MAY	May
Maritime Safety Information	MSI	MENU	Menu
Marker	MKR	MF	Medium Frequency
Master	MSTR	MIN	Minimum
Maximum	MAX	MISSING	Missing
May	MAY	MKR	Marker
Medium Frequency	MF	MMSI	Maritime Mobile Services Identity number
Medium Pulse	MP	MON	Performance Monitor
Menu	MENU	MP	Maritime Pollutant (applies to AIS)
Minimum	MIN	MP	Medium Pulse
Missing	MISSING	MSI	Maritime Safety Information
Mute	MUTE	MSTR	Master
Navigation	NAV	MUTE	Mute
Normal	NORM	MVR	Manoeuvre
North	N	N	North
North-up	N Up (See note 2)	N UP (See note 2)	North-up
November	NOV	NAV	Navigation
October	OCT	NORM	Normal
Off	OFF	NOV	November
Officer Of the Watch	OOW	NUC	Vessel Not Under Command (applies to AIS)
Offset	OFFSET	OCT	October
On	ON	OFF	Off
Out/Output	OUT	OFFSET	Offset
Own Ship	OS	ON	On
Panel Illumination	PANEL	OOW	Officer On Watch

Parallel Index Line	PI	OS	Own Ship
Passenger Vessel (applies to AIS)	PASSV	OUT	Out/Output
Performance Monitor	MON	PAD	Predicted Area of Danger
Permanent	PERM	PANEL	Panel Illumination
Person Overboard	POB	PASSV	Passenger Vessel (applies to AIS)
Personal Identification Number	PIN	PDOP	Positional Dilution Of Precision
Pilot Vessel (applies to AIS)	PILOT	PERM	Permanent
Port/Portside	PORT	PI	Parallel Index Line
Position	POSN	PILOT	Pilot Vessel (applies to AIS)
Positional Dilution Of Precision	PDOP	PIN	Personal Identification Number
Power	PWR	PL	Pulse Length
Predicted	PRED	PM	Pulse Modulation
Predicted Area of Danger	PAD	POB	Person Overboard
Predicted Point of Collision	PPC	PORT	Port/Portside
Pulse Length	PL	POSN	Position
Pulse Modulation	PM	PPC	Predicted Point of Collision
Pulse Repetition Frequency	PRF	PPR	Pulses Per Revolution
Pulse Repetition Rate	PRR	PRED	Predicted
Pulses Per Revolution	PPR	PRF	Pulse Repetition Frequency
Racon	RACON	PRR	Pulse Repetition Rate
Radar	RADAR	PWR	Power
Radius	RAD	RACON	Racon
Rain	RAIN	RAD	Radius
Range	RNG	RADAR	Radar
Range Rings	RR	RAIM	Receiver Autonomous Integrity Monitoring
Raster Chart Display System	RCDS	RAIN	Anti-clutter Rain
Raster Navigational Chart	RNC	RAIN	Rain
Rate Of Turn	ROT	RCDS	Raster Chart Display System
Real-time Kinematic	RTK	REF	Reference
Receiver	RX (See note 2)	REL (See note 3)	Relative
Receiver Autonomous Integrity Monitoring	RAIM	RIM	Vessel Restricted in Manoeuvrability (applies to AIS)

(Continued)

APPENDIX – LIST OF STANDARD TERMS AND ABBREVIATIONS (Continued)

Term	Abbreviation	Abbreviation	Term
Reference	REF	RM	Relative Motion
Relative	REL (See note 3)	RMS	Root Mean Square
Relative Motion	RM	RNC	Raster Navigational Chart
Revolutions per Minute	RPM	RNG	Range
Roll On/Roll Off Vessel (applies to AIS)	RoRo	RoRo	Roll On/Roll Off Vessel (applies to AIS)
Root Mean Square	RMS	ROT	Rate Of Turn
Route	ROUTE	ROUTE	Route
Safety Contour	SF CNT	RPM	Revolutions per Minute
Sailing Vessel (applies to AIS)	SAIL	RR	Range Rings
Satellite	SAT	RTK	Real-time Kinematic
S-Band (applies to Radar)	S-BAND	RX (See note 2)	Receiver
Scan to Scan	SC/SC	S	South
Search And Rescue Transponder	SART	SAIL	Sailing Vessel (applies to AIS)
Search And Rescue Vessel (applies to AIS)	SARV	SART	Search And Rescue Transponder
Select	SEL	SARV	Search And Rescue Vessel (applies to AIS)
September	SEP	SAT	Satellite
Sequence	SEQ	S-BAND	S-Band (applies to Radar)
Set (i.e. set and drift, or setting a value)	SET	SC/SC	Scan to Scan
Ship's Time	TIME	SDME	Speed and Distance Measuring Equipment
Short Pulse	SP	SEA	Anti-clutter Sea
Signal to Noise Ratio	SNR	SEL	Select
Simulation	SIM(See note 4)	SEP	September
Slave	SLAVE	SEQ	Sequence
South	S	SET	Set (i.e. set and drift, or setting a value)
Speed	SPD	SF CNT	Safety Contour
Speed and Distance Measuring Equipment	SDME	SIM(See note 4)	Simulation

Speed Over the Ground	SOG	SLAVE	Slave
Speed Through the Water	STW	SNR	Signal to Noise Ratio
Stabilized	STAB	SOG	Speed Over the Ground
Standby	STBY	SP	Short Pulse
Starboard/Starboard Side	STBD	SPD	Speed
Station	STN	STAB	Stabilized
Symbol(s)	SYM	STBD	Starboard/Starboard Side
Synchronization	SYNC	STBY	Standby
Target	TGT	STN	Station
Target Tracking	TT	STW	Speed Through the Water
Test	TEST	SYM	Symbol(s)
Time	TIME	SYNC	Synchronization
Time Difference	TD	T	True
Time Dilution Of Precision	TDOP	TCPA	Time to CPA
Time Of Arrival	TOA	TCS	Track Control System
Time Of Departure	TOD	TD	Time Difference
Time to CPA	TCPA	TDOP	Time Dilution Of Precision
Time To Go	TTG	TEST	Test
Time to Wheel-over Line	TWOL	TGT	Target
Track	TRK	THD	Transmitting Heading Device
Track Control System	TCS	TIME	Ship's Time
Track Made Good	TMG (See note 5)	TIME	Time
Trail(s)	TRAIL	TM	True-motion
Transceiver	TXRX (See note 2)	TMG (See note 5)	Track Made Good
Transferred Line Of Position	TPL	TOA	Time Of Arrival
Transmitter	TX (See note 2)	TOD	Time Of Departure
Transmitting Heading Device	THD	TOW	Vessel Engaged in Towing Operations (applies to AIS)
Trial	TRIAL (See note 4)	TPL	Transferred Line Of Position
Trigger Pulse	TRIG	TRAIL	Trail(s)
True	T	TRIAL (See note 4)	Trial
True-motion	TM	TRIG	Trigger Pulse

(Continued)

APPENDIX – LIST OF STANDARD TERMS AND ABBREVIATIONS (Continued)

Term	Abbreviation	Abbreviation	Term
Tune	TUNE	TRK	Track
Ultrahigh Frequency	UHF	TT	Target Tracking
Universal Time, Coordinated	UTC	TTG	Time To Go
Unstabilized	UNSTAB	TUNE	Tune
Variable Range Marker	VRM	TWOL	Time to Wheel-over Line
Variation	VAR	TX(See note 2)	Transmitter
Vector	VECT	TXRX(See note 2)	Transceiver
Very High Frequency	VHF	UHF	Ultrahigh Frequency
Very Low Frequency	VLF	UNSTAB	Unstabilized
Vessel Aground (applies to AIS)	GRND	UTC	Universal Time, Coordinated
Vessel at Anchor (applies to AIS)	ANCH	UWE	Vessel Underway Using Engine (applies to AIS)
Vessel Constrained by Draught (applies to AIS)	VCD	VAR	Variation
Vessel Engaged in Diving Operations (applies to AIS)	DIVE	VCD	Vessel Constrained by Draught (applies to AIS)
Vessel Engaged in Dredging or Underwater Operations (applies to AIS)	DRG	VDR	Voyage Data Recorder
Vessel Engaged in Towing Operations (applies to AIS)	TOW	VECT	Vector
Vessel Not Under Command (applies to AIS)	NUC	VHF	Very High Frequency
Vessel Restricted in Manoeuvrability (applies to AIS)	RIM	VID	Video
Vessel Traffic Service	VTS	VLF	Very Low Frequency
Vessel Underway Using Engine (applies to AIS)	UWE	VOY	Voyage
Video	VID	VRM	Variable Range Marker
Voyage	VOY	VTS	Vessel Traffic Service
Voyage Data Recorder	VDR	W	West
Warning	WARNING	WARNING	Warning
Water	WAT	WAT	Water
Waypoint	WPT	WOL	Wheel-over Line

West	W	WOT	Wheel-over Time
Wheel-over Line	WOL	WPT	Waypoint
Wheel-over Time	WOT	X-BAND	X-Band (applies to Radar)
X-Band (applies to Radar)	X-BAND	XTD	Cross Track Distance

List of Standard Units of Measurement and Abbreviations

Term	Abbreviation	Abbreviation	Term
cable length	cbl	cbl	cable length
cycles per second	cps	cps	cycles per second
degree(s)	deg	deg	degree (s)
fathom(s)	fin	fin	fathom(s)
feet/foot	ft	ft	feet/foot
gigahertz	GHz	GHz	gigahertz
hectopascal	hPa	hPa	hectopascal
hertz	Hz	Hz	hertz
hour(s)	hr(s)	hr(s)	hour(s)
kilohertz	kHz	kHz	kilohertz
kilometre	km	km	kilometre
kilopascal	kPa	kPa	kilopascal
knot(s)	kn	kn	knot(s)
megahertz	MHz	MHz	megahertz
minute(s)	min	min	minute (s)
Nautical Mile(s)	NM	NM	Nautical Mile(s)

Notes:

.1 Terms and abbreviations used in nautical charts are published in relevant IHO publications and are not listed here.

.2 In general, terms should be presented using lower case text and abbreviations should be presented using upper case text. Those abbreviations that may be presented using lower case text are identified in the list, e.g. 'dGNSS' or 'Rx'.

.3 Abbreviations may be combined, e.g. 'CPA LIM' or 'T CRS'. When the abbreviation for the standard term 'Relative' is combined with another abbreviation, the abbreviation 'R' should be used instead of 'REL', e.g. 'R CRS'.

(Continued)

.4 The use of the abbreviations 'SIM' and 'TRIAL' are not intended to replace the appropriate symbols listed in Annex 1.

.5 The term 'Course Made Good' has been used in the past to describe 'Track Made Good'. This is a misnomer in that 'courses' are directions steered or intended to be steered with respect to a reference meridian. 'Track Made Good' is preferred over the use of 'Course Made Good'.

.6 Where information is presented using SI units, the respective abbreviations should be used.

11.2.4 Extract from IMO Resolution A.615(15) Marine Uses of Radar Beacons and Transponders

ANNEX 1

1 INTRODUCTION

1.1 The uncontrolled provision of radar beacons and transponders could cause degradation of ships' navigational radar and ARPA displays, produce incompatibilities among devices developed for different uses, or necessitate a succession of modifications to ships' radar displays to accommodate progressive developments of radar beacons and transponders.

1.2 To avoid these possibilities, the following recommendations are made concerning the appropriate applications for radar beacons and transponders, where an operational requirement for such a device exists, and concerning measures for general administration of radar beacons and transponders.

1.3 The technical criteria and operation of radar beacons and transponders are similar. However, the terms *radar beacon (racon)* and *transponder*, as used in this recommendation, are understood to have the following meanings:

.1 *Radar beacon (racon)*: A receiver-transmitter device associated with a navigational mark which, when triggered by a radar, automatically returns a distinctive signal which can appear on the display of the triggering radar, providing range, bearing and identification information. The terms *radar beacon* and *racon* should be reserved exclusively for this use and include devices mounted on fixed structures, or on floating aids anchored at fixed positions, for navigational purposes. The racon itself is considered a separate aid to navigation, whether used alone, or mounted on another aid to navigation (such as a visible mark).

.2 *Transponder*: A receiver-transmitter device in the maritime radio determination service which transmits automatically when interrogated, or when a transmission is initiated by a local command. The transmission may include a coded identification signal and/or data. The response may be displayed on a radar PPI, or on a display separate from any radar, or both, depending upon the application and content of the signal.

2 GENERAL OPERATIONAL CHARACTERISTICS

2.1 Radar beacons

2.1.1 A radar beacon in the maritime radionavigation service is a device which will:

.1 be triggered automatically by the transmissions of any radar operating in the appropriate frequency band; and
.2 transmit a response immediately on receipt of the triggering pulse for display as part of the normal picture of the triggering radar.

2.1.2 Where a radar beacon incorporates a user-selectable mode, so that a user can control the presentation of a radar beacon response, it will also:

.1 be triggered automatically by the transmission of a suitably configured radar in the vicinity operating in the appropriate frequency band and using its own user-selectable facility; and
.2 transmit a response so that it can:
  .2.1 be shown on the radar display in a manner distinct from that used for radar information;
  .2.2 be shown on the radar display or other display without other information; or
  .2.3 not be shown on the radar display.

2.1.3 In special circumstances, a radar beacon not being used for general navigational purposes may operate exclusively in the user-selectable mode.

2.2 Transponders

A transponder is a device which can provide for:

.1 ship radar target identification and echo enhancement with the proviso that such enhancement should not significantly exceed that which could be achieved by passive means on the radar display of an interrogating ship or shore station;
.2 radar target correlation with voice or other radio transmission for identification on the radar display of an interrogating ship or shore station;
.3 user-selectable presentation of transponder responses either superimposed on the normal radar display, or free of clutter and other targets; and
.4 transfer of information pertinent to avoidance of collision or other hazards, manoeuvre, manoeuvring characteristics, etc.

3 OPERATIONAL USE

3.1 Radar beacons should be used only for radionavigational purposes (not for detection of marine craft), for example:

.1 ranging on and identification of positions on inconspicuous coastlines;
.2 identification of positions on coastlines which permit good ranging but are featureless;
.3 identification of selected navigational marks both seaborne and land-based;
.4 landfall identification;
.5 as a warning device to identify temporary navigational hazards and to mark new and uncharted dangers;
.6 bridge marking;
.7 leading lines;
.8 identification of offshore structures;
.9 marking important features in channels.

3.2 Radar beacons used at locations where clutter from land, sea, ice or weather could mask their response may, at the discretion of the Administration concerned, incorporate a user-selectable mode.

3.3 Where an operational requirement exists for a responding device, other than for radionavigational purposes, a transponder should be used. Examples of requirements suitable for transponders are:

.1 identification of certain classes of ships (ship-to-ship) and towed devices;
.2 identification of ships for VTS and other shore surveillance purposes;
.3 search and rescue operations;
.4 identification of individual ships and data transfer; and
.5 establishing positions for hydrographical purposes.

4 GENERAL ADMINISTRATION OF RADAR BEACONS AND TRANSPONDERS

4.1 All radar beacons should be authorized by an Administration or by a competent navigation authority. Before authorizing or approving the setting up of a radar beacon, account should be taken of the density of such devices in the particular area and the need to prevent degradation of ships' radar displays.

4.2 Except in the case of SAR transponders (see resolution A.530(13)), transponder systems designed to respond in a frequency band used by marine radars should be authorized by an Administration. Before giving such authorization, account should be taken of the effect such transmissions would have on ships' radars.

ANNEX 2

RECOMMENDATION ON OPERATIONAL STANDARDS FOR RADAR BEACONS

1 INTRODUCTION

1.1 Radar beacons should conform to the following minimum operational standards.

1.2 Radar beacons should be operationally compatible with navigational radar and ARPA equipment which conforms to the performance standards recommended by the Organization.

2 OPERATING FREQUENCIES

2.1 Radar beacons designed to operate on a wavelength of 3 cm should be capable of being interrogated by any navigational radar equipment operating on any frequency between 9320 MHz and 9500 MHz and respond within this frequency band.

2.2 Radar beacons designed to operate on a wavelength of 10 cm should be capable of being interrogated by any navigational radar equipment operating on any frequency between 2900 MHz and 3100 MHz and respond within this frequency band.

3 TRANSMITTER TUNING CHARACTERISTICS

The tuning characteristics of the transmitter should be such that the beacon response can appear on a radar display in a recognizable form at least once every 2 minutes.

4 OPERATING RANGE

The operating range should be compatible with the navigational requirements for the radar beacon at its location.

5 RESPONSE CHARACTERISTICS

5.1 On receipt of an interrogating signal, the radar beacon should commence its response in such time that the gap on the radar display between the radar target and the beacon response does not normally exceed approximately 100 m. In certain cases, the operational use of radar beacons may be aided by increasing this delay slightly. Under such circumstances, the delay time should be as short as practicable and the details should be shown in appropriate navigational publications.

5.2 The duration of the response should be approximately 20% of the maximum range requirement of the particular radar beacon, or should not exceed 5 NM, whichever is the lower value. In certain cases, the duration of the response may be adjusted to suit the operational requirements for the particular radar beacon.

5.3 The leading edge of the response should be sufficiently sharp to permit satisfactory range determination.

6 IDENTIFICATION CODING

6.1 Identification coding should normally be in the form of a Morse letter. The identification coding used should be described in appropriate navigational publications.

6.2 The identification coding should comprise the full length of the radar beacon response and, where a Morse letter is used, the response should be divided with a ratio of 1 dash equal to 3 dots and 1 dot equal to 1 space.

6.3 The coding should normally commence with a dash.

7 USER-SELECTABLE MODE

7.1 Radar beacons may be provided with a user-selectable mode. In this mode the radar beacon, in addition to satisfying the response characteristics set out in section 5, should be capable of transmitting a response after receipt of an interrogating signal from a suitably configured radar using its own user-selectable facility.

7.2 The characteristics of the interrogating signal and the response should conform to the appropriate CCIR Recommendations.

7.3 Radar beacons provided with a user-selectable mode should, unless operating exceptionally only in the user-selectable mode, be capable of responding to interrogations from both normal radar signals and special interrogating signals with a minimum of interruption in response to any user.

8 CONSTRUCTION

Radar beacons should be designed to provide high availability when installed permanently in a marine environment.

ANNEX 3

RECOMMENDATION ON TRANSPONDERS

.1 The design of transponder systems should ensure that there is no significant degradation in the use of radar beacons, and the response of a transponder should not be capable of being interpreted as being from a radar beacon of any type.

.2 Where a transponder is to be used with a marine navigational radar or ARPA, any modifications necessary to the radar or ARPA should not degrade its performance; they should be kept to a minimum, be simple and be compatible with a user-selectable beacon facility.

.3 Transponders should not be used to enhance the detection of marine craft, except for search and rescue or when specially authorized by Administrations for safety purposes. Transponders used for search and rescue purposes should be capable of transmitting signals which will appear on a radar or ARPA display as a series of equally spaced dots (resolution A.530(13)).

11.2.5 Extract from IMO Resolution A.802(19) Performance Standards for Survival Craft Radar Transponders for Use in Search and Rescue Operations (As Amended)

1 INTRODUCTION

The 9 GHz SAR transponder (SART), in addition to meeting the requirements of the relevant ITU-R Recommendation and the general requirements set out in resolution A.694(17), should comply with the following performance standards.

2 GENERAL

The SART should be capable of indicating the location of a unit in distress on the assisting unit's radars by means of a series of equally spaced dots (see resolution A.530(13)).

2.1 The SART should:

.1 be capable of being easily activated by unskilled personnel;

.2 be fitted with means to prevent inadvertent activation;

.3 be equipped with a means which is either visual or audible, or both visual and audible, to indicate correct operation and to alert survivors to the fact that a radar has triggered the SART;

.4 be capable of manual activation and deactivation, provision for automatic activation may be included;

Note: If an on-board test is performed using a shipborne 9 GHz radar, activation of the SART should be limited to a few seconds to avoid harmful interference with other shipborne radars and excessive consumption of battery energy.

.5 be provided with an indication of the standby condition;

.6 be capable of withstanding without damage drops from a height of 20 m into water;

.7 be watertight at a depth of 10 m for at least 5 minutes;

.8 maintain watertightness when subjected to a thermal shock of 45°C under specified conditions of immersion;

.9 be capable of floating if it is not an integral part of the survival craft;

.10 be equipped with buoyant lanyard, suitable for use as a tether, if it is capable of floating;

.11 not be unduly affected by seawater or oil;

.12 be resistant to deterioration in prolonged exposure to sunlight;

.13 be of a highly visible yellow/orange colour on all surfaces where this will assist detection;

.14 have a smooth external construction to avoid damaging the survival craft; and

.15 be provided with a pole or other arrangement compatible with the antenna pocket in a survival craft in order to comply with requirements referred to in paragraph 2.4, together with illustrated instructions.

2.2 The SART should have sufficient battery capacity to operate in the standby condition for 96 h and, in addition, following the standby period, to provide transponder transmissions for 8 h when being continuously interrogated with a pulse repetition frequency of 1 kHz.

2.3 The SART should be so designed as to be able to operate under ambient temperatures of −20°C to +55°C. It should not be damaged in stowage throughout the temperature range of −30°C to +65°C.

2.4 The height of the installed SART antenna should be at least 1 m above sea level.

2.5 Horizontal polarization or circular polarization should be used for transmission and reception.

2.6 The SART should operate correctly when interrogated at a distance of up to at least 5 NM by a navigational radar complying with resolutions A.477 (XII) and A.222(V11), with an antenna height of 15 m. It should also operate correctly when interrogated at a distance of up to 30 NM by an airborne radar with at least 10 kW peak output power at a height of 3000 ft.

3 TECHNICAL CHARACTERISTICS

Technical characteristics of the SART should be in accordance with Recommendation ITU-R M.628−2.

4 LABELLING

In addition to the items specified in resolution A.694(17) on general requirements, the following should be clearly indicated on the exterior of the equipment:

.1 brief operating instructions; and
.2 expiry date for the primary battery used.

11.2.6 Extract from IMO Resolution A.384(X) Performance Standards for Radar Reflectors

1 INTRODUCTION

1.1 Small craft referred to in paragraph 2 of this Recommendation should be fitted with radar reflectors to improve the range and probability of their radar detection.

1.2 Radar reflectors should comply with the minimum performance requirements as specified in this Recommendation.

1.3 In the following paragraphs, the echoing areas specified are those for the frequency of 9.3 GHz (corresponding to a wavelength of 3.2 cm).

2 APPLICATION

2.1 All ships of less than 100 tons gross tonnage operating in international waters and adjacent coastal areas should, if practicable, be fitted with a radar reflector.

2.2 The radar reflector should be of an approved type with an adequate polar diagram in azimuth, and an echoing area:

.1 preferably, of at least 10 m^2, mounted at a minimum height of 4 m above water level; or
.2 if this is not practicable, of at least 40 m^2, mounted at a minimum height of 2 m above water level.

3 PERFORMANCE

.1 Reflectors should be capable of performance around 360° in azimuth using a typical marine navigational radar.
.2 The echoing areas referred to in paragraph 2 correspond to the maximum values of the main lobes of the polar diagram.
.3 The azimuthal polar diagram should be such that the response over a total angle of 240° is not less than −6 dB with reference to the maxima of the main lobes and that the response should not remain below −6 dB over any single angle of more than 10°.

4 CONSTRUCTION

The reflector should be capable of maintaining its reflection performance under the conditions of sea states, vibration, humidity and change of temperature likely to be experienced in the marine environment.

5 INSTALLATION

5.1 Fixing arrangements should be provided so that the reflector can be fitted either on a rigid mount or suspended in the rigging.

5.2 If there is a preferred orientation of mounting, this should be clearly marked on the reflector. In the case of an octahedral reflector, the correct method of mounting is one corner cavity at the top and one at the bottom. Any other method might reduce its performance below that in paragraph 3.3.

11.3 IMO PERFORMANCE STANDARDS FOR OTHER RELATED EQUIPMENT

11.3.1 Extract from IMO Resolution MSC.74(69) Annex 3 Performance Standards for a Universal Shipborne Automatic Identification System (AIS)

1 SCOPE

1.1 These performance standards specify the requirements for the universal AIS.

1.2 The AIS should improve the safety of navigation by assisting in the efficient navigation of ships, protection of the environment, and operation of Vessel Traffic Services (VTS), by satisfying the following functional requirements:

.1 in a ship-to-ship mode for collision avoidance;
.2 as a means for littoral States to obtain information about a ship and its cargo; and
.3 as a VTS tool, i.e. ship-to-shore (traffic management).

1.3 The AIS should be capable of providing to ships and to competent authorities, information from the ship, automatically and with the required accuracy and frequency, to facilitate accurate tracking. Transmission of the data should be with the minimum involvement of ship's personnel and with a high level of availability.

1.4 The installation, in addition to meeting the requirements of the Radio Regulations, applicable ITU-R Recommendations and the general requirements as set out in resolution A.694(17), should comply with the following performance standards.

2 FUNCTIONALITY

2.1 The system should be capable of operating in a number of modes:

.1 an 'autonomous and continuous' mode for operation in all areas. This mode should be capable of being switched to/from one of the following alternate modes by a competent authority;

.2 an 'assigned' mode for operation in an area subject to a competent authority responsible for traffic monitoring such that the data transmission interval and/or time slots may be set remotely by that authority; and

.3 a 'polling' or controlled mode where the data transfer occurs in response to interrogation from a ship or competent authority.

3 CAPABILITY

3.1 The AIS should comprise:

.1 a communication processor, capable of operating over a range of maritime frequencies, with an appropriate channel selecting and switching method, in support of both short and long range applications;

.2 a means of processing data from an electronic position-fixing system which provides a resolution of one ten-thousandth of a minute of arc and uses the WGS-84 datum.

.3 a means to automatically input data from other sensors meeting the provisions as specified in paragraph 6.2;

.4 a means to input and retrieve data manually;

.5 a means of error checking the transmitted and received data; and

.6 built-in test equipment (BITE).

3.2 The AIS should be capable of:

.1 providing information automatically and continuously to a competent authority and other ships, without involvement of ship's personnel;

.2 receiving and processing information from other sources, including that from a competent authority and from other ships;

.3 responding to high priority and safety related calls with a minimum of delay; and

.4 providing positional and manoeuvring information at a data rate adequate to facilitate accurate tracking by a competent authority and other ships.

4 USER INTERFACE

To enable a user to access, select and display the information on a separate system, the AIS should be provided with an interface conforming to an appropriate international marine interface standard.

5 IDENTIFICATION

For the purpose of ship and message identification, the appropriate Maritime Mobile Service Identity (MMSI) number should be used.

6 INFORMATION

6.1 The information provided by the AIS should include:

.1 Static:
- **a.** IMO number (where available)
- **b.** Call sign and name
- **c.** Length and beam
- **d.** Type of ship
- **e.** Location of position-fixing antenna on the ship (aft of bow and port or starboard of centreline).

.2 Dynamic:
- **a.** Ship's position with accuracy indication and integrity status
- **b.** Time in UTC (date to be established by receiving equipment)
- **c.** Course over ground
- **d.** Speed over ground
- **e.** Heading
- **f.** Navigational status (e.g. NUC, at anchor, etc. – manual input)
- **g.** Rate of turn (where available)
- **h.** Optional – Angle of heel (where available)
- **i.** Optional – Pitch and roll (where available).

.3 Voyage-related:
- **a.** Ship's draught
- **b.** Hazardous cargo (type) (as required by competent authority)
- **c.** Destination and ETA (at master's discretion)
- **d.** Optional – Route plan (waypoints).

.4 Short safety-related messages.

6.2 Information update rates for autonomous mode

The different information types are valid for a different time period and thus need a different update rate:

- Static information: Every 6 minutes and on request
- Dynamic information: Dependent on speed and course alteration according to Table 1
- Voyage-related information: Every 6 minutes, when data has been amended and on request
- Safety-related message: As required

Ship Reporting Capacity – the system should be able to handle a minimum of 2000 reports per minute to adequately provide for all operational scenarios envisioned.

TABLE 1

Type of ship	Reporting interval
Ship at anchor	3 minutes
Ship 0–14 knots	12 seconds
Ship 0–14 knots and changing course	4 seconds
Ship 14–23 knots	6 seconds
Ship 14–23 knots and changing course	2 seconds
Ship > 23 knots	3 seconds
Ship > 23 knots and changing course	2 seconds

6.3 Security

A security mechanism should be provided to detect disabling and to prevent unauthorized alteration of input or transmitted data. To protect the unauthorized dissemination of data, IMO guidelines (Guidelines and Criteria for Ship Reporting Systems (Resolution MSC.43(64)) should be followed.

7 PERMISSIBLE INITIALIZATION PERIOD

The installation should be operational within 2 minutes of switching on.

8 POWER SUPPLY

The AIS and associated sensors should be powered from the ship's main source of electrical energy. In addition, it should be possible to operate the AIS and associated sensors from an alternative source of electrical energy.

9 TECHNICAL CHARACTERISTICS

The technical characteristics of the AIS such as variable transmitter output power, operating frequencies (dedicated internationally and selected regionally), modulation and antenna system should comply with the appropriate ITU-R Recommendations.

11.3.2 Extract from IMO Resolution MSC.112(73) Performance Standards for Shipborne Global Positioning System (GPS) Receiver Equipment Valid for Equipment Installed on or After 1 July 2003

1 INTRODUCTION

1.1 The Global Positioning System (GPS) is a space-based positioning, velocity and time system that has three major segments: space, control and user. The GPS space segment will normally be composed of 24 satellites in six orbits. The satellites operate in circular 20,200 km orbits at an inclination angle of 55° with a 12-hour period. The spacing of satellites in orbit will be arranged so that a minimum of four satellites will be in view to users worldwide, with a position dilution of precision (PDOP) of ≤ 6. Each satellite transmits on two 'L' band frequencies, L1 (1575.42 MHz) and L2 (1227.6 MHz). L1 carries a precise (P) code and coarse/acquisition (C/A) code. L2 carries the P code. A navigation data message is superimposed on these codes. The same navigation data message is carried on both frequencies.

1.2 Receiver equipment for the GPS intended for navigational purposes on ships with maximum speeds not exceeding 70 knots should, in addition to the general requirements contained in resolution A.694(17), comply with the following minimum performance requirements.

1.3 These standards cover the basic requirements of position-fixing for navigation purposes only and do not cover other computational facilities which may be in the equipment.

2 GPS RECEIVER EQUIPMENT

2.1 The words 'GPS receiver equipment' as used in these performance standards include all the components and units necessary for the system properly to perform its intended functions. The equipment should include the following minimum facilities:

.1 antenna capable of receiving GPS signals;
.2 GPS receiver and processor;
.3 means of accessing the computed latitude/longitude position;
.4 data control and interface; and
.5 position display and, if required, other forms of output.

2.2 The antenna design should be suitable for fitting at a position on the ship which ensures a clear view of the satellite constellation.

3 PERFORMANCE STANDARDS FOR GPS RECEIVER EQUIPMENT

The GPS receiver equipment should:

.1 be capable of receiving and processing the Standard Positioning Service (SPS) signals as modified by Selective Availability (SA) and provide position information in latitude and longitude World Geodetic System (WGS)-84 coordinates in degrees, minutes and thousandths of minutes and time of solution referenced to UTC (USNO). Means may be provided for transforming the computed position based upon WGS-84 into data compatible with the datum of the navigational chart in use. Where this facility exists, the display should indicate that coordinate conversion is being performed, and should identify the coordinate system in which the position is expressed;

.2 operate on the L1 signal and C/A code;

.3 be provided with at least one output from which position information can be supplied to other equipment. The output of position information based upon WGS-84 should be in accordance with international standards (see IEC 61162);

.4 have static accuracy such that the position of the antenna is determined to within 100 m (95%) with horizontal dilution of precision (HDOP) ≤ 4 (or PDOP ≤ 6);

.5 have dynamic accuracy such that the position of the ship is determined to within 100 m (95%) with HDOP ≤ 4 (or PDOP ≤ 6) under the conditions of sea states and ship's motion likely to be experienced in ships (see A694(17));

.6 be capable of selecting automatically the appropriate satellite-transmitted signals for determining the ship's position with the required accuracy and update rate;

.7 be capable of acquiring satellite signals with input signals having carrier levels in the range of −130 dBm to −120 dBm. Once the satellite signals have been acquired, the equipment should continue to operate satisfactorily with satellite signals having carrier levels down to −133 dBm;

.8 be capable of acquiring position to the required accuracy, within 30 minutes, when there is no valid almanac data;

.9 be capable of acquiring position to the required accuracy, within 5 minutes, when there is valid almanac data;

.10 be capable of re-acquiring position to the required accuracy, within 5 minutes, when the GPS signals are interrupted for a period of at least 24 hours but there is no loss of power;

.11 be capable of re-acquiring position to the required accuracy, within 2 minutes, when subjected to a power interruption of 60 seconds;

.12 generate and output to a display and digital interface (see IEC 61162) a new position solution at least once every 1 s (for high speed craft 0.5 s);

.13 have a minimum resolution of position, i.e. latitude and longitude, of 0.001 minutes;

.14 generate and output to the digital interface (see IEC 61162) course over the ground (COG), speed over the ground (SOG) and universal time coordinated (UTC). Such outputs should have a validity mark aligned with that on the position output. The accuracy requirement for COG and SOG should not be inferior to the relevant performance standards for heading and SDME (see A424(XI) and A824(19));

.15 have the facilities to process differential GPS (DGPS) data fed to it in accordance with the standards of Recommendation ITU−R M.823 and the appropriate RTCM standard. When a GPS receiver is equipped with a differential receiver, performance standards for static and dynamic accuracies (paragraphs 3.4 and 3.5 above) should be 10 m (95%); and

.16 be capable of operating satisfactorily in typical interference conditions.

4 PROTECTION

Precautions should be taken to ensure that no permanent damage can result from an accidental short circuit or grounding of the antenna or any of its input or output connections or any of the GPS receiver equipment inputs or outputs for a duration of 5 minutes.

5 FAILURE WARNINGS AND STATUS INDICATIONS

5.1 The equipment should provide an indication of whether the position calculated is likely to be outside the requirements of these performance standards.

5.2 The GPS receiver equipment should provide as a minimum:

.1 an indication within 5 s if either:

.1.1 the specified HDOP has been exceeded; or

.1.2 a new position has not been calculated for more than 1 s (for high-speed craft 0.5 s). Under such conditions, the last known position and the time of the last valid fix, with explicit indication of this state, so that no ambiguity can exist, should be output until normal operation is resumed;

.2 a warning of loss of position:

.3 differential GPS status indication of

.3.1 the receipt of DGPS signals; and

.3.2 whether DGPS corrections are being applied to the indicated ship's position;

.4 DGPS integrity status and alarm; and

.5 DGPS text message display.

11.3.3 Extract from IMO Resolution MSC.232(82) Revised Performance Standards for Electronic Chart Display and Information Systems (ECDIS) Adopted on 5 December 2006

7 DISPLAY OF OTHER NAVIGATIONAL INFORMATION

7.1 Radar information and/or AIS information may be transferred from systems compliant with the relevant standards of the Organization. Other navigational information may be added to the ECDIS display. However, it should not degrade the displayed SENC information and it should be clearly distinguishable from the SENC information.

7.2 It should be possible to remove the radar information, AIS information and other navigational information by single operator action.

7.3 ECDIS and added navigational information should use a common reference system. If this is not the case, an indication should be provided.

7.4 Radar

7.4.1 Transferred radar information may contain a radar image and/or tracked target information.

7.4.2 If the radar image is added to the ECDIS display, the chart and the radar image should match in scale, projection and orientation.

7.4.3 The radar image and the position from the position sensor should both be adjusted automatically for antenna offset from the conning position.

11.3.4 Extracts from IMO Resolution MSC 252(83) Performance Standards for Integrated Navigation Systems (INS)

1 PURPOSE OF INTEGRATED NAVIGATION SYSTEMS

1.1 The purpose of integrated navigation systems (INS) is to enhance the safety of navigation by providing integrated and augmented functions to avoid geographic, traffic and environmental hazards.

1.2 By combining and integrating functions and information the INS provides 'added value' for the operator to plan, monitor and/or control safety of navigation and progress of the ship.

1.3 Integrity monitoring is an intrinsic function of the INS. The INS supports safety of navigation by evaluating inputs from several sources, combining them to provide information giving timely alerts of dangerous situations and system failures and degradation of integrity of this information.

1.4 The INS presents correct, timely and unambiguous information to the users and provides subsystems and subsequent functions within the INS and other connected equipment with this information.

1.5 The INS supports mode and situation awareness.

1.6 The INS aims to ensure that, by taking human factors into consideration, the workload is kept within the capacity of the operator in order to enhance safe and expeditious navigation and to complement the mariner's capabilities, while at the same time compensating for their limitations.

1.7 The INS aims to be demonstrably suitable for the user and the given task in a particular context of use.

1.8 The purpose of the alert management is specified in module C.

2 SCOPE

2.1 Navigational tasks

2.1.1 An INS comprises navigational tasks such as 'Route planning', 'Route monitoring', 'Collision avoidance', 'Navigation control data', 'Navigation status and data display' and 'Alert management', including the respective sources, data and displays which are integrated into one navigation system. These tasks are described in paragraph 7.

2.1.2 An INS is defined as such if workstations provide multifunctional displays integrating at least the following navigational tasks/functions:

- 'Route monitoring';
- 'Collision avoidance';

and may provide manual and/or automatic navigation control functions.

2.1.3 Other mandatory tasks

2.1.3.1 An alert management is a part of the INS. The scope and the requirements of the alert management are specified in module C.

2.1.3.2 The presentation of navigation control data for manual control as specified in paragraph 7.5.2 of these performance standards is part of the INS.

2.1.4 Other navigational tasks/functions may also be integrated in the INS.

2.2 Task stations

2.2.1 The tasks are allocated to, and operated by the operator on, a defined set of multi-functional 'task stations'.

2.2.2 The scope of an INS may differ depending on the number and kind of tasks integrated.

2.2.3 Configuration, use, operation and display of the INS is situation-dependent on:

- shift underway, at anchor, and moored;
- manual and automatic navigation control in different waters;
- planned routine navigation and special manoeuvres.

3 APPLICATION OF THESE PERFORMANCE STANDARDS

3.1 Purpose of these standards

3.1.1 The purpose of these performance standards is to support the proper and safe integration of navigational functions and information.

3.1.2 The purpose is in particular:

- to allow the installation and use of an INS instead of stand-alone navigational equipment onboard ships; and
- to promote safe procedures for the integration process;

both for

- comprehensive integration; and
- partial integration;

of navigational functions, data and equipment.

3.1.3 These standards supplement for INS functional requirements of the individual Performance Standards adopted by the Organization.

3.2 Application to tasks

3.2.1 These performance standards are applicable to systems where functions/equipment of at least the navigational tasks mentioned in paragraph 2.1.2 are combined.

3.2.2 If further tasks are integrated, the requirements of these standards should apply to all additional functions implemented in the INS.

3.3 Modules of these standards

3.3.1 These performance standards are based on a modular concept which should provide for individual configurations and for extensions, if required.

3.3.2 These standards contain four modules:

- Module A for the requirements for the integration of navigational information;
- Module B for the operational/functional requirements for INS based on a task-related structure;
- Module C for the requirements of the Alert management; and
- Module D for the Documentation requirements.

3.4 Application of modules

These performance standards are applicable to all INS as follows:

3.4.1 Modules A, C, D and paragraphs 6, 8 to 13 of module B are applicable for any INS.

3.4.2 Additionally, for each task integrated into the INS, the INS should fulfil both:

- the requirements of the respective tasks as specified in paragraph 7 of module B; and
- the relevant modules of performance standards for stand-alone equipment as specified in Table 1.

TABLE 1

INS Tasks and functions (Para of this standard)	Additionally applicable modules of specific equipment standards for tasks integrated into the INS. The modules are specified in the appendices of these performance standards, if not specified in the equipment standards.
Collision avoidance (7.4)	Radar PS (Res. MSC.192(79)) (Modules specified in Appendix 3)
	Module A: 'Sensor and Detection'
	Module B: 'Operational requirements'
	Module C: 'Design and Technical requirements'.
Route planning (7.2)	ECDIS PS (Res. MSC.232(82)).
Route monitoring (7.3)	Module A: 'Database'
	Module B: 'Operational and functional requirements'.
Track control (7.5.3 and 8.6, 8.7)	Track Control PS Res. MSC.74 (69), Annex 2 (Modules specified in Appendix 4)
	Module B: 'Operational and functional requirements'.

3.5 Acceptance of INS as navigational equipment

3.5.1 These standards may allow for accepting INS to substitute for some carriage requirements of navigational equipment as equivalent to other means under SOLAS regulation V/19. In this case, the INS should comply with:

- these performance standards; and
- for the relevant tasks of these performance standards, with the applicable modules of the equipment performance standards as specified in Table 2.

TABLE 2

Allow for accepting the INS as	INS in compliance with	
	Tasks and functions (Para of this standard)	**Applicable modules of specific equipment standards as specified in the Appendices of the document**
Radar system	Collision avoidance (7.4)	Radar PS (Res. MSC.192(79)) (Modules specified in Appendix 3)
		Module A: 'Sensor and Detection'
		Module B: 'Operational requirements'
		Module C: 'Design and Technical requirements'
ECDIS	Route planning (7.2)	ECDIS PS (Res. MSC.232(82))
	Route monitoring (7.3)	Module A: 'Database'
		Module B: 'Operational and functional requirements'
Heading control system (HCS)	Navigation control data (7.5) or Navigation status and data display (7.7)	Res. A.342, as amended – MSC.64(67), Annex 3
Track control system (TCS)	Navigation control data and track control (7.5.3 and 8.6, 8.7)	Track Control Res. MSC.74(69), Annex 2 (Modules specified in Appendix 4) Module B: 'Operational and functional requirements'
Presentation of AIS data	Collision avoidance (7.4) Navigation control data (7.5)	MSC.74 (69), Annex 3
Echosounding system	Route monitoring (7.3)	MSC.74(69), Annex 4
EPFS	Navigation control data (7.5) or Navigation status and data display (7.7)	GPS Res. A.819(19), as amended, MSC.112(73) or GALILEO, Res. MSC.233(82) or GLONAS, Res. MSC.53(66), as amended MSC.113(73)
SDME	Navigation control data (7.5) or Navigation status and data display (7.7)	Res. MSC.96(72)

3.6 The application of the alert management is specified in module C.

3.7 OTHER RELEVANT STANDARDS

3.7.1 The workstation design, layout and arrangement is not addressed in this performance standards, but in MSC/Circ. 982.

4 DEFINITIONS

For the purpose of these standards, the definitions in Appendix 1 apply.

Module A – Integration of Information

5 REQUIREMENTS FOR INTEGRATION OF NAVIGATIONAL INFORMATION

5.1 Interfacing and data exchange

5.1.1 An INS should combine, process and evaluate data from connected sensors and sources.

5.1.2 The availability, validity and integrity of data exchange within the INS and from connected sensors and sources should be monitored.

5.1.3 A failure of data exchange should not affect any independent functionality.

5.1.4 Interfacing to, from and within the INS should comply with international standards for data exchange and interfacing as appropriate.

5.1.5 The interface(s) should comply with the interface requirements of the alert management as described in Module C of these performance standards.

5.2 Accuracy

5.2.1 INS data should comply with the accuracy and resolution required by applicable performance standards of the Organization.

5.3 Validity, Plausibility, Latency

5.3.1 Validity

5.3.1.1 Data failing validity checks should not be used by the INS for functions dependent on these data, unless for cases where the relevant performance standards specifically allow use of invalid data. There should be no side effects for functions not depending on this data.

5.3.1.2 When data used by the INS for a function becomes invalid, or unavailable, a warning should be given. When data not actually in use by the INS becomes invalid, or unavailable, this should be indicated at least as a caution.

5.3.2 Plausibility

5.3.2.1 Received or derived data that is used or distributed by the INS should be checked for plausible magnitudes of values.

5.3.2.2 Data which has failed the plausibility checks should not be used by the INS and should not affect functions not dependent on these data.

5.3.3 Latency

5.3.3.1 Data latency (timeliness and repetition rate of data) within the INS should not degrade the functionality specified in the relevant performance standards.

5.4 Consistent common reference system (CCRS)

5.4.1 Consistency of data

5.4.1.1 The INS should ensure that the different types of information are distributed to the relevant parts of the system, applying a 'consistent common reference system' for all types of information.

5.4.1.2 Details of the source and the method of processing of such data should be provided for further use within INS.

5.4.1.3 The CCRS should ensure that all parts of the INS are provided with the same type of data from the same source.

5.4.2 Consistent common reference point

5.4.2.1 The INS should use a single consistent common reference point for all spatially related information. For consistency of measured ranges and bearings, the recommended reference location should be the conning position. Alternative reference locations may be used where clearly indicated or distinctively obvious. The selection of an alternative reference point should not affect the integrity monitoring process.

5.4.3 Consistency of thresholds

5.4.3.1 The INS should support the consistency of thresholds for monitoring and alert functions.

5.4.3.2 The INS should ensure by automatic means that consistent thresholds are used by different parts of an INS, where practicable.

5.4.3.3 A caution may be given when thresholds entered by the bridge team differ from thresholds set in other parts of the INS.

5.5 Integrity monitoring

5.5.1 The integrity of data should be monitored and verified automatically before being used, or displayed.

5.5.2 The integrity of information should be verified by comparison of the data derived independently from at least two sensors and/or sources, if available.

5.5.3 The INS should provide manual or automatic means to select the most accurate method of integrity monitoring from the available sensors and/or sources.

5.5.4 A clear indication of the sensors and sources of data selected for integrity monitoring should be provided.

5.5.5 The INS should provide a warning, if integrity verification is not possible or has failed.

5.5.6 Data which fails the integrity monitoring function or data where integrity monitoring is not possible should not be used for automatic control systems/functions.

5.6 Marking of data

5.6.1 The data should be marked with the source and the results of validity, plausibility checks and integrity monitoring to enable subsequent functions to decide whether their input data complies with their requirements or not.

5.7 Selection of sensors and sources

5.7.1 INS should provide two user selectable sensor/source selection modes when multiple sensors/ sources are available; manual sensor/source selection mode and automatic sensor/source selection mode.

5.7.2 In manual sensor/source selection mode it should be possible to select individual sensors/sources for use in the INS. In the case where a more suitable sensor/source is available this should be indicated.

5.7.3 In automatic sensor/source selection mode, the most suitable sensors/sources available should be automatically selected for use in the INS. It should further be possible to manually exclude individual sensors/sources from being automatically selected.

Module B – Task related requirements for Integrated Navigation Systems

6 OPERATIONAL REQUIREMENTS

6.1 The design of the INS should ease the workload of the bridge team and pilot in safely and effectively carrying out the navigation functions incorporated therein.

6.2 The integration should provide all functions, depending on the task for which the INS is used and configured, to facilitate the tasks to be performed by the bridge team and pilot in safely navigating the ship.

6.3 Each part of the INS should comply with all applicable requirements adopted by the Organization, including the requirements of these performance standards.

6.4 When functions of equipment connected to the INS provide facilities in addition to these performance standards, the operation and, as far as is reasonably practicable, the malfunction of such additional facilities should not degrade the performance of the INS below the requirements of these standards.

6.5 The integration of functions of individual equipment into the INS should not degrade the performance below the requirements specified for the individual equipment by the Organization.

6.6 Alerts should be generated and presented according to Module C.

7 TASK AND FUNCTIONAL REQUIREMENTS FOR AN INS

7.1 General

7.1.1 The configuration of the INS should be modular and task-oriented. The navigational tasks of an INS are classified as 'Route planning', 'Route monitoring', 'Collision avoidance', 'Navigation control data', 'Status and data display' and 'Alert management'. Each of these tasks comprises the respective functions and data.

7.1.2 All tasks of an INS should use the same electronic chart data and other navigational databases such as routes, maps, tide information.

7.1.3 If Electronic Navigational Charts (ENCs) are available, they should be used as common data source for INS.

7.1.4 Paragraphs 7.2 to 7.5 and 7.7 apply, if the respective task is integrated into the INS.

7.2 Task 'Route planning'

7.2.1 ECDIS performance standards-related mandatory functions and data.

The INS should provide the route planning functions and data as specified in Modules A and B of the revised ECDIS performance standards (resolution MSC.232(82)).

7.2.2 Procedures for voyage planning

The INS should be capable of supporting procedures for relevant parts of voyage planning, as adopted by the Organization.**

7.2.3 Additional mandatory functions

The INS should provide means for:

- administering the route plan (store and load, import, export, documentation, protection);
- having the route check against hazards based on the planned minimum under keel clearance as specified by the mariner;
- checking of the route plan against manoeuvring limitation, if available in the INS, based on parameters turning radius, rate of turn (ROT), wheel-over and course changing points, speed, time, ETAs;
- drafting and refining the route plan against meteorological information if available in the INS.

7.3 Task 'Route monitoring'

7.3.1 ECDIS performance standards-related mandatory functions and data.

**Resolution A.893(21) on guidelines for voyage planning.

The INS should provide the route monitoring functions and data as specified in Modules A and B of the ECDIS performance standards.

7.3.2 Additional mandatory functions

The INS should provide capability for:

- optionally overlaying radar video data on the chart to indicate navigational objects, restraints and hazards to the own ship in order to allow position monitoring evaluation and object identification;
- determination of deviations between set values and actual values for measured under-keel clearance and initiating an under-keel clearance alarm, if fitted;
- the alphanumeric display of the present values of latitude, longitude, heading, COG, SOG, STW, under-keel clearance, ROT (measured or derived from change of heading);
- AIS reports of AtoNs, and if track control is integrated into the INS;
- it should be possible to include the planned track and to provide, monitor and display the track related and manoeuvring data.

7.3.3 Optional Functions

For navigational purposes, the display of other route-related information on the chart display is permitted, e.g.:

- tracked radar targets and AIS targets
- AIS binary and safety-related messages
- initiation and monitoring of man-over-board and SAR manoeuvres (search and rescue and man-over-board modes)
- Navtex
- tidal and current data
- weather data
- ice data.

7.3.4 Search and rescue mode

7.3.4.1 If available it should be possible to select on the route monitoring display a predefined display mode for a 'search and rescue' situation, that can be accessed upon simple operator command.

7.3.4.2 In the search and rescue mode a superimposed graphical presentation of the datum (geographic point, line, or area used as a reference in search planning), initial most probable area for search, commence search point and search pattern chosen by the operator (expanding square search pattern, sector search pattern or parallel track search pattern), with track spacing defined by him, should be presented.

7.3.5 Man-over-board (MOB) mode

7.3.5.1 If available it should be possible to select on the route monitoring display a predefined display mode for a 'man-over-board' situation, that can be accessed upon simple operator command.

7.3.5.2 In the man-over-board mode a superimposed graphical presentation of an operator selectable man-over-board manoeuvre should be presented.

7.3.5.3 The man-over-board position should be memorised by a simple operator action.

7.3.5.4 An urgency manoeuvring procedure should be available at the display, taking set and drift into consideration.

7.4 Task 'Collision Avoidance'

7.4.1 Radar performance standards related mandatory functions and data.

The INS should provide the collision avoidance functions and data as specified in Modules A and B of the Radar performance standards.

7.4.2 Additional mandatory functions

7.4.2.1 It should be possible to present less information of ENC database objects than specified in MSC.232(82) for display base.

7.4.2.2 Target association and target data integration.

If target information from multiple sensors/sources (radar and AIS; 2 radar sensors) are provided on one task station:

- the possibility of target association should be provided for mutual monitoring and to avoid the presentation of more than one symbol for the same target;
- the association of AIS and radar targets should follow the requirements of resolutions MSC.192 (79) and MSC.191(79);
- common criteria should be used for raising target related alerts, e.g., CPA/TCPA.

7.4.2.3 Target identifier

For identical targets unique and identical target identifiers should be used for presentation on all INS displays.

Where a target from more than one source can be presented on one display the identifier should be amended as required. Amended target identifiers should be used for all INS display presentations.

7.4.2.4 Combined radar signals

A display may present combined radar signals from more than one radar source. The malfunctions of this additional facility should not degrade the presentation of the radar source selected as primary.

The primary and the other source(s) should be indicated as such.

7.4.3 Optional functions

Optionally, the following information may be displayed:

- true scaled ship symbols and CPA/TCPA and bow crossing range (BCR)/bow crossing time (BCT) related to the real dimensions;
- chart data from the common database of INS: traffic-related object layers.

7.5 Task 'Navigation Control Data'

7.5.1 General

To support the manual and automatic control of the ship's primary movement, the INS navigation control task should provide the following functionality:

- display of data for the manual control of the ship's primary movement;
- display of data for the automatic control of the ship's primary movement;
- presentation and handling of external safety related messages.

7.5.2 Presentation of navigation control data for manual control

7.5.2.1 For manual control of the ship's primary movement the INS navigation control display should allow at least to display the following information:

- under keel clearance (UKC) and UKC profile
- STW, SOG, COG
- position
- heading, ROT (measured or derived from change of heading)
- rudder angle
- propulsion data
- set and drift, wind direction and speed (true and/or relative selectable by the operator), if available
- the active mode of steering or speed control
- time and distance to wheel-over or to the next waypoint
- safety related messages e.g., AIS safety-related and binary messages, Navtex.

7.5.3 Presentation of navigation control data for automatic control

7.5.3.1 For automatic control of the ship's primary movement, the INS navigation control display should allow at least and as default the display of the following information:

- all information listed for manual control;
- set and actual radius or rate of turn to the next segment.

7.5.4 The navigation control data should be presented:

- in digital and where appropriate in analogue form, e.g. mimic elements, logically arranged on and around a symbolic outline of a ship,
- if applicable, together with their 'set-values',
- if applicable and on demand together with a history presentation to indicate the trend of the parameter.

7.6 Task 'Alert management'

7.6.1 Scope, operational requirements and alert-related requirements are specified in Module C of these performance standards.

7.7 Task 'Status and data display'

7.7.1 Mandatory data display functions

The INS should provide the following data display functions:

- presentation of mode and status information
- presentation of the ship's static, dynamic and voyage-related AIS data
- presentation of the ship's available relevant measured motion data together with their 'set-values'
- presentation of received safety related messages, such as AIS safety-related and binary messages, Navtex
- presentation of INS configuration
- presentation of sensor and source information.

7.7.2 Mandatory data management functions

The INS should provide the following management functions:

- setting of relevant parameters;
- editing AIS own ship's data and information to be transmitted by AIS messages.

7.7.3 Optional data display functions

The INS may provide on demand:

- tidal and current data;
- weather data, ice data;
- additional data of the tasks Navigation control and Route monitoring and AIS target data.

APPENDIX – DEFINITIONS

Added Value	The functionality and information, which are provided by the INS, in addition to the requirements of the performance standard for the individual equipment.
Alarm	An alarm is the highest priority of an alert. Condition requiring immediate attention and action by the bridge team, to maintain the safe navigation of the ship.
Alert	Alerts are announcing abnormal situations and conditions requiring attention. Alerts are divided in three priorities: alarms, warnings and cautions.
Alert announcements	Visual and acoustical presentation of alerts.
Alert history list	Accessible list of past alerts.
Alert management	Concept for the harmonized regulation of the monitoring, handling, distribution and presentation of alerts on the bridge.
Automatic control functions	Functions that include automatic heading, and/or track and/or speed control or other navigation related automatic control functions.
Category A alerts	Alerts where graphical information at the task station directly assigned to the function generating the alert is necessary, as decision support for the evaluation the alert related condition.
Category B alerts	Alerts where no additional information for decision support is necessary besides the information which can be presented at the central alert management HMI.
Caution	Lowest priority of an alert. Awareness of a condition which does not warrant a alarm or warning condition, but still requires attention out of the ordinary consideration of the situation or of given information.
Collision avoidance	The navigational task of detecting and plotting other ships and objects to avoid collisions.
Consistent common reference system (CCRS)	A sub-system or function of an INS for acquisition, processing, storage, surveillance and distribution of data and information providing identical and obligatory reference to sub-systems and subsequent functions within an INS and to other connected equipment, if available.
Consistent common reference point (CCRP)	The Consistent Common Reference Point (CCRP) is a location on own ship, to which all horizontal measurements such as target range, bearing, relative course, relative speed, closest point of approach (CPA) or time to closest point of approach (TCPA) are referenced, typically the conning position of the bridge.
Degraded condition	Reduction in system functionality resulting from failure.
Essential functions	Indispensable functions to be available as required for the relevant operational use.
Essential information	Indispensable information to be available as required for the relevant functions.
External safety related messages	Data received from outside of the ship concerning the safety of navigation, through equipment listed in SOLAS chapter V and/or NAVTEX.
Failure analysis	The logical, systematic examination of an item, including its diagrams or formulas, to identify and analyse the probability, causes and consequences of potential and real failures.
Human factor	Workload, capabilities and limits of a user trained according to the regulations of the Organization.

Human machine interface (HMI)	The part of a system an operator interacts with. The interface is the aggregate of means by which the users interact with a machine, device, and system (the system). The interface provides means for input, allowing the users to control the system and output, allowing the system to inform the users.
Indication	Display of regular information and conditions, not part of alert management.
Integrated navigation system	An INS is a composite navigation system which performs at least the following tasks: collision avoidance, route monitoring thus providing "added value" for the operator to plan, monitor and safely navigate the progress of the ship. The INS allows meeting the respective parts of SOLAS regulation V/19 and supports the proper application of SOLAS regulation V/15.
Integrity	Ability of the INS to provide the user with information within the specified accuracy in a timely, complete and unambiguous manner, and alerts within a specified time when the system should be used with caution or not at all.
Partial integrations	Smaller integrations which are not covering the tasks "route monitoring" and "collision avoidance".
Man-over-board mode (MOB)	Display mode for operations and actions of a ship after a Man-over-board accident happened (release of safety equipment, e.g., life buoy and life belt, performance of a return manoeuvre etc.).
Multifunction display	A single visual display unit that can present, either simultaneously or through a series of selectable pages, information from more than a single function of an INS.
Mode awareness	The perception of the mariner regarding the currently active Modes of Control, Operation and Display of the INS including its subsystems, as supported by the presentations and indications at an INS display or workstation.
Navigation control data	Task that provides information for the manual and automatic control of the ship's movement on a task station.
One equipment concept	The equipment which is recognized as one type of equipment by integrating the function of mandatory equipment of SOLAS of a plural number.
Operational modes	Modes of operation depending on the sea area.
Operational/functional modules	Modules comprising the operational/functional requirements for navigational systems.
Plausibility of data	The quality representing, if data values are within the normal range for the respective type of data.
Route monitoring	The navigational task of continuous surveillance of own ships position in relation to the pre-planned route and the waters.
Safety related automatic functions	Automatic functions that directly impinge on hazards to ship or personnel, e.g., target tracking.
Search and rescue mode	Display mode for operations of a ship involved in search and rescue actions.
Sensor	A navigational aid (measuring device), with or without its own display, processing and control as appropriate, automatically providing information to operational systems or INS.
Sensor/source modules	Modules comprising the senor/source requirements.
Ship's primary movement	The longitudinal directional, lateral directional and heading-rotational movement of the ship.

(*Continued*)

APPENDIX – DEFINITIONS (Continued)

Simple operator action	A procedure achieved by no more than two hard-key or soft-key actions, excluding any necessary cursor movements, or voice actuation using programmed codes.
Single operator action	A procedure achieved by no more than one hard-key or soft-key action, excluding any necessary cursor movements, or voice actuation using programmed codes.
Situation awareness	Situation awareness is the mariner's perception of the navigational and technical information provided, the comprehension of their meaning and the projection of their status in the near future, as required for timely reaction to the situation. Situation awareness includes mode awareness.
Source	A device, or location of generated data or information (e.g. chart database), which is part of the INS automatically providing information to INS.
System alerts	Alerts related to equipment failure or loss (system failures).
System integrator	The organization responsible for ensuring that the INS complies with the requirements of this standard.
System position'	Position calculated in the INS out of at least two positioning sensors.
Task station	Multifunction display with dedicated controls providing the possibility to display and operate any navigational tasks. A task station is part of a workstation.
Track	Path to be followed over ground.
Track control	Control of the ship movement along a track.
Warning	Condition requiring no-immediate attention or action by the bridge team. Warnings are presented for precautionary reasons to make the bridge team aware of changed conditions which are not immediately hazardous, but may become so, if no action is taken.
Watchdog	System which monitors the software and Hardware well running at regular intervals.
Workstation	The combination of all job-related items, including the console with all devices, equipment and the furniture, to fulfil certain tasks. Workstations for the Bridge are specified in MSC/Circ.982.

11.3.5 Extract from IMO Resolution A.861(20) Performance Standards for Shipborne Voyage Data Recorders (VDRs) (As Amended)

1 PURPOSE

The purpose of a voyage data recorder (VDR) is to maintain a store, in a secure and retrievable form, of information concerning the position, movement, physical status, command and control of a vessel over the period leading up to and following an incident having an impact thereon. Information contained in a VDR should be made available to both the Administration and the shipowner. This information is for use during any subsequent investigation to identify the cause(s) of the incident.

2 APPLICATION

A VDR with capabilities not inferior to those defined in these performance standards is required to be fitted to ships of classes defined in SOLAS chapter V, as amended.

3 REFERENCES

3.1 SOLAS:

- 1995 SOLAS Conference, resolution 12.

3.2 IMO resolutions:

- A.662(16) Performance standards for float-free release and activation arrangements for emergency radio equipment

- A.694(17) General requirements for shipborne radio equipment forming part of the GMDSS and for electronic navigational aids
- A.824(19) Performance standards for devices to indicate speed and distance
- A.830(19) Code on Alarms and Indicators, 1995
- MSC.64(67), annex 3 Performance standards for heading control systems
- MSC.64(67), annex 4 Performance standards for radar equipment, as amended

4 DEFINITIONS

4.1 *Voyage data recorder (VDR)* means a complete system, including any items required to interface with the sources of input data, for processing and encoding the data, the final recording medium in its capsule, the power supply and dedicated reserve power source.

4.2 *Sensor* means any unit external to the VDR, to which the VDR is connected and from which it obtains data to be recorded.

4.3 *Final recording medium* means the item of hardware on which the data is recorded such that access to it would enable the data to be recovered and played back by use of suitable equipment.

4.4 *Playback equipment* means the equipment, compatible with the recording medium and the format used during recording, employed for recovering the data. It includes also the display or presentation hardware and software that is appropriate to the original data source equipment.

Note: Playback equipment is not normally installed on a ship and is not regarded as part of a VDR for the purposes of these performance standards.

4.5 *Dedicated reserve power source* means a secondary battery, with suitable automatic charging arrangements, dedicated solely to the VDR, of sufficient capacity to operate it as required by 5.3.2.

5 OPERATIONAL REQUIREMENTS

5.1 General

5.1.1 The VDR should continuously maintain sequential records of preselected data items relating to the status and output of the ship's equipment, and command and control of the ship, referred to in 5.4.

5.1.2 To permit subsequent analysis of factors surrounding an incident, the method of recording should ensure that the various data items can be co-related in date and time during playback on suitable equipment.

5.1.3 The final recording medium should be installed in a protective capsule which should meet all of the following requirements:

.1 be capable of being accessed following an incident but secure against tampering;
.2 maximize the probability of survival and recovery of the final recorded data after any incident;
.3 be of a highly visible colour and marked with retro-reflective materials; and
.4 be fitted with an appropriate device to aid location.

5.1.4 The design and construction, which should be in accordance with the requirements of resolution A.694(17) and international standards acceptable to the Organization (IEC 945), should take special account of the requirements for data security and continuity of operation as detailed in 5.2 and 5.3.

5.2 Data selection and security

5.2.1 The minimum selections of data items to be recorded by the VDR are specified in 5.4. Optionally, additional items may be recorded provided that the requirements for the recording and storage of the specified selections are not compromised.

5.2.2 The equipment should be so designed that, as far as is practical, it is not possible to tamper with the selection of data being input to the equipment, the data itself nor that which has already been recorded. Any attempt to interfere with the integrity of the data or the recording should be recorded.

5.2.3 The recording method should be such that each item of the recorded data is checked for integrity and an alarm given if a non-correctable error is detected.

5.3 Continuity of operation

5.3.1 To ensure that the VDR continues to record events during an incident, it should be capable of operating from the ship's emergency source of electrical power.

5.3.2 If the ship's emergency source of electrical power supply fails, the VDR should continue to record bridge audio (see 5.4.5) from a dedicated reserve source of power for a period of 2 h. At the end of this 2 h period all recording should cease automatically.

5.3.3 Recording should be continuous unless interrupted briefly in accordance with 6 or terminated in accordance with 5.3.2. The time for which all stored data items are retained should be at least 12 h. Data items which are older than this may be overwritten with new data.

5.4 Data items to be recorded

5.4.1 Date and time

Date and time, referenced to UTC, should be obtained from a source external to the ship or from an internal clock. The recording should indicate which source is in use. The recording method should be such that the timing of all other recorded data items can be derived on playback with a resolution sufficient to reconstruct the history of the incident in detail.

5.4.2 Ship's position

Latitude and longitude, and the datum used, should be derived from an electronic position-fixing system (EPFS). The recording should ensure that the identity and status of the EPFS can always be determined on playback.

5.4.3 Speed

Speed through the water or speed over the ground, including an indication of which it is, derived from the ship's speed and distance-measuring equipment.

5.4.4 Heading

As indicated by the ship's compass.

5.4.5 Bridge audio

One or more microphones positioned on the bridge should be placed so that conversations at or near the conning stations, radar displays, chart tables, etc. are adequately recorded. As far as practicable, the positioning of microphones should also capture intercom, public address systems and audible alarms on the bridge.

5.4.6 Communications audio

VHF communications relating to ship operations should be recorded.

5.4.7 Radar data, post-display selection

This should include electronic signal information from within one of the ship's radar installations which records all the information which was actually being presented on the master display of that radar at the time of recording. This should include any range rings or markers, bearing markers, electronic plotting symbols, radar maps, whatever parts of the SENC or other electronic chart or map that were selected, the voyage plan, navigational data, navigational alarms and the radar status data that were visible on the display. The recording method should be such that, on playback, it is possible to present a faithful replica of the entire radar display that was on view at the time of recording, albeit within the limitations of any bandwidth-compression techniques that are essential to the working of the VDR.

5.4.8 Echo-sounder

This should include depth under keel, the depth scale currently being displayed and other status information where available.

5.4.9 Main alarms

This should include the status of all mandatory alarms on the bridge.

5.4.10 Rudder order and response

This should include status and settings of autopilot if fitted.

5.4.11 Engine order and response

This should include the positions of any engine telegraphs or direct engine/propeller controls and feedback indications, if fitted, including ahead/astern indicators. This should also include status of bow thrusters if fitted.

5.4.12 Hull openings status

This should include all mandatory status information required to be displayed on the bridge.

5.4.13 Watertight and fire door status

This should include all mandatory status information required to be displayed on the bridge.

5.4.14 Accelerations and hull stresses

Where a ship is fitted with hull stress and response monitoring equipment, all the data items that have been pre-selected within that equipment should be recorded.

5.4.15 Wind speed and direction

This should be applicable where a ship is fitted with a suitable sensor. Either relative or true wind speed and direction may be recorded, but an indication of which it is should be recorded.

6 OPERATION

The unit should be entirely automatic in normal operation. Means should be provided whereby recorded data may be saved by an appropriate method following an incident, with minimal interruption to the recording process.

7 INTERFACING

Interfacing to the various sensors required should be in accordance with the relevant international interface standard, where possible. Any connection to any item of the ship's equipment should be such that the operation of that equipment suffers no deterioration, even if the VDR system develops faults.

8 DOWNLOAD AND PLAYBACK EQUIPMENT FOR INVESTIGATION AUTHORITIES

8.1 Data output interface

The VDR should provide an interface for downloading the stored data and playback the information to an external computer. The interface should be compatible with an internationally recognized format, such as Ethernet, USB, FireWire, or equivalent.

8.2 Software for data downloading and playback

8.2.1 A copy of the software programme providing the capability to download the stored data and playback the information onto a connected external laptop computer and for the playback of the data should be provided for each VDR installation.

8.2.2 The software should be compatible with an operating system available with commercial-off-the-shelf laptop computers and provided on a portable storage device such as a CD-ROM, DVD, USB-memory stick, etc.

8.2.3 Instructions for executing the software and for connecting the external laptop computer to the VDR should be provided.

8.2.4 The portable storage device containing the software, the instructions and any special (not commercial-off-the-shelf) parts necessary for the physical connection of the external laptop computer, should be stored within the main unit of the VDR.

8.2.5 Where non-standard or proprietary formats are used for storing the data in the VDR, the software for converting the stored data into open industry standard formats should be provided on the portable storage device or resident in the VDR.

11.3.6 Extract from IMO Resolution A.694(17) General Requirements for Shipborne Radio Equipment Forming Part of the Global Maritime Distress and Safety System (GMDSS) and for Electronic Navigational Aids

1 INTRODUCTION

1.1 Equipment, which:

.1 forms part of the global maritime distress and safety system; or

.2 is required by regulation V/12 of the 1974 SOLAS Convention as amended and other electronic navigational aids, where appropriate;

should comply with the following general requirements and with all applicable performance standards adopted by the Organization.

1.2 Where a unit of equipment provides a facility which is additional to the minimum requirements of this Recommendation, the operation and, as far as is reasonably practicable, the malfunction of such additional facility should not degrade the performance of the equipment specified in 1.1.

2 INSTALLATION

Equipment should be installed in such a manner that it is capable of meeting the requirements of 1.1.

3 OPERATION

3.1 The number of operational controls, their design and manner of function, location, arrangement and size should provide for simple, quick and effective operation. The controls should be arranged in a manner which minimizes the chance of inadvertent operation.

3.2 All operational controls should permit normal adjustments to be easily performed and should be easy to identify from the position at which the equipment is normally operated. Controls not required for normal operation should not be readily accessible.

3.3 Adequate illumination should be provided in the equipment or in the ship to enable identification of controls and facilitate reading of indicators at all times. Means should be provided for dimming the output of any equipment light source which is capable of interfering with navigation.

3.4 The design of the equipment should be such that misuse of the controls should not cause damage to the equipment or injury to personnel.

3.5 If a unit of equipment is connected to one or more other units of equipment, the performance of each should be maintained.

3.6 Where a digital input panel with the digits 0 to 9 is provided, the digits should be arranged to conform with relevant CCITT recommendations (CCITT E161/Q11). However, where an alphanumeric keyboard layout, as used on office machinery and data processing equipment, is provided, the digits 0 to 9 may, alternatively, be arranged to conform with the relevant ISO standard (ISO 3791).

4 POWER SUPPLY

4.1 Equipment should continue to operate in accordance with the requirements of this Recommendation in the presence of variations of power supply normally to be expected in a ship.

4.2 Means should be incorporated for the protection of equipment from the effects of excessive current and voltage, transient and accidental reversal of the power supply polarity.

4.3 If provision is made for operating equipment from more than one source of electrical energy, arrangements for rapidly changing from one source to the other should be provided but not necessarily incorporated in the equipment.

5 DURABILITY AND RESISTANCE TO ENVIRONMENTAL CONDITIONS

Equipment should be capable of continuous operation under the conditions of various sea states, ship's motion, vibration, humidity and temperature likely to be experienced in ships (IEC 92–101 and 945).

6 INTERFERENCE

6.1 All reasonable and practicable steps should be taken to ensure electromagnetic compatibility between the equipment concerned and other radio-communication and navigational equipment carried on board in compliance with the relevant requirements of chapter IV and chapter V of the 1974 SOLAS Convention (IEC 533 and 945).

6.2 Mechanical noise from all units should be limited so as not to prejudice the hearing of sounds on which the safety of the ship might depend.

6.3 Each unit of equipment normally to be installed in the vicinity of a standard compass or a magnetic steering compass should be clearly marked with the minimum safe distance at which it may be mounted from such compasses.

7 SAFETY PRECAUTIONS

7.1 As far as is practicable, accidental access to dangerous voltages should be prevented. All parts and wiring in which the direct or alternating voltages or both (other than radio frequency voltages) combine to give a peak voltage greater than 55 V should be protected against accidental access and should be isolated automatically from all sources of electrical energy when the protective covers are removed. Alternatively, the equipment should be so constructed that access to such voltages may only be gained after having used a tool for this purpose, such as spanner or screwdriver, and warning labels should be prominently displayed both within the equipment and on protective covers.

7.2 Means should be provided for earthing exposed metallic parts of the equipment, but this should not cause any terminal of the source of electrical energy to be earthed.

7.3 All steps should be taken to ensure that electromagnetic radio frequency energy radiated from the equipment shall not be a hazard to personnel.

7.4 Equipment containing elements such as vacuum tubes which are likely to cause X-radiation should comply with the following requirements:

.1 external X-radiation from the equipment in its normal working condition should not exceed the limits laid down by the Administration concerned;

.2 when X-radiation can be generated inside the equipment above the levels laid down by the Administration, a prominent warning should be fixed inside the equipment and the precautions to be taken when working on the equipment should be included in the equipment manual; and

.3 if malfunction of any part of the equipment can cause an increase in X-radiation, adequate advice should be included in the information about the equipment, warning of the circumstances which could cause the increase and stating the precautions which should be taken.

8 MAINTENANCE

8.1 The equipment should be so designed that the main units can be replaced readily, without elaborate recalibration or readjustment.

8.2 Equipment should be so constructed and installed that it is readily accessible for inspection and maintenance purposes.

8.3 Adequate information should be provided to enable the equipment to be properly operated and maintained. The information should:

.1 in the case of equipment so designed that fault diagnosis and repair down to component level are practicable, provide full circuit diagrams, component layouts and a component parts list; and

.2 in the case of equipment containing complex modules in which fault diagnosis and repair down to component level are not practicable, contain sufficient information to enable a defective complex module to be located, identified and replaced. Other modules and those discrete components which do not form part of modules should also meet the requirements of (1) above.

9 MARKING AND IDENTIFICATION

Each unit of the equipment should be marked externally with the following information which should be clearly visible in the normal installation position:

.1 identification of the manufacturer;

.2 equipment type number or model identification under which it was type tested; and

.3 serial number of the unit.

11.4 EXTRACTS FROM UK STATUTORY INSTRUMENT 1993 NO. 69, THE MERCHANT SHIPPING (NAVIGATIONAL EQUIPMENT) REGULATIONS 1993

11.4.1 Extract from Part IV, Radar Installation

Radar performance standards and interswitching facilities

19.–

.1 Every radar installation required to be provided shall comply with the performance standard adopted by the Organization and shall, in the case of a United Kingdom ship, comply with the relevant performance standard.

.2 (a) Where such a radar installation includes additional radar units and facilities for interswitching, at least one arrangement of units when used together shall comply with all the requirements of this Part of these Regulations;

(b) where two radar installations are required to be provided on a ship, they shall be so installed that each radar installation can be operated individually and both can be operated simultaneously without being dependent upon one another.

Provision of plotting facilities

20. Facilities for plotting radar readings shall be provided on the navigating bridge of every ship required to be fitted with a radar installation. In ships of 1600 tons gross tonnage and upwards constructed on or after 1st September 1984 the plotting facilities shall be at least as effective as a reflection plotter.

Radar watch

21. While a United Kingdom ship which is required to be fitted with a radar installation is at sea and a radar watch is being kept, the radar installation shall be under the control of a qualified radar observer who may be assisted by unqualified personnel.

Performance of radar installations

22. The performance of the radar installation shall be checked before the ship proceeds to sea and at least once every four hours whilst the ship is at sea and radar watch is being maintained.

Qualifications of radar observers

23. For the purposes of these Regulations, a person is a 'qualified radar observer' if he holds:

(a) a valid Radar Observer's Certificate granted by the Secretary of State; or
(b) a valid certificate of attendance granted at the conclusion of a radar simulator course which has been approved by the Secretary of State; or
(c) a valid Electronic Navigation Systems Certificate granted by the Secretary of State; or
(d) a valid Navigation Control Certificate granted by the Secretary of State; or
(e) a certificate recognized by the Secretary of State as being equivalent to any of the certificates mentioned in (a), (b), (c) or (d).

Siting of radar installation

24.—

.1 The antenna unit of the radar installation shall be sited so that satisfactory overall performance is achieved in relation to:

(a) the avoidance of shadow sectors;
(b) the avoidance of false echoes caused by reflections from the ship's structure; and
(c) the effect of antenna height on the amplitude and extent of sea-clutter.

.2 The radar display shall be sited on the bridge from which the ship is normally navigated. The siting of one of the displays shall be such that:

(a) an observer, when viewing the display, faces forward and is readily able to maintain visual lookout;
(b) there is sufficient space for two observers to view the display simultaneously.

Alignment of heading marker

25. The radar heading marker (and stern marker if fitted) shall be aligned to within 1° of the ship's fore-and-aft line as soon as practicable after the radar installation has been installed in the ship. Where inter-switching facilities are provided, the heading marker shall be aligned with all arrangements of units. The marker shall be realigned as soon as practicable whenever it is found to be substantially inaccurate.

Measurement of shadow sectors

26. The angular width and bearing of any shadow sectors displayed by the radar installation shall be determined and recorded. The record shall be shown on a diagram adjacent to the radar display and be kept up-to-date following any change likely to affect shadow sectors.

Display sizes

27. A radar installation required to be provided which is or was installed onboard a ship on or after 1st September 1984 shall provide a relative plan display having an effective diameter, without external magnification, of not less than:

(a) 180 millimetres on ships of 500 tons or over but less than 1600 tons;
(b) 250 millimetres on ships of 1600 tons or over but less than 10,000 tons;
(c) 340 millimetres in the case of one radar installation and 250 millimetres in the case of the other on ships of 10,000 tons or over.

11.4.2 Extract from Part IX, Automatic Radar Plotting Aid Installation

Automatic radar plotting aid performance standards

39. Every automatic radar plotting aid installation required to be provided shall comply with the performance standard adopted by the Organization and shall, in the case of a United Kingdom ship, comply with the relevant performance standard.

Siting and other requirements of automatic radar plotting aid installations

40.—

.1 Where the automatic radar plotting aid installation is provided as an additional unit to a radar installation it shall be sited as close as is practicable to the display of the radar with which it is associated.

.2 Where the automatic radar plotting aid installation forms an integral part of a complete radar system that radar system shall be regarded as one of the radar installations required by regulation 3(4)(b) and accordingly shall comply with the relevant requirements of Part IV of these Regulations.

.3 The automatic radar plotting aid installation shall be interconnected with such other installations as is necessary to provide heading and speed information to the automatic radar plotting aid.

Use of an automatic radar plotting aid to assist in the radar watch

41. When at any time on or after the coming into force of these Regulations, a United Kingdom ship required to be fitted with an automatic radar plotting aid is at sea and a radar watch is being kept on the automatic radar plotting aid, the installation shall be under the control of a person qualified in the operational use of automatic radar plotting aids, who may be assisted by unqualified personnel.

Qualifications of observers using an automatic radar plotting aid to assist in keeping a radar watch

42. For the purpose of regulation 41 of these Regulations, a person shall be qualified in the operational use of automatic radar plotting aids if he holds:

(a) a valid Electronic Navigation Systems Certificate granted by the Secretary of State; or

(b) a valid Navigation Control Certificate granted by the Secretary of State; or

(c) a valid Automatic Radar Plotting Aids Certificate granted by the Secretary of State; or

(d) a certificate recognized by the Secretary of State as being equivalent to any of the certificates mentioned in (a), (b) or (c).

Glossary of Acronyms and Abbreviations

AEB	area exclusion boundary
AFC	automatic frequency control
AIS	automatic identification system
ARB	area rejection boundary
ARPA	automatic radar plotting aid
ASCII	American Standard Code for Information Interchange
ATA	auto-tracking aid
BCR	bow crossing range
BCT	bow crossing time
C/A	coarse/acquisition (code)
CCITT	Comite Consultatif International Telephonique et Telegrahique (now known as ITU)
CFAR	constant false alarm rate
COG	course over ground
CPA	closest point of approach
CRT	cathode ray tube
CSR	controlled silicon rectifier
dB	decibel
DGPS	differential global positioning system
DOP	dilution of precision
DR	dead reckoning
DSC	digital scan converter (conversion)
EBI	electronic bearing indicator
EBL	electronic bearing line
EBM	electronic bearing marker
ECDIS	electronic chart display and information system
EHT	extra high tension (very high voltages)
EIA	Electronic Industry Association
ENC	electronic navigational chart
EPA	electronic plotting aid
EPFS	electronic position-fixing system
ERBL	electronic range and bearing line
ETA	estimated time of arrival
FTC	fast time constant (rain clutter control)
FTE	false target elimination
GHz	gigahertz
GIS	geographical information system
GLONASS	Russian satellite navigation system
GMDSS	global maritime distress and safety system
GNSS	global navigation satellite system
GPS	global positioning system
GRP	glass reinforced plastic (fibreglass)
HBW	horizontal beamwidth
HDOP	horizontal dilution of precision
HM	heading marker
HSC	high-speed craft
Hz	hertz
IBS	integrated bridge system
IEC	International Electrotechnical Commission
IF	intermediate frequency
IHO	International Hydrographic Organization
IKBS	intelligent knowledge-based system
IMO	International Maritime Organization
INS	integrated navigation system

ISO	International Organization for Standardization
ITOFAR	interrogated time-offset frequency-agile racon
ITU	International Telecommunication Union
kHz	kilohertz
kn	knot(s)
kW	kilowatt
LCD	liquid crystal display
LORAN	long range radio navigation (system)
LRIT	long range identification and tracking
LSB	least significant bit
LSR	least sampling rate
m	metres
mm	millimetres
M	military (code)
MCS	master control station
MHz	megahertz
MKD	minimum keyboard display
MMSI	maritime mobile service identity
ms	millisecond
MSB	most significant bit
MSC	Maritime Safety Committee (of IMO)
MTBF	mean time between failures
NM	nautical mile(s)
NMEA	National Marine Electronics Association
NUC	not under command
OAW	plotting triangle symbols
OOW	officer of the watch
P	precision (code)
PAD	predicted area of danger
PDOP	position dilution of precision
PFN	pulse-forming network
PPC	predicted point of collision
PPI	plan position indicator
PRF	pulse repetition frequency
PRP	pulse repetition period
PRR	pulse repetition rate
Radar	radio detection and ranging
RCS	radar cross section
RDF	radio direction finder
RF	radio frequency
RH	relative humidity
RS232	Recommended Standard No 232 from the EIA
s	seconds
SA	selective availability
SAR	search and rescue
SART	search and rescue transponder
SCR	silicon controlled rectifier
SDME	speed and distance measuring equipment
SENC	system electronic navigation chart
SI	statutory instrument
SOD	sector of danger
SOG	speed over ground
SOLAS	safety of life at sea
SOP	sector of preference
SOTDMA	self-organizing time division multiple access
SPS	standard positioning service
SVDR	simplified voyage data recorder
STC	sensitivity time control
TCPA	time to closest point of approach
TDMA	time division multiple access
UHF	ultra-high frequency
UK	United Kingdom
US	United States
UTC	coordinated universal time
VBW	vertical beamwidth
VDOP	vertical dilution of precision
VDR	voyage data recorder
VHF	very high frequency
VRM	variable range marker
VTS	vessel traffic service
W	watts
WGS	world geodetic system
μS	microsecond

See also Appendices of Section 11.2.2 for IMO standard terms and abbreviations.

Index

Note: Page numbers followed by '*f*' and '*t*' refer to figures and tables, respectively.

A

Acquisition of targets, 217–220, 466, 473*t*
  performance standard, 466
Active array, 59, 60*f*
Active radar reflectors, 157–158, 169–170
Active reflector, 157–158
Adaptive gain, 180–182, 188–189, 302–304
Aerial. *see* Antenna
Afterglow
  misleading effect of, 422
  trail, 13
Air traffic control (ATC) radars, 127
AIS (Automatic identification system), 32–33, 215, 255, 371–372
  benefits to shore monitoring stations, 270–271
  and bridge displays, 264–266
    Class B equipment, 266
    integration of AIS with ARPA and/or ECDIS, 265–266
    minimum keyboard display (MKD), 264
    pilot plug and display, 266
    receivers, 266
    standalone graphical display, 265
  capacity, 267–268
  carriage, 270
  dependence on GNSS, 268–269
  information transmission, 259–261
  information transmitted by Class A vessel, 259–261
  internet-based networks, 273
  long range identification and tracking (LRIT), 274–275
  messages and types, 261–264
    AIS on buoys, 263
    binary messages, 263–264
    Class A transmissions, 261
    Class B transmissions, 261
    SAR aircraft, 261
    SART, 261–263
    synthetic AIS, 263
    virtual AIS, 263
  modes, 256–259
  operational controls, 311–315
    with ARPA, 313–315
    standalone AIS equipment, 311–313
  organization of, 256–259
    assigned mode, 257
    autonomous and continuous mode, 265–266
    Class B transmissions, 258–259
    polling mode, 258
  participation by vessels, 269
  performance standards, 466–467, 503–505
  pilot plug, 266
  radar, comparison with, 271–273, 468
  Radar/ARPA, comparison with, 271–273
  satellite AIS, 273–274
  signals transmission around large land features, 267
  symbols, 477, 480
  target swap, 267
  vulnerability to false reports, spoofing and jamming, 269–270
Alarms. *see* Warnings, ARPA
Amplifier. *see also* Receiver
  intermediate frequency, 66–67
  linear, 72
  logarithmic, 67, 71–74, 73*f*
  noise. *see* Thermal noise
  radio frequency, 69–70
  video, 80
Amplitude modulation, 43
Analog processing, 32
Analog signals, 63–64
Analog-to-digital conversion, 77–79, 78*f*
Angle of cut, 379
Anode, 44
Antenna, 30–32
  angle, 11
  array. *see* Array antennas
  beamwidth and sidelobes, 47–48
  characteristics, 140–141
  gain and directivity, 48
  height, 132
  isotropic source, 45–46
  performance standard, 471
  polarization, 48–49
  power density, 46
  principles, 45–61
  radiation intensity, 46–47
  radiation pattern, 47
  receiving characteristics, 48
  rotation mechanism, 60–61
  siting, 515
Antenna location, choice of, 136
Antenna siting, 128–133, 130*f*
  antenna height, 132
  blind and shadow sectors, 128–132, 129*f*
  damage, susceptibility to, 132–133
  indirect echoes, 132
  radiation hazard, 133
ARB (Area rejection boundaries), 219–220
ARPA
  accuracy and errors, 407, 422–423, 467
  see also Errors associated with ARPA
  AIS integrated with, 313–315

ARPA (*Continued*)
alarms and warnings. *see* Warnings, ARPA
carriage requirements, 216–217
data brilliance control, 237
definition of terms, 473*t*
display modes, 236–237
display sizes, 464
integral/stand-alone, 216
malfunction of, 470
observer qualification, 515
over-reliance on, 422–423
performance standards – IMO, 217, 460–503
simulation facilities, 471
target trails and past positions, 229–232
test programmes, 244–245, 477
test scenarios, 407
Array
antenna phased, 49–60, 52*f*
multiple radiating elements, 49–53
parallel fed and active arrays, 59–60, 59*f*
practical considerations, 53–55
sidelobes, 53
slotted waveguide, 55–57, 56*f*
vertical beamwidth, 57–59
radar reflectors
dihedral, 155
double pentagonal, 155
octahedral, 155
pentagonal, 155
A-scan display, 5, 5*f*
Aspect, 320
target, 143–144
contribution to response, 146–147
changes as vessel steams, 208
variable, 145
vessel
determination by plot, 320
basis for collision avoidance strategy, 320
ATA (automatic tracking aid), 215
Atmospheric conditions, 191–194
ducting, 196–197
standard, 191–194
sub-refraction, 194–195
super-refraction, 195–196
ATT (Automatic Target Tracker), 217
Attenuation due to precipitation, 184–185, 190
Autocorrelation, 114, 119*f*
Automatic Identification System. *see* AIS (Automatic identification system)
Automatic radar plotting aid (ARPA), 62–63, 255, 371
integration of AIS with, 265–266
Auto-tracking aid (ATA), 360–361
Azimuth cell
defining, 82–84
increment, 5–6
Azimuth encoder, 61
'Azimuth pulses', 10–11
Azimuth quantization error, 410

B
Backlash, in gearing, 409
Bandwidth, 68–69
Barker code, 120, 121*t*
Baseband processing, 67
Baseband signal, 74
Beacon band operation, 163
Beam processing, 206
Beamwidth. *see* Antenna
Bearing
accuracy, 300–301
discrimination, 49, 103–105, 104*f*
errors, 409–410
facilities for measurement of, 237, 295–301
measurement, 11–12, 12*f*
performance standard, 463–464
Beat frequency, 67
Bediou navigation satellite system, 441
Bels, 45
Binary messages, AIS, 263–264
Bit, 77
Black box, 454
Blind and shadow sectors, 128–132, 129*f*
Buoy
AIS on, 263
patterns, 155–157

C
Calibration levels, 293
Cathode, 44
Cathode ray tube (CRT), 6, 33, 89
Cavity magnetron, 44, 44*f*
Circulator, 65
Class A transmissions, 261
Class B transmissions, 261
Closed corner, 153
Close-quarters situation, 363–364
Closest point of approach (CPA), 27, 227, 233, 235
accuracy of, 229
determination of, 318
safe limits, 239, 468
Clutter, 32–33, 84–87
rain, 86–87
sea, 84–86
Clutter, precipitation, 183–191
attenuation in precipitation, 184–185
combating the attenuation caused by precipitation, 190
detection of targets beyond, 190, 304–305
detection of targets in, 183–191
effect of precipitation type, 185–186
exploiting ability to detect, 190–191
nature of, 183–184
performance standard, 461, 463
rain clutter circuit, 187–188
rain clutter control, 304
suppression of (practice), 304–305
suppression of (theory), 186–190
Clutter, sea, 84–86, 170–183, 301–304
control, 176–177
see also Adaptive gain; Rotation-to-rotation correlation
detection of targets in, 170–183
nature of, 170–175
performance standard, 143, 461
suppression of (theory), 175–176
suppression of, 301–304
Coaxial cable, 29–30, 31*f*
Coherent radar, 1, 110–112
concepts, 110–114
principles, 34–35
Collision, action to avoid, 364–365
Collision avoidance
display, 8
radar/ARPA and AIS comparison for, 271–273
regulations. *see* Regulations for preventing collisions at sea
Colour, use of, 93–94
Compass, 441
safe distances, 134–135
stabilization, 15
Composite shadowing, 376
Consistent Common Reference Point (CCRP), 129–131

Constant false alarm rate (CFAR), 302–303
Controls, for suppression of unwanted responses, 301
Correlated video, 205–206
Correlation, 75, 114–116
  example of, 116–120
Course Up orientation mode, 17, 18*f*, 98–99
CPA. *see* Closest point of approach (CPA)
Cross-correlation, 114
CSTDMA, 258–259
Cumulative turn, 365–366
Cursor, 465
  see also Perspex Cursor

D

'Danger circle', 389
Data (ARPA)
  accuracy of displayed, 407–408, 422–423
  alphanumeric display of, 237–238, 468
  analysis, 223–229
  fusion, 314
  storage of, 227, 414–415, 419*f*
Default conditions, 281
Demodulator, 74
Detection of targets. *see* Target, detection of
Detector, 67, 74, 74*f*
Differentiator. *See* Clutter, precipitation: rain clutter circuit
Diffraction, 128–129
Digital display, 92–94
  colour, use of, 93–94
Digital processing, 33
Digital signals, 63–64
Digital storage, 225
Digitization, 77–79
Dihedral array, 155
Direct drive motor, 60–61
Directional transmission and reception, 4
Directive gain, 48
Directivity, 48
Discrimination, 373–374
  bearing, 103–105
  performance standard, 463
  range, 103
Display, 5–10, 9*f*, 89–107
  see also Cathode ray tube (CRT); Liquid crystal display (LCD) technology
  alphanumeric data, 237–238, 480
  ARPA, performance standard, 473*t*
  A-scan, 5, 5*f*
  digital display, 92–94
  flat panel display technology – LCDs, 94–96
  formatting, 96–100
  graphical data, 461, 470
  history. *see* History, ARPA tracking
  horizontal reference mode, 100
  motion modes, 99–100
  orientation modes, 98–99
  plan position indicator, 5–8
  of radar echoes, 100–105
  radial scan PPI, 90–92, 90*f*
  raster-scan. *see* Raster-scan display
  relative-motion. *see* Relative-motion presentation
  requirements, 97–98
  setting up the radar, 278–288
  size, 236, 464, 466
  target trails, 8–10, 10*f*, 15*f*
  true motion. *see* True-motion display
  and user interface, 33–34, 105–107
  vector. *see* Vector
Display unit, 134
Doppler effect, 109
Double pentagonal array, 155
Down-conversion, 63–64
Down-converted resultant signal, 67
Dual-axis Doppler log, in ground track mode, 309–310
'Dual-fuel' system, 444
Dwell time, 40

E

'E' plot, 358–359
ECDIS. *see* Electronic chart display and information system (ECDIS)
Echo box, 288
  principle, 288–290
  siting, 290–292
Echo paint, 100–102, 101*f*
Echo principle, 2–3, 3*f*
Echo reference. *see* Stabilization, ground
Echo stretch, 305, 372
Echoes, unwanted. *see* False echoes
'Effective energy', 40–41
Electronic bearing line (EBL), 11–12, 297–298, 373
Electronic Chart Display and Information System (ECDIS), 89, 396, 442–447, 469, 507
  comparison
    electronic and paper, 444–445
    raster and vector, 445–446
  integration of AIS with, 265–266
  publications associated with, 446–447
  radar, integration with, 447
  raster, 444
  unapproved data, 446
  vector, 443
Envelope of received signal, 67
EPA (electronic plotting aid), 359–360
  and reflection plotter, comparison between, 359–360
Equivalent flat plate area, 148
Errors associated with ARPA
  displayed data in, 413–420
  due to incorrect data input, 415–417
  effect of random gyro compass errors, 411–412
  interpretation in, 420–423
  quantization, 410–411
  in range measurement, 410–411
  sources of, 408
  target swap, 223, 413
  tracking errors, 413–415
Errors associated with manual plotting, 347–351
Errors associated with the true-motion presentation, 351–353
Exclusion areas. *see* ARB (Area rejection boundaries)

F

False alarms, 302–303
False echoes, 197–213
  due to power cables, 210–213
  indirect, 197–198
  multiple, 201–202
  radar to radar interference. *see* Interference
  second-trace, 206–210
  side, 202–204
Fast time constant (FTC) control, 86, 187
Field effect transistor (FET), 43

Fixed range rings, 296–297
Fixed target, 241–242
Fjord effect, 267, 268*f*
Flat panel display technology – LCDs, 94–96
Formatting the display, 96–100
Fourier analysis, 68–69
Fourier transformation, 42
'Frame grabbing', 456
Frequency, 35–36, 461
  choice of, 135–136
  pulse repetition, 38–39, 39*t*
Frequency modulated coherent radar, 111–112
Frequency modulated continuous wave (FMCW) pulse compression radars, 113–114, 114*f*
Frequency modulated pulse, 110–111
Frequency shift keying (FSK), 114–115
FTC. *See* Clutter, precipitation: rain clutter circuit
FTE (False target elimination), 210

G

Gain
  control, 284–286
  overall, 463
Galileo system, 441
Gate, tracking. *see* Tracking
Geo-graphical information system (GIS), 442
Glass-reinforced plastic (GRP), 56–57
Glint, 409
GLONASS satellites, 440–441
GMDSS (Global marine distress and safety system), 519–520
GNSS (global navigation satellite system), 17–19, 242, 309, 371
  see also GPS; GLONASS satellites; Galileo system
  accuracy and errors, 432–437
  AIS dependence on, 268–269
  Bediou navigation satellite system, 441
  global positioning system, 425–426
  inter-relationship with ECDIS, 442
  inter-relationship with radar, 442
  position fixing, 430
  radar, integration with, 399–400
  range and time, measurement of, 428–430
  regional satellite navigation systems, 441–442
  user equipment and display of data, 430–432
GPS (global positioning system), 425–428
  DGPS (differential GPS), 435–437
  DOP (dilution of precision), 435
  improvements to, 437–438
  performance standards, 505–506
  ranging principle, 429*f*
Grey-level, 286
Grids, 44
Ground lock, 306
Growlers, 151
Guard rings and zones, 219–220, 238–239
Guard zones, 311

H

Hazards, high-voltage, 135
Head Up orientation mode, 98
Heading and speed input data, 306–308
Heading marker, 10–11
  alignment of, 515
  performance standard, 470
  setting-up, 282
Head-up orientation, 13–15, 14*f*
Hertz, Heinrich, 1
High-voltage hazards, 135
History, ARPA tracking, 364
Hit matrix, 224–225
Horizontal beamwidth, 4, 53
Horizontal reference mode, 100
Horizontal shadowing, 376, 385*f*
Hülsmeyer, Christian, 1

I

IBS (Integrated bridge systems), 265, 447–448
IEC 61162, 98–99
IKBS (intelligent knowledge-based systems), 367–369
IMO, 255–256
  abbreviations, 484–498
  AIS (Automatic identification system), 466–467
  ECDIS (Electronic chart display and information system), 507
  general requirements for radio equipment, 519–520
  GPS (Global positioning system), 505–506
  INS (Integrated navigation systems), 507–515
  radar beacons and transponders, 499–502
  radar performance standards, 255–256, 264, 460–503
  radar reflectors, 503
  shipborne navigational displays, 477–483
  survival craft transponders (incl. SARTs), 502–503
  symbols for navigation displays, 484–498
  terms for navigation displays, 484–498
  voyage data recorders, 516–519
In-band interference, 75–76
Indexing target, 383–384
Indirect echoes, 132
Information overload, 265, 453
INS (Integrated navigation systems), 448
  performance standards, 507–515
Integrated circuits (IC), 66–67
Integrated navigation concepts (INS), 135
Integrated video, 303
Integration. *see also* System integration
  AIS/radar/ARPA/ATA comparison, 271–273
  ECDIS with radar and ARPA/ATA, 447
  GNSS with ECDIS, 442
  GNSS with radar, 442
Interference, 75–77
  in-band, 75–76
  out-of-band, 76
  radar to radar, 204–206, 461, 463
  solid-state coherent radars, 76–77
  standards, 513
  suppression of, 305
Intermediate frequency (IF) amplifier, 66–67
International Maritime Organization (IMO) Performance Standards, 30
Interrogated time-offset frequency-agile racons (ITOFAR), 163
Interswitching, 43*f*, 135–137, 140, 374, 515
  antenna location, choice of, 136

combination of radar data from multiple antennas, 137
frequency, choice of, 135–136
master and slave controls, 136–137
partial interswitching, 136
picture orientation and presentation, choice of, 136
switching protection, 136
IQ demodulation, 113
Isotropic source, 45–46

J
Joystick, 218, 298–299

K
King's Bank, 160

L
Landscape mode, 33
LCD technology. *see* Liquid crystal display (LCD) technology
Limiter, 65–66
Line-to-line correlation, 205
Liquid crystal display (LCD) technology, 89, 94–96, 95*f*
Local oscillator (LO), 66–68
Log errors, 408–409, 412–413
Logarithmic amplifier, 67, 71–74, 73*f*
Long range identification and tracking (LRIT), 274–275
Long range target identification, 372–373
Lookout, proper, 239, 362
'Low-noise RF amplification', 66
Lunenburg lens, 155–157

M
Magnetron
-based marine radar, 37
-based transmitter, 42–43
operation, 44–45
Manoeuvre
to avoid collision, 364
avoidance, 364
characteristics, 328–338
forecasting effect of, 328, 364–365
resumption, 365
target, effect of a, 326–328
trial. *see* Trial manoeuvre
Manual plotting, 347–351
Manual speed, 306
Map presentation, 244*f*
Margin lines, 389–390, 391*f*
Margin-of-safety lines, 402
Marine radar, 30*f*, 108–109
Master display, 136–137
Material, reflecting properties, 144–145
Minimum detectable signal, 41–42
Minimum keyboard display (MKD), 264
Mixer, 66–67
Mixer principle, 67–68
Modern radar display, 97–98, 98*f*
Modulator unit, 43*f*
Motion and stabilization modes, 17–28
relative-motion presentation, 19–22, 20*f*
true-motion presentation, 22–26
Motion modes of display, 99–100
Mouse, 105–107
see also Joystick
Moving Target Indication (MTI) radars, 127
Multiple antennas, combination of radar data from, 137
Multiple Echoes, 201–202
see also False Echoes
Multiple installations. *see* Interswitching

N
Navigation lines, 242–244, 400–402, 469
Near-field effect, 52–53
Nematic phase, 94–95
NMEA 0183, 449–452
Noise. *see* Thermal noise
Non-coherent radar, 110
North Up orientation mode, 15–17, 16*f*, 98–99
NT (New Technology) Radar. *see* Solid-state radars
Nyquist rate, 82
Nyquist sampling theorem, 77

O
Octahedral array, 155
Off-centring, 99–100
On Board ship units, 128–137
antenna siting, 128–133, 130*f*
compass safe distances, 134–135
display unit, 134
exposed and protected equipment, 135
high-voltage hazards, 135
interswitching. *see* Interswitching
power supplies, 135
transceiver unit, 133–134
Operational controls, 277
Operational display area, 6
Operational warnings. *see* Warnings, ARPA
Orientation, 12–17
choice of, 17
course-up, 17, 18*f*
head-up, 13–15, 14*f*
north-up, 15–17, 16*f*
Orientation modes of display, 98–99
Orientation of picture, 282
course-up, 282
head-up, 282
north-up, 282
Orthogonal fields, 48–49
Out-of-band interference, 76
Output monitor, 290

P
PAD (Predicted area of danger), 250–253
errors of interpretation, 421–422
plotting and construction of, 339–341
Parallel fed array, 59, 59*f*
Parallel indexing, 380–405
anti-collision manoeuvring, 402–405
collision avoidance while, 402–405
complex series of, 388*f*, 389
cursor, 355, 386
facilities for, 381–382
margin lines, 389–390, 391*f*
modern radar navigation facilities, 396–402
preparations and precautions, 382–385
progress monitoring, 392–394
technique, 385–392
true-motion display, on, 382, 394–396, 400*f*
true-motion radar presentation for, 394–396
unplanned, 402
unplanned parallel indexing, 402
wheel-over position, 390–391, 395*f*
Partial interswitching, 136
Past positions. *see* History, ARPA tracking
Pattern matching, 127–128

Pentagonal array, 155
Performance monitor signal, 288–289
Performance monitoring, 288–294
calibration levels, 293
echo box
principle, 288–290
siting, 290–292
modern trends, 294
performance check procedure, 293–294
power monitors, 292
transponder performance monitors, 292–293
Performance standards – IMO:. *see* IMO
Perspex cursor, 355
Phase modulation, 115–116, 116*f*
Picture. *see also* Display
build-up of, 11*f*
optimum, 284–288
storage of, 224–225
synthetic, 89
Picture orientation and presentation, choice of, 136
Pilotage situations targets, 378
Plan Position Indicator (PPI), 5–8, 89, 373–374
Plotting, 317
accuracy and errors, 347–351
aids, 353–361
avoiding action, 364–365
facilities for, 521–522
need for, 361
PADs, 250–253, 339–340
PPCs, 245–250, 338–339
practicalities of, 321–323
relative, 317–324
SODs, 341–343
SOPs, 343–344
stopping distances, 331–334
in tide, 345–347
true, 324–325
turning circles, 328–329, 332, 333*f*
when ownship only manoeuvres, 328–329
when ownship resumes, 335–336
when target only manoeuvres, 326–328
Plume, 290
Pointing devices, 105
Polarization, 48–49, 161–165, 175, 189, 463
Portrait mode, 33
Position fixing, 378–380, 430, 432*f*
selection of targets, 379
types of position line, 379–380
Power density, 46
Power divider network, 59
Power monitors, 292
Power supplies, 135
PPC (Predicted point of collision), 245–250
error due to incorrect sensor input, 418–420
plotting of, 338–339
Precipitation. *see* Clutter, precipitation
Presentation of picture
choice of, 26–28
ground-stabilization. *see* Stabilization
relative motion, 19–22
sea-stabilization. *see* Stabilization
true-motion, 22–26
Prime register, 224
Prioritization, 219
Probability of detection, 41
Probability of false alarm, 41
Processor, 225
role of, 62–64
Progress monitoring, 392–394
Pseudo-random number generation, 39
Pulse
compression, 34–35, 108–109, 112*f*, 126–128
compression radar, 112–113
jitter, 39
length, 36–38, 39*t*, 81*f*
repetition frequency, 38–39
shape, 41–42, 41*f*
Pulse repetition frequency (PRF), 4

Q

Qualifications
ARPA observer, 516
radar observer, 515

R

Racon. *see* Radar beacons and transponders
Radaflare, 166
Radar, 1, 2*f*
Radar beacons and transponders
chart symbols, 165–166
clutter, 160, 163
operation of radar with, 463
performance standards for. *see* IMO
racon, 157–158, 499
sources of information, 165–166
survival craft, for. *see* SARTs
Radar bearings, 380
Radar cross-section of target, 148–150
Radar cross-sectional area (RCS), 62
Radar display, setting up, 278–288
Radar echoes, display of, 100–105
bearing discrimination, 103–105, 104*f*
practical considerations, 102–103
range discrimination, 103
Radar equation, 61–62
Radar equipment for new ships, performance standards for, 460–476
Radar horizon, 191–197
Radar installation, 520–521
errors generated in, 408–413
Radar log, 151–152
Radar maintenance, 514
Radar processor, 32–33
Radar range equation, 139
Radar ranges, 379–380
Radar reflectors, 143, 152, 271, 503
arrays of, 154–155
corner, 152–154
performance standards, 459, 503
Radar signal, 34–42
choice of frequency, 35–36
fundamental considerations, 34–35
power considerations, 39–41
pulse length, 36–38
pulse repetition frequency, 38–39
pulse shape, 41–42
Radar system
characteristics, 140–143
obligation to use, 363
performance standards, 264, 266, 460–503
requirement to carry, 461
siting of, 515
technical principles, 29
use in clear weather, 317, 364
Radar transmitter, 42–45
magnetron-based transmitter, 42–43
magnetron operation, 44–45
Radar-conspicuous, 151–152
Radar-conspicuous targets, 376–378
Radial scan plan position indicator (PPI), 89
Radial scan PPI, 90–92, 90*f*

Radiation hazard, 133
Radiation intensity, 46–47
Radiation pattern, 46–47, 47*f*, 48*f*
Radio frequency (RF), 29–30
Radio waves, 1
Rain clutter, 86–87, 183
see also Clutter, precipitation
Ramark. *see* Radar beacons and transponders
Range
accuracy, 299–300
discrimination, 36–38, 37*f*, 103, 373–374, 463
facilities for measurement of, 237, 295–301
as a function of time, 3–4
maximum detection, 139, 191–193
minimum detection, 141–143, 463
performance standard, 458
rings, 6
sampling example, 81–82
sidelobes, 120–126
Range cell increment, 5–6
Range scales
changing, 237, 294–295, 355
choice of, 323, 384
'Range-dependent' pattern, 52–53
Ranging principle, 429*f*
Raster charts, 444–446
Raster electronic chart version (ARCS), 445–446
Raster-scan display, 357–358
see also Cathode ray tube (CRT); Liquid crystal display (LCD) technology
Rate aiding, 221, 413
Rayleigh roughness criterion, 64–67, 150
see also Amplifier
Receiver, 32, 67–74
bandwidth, 68–69
characteristics, 141
logarithmic amplifier, 71–74, 73*f*
mixer and intermediate frequency amplifier, 66–67
mixer principle, 67–68
noise, 206, 288
protection, 64–66
on raster-scan displays, 357–358
RF amplification, 66
role of, 62–64
second detector, 74, 74*f*
sensitivity time control (STC), 71–74, 85*f*
signal amplification, 69–70
signal detection and baseband processing, 67
thermal noise, 70–71
tuning of, 68
Reception, 4
Reflected echoes. *see* Indirect echoes
Reflection plotters, 354–358
and electronic plotting aid, comparison between, 359–360
Reflectors. *see* Atmospheric conditions
Refraction. *see* Atmospheric conditions
Regional satellite navigation systems, 441–442
Regulations for preventing collisions at sea
application of, 361–367
summary, 366–367
Relative motion mode, 99–100
Relative motion true-tracks (RMT), 284
Relative/true vector mode, selecting, 308
Relative-motion presentation, 19–22, 20*f*
Reserve gain, 153
Resolution cell, 102–103, 146–147, 146*f*, 224–225, 372–373
'Resonant' effect, 56
Restricted visibility, 364–365
RF amplification, 66
Rise time, 42
Rotation-to-rotation correlation, 182–183, 303–304
Route monitoring display, 8
'Rule of thumb' method, 403

S

Safe limits, 239, 310
Safe speed, 362
Sailing directions, 373
Sample rate, 63–64
SAR aircraft, 261
SARTs (search and rescue transponders), 256, 261–263
Satellite AIS, 273–274
Saturation state, 79
S-band characteristics, 174, 184
S-band radars, 35–36, 36*t*, 55
S-band vertical pattern, 57–58
Scanner. *see* Antenna
Scanty radar information, 363
Scintillation, 62
Screen marker, 218, 237, 298–299
SDME (speed and distance measuring equipment), 217
Sea clutter. *see* Clutter, sea
Sea stabilization. *see* Stabilization
Search and rescue transponder (SART), 166–169, 463, 502–503
Searching, 177
Sea-stabilized true vectors, 308–310
Second detector, 74, 74*f*
Second time around effect, 38
Second trace echoes
effect, 38
elimination, 305
Sensitivity time control (STC), 71–74, 85*f*, 86–87, 177
Sensor errors, 408
Setting-up the radar display
coherent radar, 288
default conditions and startup, 279–281
gain, 284–286
heading marker, 282
orientation of picture, 282
presentation of radar, 282–284
screen brilliance setting, 279–281
switching on, 278–279
tuning, 286–288
Shadow areas, 374–376
composite shadowing, 376
horizontal shadowing, 376, 385*f*
pilotage situations, 378
radar-conspicuous targets, 376–378
tide, rise and fall, 376
vertical shadowing, 374–376, 375*f*
Shadow sectors, 41*f*, 128–132, 129*f*
Shipborne navigational displays, 477–483
Shore monitoring stations, benefits of AIS to, 270–271
Side echoes, 202–204
Side lobe suppression, 163
Sidelobes
of antenna, 47–48
of array antennas, 53
Signal amplification, 69–70
Signal detection, 67
Signal processing, 62–63, 181–182, 463
'Signal threshold gain', 80
Siting of units, 515
Slave display, 136–137

Sloping surface, effect on target response, 147
Slotted waveguide array antenna, 55–57, 56*f*
SOD (sector of danger), 341–343
SOLAS vessels, 255–256
Solid-state coherent radars, 76–77
Solid-state radars, 108–128, 279
  coherent radar concepts, 110–114
  correlation, 114–120
  pulse compression, 126–128
  range sidelobes, 120–126
SOP (sector of preference), 343–344
SOTDMA (self-organizing time division multiple access), 257, 257*f*
Speed and Distance Measuring Equipment (SDME), 99
Spot size, 374
Spurious echoes, 197
Stabilization
  azimuth, 465
  ground, 24–26, 241–242, 308–310, 396–405, 465, 473*t*
  sea, 22–24, 308–310, 465, 473*t*
Stabilized head-up, 13
Standalone graphical display, 265
Stand-by condition, 279, 295
Statutory instrument No. 69 of 1993, extracts, 520–522
Surface texture (of target), 144
Swept gain, 177
Switching on, 278–279, 306
Switching protection, 136
Symbols for radar and navigation, 484–498
Syncro transmitter, 61
Synthetic AIS transmissions, 263
Synthetic picture. *see* Display
System integration
  advantages of, 453
  connectivity and interfacing, 449–453
  dangers of data overload, 453
  data logging. *see* VDR (voyage data recorders)
  data monitoring and logging, 454
  integrated bridge systems (IBS), 447–448
  integrated navigation systems (INS), 448
  sensor errors and accuracy of integration, 453–454
  system cross-checking, warnings and alarms, 453
  typical systems, 448
Systematic observation of detected echoes, 317–318, 362
  see also Plotting

T

Target
  aspect, 143–144, 319, 320*f*
  characteristics, 143–152
  detection, minimum range, 141–143
  detection, minimum standards, 463
  detection in clutter, 170–183
  detection of, 139, 461
  enhancement, active, 157–170
  enhancement, passive, 152–157
  equivalent flat plate area, 148
  identification of, 372–379
    long range, 372–373
  loss of, 222–223, 239–240, 486*t*
  material, 144–145
  radar conspicuous, 376–378
  radar cross-section, 148–150
  responses from specific, 151–152
  shape, 145
  simulation facility, 244–245
  size, 145–150
  sloping surface, 147
  surface texture, 144
  swap, 223, 267, 413, 466, 473*t*
  tracking, 32, 87–89, 88*f*, 466
  trails, 8–10, 10*f*, 15*f*
TCPA. *see* CPA
TFT (Thin film transistor). *see* Liquid crystal display (LCD) technology
THD (transmitting heading device), 217
Theoretical minimum detection range, 142
Thermal noise, 70–71, 141, 218
Thin film transistor (TFT), 96
Threat assessment markers, 354
Tide, rise and fall, 376
Tide/current, manual estimate of, 309
Time, range as a function of, 3–4
Time sidelobes, 123
Track file, 225
Tracker
  'full' warning, 240–241
  ball, 105–107, 299
  see also Joystick
  philosophy, 224–227
Tracking, 220–221
  accuracy, 407
  course errors, 353
  errors, 413–415
  gate, 218, 221, 226, 415
  number of targets, 220–232
  performance standards, 460, 466
  preparation for, 310–311
  speed errors, 353
Transceiver unit, 133–134
Transmission, 4
Transmitter, 29–30
  characteristics, 140
  magnetron-based, 42–43
Transponder. *see also* Radar beacons and transponders; AIS (Automatic identification system)
  performance monitors, 292–293
Trend of targets motion, 223, 407, 466
Trial manoeuvre, 216, 234–236, 338
  alarm, 240
  performance standard, 469, 473*t*, 481
Triggers, 42–43
True motion mode, 99
True-motion constant display (TMCD), 284
True-motion display, 394–396
True-motion presentation, 22–26
  choice of presentation, 26–28
  errors associated with, 351–353
  ground-stabilized, 24–26, 25*f*
  sea-stabilized, 22–24, 23*f*
True-motion radar presentation for parallel indexing, 394–396
Tuning control, 68
  purpose, 293–294
  practical operation, 293–294
  performance check, 288
Turning
  mechanism/gear, 60–61
  rate, 60–61

U

United kingdom statutory instruments navigation
  radar installation, 520–522
User interface, 105–107
  display and, 33–34

V

Variable range marker (VRM), 6, 10*f*, 297

VDR (Voyage data recorder), 454–457
  data capsule, 454–455
  equipment, 454–455
  future, 457
  performance standards, 516–519
  playback, 456–457, 510
  radar data, 455–456, 511–512
Vector, 232–236
  AIS, 467
  chart, 443, 445–446
  display, 481
  mode, 473*t*
  relative, 233
  true, 233
Vector time control, setting, 308
Vertical lobing pattern, 57, 58*f*
Vertical shadowing, 374–376, 375*f*
Vessel traffic service (VTS), 255
Video
  levels, 285–286
  signal, 74
Virtual AIS transmissions, 263
VTS stations, 270

W

Warnings, ARPA
  anchor watch, 240
  CPA/TCPA, 239, 468
  guard rings/zones, 238–239
  malfunction, 470, 514
  operational, 468–469
  performance standards, 468
  performance tests and, 241
  safe limits, 468
  safe limit vector suppression, 241
  target lost, 239–240, 469
  time to manoeuvre, 240
  track change, 240
  tracks full, 240–241
  trial alarm, 241
  wrong request, 241
Water lock, 306
Water speed input, 306–308
Waveguide, 29–30, 31*f*
Wheel-over position, 390–391, 395*f*
White noise, 70
Words, 77

X

X-band characteristics, 174, 184
X-band radars, 35–36, 36*t*, 55

Y

Yaw, effect of, 15, 15*f*

Printed in the United States
By Bookmasters